# Applications of Fluorescence in the Biomedical Sciences

# Applications of Fluorescence in the Biomedical Sciences

Proceedings of a Meeting Held in Pittsburgh, Pennsylvania
April 12–15, 1985


## Editors

**D. Lansing Taylor**

**Alan S. Waggoner** **Frederick Lanni**

**Robert F. Murphy** **Robert R. Birge**

Departments of Biological Sciences and Chemistry
Center for Fluorescence Research in Biomedical Sciences
Carnegie-Mellon University
Pittsburgh, Pennsylvania



**Alan R. Liss, Inc. • New York**

**Address all Inquiries to the Publisher**
**Alan R. Liss, Inc., 41 East 11th Street, New York, NY 10003**



**Printed in the United States of America**



**Library of Congress Cataloging-in-Publication Data**
Main entry under title:

Applications of fluorescence in the biomedical sciences.

Includes index.
1. Fluorescence spectroscopy—Congresses.
2. Fluorescence microscopy—Congresses. 3. Flow cytometry—Congresses. I. Taylor, D. Lansing. [DNLM: 1. Cell Separation—congresses. 2. Flow Cytometry—congresses. 3. Microscopy, Fluorescence—congresses. 4. Spectrometry, Fluorescence—congresses. QH 324.9.F4 A652 1985]
QH324.9.F4A67 1986 574'.01'53535 85-23713
ISBN 0-8451-4210-0

# Contents

# Contributors

**George K. Alameda,** Department of Biological Sciences, Center for Fluorescence Research in Biomedical Sciences, Carnegie-Mellon University, Pittsburgh, PA 15213 **[203]**

**Randall S. Alberte,** Department of Molecular Genetics and Cell Biology, University of Chicago, Chicago, IL 60637 **[91]**

**R.R. Alfano,** Department of Physics and Institute for Ultrafast Spectroscopy and Lasers, City College of the City University of New York, New York, NY 10031 **[69]**

**Philip A. Amato,** Department of Biological Sciences, Center for Fluorescence Research in Biomedical Sciences, Carnegie-Mellon University, Pittsburgh, PA 15213 **[347]**

**Daniel Axelrod,** Department of Physics, Biophysics Research Division, University of Michigan, Ann Arbor, MI 48109 **[461]**

**Robert R. Birge,** Department of Chemistry, Center for Fluorescence Research in Biomedical Sciences, Carnegie-Mellon University, Pittsburgh, PA 15213 **[xv, 105, 203]**

**John J. Birmingham,** Department of Biochemistry, Dundee University, Dundee DD1 4HN, Scotland, United Kingdom **[245]**

**A.A. Brian,** Department of Chemistry, Stanford University, Stanford, CA 94305; present address: Department of Chemistry, University of California at San Diego, La Jolla, CA 92093 **[499]**

**Gary R. Bright,** Department of Biological Sciences, Center for Fluorescence Research in Biomedical Sciences, Carnegie-Mellon University, Pittsburgh, PA 15213 **[257]**

**Laura J. Brvenik,** Department of Chemistry, Syracuse University, Syracuse, NY 13210 **[141]**

**R.H. Callender,** Department of Physics and Institute for Ultrafast Spectroscopy and Lasers, City College of the City University of New York, New York, NY 10031 **[69]**

**Sarina A. Cavalier,** Department of Chemistry and Institute of Molecular Biology, University of Oregon, Eugene, OR 97403 **[159]**

**Thomas M. Chused,** Laboratory of Microbial Immunity, N.I.A.I.D., N.I.H., Bethesda, MD 20205 **[531]**

**Lawrence B. Cohen,** Department of Physiology, Yale University School of Medicine, New Haven, CT 06510 **[423]**

**Alison Cooney-Freed,** Department of Chemistry and Institute of Molecular Biology, University of Oregon, Eugene, OR 97403 **[159]**

The number in brackets is the opening page number of the contributor's article.

**Laurence Corash,** Department of Laboratory Medicine, University of California at San Francisco, San Francisco, CA 94143 **[567]**

**Richard T. Cummings,** Department of Chemistry, Syracuse University, Syracuse, NY 13210 **[141]**

**A.G. Doukas,** Department of Physics and Institute for Ultrafast Spectroscopy and Lasers, City College of the City University of New York, New York, NY 10031 **[69]**

**Jeffrey L. Evelhoch,** Department of Chemistry, Center for Fluorescence Research in Biomedical Sciences, Carnegie-Mellon University, Pittsburgh, PA 15213 **[203]**

**Leonore A. Findsen,** Department of Chemistry, Center for Fluorescence Research in Biomedical Sciences, Carnegie-Mellon University, Pittsburgh, PA 15213 **[105]**

**Graham R. Fleming,** Department of Chemistry, University of Chicago, Chicago, IL 60637 **[91]**

**Robert M. Fulbright,** Department of Physics, Biophysics Research Division, University of Michigan, Ann Arbor, MI 48109 **[461]**

**Ruth H. Furukawa,** Department of Chemistry, Syracuse University, Syracuse, NY 13210 **[141]**

**Peter B. Garland,** Department of Biochemistry, Dundee University, Dundee DD1 4HN, Scotland, United Kingdom; present address: Unilever Research, Colworth Laboratory, Sharnbrook, Bedford MK44 1LQ, England, United Kingdom **[245]**

**H.E. Gaub,** Department of Chemistry, Stanford University, Stanford, CA 94305; present address: Department of Physics, Technische Universität, München, Garching D8059, Federal Republic of Germany **[499]**

**David Gross,** Department of Applied and Engineering Physics, Cornell University, Ithaca, NY 14853 **[405]**

**Robert J. Gulotty,** Department of Chemistry, University of Chicago, Chicago, IL 60637 **[91]**

**Danni L. Harris,** Department of Chemistry and Institute of Molecular Biology, University of Oregon, Eugene, OR 97403 **[159]**

**Edward H. Hellen,** Department of Physics, Biophysics Research Division, University of Michigan, Ann Arbor, MI 48109 **[461]**

**Yasushi Hiraoka,** Department of Biophysics, Faculty of Science, Kyoto University, Kyoto 606, Japan; present address: Department of Biochemistry and Biophysics, University of California at San Francisco, San Francisco, CA 94143 **[321]**

**Bruce S. Hudson,** Department of Chemistry and Institute of Molecular Biology, University of Oregon, Eugene, OR 97403 **[159]**

**Dennis E. Koppel,** Department of Biochemistry, University of Connecticut Health Center, Farmington, CT 06032 **[477]**

**Grant A. Krafft,** Department of Chemistry, Syracuse University, Syracuse, NY 13210 **[141]**

**Joseph R. Lakowicz,** Department of Biological Chemistry, University of Maryland School of Medicine, Baltimore, MD 21201 **[29, 225]**

**Frederick Lanni,** Department of Biological Sciences, Center for Fluorescence Research in Biomedical Sciences, Carnegie-Mellon University, Pittsburgh, PA 15213 **[xv, 449, 505]**

**Leslie M. Loew,** Department of Physiology, University of Connecticut Health Center, Farmington, CT 06032 **[423]**

**Jill A. London,** Department of Physiology, Yale University School of Medicine, New Haven, CT 06510 **[423]**

**Katherine Luby-Phelps,** Department of Biological Sciences, Center for Fluorescence Research in Biomedical Sciences, Carnegie-Mellon University, Pittsburgh, PA 15213 **[347]**

**Richard D. Ludescher,** Department of Chemistry and Institute of Molecular Biology, University of Oregon, Eugene, OR 97403 **[159]**

**Richard A. Mathies,** Department of Chemistry, University of California, Berkeley, CA 94720 **[129]**

**Seiji Matsumoto,** Department of Biophysics, Faculty of Science, Kyoto University, Kyoto 606, Japan **[321]**

**H.M. McConnell,** Department of Chemistry, Stanford University, Stanford, CA 94305; present address: Stauffer Laboratory for Physical Chemistry, Stanford University, Stanford, CA 94305 **[499]**

**Paul L. McNeil,** Department of Biological Sciences, Center for Fluorescence Research in Biomedical Sciences, Carnegie-Mellon University, Pittsburgh, PA 15213 **[347]**

**Laurens Mets,** Department of Molecular Genetics and Cell Biology, University of Chicago, Chicago, IL 60637 **[91]**

**Yuen Mok,** Department of Laboratory Medicine, University of California at San Francisco, San Francisco, CA 94143 **[567]**

**Kosuke Morikawa,** Department of Biophysics, Faculty of Science, Kyoto University, Kyoto 606, Japan **[321]**

**Robert F. Murphy,** Department of Biological Sciences, Center for Fluorescence Research in Biomedical Sciences, Carnegie-Mellon University, Pittsburgh, PA 15213 **[xv, 525, 545]**

**Lionel P. Murray,** Department of Chemistry, Center for Fluorescence Research in Biomedical Sciences, Carnegie-Mellon University, Pittsburgh, PA 15213; present address: Laboratory of Chemical Physics, National Institute of Arthritis, Diabetes, Digestive and Kidney Diseases, N.I.H., Bethesda, MD 20205 **[105]**

**Diana Gold Myles,** Department of Physiology, University of Connecticut Health Center, Farmington, CT 06032 **[477]**

**Shun Okada,** Department of Biophysics, Faculty of Science, Kyoto University, Kyoto 606, Japan **[321]**

**Harry S. Orbach,** Division of Biology, California Institute of Technology, Pasadena, CA 91125 **[423]**

**Brian M. Pierce,** Department of Chemistry, Center for Fluorescence Research in Biomedical Sciences, Carnegie-Mellon University, Pittsburgh, PA 15213; present address: Advanced Technology Projects Laboratory, Support Systems, Hughes Aircraft Company, Long Beach, CA 90810 **[105]**

**J.S. Ploem,** Department of Histochemistry and Cytochemistry, Leiden University, AL Leiden 2333, The Netherlands **[289]**

**Paul Primakoff,** Department of Physiology, University of Connecticut Health Center, Farmington, CT 06032 **[477]**

**Margaret Rheinschmidt,** Department of Laboratory Medicine, University of California at San Francisco, San Francisco, CA 94143 **[567]**

**Mario Roederer,** Department of Biological Sciences, Center for Fluorescence Research in Biomedical Sciences, Carnegie-Mellon University, Pittsburgh, PA 15213 **[545]**

**Anthony Ruggiero,** Department of Chemistry and Institute of Molecular Biology, University of Oregon, Eugene, OR 97403 **[159]**

**E.D. Salmon,** Department of Biology, University of North Carolina at Chapel Hill, Chapel Hill, NC 27514 **[377]**

**G. Alan Schick,** Department of Chemistry, Center for Fluorescence Research in Biomedical Sciences, Carnegie-Mellon University, Pittsburgh, PA 15213 **[203]**

**Bruce E. Seligmann,** Laboratory of Clinical Investigation, N.I.A.I.D., N.I.H., Bethesda, MD 20205; present address: Department I.A.R., Ciba-Geigy Corporation, Ardsley, NY 10502 **[531]**

**John A. Steinkamp,** Experimental Pathology Group, Los Alamos National Laboratory, Los Alamos, NM 87545 **[585]**

**Anita P. Stevenson,** Experimental Pathology Group, Los Alamos National Laboratory, Los Alamos, NM 87545 **[585]**

**Carleton C. Stewart,** Experimental Pathology Group, Los Alamos National Laboratory, Los Alamos, NM 87545 **[585]**

**Lubert Stryer,** Department of Cell Biology, Sherman Fairchild Center, Stanford University School of Medicine, Stanford, CA 94305 **[129]**

**Lerson Tanasugarn,** Department of Biological Sciences, Center for Fluorescence Research in Biomedical Sciences, Carnegie-Mellon University, Pittsburgh, PA 15213 **[347]**

**D. Lansing Taylor,** Department of Biological Sciences, Center for Fluorescence Research in Biomedical Sciences, Carnegie-Mellon University, Pittsburgh, PA 15213 **[xv, 257, 347]**

**Roger Tsien,** Department of Physiology-Anatomy, University of California, Berkeley, CA 94720 **[531]**

**Tadashi Uemura,** Department of Biophysics, Faculty of Science, Kyoto University, Kyoto 606, Japan **[321]**

**Pat Wadsworth,** Department of Biology, University of North Carolina at Chapel Hill, Chapel Hill, NC 27514 **[377]**

**Alan S. Waggoner,** Department of Biological Sciences, Center for Fluorescence Research in Biomedical Sciences, Carnegie-Mellon University, Pittsburgh, PA 15213 **[xv, 3]**

**John E. Wampler,** Department of Biochemistry, Instrument Design Group, University of Georgia, Athens, GA 30602 **[301]**

**Bennie R. Ware,** Department of Chemistry, Syracuse University, Syracuse, NY 13210 **[141]**

**T.H. Watts,** Department of Chemistry, Stanford University, Stanford, CA 94305 **[499]**

**Watt W. Webb,** Department of Applied and Engineering Physics, Cornell University, Ithaca, NY 14853 **[405]**

**Gregorio Weber,** Department of Biochemistry, School of Chemical Sciences, University of Illinois at Champaign-Urbana, Urbana, IL 61801 **[601]**

**H. Alexander Wilson,** Laboratory of Microbial Immunity, N.I.A.I.D., N.I.H., Bethesda, MD 20205 **[531]**

**Mitsuhiro Yanagida,** Department of Biophysics, Faculty of Science, Kyoto University, Kyoto 606, Japan **[321]**

**Dejan Zecevic,** Department of Physiology, Yale University School of Medicine, New Haven, CT 06510 **[423]**

# Preface

The application of luminescence techniques to investigate the chemical and molecular details of life processes has become very exciting in the last few years. Fluorescence spectroscopy has emerged as one of the powerful luminescence methods in the biomedical sciences because of its sensitivity, specificity, temporal resolution, potential for yielding spatial information, and finally because of the power of spectroscopic measurements. Fundamental problems in the life sciences can now be addressed with techniques involving fluorescence spectroscopy. This book is designed to introduce the general biomedical research scientist to the potential of fluorescence techniques. We have attempted to integrate the three principal areas of fluorescence technology as practiced today: fluorescent probes and solution spectroscopy; quantitative fluorescence microscopy and imaging techniques; and flow cytometry. General review chapters are included along with selected research papers in these three areas.

This volume evolved out of an international meeting held in Pittsburgh in the spring of 1985. The meeting and this book reflect the philosophy of the editors in approaching the use of fluorescence spectroscopy in the life sciences. Major advances in the technologies and in the understanding of specific biomedical problems could occur rapidly when the heretofore distinct fields of chemistry, biology, physics, computer science, engineering and medicine are integrated. This philosophy is encouraged at Carnegie-Mellon University through the leadership of its President, Richard M. Cyert. Dr. Cyert's vision made possible the creation of the Center for Fluorescence Research in the Biomedical Sciences at Carnegie-Mellon in 1982. We respectfully dedicate this volume to him. In addition, we thank the members of the science and engineering departments for their support and encouragement.

D. Lansing Taylor
Alan S. Waggoner
Robert F. Murphy
Frederick Lanni
Robert R. Birge

# Acknowledgments

The International Meeting and thus this volume were made possible by the generous support of the following:

Corporate Sponsors and Exhibitors:

Abbott Laboratories
Abbott Park
North Chicago, IL 60064

Becton Dickinson
Immunocytometry Systems
2375 Garcia Drive
Mountain View, CA 94043

Colorado Video, Inc.
17 Amble Road
Chelmsford, MA 01824

Coulter Corporation
Epics Division
P.O. Box 4486
Hialeah, FL 34014

Dage/MTI, Inc.
208 Wabash Street
Michigan City, IN 46360

Interactive Video
Systems, Inc.
358 Baker Avenue
Concord, MA 01742

Iss, Inc.
2010 Crescent Drive
Champaign, IL 61821

Nikon, Inc.
623 Stewart Avenue
Garden City, NY 11530

Omega Optical, Inc.
3 Grove Street
P.O. Box 573
Brattleboro, VT 05301

Pandex Laboratories, Inc.
909 Orchard Street
Mundelein, IL 60060

Photochemical Research
Associates, Inc.
45 Meg Drive
London, Ontario, Canada N6E 2V2

Photonic Microscopy, Inc.
2625 Butterfield Road
Suite 204 South
Oak Brook, IL 60521

Quantex Corporation
252 North Wolfe Avenue
Sunnyvale, CA 94086

Spex Industries, Inc.
3880 Park Avenue
Edison, NJ 08820

Vicom Systems, Inc.
2520 Junction Avenue
San Jose, CA 95134

Carl Zeiss, Inc.
One Zeiss Drive
Thornwood, NY 10594

Corporate Donors:

Applied Biosystems, Inc.
850 Lincoln Center Drive
Foster City, CA 94404

Eastman Kodak Company
343 State Street
Rochester, NY 14650

E.I. Du Pont de Nemours & Co.
Biomedical Products Dept.
549 Albany Street
Boston, MA 02118

Alan R. Liss, Inc.
Scientific, Medical and
Scholarly Publications
41 East 11th Street
New York, NY 10003

Molecular Probes, Inc.
24750 Lawrence Road
Junction City, OR 97448

Perkin-Elmer Corporation
Oak Brook Instrument Division
2000 York Road
Oak Brook, IL 60521

Smith, Kline Beckman Corp.
Corporate Science &
Technology Center
One Franklin Plaza
P.O. Box 79929
Philadelphia, PA 19101

Grant Support:

National Science Foundation
Argonne University Trust

# PART I
# Fluorescent Probes and Spectroscopy

*Applications of Fluorescence in the Biomedical Sciences, pages 3–28*


# 1 Fluorescent Probes for Analysis of Cell Structure, Function, and Health by Flow and Imaging Cytometry

*Alan S. Waggoner*

The use of fluorescent probes is a "sensitive method" to get information about the structure, function, and health of cells. Actually, the method can be sensitive in two ways. First, it is possible to *detect* very few fluorescent molecules in a cell. Barak and Webb [1981] demonstrated that less than 50 fluorescent lipid analogs associated with the low density lipoprotein (LDL) receptor could be seen with a SIT camera on a fluorescence microscope. A key factor for the sensitivity of this experiment was the localization of the fluorescent lipid analogs to an area less than 0.1 $\mu$m in diameter, so that the emitted photons could be focussed on a small region of the photocathode of the camera. Concentration of the probe signal in just a few pixels is important for detection because the number of "signal photons" per pixel must be larger than the number of "noise photons" that come from the combination of camera-associated electrical noise, photon noise from lense glare and filter leakage, and autofluorescence from endogenous cell chromophores. These sources of background are apparently small when integrated over the few pixels containing the image of an LDL receptor which has 50 probe molecules, so the probe signal stands out. Detection sensitivity is less in flow cytometry because one generally visualizes fluorescence from a whole cell and at least several thousand probes must usually be on the cell before the signal is larger than the noise. Mathies and Stryer, and Bright and Taylor in this volume discuss detection sensitivity more fully.

Fluorescent chromophores for labeling antibodies, DNA, and lipids are designed for high detection sensitivity. They have large extinction coeffi-

Department of Biological Sciences, Center for Fluorescence Research in Biomedical Sciences, Carnegie-Mellon University, Pittsburgh, Pennsylvania 15213.

cients, high quantum yields, and are *insensitive* to pH and the polarity of their local molecular environments.

The second way fluorescence can be sensitive is in the ability of probe molecules to respond spectroscopically to subtle changes in their molecular-environment. The dependence of absorption wavelength, extinction, emission wavelength, quantum yield, excited-state lifetime, and emission polarization on fluorescent probe microenvironment provides a powerful method for understanding the behavior of cells. One of the key requirements for exploiting the sensitivity of probes of microenvironment is the careful design of the fluorophore structure so that it can be targeted to the site of interest in the cell—for example, plasma membrane vs. endoplasmic reticulum, or mitochondria vs. cytoplasm, or one class of lysosomes vs. another. Therefore, the skills of the organic chemist must be blended with those of the spectroscopist, biochemist, and cell biologist to optimally exploit this potential of fluorescent probes.

## IDEAL PROPERTIES OF FLUORESCENT PROBES

### Large Extinction Coefficient ($\epsilon$) at the Region of Excitation

The value of $\epsilon$ should be at least as high as that for fluorescein (60,000 liters/mol cm) and ideally as high as the multichromophore phycobiliproteins ($2 \times 10^6$ liters/mol cm) [Glazer and Stryer, 1984] (Mathies and Stryer, this volume) if the probe is to be used as a label in a sensitive protein or nucleotide detection experiment.

### High Quantum Yield ($\Phi$)

$\Phi$ should be greater than 0.3 when the probe is bound to the target and is in the solvent environment where the fluorescence measurement is to be made. For example while various fluorescein-labeled proteins have quantum yields in the range of 0.2 to 0.7 at pH 8, the $\Phi$ drops off rapidly with decreasing pH. Rhodamine, on the other hand, is much less sensitive to pH and is still fluorescent in acidic compartments of cells or in the presence of acidic fixatives.

### Optimal Excitation Wavelength

Cells excited at wavelengths below 500 nm produce considerable autofluorescence that arises mainly from flavins, flavoproteins, NADH, etc. [Aubin, 1979; Benson et al, 1979]. In situations where autofluorescence can swamp the probe fluorescence, it is useful to have the probe absorb at wavelengths above 500 nm.

### Photostability

Fluorescein can survive between $10^4$ and $10^5$ excitations before decomposing (Mathies and Stryer, this volume). As Mathies and Stryer discuss, photostability is important for detecting a small number of probes in solutions. Photostability is also a concern for visualization of tagged materials with conventional fluorescence microscopes. For example, the fading of fluorescein fluorescence under intense microscope illumination is well known and numerous chemical approaches have been taken with various degrees of success to reduce fading [Picciolo and Kaplan, 1984; Bock et al., 1985]. Removal of oxygen is the most effective step that can be taken but this is impossible in studies of most living cells. Photofading is generally not a problem in flow cytometry because the stained sample only remains in the laser beam a short time.

### Excited-State Lifetime ($\tau$)

In flow cytometry the greatest sensitivity can be achieved with a combination of high laser power and probes with short fluorescence lifetimes. Under saturation conditions, the strong illumination of a focused laser spot can excite dye molecules immediately after they have fluoresced or relaxed from a previous excitation. Thus, as a cell passes through the beam the greatest number of excitations can occur when a dye has a very short $\tau$. Most highly fluorescent molecules don't have lifetimes much shorter than a nanosecond. Fluorescein and pyrene have emission lifetimes of about 4 and 100 nsec, respectively. At the other extreme, very long lifetimes can also be valuable in high-sensitivity detection. In these measurements emission is detected with a gated photomultiplier tube well after the decay of a short-excitation pulse. Thus the scattered light and background fluorescence (autofluorescence) is eliminated from the signal. Europiun chelates in special micellar solvents have lifetimes as long as 100 $\mu$sec [Hemmilä et al., 1984]. The sensitivity of this approach has been exploited commercially in immunoassay systems (LKB incorp).

### Minimal Perturbation by Probe

The probe should not perturb the function of the cell, organelle, or target molecule by reacting with key groups in active sites or by causing nonspecific binding of a labeled protein, or by causing steric perturbations because of its size. The probe should also not be phototoxic.

Few fluorescent probes have all the ideal properties but a number have proved to be very valuable and are widely used.

## FLUORESCENT PROBES USED AS COVALENT LABELS

In addition to the ideal properties mentioned above, fluorescent chromophores that are to be used to tag proteins, polynucleotides, lipids, or other biological molecules must have the following properties.

The labeling reagent must have the proper *reactivity* and *selectivity*. If too reactive, the tagging reagent hydrolyzes before binding to the protein. Only a few functional groups have the appropriate reactivity. Isothiocyanates, chlorotriazinyl derivatives, and hydroxysuccinimido active esters are the most common functional groups that allow chromophores to be attached covalently to primary amino groups of proteins. Generally, the reactions are carried out in aqueous solutions at pH 8.5–9.0, at which pH amino groups spend a significant time in the uncharged (and reactive) form. Iodoacetamido and maleimido functional groups on chromophores can be used to form linkages to protein sulfhydryl groups. This chemistry has been reviewed by Haugland [1983].

Tagging reagents should be soluble in aqueous solutions used for labeling.

And finally, the fluorescence of the label should be insensitive to solvent pH and polarity.

Certainly the most commonly used fluorescent tag is fluorescein isothiocyanate. This molecule is mainly selective for labeling primary, neutral amino groups. It has been attached to antibodies, lectins, avidin hormones, biological macromolecules, and lipids (see Haugland [1983] for general review of fluorescent probes). Properties that make fluorescein a valuable tag are its reasonably large extinction coefficient, high quantum efficiency, and its water solubility, which makes the labeling procedure easy. Antibodies, lectins, and avidin can be labeled with two to five fluorescein molecules before quenching begins to occur as a result of close proximity of tags on the macromolecule surface.

In spite of its general usefulness, there are cases where fluorescein has disadvantages. It is not particularly photostable, its absorption spectrum falls in a region where cell autofluorescence is produced, and its fluorescence is diminished as the pH drops below 8.

Other fluorescent probes have been developed to either improve upon the properties of fluorescein or to provide additional reagents which absorb and emit at different wavelengths that can be detected simultaneously. A number of these reagents are listed in Figure 1. Rhodamine derivatives are generally more stable than fluorescein, excite in spectral regions where less autofluorescence is produced, and are not sensitive to pH in the physiological range. However, problems with the limited water solubility of rhodamine sometimes make this reagent less attractive for labeling macromolecules. Lissamine rhodamine and Texas Red are more water soluble and absorb further to the red.

The phycobiliproteins are unique in their strong absorbing and fluorescing capabilities [Oi et al, 1982; Glazer and Stryer, 1984; Mathies and Stryer, this volume]. Methods were developed for coupling these large protein-chromophore complexes to antibodies and the phycobiliprotein-tagged antibodies have become very popular for flow cytometric studies. Of particular value

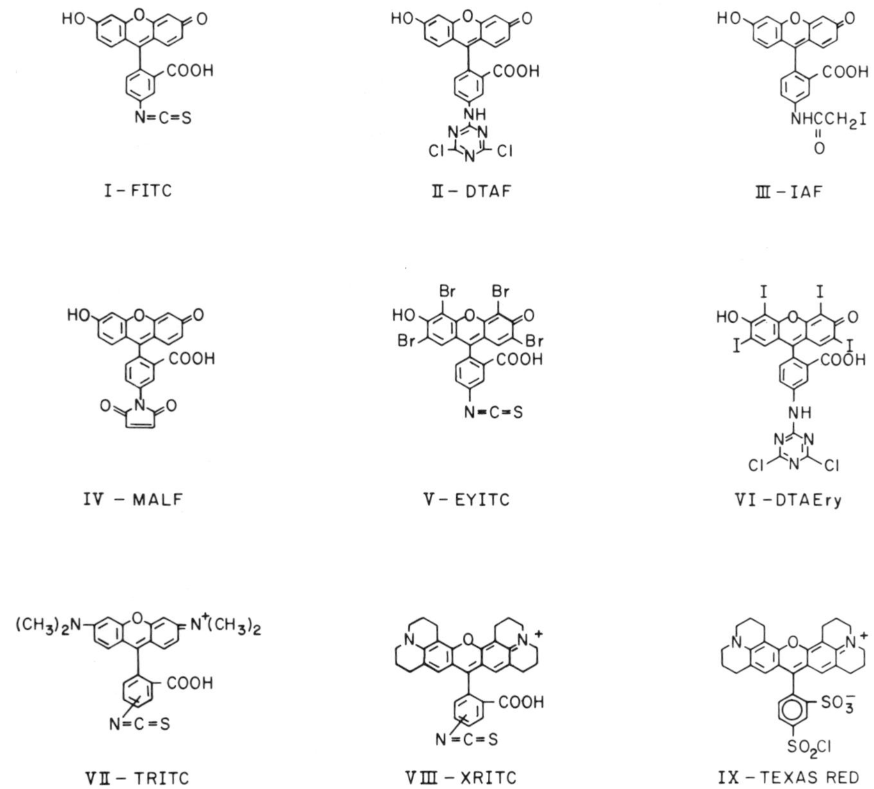


Fig. 1. Fluorescent labeling probes. I, fluorescein isothiocyanate; II, dichlorotriazinylaminofluorescein; III, iodoacetamidofluorescein; IV, maleimidofluorescein; V, eosin Y isothiocyanate; VI, dichlorotriazinylaminoerythrosin; VII, tetramethylrhodamine isothiocyanate; VIII, rhodamine isothiocyanate analog; IX, Texas Red, a trademark of Molecular Probes, Inc.

are phycoerythrin-tagged antibodies which can be used simultaneously with fluorescein-tagged antibodies for two-color analysis of cells [Kronick and Grossman, 1983]. Both tags can be excited at 488 nm but the phycoerythrin has a large Stokes shift and its red orange emission can be detected independently of the yellow green fluorescein emission.

More labelling reagents need to be developed. Photostable and environment-insensitive chromophores are needed to improve the power and versatility of flow cytometry and fluorescence microscopy. Probes that absorb and emit further to the red (even into the near infrared) than those available now would be valuable for a number of reasons. For example, it is conceivable that fluorescence microscopes equipped with red-sensitive intensified cameras or charge-coupled device (CCD) cameras could quantify the fluorescence of six to eight different-colored probes on a single preparation just by sliding in this many filter combinations. There is no reason to exclude probes that absorb as far to the red as 800–850 nm. They could even be used with flow cytometers "built on chips" which incorporate laser diodes for excitation and solid-state detectors.

Fluorescent probes can also be used to localize and monitor interesting biological processes in living cells. Endocytic processing of receptors and ligands, e.g., insulin and EGF, has been studied (see Murphy, this volume). Taylor and his associates as well as others have followed the movement, localization, and association of tagged cytostructural proteins which have been inserted into living cells (see Taylor et al., this volume).

Detection of specific genes in situ with biotin-derivatized DNA probes and fluorescein-tagged avidin first occurred in 1981 [Langer et al.]. Subsequently, hapten molecules were attached to DNA probes so that they could be detected with fluorescent monoclonal antibodies. The result is a powerful technology for detection of infectious agents, oncogenes, and genetic defects in tissues, cells, and fluids [Brigati et al., 1983]. Another important advance was the development of fluorescein-tagged antibodies to quantify incorporation of bromodeoxyuridine into DNA [Dolbeare et al., 1983]. This reagent is valuable for identifying S-phase cells in cell kinetic analysis.

There is no doubt that there is a bright future for the application of fluorescent probes for analysis of antigens, genes, and biochemical processes in living and fixed cells and tissues. Still needed is further development of fluorescent, phosphorescent, and chemiluminescent chromophores with extended spectroscopic properties which can be covalently attached to biological molecules without interfering with their function. Interested organic and inorganic chemists, take note!

## NONCOVALENTLY ASSOCIATING FLUORESCENT PROBES

Because of their particular molecular composition, fluorescent probes in this class associate noncovalently with special structures in cells. Structures that can be visualized with these probes include DNA, RNA, lipid, electrically negative or positive compartments like mitochondria, and compartments with a low pH.

### DNA and RNA Probes

Some DNA probes are known to intercalate between the base pairs. Ethidium and propidium are the classical examples (Fig. 2). Biophysical studies have elucidated the binding and explained the absorption shifts and quantum yield changes that occur upon intercalation [LePecq and Paoletti,

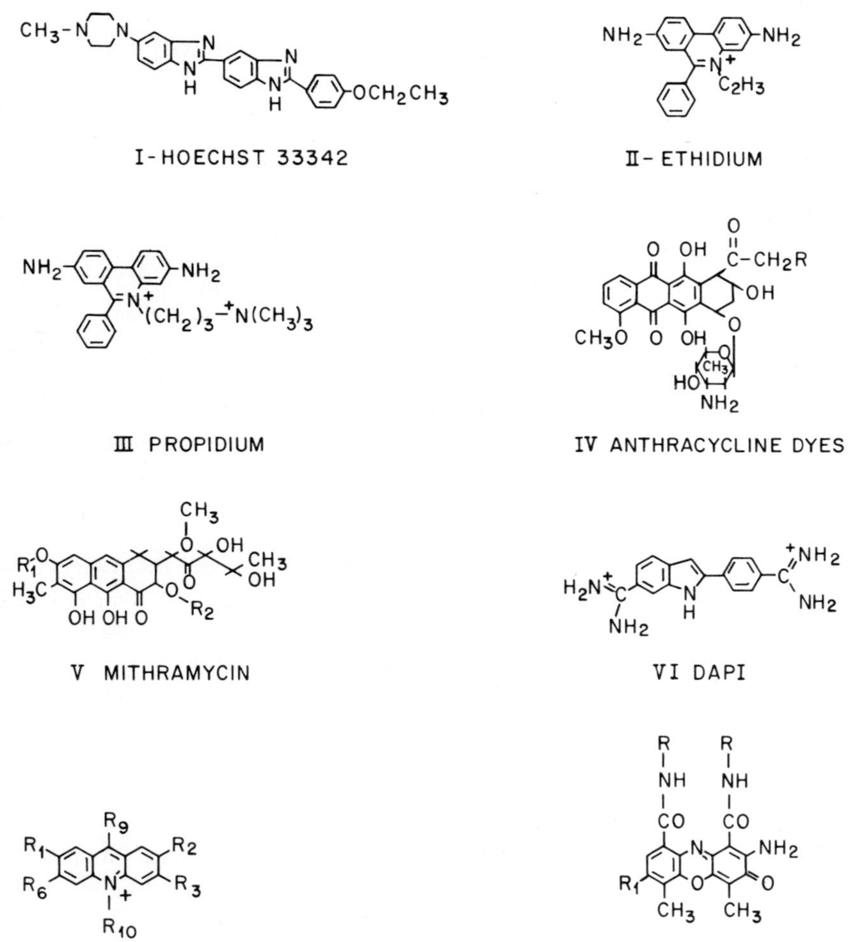


Fig. 2. Fluorescent probes that bind noncovalently to DNA. For the anthracycline dyes (IV) the R-group is −H on Daunomycin and −OH on Adriamycin. The R-groups of mithramycin (V) are 2- and 3-unit oligosaccharides. The acridines (VII) have a wide variety of R-groups including alkyl, amino, alkylamino, methoxy, and halogen substituents. The R-groups on Actinomycin D (VIII) are cyclic polypeptide chains and $R_1$ is a hydrogen.

1967]. Propidium is frequently used to quantify the DNA in cell by flow cytometry [Muirhead et al., 1985]. It does not readily cross cell membranes; therefore, permeabilization procedures such as fixation are necessary to provide access to the nucleus. (Propidium readily crosses the membranes of "dead" cells to make them highly fluorescent. This latter property is useful in immunology for cell killing assays [Muirhead et al., 1985].)

Hoechst 33342, unlike propidium, slowly crosses intact membranes and is useful for DNA analysis in live cells. Thus living cells in different stages of the cell cycle can be sorted as well as analyzed on the basis of their DNA content as measured by Hoechst 33342 [Arndt-Jovin and Jovin, 1977]. Hoechst 33342, DAPI, propidium, and acridine orange have been extensively used in flow cytometry to determine the number of aneuploid cells in dispersed tumor samples [Muirhead et al., 1985]. The degree of aneuploidy may have prognostic value in cancer treatment. Many other uses of DNA-specific probes have been reviewed elsewhere [Muirhead et al., 1985; Shapiro, 1985; Latt et al., 1984; Steinkamp, 1984; Darzynkiewicz et al., 1984].

Precise measurements of DNA content often require that RNA be removed enzymatically because of the affinity of certain probes for double-stranded RNA. This is particularly true for analysis using acridine orange. It is interesting, however, that acridine orange binds differently to single-stranded RNA and to DNA. The green fluorescence of acridine orange intercalating in DNA can be readily measured separately from the red fluorescence of acridine orange molecules associated with single-stranded RNA; hence, DNA and RNA can be quantified simultaneously. Darzynkiewicz and his colleagues have exploited this property to study the production of RNA during the cell cycle [Darzynkiewicz et al., 1980]. They found evidence for new cell cycle compartments by flow cytometric analysis of acridine-orange-stained cells. These investigators have also contributed to our knowledge of the mechanism of the association of acridine orange with RNA to produce red fluorescence [Kapuscinski et al., 1982].

In a similar approach Shapiro has used Pyronin Y as an RNA probe along with methyl green or Hoechst 33342 to block Pyronin Y binding to DNA. This author describes his method and provides insights into other approaches in his recent text on flow cytometry [Shapiro, 1985]. As an extension of this work, Shapiro and Stephens [1986] have discovered helium-neon-laser-excitable laser dyes that can be useful for quantifying DNA. This advance leaves open shorter-excitation wavelengths for measurement of other cell parameters with flow cytometers.

## Membrane Potential Probes

Membrane potential is an important property of cells and some organelles. The electrical potential difference across the plasma membrane is directly involved in the transport of ions and nutrients into and out of the cell. Modulation of the cell membrane potential is the mechanism of nerve conduction and a part of sensory transduction. The electrical potential difference works together with the pH gradient across the mitochondrial membrane to couple energy released during electron transport to the phosphorylation of ADP to form ATP. Quantification of membrane potentials can be accomplished with microelectrodes in large cells, but in small cells and organelles, membrane potentials are estimated from the transmembrane distribution of membrane-permeant cations and anions that are either radioactive or fluorescent. Cyanine dyes (Fig. 3) were the first fluorescent membrane potential probes to have their mechanism determined [Sims et al., 1974] and they have been used extensively since [Waggoner, 1979, 1985]. The cationic cyanine dyes are accumulated in electrically negative compartments such as cell interiors and mitochondria. Depolarization of the cell membrane or uncoupling of mitochondria produces loss of fluorescence as cyanine dye leaks out. Rhodamine 123 is also a membrane potential stain exploited by Chen and his colleagues for visualization of mitochondria in living cells under the fluorescence microscope [Johnson et al., 1981]. Rhodamine 123 has been termed a mitochondria-specific dye because the mitochondria are so brightly stained. Cyanine dyes also yield bright mitochondria in healthy cells. Undoubtedly, the high fluorescence of mitochondria arises because they have a negative membrane potential. Uncoupling agents and inhibitors reduce mitochondrial fluorescence. The large membrane surface area in the mitochondrial matrix may contribute to the staining by binding large amounts of accumulated probe. Rhodamine 123 retention by tumor cell lines has been studied and related to anticancer drug sensitivity [Lampidis et al., 1982]. Other uses of this mitochondria stain have been reviewed recently [Weiss and Chen, 1984].

Cyanine dyes and other membrane potential probes have been used in flow cytometry experiments to study changes in neutrophils, platelets, and lymphocytes following stimulation of the cells with formylated peptides, platelet release factors, and mitogen, respectively. The fluorescent changes are substantial but the mechanisms are not well understood [Waggoner, 1985]. A major complication is the possible contribution of mitochondria to the total cell fluorescence change. Chused et al. (this volume) show that anionic oxonol dyes may provide answers to this question. The large fluorescence changes undoubtedly reflect important physiological changes taking place in cells. It will be interesting to eventually determine the nature of these changes and how probe fluorescence is being altered. Exciting advances have been made in the use of fluorescent membrane potential probes to study excitable membrane behavior and to map neural circuits. This work has been pioneered by Cohen, who describes the work in this volume (London et al.).

## Probes for Visualization of Membranes and Lipid Compartments

Fluorescent probes for lipid bilayers (Fig. 4) have been available for some time [Waggoner and Stryer, 1971]. Many are uncharged hydrocarbons like diphenyl hexatriene (DPH). Others are charged but have long hydrocarbon

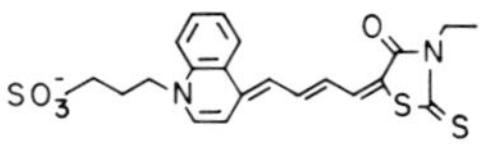

WW375

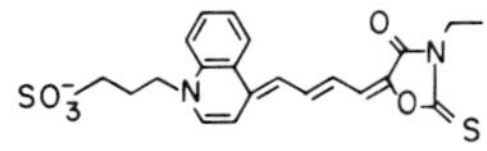

NK 2367

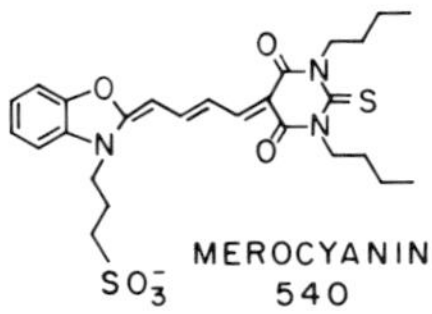

MEROCYANINE 540

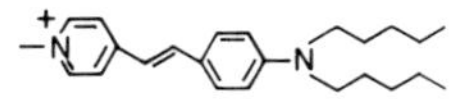

di-$C_5$-ASP

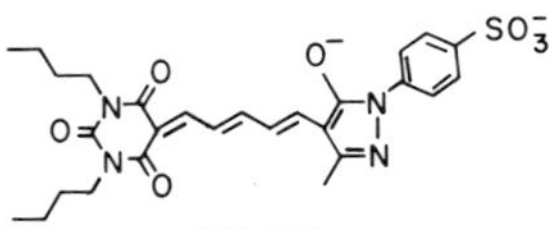

WW781

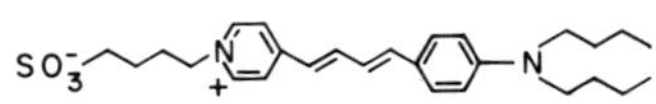

RH 160

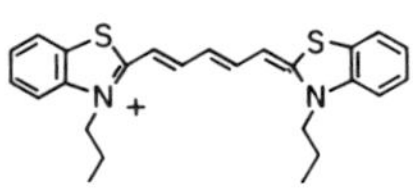

diS-$C_3$-(5)

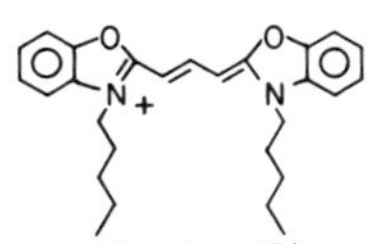

diO-$C_5$-(3)

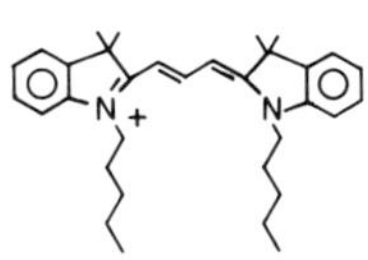

diI-$C_5$-(3)

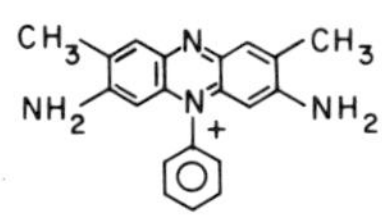

SAFRANIN O

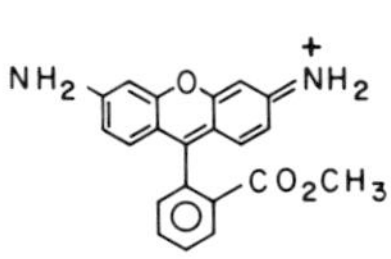

RHODAMINE 123

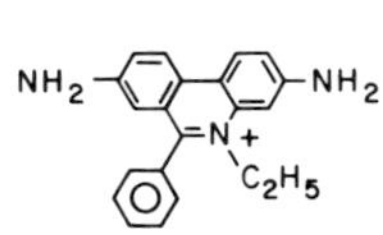

ETHIDIUM

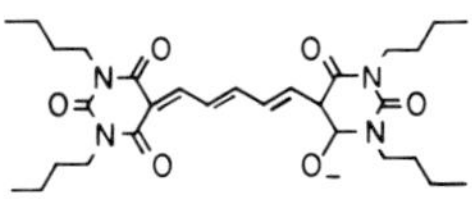

diBA-$C_4$-(5)

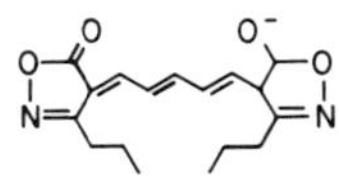

OX-VI

chains attached [e.g., dioctylindocarbocyanine, $diIC_{18}$ (3)]. Since not all lipid membranes in a cell are alike it is not surprising that fluorescent lipid probes with different structures partition between the different membranes in interesting ways [Hudson et al., this volume]. Pagano and his associates have made extensive use of fluorescent lipid analogs to trace pathways of membrane movement during endocytosis and recycling [see review by Pagano and Sleight, 1985]. Rhodamine- and Lucifer-yellow-labeled gangliosides have been used as probes for the retention and organization of fibronectin by ganglioside-deficient mouse cells [Spiegel et al., 1985]. Cyanine-dye-based lipid analogs seem to localize preferentially in fluid or gel regions of lipid bilayers depending on the length of the long alkyl chains attached to the dye structure [Klausner and Wolf, 1980; Packard and Wolf, 1985]. The hydrophobic dye Nile Red [Greenspan et al., 1985] has been shown to be highly specific for lipid oil droplets in cells. It has recently been reported that cyanine dye, $diOC_5(3)$ is useful for visualizing endoplasmic reticulum [Terasaki et al., 1984]. This staining is probably related to the lipid solubility of this dye rather than to its membrane potential sensitivity.

## pH Probes That Partition Between Compartments With Different pH

There are two classes of pH probes (Fig. 5). One class of fluorophores have *spectral properties* that change with pH, e.g., fluorescein. These fluorophores will be discussed in the next major section, which centers on environment-sensitive fluorescent probes.

A second class of pH probes *partition* differently into cell compartments with different internal pH. The pH of each compartment can be estimated by the amount of fluorescent weak acid or weak base that distributes into that compartment. Weak bases cross membranes as the conjugate base (usually the whole molecule is neutral) but become trapped in acidic compartments when protonated. The converse is true for weak acids. Acridine orange is accumulated in neutrophil granules [Abrams et al., 1983] and acidic lysosomal compartments presumably for this reason. Unfortunately, the lysosomal fluorescence is masked or complicated by acridine orange fluorescence from RNA and DNA; 9-amino acridine also is accumulated in acidic compartments

---

Fig. 3. Membrane potential probes. The top six structures are "fast dyes" that respond to membrane potential changes that take place in microseconds. These probes are membrane impermeant and are useful for detecting action potentials in excitable membranes. The work of London et al. in this volume illustrates the use of these probes. The bottom eight structures are "slow dyes" that redistribute across membranes over a period of milliseconds to seconds and accumulate in cells or cellular compartments according to the membrane potential difference. Chused et al. discusses the application of several of these probes in this volume. These membrane potential probes are also discussed by Waggoner [1985].

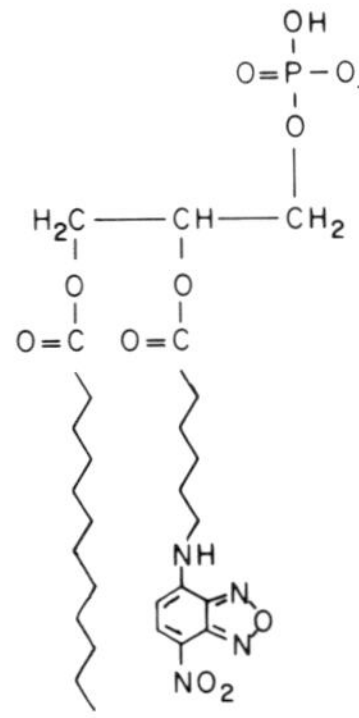

C$_6$ NBD PA

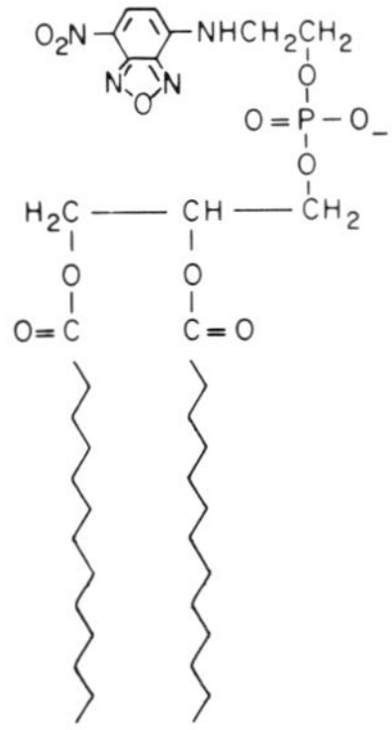

NBD PE

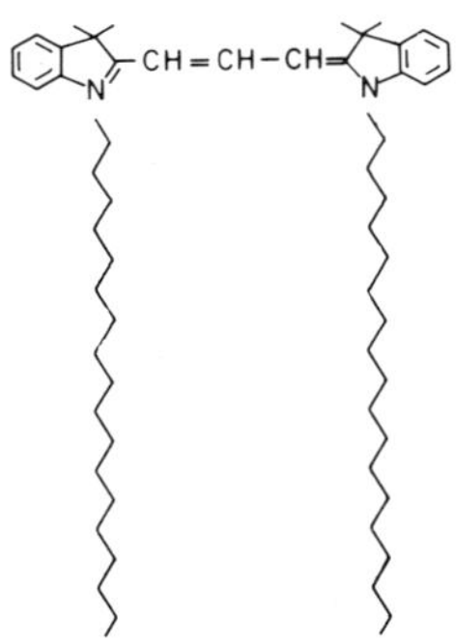

diI C$_{18}$-(3)

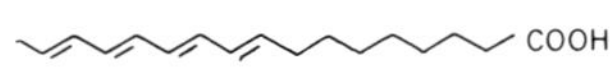

TRANS-PARANARIC ACID

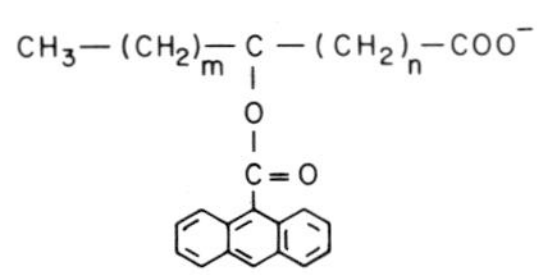

(m,n)-ANTHROYLSTEARATE

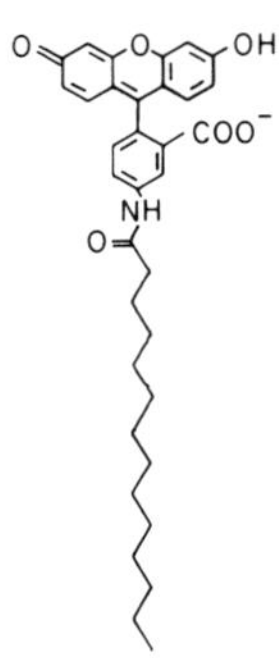

ACYL AMINOFLUORESCEIN

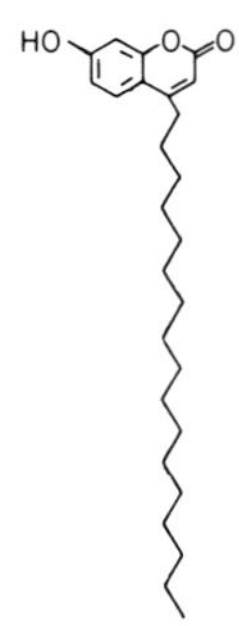

4-ALKYL UMBELLIFERONE

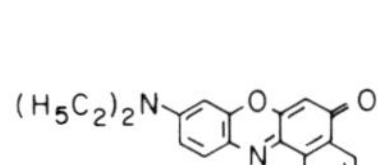

NILE RED

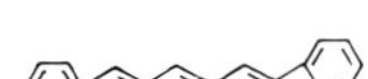

DPH

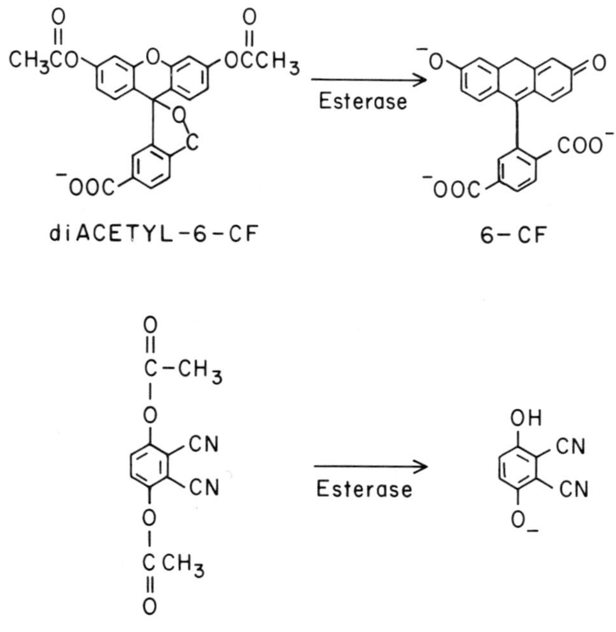


Fig. 5. pH probes. Two reactions illustrate the conversion, by cellular esterase enzymes, of membrane-permeant ester molecules into fluorescent pH probes that are membrane impermeant because of their increased charge. The trapped products are 6-carboxyfluorescein (top) and 2,3-dicyano-1,4-dihydroxybenzene (bottom).

[Deamer et al., 1972] but unless it is used at extremely low concentrations the fluorescence of accumulated 9-amino acridine molecules is actually reduced as a result of concentration quenching inside the compartment. It would be valuable if there were specific fluorescent probes for acidic and basic compartments. This should provide another area for collaboration of organic chemists and spectroscopists.

---

Fig. 4. Lipid probes. The top two probes have been exploited by Pagano and his colleagues [Pagano and Sleight, 1985]. Barak and Webb [1981] used $diIC_{18}$-(3) to study the LDL receptor in cultured cells. Paranaric acid is discussed by Hudson in this volume. The anthroylstearates have been used to study lipid fluidity and organization as a function of depth in bilayer membranes [Blatt and Sawyer, 1985]. The pH-sensitive fluorescein and umbelliferone chromophores with long alkyl chain substituents can be used to study the effects of membrane surface potential on the pH of aqueous regions close to the membrane surface [e.g., Fromhertz and Masters, 1972]. Nile Red is a lipid droplet probe [Greenspan et al., 1985]. Diphenylhexatriene (DPH) is a membrane lipid fluidity probe discussed by Lakowicz in this volume.

## PROBES THAT ARE SENSITIVE TO THEIR MICROENVIRONMENT

Fluorescent chromophores can be used to estimate the following properties of their local environment: pH, calcium concentration redox potential, polarity, fluidity, and the presence of other ions or molecules that can deactivate the excited state of the chromophore by energy transfer or charge transfer. The major challenge in making use of these abilities is finding a method, by synthetic modification of the probe or otherwise, to target the probe to the region of interest within the cell.

### pH Probes That Respond With Spectral Changes

The binding of hydrogen ions or metal ions changes the electronic structure of a number of dyes. The absorption properties, the fluorescence properties, or both, of the molecule can be affected depending on whether the hydrogen or metal ion concentration is near the dissociation constant of the ground state or the excited state, respectively [Schulman, 1976]. The rates of complexation and decomplexation relative to the excited-state lifetime of the probe also play a role in the sensitivity of fluorescence to ion binding. An example of a probe with an absorption spectrum sensitive to pH changes in the physiological range is 6-carboxy-fluorescein, which has a ground-state pKa near 6.5. The absorption spectrum shifts from the 440–450 region at more acid pH to near 490 as the pH is raised [Thomas et al., 1979]. Emission is not especially affected for reasons that have been discussed by Martin and Lindquist [1975]. Therefore, pH measurements are made by calculating the ratio of the emission produced with 496 excitation and the emission produced with 452 excitation. Ohkuma and Poole [1978] used a standard fluorometer to measure pH in endosomes. Heiple and Taylor [1980] and Tanasugarn et al. [1984] have demonstrated how this method can be used to determine cellular pH with a fluorescence microscope. Murphy et al. [1984] used an interesting variation of this technique to measure endosomal pH of single cells with a dual laser excitation flow cyctometer. Instead of exciting fluorescein at two wavelengths, a mixture of endocytosed fluorescein and rhodamine-labeled insulin molecules were excited. pH-insensitive rhodamine fluorescence provided a measure of the amount of the both ligands in each cell. The ratio of fluorescein emission to rhodamine emission gave the pH of the cell. In this volume, Taylor et al. more fully describe ratio imaging methods for estimating spatial distributions of pH in cells. Murphy, also in this volume, describes pH measurements by flow cytometry. Both of these authors have covalently attached fluorescein to biological molecules of interest in order to target the pH measurement to an interesting region.

There is also a useful pH probe that changes its *emission* properties in the physicological pH range. Valet et al. [1981] have shown that 2,3-dicyano

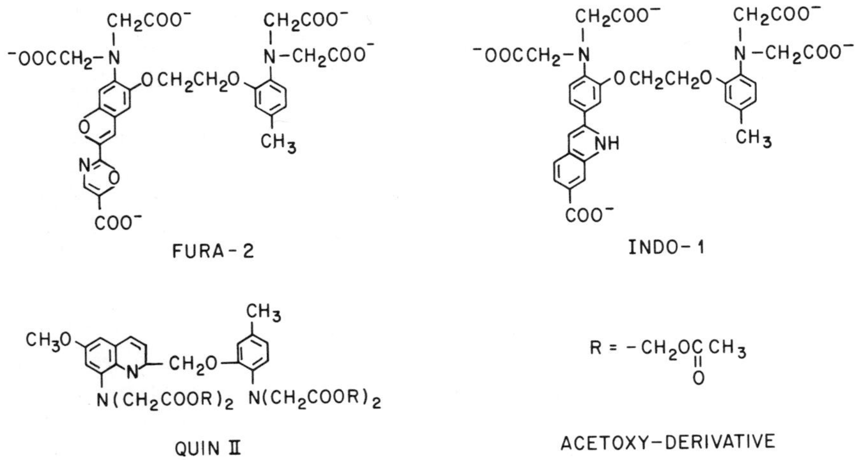


Fig. 6. Calcium probes; fura-2 and indo-1 can be obtained as acetoxyester derivatives, similar to the derivatized form of Quin II shown at the bottom left of this figure. The ester derivatives are membrane permeant. Cellular esterases convert the derivatives into the free-acid forms of the probes, which are trapped inside cells and are able to chelate calcium.

1,4-dihydroxy benzene can be excited at a single wavelength and that the emission spectrum shifts from 450 nm to 483 nm as pH increase [Kurtz and Balaban, 1985]. The excited-state pKa of this molecule is in the physiological range and proton exchange can occur within its excited-state lifetime. The probe is loaded into cells in the diacetyl form. Nonspecific esterases expose the pH-sensitive hydroxyl groups. Thus, this probe is expected to report the pH of esterase-rich regions including the cytoplasmic compartment. The ester-loading trick can be used with 6-carboxy fluorescein diacetate, also [Thomas et al., 1979]. Unfortunately, these probes may not be limited to any one compartment of most cells.

Hydroxypyrene trisulfonate [Clement and Gould, 1981] is pH sensitive in the physiological range. The molecule has the advantage that it does not leak out of cells but has the disadvantage that microinjection methods or other loading methods are required to get the probe into cells.

## Calcium Probes

Calcium, because of its important role in regulating many cellular processes, has received considerable attention in recent years [Tsien et al., 1984; Campbell, 1983]. The indicator Quin II, developed by Tsien [1980], has played an important role. The use of Quin II provides the first fluorescence method for detecting changes in cytoplasmic calcium [Tsien et al., 1984]. The structure of Quin II is shown in Figure 6. Its acetoxy ester precursor is membrane permeant. Cytoplasmic esterase activity liberates the carboxylate groups, which limit the escape of the probe into the medium and provide chelation groups for binding the calcium. Bound calcium shifts the absorption peak toward the red and measurements are usually made with excitation near 340 nm. Calcium determinations with Quin II, and its improved analog, fura-2 [Grynkiewicz et al., 1985], are best made by a ratio technique where emission is measured at the two optimal excitation wavelengths for probe with and without bound calcium. Quin II is sufficiently nonfluorescent so that cells must be loaded with 0.1–1.0 mM probe. The large Quin II concentration has worried some investigators because the probe is such a strong calcium buffer. The dissociation constant of the probe calcium complex is below $10^{-6}$ M. Thus, "makeup" calcium must enter the cytoplasm, which normally contains less than $10^{-6}$ M calcium, before the cytoplasmic Quin II is at equilibrium and calcium measurements can begin. Makeup calcium is provided in the medium during experiments. Several recent reports describing internal calcium concentration changes in platelets [Johnson et al., 1985] and 3T3 fibroblasts [McNeil et al., 1985; Moolenaar et al., 1984] suggest that Quin II signals may not be accurately indicating the magnitude or kinetics of the changes. Thus Quin II may be best for qualitatively detecting calcium level changes. The improved probe from Tsien's laboratory, fura-2, is more sensitive and can be used at lower concentrations [Grynkiewicz et al., 1985]. Another advance is the synthesis of indo-1, which shows an emission shift upon calcium binding [Grynkiewicz et al., 1985]. In flow cytometry this is a definite advantage since indo-1-stained cells can be excited with a single laser line and calcium concentration determined from the ratio at two emission wavelengths (see Chused et al., this volume). Care must be exercised in applying and interpreting the fluorescence signals of the new generation of fluorescent calcium indicators.

**Fluorescent Redox Potential Probes**

Neutrophils release superoxide and hydrogen peroxide when they are stimulated with chemotactic factors. Several fluorescent probes have been used to monitor the kinetics of this process in normal and diseased cells. Dichlorofluorescin (Fig. 7) can be loaded into cells in its acetylated form and trapped as a result of esterase activity. Bass and his associates [Bass et al., 1983] have shown that $H_2O_2$ released from stimulated neutrophils converts the nonfluorescent reduced form into dichlorofluorescein, thereby making stimulated cells fluorescent. Chemiluminescent methods have recently become popular for detecting peroxides [Campbell et al., 1985]. For instance, luminol was used as a luminescent probe to measure production of reactive oxygen metabolites from neutrophils upon stimulation with complement components [Campbell and Morgan, 1985]. Cellular NADH and NADPH

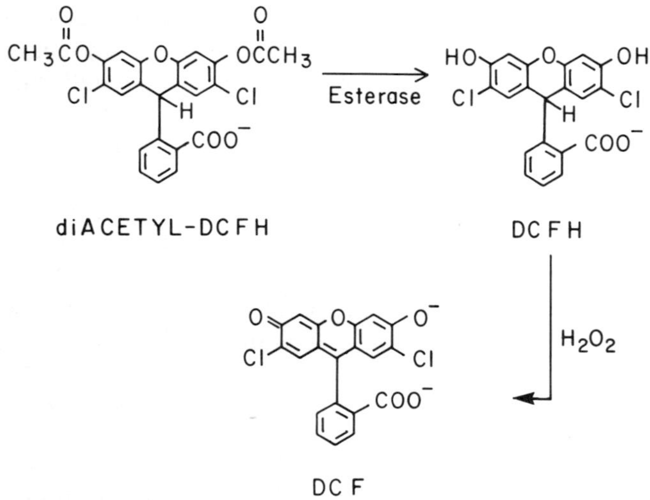


Fig. 7. Peroxide probe. Membrane-permeant diacetyl-dichlorofluorescin is converted by cellular esterases to dichlorofluorescin. Peroxides released upon neutrophil stimulation oxidize the nonfluorescent dichlorofluorescin to the fluorescent dichlorofluorescein. Details can be found in Bass et al. [1983].

are intrinsic fluorescent probes of the redox state of cells and have also been used to monitor the energy state of cells and tissues [Thorell, 1983].

## Fluorescent Probes for Measuring Enzyme Activities

Fluorogenic enzyme substrates are chromophores that are converted by specific enzymes into products that have either increased fluorescence or shifted spectra. These substrates can therefore be said to be sensitive to the presence of certain enzymes in their microenvironment. Dolbeare [1981] has provided an excellent review of the use of fluorogenic substrates.

Hydrolase enzymes play crucial roles in cell function. Lysosomal proteases, for example, are responsible for breakdown of intracellular proteins and peptides as well as polypeptide material that has been endocytosed [Barret, 1979; Holtzman, 1976]. Other proteases are bound to the external surfaces of cells or are secreted in soluble form. The latter enzymes break down extracellular protein molecules [Struli et al., 1980; Zucker-Franklin et al., 1981; Poole and Mort, 1981; Pietras et al., 1981]. The breakdown of extracellular macromolecules is required for movement of phagocytic cells like macrophages and neutrophils through tissues [Herscowicz et al., 1981; Wright, 1982]. It is also thought that the extracellular proteases play a significant role in tumor invasion, rheumatoid arthritis, and other disease states [Struli et al., 1980]. Thus, because of their significant roles in cell function it is important to be able to measure activity and localization of enzymes such as proteases, glycosidases, esterases, phosphodiesterases, phosphatases, sulfates, etc., in tissues, cells, organelles, and fluids.

Antibodies against proteases have come into wide use in the past 5 years for localizing and quantifying the enzymes in fixed tissues and cells and in serum. However, antibodies do not indicate whether the enzymes they bind to are functional (since the epitope on the enzyme is usually not associated with the active site of the enzyme) nor are antibodies sensitive to features involved in allosteric regulation of enzyme activity.

Most fluorogenic substrates have been constructed from chromophores that have either an amino or a hydroxyl group which participate in the electronic conjugation (Fig. 8). When an appropriate amino acid sequence is attached by an amide bond to an amino-chromophore, the molecule usually becomes much less fluorescent. Cleavage of the amide bond of the substrate by a peptidase that will act upon the particular chromophore-peptide complex releases the more fluorescent primary amine chromophore. Naphthylamine, amino coumarins, amino quinolines, and rhodamine have served as chromophore bases for fluorogenic peptidase substrates [Dolbeare, 1981; Leytus et al. 1983].

Esterase, glycosidase, sulfatase, phosphatase, and phosphodiesterase substrates can be made by attaching the appropriate functional group to a hydroxyl-containing chromophore as shown in Figure 8. Hydroxy coumarins and fluorescein have most often been used for construction of these substrates.

Fluorogenic substrates are now routinely used to measure activities of enzymes in fluids such as blood plasma [Smith, 1983]. However, these substrates have found little use in studies of enzyme activities in intact cells because they do not usually diffuse readily into cells and the fluorescent products rapidly leak out of the cells [Dolbeare and Vanderlaan, 1979; Dolbeare, 1981]. A variety of coupling reagents have been developed which convert products into insoluble colored substances that can be measured by absorbance techniques, but only one coupling reaction is available that results in an insoluble fluorescent product that stays within the cell. The later involves coupling of nitrosalisaldehyde to the methoxynaphthylamine product that is liberated by peptidase activity. This reaction, developed by Dolbeare and Smith [1977], is illustrated in Figure 8. It is surprising that little use has been made of this coupling reaction since it provides a way to produce fluorescence in cells that is proportional to enzyme activity.

A second disadvantage of presently available substrates is that most need to be excited with UV light and hence their signals may be obscured by cell autofluorescence. The most commonly used peptidase substrates are based on the coumarin and naphthylamine chromophores that are excited near 365 nm. A recent and significant improvement was made when peptidase sub-

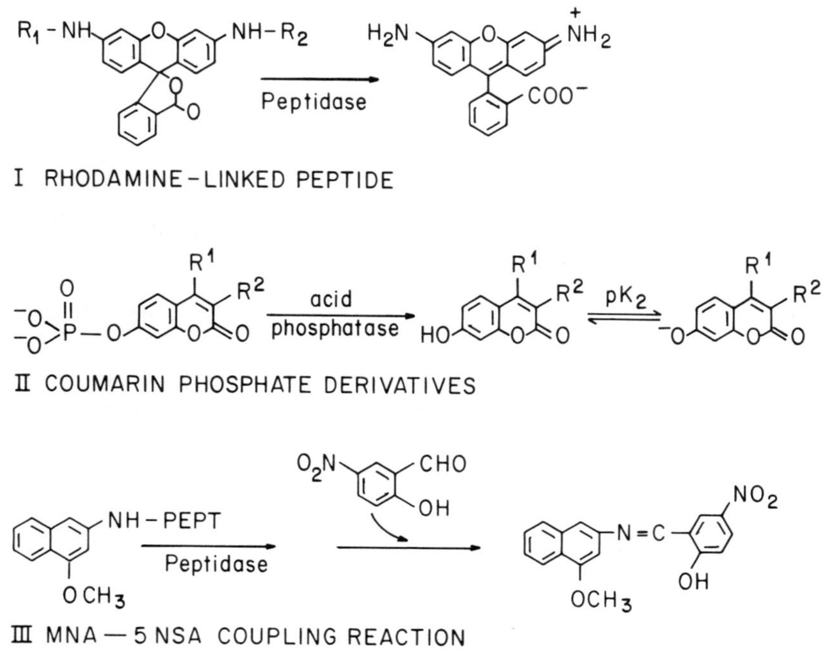


Fig. 8. Fluorogenic enzyme substrates. I, rhodamine-linked peptides developed by Leytus et al. [1983]. The R-groups are peptides specific for the enzyme of interest. II, acid phosphatase derivatives based on the coumarin structure. The R-groups are chosen to shift the fluorescence of these derivatives toward the visible region of the spectrum [Koller and Wolfbeis, 1985]. For I and II the products are more fluorescent than the substrates. III, the coupling reaction devised by Dolbeare and Smith [1977]. The free amino group, which is made available when the peptide bond of peptidyl-methoxynaphthylamine is hydrolyzed by a peptidase, couples with 5-nitrosalicaldehyde, which is present during the reaction, to form the insoluble red-orange-fluorescing product.

strates based on the rhodamine chromophore were developed by Leytus et al. in 1983. These substrates are excited at wavelengths near 490 nm. A limitation of presently available fluorogenic substrates is that none can be attached to peptide hormones or to phagocytosable particles like viruses, bacteria, and parasites so that the kinetics of lysosomal processing of these materials by living cells can be followed by sensitive fluorescent methods.

Certainly if the technical problems with fluorogenic substrates can be overcome there is potential for their wide use. Many hydrolases exist in cells and substrates need to be developed for the measurement of their activities in fluids, tissues, and cells.

## Energy Transfer

Fluorescent chromophores in the excited state are sensitive to the presence of nearby chromophores that can act as acceptors for resonance energy transfer. Energy transfer can occur between two chromophores separated by distances of tens of angstroms provided that the donor fluorescence spectrum overlaps significantly with the acceptor absorption spectrum. The relative orientation of the optical transition moments of the two chromophores also affects the efficiency of transfer. Energy transfer is a nonradiative process. That is to say, there is not absorption by the acceptor of a photon emitted by the donor. If the acceptor is fluorescent, however, it can emit a photon as a result of excitation of the donor. The theory and early uses of energy transfer to study protein and membrane structure have been reviewed elsewhere [Stryer, 1978; Yguerabide and Foster, 1981]. More recently the technique has been used to detect aggregation of cell surface components which were separately labeled with fluorescein (donor) and rhodamine (acceptor) [Fernandez and Berlin, 1976; Szollosi et al., 1984]. Energy transfer could also be used to look at self-assembly or binding events even inside living cells [Taylor et al., 1981].

Energy transfer can be a nuisance as well. For example, in experiments designed to quantify cell surface markers with antibodies tagged with different colored fluorophores, extensive capping or patching may, depending on the extent of tagging and the size of the antibody antigen complex, result in energy transfer. The quantity of marker tagged with the energy donor would be underestimated in this circumstance. This possibility is often overlooked in multicolor flow cytometry experiments.

In addition to energy transfer, it is possible for certain molecules and ions to quench fluorescence directly by deactivating the excited state. Direct quenching of probe fluorescence can occur in the presence of oxygen, heavy atoms, and molecules that form charge transfer complexes [Lakowicz, 1983]. The latter processes can in principle be used to learn something about biophysical properties of cells—for example, accessibility of dissolved oxygen to structures containing fluorescent probes.

## Solvent Polarity

Almost all fluorescent molecules are sensitive to some degree to the dynamic polarity of their solvent environment. It has been known for decades that the quantum yield of anilino naphthaline sulfonate (ANS) is high in nonpolar solvents but more than 30-fold lower in water [Weber and Laurence, 1954]. A large shift in the emission wavelength of ANS also occurs in nonviscous polar solvents. This happens because energy of the excited state is lost to solvent when polar solvent molecules reorient around the more dipolar structure of excited ANS shortly after excitation. Reduction of the distance between the excited and ground state means emitted photons will have lower energy, i.e., be red shifted.

Quantum yield changes and wavelength shifts of fluorescent probes attached selectively to specific sites of perturbed or functionary protein molecules have been used to support models involving conformational changes. Much of this work has been reviewed elsewhere (Lakowicz, this volume).

Sensitivity to solvent polarity is an important component of the mechanism of certain probes. For example, propidium is more fluorescent when protected from water by DNA intercalation. Also, many potential sensitive dyes are more fluorescent in hydrocarbon environments like membranes and therefore show fluorescence changes when driven by membrane potential changes to compartments of different polarity [Waggoner, 1985].

Sometimes it is better if a probe is not sensitive to solvent polarity. Quantification of the binding of fluorescent-labeled antibodies may be difficult if the probe is sensitive to the polarity of the cellular environment near the antigen or to different aqueous and nonaqueous preserving agents or to air drying. It complicates physiological interpretations when pH probes, calcium probes, energy transfer probes, and distribution-type membrane potential probes also change their spectral properties because of their solvent environment changes rather than because of a change in the property they are intended to monitor.

## Resistance to Rotational and Translational Diffusion

There is a direction defined on each chromophore, the transition moment, for which polarized light will be optimally absorbed or emitted. Absorption and emission transition moments are not always in the same direction but often they are. This property can be used to find the orientation of probes bound to biological structures illuminated with polarized light. The property can also be utilized to determine the rotational mobility of probes and therefore determine the microviscosity of their environment or the binding of the probe. The most common method is to measure the rate of depolarization of fluorescence from probe molecules excited with polarized light. Why should fluorescence be polarized in the first place? Excitation, whether continuous or pulsed, sets up a population of excited chromophores with their transition moments oriented preferentially in the direction of polarization. In a viscous solvent the molecules will remain essentially fixed during the excited-state lifetime so that emitted light will be polarized. The degree of polarization can be assessed by measuring fluorescence passed through analyzers oriented parallel and perpendicular to the direction of the polarized excitation light. At the other extreme, molecules excited in a fluid environment will rotate during the excited-state lifetime and the emission transition moments will tend to randomize. The greater the randomization, the greater the depolarization of fluorescence. Methods for analyzing fluorescence depolarization and examples of its usefulness in studying segmental flexibility of tagged proteins and rotational mobility of lipid fluidity probes are discussed by Lakowicz in this volume. Triplet state absorption and phosphorescence methods for quantifying the rotational mobility of probes are also discussed in this volume by Garland.

Whereas spectroscopic methods that quantify the rotational diffusion rate of a probe allow one to put a number on the "microviscosity" of the local environment of the probe [which may be unrelated to bulk viscosity (Lakowicz, this volume)], measurements of translational diffusion can also be made by fluorescent methods and provide a measure of restrictions to diffusion of the probe. Diffusional mobility measurement of free probes or tagged macromolecules, polymers, or lipids can be readily measured by photobleaching methods and such measurement is described in this volume by Lanni, and by Ware et al. The restrictions to probe diffusion may be completely unrelated to the viscosity or microviscosity, which would be determined by bulk method or by determining rotational mobilities. For example, lipids and proteins may freely diffuse rotationally and translationally in the fluid lipid bilayer but others may not move because of attachment to cytostructure. Small labeled protein and polymer molecules may diffuse more slowly [Luby-Phelps et al., 1986; Jacobson and Wojcieszyn, 1984].

No doubt many interesting structural and functional properties of membranes and cytoplasm will be learned from these approaches.

## REFERENCES

Abrams WR, Diamond LW, Kane AB (1983): A flow cytometric assay of neutrophil degranulation. J Histochem Cytochem 31:737–744.

Arndt-Jovin DJ, Jovin TM (1977): Analysis and sorting of living cells according to DNA content. J Histochem Cytochem 25:585–000.

Aubin J (1979): Autofluorescence of viable cultured mammalian cells. J Histochem Cytochem 27:36–43.

Barak LS, Webb WW (1981): Fluorescent low density lipoprotein for observation of dynamics of individual receptor complexes on cultured human fibroblasts. J Cell Biol 90:595–604.

Barrett AJ (1979): "Proteinases in Mammalian Cells and Tissues." Amsterdam/New York: Elsevier/North Holland.

Bass DA, Parce JW, Dechatelet LR, Szejda P, Seeds Mc, Thomas M (1983): Flow cytometric studies of oxidative formation by neutrophils: A graded response to membrane stimulation. J Immunol 130:1910–1917.

Benson R, Meyer RA, Zaruba M, McKhann G (1979): Cellular autofluorescence. Is it due to flavins? J Histochem Cytochem 27:44–48.

Blatt E, Sawyer WH (1985): Depth-dependent fluorescence quenching in micelles and membranes. Biochim Biophys Acta 822:43–62.

Bock G, Hilchenbach M, Schauenstein K, Wick G (1985): Photometric analysis of antifading reagents for immuno-fluorescence with laser and conventional illumination sources. J Histochem Cytochem 33:699–705.

Brigati DJ, Myersow D, Leary JJ, Spalholz B, Travis SZ, Fong CKY, Hsiung GD, Ward DC (1983): Detection of viral genomes in cultured cells and parofin embedded tissue sections using biotin-labeled hybridization probes. Virology 126:32–50.

Busa WB, Nuccitelli R (1984): Metabolic regulation via intracellular pH. Am J Physiol 246:R409–R483.

Campbell AK (1983): "Intracellular Calcium." New York: John Wiley and Sons.

Campbell AK, Morgan BP (1985): Monoclonal antibodies demonstrate projection of polymorphonuclear leukocytes against complement attack. Nature 317:164–166.

Campbell AK, Holt M, Patel A (1985): Chemiluminescence in medical biochemistry. Rec Adv Clin Biochem 3:1–30.

Clement NR, Gould MJ (1981): Pyranine (8-hydroxy-1,3,6-pyrenetrisulfonate) as a probe of internal aqueous hydrogen ion concentration of phospholipid vesicles. Biochemistry 20:1534–1538.

Darzynkiewicz Z, Traganos F, Malamed M (1980): New cell cycle compartments identified by multiparameter flow cytometry. Cytometry 1:98–108.

Darzynkiewicz Z, Traganos F, Kapuscinsky J, Staiano-Coico L, Melamed M (1984): Accessibility of DNA in situ to varius fluorochromes: Relationship to chromatin changes during erythroid differentiation of Friend leukemic cells. Cytometry 5:355–363.

Deamer DW, Prince RC, Crofts AR (1972): The response of fluorescent amines to pH gradients across liposome membranes. Biochim Biophys Acta 274:323–335.

Dolbeare F (1981): Fluorometric quantification of specific chemical species in single cells. In Wehry EL (ed): "Modern Fluorescence Spectroscopy." New York: Plenum Press, pp 251–293.

Dolbeare FA, Smith RE (1977): Flow cytometric measurement of peptidases with use of 5-nitro salicylaldehyde and 4-methoxy-B-naphthylamine derivatives. Clin Chem 23:1485–1491.

Dolbeare F, Vanderlaan M (1979): A fluorescent assay of proteinases in cultured mammalian cells. J Histochem Cytochem 27:1493–1495.

Dolbeare F, Gratzner H, Pallovicini MG, Gray JW (1983): Flow cytometric measurement of total DNA content and incorporation of bromodeoxyuridine. PNAS 80:5573–5577.

Fernandez SM, Berlin RD (1976): Cell surface distribution of lectin receptors determined by resonance energy transfer. Nature 264:411–415.

Fromhertz P, Masters B (1974): Interfacial pH at electrically charged lipid monolayers investigated by the lipoid pH-indicator method. Biochim Biophys Acta 356:270–275.

Glazer AN, Stryer L (1984): Phycofluor probes. TIBS (Oct). 423–427.

Greenspan P, Mayer EP, Fowler SD (1985): Nile Red: A selective fluorescent stain for intracellular lipid droplets. J Cell Biol 100:965–973.

Grynkiewicz G, Poenie M, Tsien RY (1985): A new generation of $Ca^{+2}$ indicators with greatly improved fluorescence properties. J Biol Chem 260:3440–3450.

Haugland RP (1983): Covalent fluorescent probes. In Steiner RF (ed): "Excited States of Biopolymers." New York: Plenum Press, pp 29–58.

Heiple JM, Taylor DL (1980): Intracellular pH in single motile cells. J Cell Biol 86:885–890.

Hemmilä I, Dakubu S, Mukkala V-M, Siitari H, Lovgren T (1984): Europium as a label in time-resolved immunofluorometric assays. Anal Biochem 137:335–343.

Herscowicz H, Holden H, Bellante J, Ghaffor A (1981): "Manual of Macrophase Methodology." New York: Marcel Dekker.

Holtzman E (1976): "Lysosomes: A Survey." New York: Springer-Verlag.

Jacobson K, Wojcieszyn (1984): The translational mobility of substances within the cytoplasmic matrix. Proc Natl Acad Sci USA 81:6747–6751.

Johnson LV, Walsh ML, Bockus BJ, Chen LB (1981): Monitoring of relative mitochondrial

membrane potential in living cells by fluorescence microscopy. J Cell Biol 88:526–535.
Johnson PC, Ware JA, Cliveden PB, Smith M, Dvorak AM, Salzman EW (1985): Measurement of ionized calcium in blood platelets with the photoprotein aquorin. Comparison with quin II. J. Biol Chem 260:2069–2076.
Kapuscinski J, Darzynkiewicz Z, Melamed M (1982): Luminescence of the solid complexes of acridine orange with RNA. Cytometry 2:201–211.
Klausner RD, Wolf DE (1980): Selectivity of fluorescent lipid analogues for lipid domains. Biochemistry 19:6199–6203.
Koller E, Wolfbeis OS (1985): Synthesis and spectral properties of long wavelength absorbing and fluorescing substrates for direct and continuous kinetic assay of carboxyesterases, phosphatases, and sulfatases. Monatschefte Chemie 116:65–75.
Kronick MN, Grossman PD (1983): Immunoassay techniques with fluorescent phycobiliprotein conjugates. Clin Chem 29:1582–1586.
Kurtz I, Balaban RS (1985): Fluorescence emission spectroscopy of 1,4 dihydroxyphthalonitrile. Biophys J 48:499–508.
Lakowicz JR (1983): "Principles of Fluorescence Spectroscopy." New York: Plenum Press.
Lampidis TJ, Bernal SD, Summerhays IC, Chen LB (1982): Rhodamine 123 is selectively toxic and preferentially retained in carcinoma cells in vitro. Ann NY Acad Sci 397:299–302.
Langer PR, Waldrop AA, Ward D (1981): Enzymatic synthesis of biotin labeled polynucleotides: Novel nucleic acid affinity probes. Proc Natl Acad Sci USA78:6633–6637.
Latt SA, Marino M, Lalande M (1984): New fluorochromes, compatible with high wavelength excitation for flow cytometric analysis of cellular nucleic acids. Cytometry 5:339–347.
LePecq JB, Paoletti C (1967): A fluorescent complex between ethidium bromide and nucleic acids. Physical-chemical characterization. J Mol Biol 27:87–000.
Leytus SP, Patterson WL, Mangel WF (1983): New clan of sensitive and selective fluorogenic substrates for serine proteinases. Biochem J 215:253–260.
Luby-Phelps K, Taylor DL, Lanni F (1986): Probing the structure of cytoplasm (in press).
Martin MM, Lindquist L (1975): The pH dependence of fluorescein fluorescence. J Luminescence 10:381–390.
McNeil PL, McKenna MP, Taylor DL (1985): A transient rise in cytosolic calcium follows stimulation of quiescent cells with growth factors and is inhibitable with phorbol myristate acetate. J Cell Biol 101:372–379.
Moolenaar WH, Tertoolen LGJ, deLaat SW (1984): Growth factors immediately raise cytoplasmic free $Ca^{++}$ in human fibroblasts. J Biol Chem 259:8066–8069.
Muirhead K, Horan PK, Poste G (1985): Flow cytometry: Present and future. Bio/Tech 3:337–356.
Murphy RF, Powers S, Cantor CR (1984): Endosome pH measurement in single cells by dual fluorescence flow cytometry: Rapid acidification of insulin to pH 6. J Cell Biol 98:1757–1762.
Ohkuma S, Poole B (1978): Fluorescence probe measurement of the intralysosomal pH in living cells and the perturbation of pH by various agents. Proc Natl Acad Sci USA 75:3327–3331.
Oi V, Glazer AN, Stryer L (1982): Fluorescent phycobiliprotein conjugates for analysis of cells and molecules. J Cell Biol 93:981–986.
Packard BS, Wolf DE (1985): Fluorescence lifetimes of carbocyanine lipid analogues in phospholipid bilayers. Biochemistry 24:5176–5181.
Pagano RE, Sleight RG (1985): Defining lipid transport pathways in animal cells. Science 229:1051–1057.
Picciolo GL, Kaplan DS (1984): Reduction of fading of fluorescent reaction product for

microphotometric quantitation. Adv Appl Microbiol 30:197–234.

Pietras RJ, Szego CM, Roberts JA, Seelser BJ (1981): Lysosomal cathepsin B-like activity: Mobilization in prereplicative and neoplastic epithelial cells. J Histochem Cytochem 29:440–450.

Poole A, Mort J (1981): Biochemical and immunological studies of lysosoma land related proteinases in health and disease. J Histochem Cytochem 29:494–500.

Schulman SG (1976): Acid-base chemistry of excited singlet states. In Wehry EL (ed): "Modern Fluorescence Spectroscopy." Vol. 2. New York: Plenum Press, pp 239–275.

Shapiro H (1985): "Practical Flow Cytometry." New York: Alan R. Liss.

Shapiro HM, Stephens S (1986): Flow cytometry of DNA content using oxazine 750 or related laser dyes with 633 nm laser excitation. Cytometry (in press).

Sims PJ, Waggoner AS, Wang CH, Hoffman JF (1974): Biochemistry 13:3315–3330.

Smith RE (1983): Contributions of histochemistry to the development of the proteolytic enzyme detection system in diagnostic medicine. J Histochem Cytochem 31:199–209.

Spiegel S, Yamada KM, Hom BE, Moss J, Fishman PH (1985): Fluorescent gangliosides as probes for the retention and organization of fibronectin by ganglioside-deficient mouse cells. J Cell Biol 100:721–726.

Steinkamp JA (1984): Flow cytometry. Rev Sci Instrum 55:1375–1400.

Struli P, Barrett AJ, Bauci A (1980): "Proteinases and Tumor Invasion." New York, Raven Press.

Stryer L (1978): Fluorescence energy transfer as a spectroscopic ruler. Ann Rev Biochem 47:819–846.

Szollosi J, Trons L, Damjanoviek S, Hellewell SH, Arndt-Jovin D, Jovin TM (1984): Fluorescence energy transfer measurements on cell surfaces: A critical comparison of steady state fluorimetric and flow cytometric methods. Cytometry 5:210–216.

Tanasugarn L, McNeil P, Reynolds GT, Taylor DL (1984): Microspectro-fluorometry by digital image processing: Measurement of cytoplasmic pH. J Cell Biol 98:717–724.

Taylor DL, Reidler J, Spudich J, Stryer L (1981): Detection of actin assembly by fluorescence energy transfer. J Cell Biol 89:362–367.

Terasaki M, Hendrickson S, Stiles CD, Chen LB (1984): Changes in the organization of the endoplasmic reticulum in serum stimulated 3T3 cells. J Cell Biol 99:358a.

Thomas JA, Buschbaum RN, Zimmick A, Racker E (1979): Intracellular pH measurements in Ehrlich as cites tumor cells utilizing spectroscopic probes generated in situ. Biochemistry 18:2210–2218.

Thorell B (1983): Flow cytometric monitoring of intracellular flavins simultaneously with NAD(P)H levels. Cytometry 4:61–65.

Tsien RY (1980): New calcium indicators and buffers with high selectivity against magnesium and protons: Design, synthesis, and properties of prototype structures. Biochemistry 19:2396–2404.

Tsien RY, Pozzan T, Rink TJ (1984): Measuring and manipulating cytosolic $Ca^{2+}$ with trapped indicators. Trends Biochem Sci June:263–266.

Valet G, Raffael A, Moroder L, Wunsch E, Ruhenstroth-Bauer G (1981): Fast intracellular pH determination in single cells by flow cytometry. Naturwiss 68:265–266.

Waggoner AS (1979): Dye indicators of membrane potential. Ann Rev Biophys Bioeng 8:47–68.

Waggoner AS (1985): Dye probes of cell organelle, and vesicle membrane potentials. In Martonosi A (ed): "The Enzymes of Biological Membranes." New York: Plenum Press, pp 313–331.

Waggoner AS, Stryer L (1970): Fluorescent probes of biological membranes. Proc Natl Acad

Sci USA 67:579–589.
Weber G, Laurence JR (1954): Fluorescent indicators of adsorption in aqueous solution and on solid phases. Biochem J 56:31.
Weiss MJ, Chen LB (1984): Rhodamine 123: A lipophilic mitochondrial-specific vital dye. Kodak Lab Chem Bull 55:1–4.
Wright DG (1982): The neutrophil as a secretory organ in host defense. In Gallin JI, Fauci AS (eds): "Host Defense Mechanisms." Vol. 1. New York: Raven Press, pp 75–110.
Yguerabide J, Foster M (1981): Fluorescence spectroscopy of biological membranes. Mol Biol Biochem Biophys 31:199–269.
Zucker-Franklin D, Lavi D, Franklin E (1981): Demonstration of membrane-bound proteolytic activity on the surface of mononuclear leukocytes. J Histochem Cytochem 29:451–456.

*Applications of Fluorescence in the Biomedical Sciences, pages 29–67*


# 2 A Review of Photon-Counting and Phase-Modulation Measurements of Fluorescence Decay Kinetics

*Joseph R. Lakowicz*

The fluorescence lifetime of a substance represents the average amount of time a molecule remains in the excited state prior to its return to the ground state. An analysis of the fluorescence decay kinetics can reveal the interactions of the fluorophore with its environment. For example, multiple decay constants can be the result of a fluorophore being in several distinct environments, the result of an excited state process, or due to intrinsic heterogeneity in the decay of the fluorophore. Lifetime measurements can also reveal the frequency of collisional encounters with quenching agents, the rate of energy transfer, and the rates of excited-state reactions. Measurement of fluorescence decays is difficult because these values are typically near 10 nsec, necessitating the use of high-speed electronic devices and detectors. Additionally, one requires a light source whose intensity varies at a rate which is rapid relative to the intensity decay. The light source may be pulsed or intensity modulated.

There are two general methods for the measurement of fluorescence lifetimes. These are the pulse method and the phase-modulation (harmonic) method. In the pulse method the sample is excited with a brief pulse of light. The time-dependent decay of fluorescence intensity is measured and is used to estimate the decay law. In the phase-modulation method the sample is

Department of Biological Chemistry, University of Maryland School of Medicine, Baltimore, Maryland 21201.

excited with intensity-modulated light. The phase shift and demodulation of the emission, relative to the incident light, are used to calculate the decay law. During the past several years the capabilities of methods have dramatically improved. The capabilities of pulse fluorometry have increased as the result of improved flash lamps with good stability and pulse widths near 1 nsec. Additionally, dramatic improvements in resolution are obtained using mode-locked and cavity-dumped laser sources, which provide 10-psec pulses, repetition rates as high as 4 MHz, and high intensity. The capabilities of phase-modulation fluorometry have been increased by the recent possibility of variable-frequency light modulation. Current instruments operate from 1 to 200 MHz and provide resolution which is at least equivalent to that obtainable with a pulsed laser source and single photon counting. Each measurement, time-domain, or frequency-domain possesses its own advantages and disadvantages. In this chapter we describe the principles of each measurement, its past and present status, and the advantages and disadvantages. Additional detail on pulse fluorometry can be found in many books and reviews [Demas, 1983; Lakowicz, 1983; Ware, 1983]. For additional details on variable-frequency phase-modulation fluorometry see Lakowicz and Maliwal [1985], Lakowicz et al. [1984a,b], and Gratton and Limkemann [1983].

## INTRODUCTION TO PULSE AND PHASE-MODULATION FLUOROMETRY

The objective of both techniques is to determine the decay law of the sample. This decay is best visualized by considering excitation of the sample with an infinitely short pulse of light, resulting in an initial population $N_0$ of fluorophores in the excited state. If the population of the excited molecules $[N(t)]$ is homogeneous, then the decay is often described as a single exponential

$$N(t) = N_0 e^{-t/\tau} \tag{1}$$

where $\tau = (\gamma + k)^{-1}$ is the lifetime of the excited state, $\gamma$ is the emission rate, and $k$ is the rate of nonradiative decay (Fig. 1). The fluorescence intensity $[F(t)]$ is proportional to the emission rate ($\gamma$) and to the excited-state population $[F(t) = C\,\gamma\,N(t)]$, where $C$ is a constant. The fluorescence lifetime is generally equated with the time required for the intensity or population to decay to $1/e$ of its initial value, which is $(\gamma + k)^{-1}$. More complex considerations are needed when the decay is more complex than a single exponential.

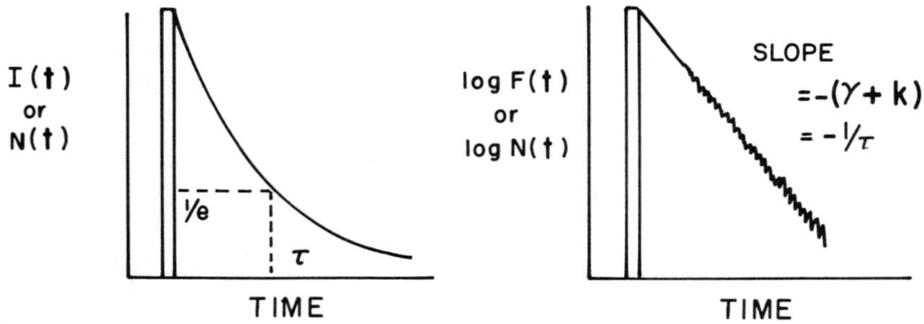


Fig. 1. Schematic description of pulse lifetime measurements [from Lakowicz, 1983].

The lifetime may also be considered to be the average amount of time a fluorophore spends in the excited state. This average is given by

$$\langle t \rangle = \frac{\sum_i t_i N_i(t)}{\sum_i N_i(t)} \tag{2}$$

For a large number of fluorophores and small time intervals this sum becomes

$$\langle t \rangle = \frac{\int_0^\infty t\, N(t)\, dt}{\int_0^\infty N(t)\, dt} = \frac{\int_0^\infty te^{-t/\tau}dt}{\int_0^\infty e^{-t/\tau}dt} \tag{3}$$

Hence, for an exponential decay the average time a fluorophore remains in the excited state is equal to the lifetime,

$$\langle t \rangle = \tau \tag{4}$$

It is important to note that equation 4 is not true for more complex decay laws, such as multi- or nonexponential decays. By using the observed decay law, an average lifetime can always be calculated by using equation 3. However, this apparent value can be a complex function of the parameters describing the actual decay (equation 11).

The second method by which lifetimes are routinely measured is the phase-modulation method. Instead of pulsed excitation the sample is excited with light whose intensity is modulated sinusoidally (Fig. 2). The emission is a forced response to the excitation, and therefore the emission is modulated at the same circular frequency ($\omega = 2\pi \times$ frequency in Hz) as the excitation. Because of the finite lifetime of the excited state, the modulated emission is

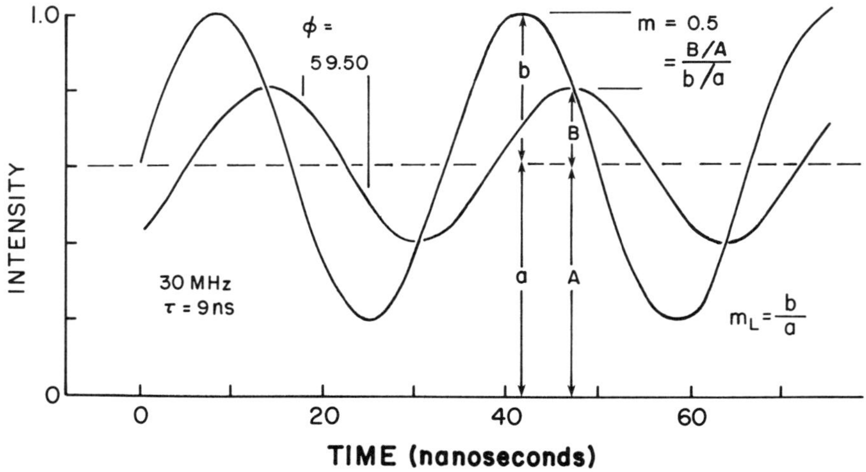


Fig. 2. Schematic description of phase-modulation fluorometry [from Lakowicz, 1983].

delayed in phase by an angle $\phi$ relative to the excitation. Furthermore, the emission is less modulated (demodulated) relative to the excitation. That is, the relative amplitude of the variable portion of the emission ($B/A$) is smaller for the emission than for the excitation ($b/a$). The phase angle ($\phi$) and demodulation factor ($m = Ba/bA$) are both measured and used to calculate the phase ($\tau_p$) and modulation ($\tau_m$) lifetimes by using

$$\tan \phi = \omega\tau_p, \qquad \tau_p = \omega^{-1} \tan \phi \tag{5}$$

$$m = [1 + \omega^2 \tau_m{}^2]^{-1/2}, \qquad \tau_m = \omega^{-1} \left[\frac{1}{m^2} - 1\right]^{1/2} \tag{6}$$

These expressions (equations 5 and 6) are only correct if the decay is a single exponential. If the decay is multi- or nonexponential then the lifetimes ($\tau_p$ and $\tau_m$) are only apparent values. The emission (Fig. 2) was drawn by assuming a modulation frequency of 30 MHz and a lifetime of 9 nsec. Using equations 5 and 6 one can readily calculate that the phase angle is 59.5° and the demodulation factor is 0.5. Since phase-modulation fluorometry is somewhat less well understood than pulse fluorometry the observable quantities are illustrated in Figure 3, where we plotted the dependence of $m$ and $\phi$ on the fluorescence lifetime. These values are presented for modulation frequencies of both 10 and 30 MHz. At a given frequency, the phase angles increase and the demodulation factors decrease as the lifetime increases. The modu-

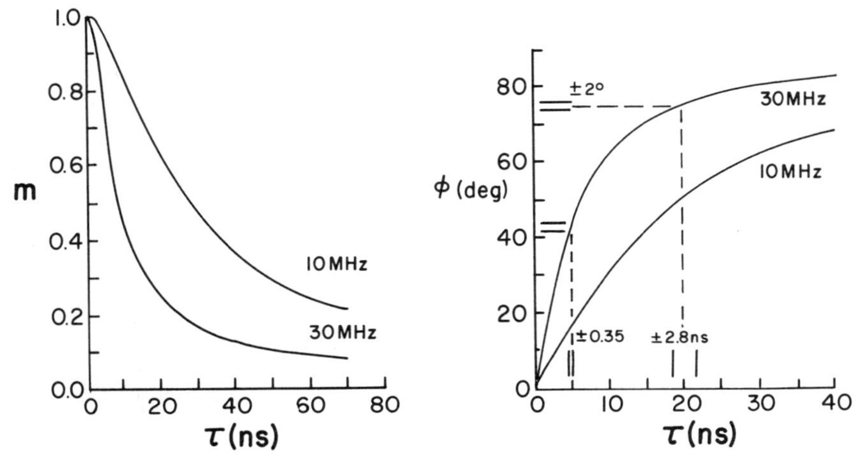


Fig. 3. Dependence of phase shifts and demodulation factors on the fluorescence lifetimes [from Lakowicz, 1983].

lation frequency must be comparable to the decay rate to obtain measurable values of $\phi$ and m. For a constant lifetime, increasing the modulation frequency results in larger phase shifts and greater degrees of demodulation (Fig. 3).

Comparison of Figures 1 and 2 reveals the ideal conditions needed for pulse and phase fluorometry. For pulse measurements one requires the shortest possible flash of light and a fast detector. In the frequency-domain the source and detector must allow excitation and detection at a large number of frequencies. Variable frequency light modulation is required for reliable resolution of complex decays by phase-modulation fluorometry. Until several years ago the resolution of both methods was limited by either the width of the excitation pulse or by the use of only a few modulation frequencies. These restrictions are no longer present in state-of-the-art instrumentation. In the following sections on pulse and phase-modulation fluorometry, first the standard instrumentation and then the present state of the art will be described. The two methods will be referred to as time-resolved and frequency-domain fluorometry.

## TIME-RESOLVED FLUOROMETRY

### Flash Lamps

For time-resolved fluorescence the most commonly used excitation source is the flash lamp. These devices are relatively inexpensive, provide reasonable intensities, and provide pulse widths ranging from one to several nanoseconds. These lamps may be filled with nitrogen, hydrogen, or other gases yielding different spectral outputs. For example, nitrogen provides high intensity, but the output is limited to discrete wavelengths (Fig. 4). Hydrogen and deuterium provide continuous output from the UV through the visible, but the intensity is less at each individual wavelength. It is informative to describe the intensities in operational terms. Time-resolved fluorescence is generally measured as single photon events (next section). Hence, a light source is of adequate intensity if a single photon from the sample is detectable following pulsed excitation. For a typical sample with an optical density near 0.05 at the excitation wavelength and a reasonable quantum yield, a single photon can almost always be detected by using nitrogen at one of its peak wavelengths (Fig. 4). This includes the use of monochromators for wavelength selection for both excitation and emission. When hydrogen or deuterium is used the intensity is just adequate. For samples with high quantum yields both excitation and emission monochromators may be used. For less-intense samples it is often necessary to remove the emission monochromator to obtain adequate intensity, especially if polarizers are used in the light paths. Frequently, hydrogen and deuterium are used for excitation of protein fluorescence because of the output in the UV. Since protein fluorescence may be weak, one often encounters difficulties in obtaining adequate intensities for measurement of protein fluorescence. Substitution of the emission monochromator with interference or broad-band filters usually results in adequate signal levels.

The presently available flash lamps are superior to those available just a few years ago. Typical time profiles for nitrogen and hydrogen are shown in Figure 5. These profiles are from a lamp built by Edinburgh Instruments [Birch and Imhof, 1981] and no special precautions were taken to minimize the pulse width. For hydrogen the full width at half-maximum intensity (FWHM) is about 1 nsec. For nitrogen the width is about 2.3 nsec. Typically, nitrogen displays a small afterglow, which can be diminished by the use of a slightly elevated gas pressure (1.5 atm). Importantly, the pulse profile of this lamp is stable for hours. This stability allows measurement of subnanosecond lifetimes or the resolution of multiexponential decays. This is possible with a stable lamp profile because the lamp profile and fluorescence response can be subjected to iterative reconvolution analysis (equation 10). The nsec pulse widths of all flash lamps are more troublesome for the measurement of rapid anisotropy decays, which often decay during the lamp profile. This will be evident when we discuss the data in Figure 10.

### Single-Photon Counting

Following excitation it is necessary to quantify the time-dependent emission. Several methods have been described, including pulse-sampling meth-

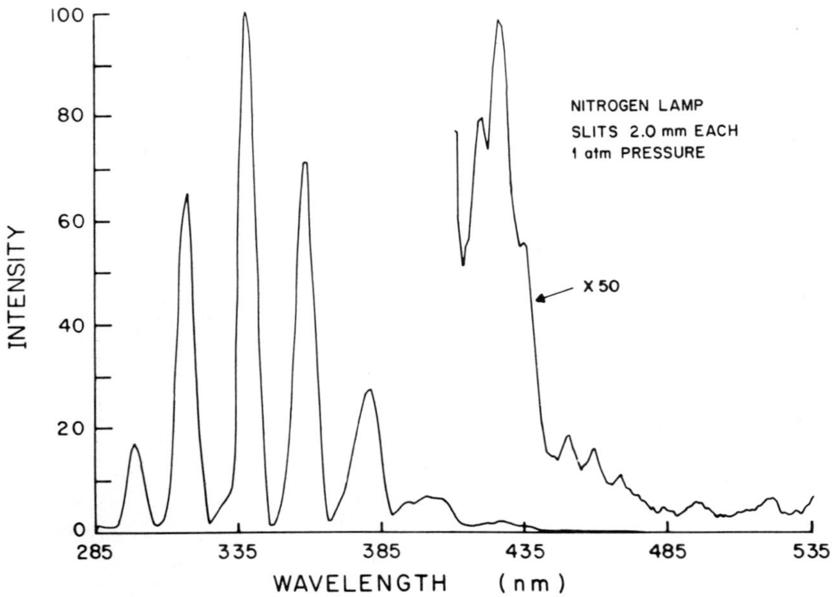


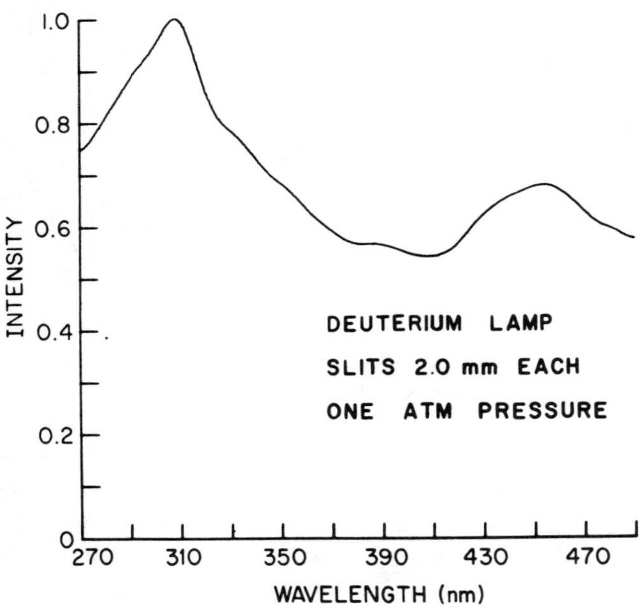


Fig. 4. Spectral output of nitrogen and deuterium flash lamps [from Badea and Brand, 1971].

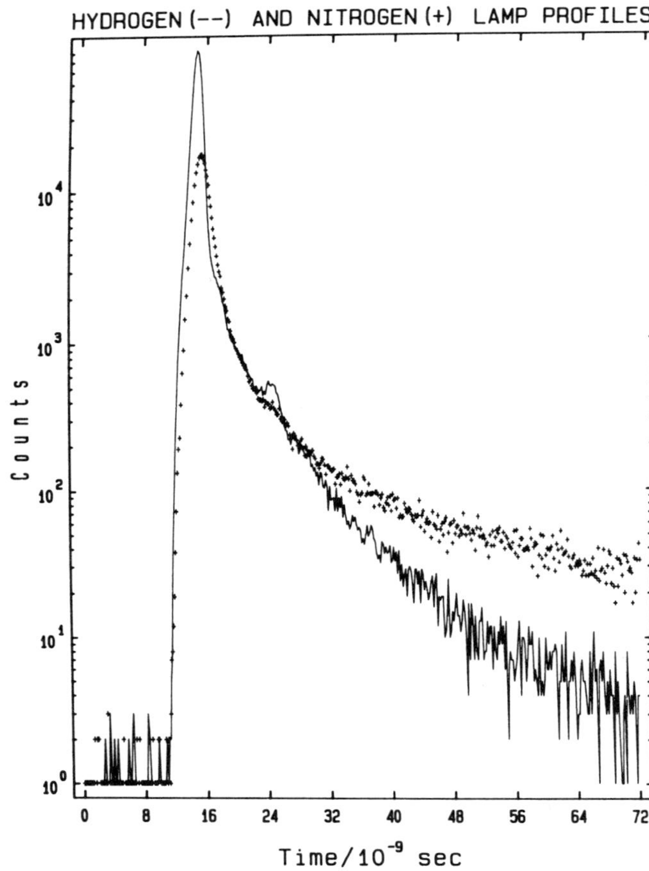


Fig. 5. Typical time-intensity profiles for nitrogen and hydrogen flash lamps. Hydrogen, 0.5 atm; Nitrogen, 1.5 atm. Profiles were obtained by using a system 199 from Edinburgh Instruments, Inc. The flash lamp is that described by Birch and Imhof [1981]. The nitrogen (1.5 atm) pulse (+) displays more intensity at longer times than the $H_2$ profile (——).

ods and the use of sampling oscilloscopes [Bennet, 1960; Ware et al., 1970; Hundly et al., 1967]. In general, the sampling methods were used because the time-response of the detectors or the associated electronics have not been adequate to accurately record the time-dependent emission. The availability of faster photomultiplier tubes and amplifiers may result in recording of the entire time response following each excitation pulse. At present, the preferred method is single-photon counting. The sample is excited with the lamp flash and the time interval is measured between the lamp flash and the arrival of the single photon-electron pulse at the photomultiplier tube (PMT) anode.

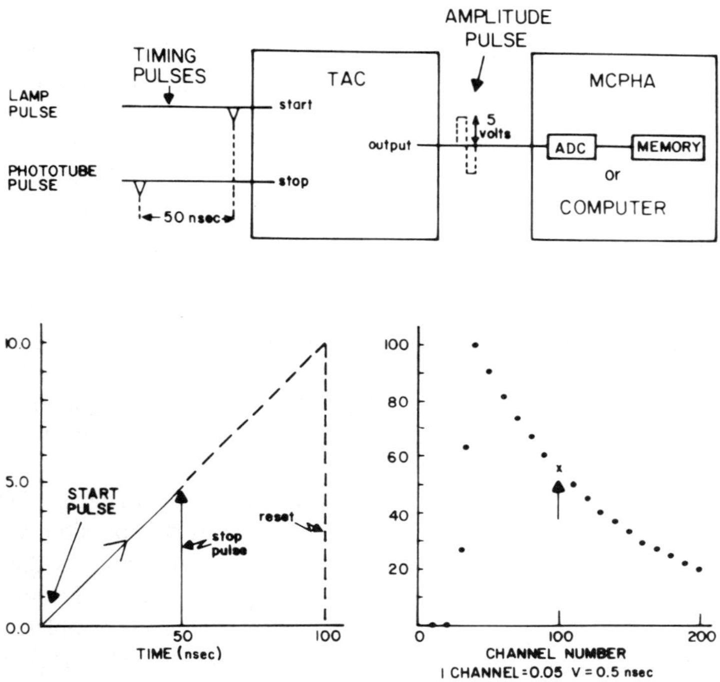


Fig. 6. Measurement of time-resolved fluorescence by single-photon counting [from Badea and Brand, 1971]. ADC, analog-to-digital converter.

The flash intensity or emission sensitivity is adjusted so that this pulse is due to a single photon incident upon the photocathode. Fluorescence emission is a random event. If only a single photon is detected, the probability of this event at various times is the same as the time-resolved decay for a large number of fluorophores. In the single-photon counting method the intensity level must be adjusted so that one photon or less is observed for each of 20 pulses. If there are multiple photons per flash the photon-counting circuit detects only the first event. Consequently, the measured decay is skewed toward lifetimes which are shorter than the actual value.

The time interval between the lamp pulse and the anode pulse is measured with a time-to-amplitude converter (TAC). The operation of this device is illustrated in Figure 6. The lamp pulse starts the TAC, which generates a voltage which increases linearly with time. The anode pulse, due to the single detected photon, stops the TAC. This device now contains a voltage which is proportional to the time between these two events. If no photon (stop pulse) is detected the device is reset to zero. The detected pulses are recorded in a digital manner in a multichannel pulse height analyzer (MCPHA). A histogram is generated in which the height represents the number of counts and the x-axis represents time (voltage generated by the TAC). For example, if a photon is detected 50 nsec after the lamp pulse, a single count is stored in the 50-nsec location of the MCPHA. This position is indicated by the arrow in Figure 6. This procedure is repeated until a large number of photons have been detected, typically $10^6$ or more. The number of times each voltage was generated by the TAC is presented in histogram form by the MCPHA. These data represent the time-resolved decay, which of course is distorted by the width of the lamp pulse. Using the example presented in Figure 6 we see that no pulses were measured for channel numbers 0–10, because the lamp has not yet fired. A maximum number of counts is seen near channel 30, which corresponds to the approximate time of the excitation pulse. The time $t = 0$ is not precisely defined for pulses other than $\delta$-pulses. At higher channel numbers, which represent progressively longer times, the number of counts is lower.

The single-photon counting method has several advantages. The time resolution is quite good, about 0.2 nsec. This resolution is possible since the detector and electronics only need to identify the rising edge of the single photon pulse. This is generally easier than detecting the detailed shape of the entire pulse. High-quality electronic components, and even complete instruments with analysis software, are commercially available. The single photon counting method is highly sensitive; only dilute solutions are needed and the low excitation intensity generally eliminates problems due to photodecomposition. And finally, for measurement of time-resolved intensities the times needed for data acquisition are not excessively long. For example, with a flash repetition rate of 20 kHz and a photon collection rate of 1,000 cps (5%), the accumulation of $10^6$ counts requires about 17 minutes. The time for data acquisition becomes longer when one requires $10^7$ counts or more to resolve closely spaced decay times or fluorescence anisotropy decays.

## Analysis of Time-Resolved Decays of Fluorescence

The objective of time-resolved measurements is to determine the impulse response function, $F(t)$, which would be observed for an infinitely sharp excitation pulse, or for a pulse width much shorter than the decay time of the sample. Unfortunately, pulse widths are frequently comparable to the decay times, as is illustrated for the decay of anthracene in ethanol (Fig. 7). Because of the width of the lamp pulse the zero time point is not defined, and one cannot simply use the $1/e$ point to determine the lifetimes. Furthermore, the emission from a sample may not be represented by a single exponential decay, again limiting the meaning of the $1/e$ point.

Because the width of the lamp pulse is comparable to the fluorescence decay times it is necessary to correct for the pulse width. This requires measurement of the lamp pulse [$L(t)$]. Since the lamp profiles may change

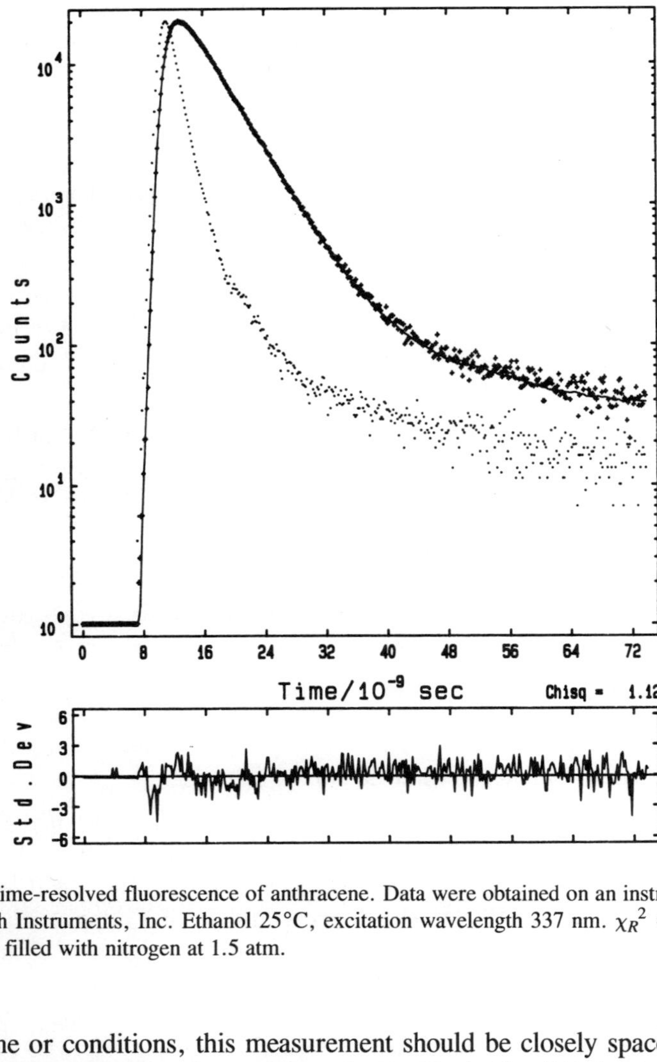


Fig. 7. Time-resolved fluorescence of anthracene. Data were obtained on an instrument from Edinburgh Instruments, Inc. Ethanol 25°C, excitation wavelength 337 nm. $\chi_R^2 = 1.12$. The lamp was filled with nitrogen at 1.5 atm.

with time or conditions, this measurement should be closely spaced in time to measurement of the fluorescence decay. Typically, $L(t)$ is measured by using a solution which scatters light. The observed decay of fluorescence $[R(t)]$ is given by the convolution of the lamp pulse with the impulse response of the sample,

$$R(t) = \int_0^\infty L(t')\, F(t - t')\, dt' \tag{7}$$

The lamp pulse can be considered as a large number of narrow pulses at times $t_i$ with equal widths $\Delta t$, where $\Delta t$ is much smaller than the decay time of the sample [Badea and Brand, 1971; Yguerabide, 1972]. Each individual pulse [$L(t_i)$] induces an impulse response in the sample [$F(t - t_i)$]. Hence,

$$R_i(t) = L(t_i)\,\Delta t\,F(t - t_i) \quad (t > t_i) \tag{8}$$

The term $(t - t_i)$ appears because the time-resolved decay must be relative to the time of the excitation, which is $t_i$. For $t < t_i$ it is understood that $R_i(t) = 0$. The measured decay is a summation of all these individually induced responses.

$$R(t) = \sum_{t=0}^{t=t_i} L(t_i)\,F(t - t_i)\,\Delta t \tag{9}$$

As $\Delta t$ approaches zero this becomes the convolution integral, equation 7. For convenience the variable is generally changed so that the term $(t - t_i)$ does not appear in the impulse response function. Using $t' = t - \eta$ equation 8 becomes

$$R(t) = \int_0^t L(t - \eta)\,F(\eta)\,d\eta \tag{10}$$

An analysis method is needed to determine the impulse response function from the convolved data.

## Least-Squares Analysis of Time-Resolved Decays

A variety of methods have been proposed for estimation of the impulse response function $F(t)$ from the measured decay curve $R(t)$ and the lamp profile $L(t)$. These include the method of least-squares [Grinvald and Steinberg, 1974], moments [Small and Isenberg, 1983], and Laplace transformation [Gafni et al., 1975]. Considerable controversy has accompanied the choice of analysis methods, particularly with regard to moments and least squares [Cundall and Dale, 1983]. At present we see no reason to doubt the conclusions of O'Connor et al. [1979] and Grinvald [1976], who claim that the method of nonlinear least squares provides the most generally reliable results. There may be specific circumstances for which an alternative method is superior, and one cannot help but be impressed by the resolution obtained by E. Small using the method of moments (Fig. 13). However, no analysis method has been demonstrated to be consistently superior to nonlinear least squares. This method is used in commercially available software.

Assume the impulse response can be described by a sum of exponentials

$$F(t) = \sum_i \alpha_i\,e^{-t/\tau_i} \tag{11}$$

The basis of the least-squares method is calculation of the expected value of $R(t)$ given assumed values of $\alpha_i$ and $\tau_i$. Specifically, the convolution integral (equation 10) is calculated by using the assumed values of $\alpha_i$ and $\tau_i$ and the measured time-profile of the lamp pulse. The calculated values $[R_c(t)]$ are compared with the observed values $R(t)$. The $\alpha_i$ and $\tau_i$ values are varied until the best fit is obtained. The goodness of fit ($\chi^2$) is calculated from

$$\chi^2 = \sum_{i=1}^{n} \omega_i \, [R(t) - R_c(t)]^2 \qquad (12)$$

where $\omega_i = 1/R(t)$ is a statistical weighting factor to account for the expected error in each value of $R(t)$. It is essential to use the proper weighting factor; otherwise serious errors can arise. This procedure is frequently referred to as deconvolution, which is inaccurate. In reality an assumed decay law is convoluted with $L(t)$ and the results are compared with $R(t)$. More properly this procedure is one of iterative reconvolution. Of course, numerous calculations must be performed and the use of a computer is essential.

The weighting factor in equation 12 is often a source of confusion because $\chi^2$ is a dimensionless number, and casual inspection of equation 12 indicates that the units are counts/sec. The value of $\chi^2$ is the sum of the squared deviations between the measured $[R(t)]$ and expected values $[R_c(t)]$, each divided by the amount of deviation expected for randomly distributed deviations. For photon counting data the expected deviations are given by $\sqrt{R(t)}$. For instance, assume a channel contained $10^4$ counts. Then the expected deviation for this measurement is 100 counts. If the assumed model accounts for the measured data then the numerator and denominator of equation 12 are both $(10^2)^2$, and this data point contributes 1.0 to the value of $\chi^2$. Hence, for random errors, $\chi^2$ is expected to be approximately equal to the number of data points (channels).

It is inconvenient to interpret the values of $\chi^2$ because $\chi^2$ depends on the number of data points. For this reason one uses the value of reduced $\chi^2$,

$$\chi_R^2 = \frac{\chi^2}{N - p} \qquad (13)$$

where $N$ is the number of data points and $p$ is the number of floating parameters. If only random errors contribute to $\chi_R^2$ then this value is expected to be near unity because the average $\chi^2$ per data point should be about one. If the model does not fit the data then the individual values of $\chi^2$ and $\chi_R^2$ are both larger than expected for random errors.

The minimum value of $\chi_R^2$ indicates the best fit possible for an assumed model. The values of the parameters ($\alpha_i$ and $\tau_i$) are found by using any of a number of search procedures [Bevington, 1969]. Since the value of $\chi_R^2$ can vary due to the random distribution of the errors one uses statistical considerations to accept or reject an assumed model. For example, assume there are 203 data points, and three degrees of freedom (a double exponential decay). Then a value of $\chi_R^2 = 1.25$ would only be observed 1% of the time for random errors. Hence, values this large could be used to reject the model. In this laboratory we use a more conservative approach. Values of $\chi_R^2 = 2$ or larger are taken to indicate a poor fit. Values of $\chi_R^2 = 1.2$ or less are taken to represent a good fit, especially if the fit is not improved by the addition of an additional component in the decay law. Intermediate values are judged on the basis of our knowledge about the sample.

The appearance of the analyzed data is illustrated in Figure 7 for anthracene. In this case the emission is strongly convoluted with the lamp because the decay time of anthracene is near 4.2 nsec. The half-width of the lamp profile is about 1.5 nsec. The solid line represents the calculated values $R(t)$ [$R_c(t)$] for an assumed single exponential decay with a lifetime of 4.2 nsec. The calculated response overlaps precisely with the measured data, and the deviations between the measured and calculated values are randomly distributed around zero (Fig. 7). The weighted residuals shown in the lower portion of the figure are calculated by using

$$\text{Residual}_i = \frac{R(t_i) - R_c(t_i)}{\sqrt{R(t_i)}} \tag{14}$$

Nonrandom deviations of the residuals from zero indicate that the data are not adequately described by the assumed model. An additional indicator of a good fit is the reduced value of $\chi_R^2$, which is near unity. In this case the value of $\chi_R^2$ could not be improved by assuming a double exponential decay.

Now consider a sample which displays a multiexponential decay of intensity. For illustrative purposes we used a mixture of anthracene (4.2 nsec) and 9-vinylanthracene (7.6 nsec). The inherent difficulty in resolving such closely spaced lifetimes is illustrated in Figure 8, which shows decay curves for each individual compound and for a mixture. Visually, the decay of the mixture (central curve) is almost indistinguishable from the individual single exponential decays. The distinction between the single and double exponential decays is only evident following reconvolution analysis (Fig. 9). It was not possible to fit the data for the mixture by using a single exponential decay law (Fig. 9, left). This is evident from the lack of agreement between the measured (. . .) and calculated (——) data, the systematic deviations seen in the weighted residuals, and in the elevated value of $\chi_R^2$ (8.88).

The right side of Figure 9 shows the same data, except that the assumed decay law was doubly exponential,

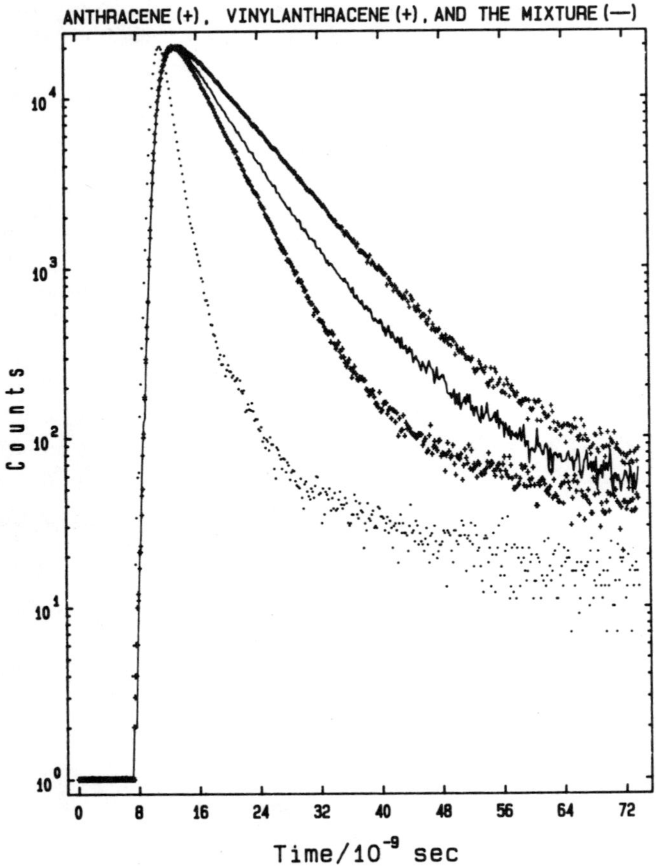


Fig. 8. Time-resolved decays of anthracene, 9-vinylanthracene, and a mixture. On the trailing side of the response the curves, from left to right, are the lamp profile, anthracene, the anthracene-vinylanthracene mixture, and vinylanthracene, respectively.

$$F(t) = \alpha_i\, e^{-t/\tau_1} + \alpha_2\, e^{-t/\tau_2} \tag{15}$$

The fitting algorithm calculates

$$R_c(t) = \int_0^t L(t - \eta)\, [\alpha_1 e^{-\eta/\tau_1} + \alpha_2 e^{-\eta/\tau_2}]\, d\eta \tag{16}$$

and compares these values with $R(t)$. $\alpha_1$, $\alpha_2$, $\tau_1$ and $\tau_2$ are all varied until ${\chi_R}^2$ reaches a minimum value. The right-hand side shows the fit obtained

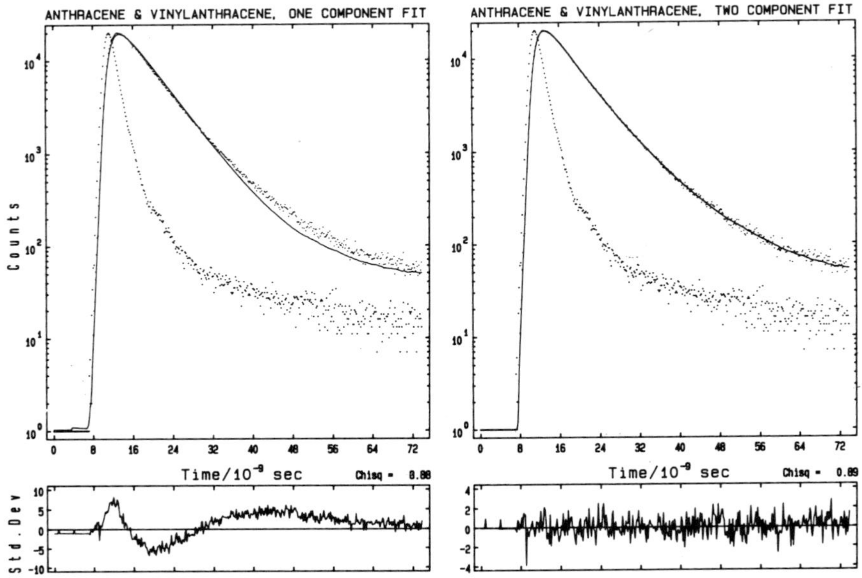


Fig. 9. Time-resolved fluorescence of a mixture of anthracene and 9-vinylanthracene. Ethanol 25°C, nm excitation, nm emission. Left, single exponential fit $\chi_R^2 = 8.88$; right, double exponential fit, $\chi_R^2 = 0.89$. For the two-component fit the recovered lifetimes were 4.08 ± 0.07 nsec 7.74 ± 0.04 nsec, in excellent agreement with the independently measured values for anthracene (4.20 nsec) and vinylanthracene (7.56 nsec).

with a double exponential model (equations 15 and 16). In this case a clear improvement is noted. The residuals now show a random distribution of errors and $\chi_R^2 = 0.89$. The improved fit indicates that at least a double exponential model is needed to describe the sample. In contrast to this mixture, a more complex decay law did not improve the fit to the data for anthracene (Fig. 7). Such lack of improvement is taken to indicate that the decay of anthracene is adequately described by a single exponential decay.

We note that an improved fit can often be obtained with a more complex model even if the more complex model is not appropriate for the sample. For the mixture the origin of the double exponential decay was clearly due to heterogeneity. However, for many fluorophores the decays are not single exponentials, even if the fluorophore is present in a homogeneous environment. Decays can be multiexponential or nonexponential, but the data can almost always be fit by using a multiexponential model. In all such cases one must exercise caution when interpreting the recovered decay times in terms of discrete molecular entities within the sample. Additionally, $\chi_R^2$ can be large because of systematic errors in the data. Perhaps the most common error is a change in the lamp profile during the data acquisition. Such changes yield nonrandom deviations of the residuals which are most evident near the peak of the lamp pulse. It is not correct to compensate for such errors using a more complex model.

Most least-squares analysis methods provide an estimation of the uncertainties in the parameter values. Two such estimates are given in Figure 9. These estimates are usually based on the assumption that the parameters are not correlated. This assumption is not true for multiexponential decays, and the $\alpha_i$ and $\tau_i$ values become more strongly correlated as the decay times become more closely spaced. The effect of correlation between parameters is that any one parameter can be varied with a compensating change in another parameter while the value of $\chi_R^2$ remains unchanged. Hence, the actual uncertainties in the parameters are larger than those returned by the least-squares algorithm. If the lifetimes are widely spaced then the effects of parameter correlation are minimal. Given the need to reliably interpret the parameters it seems logical that the effects of correlation be included in all analyses, as suggested by M. Johnson [1983].

## Anisotropy Decays

The pulse width and low repetition rate of flash lamps are troublesome for measurement of anisotropy decays. For anisotropy measurements the needed information is contained in the time-dependent difference between the parallel [$R_{\|}(t)$] and perpendicular [$R_{\perp}(t)$] components of the emission. Representative data are shown in Figure 10 for tyrosine fluorescence from calmodulin. At the pH and ionic strength of the measurements the calmodulin is probably in an expanded form with considerable segmental freedom. Notice that there are relatively few counts in the difference, and that the difference decays within the pulse width of the lamp. These difficulties result from the low repetition rate and long pulse width, respectively, of the flash lamps. These limited data do not sustain detailed analysis of the anisotropy decay law, especially in the presence of rapid segmental motions of the fluorophore. If the correlation times are somewhat longer then reliable information about the anisotropy decay can be obtained. This is illustrated in Figure 11 for tyrosine in glycerol at 20°C. In this case the correlation time can be easily estimated to be 11 nsec. However, even after 2 hours of data acquisition the anisotropy data could probably not sustain a multiexponential analysis.

What factors limit the resolution of multiexponential decays using single photon counting? With flash lamp sources the maximum possible repetition rates are near 30 kHz, restricting the maximum photon counting rates to about 2% of this value, or 600 Hz. Under these circumstances the resolution is usually limited by the photon statistics. It is then difficult to collect enough counts to resolve closely spaced lifetimes or complex decays of fluorescence

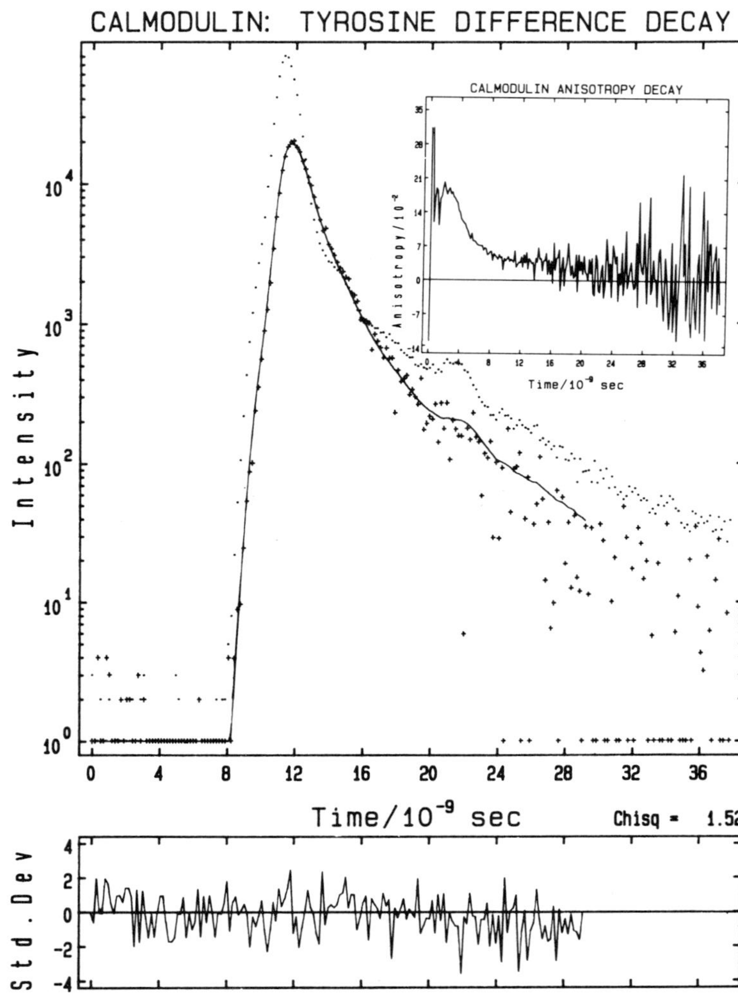


Fig. 10. Time-resolved decay of the difference between the parallel and perpendicular components of the tryosine emission from calmodulin. The solid line shows the fitted function through the difference decay (+ + +). The dots are the lamp profile. These data required about 2 hours of data acquisition. Calmodulin was provided by R.F. Steiner, University of Maryland, Baltimore County, Department of Chemistry.

anisotropy. The number of counts can be increased by acquiring data for longer periods of time. However, this is usually impractical because of the need to examine a number of samples on the same day, because of the instability of the sample, or because of the instability in the lamp profile. Also, with weakly fluorescent samples and/or the low intensity of some flash

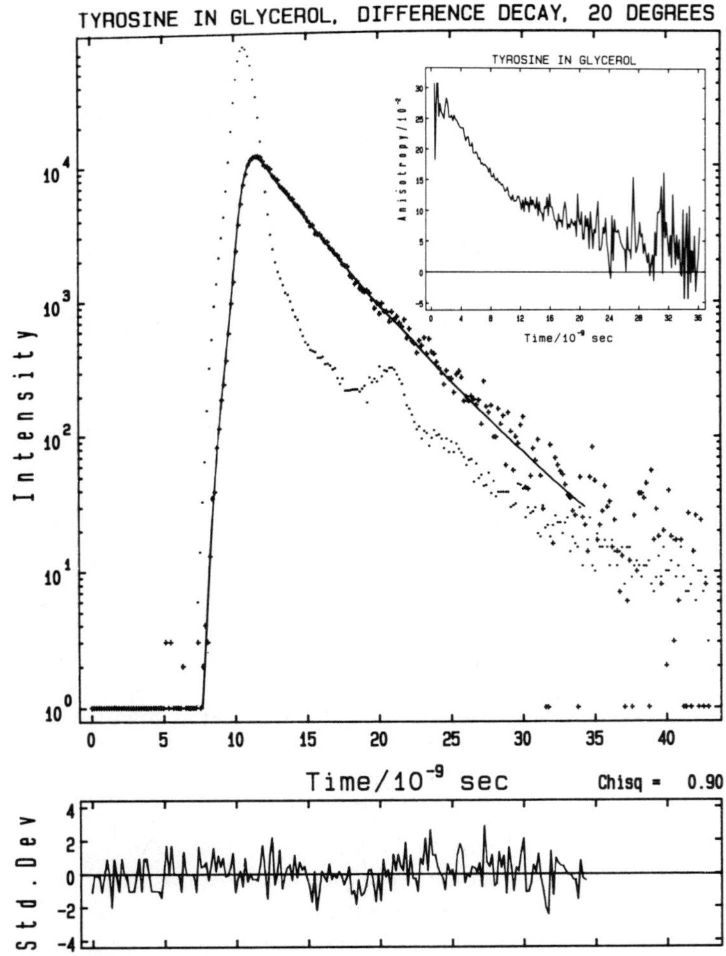


Fig. 11. Difference decay and anisotropy decay for tyrosine in glycerol at 20°C. $H_2$ lamp, 2 hours of data acquisition. The correlation time was 11.0 nsec and the apparent anisotropy at $t = 0$ was 0.229, $\chi_R^2 = 0.9$.

lamps, it can be difficult to obtain even a 600-Hz count rate. These limitations can be overcome by the use of pulsed laser sources, but at the cost of considerable expense and complexity.

## Pulsed Laser Sources

During the past 5 years many laboratories have described photon-counting instruments which use pulsed lasers as the excitation source [Small et al., 1984; Koester and Dowben, 1978; van Hoek et al., 1983; Zimmerman et al.,

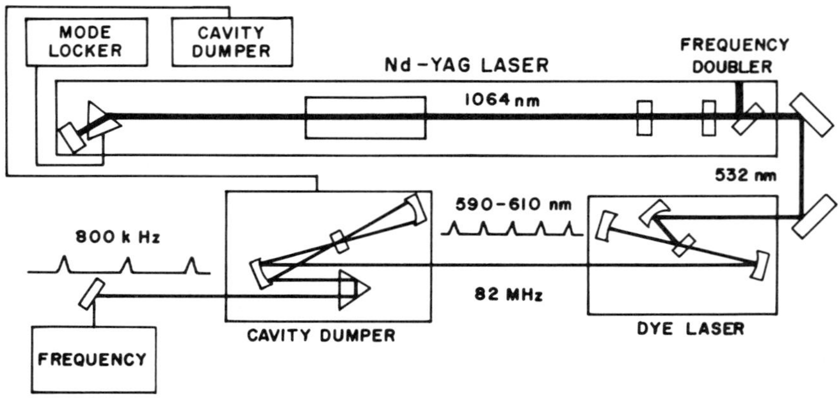


Fig. 12. Schematic description of a mode-locked synchronously pumped and cavity-dumped dye laser. Redrawn from literature distributed by Spectra Physics.

1982]. Most pulsed laser sources in current use consist of a similar grouping of components (Fig. 12). The main source is either a mode-locked and frequency-doubled yttrium aluminum garnet (Nd-YAG) laser or a mode-locked argon ion laser. For either laser the output is a pulse train at approximately 82 MHz, at wavelengths near 520 nm, with a pulse width near 100 psec. This pulse train is used to excite a dye laser, resulting in an 82-MHz pulse train at the wavelength governed by the dye and tuning elements within the laser cavity. For rhodamine 6G, a commonly used laser dye, the wavelengths range from 570 to 630 nm. The pulse widths of the dye laser output are typically near 10 psec, which is 100-fold shorter than the typical pulse width of flash lamps. At 82 MHz the pulses are spaced only 12 nsec apart. This is too fast for most samples because the intensity decay will not be complete prior to the arrival of the next pulse. The repetition rate is decreased by using the cavity-dumped output of the dye laser. An acousto-optic device (Bragg cell) is activated by a brief pulse of RF power. The traveling wave in the Bragg cell, typically made of quartz, deflects a single pulse out of the cavity. In this way the repetition rate can be chosen as desired, typically from single shot to 4 MHz. Power accumulates in the dye laser between cavity-dumped pulses. Hence, the average power from a cavity-dumped dye laser is not greatly different from the power at 82 MHz. This is fortunate because it is usually necessary to frequency double the dye laser output to obtain wavelengths useful for the excitation of biological samples. Frequency doubling of R6G yields 285–315 nm.

There are several advantages to this type of light source. The intensity is high, so that one is usually not limited by the quantum yield or concentration of the sample, but rather by background fluorescence. The fast repetition rate allows intensity decays to be accumulated with $10^6$ counts in the peak with only minutes of data acquisition. Since the time needed for data acquisition is brief (minutes) it is practical to also measure the signal due to background and to subtract this component. The larger number of counts improves the statistical accuracy of the data, allowing the resolution of more closely spaced decay times. Also, the short acquisition time frequently improves the results because of a decreased probability of instrumental drift.

The detectors used for time-resolved fluorescence have also improved. At present, the microchannel plate photomultiplier tubes are the detector of choice. For example, pulse widths for a single photoelectron are typically near 0.3 nsec, as compared to 1.5–2.5 nsec for fast PMTs. There are some difficulties with these devices, such as a count-rate dependent time-response and limited life span.

The remarkable improvement possible in the time-resolved data is evident from the data in Figure 13 (from E. Small). These data are for a four-component decay of tyrosine fluorescence, with a maximum of 4 nsec for the longest lifetime. Even for the shortest of the four lifetimes (0.38 nsec) the scatter profile is significantly shorter than the fluorescence decay. Additionally, there are $10^6$ counts in the peak channel. It is easily possible to have $10^7$ counts in the decay with only minutes of data acquisition.

The combination of increased count rate and increased time resolution of the detection system allows the easy resolution of two- and three-component decays. In this case the analysis was accomplished by using the method of moments [Small and Isenberg, 1977; Isenberg and Small, 1982]. To the best of my knowledge this resolution represents the most difficult sample which has been analyzed by using time-resolved fluorescence.

The advantages of the pulsed laser source are further illustrated by the anisotropy difference decay for rhodamine B in water (Fig. 14). There are nearly $10^5$ counts in the peak channel, and the laser pulse decays more rapidly than the difference decay. In contrast to the data obtained by using flash lamps (Fig. 10, 11), these data are likely to support a detailed analysis of the anisotropy decay law.

## Advances in Analysis and Experimental Techniques

In addition to advances in instrumentation there have been substantial advances in the methods of data analysis. The most notable of these are the global analysis methods described by Brand and co-workers [Beechem et al., 1985; Knutson et al., 1983]. This approach is illustrated by measurements of a mixture of fluorophores. Suppose the measurements are performed at multiple emission wavelengths. Then the $\alpha_i$ values are variable (equation 11), now given by $\alpha_i(\lambda)$, where $\lambda$ is the wavelength. However, the lifetimes are not wavelength dependent. When decays at various wavelengths are analyzed

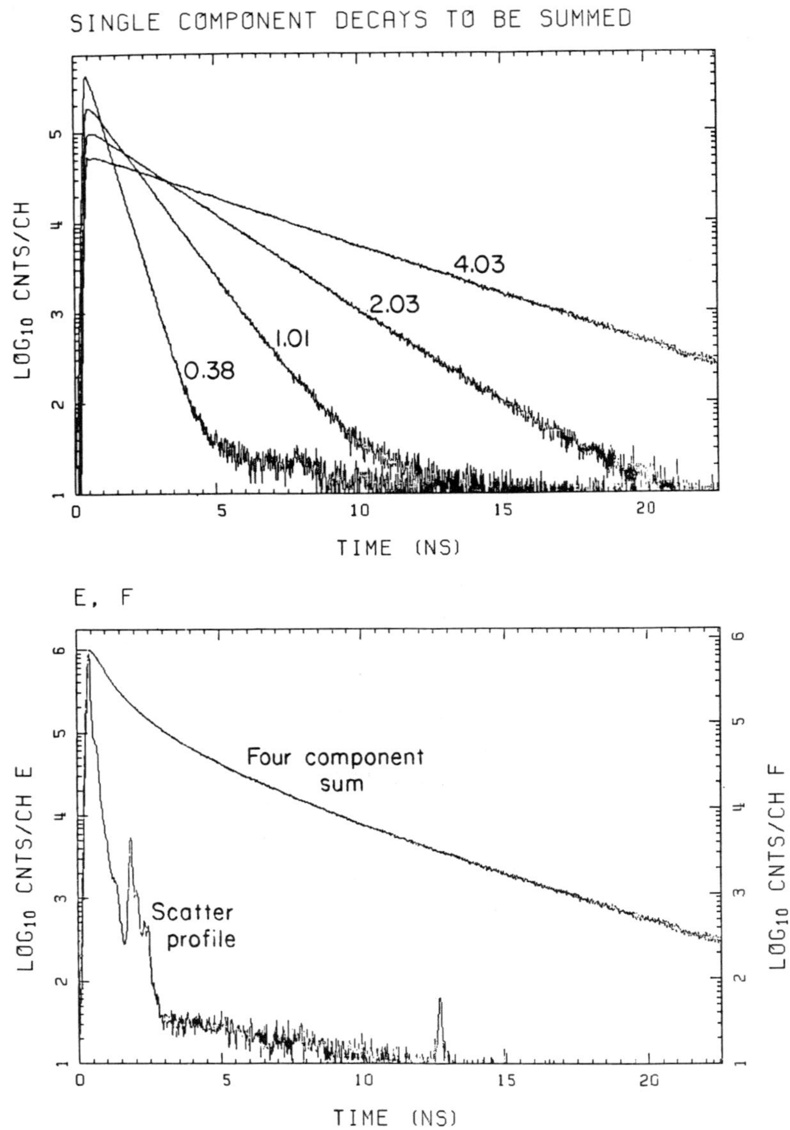


Fig. 13. Four-component decay of tyrosine fluorescence. The data were provided by Dr. Enoch Small, University of Oregon. The source was described previously [Small et al., 1984] except for the addition of a microchannel plate photomultiplier tube (PMT). Four single-component decays with lifetimes of 0.38, 1.01, 2.03, and 4.03 nsec are shown in the upper panel. These data sets were summed to give the four-component decay in the lower panel. The samples were excited at 280 nm and the emission measured through a narrow band-pass filter at 300 nm. CNTS, counts; CH, channel.

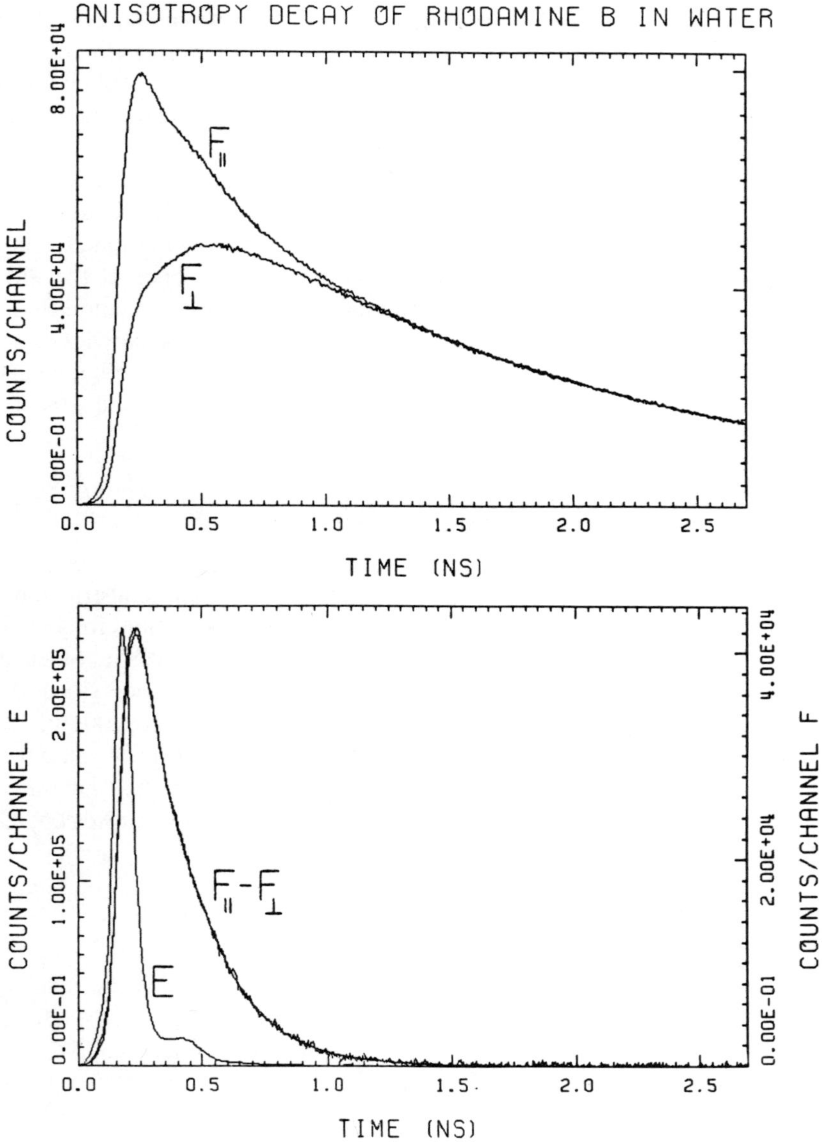


Fig. 14. Time-resolved difference decay $[(R_{\parallel}(t) - R_{\perp}(t)]$ for rhodamine B in water at 20°C. These data were provided by Dr. Enoch Small, University of Oregon. The upper panel shows the decays obtained with the emission polarizer either parallel or perpendicular to the excitation polarizer direction. The emission was measured through a Corning CS 3–66 cutoff filter. The lower panel shows the lamp profile for 556-nm excitation and the difference decay $(F_{\parallel} - F_{\perp})$.

together, the additional data greatly stabilize the analysis, and allow more closely spaced decay times to be resolved. Such methods can be extended to include multiple samples, say at different temperatures or ligand concentrations. Global methods have already been extended to phase-modulation fluorometry [Beechem et al., 1983] and to frequency-swept phase-modulation fluorometry [Lakowicz and Maliwal, 1985].

Another advance is the use of reference samples in place of scattered light for quantitation of the lamp profile [Libertini and Small, 1984; Zuker et al., 1985]. This procedure can improve the results by minimizing the effects of wavelength or geometric considerations on the time-response of the PMT. The use of standards has become commonplace in phase-modulation fluorometry where systematic errors are more easily detected [Lakowicz et al., 1981]. Additionally, the analysis of single exponential reference compounds should be considered an essential control in any analysis of multiexponential decays [Lampert et al., 1983; Grinvald, 1976].

## PHASE-MODULATION FLUOROMETRY

The capability of phase-modulation fluorometry has advanced dramatically in the past 2 years. The major improvement has been the construction of instruments which operate over a wide range of modulation frequencies [Gratton and Limkemann, 1983; Lakowicz and Maliwal, 1985]. Since the method of frequency-domain fluorometry is less widely understood than pulse methods we will describe the instrumentation, the methods of data analysis for multiexponential decays of fluorescence, and briefly some measurements of anisotropy decays. In a separate chapter in this volume we describe the application of frequency-domain fluorometry to anisotropy decays and time-resolved emission spectra (see Lakowicz, this volume).

### Fixed-Frequency Phase-Modulation Fluorometry

In phase-modulation fluorometry the sample is excited with intensity-modulated light (Fig. 2). Because of the time lag between absorption and emission, the emission is delayed in phase and demodulated relative to the incident light. The phase delay ($\phi$) and demodulation factor ($m$) can be used to calculate the fluorescence lifetime (equations 5 and 6). Historically, the phase method preceded the pulse method, the first such instrument being constructed by Gaviola in 1926. In subsequent years a number of phase fluorometers have been constructed [Muller et al., 1965; Bailey and Rollefson, 1953; Bonch-Breuvich et al., 1959; Birks and Dyson, 1961]. Most current and commercially available instruments are based upon the design described by Spencer and Weber [1969]. This instrument uses an ultrasonic modulator [Debye and Sears, 1932; Bergman, 1938] to obtain modulated light and employs cross-correlation detection. In this application cross-correlation consists of modulating the gain of the detector at a frequency slightly different than that used for light modulation. The low-frequency difference signal contains the same phase and modulation information as the original higher-frequency signals. This procedure simplifies the design and use of these instruments, and improves precision and reliability. Until recently, technical factors have limited the instrumentation to two or three modulation frequencies. Under favorable circumstances, the $\phi$ and $m$ values measured at these frequencies can be adequate to obtain the components of a heterogeneous decay. However, with this limited amount of data it has been difficult to obtain reliable resolution of multiexponential decays [Jameson and Gratton, 1983].

Phase-modulation methods, even at fixed modulation frequencies, have a number of advantages. Short (subnanosecond) lifetimes are easily measured by the phase method. These shorter lifetimes are difficult to measure by the pulse method since lamp pulses are usually 2 nsec wide. Phase and modulation measurements are rapid, requiring only seconds of data acquisition. Longer acquisition times are needed to determine the time-resolved decays. Finally, the technique of phase-sensitive detection of fluorescence increases the usefulness of phase-modulation methods in the analysis of heterogeneous samples. By using this technique with a single modulation frequency it is possible to determine the lifetimes and directly record the individual emission spectrum of each fluorophore from a mixture of two fluorophores [Lakowicz and Cherek, 1981a,b]. More recently, it has become possible to resolve three-component mixtures by using phase-sensitive detection [Keating-Nakamoto et al., 1985].

The resolution possible by using phase-sensitive detection was illustrated in the resolution of tyrosine and tryptophan emissions [Lakowicz and Cherek, 1981b]. This technique allowed the resolution of these closely spaced lifetimes, only 1 nsec apart, even in the UV where sensitivity is limited. Apparently the reason this procedure has been successful is that it uses data obtained over the entire emission spectrum. In this respect it is comparable to the global method of Brand and co-workers. Recently the capabilities of this method have been extended to allow three components to be resolved at a single modulation frequency [Keating-Nakamoto et al., 1985].

Several groups have attempted to improve on this technique. Gratton and Jameson [1985] developed a numerical procedure for calculating the spectra based on known lifetimes. This technique may be of some use when the lifetimes are known prior to the analysis, which is not usually the case. Also, these authors examined samples with widely spaced lifetimes, as compared to the closely spaced lifetimes resolved by Lakowicz et al. [1981a,b].

The main limitation in phase-modulation fluorometry has been the lack of variable-frequency modulation. The Debye-Sears modulators used in commercial instrumentation operate only at fixed frequencies. Also, the ethanol-water mixtures used in the Debye-Sears modulators may absorb the RF energy above 50 MHz, and generally do not provide stable modulation at any frequency. Because of the recent development of the variable-frequency instruments it seems probable that additional frequencies will be added to the present fixed-frequency fluorometers. In my opinion, a limited number of fixed frequencies, say ten or less, will not be adequate for the resolution of complex intensity or anisotropy decays. This is because some of the frequencies will be too fast or slow to provide useful information, and the remaining four or five frequencies will simply not contain enough information to resolve two- or three-component decays. In our experience a reasonable number of frequencies (20–30) allows for the confident distinction between competing models.

## Analysis of Phase and Modulation Data at Fixed Frequencies

A number of methods have been described to analyze heterogeneous emission by using phase-modulation data. Weber described a direct closed-form procedure to calculate the $\alpha_i$ and $\tau_i$ values in an N-component decay from the phase and modulation values obtained at N-modulation frequencies [Weber, 1981]. This procedure was used to resolve the pH-dependent decay times of tryptophan fluorescence [Jameson and Weber, 1981]. At pH values near neutrality, the zwitterionic form of tryptophan has a lifetime of 3 nsec. At pH values near 10, the negatively charged form displays a lifetime of about 9 nsec. At intermediate pH values both species are present. These species do not interchange during the lifetime of the excited state and hence these tryptophan solutions display ground-state heterogeneity. From the phase and modulation data obtained at two frequencies, Jameson and Weber were able to calculate the correct values of $\alpha_i$ and $\tau_i$. The errors in these calculated values were about tenfold larger than in the measured values. Hence, highly accurate data are necessary. It seems unlikely that this method will provide reliable resolution of more closely spaced decay times. This is not because the method is flawed, but rather because of the limited information (two to three phase and modulation values) which is used to determine the parameters. A small error in any of the values can lead to misleading results. It is known that commercial fixed-frequency instruments have substantial systematic errors [Lakowicz et al., 1981]. Such errors in any one data point will also invalidate the least-squares analysis of the fixed-frequency data proposed by Matayoshi [1980].

## Analysis of Variable-Frequency Phase and Modulation Data

Since variable-frequency instruments are now available it is appropriate to describe the methods used for analyzing these data. We developed a procedure by using nonlinear least squares [Lakowicz et al., 1984], which is analogous to the analysis of time-resolved data. It is possible to predict the phase and modulation value for any assumed decay law, just as it is possible to predict the time-resolved fluorescence. In the frequency-domain, convolution with the lamp profile is replaced with sine and cosine transformations of the impulse response (equation 11).

$$N_\omega = \frac{\int_0^\infty F(t) \sin \omega t \, dt}{\int_0^\infty F(t) \, dt} \tag{17}$$

$$D_\omega = \frac{\int_0^\infty F(t) \cos \omega t \, dt}{\int_0^\infty F(t) \, dt} \tag{18}$$

At each modulation frequency ($\omega = 2\pi \times$ frequency in cycles per second) the phase angle and modulation can be predicted from

$$\phi_{c\omega} = \text{arc tan } (N_\omega / D_\omega) \tag{19}$$

$$m_{c\omega} = (N_\omega^2 + D_\omega^2)^{1/2} \tag{20}$$

The predicted values are compared with the measured values ($\phi_\omega$ and $m_\omega$) and the parameters describing the decay law are adjusted to yield the best fit (minimum $\chi_R^2$). The detailed expressions are as follows. Assume the decay is described by equation 11. Then

$$N_\omega J = \sum_i \frac{\alpha_i \omega \tau_i^2}{1 + \omega^2 \tau_i^2} \tag{21}$$

$$D_\omega J = \sum_i \frac{\alpha_i \tau_i}{1 + \omega^2 \tau_i^2} \tag{22}$$

and

$$J = \sum_i \alpha_i \tau_i \tag{23}$$

The goodness-of-fit between the measured values and the assumed model is judged by the value of $\chi^2$, which is given by

$$\chi^2 = \sum_{\omega} \frac{1}{\sigma_{\phi\omega}^{\;2}} (\phi_\omega - \phi_{c\omega})^2 + \sum_{\omega} \frac{1}{\sigma_{m\omega}^{\;2}} (m_\omega - m_{c\omega})^2 \tag{24}$$

In this expression $\sigma_{\phi\omega}$ and $\sigma_{m\omega}$ are the errors of the measured phase and modulation values under the chosen experimental conditions. The goodness-of-fit is usually judged by the value of reduced $\chi^2$,

$$\chi_R^2 = \frac{\chi^2}{\nu} = \frac{\chi^2}{2N - p} \tag{25}$$

where the number of degrees of freedom is given by $\nu = 2N - p$. $N$ is the number of modulation frequencies and p is the number of floating parameters. If the assumed model is appropriate for the sample, and if the errors are random and properly estimated by $\sigma_{\phi\omega}$ and $\sigma_{m\omega}$, then $\chi_R^2$ is expected to be near unity.

The appearance of phase and modulation data is illustrated in Figure 15, which shows simulated data for a two-component decay, 5 and 20 nsec (•). The phase angles increase and the modulation decreases with increasing modulation frequency. The dashed line shows the best possible fit through the simulated data with a single-component decay law. This fit is clearly unacceptable. The systematic deviations between the "measured" and calculated (–○–) values (lower panels) and the high value of $\chi_R^2$ also indicate the fit is not acceptable. Inclusion of a second component in the decay law yields a value of $\chi_R^2 = 1.0$ and random deviations (•). Additionally, the calculated values (solid line, top panel) now overlap with the stimulated data. We chose this method of presentation to be analogous to that used in time-resolved fluorometry (Figs. 7, 9).

The nature of the $\chi_R^2$ surface determines the decay times which can be resolved. Cross sections of one such surface are shown in Figure 16. If the lifetimes are widely spaced (5 and 20 nsec, ——) then $\chi_R^2$ displays a sharp minimum near the correct values. If one of these lifetimes is held fixed at a slightly incorrect value (say 21 nsec), then $\chi_R^2 = 7$, even after adjustment of the other variable parameters [Lakowicz et al., 1984]. Hence, the correct values are easily identified. If the lifetimes are more closely spaced, then the $\chi_R^2$ surface displays broader minima. When the lifetimes differ by only 25% (20 and 15 nsec) the $\chi_R^2$ minima are quite flat, resulting in less certain values of the resolved lifetimes. With a single wavelength (nonglobal methods) our present resolution allows lifetimes which are different by 40% to be reliably recovered.

The lower panel shows the effects of a higher level of random noise, which is comparable to a lower number of counts in time-resolved fluorometry. An

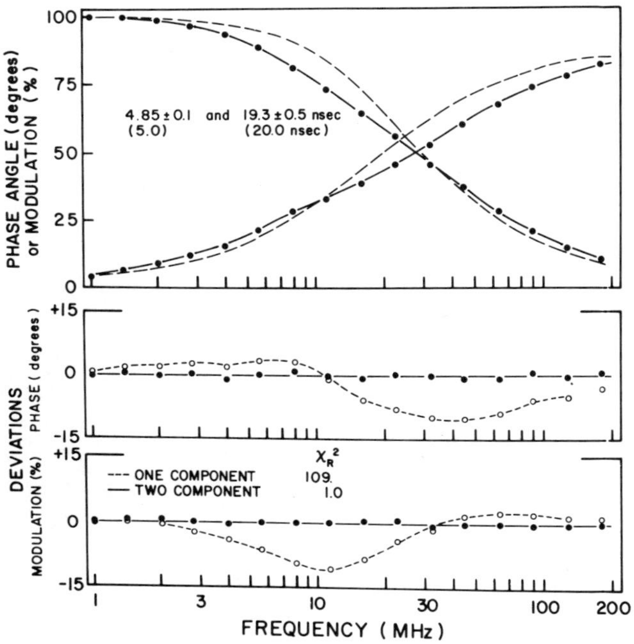


Fig. 15. Simulated phase and modulation data for a two-component decay, 5 and 20 nsec. The fractional intensities of the components were assumed equal. The added levels of Gaussian noise where $\sigma_p = 0.5°$ and $\sigma_m = 0.005$. From Lakowicz et al. [1984].

increased level of uncertainty in the measured values results in increased uncertainty in the calculated parameter. Similar surfaces could be constructed for time-resolved data.

## A Frequency-Domain Phase-Modulation Fluorometer

The technique of frequency-domain is relatively new, and it seems valuable to describe the instrument. Additional details may be found elsewhere [Lakowicz and Maliwal, 1985]. A schematic description of our instrument is shown in Figure 17. This instrument can be easily constructed by using commercially available components. The output of a CW-laser is intensity modulated in an approximate sinusoidal manner at a frequency $F$. The intensity-modulated light excites the sample, and the emission is observed by using a photomultiplier. Another PMT is placed near a quartz beam splitter and is used as a phase reference. This PMT can also be positioned in the T-format location for differential polarized phase measurements. Measurement of the phase angle difference between the sample and reference emission is accomplished by rotation of a turret. The phase angle difference ($\phi_\omega$) and the

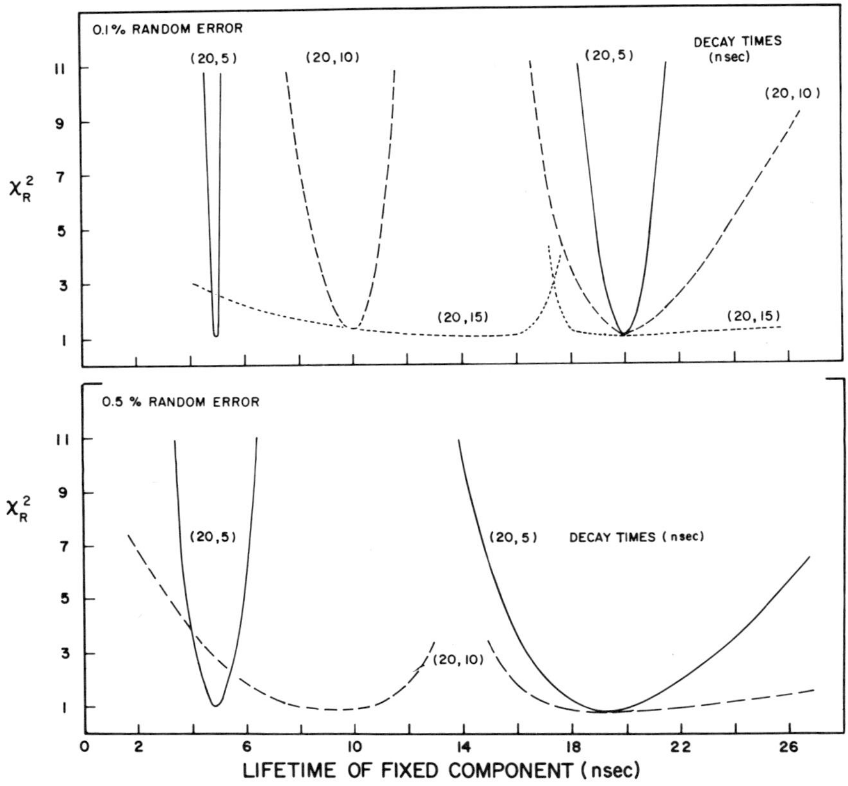


Fig. 16. Dependence of $\chi_R{}^2$ on the lifetime values used to fit a two-component decay. Upper panel, $\sigma_p = 0.1°$ and $\sigma_m = 0.001$; lower panel, $\sigma_p = 0.5°$ and $\sigma_m = 0.005$. From Lakowicz et al. [1984].

relative modulation of the sample ($m_\omega$) can be measured relative to either scattered light or a reference fluorophore. The subscript $\omega$ refers to the modulation frequency, expressed in radians per second ($\omega = 2\pi F$). Detection of the emission is performed by using the cross-correlation method. Specifically, the gains of the PMTs are modulated at the frequency $F + 25$ Hz, where $F$ is the modulation frequency of the incident light. The phase and modulation are measured at the 25 Hz cross-correlation frequency by a time-interval counter and a ratio digital voltmeter, respectively. The two closely spaced frequencies are provided by two frequency synthesizers which are each driven by the same crystal reference source (phase locked). The data are averaged and stored by an interfaced computer. Measurement at 20 modulation frequencies typically requires 1 hour. At present the data acqui-

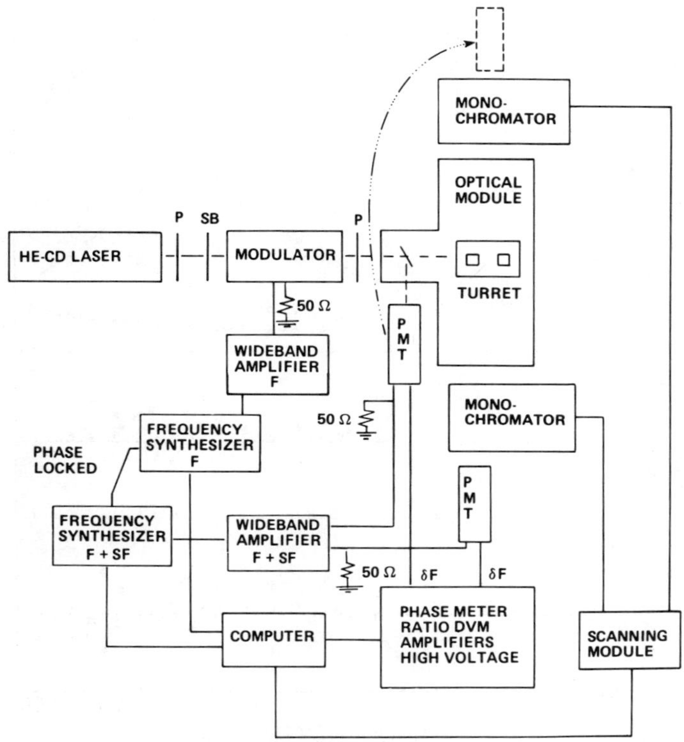


Fig. 17. Schematic of the variable-frequency phase-modulation fluorometer. P, polarizer; SB, Soleil-Babinet compensator; F, frequency; δF, cross-correlation frequency; PMT, photomultiplier tube. From Lakowicz and Maliwal [1985].

sition is not automated, and most of this time is spent in instrument manipulation rather than data acquisition.

Intensity modulation of the laser light is accomplished by using an electro-optic modulator. In response to an applied electric field these devices change linearly polarized light into circularly polarized light. The theory and operation of these devices are described elsewhere [Wilson and Hawkes, 1983; Kaminov, 1974]. The electro-optic modulator is placed between crossed polarizers. If an RF voltage is applied to the modulator, and no optical or electrical bias is applied, then the transmitted light is modulated at twice the applied frequency. We use the modulators in the approximate linear mode. This can be accomplished by using either a voltage bias or an optical bias. If the bias compensator is adjusted to the λ/4 position then the light incident on the modulator is circularly polarized. When a voltage is applied the polari-

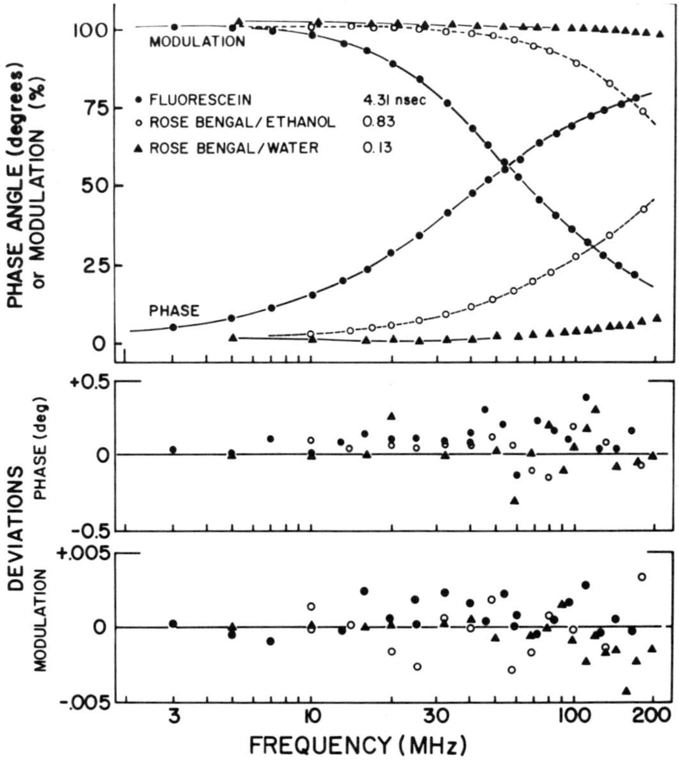


Fig. 18. Phase and modulation data for single-component solutions. Data are shown for fluorescein in water (0.02 M tris, pH = 8) (●), rose bengal in ethanol (○), and in water (▲). The lower panel shows the deviations between the measured and calculated values. From Lakowicz and Maliwal [1985].

zation is rotated further. Hence, the transmission is varied, and the incident light is modulated at the applied frequency.

## Results From Frequency-Domain Fluorometry

The data in Figure 18 illustrate the use of frequency-domain data for measurement of single exponential decay times. The data are for fluorescein in water (4.3 nsec) and for rose bengal in ethanol (0.83 nsec) and in water (0.13 nsec). At any given frequency larger phase angles and smaller modulations are seen for the longer lifetimes. The phase angles increase and the modulation values decrease with increasing modulation frequency. The solid lines (top panel) represent the theoretical curves calculated for the lifetimes which yielded minimum values of $\chi_R^2$. Obviously, the calculated values for a single exponential decay accurately reproduce the measured values (•, ○, ▲). The agreement between the measured and calculated data is further illustrated in the lower panel of Figure 18, which shows the deviations between the measured data and the calculated values. The deviations are randomly distributed around zero. This indicates that the single exponential model is adequate to account for the data, and that the measurements are free of systematic errors. The average deviations are near 0.2 degrees for phase and 0.002 for modulation, when the instrument in this laboratory is used. We found these random errors, or smaller, to be the apparent level of random error for most of our measurements. These values were used for the calculation of ${\chi_R}^2$. In contrast to photon counting data, the random errors in phase and modulation cannot be estimated on the basis of counting statistics. Rather, we estimate the level of random error by comparison of the data with the calculated values for cases in which the assumed decay law is appropriate. The fact that ${\chi_R}^2$ are often less than unity indicates that the actual level of random error is less than 0.2 degrees of phase and less than 0.002 in modulation.

We attempted to fit the data for fluorescein and rose bengal to double exponential decays of intensity. The values of ${\chi_R}^2$ did not substantially decrease and the amplitude of the assumed second component was negligible. Such results demonstrate that inclusion of an additional component is not justified from the data. These results illustrate the precision and accuracy of our measurements. There do not appear to be any significant systematic errors which result in the apparent detection of small additional components in the decay law. We note that these measurements were performed without use of standard fluorophores, indicating the absence of significant color or geometric artifacts in the time response of the photomultiplier tubes.

In our opinion the major challenge is not measurement of the shortest lifetimes. In biochemical research it is more important to resolve closely spaced decay times. Since the decay times ($\tau_i$) and the preexponential factors ($\alpha_i$) are correlated parameters, one requires precise measurements to obtain reliable estimates of $\alpha_i$ and $\tau_i$. To illustrate the resolution obtainable from the frequency-domain measurements we examined several two-component mixtures. One such resolution is shown in Figure 19 for a mixture of anthracene (4.08 nsec) and 9,10-diphenylanthracene (6.31 nsec). Obviously, a good fit to the measured data (•) is provided by an assumed two-component decay law (——).

The lower panels show the deviations between the measured and calculated values for the best fit obtainable by using a single (-○-) and double (—•—) exponential decay law. For this mixture, in which the decay times differ by 40%, the experimental deviations found by using a single decay time are about ten- to twentyfold larger than the random error level of the measure-

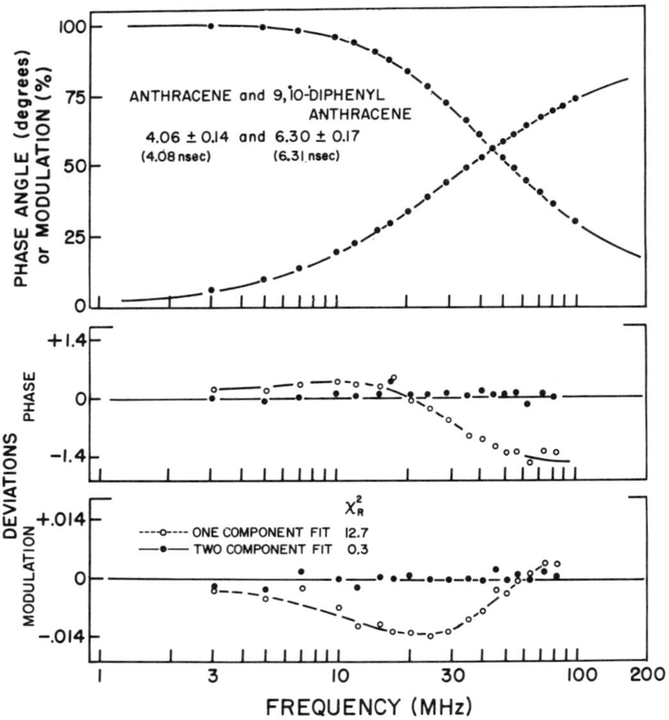


Fig. 19. Phase and modulation data for a mixture of anthracene and 9,10-diphenylanthracene. The dashed line indicates the best one-component fit, and the solid line the best two-component fit. The values not in parentheses are the experimental values and those in parentheses are from measurements on the pure components. From Lakowicz and Maliwal [1985].

ments. For the single-component fit the deviations vary systematically with the frequency, indicating that the deviations are not due to random error and are probably due to an inappropriate decay law. Additionally, the value of $\chi_R^2 = 12.7$ is clearly unacceptable. The inclusion of a second component in the decay law results in a fortyfold decrease in $\chi_R^2$ to 0.30, and the deviations become small and randomly distributed around zero. Importantly, the calculated decay times and fractional intensities are in good agreement with the expected values.

We note that lifetimes more closely spaced than 4 and 6 nsec can also be resolved. By using a method analogous to the global method of Brand and coworkers, we resolved a two-component decay with lifetimes differing by only 8%, 4.1 and 4.4 nsec [Lakowicz and Maliwal, 1985]. To the best of our knowledge these are the most closely spaced lifetimes which have been resolved to date, excluding those resolved by phase-sensitive detection in

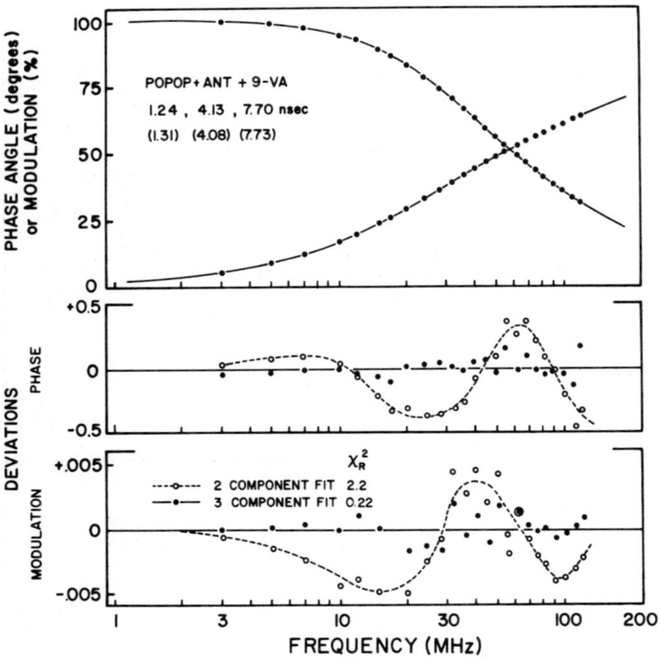


Fig. 20. Phase and modulation data for a three-component mixture of POPOP, anthracene and 9-vinylanthracene. The fractional intensities of the three components are equal ($f_1 = f_2 = f_3 = 0.33$). From Lakowicz and Maliwal [1985].

which the emission spectra of the components were known. Irrespective of the use of global methods, a 40-fold decrease in $\chi_R^2$ from a one- to a two-component fit (Fig. 19) illustrates the substantial resolution possible with frequency-domain fluorometry. A 40-fold decrease in $\chi_R^2$ is more than is required, which indicates that more closely spaced lifetimes could be resolved. (We note that in our analysis the absolute values of $\chi_R^2$ are not always near unity.)

It is well known that the resolution of a three-component decay is considerably more difficult than a two-component decay [Grinvald and Steinberg, 1974]. Hence, it was of interest to test our instrument by using a three-component mixture. We examined three-component mixtures of POPOP (1.32 nsec), anthracene (4.1 nsec), and 9-vinylanthracene (7.7 nsec). The resolution of a three-component mixture with a sixfold range of lifetimes is a difficult problem. Figure 20 shows the two- and three-component fit. The two-component fit is easily judged to be unsuccessful because of the systematically varying deviations and the large value of $\chi_R^2$. The two-component fit

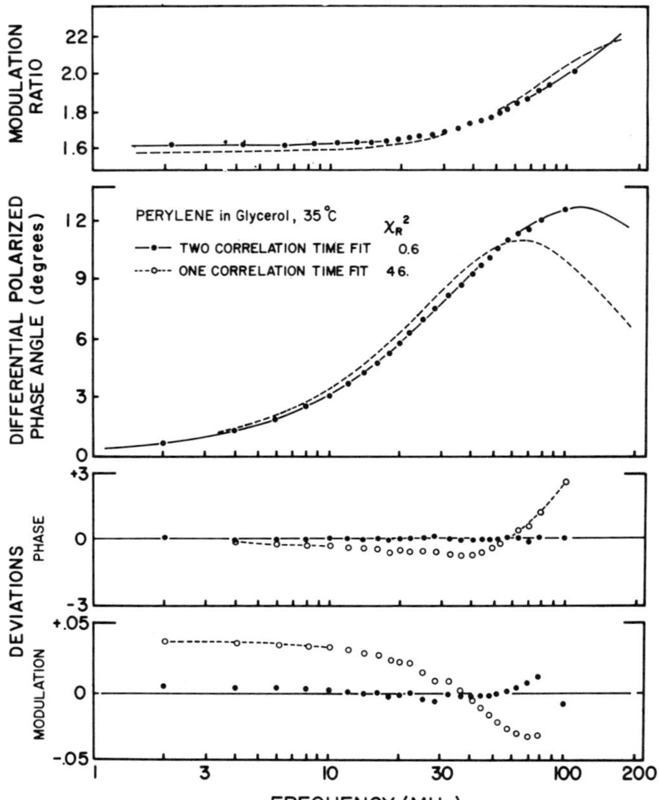


Fig. 21. Anisotropy decay of perylene in glycerol. The correlation times for the two-component fit are 1.5 and 10.8 nsec. From Lakowicz and Maliwal [1985].

displays deviations which are about fivefold larger than the expected level of random error. Inclusion of the third component yields a tenfold decrease in $\chi_R^2$ and more random deviations.

## Anisotropy Decay From Frequency-Domain Measurements

Phase-modulation methods are uniquely suited for the measurement of rapid and/or complex decays of fluorescence anisotropy. This is because the signal resulting from rotational diffusion is measured directly. One representative case is shown in Figure 21 for perylene in glycerol. This molecule displays at least two rotational correlation times, with the in-plane rotation being more rapid. The procedures used for measurement and analysis of anisotropy decays are described elsewhere in this volume in more detail (see Lakowicz, this volume). The important point is that the frequency-domain data provided a distinction between one and two rotational correlation times with nearly an 80-fold decrease in $\chi_R^2$. We have also demonstrated that rotational correlation times as short as 47 psec can be measured with the present instrumentation. These results demonstrate that frequency-domain fluorometry provides remarkable resolution of rapid and/or complex decays of fluorescence anisotropy.

## SUMMARY

Technological advances have resulted in substantially improved instrumentation for both time-resolved and frequency-domain fluorescence. By using either method it is possible to obtain reliable three-component resolutions. This ability to obtain reliable estimates of complex decay laws will facilitate the interpretation of the decay data in terms of the molecular properties of the biological macromolecules.

## ACKNOWLEDGMENTS

This work was supported by grants from the National Science Foundation (DMB-8502853, PCM 82-108778, and 81-20068), from the Department of Defense (DAAG 29-85-G-0017), and from the National Institute of Health (GM-29318). The author thanks Dr. Badri P. Maliwal for measuring the time-resolved decays of anthracene and 9-vinylanthracene. My special thanks goes to Dr. Enoch Small for providing the data on tyrosine decays and rhodamine B.

## REFERENCES

Badea MG, Brand L (1971): Time-resolved fluorescence measurements. Methods Enzymol 61:378–425.

Bailey EA, Rollefson GK (1953): The determination of the fluorescence lifetimes of dissolved substances by a phase shift method. J Chem Physiol 21:1315–1326.

Beechem JM, Knutson JR, Ross JAB, Turner BJ, Brand L (1983): Global resolution of heterogeneous decays by phase/modulation fluorometry: Mixtures and Proteins. Biochemistry 22:6054–6058.

Beechem JM, Ameloot M, Brand L (1985): Analysis of fluorescence intensity and anisotropy decay surfaces. In Szabo A, Marsotti L (eds): "Excited State Probes in Biochemistry and Biology." New York: Plenum (in press).

Bennett RG (1960): Instrument to measure lifetimes in the millisecond region. Rev Sci Instrum 31:1275–1279.

Bergman L (1938): "Ultrasonics and Their Scientific and Technical Applications." London: G. Bell and Sons, Ltd.

Bevington PR (1969): "Data Reduction and Error Analysis for the Physical Sciences." New York: McGraw-Hill.

Birch DJS, Imhof RE (1981): Coaxial nanosecond flashlamp. Rev Sci Instrum 52:1206–1212.

Birks JB, Dyson DJ (1961): Phase and modulation fluorometer. J Sci Instrum 38:282–285.
Bonch-Breuvich AM, Kazarin IM, Molchanov VA, Shirkov IV (1959): An experimental model of a phase fluorometer. Instrum Exp Tech 2:231–236.
Cundall RB, Dale RE (1983): "Time-Resolved Fluorescence Spectroscopy in Biochemistry and Biology." New York: Plenum, pp 115–296.
Debye P, Sears FW (1932): On the scattering of light by supersonic waves. Proc Natl Acad Sci USA 18:409–414.
Demas JN (1983): "Excited State Lifetime Measurements." New York: Academic Press.
Gafni A, Modlin RL, Brand L (1975): Analysis of fluorescence decay curves by means of the Laplace transformation. Biophys J 15:263–280.
Gaviola Z (1926): Ein Fluorometer. Apparat zur Messung von Fluoreszenzabklingungszeiten. Zeitschrift Physik 42:853–861.
Gratton E, Limkemann M (1983): A continuously variable frequency cross-correlation phase fluorometer with picosecond resolution. Biophys J 44:315–324.
Gratton E, Jameson DM (1985): A new approach to phase and modulation resolved spectra. Anal Chem (in press).
Gratton E, Limkemann M, Lakowicz JR, Maliwal BP, Cherek H, Laczko G (1984): Resolution of mixtures of fluorophores using variable-frequency phase and modulation data. Biophys J 46:479–486.
Grinvald A (1976): The use of standards in the analysis of fluorescence decay data. Anal Biochem 75:260–280.
Grinvald A, Steinberg IZ (1974): On the analysis of fluorescence decay kinetics by the method of least-squares. Anal Biochem 59:583–598.
Hundly L, Coburn T, Garwin E, Stryer L (1967): Nanosecond fluorometer. Rev Sci Instrum 38:488–492.
Isenberg I, Small EW (1982): Exponential depression as a test of estimated decay parameters. J Chem Physiol 77:2799–2805.
Jameson DM, Weber G (1981): Resolution of the pH-dependent heterogeneous fluorescence decay of tryptophan by phase and modulation measurements. J Phys Chem 85:953-958.
Jameson DM, Gratton E (1983): Analysis of heterogeneous emissions by multifrequency phase and modulation fluorometry. In Eastwood D (ed): "New Directions in Molecular Luminescence." Philadelphia: ASTM, pp 67–81.
Johnson ML (1983): Evaluation and propagation of confidence intervals in nonlinear, asymmetrical variances spaces; analysis of ligand binding data. Biophys J 44:101–106.
Kaminov IP (1974): "An Introduction to Electro-Optic Devices." NY: Academic Press.
Keating-Nakamoto S, Cherek H, Lakowicz JR (1985): A new method for resolution of two and three component mixtures of fluorophores by phase sensitive detection of fluorescence. Anal Biochem 148:349–356.
Knutson JR, Beechem JM, Brand L (1983): Simultaneous analysis of multiple fluorescence decay curves: A global approach. Chem Phys Lett 1-2:501–507.
Koester VJ, Dowben RM (1978): Subnanosecond single photon counting fluorescence spectroscopy using synchronously pumped tunable dye laser excitation. Rev Sci Instrum 49:1186–1191.
Lakowicz JR (1983): "Principles of Fluorescence Spectroscopy." New York: Plenum.
Lakowicz JR, Cherek HC (1981a): Phase-sensitive fluorescence spectroscopy. A new method to resolve fluorescence lifetimes or emission spectra of components in a mixture of fluorophores. J Biochem Biophys Methods 5:19–35.
Lakowicz JR, Cherek H (1981b): Resolution of heterogeneous fluorescence from proteins and aromatic amino acids by phase-sensitive detection of fluorescence. J Biol Chem 256:6348–6353.

Lakowicz JR, Maliwal BP (1985): Construction and performance of a variable-frequency phase-modulation fluorometer. Biophys Chem 21:61–78.

Lakowicz JR, Cherek H, Balter A (1981): Correction of timing errors in photomultiplier tubes used in phase-modulation fluorometry. J Biochem Biophys Methods 5:131–146.

Lakowicz JR, Cherek H, Maliwal BP, Gratton E (1984a): Time-resolved fluorescence anisotropies of diphenylhexatriene and perylene in solvents and lipid bilayer obtained from multifrequency phase-modulation fluorometry. Biochemistry 24:376–383.

Lakowicz JR, Laczko G, Cherek H, Gratton E, Limkemann M (1984b): Analysis of fluorescence decay kinetics from variable-frequency phase shift and modulation data. Biophys J 46:463–477.

Lampert RA, Chewter LA, Phillips D, O'Connor DV, Roberts AJ, Meech SR (1983): Standards for nanosecond fluorescence decay measurements. Anal Chem 55:68–73.

Libertini LJ, Small EW (1984): F/F deconvolution of fluorescence decay data. Anal Biochem 138:314–318.

Matayoshi ED (1980): Distribution of shape-changing compounds across the red cell membrane. Biochemistry 19:3414–3422.

Muller A, Lumry R, Kokubun H (1965): High performance phase fluorometer constructed from commercial subunits. Rev Sci Instrum 36:1214–1226.

O'Connor DVO, Ware WR, Andre JC (1979): Deconvolution of fluorescence decay curves. A critical comparison of techniques. J Phys Chem 83:1333–1343.

Small EW, Isenberg I (1977): On moment index displacement. J Chem Physiol 66:3347–3451.

Small EW, Isenberg I (1983): Fluorescence decay analysis by the method of moments. In Cundall RB, Dale RE (eds): "Time-Resolved Fluorescence Spectroscopy in Biochemistry and Biology." NY: Plenum Publishing Co.

Small EW, Libertini LJ, Isenberg I (1984): Construction and tuning of a monophoton decay fluorometer with high resolution capabilities. Rev Sci Instrum 55:879–885.

Spencer RD, Weber G (1969): Measurement of subnanosecond fluorescence lifetimes with a cross-correlation phase fluorometer. Ann NY Acad Sci 158:361–376.

van Hoek A, Vervoort J, Visser AJWG (1983): A subnanosecond resolving spectrofluorometer for the analysis of protein fluorescence. J Biochem Biophys Methods 7:243–254.

Ware WR (1983): Technique of pulse fluorometry. In Cundall RB, Dale RE (eds): "Time-Resolved Fluorescence Spectroscopy in Biochemistry and Biology." New York: Plenum.

Ware WR, Lee SK, Brant GJ, Chow PO (1970): Nanosecond time-resolved emission spectroscopy: Spectral shifts due to solvent-excited solute relaxation. J Phys Chem 54:4729–4737.

Weber G (1981): Resolution of the fluorescence lifetimes in a heterogeneous system by phase and modulation measurements. J Phys Chem 85:949–953.

Wilson J, Hawkes JFB (1983): "Optoelectronics: An Introduction." New Jersey: Prentice-Hall, pp 85–125.

Yguerabide J (1972): Nanosecond fluorescence spectroscopy of macromolecules. Methods Enzymol 26:498–578.

Zimmerman HE, Penn JH, Carpenter CW (1982): Evaluation of single-photon-counting measurements of excited state lifetimes. Proc Natl Acad Sci USA 79:2128–2132.

Zuker M, Szabo AG, Bromall L, Krvzcarski DT, Selinger B (1985): The delta function deconvolution method (DFCM) for fluorescence decay experiments. Submitted for publication.

*Applications of Fluorescence in the Biomedical Sciences, pages 69–89*


# 3 Ultrafast Photophysical Processes in Visual Pigments

*A.G. Doukas, R.H. Callender, and R.R. Alfano*

Rhodopsins, the name generally given to visual pigments, consist of a single chromophore, 11-*cis* retinal, covalently bound in the form of a protonated Schiff base to the ϵ-amino group of a lysine in the apoprotein, opsin. Vision is initiated when the chromophore absorbs a photon, and rhodopsin is transformed to its primary photoproduct, bathorhodopsin [for recent reviews, see Ottolenghi, 1980; Birge, 1981; Balogh-Nair and Nakanishi, 1982]. There are a number of remarkable properties in the primary photophysics of visual proteins. The formation quantum yield of bathorhodopsin from rhodopsin (0.67) is very high, indicative of an extremely efficient process. Also, some two-thirds of the incident photon's energy is converted to chemical energy in the rhodopsin to bathorhodopsin phototransition since the enthalpy of bathorhodopsin is some 35 kcal/mol higher than that of rhodopsin's [Cooper, 1979]. This chemical energy is used to drive subsequent thermal reactions, ultimately resulting in visual perception.

The rhodopsin-bathorhodopsin transformation has been extensively studied. The in situ retinal chromophore isomerizes, being converted from an 11-*cis* isomer to a *trans*-like species. This structural change and the fact that the Schiff base remains protonated in the rhodopsin-bathorhodopsin photoreaction have been characterized in fair detail by resonance Raman [for reviews, see Callender and Honig, 1977; Mathies, 1977; also see Aton et al., 1980; Eyring et al., 1982], fast Fourier infrared difference spectroscopy [Bagley et al., 1985], and two-photon absorption spectroscopy [Birge et al., 1985]. The bathorhodopsin formation time is very fast, less than 6 psec at room temper-

Department of Physics and Institute for Ultrafast Spectroscopy and Lasers, City College of the City University of New York, New York, New York 10031.

ature, being as yet unresolved in picosecond absorption spectroscopic measurements at physiological temperatures. At much lower temperatures near liquid helium, the formation time increases and is resolved at around 20°K. Peters et al. [1977] found that the rise time of bathorhodopsin followed a non-Arrhenius temperature dependence and increased markedly when measured for samples suspended in $D_2O$. These results prompted their suggestion that a proton(s) translocation process is the rate-limiting step in bathorhodopsin formation. Thus, there are at least two steps in the rhodopsin-bathorhodopsin transformation, isomerization of the retinal chromophore and proton motion of an unspecified, but exchangeable, opsin proton.

For many years, there has been some question as to the identity of the photoproduct of rhodopsin. Low-temperature conventional absorption spectroscopy suggested two possibilities. At 77°K, rhodopsin was shown to photochemically convert to a species absorbing to the red of the primary pigment which was called bathorhodopsin after the bathochromic shift in the absorption maximum [Yoshizawa and Wald, 1963]. Upon warming, bathorhodopsin was found to decay thermally through a series of steps. This is called bleaching since the absorption maximum of the final product absorbs in the near UV and is more or less colorless to the eye. Similar measurements performed near liquid helium temperatures result in the photochemical formation of a blue-shifted species, hypsorhodopsin, in addition to bathorhodopsin [Yoshizawa and Horiuchi, 1973]. Hypsorhodopsin has been found to decay thermally to bathorhodopsin; and this raised the possibility that hypsorhodopsin is a precursor to bathorhodopsin. Kinetic measurements, as will be described in detail below, have yielded conflicting results. However, recent careful work by Yoshizawa and his collaborators suggest that hypsorhodopsin is formed by two-photon reactions [Shichida et al., 1984; Matuoka et al., 1984]. Thus, under normal light conditions, hypsorhodopsin appears to play no physiological role.

There are a number of review articles on picosecond measurements [Applebury, 1980; Alfano, 1982]. In this work we review below our current knowledge of the fast kinetics of visual pigments. We have divided the article into several sections. We first describe the techniques that permit measurement of changes in absorption and fluorescence emission upon excitation on the picosecond time scale. We then describe results obtained with picosecond absorption spectroscopy. Next we discuss the very recent picosecond fluorescence studies from our laboratory. Finally, we describe a simple model that can be used to qualitatively understand the primary photophysics of visual pigments.

## PICOSECOND TIME-RESOLVED TECHNIQUES

The time-resolved techniques that have been developed make use of the physical processes of absorption, fluorescence, and light scattering. The two techniques we will discuss here are picosecond absorption spectroscopy and picosecond fluorescence spectroscopy with a streak camera. A number of reviews are available on picosecond techniques [Rentzepis, 1978; Doukas et al., 1982; Rosen et al., 1984].

The basic method behind the picosecond absorption experiments is the "excite and probe technique". The sample is first excited by an intense pulse and subsequently is probed by a weak pulse as the material relaxes to its original state or moves through some photochemical reaction. The second harmonic at 530 nm from an Nd:YAG or Nd:glass laser is commonly used as a pump pulse. This wavelength is well situated for the study of visual pigments. For the probe pulse a broad-band source is required, so that a large region of the spectrum can be probed at one time. Such a spectrally white ultrafast pulse is available. It is known as the supercontinuum [Alfano and Shapiro, 1970]. When a very strong laser pulse, the Nd:glass 1060 nm in this case, is focused in $CCl_4$ (or any condensed material or gas for that matter), it generates a broad-band pulse by self phase modulation and four-photon mixing covering the entire visible and IR regions. The duration of the continuum is approximately the same as the generating laser pulse. Figure 1 shows a schematic diagram of the picosecond absorption apparatus. The pump pulse travels a variable path length through a delay prism. The kinetics of the absorption changes can be followed point by point by moving the delay prism across the time domain of interest. At the sample site, the probe beam is divided into two pulses of approximately equal intensity $I^e(\tau)$and $I^r(\tau)$, where $\tau$ is the delay time between pump and probe pulses. $I^e(\tau)$ is transmitted through the area of the sample which is excited by the pump pulse. The intensity of the transmitted probe beam depends on the photochemical changes initiated by the exciting pulse. $I^r(\tau)$ is the reference beam. The change in the optical density is obtained as:

$$\Delta\mathrm{OD}(\tau) = -\log\left\{ \left[\frac{I^e(\tau)}{I^r(\tau)}\right] \left(\frac{I^r}{I^e}\right)\right\} \tag{1}$$

where the ratio $I^r/I^e$ is the normalization factor for the two laser pulses before the sample. The resolution of the picosecond absorption technique is limited only by the duration of laser pulse and the dispersion of the supercontinuum.

The streak camera experimental arrangement is shown in Figure 2. The laser pulse is divided by a beam splitter. Part of the beam is directed to a photodiode to trigger the streak camera. The rest of the beam excites the sample. A wedge in front of the sample is used to reflect a small fraction of the pulse directly into the streak camera. This prepulse is used for the calibration of the zero time. The collected fluorescence from the sample is passed through a cutoff filter to eliminate the excitation pulse and focused onto the slit of the streak camera.

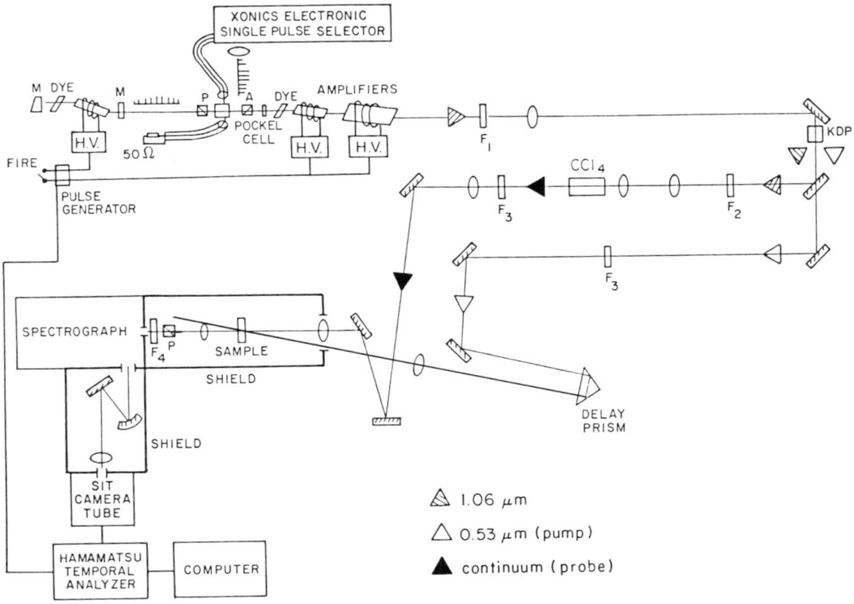


Fig. 1. Schematic diagram of the apparatus used to measure absorption kinetics. A single pulse is selected and amplified. Filter $F_1$ eliminates the scattered light from the flash lamp. The beam is passed through a KDP crystal to produce 530-nm pulses and 1,060 and 530 nm are separated by a specially coated mirror. Filter $F_2$ eliminates the 530-nm pulse. A set of lens collimates and focuses the 1,060-nm pulse into the $CCl_4$ cell to produce the "continuum." Filter $F_3$ eliminates the 1,060-nm pulse and filter $F_4$ eliminates the scattered light from the pump pulse. The probe beam is analyzed by a spectrograph and projected onto the detector.

At the streak camera the time information of the luminous event is converted into spatial information (for a detailed description of the streak camera see Schiller and Alfano [1982], Tsuchida [1984]). Photons striking the photocathode of the streak tube produce an emission of electrons that is proportional to the intensity of the incident light. The photoelectrons are deflected by an applied voltage ramp and impinged across a phosphorescent screen. Photoelectrons released at different times from the photocathode strike the phosphorescent screen at different positions, causing a track with a spatial intensity profile directly proportional to the incident temporal intensity profile of the fluorescence. The phosphorescent track is sampled by an optical multichannel analyzer.

In all picosecond measurements of visual pigments, a possible source of systematic error can arise from the photochemistry of rhodopsin. Rhodopsin can be converted to bathorhodopsin and subsequently to isorhodopsin by

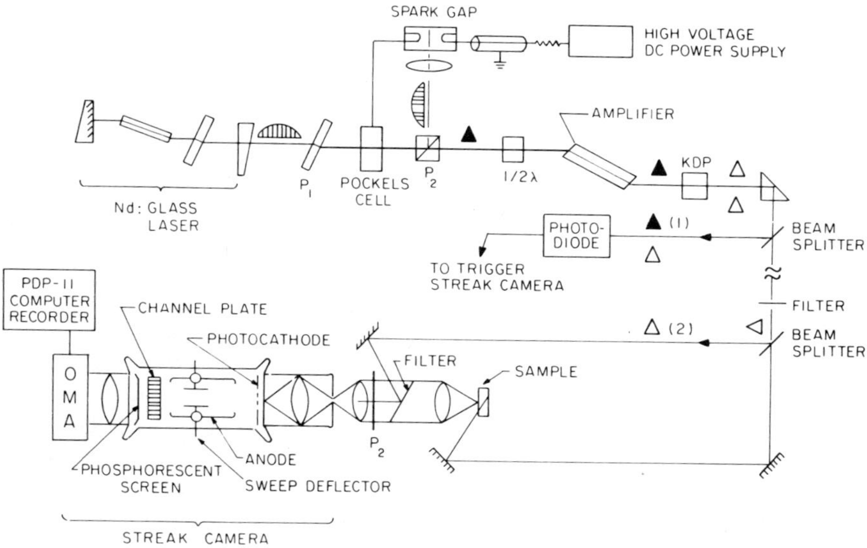


Fig. 2. Schematic diagram of the apparatus used to measure fluorescence kinetics by a streak camera. A single pulse is selected, amplified, and passed through a KDP crystal to produce the second harmonic. A beam splitter provides two side beams. Beam 1 triggers the steak camera. Beam 2 arrives at the streak camera at an earlier time to provide a calibration pulse. The main 530-nm pulse excites the fluorescence from the sample. The streak produced at the phosphorescent screen is recorded by an optical multichannel analyzer (OMA).

light. Under high-intensity radiation, rhodopsin, bathorhodopsin, and to a lesser extent, isorhodopsin, are present. The kinetics studied are, in the higher-intensity regime, the resultant of many different and not well-known photoreactions. On the other hand, decreasing the intensity of the laser pulses is not always possible because the observed signals are normally very weak. It is thus extremely important, in all kinetic studies of visual pigments, to accurately report the parameters characterizing the excitation pulse and to characterize the formation of unwarranted photoproducts.

## PIGMENT TIME-RESOLVED ABSORPTION SPECTROSCOPY

In 1972, the formation kinetics of bathorhodopsin formed photochemically from rhodopsin was probed by picosecond absorption spectroscopy in a pioneering experiment by Bush et al. [1972]. They found bathorhodopsin formation occurred faster than their instrumental resolution of 6 psec.

In Figure 3, we show typical kinetic data, where the absorptive changes at a wavelength near the absorption maximum of bathorhodopsin are monitored, showing the formation of bathorhodopsin after squid rhodopsin was excited. In Figure 4, we show the changes in sample absorption during the excitation pulse (zero time) and 100 psec after excitation in the visible spectral region for experiments performed on cattle rhodopsin. Note the decrease of rhodopsin present in the sample due to excitation resulting in a negative optical absorbance signal, centered near 490 nm, and the simultaneous increase in absorbance centered near 570 nm due to bathorhodopsin formation. Such a fast process prompted some speculation as to whether a major rotation about the retinal chromophore C11-C12 double bond could occur within 6 psec, as required by the 11-*cis* to *trans* isomerization model for the rhodopsin-bathorhodopsin transformation. It is now evident theoretically that such fast rotations upon the absorption of a photon are possible [Warshel, 1976; Birge and Hubbard, 1980; Doukas et al., 1984], even likely, for isomers of protonated Schiff bases. Isorhodopsin has been of special value to the study of the photophysics of visual pigments; it has the property of photochemically forming the same bathorhodopsin as rhodopsin. The earliest experimental

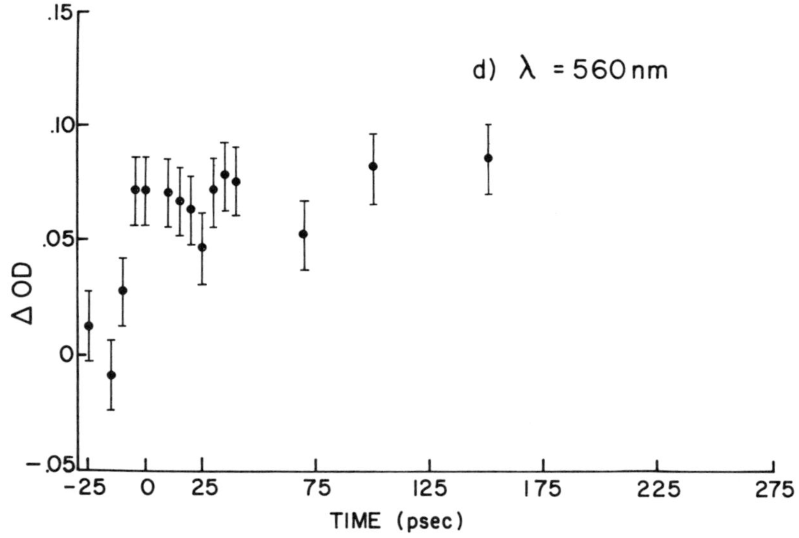


Fig. 3. Laser-induced absorbance changes of solubilized squid rhodopsin (*Todarodes pacificus)* at 560 nm as a function of the delay time between pump and probe pulses at room temperature. Negative time indicates that probe pulse arrives before the pump pulse. The zero time is located by using $CS_2$ Kerr optical shutter at the sample site. The rise time of the optical absorption is unresolved, less than 7 psec [unpublished data from our laboratory].

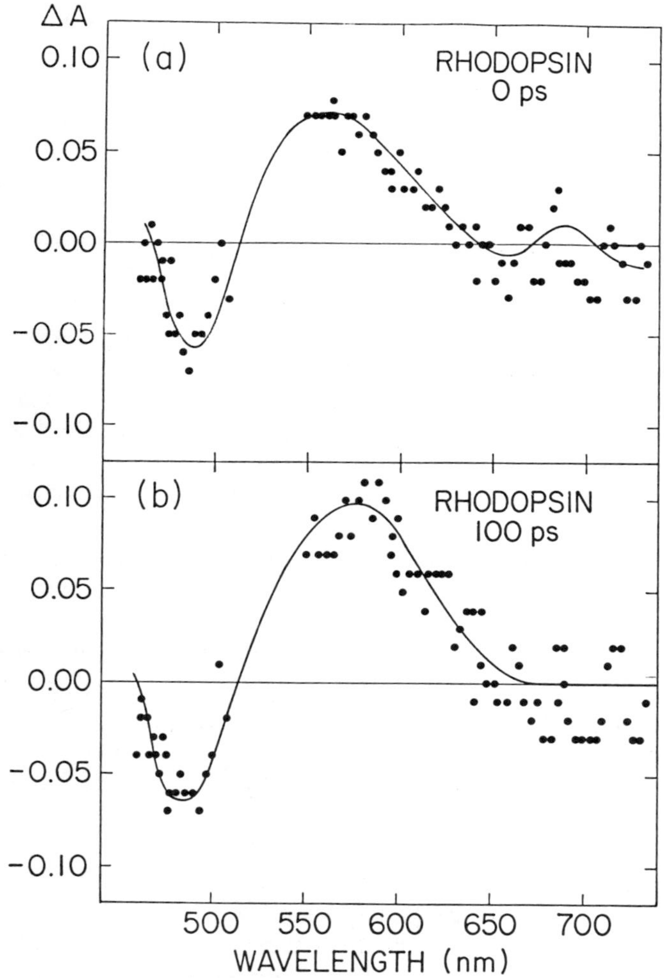


Fig. 4. Laser-induced difference spectra for the formation of bathorhodopsin in detergent-solubilized bovine rhodopsin at room temperature. a. Absorbance changes measured during laser excitation (zero time). b. Absorbance changes measured 100 psec after laser excitation. The small absorption band indicated near 670 nm in the zero time difference spectra is questionable. The isosbestic point near 460 nm indicates that additional positive absorbance changes are generated for $\lambda < 460$ nm [see Monger et al., 1979].

evidence that bathorhodopsin is an isomerized photoproduct of rhodopsin arises from the observation that bathorhodopsin is a common photoproduct of two different *cis* isomers, rhodopsin (11-*cis*) and isorhodopsin (9-*cis*) [Hubbard and Kropf, 1958; Yoshizawa and Wald, 1963]. Green et al. [1977] found that bathorhodopsin formed in less than 6 psec from isorhodopsin and confirmed the results of Bush et al. [1972] for rhodopsin. They argued that since the common bathorhodopsin species was formed from two different *cis* isomers, both in less than 6 psec, a very fast rotation must occur in this system. By deconvoluting the actinic and probe pulse shapes, Green et al. [1977] also obtained the result that bathorhodopsin was formed in approximately 3 psec.

A particularly important series of experiments was performed by Rentzepis and co-workers. They [Peters et al., 1977] found that the formation kinetics of bathorhodopsin could be resolved at very low temperatures and was markedly affected by suspending samples in $D_2O$. The data are summarized in Table 1. The non-Arrhenius temperature dependence and the large isotope effect was interpreted to indicate that the rate-limiting step in bathorhodopsin formation involved proton(s) tunneling or proton(s) translocation. Proteins contain hundreds of exchangeable protons. Since there is little change in the protein structure in the rhodopsin-bathorhodopsin transformation, it would seem reasonable to conclude that only protons in the vicinity of the chromophore participate. It has been suggested [Peters et al., 1977] that the retinal chromophore's Schiff base proton (the only exchangeable proton of the chromophore) translocated during bathorhodopsin formation, but this suggestion appears to be inconsistent with resonance Raman studies [Elyring and Mathies, 1979; Aton et al., 1980; Narva and Callender, 1980]. It seems clear that bathorhodopsin formation involves two events; the ordering of those two processes appears to have been resolved by recent picosecond fluorescence studies, discussed below, which suggest that isomerization occurs first.

As mentioned in the beginning of this chapter, a blue (ca. 430 nm) absorbing pigment, hypsorhodopsin, has been observed as a stable photoproduct of rhodopsin at liquid helium temperatures which, upon sample warming, thermally decays to bathorhodopsin [Yoshizawa and Horiuchi, 1973]. This prompted the early suggestion that hypsorhodopsin, rather than bathorhodopsin, is the primary photoproduct of rhodopsin. However, hypsotype intermediates have not been observed for all visual pigments [Yoshizawa

**TABLE 1. The Formation Time of Bathorhodopsin at Low Temperature**

| Temperature (°K) | Rhodopsin in $H_2O$ (psec) | Rhodopsin in $D_2O$ (psec) |
|---|---|---|
| 50 | | <6 |
| 40 | | 17 ± 2 |
| 30 | <6 | 51 ± 5 |
| 20 | 9 ± 2 | 175 ± 13 |
| 10 | 29 ± 2 | |
| 4 | 36 ± 2 | 257 ± 28 |

and Horiuchi, 1973]; they appear apparently in much greater yield for the cephalopod pigments of squid and octopus rhodopsins than for cattle [Shichida et al., 1979; Tsuda et al., 1980]. Moreover, it appears that cattle rhodopsin [Yoshizawa and Horiuchi, 1973; Shichida et al., 1979] was converted to bathorhodopsin without forming detectable amounts of hypsorhodopsin when irradiated at liquid helium temperatures and similarly for octopus rhodopsin [Tsuda et al., 1980]. Picosecond kinetic studies that measure the time course of bathorhodopsin and hypsorhodopsin formation have given ambiguous results. In some cases, hypsorhodopsin was observed [Shichida et al., 1977; 1978; Kobayashi, 1979] but not in others [Sundstrom et al., 1977; Doukas et al., 1980]. This may be a result of differences in experimental conditions (e.g., use of rhodopsin of different species, varying sample pH or detergent used in solubilizing the pigment, different actinic pulse wavelength). However, the most plausible explanation for the differing kinetic results has come from the laboratory of Yoshizawa and his co-workers [Shichida et al., 1984; Matuoka et al., 1984]. They found significant hypsorhodopsin formation for actinic pulses of high intensity and none for lower power levels. They suggest that hypsorhodopsin is formed by a two-photon reaction. Although of major interest in understanding the photophysics of visual pigments, these studies thus suggest that hypsorhodopsin plays no role in the visual transduction process.

There has been some investigation of whether the bathorhodopsin formed from rhodopsin is the same as that formed from isorhodopsin. As pointed out above, steady-state absorption spectroscopy yields an identical bathorhodopsin spectrum whether photochemically produced from rhodopsin or bathorhodopsin. Picosecond absorption experiments have been performed in two different laboratories to probe whether the "two" bathorhodopsins are the same soon after either rhodopsin or isorhodopsin is excited. The results have been inconclusive in that one laboratory reports a small spectroscopic difference [Spalink et al., 1983] while the other reports that the batho intermediate formed from rhodopsin is the same as that formed from isorhodopsin [Rudzki and Peters, 1984]. Although the experiments differ in some of their attributes, the fundamental accuracy of the two reported studies is quite similar.

A very recent study [Shichida et al., 1984] suggests that the kinetics of bathorhodopsin formation is slightly more complicated than the picture given above. A series of visible difference spectra at various times after the actinic pulse of the type given in Figure 2 were obtained. It was found that a photoproduct, whose absorption maximum lay at a slightly longer wavelength (ca. 10 nm) than that of bathorhodopsin, was produced within the 25-psec half-width of their actinic pulse. They determined that this photoproduct, which they called photorhodopsin, decayed to bathorhodopsin in 40 psec in the case of cattle rhodopsin and in several hundred picoseconds in the case of squid and octopus rhodopsins. Such small changes would not have been observed in many of the studies, e.g., in those giving rise to Figure 2. Shichida et al. [1984] speculated that photorhodopsin is likely a ground-state precursor to bathorhodopsin. This view is strongly reinforced by an important resonance Raman picosecond experiment performed by Stryer and his colleagues [Hayward et al., 1981]. Resonance Raman steady-state measurements [Oseroff and Callender, 1974] have shown that the resonance Raman spectrum of cattle bathorhodopsin contains unique marker bands at 853, 874, and 971 $cm^{-1}$. Hayward et al. [1981] found that these bands appeared within the resolution (35 psec) of their kinetic Raman spectrometer. Taken together, the two picosecond experiments suggest some sort of relaxation process of the protein cavity which surrounds the chromophore. It is easily imagined that this occurs in response to the isomerization to the retinal chromophore from 11-*cis* to distorted *trans*.

## PICOSECOND FLUORESCENCE SPECTROSCOPY

In recent years a second spectroscopic technique, namely, picosecond time-resolved fluorescence spectroscopy, has been used to investigate the primary events in vision. Fluorescence kinetic measurements examine excited-state relaxation processes. Fluorescence measurements on visual pigments have been hampered by the small quantum yield of the intrinsic chromophore fluorescence that is easily masked in the presence of fluorescing impurities in the samples. The first measurements of steady-state fluorescence were made by Guzzo and Pool [1968]. They obtained a quantum yield of $5 \times 10^{-3}$. More recently, we have examined the fluorescence kinetics on the picosecond time scale [Doukas et al., 1981, 1984, 1985]. Two emission components are generally observed. One component is resolution limited (less than 12 psec), and the second one has a lifetime of approximately 250 psec. Upon bleaching of the sample (so that the visible absorption band of the chromophore moves to the near UV) in the presence of hydroxylamine (a chemical which binds the free chromophore), the "fast" component disappears, leaving only the "slow" component. Thus, we have assigned the "fast" component to the intrinsic fluorescence of rhodopsin or isorhodopsin and the "slow" component to impurities of the sample. The assignment of the "slow" component to fluorescing impurities of the sample is strengthened by the fact that the characteristics of the slow component vary considerably with the sample preparation but not in any consistent way. In general, fresh samples tend to have a lower background than older samples. Figure 5 shows the fluorescence kinetic from bovine rhodopsin and the fluorescence background of the bleached sample. The fluorescence background in the figure is rather low.

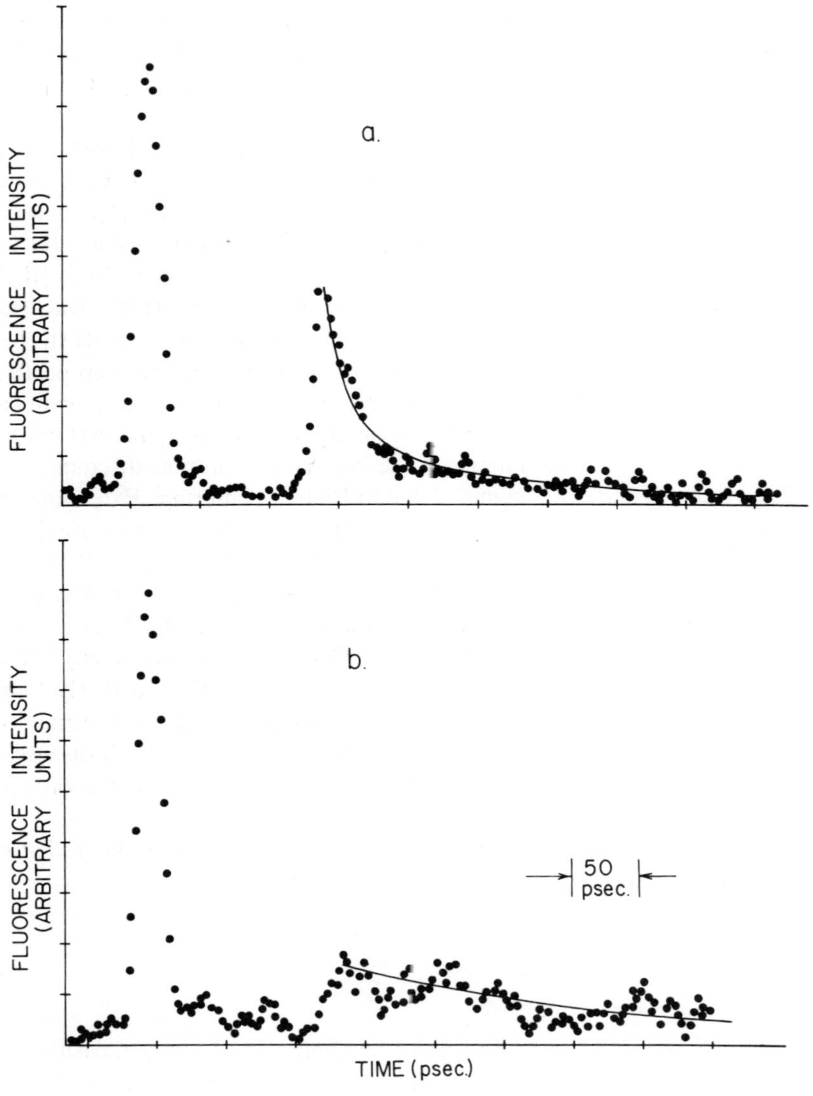


Fig. 5. Fluorescence kinetics integrated from 560 to 800 nm of (a) bovine rhodopsin in suspension and (b) bleached rhodopsin in the presence of hydroxylamine. Both spectra were produced by exciting with a 527-nm laser pulse. The sharp pulse on the left in both (a) and (b) is the calibration pulse. Curve (a) is fitted to a double exponential (solid line), a "fast" component of 12 psec (resolution limited) and a "slower" component of 250 psec. Curve (b) is fitted to a single exponential (solid line) of 250 psec.

Even so, the integrated signal from the background is of the same order of magnitude as the fast component. It is clear to see that the steady-state fluorescence measurements of rhodopsin are almost impossible in any but the best preparations.

The fluorescence quantum yields of bovine and squid rhodopsins (after subtracting out the uninteresting slow impurities component) were found to be $(1.2 \pm 0.5) \times 10^{-5}$ and $(1.3 \pm 0.5) \times 10^{-5}$, respectively, at room temperature [Doukas et al., 1984]. These are in good agreement with the values measured by Birge and co-workers of $(7.8 \pm 2.3) \times 10^{-6}$ [R.R. Birge, J.A. Bennett, and L.P. Murray, private communication], who employed nanosecond techniques. These extremely low quantum yields suggest a fast and efficient excited-state process. The fluorescence emission profiles for both pigments are broad, approximately 60 nm full width at half maximum, peaking at approximatley 620 nm for squid rhodopsin and 600 nm for bovine rhodopsin. The broadness of the observed emission discounts any possibility that the observed signal is due to Raman scattering. Extending the room temperature studies, the fluorescence kinetics of rhodopsin were determined at low temperatures in the range 5–40°K [Doukas et al., 1985]. The kinetics remain unresolved even at 5°K. In addition, the fluorescence quantum yield remains unchanged within the statistical variations. What is more important is that substitution of $H_2O$ with $D_2O$, does not cause any effect even at 5°K. The quantum yields of rhodopsins in $D_2O$ and $H_2O$ are identical. These observations are in marked contrast with the formation kinetics of bathorhodopsin at low temperatures [Peters et al. 1977] discussed above. The implication is that the proton translocation is not the process competing with the fluorescence.

The quantum yield, $\phi$, the nonradiative lifetime $\tau_n$, and the radiative lifetime $\tau_r$ are related through the well-known equation,

$$\phi = \frac{\tau_n}{\tau_r} \tag{2}$$

Therefore, though the fluorescing lifetime cannot be directly measured, it can be estimated from the quantum yield. Using an average quantum yield for bovine and squid rhodopsin of $(1.2 \pm 0.5) \times 10^{-5}$ and a radiative lifetime of 10 nsec (see below) we find that the rate constant of the process that competes with the fluorescence must be of the order of $\tau_n = 0.1$ psec. Given the large quantum yield of isomerization of rhodopsin's retinal chromophore, this process is likely to involve the motion along the retinal's C11-C12 torsional coordinate that leads to *cis-trans* isomerization.

It is possible to test whether isomerization plays a role in determining the fluorescence characteristics by measuring emission from artificial rhodopsins where the 11-*cis* retinal chromophore is replaced by well-chosen analogues.

The two artificial pigments that we have studied, so far, are isorhodopsin (9-*cis* retinal) and a rhodopsin analogue with a fixed 11-ene (Rh7) [Akita et al., 1980]. In the latter, the 11-*cis* retinal chromophore has been replaced by an 11-*cis* analogue where 10C and 13C of the retinal polyene backbone have been bridged by a three-carbon alkyl group, forming a seven-membered ring. The three-carbon alkyl bridge prevents isomerization about the C11-C12 double bond, although some rotation is possible. Preliminary measurements of the isorhodopsin emission quantum yield give a value of $(2.0 \pm 0.6) \times 10^{-5}$ [Doukas, Alfano, Callender, unpublished data], while in the case of Rh7 [Buchert et al., 1983] the quantum yield has been measured to be $4 \times 10^{-4}$, more than one order of magnitude larger than rhodopsin. The somewhat large quantum yield of isorhodopsin can be rationalized by noting that the rotational moment of inertia about the *cis* bond is larger for a 9-*cis* isomer than for an 11-*cis* isomer. As seen below, the rotational motion can be directly related to $\tau_n$ so we would expect more emission from isorhodopsin than rhodopsin, as is observed. Although some rotational motion about the C11-C12 double bond is possible in the seven-membered ring by using retinal analogues, such motion is greatly restricted, and this would result in a marked increase in fluorescence emission as observed. It thus seems quite clear that the fluorescence properties of the retinal chromophore in rhodopsin are related to the rotation about retinal's C11-C12 double bond.

Relating $\tau_n$ to rotation about the C11-C12 double bond requires a construction of the potential surfaces along this coordinate. As it turns out, the ground-state potential surface is quite well known. Bathorhodopsin (*trans* or $\theta = 180°$ in Fig. 6) lies some 35 kcal/mol above rhodopsin ($\theta = 0°$) [Cooper, 1979]. There is some information on the thermal reactions, from rhodopsin to bathorhodopsin and vice versa, so that the barrier between the two species can be determined to be $E^a \sim 5$ kcal/mol within a few kcal/mol [see Doukas et al., 1984; Cooper, 1979].We model the excited-state potential surface as a barrierless potential well, as shown in Figure 6. There is good reason to believe this is so [Rosenfeld et al., 1977]; and, moreover, it would be difficult to understand the extremely fast $\tau_n = 0.1$ psec and temperature-independent rate should there be any significant barrier, as will become clear below.

Using this simple picture and ignoring any other coordinates, we can see that following vertical excitation, the chromophore would rotate about the torsional angle as shown in Figure 6. Assuming that the dipole moment matrix between ground and excited states is not dependent on the torsional angle and noting the radiative lifetime as a function of $\theta$ changes dramatically due to the $\nu^{-3}$ dependence of the Einstein coefficient for spontaneous emission, most of the emission will occur for small angles because the ground-state surface is so steeply rising for $\theta > 20°$. It is easy to calculate that the radiative lifetime at $\theta = 0°$ is two times faster than $\theta = 22°$ and ten times faster than $\theta = 45°$. This implies that the fluorescence intensity decreases by

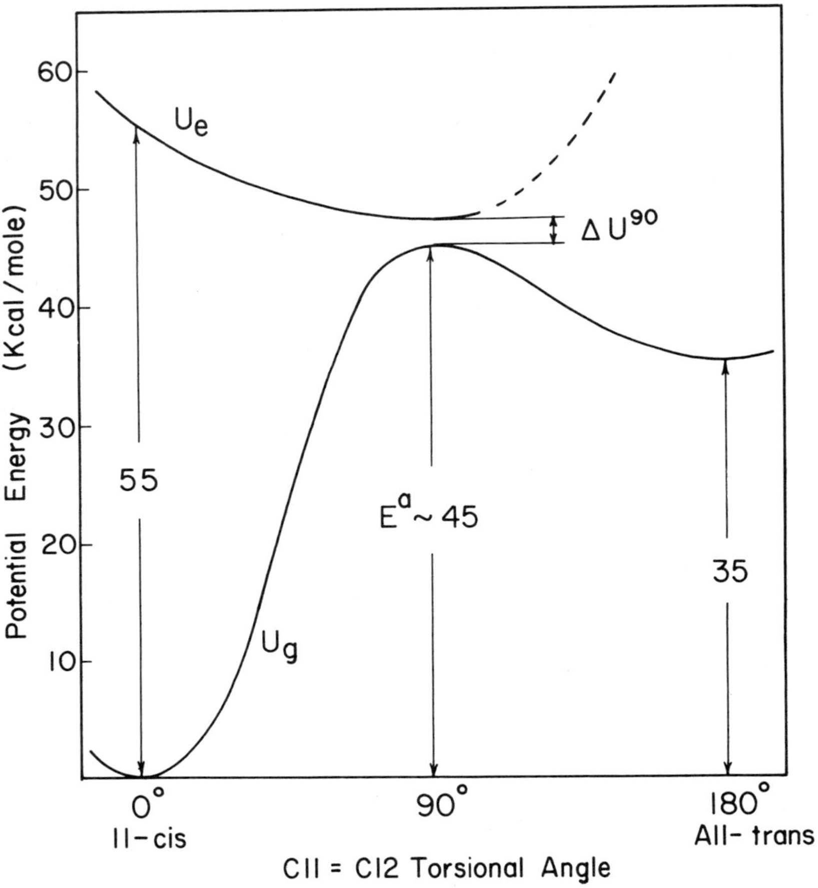


Fig. 6. A scheme of the ground- and excited-state potential surfaces along the chromophore C11-C12 torsional angle. The ground-state potential can be represented by $U_g = E^a \sin^2\theta$ in the region from 0° to 90°. The excited-state potential is represented by a parabola with a minimum at 90° of the form $U^e = E^a + \Delta U^{90} + (k^e/2)(\theta - 90)^2$. $k$ is the torsional force constant associated with the excited electronic state. The value of $E^a$ of 45 kcal/mol is derived from the value of rhodopsin-bathorhodopsin enthalpy increase and the activation of the back-reaction bathorhodopsin → iodopsin [Cooper 1979; Doukas et al., 1984].

an order of magnitude during the rotation of the C11-C12 double bond from $\theta = 0°$ to $\theta = 45°$. We have called the region between the torsional angles $0° \leqq \theta \leqq 45°$ the "fluorescing region" because most of the fluorescence occurs during the chromophore rotation from 0° to 45°. The radiative lifetime at $\theta = 0°$ can be calculated from integrating the absorption spectrum; it was found that $\tau_r(0°) = 5$ nsec [Doukas et al., 1981]. We have used the value of $\tau_r = 10$ nsec as an "average" value when applying $\phi = \tau_n/\tau_r$ since this is the value of $\tau_r(\theta)$ for $\theta = 22°$, in the middle of the fluorescing region. As estimated, the nonradiative lifetime, $\tau_n$, is then the time it takes the molecule to follow a trajectory in the excited state from $\theta = 0°$ to $\theta = 45°$.

This time may be estimated by the following calculation. Assuming the excited-state surface is a parabola, we have the equation

$$\left(\frac{\pi}{2} - \theta\right) = \frac{\pi}{2} \cos\left(\frac{2\pi t}{T_e}\right) \tag{3}$$

where the period $T_e$ is related to the rotational moment of inertia, $I$, and the rotational force constant, $k$, by

$$T_e = 2\pi\sqrt{\frac{I}{k}} \tag{4}$$

When $\theta = 45°$, $t$ corresponds to the nonradiative lifetime, $\tau_n$. By substituting $\tau_n = 0.12$ and $\theta = 45°$ into equation (3) we obtain the period of oscillation $T_e = 9 \times 10^{-13}$ sec, and from equation 4 we derive the moment of inertia $I = 100\ m_p\text{Å}^2$ where $m_p$ is the mass of the proton assuming a value of the rotational force constant, $k$, corresponding to $\Delta V^{90} = 0$. This latter assumption amounts to taking the steepest parabolic excited-state potential surface such that the ground and excited-state surfaces just "touch." (Other values of the parameter can be considered; see Doukas et al. [1984] for more details.) It is easy to calculate the moment of inertia of the 11-*cis* retinal rotating as a rigid body around the C11-C12 double bond. If we include the Schiff base terminus of the retinal plus two carbons of the lysine to which it is attached, we obtain a moment of inertia of $I = 400\ m_p\text{Å}^2$. Our measurements so far seem to imply that the isomerization of the 11-*cis* retinal chromophore based on the moment of inertia corresponds to a rotation that is less than rigid. It is also possible that the excited-state surface near $\theta = 0°$ is much steeper than our model. The effect of this would be to permit a larger moment of inertia for the same $\tau_n$. It should be pointed out, however, that the values given above are rough estimates. They demonstrate how measurements of fluorescence quantum yield could provide information about the actual pathway of isomerization of the chromophore.

It is interesting to compare the isomerization of the retinal chromophore in the opsin matrix to that of 11-*cis* protonated Schiff base in solution. Huppert et al. [1977] measured the isomerization of 11-*cis* to all-*trans* retinal in solution to be 11 nsec, almost five orders of magnitude slower than retinal in opsin's active site. In addition the isomerization quantum yield of the free chromophore is 0.005 to 0.25 depending on the excitation wavelength [Rosenfeld et al., 1977] as opposed to 0.67 for rhodopsin. It is remarkable that the protein matrix allows and even facilitates a significant displacement of atoms in such a short time. How this is accomplished and what kind of chromophore-protein interactions mediate the observed behavior become a central question in our understanding of the primary event.

## DISCUSSION

The picosecond experimental work we described in the previous sections seems to confirm the mechanism of light energy transduction proposed by Honig andco-worker [Honig et al., 1979]. An 11-*cis* to *trans* isomerization of the retinal chromophore is taken to be the first stage of the primary event. This isomerization, as we have discussed above, takes place in the excited state of the molecule. After photoisomerization has occurred, the chromophore finds itself in a new conformation in which the positively charged nitrogen has been removed from the vicinity of an as-yet-unidentified negatively charged counter-ion which is anchored and belongs to the apoprotein surrounding the chromophore (Fig. 7). The charge separation between the proton of the Schiff base and the counter-ion of the protein increased from 3 Å center-to-center distance of the salt bridge to a maximum of 7 Å, if a rigid rotation is assumed. This separation of charge represents a large amount of electrostatic potential energy, as can be calculated from electrostatic energy $= -e_1e_2/\epsilon r_{12}$where $e_1$ and $e_2$ are the two charges, $r_{12}$ the internuclear separation, and $\epsilon$ the dielectric constant. (It is unclear how to model the chromophore site, and, hence, there is a difficulty in properly treating how the charges are screened. These efforts are probably inadequately represented by $\epsilon$.) In Honig et al. [1979], the center to center distance between the Schiff base and its counter-ion was assumed to be only 5 Å, a charge of 0.5 was used for the Schiff base nitrogen, and $\epsilon$ was taken as $\epsilon = 1$ before isomerization and $\epsilon = 2.5$ after isomerization. The change in electrostatic potential energy was calculated to be 40 kcal/mol. Clearly large amounts of light energy can be converted to chemical energy in this way. As noted above, bathorhodopsin is a distorted *trans* species. It is also thus possible that some of bathorhodopsin's enthalpy of 36 kcal/mol consists of torsional strain energy.

From the measurements described above, the rate-limiting step in bathorhodopsin formation, which involves proton translocation [Peters et al., 1977], occurs after photoisomerization has occurred. This is depicted in Figure 7 as motion of one or more protons associated with amino acid residues of the apoprotein presumably near the chromophore. It would seem

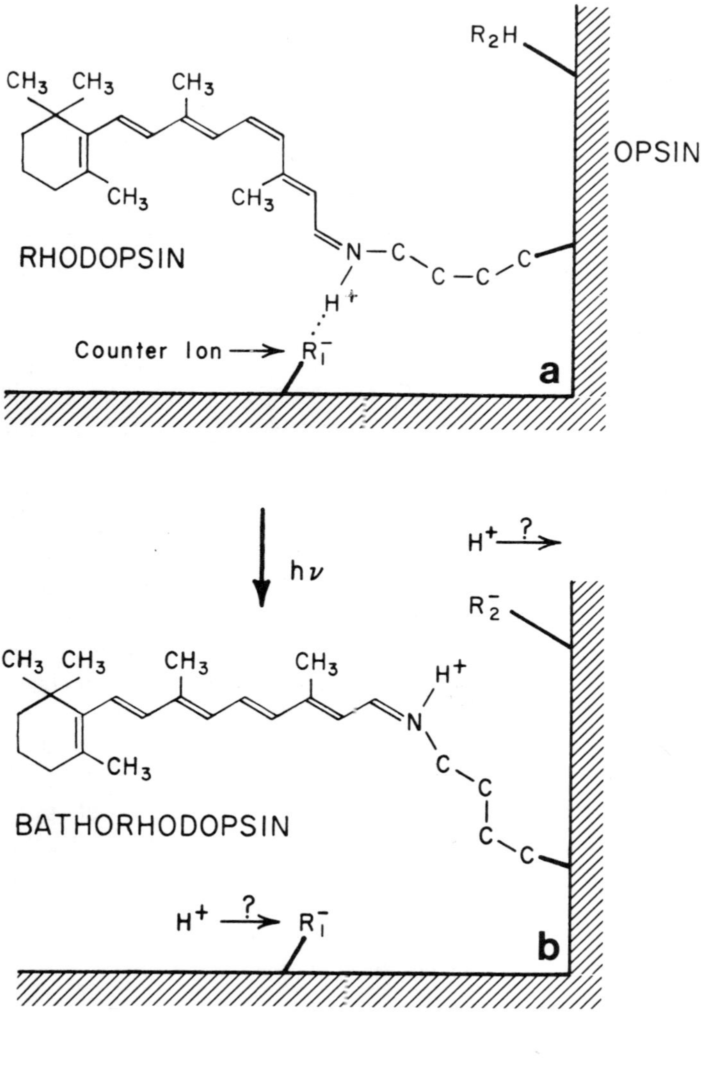


Fig. 7. Model for the early events in visual transduction. The 11-*cis* chromophore is shown with its Schiff base forming a salt bridge with a negative counter-ion. The isomerization of the retinal chromophore around the C11-C12 double bond brings the chromophore to a translike state and produces a charge separation as shown in (b). The pk value of the Schiff and neighboring groups like $R_1$ and $R_2$ is considerably changed because the salt bridge is broken. A variety of proton translocations are possible presumably associated with some kind of relaxation of the protein matrix as it adjusts to the new conformation of the chromophore.

likely those motions occur in the ground state as a result of the severe internal disruption caused by the isomerization and consequent charge separation. Other ground-state relaxation processes associated with the apoprotein appear also to occur subsequently to the proton translocation step as measured by Shichida et al. [1984], and their identification of photorhodopsin.

There has been a great deal of progress in molecular dynamical calculation of the photochemical isomerization. In the earliest calculation, Warshel [1976] calculated that the formation time of bathorhodopsin could be on the order of 0.2 psec if the rotation involves a concerted "bicycle-pedal" motion, where both the C11-C12 and C15-C16 rotate simultaneously. More recently, Birge and Hubbard [1980, 1981] have carried out extensive simulation studies on the isomerization dynamics. An assumption of these calculations is that bathorhodopsin is formed via a one-bond C11-C12 isomerization. This leads to a bathorhodopsin formation time of ~2.2 psec. These latter studies also predict a high bathorhodopsin quantum yield, $\phi_{calc} \approx 0.6$, as is observed. To make contact with the nonradiative lifetime, $\tau_n$, ≈0.1 psec obtained in the picosecond fluorescence quantum yield measurements, one must ask the time for the molecular trajectory after excitation to reach angles of 20–30°. This takes on the order of 0.4–0.8 psec in the calculations of Birge and coworkers [private communication], depending on the assumptions used in modeling the active site and chromophore electronic properties. This must certainly be considered very good agreement with the experiment considering the detailed nature of the calculations and the imprecision in the interpretation associated with the measured value of $\tau_n$.

The spectroscopic and theoretical studies of the visual pigments in the last 20 years have culminated in a simple and elegant model for the visual transduction mechanism. The basic premise of this model is a *cis-trans* isomerization in the excited state followed by a protein relaxation in the ground state. The model explains very well the thermal stability of the system and the high efficiency and sensitivity when light is absorbed. In addition it illustrates the way photon energy is stored in the protein. This energy is available to drive the subsequent thermal reactions.

## ACKNOWLEDGMENTS

We thank our collaborators, Prof. K. Nakanishi, Prof. B. Honig, and Prof. V. Balogh-Nair, for many discussions. This work was supported by NIH grants EY02515 and EY03142 and by a City University Faculty Research Award Program (PSC-BHE) grant.

## REFERENCES

Alfano RR, Shapiro SL (1970): Emission in the region 4000 to 7000 Å via four-photon coupling. Phys Rev Letts 24:584–587,

Alfano RR (ed) (1982): "Primary Events in Biology Probed by Ultrafast Laser Spectroscopy," New York: Academic Press

Akita H, Tanis SP, Adams M, Balogh-Nair V, Nakanishi K (1980): Non-bleachable rhodopsin retaining the full natural chromophore. J Am Chem Soc 102:6370–6372.

Applebury ML (1980): The primary processes of vision: A view from the experimental side. Photochem Photobiol 32:425–431.

Aton B, Doukas AG, Narva D, Callender RH, Dinur U, Honig B (1980): Resonance Raman studies of the primary photochemcial event in visual pigments. Biophys J 29:79–94.

Bagley KA, Balogh-Nair V, Croteau AA, Dollinger G, Ebrey TG, Eisenstein L, Hong MK, Nakanishi K, Vittitow J (1985): Fourier-transform infrared difference spectroscopy of rhodopsin and its photoproducts at low temperature. Biochemistry 24:6055–6071.

Balogh-Nair V, Nakanishi K (1982): The stereochemistry of vision. In Tamm C (ed): "New Comprehensive Biochemistry." Amsterdam: Elsevier Biomedical Press, Vol. 3, pp 283–334.

Birge RR (1981): Photophysics of light transduction in rhodopsin and bacteriorhodopsin. Ann Rev Biophys Bioeng 10:315–354.

Birge RR, Hubbard LM (1981): Molecular dynamics of trans-cis isomerization in bathorhodopsin. Biophys J 34:517–534.

Birge RR, Hubbard LM (1980): Molecular dynamics of cis-trans isomerization in rhodopsin. J Am Chem Soc 102:2195–2204.

Birge RR, Murray LP, Pierce BM, Akita H, Balogh-Nair V, Findsen LA, Nakanishi K (1985): Two-photon spectroscopy of locked-11-cis rhodopsin. Evidence for a protonated Schiff base in a neutral protein binding site. Proc Natl Acad Sci USA 82:4117–4121.

Buchert J, Stefancic V, Doukas AG, Alfano RR, Callender RH, Pande J, Akita H, Balogh-Nair V, Nakanishi K (1983): Picosecond kinetic absorption and fluorescence studies of bovine rhodopsin with a fixed 11-ene. Biophys J 43:279–283.

Busch GE, Applebury ML, Lamola AA, Rentzepis P (1972): Formation and decay of prelumirhodopsin at room temperatures. Proc Natl Acad Sci USA 69:2802–2806.

Callender RH, Honig B (1977): Resonance Raman studies of visual pitments. Ann Rev Biophys Bioeng 6:33–55.

Cooper A (1979): Energy uptake in the first step of visual excitation. Nature 282:531-533.

Doukas AG, Sefancic V, Suzuki T, Callender RH, Alfano RR (1980): Squid bathorhodopsin forms within 10 picoseconds. Photobiochem Photobiophys1:305–308.

Doukas AG, Lu PY, Alfano RR (1981): Fluorescence relaxation kinetics from rhodopsin and isorhodopsin. Biophys J 35:547–550.

Doukas AG, Buchert J, Alfano RR (1982): Picosecond laser techniques and design. In Alfano RR (ed): "Biological Events Probed by Ultrafast Laser Spectroscopy." New York: Academic Press, pp 387–415.

Doukas AG, Junnarkar MR, Alfano RR, Callender RH, Kakitani T, Honig B (1984): Fluorescence quantum yield of visual pigments: Evidence for subpicosecond isomerization rates. Proc Natl Acad Sci USA 81:4790–4794.

Doukas AG, Junnarkar MR, Alfano RR, Callender RH, Balogh-Nair V (1985): The primary event in vision investigated by time-resolved fluorescence spectroscopy. Biophys J 47:xxx.

Eyring G, Mathies R (1979): Resonance Raman studies of bathorhodopsin: Evidence for a protonated Schiff base linkage. Proc Natl Acad Sci USA76:33–37.

Eyring G, Curry B, Brock A, Lugtenburg J, Mathies R (1982): Assignment and interpretation of hydrogen-out-of plan vibration in the resonance Raman spectra of rhodopsin and bathorhodopsin. Biochemistry 21:384–393.

Green BH, Monger TG, Alfano RR, Aton B, Callender RH (1977): Cis-trans isomerization in rhodopsin occurs in picoseconds. Nature 259:179-180.

Guzzo AV, Pool GL (1968): Visual pigment fluorescence. Science 159:312–314.

Hayward G, Carlsen W, Siegman A, Stryer L (1981): Retinal chromophore of rhodopsin photoisomerizes within picoseconds. Science 211:942–944.

Honig B, Ebrey T, Callender RH, Dinur U, Ottolenghi M (1979): Photoiosmerization and charge separation. A model for light energy transduction in visual pigments and bacteriorhodopsin. Proc Natl Acad Sci USA 76:2503–2507.

Hubbard R, Kropf A (1958): The action of light on rhodopsin. Proc Natl Acad Sci USA 44:130–139.

Huppert D, Rentzepis PM, Kliger DS (1977): Picosecond and nanosecond isomerization kinetics of protonated 11-cis retinylidene Schiff bases. Photochem Photobiol 25:193–197.

Kobayashi T (1979): Hypsorhodopsin: The first intermediate of the photochemical process in vision. FEBS Letts 106:313-316

Mathies R (1979): Biological applications of resonance Raman spectroscopy in the visible, and ultraviolet: Visual pigment, purple membrane, and nucleic acids. In Moore CB (ed): "Chemical and Biological Applications of Lasers," Vol. 4. New York: Academic Press, pp 55–99.

Matuoka S, Shichida Y, Yoshizawa T (1984): Formation of hypsorhodopsin at room temperature by a green picosecond pulse. Biochim Biophys Acta 765:38–42.

Monger TG, Alfano RR, Callender RH (1979): Photochemistry of rhodopsin and isorhodopsin investigated on a picosecond time scale. Biophys J 27:105–115.

Narva D, Callender RH (1980): On the state of chromophore protonation in rhodopsin: Implication for primary photochemistry in visual pigments.Photochem Photobiol 32:273–276.

Oseroff AR, Callender RH (1974): Resonance Raman spectroscopy of rhodopsin in retinal disk membranes. Biochemistry 13:4243–4288.

Ottolenghi M (1980): The photochemistry of rhodopsin. In Pitts JN, Hammond GS (eds): "Advances of Photochemistry." New York: John Wiley and Sons, Vol. XII, pp 97–200.

Peters K, Applebury ML, Rentzepis PM (1977): Primary photochemical event in vision: Proton translocation. Proc Natl Acad Sci USA 74:3119–3123.

Rentzepis PM (1978): Picosecond chemical and biological events. Science 202:174–182.

Rosen DL, Doukas AG, Katz A, Budansky Y, Alfano RR (1984): Techniques in time-resolved luminescence spectroscopy. In Alfano RR (ed):"Semiconductors Probed by Ultrafast Laser Spectroscopy." New York: Academic Press, Vol. II, pp 393–407.

Rosenfeld T, Honig B, Ottolenghi M, Hurley J, Ebrey TG (1977): Cis-trans isomerization in the photochemistry of vision. Pure Appl Chem 49:341–351.

Rudzki JE, Peters KS (1984): Picosecond absorption studies on rhodopsin and isorhodopsin in detergent and native membrand. Biochemistry 23:3843–3848.

Schiller NH, Alfano RR (1982): The streak camera. Laser Focus Magazine (August), pp 43–45.

Shichida Y, Yoshizawa T, Kobayashi T, Ohtani H, Nakagura S (1977): Squid hypsorhodopsin and bathorhodopsin by a picosecond laser photolysis. FEBS Letts 80:214–216.

Shichida Y, Kobayashi T, Ohtani H, Yoshizawa T, Nakagura S (1978): Picosecond laser photolysis of squid rhodopsin at room and low temperature. Photochem Photobiol 27:335–341.

Shichida Y, Tokunaga F, Yoshizawa T (1979): Squid hypsorhodopsin. Photochem Photobiol 29:343–351.

Shichida Y, Matuoka S, Yoshizawa T (1984): Formation of photorhodopsin, a precursor of bathorhodopsin detected by picosecond laser photolysis at room temperature. Photobiochem Photobiophys 7:221–228.

Spalink JD, Reynolds AH, Rentzepis PM, Spelring W, Applebury ML (1983): Bathorhodopsin intermediates from 11-cis-rhodopsin and 9-cis-rhodopsin. Proc Natl Acad Sci USA 80:1887–1891.

Sundstrom V, Rentzepis PM, Peters K, Applebury ML (1977): Kinetics of rhodopsin at room temperature measured by picosecond spectroscopy. Nature 267:645–646.

Suzuki K, Kobayashi T, Ohtani H, Yesaka H, Nakagura S, Shichida Y, Yoshizawa T (1980): Observation of the picosecond time-resolved spectrum of squid bathorhodopsin at room temperature. Photochem Photobiol 32:809–811.

Tsuchiya Y (1984): Advances in streak camera instrumentation for the study of biological and physical processes. In Alfano RR, Doukas AG (eds): "Special Issue on Lasers in Biology and Medicine." IEEE J Quantum Electron QE-20:1516–1528.

Tsuda, M, Tokunaga F, Ebrey TG, Yue KT, Marque J, Eisenstein (1980): Behavior of octopus rhodopsin and the photoproducts at very low temperatures. Nature 287:461–462.

Wald G (1968): Molecular basis of visual excitation. Science 162:230–239.

Warshel A (1976): Bicycle-pedal model for the first step in the vision process. Nature 260:679–683.

Yoshizawa T, Wald G (1963): Prelumirhodopsin and the bleaching of visual pigments. Nature 197:1279–1286.

Yoshizawa T, Horiuchi S (1973): Studies of intermediates of visual pigments by absorption spectra at liquid helium and circular dichroism at low temperatures. In Langer H (ed): "Biochemistry and Physiology of Visual Pigments." New York: Springer-Verlag, pp 69–81.

*Applications of Fluorescence in the Biomedical Sciences, pages 91–104*


# 4 Picosecond Fluorescence Studies of Excitation Dynamics in Photosynthetic Light-Harvesting Arrays

*Robert J. Gulotty, Laurens Mets, Randall S. Alberte, and Graham R. Fleming*

The light-harvesting process, the process by which the energy absorbed by an array of chlorophyll molecules is transferred to the special chlorophyll molecules which perform the primary chemical act of photosynthesis, is quite phenomenally efficient. Once absorbed the energy from a photon has perhaps a 95% chance of being converted into chemical energy. Not merely is the efficiency of solar energy collection high in a static sense, but plants have evolved regulation mechanisms that respond to changes in light intensity and spectral distribution. The efficiency undoubtedly stems from the fact that the antenna molecules are held in proteins, with separations and perhaps orientations designed to facilitate energy transfer. Since the processes competing with transfer and final trapping of the energy (fluorescence and intersystem crossing) have relatively fixed time scales it is clear that the optimized light-harvesting system will have very rapid energy transfer. From the physical point of view the details of the energy transfer and trapping are rather poorly understood at present. Understanding of the regulation mechanisms is even poorer, although a number of intriguing proposals exist [Bennett et al., 1980; Horton, 1983].

Department of Chemistry (R.J.G., G.R.F.) and Department of Molecular Genetics and Cell Biology (L.M., R.S.A.), The University of Chicago, Chicago, Illinois 60637.

Measurement of the decay of excited-state populations in the light-harvesting arrays of chloroplasts holds considerable promise for the development of a comprehensive picture of the dynamics and regulation of energy flows in the light-harvesting process. Picosecond-time-resolved fluorescence spectroscopy provides a sensitive and precise method of studying the population dynamics and has been exploited by several groups. The technique is sufficiently sensitive that very low light intensities may be used, thus avoiding the excitation annihilation artifacts which plagued earlier picosecond measurements. The form of the fluorescence decay curve is complex and recent studies have led to the conclusion that at least three exponential decay components are necessary to fit the decay curve [Gulotty et al., 1982; Haehnel et al., 1982; Nairn et al., 1982; Karukstis and Sauer, 1984; Holzwarth, 1985]. The relationship of these decay components to specific structural components has been rather speculative for a number of reasons. First, the higher plant chloroplast has a complex constitution containing two reaction centers (*PSI* and *PSII*) one of which (*PSII*) may exist in two forms with differing properties, and a number of chlorophyll-protein complexes some of which are more closely associated with one type of reaction center than the other [Melis and Duysens, 1979]. Thus decays may represent complex averages of the underlying processes. Second, it is not obvious that the decay of excitation in such a complex system should have exponential or sum of exponential form. For example, Pearlstein [1984] has pointed out that the fluorescence decay may be nonexponential if the excitation trapping is diffusion controlled rather than trap limited.

We have studied the fluorescence decay behavior of a series of photosynthetic mutants of *Chlamydomonas reinhardii* in order to simplify the observed decays and buildup a physical description of the energy transfer dynamics [Gulotty et al., 1985; Gulotty et al., 1986]. The use of mutants has a major advantage over the standard biochemical isolation procedures, in preserving the integrity of light-harvesting array, and enables a definitive assignment of the components observed in the wild-type decay. After a brief discussion of our experimental methods, the remainder of the paper is devoted to a discussion of three major topics: (1) the origin of the decay components of the wild-type chloroplast, (2) the dynamics of energy transfer in the *PSI* core chlorophyll-protein and (3) the dynamics of energy transfer in the *PSII*-lacking Chl *a*/*b* protein-deficient mutant A4-d.

## EXPERIMENTAL METHODS

The *C. reinhardii* wild-type and mutant strains were grown heterotrophically in low light on Tris-acetate/phosphate buffer. The biochemical characterization of the various mutants is described elsewhere [Gulotty et al., 1985].

The time-resolved fluorescence decay measurements were carried out by using the time-correlated single-photon counting method. A review article describing our apparatus will appear elsewhere [Chang et al., 1985], and only brief details are given here. A diagram of the apparatus is shown in Figure 1. The light source was a synchronously pumped, cavity-dumped dye laser generating 10-ps pulses at 75-kHz repetition rate. The laser wavelength was generally kept at 610 nm. Fluorescence was detected by a Hamamatsu R1645-01u micro-channel plate photomultiplier amplified by a Hewlett-Packard PH 8447f amplifier. Fluorescence was collected at 90° from the flowing sample through a Ditric Optics 3 plate filter at 680 nm. Dark-adapted conditions were maintained by flowing the sample at 1 liter/min. The instrument response function of the single-photon counting apparatus was measured by scattering from a suspension of nondairy creamer in water placed

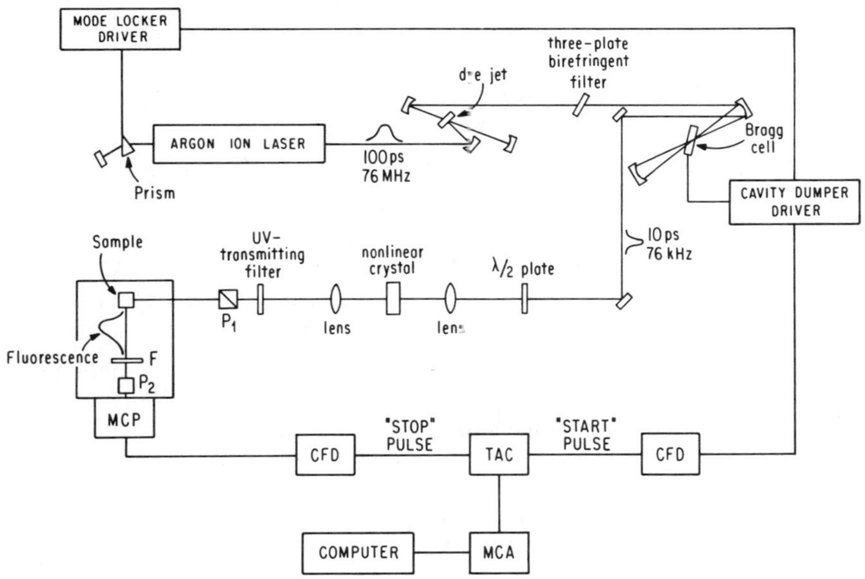


Fig. 1. Experimental arrangement for time-correlated single-photon counting. MCP: multichannel plate; CFD: constant fraction discriminator; TAC: time-to-amplitude converter. The case of ultraviolet excitation is shown above. If the pulses from the cavity dumper are vertically polarized, a λ/2 plate must be inserted in the laser beam before the nonlinear crystal (usually $LiIO_3$ or KDP) in order to produce vertically polarized UV. The polarizer, $P_1$, "cleans up" the polarization of the excitation beam. F is either an interference or a cutoff filter to select a portion of the fluorescence spectrum and to discriminate against scattered light. $P_2$ is a polarizer whose transmission axis is set either parallel or perpendicular or at 54.7° to the polarization of the excitation.

in the flow-cell. The instrument function had a FWHM of 130 ps and a full width at tenth maximum of 250 ps. For a single exponential decay the time resolution of the apparatus is 20–30 ps.

The data were analyzed as sums of exponentials by the standard method of iterative convolution. Single exponential standards of appropriate lifetime were run before and after each set of measurements. As discussed by Cross and Fleming [1984] and Chang et al. [1985] the single-photon counting method can give data in which only statistical (counting) error is detectable. This does not, of course, mean that arbitrary numbers of parameters may be extracted from a decay curve. In simulations in which four exponential components were convoluted with a measured instrument function and appropriate Gaussian noise added, it was found that for $10^4$ peak counts only three components could be reliably recovered, whereas for $10^5$ peak counts the input parameters were accurately extracted from the fit. In other words the "information content" of the $10^4$ peak count curve is insufficient to enable extraction of four decay times and four pre-exponential factors.

## RESULTS AND DISCUSSION

### Origin of the Decay Components in the Wild-Type Chloroplast

Figure 2 shows typical single-photon counting fluorescence decay data for the *C. reinhardii* wild-type strain 2137 (curve a), *PSI* mutant strain 12-7 (curve b), *PSII* mutant strain 8-36c (curve c), and the *PSI-PSII* mutant strain C2 (curve d). (Note that *PSI* mutant refers to a mutant lacking *PSI*, i.e., containing *PSII*.) The figure illustrates the nonexponential decay kinetics of the strains 2137, 12-7, and 8-36c, all of which have reaction centers and the nearly single exponential behavior of the *PSI-PSII* mutant C2. The *PSI* and *PSII* mutant strains show an increased weight of long-lifetime component compared with the wild type. This component has a similar lifetime to the C2 mutant which lacks both reaction centers; i.e., there are excitations with "nowhere to go" in both the *PSI* and *PSII* mutants. The mutant lacking *PSII* has a much larger weight of this component, as might be expected since most of the Chl *a*/*b* protein in the wild type is presumed to be connected to *PSII*. The short time behavior of the *PSI* mutant ($\tau_1$ = 155 ps) shows a longer decay than both the wild-type ($\tau_1$ = 89 ps) and the *PSII* mutant ($\tau_1$ = 53 ps).

Figure 3 shows a typical fluorescence decay of the wild-type strain 2137 of *C. reinhardii*. The smooth curve is the best fit of the data, f(t), using a sum of three exponential components convoluted with the instrument function. The residuals shown above the decay curve show that a satisfactory fit is only achieved when three components are used. This was also true for the *PSI* and *PSII* mutants and the data are collected in Table 1.

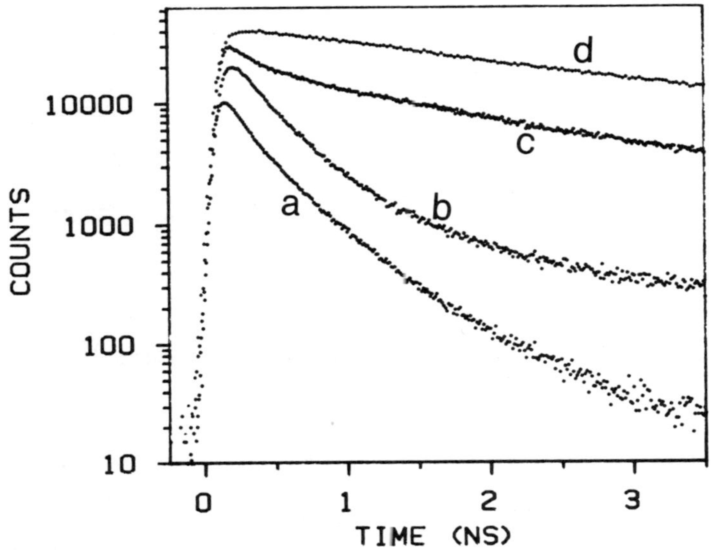


Fig. 2. Fluorescence decay curves of *C. reinhardii* strains (a) wild-type WT2137, (b) *PSI* mutant 12-7, (c) *PSII* mutant 8–36C, and (d) *PSI-PSII* mutant C2. the excitation wavelength is 610 nm and emission was collected at 680 ± 5 nm.

At this point it is important to address the question, Is the true fluorescence decay really a triple exponential or is it a different (for example more complex) function that due to the limitations of real data (e.g., finite time resolution, dynamic range, etc.) can be statistically well fit as a triple exponential decay? We address this question in two stages: we first describe a simulation of wild-type decay from the decays of the three mutants; second, a discussion of the use of sums of exponentials is given two sections later in this discussion.

Figure 4 summarizes the results of a simulation of the wild-type fluorescence in which we sum the decay properties of the *PSI* and *PSII* mutants which correspond to the photochemical paths and use the average lifetime of the *PSI-PSII* mutant to represent the lifetime of decoupled Chl *a*/*b* protein. The curves in Figure 4 were constructed according to

$$f(t) = \alpha f(t)_{PSI} + \beta f(t)_{PSII} + \gamma f(t)_{\text{decoupled antenna}} \quad (1)$$

The simulated curves were convoluted with a real instrument function, had Gaussian noise added (not shown in Fig. 4 for clarity), and fit as if they were experimental results. Curves (a) and (c) illustrate the range of the

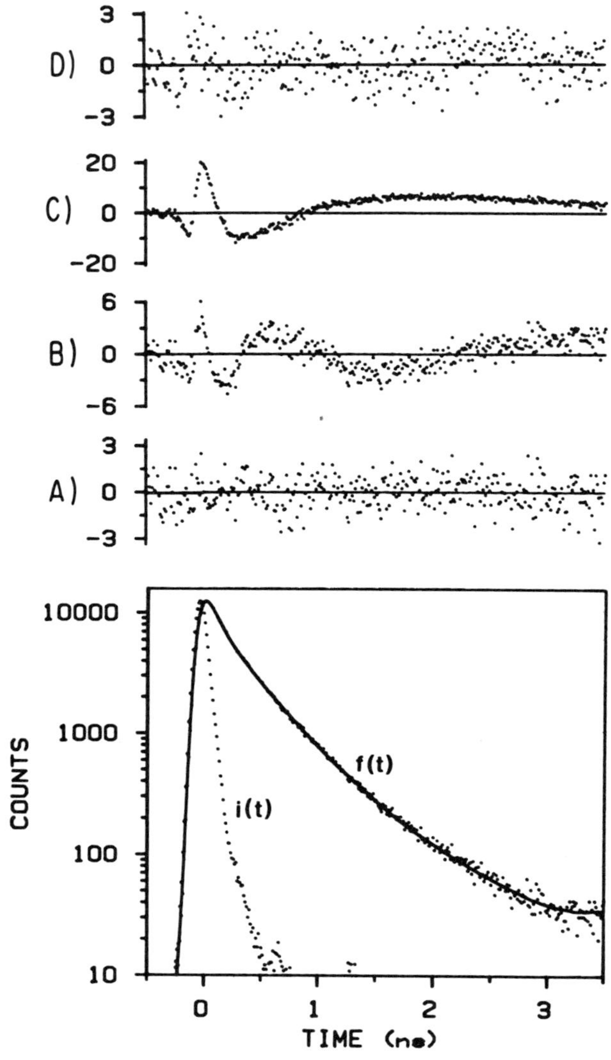


Fig. 3. Fluorescence decay curve of wild-type *C. reinhardii* strain WT2137, f(t)(dots) instrument response function i(t)(dots) and fit of the data f(t)(curve) with a sum of three exponential decay components, 0.544 exp(−t/95 ps) + 0.439 exp(−t/371 ps) + 0.018 exp(−t/2277 ps). Shown above are normalized residuals as in eq. 2 for fits of the data f(t)(dots) to (A) three, (B) two, and (C) one exponential decay components. The residuals (D) are calculated for a single exponential fit to oxazine 725 in water where the $\chi_r^2$ is 1.0 and the lifetime is 490 ps. The excitation wavelength is 610 nm, and the emission was collected at 680 ± 5 nm.

**TABLE 1. Fluorescence Decay Parameters[1] of *C. reinhardii* Wild-Type and Mutant Strains**

| Sample type | Strain | $\tau_1$ | $\tau_2$ | $\tau_3$ | $A_1$ | $A_2$ | $A_3$ | $\chi_r^2$ | Time scale (ps/channel) |
|---|---|---|---|---|---|---|---|---|---|
| Wild type | WT2123 | 86 | 387 | 1,395 | 0.586 | 0.395 | 0.019 | 1.0 | 11.75 |
| *PSII* mutant | 8-36C | 53 | 424 | 2,197 | 0.503 | 0.191 | 0.306 | 1.4 | 11.75 |
| *PSI* mutant | 12-7 | 152 | 424 | 2,924 | 0.598 | 0.378 | 0.024 | 1.1 | 11.75 |
| *PSI-PSII* mutant | C2 | | 839 | 2,720 | | 0.167 | 0.833 | 1.4 | 23.5 |
| | | | | 2,561 | | | 1.0 | 2.6 | 23.5 |
| *PSII* lacking, Chl *a/b* deficient mutant | A4-d | 52 | 361 | 1,690 | 0.64 | 0.28 | 0.08 | 1.0 | 11.75 |
| P700-dual cylinder | | 52 | 272 | | 0.56 | 0.44 | | 0.0[2] | — |
| calculation | | 80 | | | 1.0 | | | 11,043.0 | — |
| P680-dual cylinder | | 153 | 368 | | 0.20 | 0.80 | | 0.0[2] | — |
| calculation | | 276 | | | 1.0 | | | 0.28 | — |

[1]Fluorescence emission collected at 680 ± 5nm, excitation 610 nm.
[2]No noise in calculated curves.

simulation, whereas curve (b) follows the real data very closely. Curve (b) was constructed with $\alpha = 0.6$, $\beta = 0.3925$, and $\gamma = 0.075$. While all values of $\alpha$, $\beta$, and $\gamma$ resulted in curves that could be fit to three exponential decay components (despite the four input components) with a $\chi_r^2$ of 1.0–1.2 (for $10^4$ peak counts), only the convolutions with approximately equal weights of *PSI* and *PSII* fit the wild-type data. Since the intermediate lifetime in both mutants is 424 ps (Table 1), the major conclusion of the simulation is that the short component lifetime of 89 ps in the wild-type *C. reinhardii* is an average of a 53 ps due to energy transfer and trapping at *PSI* and 152-ps lifetime due to energy transfer trapping at *PSII*.

## Excitation Dynamics in *PSI*

We can use the 53-ps component assigned to energy transfer and trapping within the core antenna of *PSI*, in combination with biochemical and spectral data to provide estimates of two key quantities: the single step transfer time between communicating Chl molecules and the number of visits made to the reaction center before final trapping occurs. The starting point is Pearlstein's [1982] expression for the excitation lifetime, $M_o$.

$$M_o = [1 + (F_d/F_t)(N - 1)]k_p^{-1} + [(qF_t)^{-1} - (qF_a)^{-1}][(N - 1)^2/N] + \alpha N F_a^{-1} \quad (2)$$

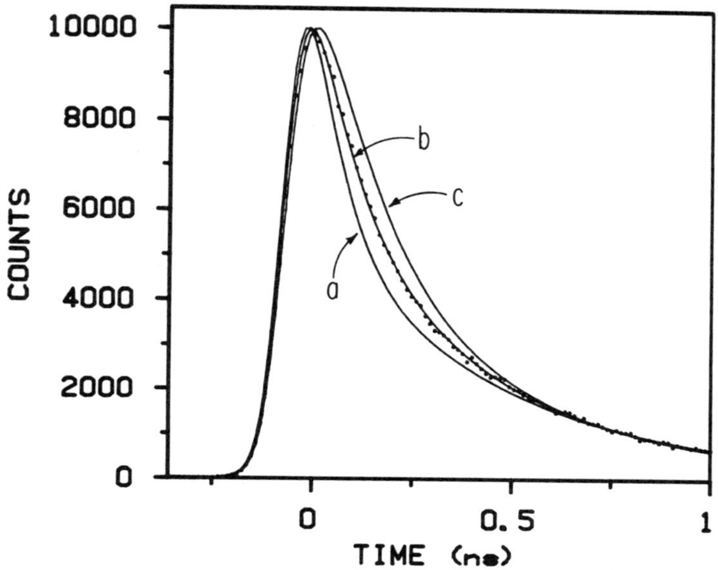


Fig. 4. Simulation of *C. reinhardii* wild-type fluorescence decay data. a. f(t) = 0.9925 *PSII* mutant decay + 0.0075 *PSI-PSII* mutant decay, (b)(curve) f(t) = 0.6 *PSII* mutant decay + 0.3925 *PSI* mutant decay + 0.0075 *PSI-PSII* mutant decay, (b)(dots) f(t) = wild-type decay, (c) f(t) = 0.9925 *PSI* mutant decay + 0.0075 *PSI-PSII* mutant decay. f(t) was convoluted with typical instrument response function to yield the curves and dots (a–c).

Here $F_t$ and $F_d$ are the average Forster rate constants (forward and reverse) for excitation transfer between antenna chlorophyll and the reaction center, $F_a$ is the average Forster rate constant for excitation transfer between antenna molecules, $N$ the array size, and $k_p$ is the primary electron transfer rate constant in the reaction center. The parameter $q$ is the coordination number of the lattice and is a numerical constant that depends on the array type [Hemenger et al., 1972]. An important feature of eq. 2 is that it accounts for multiple visits to the reaction center without requiring any assumption about the actual number of visits.

Equation 2 contains a number of parameters of which only $M_o$ and $N$ are accessible from our data. However, enough spectral and kinetic data are available to make reasonable estimates of $F_d/F_t$, $F_t/F_a$, and $k_p$. In our calculation we closely follow the assumptions used by Shipman [1980]. We assume $F_d/F_t$ is 0.107, which is simply the Boltzmann factor between a 700-nm trap and the average antenna absorption of 678 nm. Assuming the P700 is a dimer and using the same Boltzmann factor we set $F_t/F_a = 1.56$. The value for $k_p = (3 \text{ ps})^{-1}$ is based on the measurements of Holten et al. [1980] for the rate of primary electron transfer in a bacterial reaction center. Our calculation shows that single-step transfer times are quite insensitive to the value of $k_p$ in *PSI* systems. From our data for the *PSII* mutant 8–36C we let $M_o = \tau_1 = 53$ ps and $N = 111$. The remaining parameters in eq. 2 are $q$ and $\alpha$. We carried out calculations for both a square and a simple cubic lattice for which $q = 4$, $\alpha = 0.429$, and $q = 6$, $\alpha = 0.262$, respectively [Hemenger, 1972].

Putting all the above parameters into eq. 2 leads to a single-step transfer time ($1/qF_a$) of approximately 0.1 ps for both lattice types. The calculation yields the trapping and detrapping times $1/qF_t \simeq 70$ fs and $1/qF_d \simeq 0.6$ ps for both lattice types. If the short lifetime is due only to *PSI* core Chl *a* molecules, then $N = 40$–$60$ [Vierling and Alberte, 1983] and the single-step transfer time is 0.4–0.7 ps.

The second quality of interest is the ratio of the lifetime $M_o$ to the time of first passage $\tau_{FPT}$ (i.e., the time between excitation and first arrival at the reaction center. If $M_o/\tau_{FPT}$ is 1, the photochemical event is "diffusion controlled." If $M_o/\tau_{FPT} > 1$ multiple visits occur and the photochemical event is reaction controlled. $\tau_{FPT}$ is calculated from Pearlstein [1982].

$$\tau_{FPT} = N/qF_t + [\alpha - (1/q)]N/F_a \tag{3}$$

Using the values calculated above in eq. 3 with $N = 111$ we find $M_o/\tau_{FPT} = 3.7$ for both the square and cubic lattices and conclude that the excitation makes about four visits to the reaction center before the photochemical event occurs. However, if we assume $N = 40$–$60$ then $M_o/\tau_{FPT} = 1.4$–$1.7$ and the kinetics are closer to diffusion controlled.

## The A4-d Mutant—Comparison of the Fluorescence Decay Curve With a Simple Model

The *C. reinhardii* strain A4-d mutant is produced by crossing the *PSII* mutant B1 with a Chl *a/b* protein deficient strain DS521 [Gulotty, 1984]. A4-d has 80–110 Chl/P700 and a Chl *a*/Chl *b* ratio of 4–5:1. The fluorescence decay (see Table 1) contains significantly less long component than the *PSII* mutants (such as 83-C or B1) that contain substantial amounts of Chl *a/b* protein. Despite the significantly reduced complexity the decay still remains nonexponential.

We have attempted to model the excitation dynamics in A4-d and to address the sum of exponential fitting of the observed decays by considering the behavior of excitations in cylindrical arrays of chlorophyll molecules. The method of calculation, described in detail elsewhere [Gulotty et al., 1986], involves numerical integration of the relevant master equations [Shipman, 1980]. In the previous section we suggested that even in *PSI* core (and certainly in *PSII* core) trapping was reaction controlled. We have previously suggested that the longer-lifetime components in *PSI* particles or *PSI*-containing mutants arise from Chl molecules not well coupled to the *PSI* core. In order to investigate this we have studied the excitation dynamics in two coparallel cylinders, one of which contains a reaction center and one of which contains only Chl *a*/*b* protein. The use of the separate cylinder model for the light-harvesting array of higher plants is motivated by analogy with the bacterial light-harvesting structure [Miller, 1982] and by electron microscopy data on higher plants in which the light-harvesting array appears to increase in units of 30 $Å^3$ from 90–180 $Å^3$ [Staehelin and DeWit, 1984]. The model may be used to discuss the changing association of light harvesting Chl proteins with *PSI* and *PSII* due to photophosphorylation [Bennett et al., 1980], by changing the intercylinder distance. In addition the transfer of excitations between Chl in different thylakoid membranes, as in grana stacks, may be modeled in this way.

The A4-d system was modeled by means of two spatially separated cylinders containing randomly placed chlorophyll molecules. One cylinder contains a trap whereas the other contains only light harvesting Chl. The total number of molecules (90) and the spectral forms where chosen to model the Chl content of the A4-d. A cylinder 50 Å high with a radius of 36 Å containing a trap at 700 nm and 59 Chl *a* molecules (678 nm) was chosen so that the Chl density was 3,400 $Å^3$/Chl. A coparallel cylinder contained 30 molecules (15 Chl *a*, 671 nm and 15 Chl *b*, 653 nm) had the same height but a radius of 25.5 Å in order to keep the density the same as in the first cylinder. This model is consistent with the structure of the bacterial reaction center and light harvesting BChl as determined by electron microscopy by Miller [1982] in which six smaller disks containing light-harvesting BChl surround a larger disk containing a bacterial reaction center. In placing the Chl molecules a constraint of $R_{ij} > 12A$ was used to avoid concentration quenching [Beddard and Porter, 1976]. The two cylinders were separated by 20 A.

Figure 5 shows the result of the numerical integration of the master equations for two cases: (a) where the trap is P700 and (b) where the trap is P680. Both curves fit perfectly to a double exponential decay. The parameters are listed in Table 1. Comparison shows that the short components in *PSII* has about a threefold-longer lifetime. This ratio is similar to the experimental values deduced by Gulotty et al. [1985] and Holzwarth et al. [1985]. The factor of three between *PSI* and *PSII* has two sources in our calculation. The major contribution (about a factor of two) comes from the difference in trap depth. The remaining contribution comes from our assumption of a dimeric trap in *PSI* and a monomeric trap in *PSII*. Before our calculation is accepted

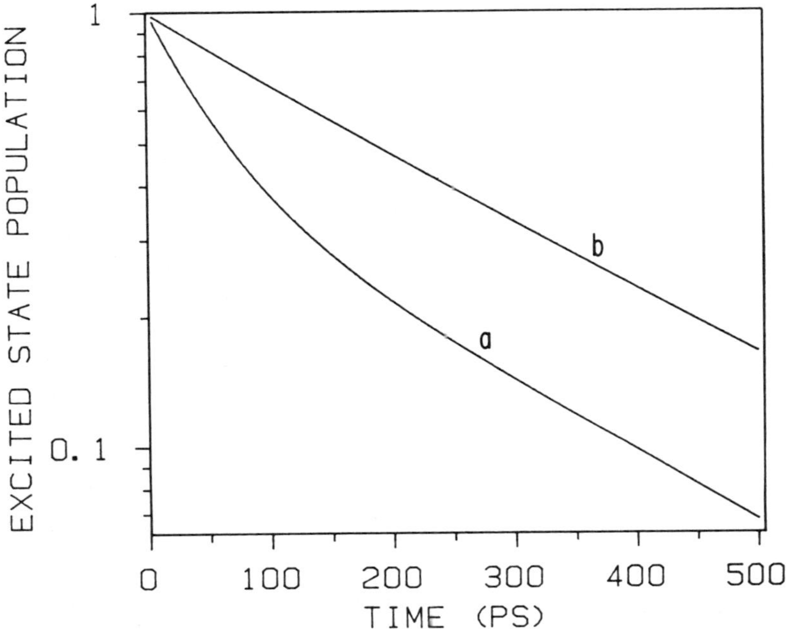


Fig. 5. Population decay calculated for equal excitation probability of chlorophyll molecules in two separate cylinders, one of which contains a *PSI* trap (curve a) or a *PSII* trap (curve b); the other contains Chl *a*/*b* protein with no trap. The intercylinder separation is 20 Å.

as support for monomeric *PSII* and dimeric *PSII* traps it must be noted that the core array size for *PSII* is not well known, and there may well be a difference in the primary electron transfer rate in the two centers [here assumed = (3 ps)$^{-1}$ in both]. Our calculation suggests that the decay time of *PSI* is more sensitive to the monomer/dimer question whereas the *PSII* decay is more sensitive to the electron transfer rate. This reflects the higher probability of detrapping in *PSII*.

Curve (a) is very similar to that of the A4-d mutant, in both lifetimes and weights of the components. The agreement in lifetime is largely due to the choice of Chl density whereas the agreement in relative weights is due to the choice of intercylinder distance. The decomposition of the mutant decay into a short (50 ps) component originating from excitation of core Chl *a* molecules and a longer (425 ps) component originating from excitation of a light-harvesting array relatively poorly coupled to the *PSI* core is supported by the decomposition shown in Figure 6. Curve (a) is the population decay when the reaction center is excited directly. Curve (b) shows the decay when only the 678 nm molecules (core Chl *a*) are excited directly. A 50-ps single exponential decay is observed. Curves (c) and (d), which are almost indistinguishable, give the decay when the 671 nm or 653 nm Chl are directly excited. In this case the population decays nonexponentially for about 200 ps and then decays exponentially with a time constant of about 300 ps. The slow initial decay in the light-harvesting cylinder is expected since it will take some time for the excitation to find the trap-containing cylinder. It is somewhat surprising, perhaps, that when all the Chl molecules have equal excitation probability the curves (a–d) in Figure 6 give the perfect double exponential of Figure 5 curve (a). Thus our calculation strongly supports a multiple exponential fitting of experimental decay curves.

## CONCLUSIONS

Photosynthetic mutants appear to be very valuable for unraveling the complexity of the light-harvesting mechanism in green plants. Our study of *C. reinhardii* mutants has led us to conclude that (1) fluorescence associated with the presence of *PSI* reaction centers must be included in analyses of

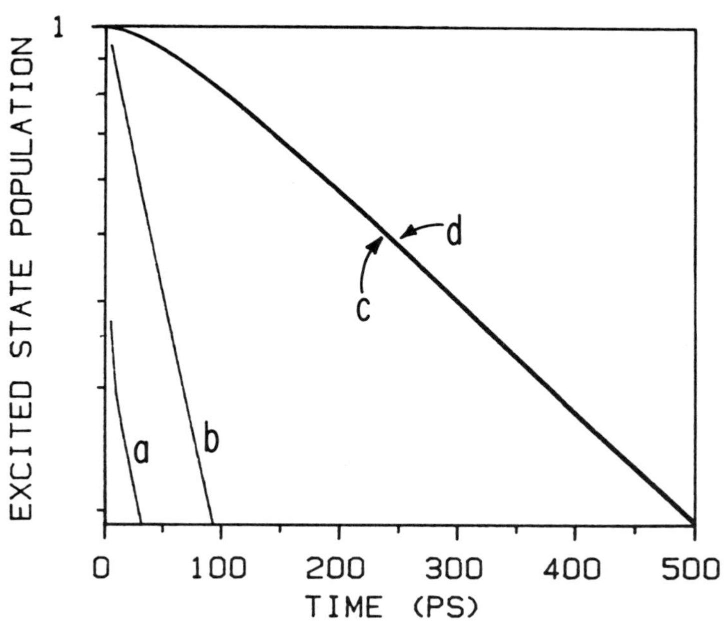


Fig. 6. Excited-state population decay of spectrally heterogeneous array in separate cylinders (intercylinder distance is 20 Å, 3,400 Å$^3$/Chl) with different initial excited-state population distributions: (a) P700 reaction center only, (b) Chl *a* 678 nm only, (c) Chl *a* 671 nm only, and (d) Chl *b* 653 nm only.

wild-type chloroplast decays. (2) The true wild-type decay is considerably more complex than the sum of three exponential components. Thus an attempt to model photophysical behavior in chloroplasts in terms of three components is an unacceptable oversimplification. (3) Calculations suggest a single-step transfer time in the *PSI* core to be about 200 fs. We also find that an excitation makes between two to four visits to *PSI* before final trapping. (4) Numerical simulations justify a multiexponential analysis and suggest the threefold difference between *PSI* and *PSII* lifetime largely arises from the greater detrapping probability in the latter.

## ACKNOWLEDGMENTS

This work was supported by a grant from the USDA to G.R. Fleming and R.S. Alberte. R.J. Gulotty was a predoctoral fellow on NIH training grant GMO7183.

## REFERENCES

Beddard GS, Porter G (1976): Concentration quenching in chlorophyll. Nature 260:366–368.

Bennett J, Steinback KE, Arntzen CJ (1980): Chloroplast phosphoproteins: Regulation of excitation energy transfer by phosphorylation of thylakoid membrane polypeptides. Proc Natl Acad Sci USA 77:5253–5257.

Chang MC, Cross AJ, Courtney SH, Gulotty RJ, Petrich JW, Fleming GR (1985): Time correlated single photon counting with microchannel plate detectors. Anal Inst (in press).

Cross AJ, Fleming GR (1984): Analysis of time-resolved fluorescence anisotropy decays. Biophys J 46:45–56.

Gulotty RJ, Fleming GR, Alberte RS (1982): Low-intensity picosecond fluorescence kinetics and excitation dynamics in barley chloroplasts Biochim Biophys Acta 682:322–331.

Gulotty RJ, Mets L, Alberte RS, Fleming GR (1985): Fluorescence decay kinetics of in vivo chl a/b protein. Photochem Photobiol 41:487–492.

Gulotty RJ (1984): Excitation Transfer and Trapping in Higher Plant Photosynthesis. Ph.D. thesis, The University of Chicago.

Gulotty RJ, Owens T, Alberte RS, Mets L, Fleming GR (1986): (in preparation).

Haehnel W, Nairn JA, Reisberg P, Sauer K (1982): Picosecond fluorescence kinetics and energy transfer in chloroplasts and algae. Biochim Biophys Acta 680:161–173.

Hemenger PR, Pearlstein RM, Lakatos-Lindenberg K (1972): Incoherent exciton quenching on lattices. J Math Phys 13:1056–1063.

Holten D, Haganson C, Windsor MW, Schenck CC, Parson WW, Migus A, Fork RL, Shank CV (1980): Subpicosecond and picosecond studies of electron transfer intermediates in *Rhodopseudomonas sphaeroides*. Biochim Biophys Acta 592:461–477.

Holzwarth AR (1985): Excited state kinetics of chlorophyll antenna pigments. In Staehelin LA, Arntzen CJ (eds): "Photosynthetic Membranes, Encyclopedia of Plant Physiology." Berlin: Springer-Verlag (in press).

Holzwarth AR, Wendler J, Haehnel W (1985): Time resolved picosecond fluorescence spectra of the antenna chlorophylls in *Chlorella vulgaris*: Resolution of photosystem: I. fluorescence. Biochim Biophys Acta 807:155–167.

Horton P (1983): Control of chloroplast electron transport by phosphorylation of thylakoid proteins. FEBS Lett 152:47–56.
Karukstis KK, Sauer K (1983): Fluorescence decay kinetics of chlorophyll in photosynthetic membrane. J Cell Biochem 23:131–158.
Melis A, Duysens LNM (1979): Biphasic energy conversion kinetics and absorbance difference spectra of photosystem II of chloroplasts. Photochem Photobiol 29:373–382.
Miller KR (1982): Three-dimensional structure of a photosynthetic membrane. Nature 300:53–55.
Nairn JA, Haehnel W, Reisberg P, Sauer K (1982): Picosecond fluorescence kinetics in spinach chloroplasts at room temperature, effects of $Mg^{2+}$. Biochim Biophys Acta 682:420–429.
Pearlstein RM (1982): Excitation migration and trapping in photosynthesis. Photochem Photophys 35:835–844.
Pearlstein RM (1984): Photosynthetic exciton migration and trapping. In Sybesma C (ed): "Advances in Photosynthesis Research," Vol. 1. Martinus Nijhoff/Dr. Junk W. The Hague, pp 13–20.
Shipman LL (1980): A theoretical study of excitons in chlorophyll *a* photosystems on picosecond time scale. Photochem Photobiol 31:157–167.
Staehelin LA, DeWit M (1984): Correlation of structure and function of chloroplast membrane at the supramolecular level. J Cell Biochem 24:261–269.
Vierling E, Alberte RS (1983): P700 chlorophyll *a*-protein. Plant Physiol 72:625–633.

*Applications of Fluorescence in the Biomedical Sciences, pages 105–127*


# 5 The Nature of the Protein Binding Site of Rhodopsin Based on Two-Photon Spectroscopy and Molecular Orbital Theory

*Lionel P. Murray, Leonore A. Findsen, Brian M. Pierce, and Robert R. Birge*

The molecular mechanism of light transduction in the visual pigment, rhodopsin, is a subject of considerable interest in contemporary biophysics. Rhodopsin, a generic name assigned to all vertebrate and invertebrate visual pigments, consists of a single polyene chromophore covalently bound to an apoprotein called opsin [Birge, 1981, 1982; Honig, 1982; Hargrave et al., 1984; Ottolenghi, 1980]. Rhodopsin's complete amino acid sequence has recently been determined [for a review see Hargrave et al., 1984], but the secondary and tertiary structure is known only approximately. While it is generally believed that seven α-helical segments span the membrane [Hargrave et al., 1984], the precise assignments of which amino acids participate in the formation of the α-helices, and the relative orientation of the helical segments, are not known. A plausible assignment is shown in Figure 1. These uncertainties lead to a corresponding uncertainty in assigning the nature of the chromophore binding site.

The importance of understanding the protein binding site to the elucidation of the molecular events associated with rhodopsin energy transduction has prompted extensive investigation [Akita et al., 1980; Arnaboldi et al., 1979; Birge, 1981, 1982; Buchert et al., 1983; Callender and Honig, 1977; Cooper, 1979a,b, 1982; Doukas et al., 1985; Eyring and Mathies, 1979; Honig, 1982; Honig et al., 1979a; Mao et al., 1980, 1981; Mateescu et al., 1981, 1983;

Department of Chemistry, Center for Fluorescence Research in Biomedical Sciences, Carnegie-Mellon University, Pittsburgh, Pennsylvania 15213.

Mathies, 1982; Monger et al., 1979; Oseroff and Callender, 1974; Ottolenghi, 1980; Peters et al., 1977; Rothschild et al., 1983; Shichida et al., 1977, 1978; Shriver et al., 1977; Yoshizawa and Wald, 1963; 1967; Yoshizawa, 1972] [for reviews see Birge, 1981, 1982; Honig, 1982; Ottolenghi, 1980].

Despite the efforts of many researchers, well-ordered three-dimensional crystals of rhodopsin have not been prepared precluding structural analysis using high-resolution x-ray diffraction. Thus, the majority of experimental studies have used electronic absorption,[¶] resonance Raman,[‖] nuclear magnetic resonance,[**] photocalorimetry,[†] Fourier-transform IR,[a] or picosecond spectroscopy.[b] The above experimental, as well as theoretical,[c] investigations are not in general agreement concerning the state of protonation of the chromophore[d] nor the number and location of the counterions inside the binding site.[e] We report here the analysis of the two-photon spectrum of rhodopsin [Birge et al., 1985] with the goal of resolving some of these issues.

---

¶For recent contributions [see: Akita et al., 1980; Arnaboldi et al., 1979; Honig et al., 1979a; Mao et al., 1980, 1981; Yoshizawa and Wald, 1963, 1967; Yoshizawa, 1972].

‖For recent contributions [see: Callender and Honig, 1977; Eyring and Mathies, 1979; Mathies, 1982; Oseroff and Callender, 1974].

**For recent contributions [see: Mateescu et al., 1981, 1983; Shriver et al., 1977].

†For recent contributions [see: Cooper, 1979a,b, 1982].

aFor recent contributions [see: Bagley et al., 1983; Rothschild et al., 1983; Siebert et al., 1983].

bFor recent contributions [see: Buchert et al., 1983; Monger et al., 1979; Peters et al., 1977; Shichida et al., 1977, 1978].

cFor recent contributions [see: Birge et al., 1982b; Birge and Hubbard, 1980, 1981; Doukas et al., 1984; Favrot et al., 1979; Harosi et al., 1978; Honig et al., 1979b, 1980; LeClercq et al., 1982; LeClercq and Sandorfy, 1981; Warshel, 1976; Warshel and Barboy, 1982; Weiss and Warshel, 1979].

dFor recent contributions [see: Akita et al., 1980; Arnaboldi et al., 1979; Birge, 1981; Buchert et al., 1983; Callender and Honig, 1977; Cooper, 1979b, 1982; Doukas et al., 1984, 1985; Eyring and Mathies, 1979; Honig, 1982; Honig et al., 1979a,b, 1980; LeClercq et al., 1982; LeClercq and Sandorfy, 1981; Mateescu et al., 1981, 1983; Mathies, 1982; Oseroff and Callender, 1974; Ottolenghi, 1980; Shriver et al., 1977].

eFor recent contributions [see: Akita et al., 1980; Arnaboldi et al., 1979; Birge, 1981; Birge and Hubbard, 1980, 1981; Buchert et al., 1983; Favrot et al., 1979; Harosi et al., 1978; Honig et al., 1979a; LeClercq et al., 1982; LeClercq and Sandorfy, 1981].

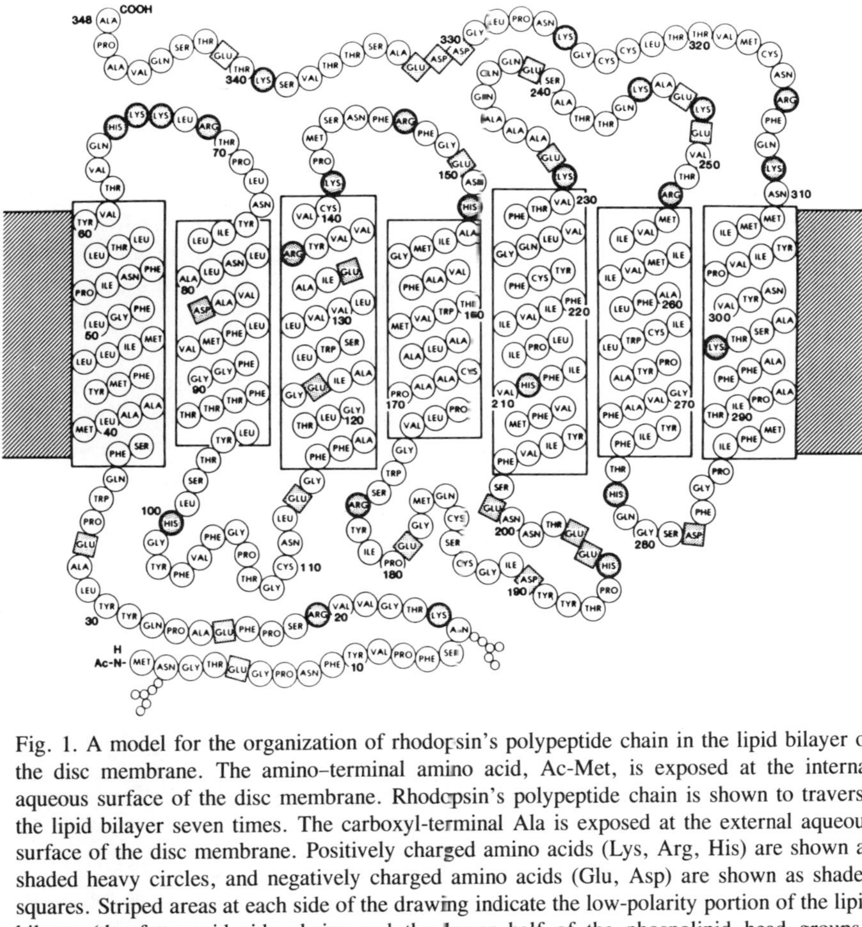


Fig. 1. A model for the organization of rhodopsin's polypeptide chain in the lipid bilayer of the disc membrane. The amino-terminal amino acid, Ac-Met, is exposed at the internal aqueous surface of the disc membrane. Rhodopsin's polypeptide chain is shown to traverse the lipid bilayer seven times. The carboxyl-terminal Ala is exposed at the external aqueous surface of the disc membrane. Positively charged amino acids (Lys, Arg, His) are shown as shaded heavy circles, and negatively charged amino acids (Glu, Asp) are shown as shaded squares. Striped areas at each side of the drawing indicate the low-polarity portion of the lipid bilayer (the fatty acid side chains and the lower half of the phospolipid head groups.) [Reprinted with permission from *Vision Res.*, 24, Hargrave et al., "Rhodopsin's Protein and Carbohydrate Structure: Selected Aspects," Copyright 1984, Pergamon Press, Ltd.]

The unique diagnostic capabilities of two-photon spectroscopy for studying the binding site of rhodopsin derive in part from the unusual electronic properties of polyenes. (The polyene chromophore of rhodopsin is 11-*cis* retinal bound to the opsin protein via a covalent linkage with the $\epsilon$-amino nitrogen of lysine-296 [Hargrave et al., 1984]). It is now generally recognized that polyenes have two low-lying $\pi\pi^*$ singlet states, a strongly allowed $^1B_u^{*+}$ (or $^1B_u^{*+}$-like symbolized by "$^1B_u^{*+}$") state responsible for the intense $\lambda_{max}$ electronic absorption band and a nearby "forbidden" $^1A_g^{*-}$ (or $^1A_g^{*-}$-like symbolized by "$^1A_g^{*-}$") state [Birge, 1981; Ottolenghi, 1980].

Although the latter state is known to be the lowest-lying excited singlet in long chain linear polyenes, this level ordering does not necessarily apply to polar polyenes such as the visual chromophore, retinal, or the Schiff bases and protonated Schiff bases of retinal [Birge et al., 1978, 1982a, 1984, 1985; Murray and Birge, 1985]. Because the "$^1B_u^{*+}$" and "$^1A_g^{*-}$" $\pi\pi^*$ states respond to environmental changes with substantially different sensitivities (see below), a knowledge of the location of both states provides a spectroscopic window into the electrostatic and dispersive characteristics of the environment of the location and photophysical properties of these two low-lying states provides a unique probe of the counterion environment in the vicinity of the chromophore.

Assigning the Franck-Condon maximum of the "$^1B_u^{*+}$" state is easily accomplished by using conventional electronic absorption spectroscopy. Assigning the location of the "$^1A_g^{*-}$" state is more difficult. The absorptivity of this state is usually too low to permit observation by using one-photon spectroscopy if this state is near the "$^1B_u^{*+}$" state and the latter state has a broad absorption band. Birge and co-workers have demonstrated that two-photon spectroscopy is the most reliable method for identifying the "$^1A_g^{*-}$" state in molecules, such as the visual chromophores, which display severe inhomogeneous broadening [Birge, 1984; Birge et al., 1978, 1982a, 1984, 1985; Murray and Birge, 1985]. A further complication associated with the analysis of the "$^1A_g^{*-}$" state in the chromophore of rhodopsin is associated with the irreversible bleaching that occurs with a high quantum yield [~0.67, Hurley et al., 1977] upon exposure of the protein to light. This characteristic is obviously desirable for a visual receptor protein and can be accommodated for most one-photon experiments due to the small light intensities required. The high light intensities and large number of measurements required to generate a nonlinear laser excitation spectrum [Birge, 1983], however, prevent the use of optically labile samples in two-photon spectroscopy. Two solutions to this problem have been proposed [Birge et al., 1984, 1985]: a two-photon double resonance technique can be used to study rhodopsin at liquid nitrogen temperatures, or a nonbleachable rhodopsin analog can be studied at ambient temperatures by using the two-photon thermal lens technique. Both methods have been investigated for rhodopsin [Birge et al., 1984, 1985]. The latter method, which is described in this paper, proved to be more reliable and is based on the previous synthetic work of Akita et al. [1980]. The incorporation of the locked-11-*cis* chromophore (shown in Fig. 2) into opsin produces a nonbleachable rhodopsin analog that can withstand the high light fluxes used in two-photon spectroscopy [Birge et al., 1984, 1985; Mao et al., 1980, 1981]. There is compelling evidence to suggest that the "locked" chromophore occupies the same binding site as the native 11-*cis* chromophore, and the spectroscopic (one-photon absorption and CD)

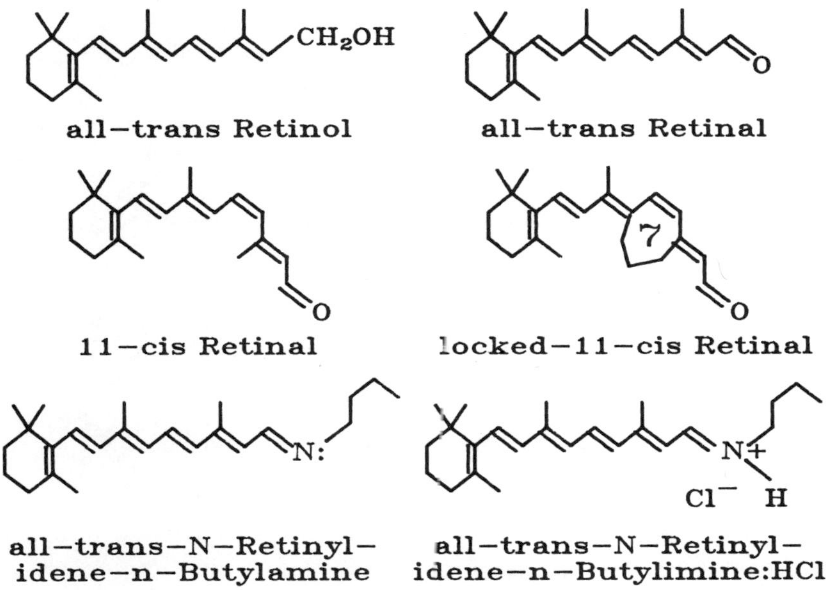


Fig. 2. Structures of the visual chromophores and visual chromophore analogs [Reprinted with permission from Birge et al., 1985.]

similarities suggest that the counterion environment is unperturbed [Akita et al., 1980].

The following sections describe the generation of the two-photon thermal lens spectrum and the theoretical analysis of the spectroscopic data. Our principal goal is to assign the state of protonation of the chromophore and determine the nature of the local counterion environment. Because we report the first two-photon spectrum of a protein an additional goal of this paper is to demonstrate the general utility of this spectroscopic technique. This method has the potential of probing the electrostatic environments of a wide variety of protein-bound chromophores.

## MATERIALS

The locked 11-cis chromophore shown in Figure 2 was synthesized following the procedures outlined in Akita et al. [1980]. Cattle rod outer segment (ROS) membranes were prepared according to the procedure of Papermaster and Dreyer [1974]. Locked 11-*cis* rhodopsin was generated following the procedures outlined in Birge et al. [1985].

## METHODS

All one-photon and two-photon spectra were taken by using ambient-temperature samples. One-photon (near-IR) spectra were taken by using a Varian Cary 17D with $10^{-5}$ M samples. Two-photon spectra were taken by using the apparatus described below with $8 \times 10^{-5}$ M samples to obtain an optical density at $\lambda_{max}$ of 3.3 O.D. (1-cm pathlength).

The optical configuration of our two-photon thermal lens spectrometer is shown in Figure 3. The 337.2-nm output of a Molectron UV-24 900-kW nitrogen laser (NL) is used to pump a Molectron DL-14 tunable dye laser (DL) with an optical amplifier (OA). The linearly polarized output of the tunable dye laser is combined with the 632.8-nm output beam of a Helium-Neon laser (GL) (Hughes 3222H-PG selected for high stability) and is adjusted for precise colinearity at three locations along the optical path. The colinear beams are focussed slightly in front of the sample cell by using a multiple element quartz achromatic lens (FL) (f = 85 mm). This configuration produces a diverging thermal lens when the sample simultaneously absorbs two photons. Approximately two meters down-field from the sample, the center of the diverging beam passes through a 10 $\mu$m aperture (AP) and three 632.8-nm interference filters (IFS). The latter remove the dye laser excitation and permit the He-Ne laser beam to illuminate a Hamamatsu 1P28 photomultiplier tube (PM2). The thermal lensing signal is monitored by using an EG&G PAR model 113 preamplifier (AC coupled, 1,000-fold gain) to amplify the signal from PM2 prior to measurement with an EG&G PAR model 165 gated integrator and an EG&G PAR model 162 boxcar averager.

The thermal lens signal is extremely "noisy" and is sensitive to irrelevant external events such as people opening doors in other parts of the building. To minimize these problems, most of the two-photon data were collected between 11 P.M. and 4 A.M. The successful collection of the reported spectrum was due in large part to the use of the automatic baseline subtraction feature of the model 165 gated integrator. This feature provides for alternating measurement of the signal and a time-delayed baseline with differential output of signal minus baseline. We found that this differential remained relatively stable whereas the absolute signal could change by ~10% over a 30-minute period. We recommend the use of automatic baseline subtraction methods when measuring weak excitation signals using the thermal lens method.

A separate 165/162 boxcar integrator/averager system was used to monitor the intensity of the tunable dye laser. A wavelength-calibrated Hamamatsu R955 photomultiplier (PM1) was used with calibration provided in a separate series of experiments with a Molectron J3 Joule meter.

The two-photon thermal lens spectrum was collected over the wavelength range from 702 to 1,000 nm. Laser excitation wavelengths from 702 to 930

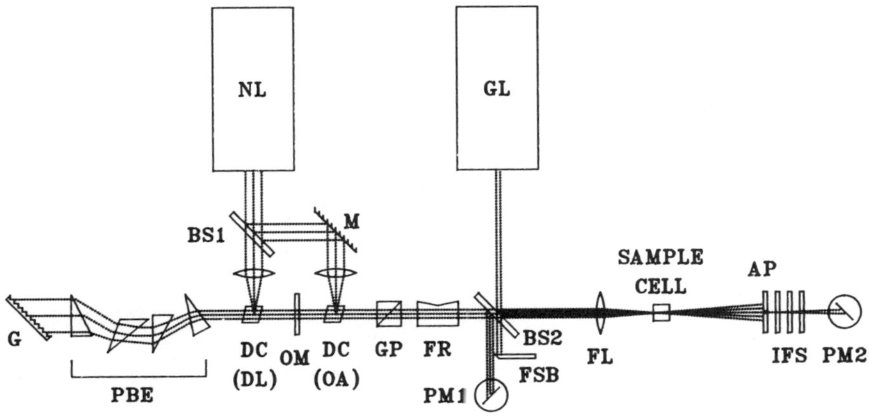


Fig. 3. Two-photon thermal lens spectrometer. A 900-kW nitrogen laser (NL) is used to simultaneously pump two dye cells (DC) to produce optically amplified tunable dye laser excitation. A beam splitter (BS2) is used to direct a small fraction (~4%) of the pulsed dye laser ouptut to a calibrated photomultiplier (PM1) for reference energy measurement. The same beam splitter combines the output of a continuous wave He-Ne gas laser (GL) with the dye laser output to provide a colinear beam that is focussed (using an achromatic f = 85-mm focussing lens, FL) slightly in front of the sample cell. A 10-$\mu$m-diameter aperture (AP) approximately 180 cm downfield passes a small fraction of the laser light. Three 632.8-nm interference filters (IFS) pass the He-Ne laser output to a 1P28 photomultiplier (PM2). The remaining symbols are G, diffraction grating, PBE, prism beam expander; DC(DL), dye cell inside dye laser cavity; BS1, beam splitter; M, dichroic mirror; OM, output mirror; DC(OA), dye cell of optical amplifier; GP, Glan polarizer; FR, Fresnel rhomb; FSB, field stop baffle (He-Ne laser beam dump).

nm were obtained by using the dyes (or dye mixtures) and solvent environments reported by Pierce and Birge [1982]. Laser excitation wavelengths from 940 to 1,000 nm were obtained by using IR 140 (3 mM in DMSO) (930–945 nm) and the binary dye mixture IR 140/DATTC (3 mM each in DMSO) (950–1,000 nm). Each spectral point of the two-photon thermal lens spectrum was obtained by collecting at least 12,000 measurements of the thermal lens signal (sample) as a function of laser energy (reference) by using a Hewlett-Packard 3497A data acquisition system controlled by an HP9826 computer. Neutral density filters (laser transmittances of 80%, 61%, 48%, 38%, and 12%) were used to attenuate the dye laser output, enabling one to remove the one-photon component inherent in the thermal lensing signal computationally with the regression techniques described by Birge [1983, pp. 165–167]. The ratios of the one-photon component of the signal to the two-photon component of the signal ($a_1/a_2$ from equation 66 of Birge, 1983) ranged from 0.12 to 0.73 although two data sets (marked with asterisks in Fig. 4) displayed ratios of ~0.95. We considered the latter two data sets to be suspect, although their inclusion in the spectrum had no significant effect (see below). Potential sample degradation was monitored by repeatedly scanning the one-photon absorption spectrum during the experiment and verifying that the optical density of the sample did not change. These tests indicated a maximum degradation of $<3\%$ over the course of the experiment. Nevertheless, potential spectroscopic error due to sample degradation was minimized by collecting data at "randomized" laser wavelengths rather than scanning monotonically in one direction.

## RESULTS

The two-photon thermal lens spectrum plotted as a function of laser excitation frequency is shown in Figure 4a. Comparison of this spectrum with the (one-photon) near-IR spectrum (bottom of Fig. 4a) indicates that the one-photon component of the thermal lensing signal has been adequately removed. The successful separation of the one-photon and two-photon components was possible due in large part to the use of $D_2O$, rather than $H_2O$, as the solvent environment. Prior attempts to measure the two-photon spectrum of the pigment suspended in $H_2O$ failed because of interference due to OH stretching overtone which overwhelmed the two-photon component of the thermal lensing signal.

Because two identical photons are absorbed in accessing the final electronic state, it is necessary to plot the two-photon spectrum as a function of laser frequency times two to properly represent the final-state vibronic distribution. The resulting spectrum is shown in Figure 4b where it is compared with the corresponding one-photon absorption spectrum. Despite the considerable random error in the individual data points (the height of the vertical error bars correspond to twice the standard deviation in the regression calculation of $S/R^2$), it is clear that the two-photon thermal lensing maximum lies above the one-photon absorption maximum in energy. A statistical analysis of the Franck-Condon separation yields $2{,}250 \pm 850$ $cm^{-1}$. In spite of the large uncertainty in the energy separation, the information content of this single measurement is extremely high. As we demonstrate below, this energy separation indicates that the protein-bound chromophore is protonated and occupies a neutral (uncharged) binding site. Comparison of theory with experiment also provides us with new insights into possible geometries of the local counterion environment.

## DISCUSSION

### Environmental Effects on Excited-State Level Ordering

The effects of conformation, substituents, and solvent environment on the photophysical properties of polyenes have been the subject of extensive

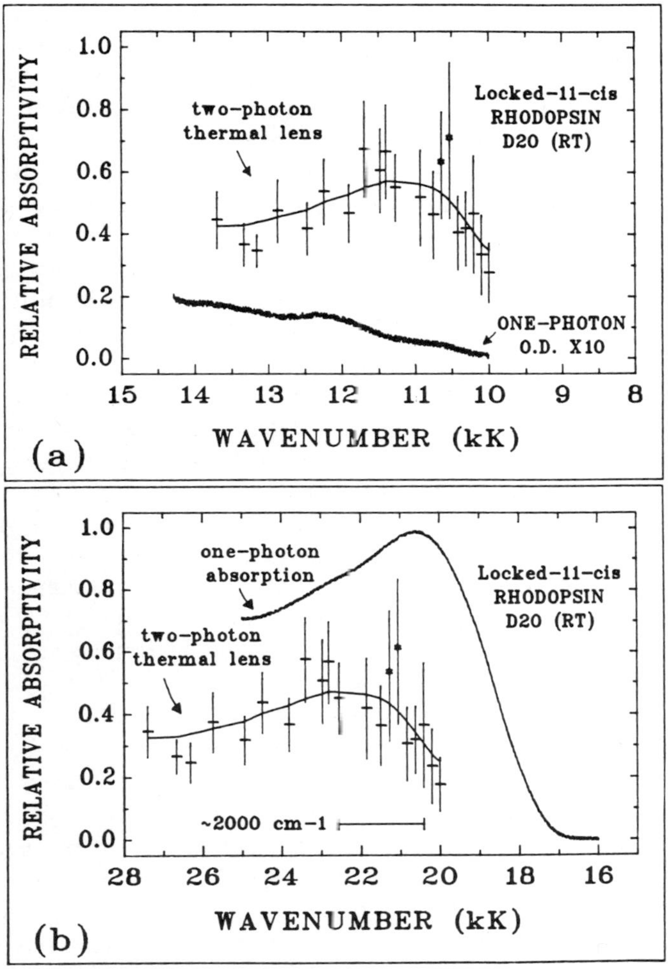


Fig. 4. Comparison of the two-photon thermal lens spectrum of locked-11-*cis* rhodopsin ($D_2O$, room temperature) with the one-photon (near-IR) absorption spectrum in the laser wavelength region (a) and with the one-photon (electronic) absorption spectrum in the ($\lambda/2$) region corresponding to the final-state vibronic distribution accessed via the combined energy of the two photons (b). Each data point represents a linear average of >12,000 measurements of sample/(reference squared) and the height of the vertical error bar equals two standard deviations (see text). The two data points marked with asterisks displayed a significant deviation (>68% confidence limit) of the two-photon slope from 2 (i.e., these two data points have a significant one-photon component that diminishes their reliability). The smooth curve drawn through the two-photon data was calculated by using a triangular slit function with an effective spectral bandwidth of 500 $cm^{-1}$. A statistical analysis of the separation between the Franck-Condon maxima of the one-photon absorption and two-photon thermal lens spectra yields 2,250 $\pm$ 850 $cm^{-1}$ (68% confidence limit). Note that the near-IR absorption spectrum in Figure 4a is multiplied by ten and that the vertical axis represents the optical density [Reprinted with permission from Birge et al., 1985.]

investigation [Birge, 1981, 1982; Birge and Pierce, 1979; Christensen and Kohler, 1973, 1974; Corsetti and Kohler, 1977; Das et al., 1979a,b; Dinur et al., 1980; Granville et al., 1979, 1980; Hargrave et al., 1984; Honig, 1982; Hudson and Kohler, 1973; Hug and Becker, 1976; Kliger, 1979; Liu et al., 1979, 1981, 1983; Liu and Asato, 1982; Muthukumar and Weimann, 1978; Ottlenghi, 1980; Schaffer et al., 1974a,b; Takemura et al., 1976, 1978; Waddell et al., 1978; Waddell and West, 1980]. In the case of long-chain (nonpolar) linear polyenes, the two-lying "$^1B_u^{*+}$" and "$^1A_g^{*-}$" $\pi\pi^*$ states respond to changes in solvent refractive index (polarizability) with dramatically different sensitivities [Hudson and Kohler, 1973]. The differences are due to the origin of the dispersive solvent shift which is proportional to the oscillator strength of the electronic transition times the polarizability of the environment [Corsetti and Kohler, 1977]. The oscillator strength of the allowed $^1B_u^{*+} \leftarrow S_o$ transition is many orders of magnitude greater than that of the forbidden "$^1A_g^{*-}$" $\leftarrow S_o$ transition in linear polyenes. The differences in oscillator strengths of the analogous transitions in an unsymmetrical, polar polyene such as all-*trans* retinal are much smaller due to symmetry breakdown. Nevertheless, the magnitude of the dispersive shift of the "$^1B_u^{*+}$" $\leftarrow S_o$ transition relative to the "$^1A_g^{*-}$" $\leftarrow S_o$ transition in all-*trans* retinal is greater by a factor of ~ 15 [Birge et al., 1982a].

Excited-state energy shifts due to electrostatic interactions are more difficult to predict because they are determined by bulk properties of the solvent (or protein) environment as well as specific interactions between the solute and neighboring polar molecules. In general, each solute must be evaluated by using theoretical procedures which are capable of accurately describing explicit intermolecular interactions. Our use of intermediate neglect of differential overlap–partial single and double configuration interaction (INDO-PSDCI) molecular orbital theory to theoretically predict the effect of counterion environment on the electronic transitions of the protein-bound chro mophore will be described below. More general statements regarding the relative sensitivity of "$^1A_g^{*-}$" and "$^1B_u^{*+}$" states to changes in electrostatic environment can be derived from an analysis of the valence bond origins of these two states.

The "$^1A_g^{*-}$" state is frequently referred to as a "covalent" state, a reference to a valence bond description involving linear combinations of bonding diagrams that have small amounts of charge separation. In contrast, the "$^1B_u^{*+}$" state is an "ionic" state that is described in terms of bonding structures with higher degrees of charge separation. The "$^1A_g^{*-}$" state is characterized by a highly correlated electron configuration in which the electrons "avoid each other," effectively reducing electron repulsion. The "$^1B_u^{*+}$" state is less correlated in the sense that electrons tend to "pile up" in various locations along the $\pi$ system.

Protonation of the Schiff base of retinal produces a significant reorganization of charge within the polyene $\pi$-system which preferentially stabilizes "ionic" states (like the "$^1B_u^{*+}$" state), and destabilizes "covalent" states (like the "$^1A_g^{*-}$" state) [Birge, 1981; Birge and Hubbard, 1980, 1981; Birge and Pierce, 1979]. Adding a single, negatively charged counterion near the protonated imine moiety restores neutrality of the $\pi$-system environment and therefore destabilizes "ionic" states and stabilizes "covalent" states. The separation between the low-lying "$^1B_u^{*+}$" and "$^1A_g^{*-}$" states is extremely sensitive to the number and location of the external counterions, a sensitivity that has significant diagnostic utility in helping to characterize the counterion environment around the protein-bound chromophore of rhodopsin (see below).

The effect that protonation and external counterions have on the relative electronic energies of the "covalent" "$^1A_g^{*-}$" state and the "ionic" "$^1B_u^{*+}$" state can be examined from a different perspective by using particle-in-a-box wave functions. Illustrative results are shown in Figure 5 where the "ionic" excited state is represented as a singlet with quantum numbers of $m = 3$ and $n = 2$ and the "covalent" excited state is represented as a spin-coupled triplet of identical spatial quantization (only one of the coupled wave functions is shown). The contour heights are proportional to the square of the wave functions of one electron assuming the other is localized in the corner of the box as shown in Figure 5a. These calculations illustrate two important characteristics of "ionic" and "covalent" states. First, "covalent" states have a lower electron-electron repulsion energy because the electron motion is correlated (i.e., there is a low probability that the two electrons will occupy the same location within the box). Second, electron correlation invariably leads to a smaller stabilization energy associated with external changes. This latter observation follows from the fact that correlated electronic structures are intrinsically optimized to avoid electron-electron repulsion and hence any external field will partially negate the relative advantages of this correlated structure.

## Two-Photon Spectra of Model Chromophores

The Franck-Condon maxima of the "$^1A_g^{*-}$" state in a variety of retinyl polyenes have been assigned by using two-photon spectroscopy [Birge et al., 1978, 1982a, 1984, 1985; Murray and Birge, 1985]. The resulting level-ordering assignments for all-*trans* retinol [Birge et al., 1978], all-*trans* retinal [Birge et al., 1982a], all-*trans* retinal Schiff base [Murray and Birge, 1985], and the protonated Schiff base [Birge et al., 1984] are shown in Figure 6. The structures are shown in Figure 2. The effect of protonation on the relative level ordering in the Schiff base compounds is pronounced. The origin of the level-ordering reversal was described in the previous section and is correctly predicted by INDO-PSDCI molecular orbital theory [Birge and Hubbard, 1980, 1981]. Comparison of the model chromophore level

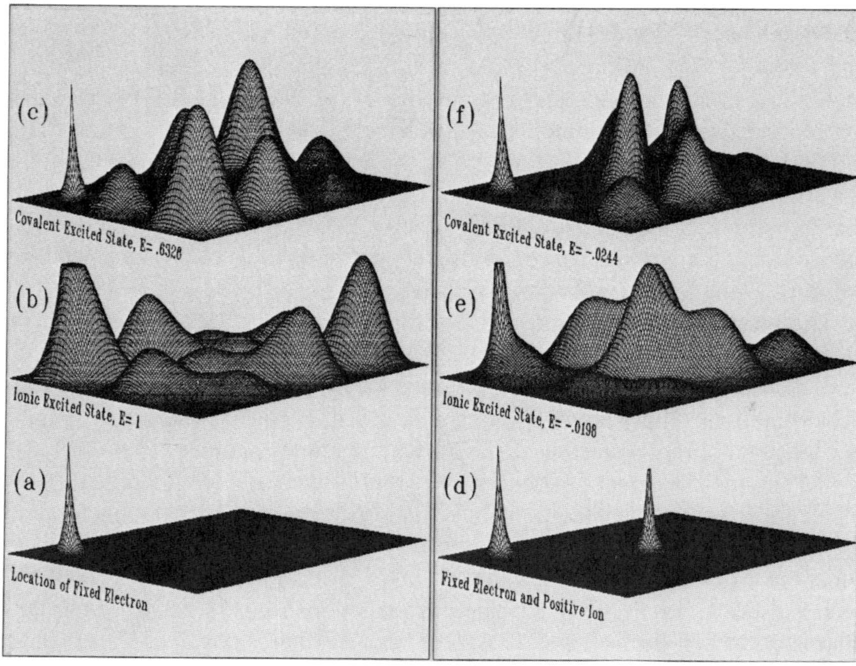


Fig. 5. Particle-in-a-box wave functions for "covalent" and "ionic" excited states for two electron systems in the absence (a–c) of and in the presence (d–f) of a positive charge localized in the center of the box. The contours shown in (b), (c), (e), and (f) represent the wave function squared for one electron when the other electron is arbitrarily fixed in the upper left-hand corner of the box (a). The ionic state is represented as an excited singlet [m = 3 (m = x axis quantum number) and n = 2 (n = y axis quantum number)]. The "covalent" state is represented as a spin-coupled triplet of identical spatial quantization. Note that the addition of a positive charge in the center of the box stabilizes the "ionic" state by 1.0198 energy units and the covalent state by 0.667 energy units.

orderings with that observed in locked-11-*cis* rhodopsin (Fig. 6) directly indicates that the chromophore in rhodopsin is protonated. Neither dispersive nor electrostatically induced shifts are realistically capable of inverting the "$^1B_u^{*+}$" and "$^1A_g^{*-}$" levels of an unprotonated Schiff base. While this observation follows from our earlier discussion, it should be noted that numerous test calculations utilizing INDO-PSDCI theory substantiate this conclusion. A second observation, which is somewhat more subtle, is that the binding site is neutral. As shown in Figure 7, INDO-PSDCI calculations demonstrate that the level ordering observed for the protein-bound chromophore can only be reproduced by using a single external negative charge (or a distribution of external charges with a total charge of −1 electron units). If

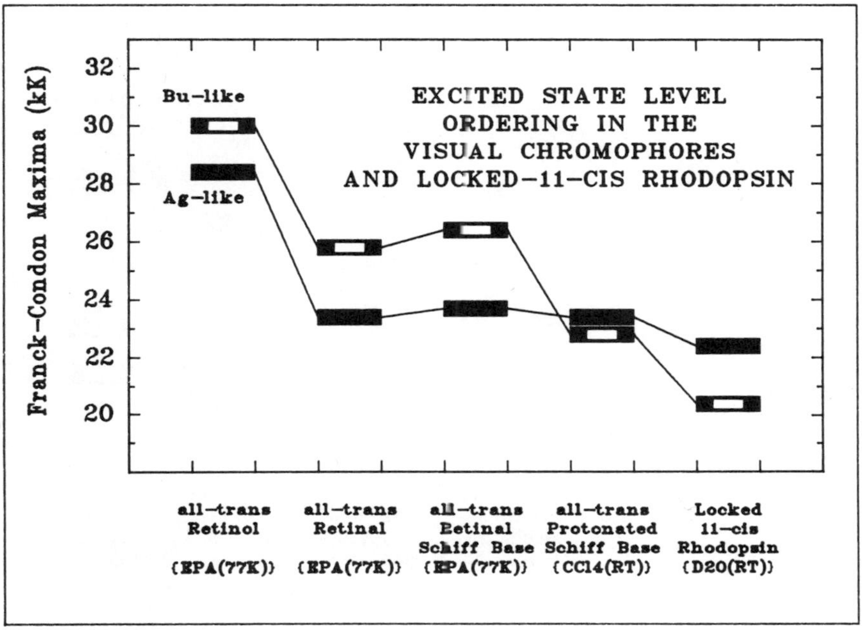


Fig. 6. Comparison of the Franck-Condon maxima associated with excitation into the "$^1B_u^{*+}$" state (one-photon absorption maxima) and the "$^1A_g^{*-}$" $\pi\pi^*$ state (two-photon absorption maxima) in selected visual chromophores and locked-11-*cis* rhodopsin. The environmental conditions are described in brackets below each compound (EPA = diethyl ether:isopentane:Ethyl alcohol; 5:5:2 v/v). [Adapted with permission from Birge et al., 1985.]

two negatively charged groups (total external charge of −2 electron units) are placed near the chromophore, the level separation increases (an "ionic" environment is produced) and more interestingly, the lowest excited "$^1B_u^{*+}$" singlet state is more strongly allowed under both one-photon and two-photon absorption relative to the higher energy "$^1A_g^{*-}$" state (see Fig. 7). While the latter observation was unexpected, the level-ordering separation arguments follow from qualitative considerations noted in the previous section. (Explicit calculations of the various binding site geometries involving two negatively charged counterions as shown in Table 2 of Kakitani et al. [1985] were also carried out. These calculations yielded similar failure in that the lowest excited $\pi\pi^*$ singlet state was the more two-photon allowed and the "$^1A_g^{*-}$"–"$^1B_u^{*+}$" splittings exceeded 7,500 $cm^{-1}$.) Two important conclusions are therefore possible: (1) the chromophore of rhodopsin is protonated, and (2) the chromophore binding site is neutral. Proceeding from this premise, we can use INDO-PSDCI procedures to simulate the effect of counterion geometry on the location of the "$^1A_g^{*-}$" and "$^1B_u^{*+}$" $\pi\pi^*$ excited states.

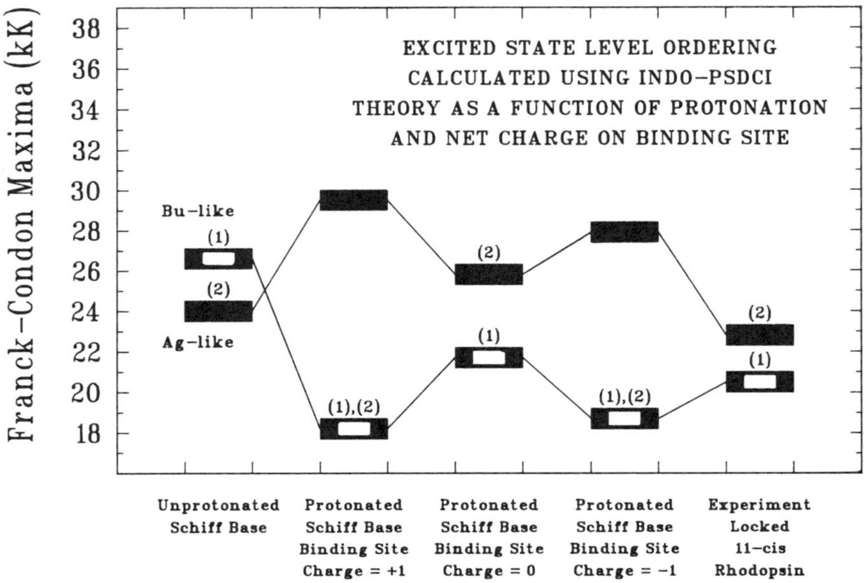


Fig. 7. Comparison of calculated (INDO-PSDCI) excited-state level orderings as a function of binding site model with the observed Franck-Condon maxima (locked-11-*cis* rhodopsin). The more one-photon-allowed excited state is indicated with the symbol "(1)" and the more two-photon-allowed excited state with the symbol "(2)". Although a number of possibilities were investigated for each binding site model, results are reported for the calculation that yielded the closest agreement with the experiment. The observation of a lowest-lying "$^1B_u^{*+}$" state indicates that the chromophore is protonated. The prediction that the lowest-lying "$^1B_u^{*+}$" state is more allowed under both one-photon and two-photon selection rules for charged binding sites indicates that the binding site is neutral.

## The Nature of the Rhodopsin Binding Site

The model chromophore studies described in the previous section indicate that the chromophore in rhodopsin is protonated and occupies a neutral binding site (chromophore charge = +1, external residue charge = −1). This observation might appear to conflict with the external point charge model proposed by Honig et al. [1979a]. This model proposed the existence of two external charged species, one near the protonated nitrogen, and the second near the $C_{12}$ and $C_{14}$ atoms [see Fig. 3 of Honig et al., 1979a]. It was presumed in the original model, however, that the latter counterion could be a member of a charged pair of a salt bridge or the negatively charged end of a neutral dipolar group. Accordingly, our observation of a neutral binding site does not conflict with the external point charge model of Honig et al. [1979a]. Our observation, however, does restrict the possibilities.

In order to explicitly investigate the nature of the counterion environment of the opsin binding site, INDO-PSDCI molecular orbital theory [Birge et al., 1982a,b; Birge and Hubbard, 1980, 1981; Murray and Birge, 1985] was used to model the effect that external counterions have on the electronic spectroscopy of the native chromophore as well as the opsin-bound hydroretinals studied previously [Arnaboldi et al., 1979; Honig et al., 1979a]. Our goal was to satisfactorily accommodate all of the available spectroscopic data. Our calculations included all single excitations below 15 eV ($\sim$150) and all double excitations below 20 eV ($\sim$250). The entire retinyl protonated Schiff base framework was included, but an ethyl moiety was substituted for the lysine residue. All other details of the calculation conform with those outlined previously [Birge et al., 1982b; Birge and Hubbard, 1980, 1981].

The semiempirical INDO-PSDCI procedures were parameterized previously to reproduce the one-photon and two-photon spectroscopic properties of retinyl polyenes [Birge et al., 1982a,b; Birge and Hubbard, 1980, 1981]. The transition energies of the low-lying strongly allowed "$^1B_u^{*+}$" states are reproduced within 0.3 eV for all-*trans* retinal [3.9 eV (3.7 eV obsvd)], all-*trans* retinal [3.5 eV (3.2 eV obsvd)], and all-*trans* retinal Schiff base [3.6 eV (3.3eV obsvd)]. Moreover, the trends are accurately reproduced. The INDO-PSDCI calculations also reproduce reasonably well the energy splitting between the two lowest-lying "$^1B_u^{*+}$" and "$^1A_g^{*-}$" states: all-*trans* retinol [0.4 eV (0.2 eV obsvd)], all-*trans* retinal [0.3 eV (0.3 eV obsvd)], all-*trans* retinal Schiff base [0.4 eV (0.4 eV obsvd)]. These data indicate that our procedures, although reliable in predicting hexaene splittings, overestimate pentaene splittings. This problem is associated with the inability of molecular orbital procedures limited to single and double CI to accurately predict trends in covalent-ionic splittings as a function of polyene size [Dinur et al., 1980]. Our procedures were parameterized for hexaenes [Birge et al., 1982a,b; Birge and Hubbard, 1980, 1981], and should be adequate for theoretically predicting the electronic spectra of the protein-bound chromophore in rhodopsin. We estimate our procedures are reliable within 0.4 eV for absolute transition energies and 0.2 eV for excited-state "$^1B_u^{*+}$"–"$^1A_g^{*-}$" splittings.

The INDO-PSDCI charge distributions in the ground and lowest two excited $\pi\pi^*$ singlet states were used to generate electrostatic contour maps predicting the (first order) effect of a single external negative charge on the energy separation between the "$^1A_g^{*-}$" and the "$^1B_u^{*+}$" $\pi\pi^*$ states (Fig. 8a) and on the energy of the "$^1B_u^{*+}$" $\leftarrow$ $S_o$ transition (Fig. 8b). It is important to note that these contour plots are qualitative because they are

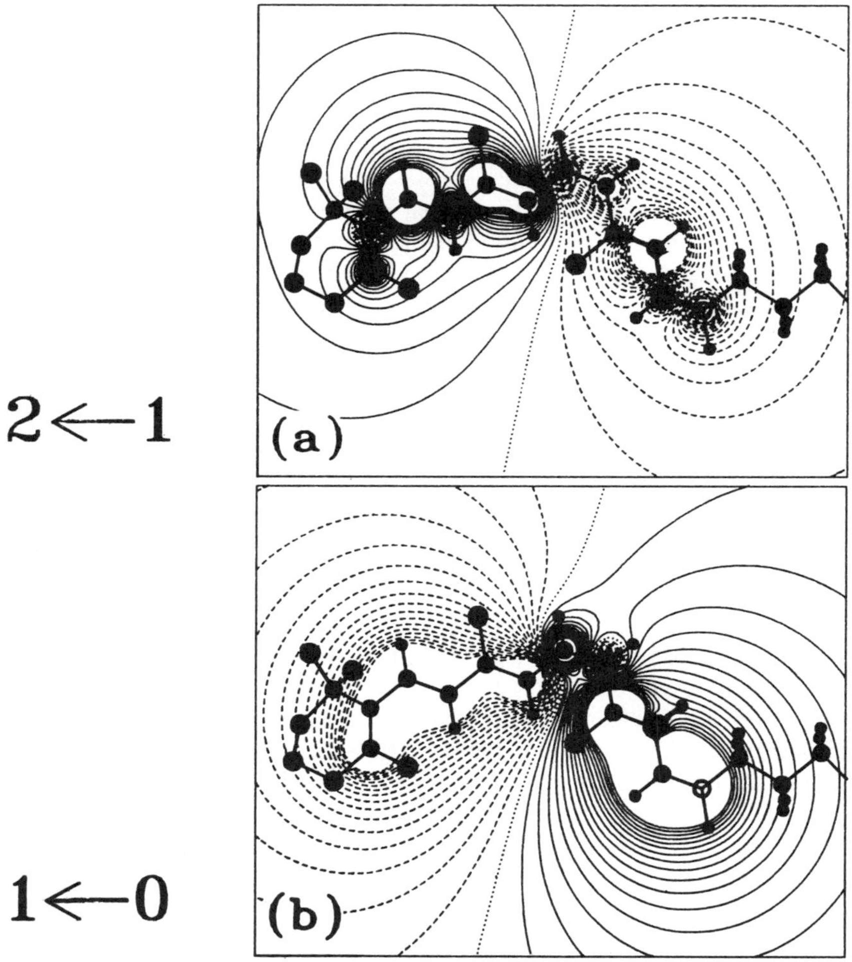


Fig. 8. Comparison of the Franck-Condon maxima for the energy splitting between the "$^1A_g{}^{*-}$" and the "$^1B_u{}^{*+}$" $\pi\pi^*$ states ("2 ← 1", a) and those associated with excitation into the "$^1B_u{}^{*+}$" $\pi\pi^*$ ("1 ← 0", b). Each contour level is drawn at increments of $\pm 800$ cm$^{-1}$ (to a maximum shift of $\pm 12{,}800$ cm$^{-1}$) where dashed lines represent red shifts (decreasing energy separation), solid lines represent blue shifts (increasing energy separation), and the dotted line represents the zero energy shift contour. The calculations were carried out by using INDO-PSDCI wave functions ($\sim 150$ single, $\sim 250$ double excitations) on the protonated Schiff base chromophore (the nitrogen atom is indicated with an open circle). Geometric and formalistic details may be found in Birge and Hubbard [1980, 1981]. The transition energies in the absence of a counterion are E("$^1B_u{}^{*+}$" ← $S_o$) = 17,100 cm$^{-1}$ and E("$^1A_g{}^{*-}$" ← $S_o$) = 29,450 cm$^{-1}$ [ΔE("$^1A_g{}^{*-}$"–"$^1B_u{}^{*+}$") = 12,350 cm$^{-1}$]. These contour plots are only accurate to first order and were used to predict possible counterion locations. INDO-PSDCI calculations were then carried out to determine the transition energies of the "supermolecule" containing the chromophore plus counterions (Fig. 9).

constructed assuming no redistribtuion of polyene electronic charge in response to external counterion location. Nevertheless, they were very useful in predicting possible locations of the counterions prior to carrying out the time-consuming [central processing unit (CPU) intensive] INDO-PSDCI calculations on the "supermolecule" comprising the chromophore plus counterion(s).

The isolated protonated Schiff base (no counterion) is calculated to have a lowest-lying "$^1B_u^{*+}$" ← $S_o$ transition at ~17,100 cm$^{-1}$ (~585 nm) and a "$^1A_g^{*-}$"–"$^1B_u^{*+}$" energy splitting of ~12,400 cm$^{-1}$. The corresponding experimental values for the chromophore of locked 11-*cis* rhodopsin are ~20,400 cm$^{-1}$ (490 nm) and ~2,000 cm$^{-1}$ (Figs. 4, 6). Analysis of Figure 8 indicates that one or more counterions must be placed near the $C_{11} \cdots N_{15}$ region of the chromophore in order to blue shift the "$^1B_u^{*+}$" ← $S_o$ transition and decrease the "$^1A_g^{*-}$"–"$^1B_u^{*+}$" splitting. By using the contour plots to estimate counterion location, a number of models were tested with the goal of reproducing the observed state locations and separations as well as the trends observed for the 11-*cis* hydroretinals shown in Figure 1 of Honig et al. [1979a]. The two models shown in Figure 9 accommodate all of the above spectroscopic data within experimental error. Although the INDO-PSDCI calculations overestimate the observed transition energies by ~0.3 eV, this error is consistent with that observed in the model chromophore calculations (see above). The single counterion model (Fig. 9a) is interesting because it provides two "external point charges" with a single aspartic (or glutamic) acid residue. In terms of simplicity, it is the more attractive model. A similar model has been proposed independently by Kakitani et al. [1985]. The two-counterion model (Fig. 9b) may have more chemical relevance, because it can be used to rationalize the origin of the deuterium isotope effect. Note that the upper counterion in Figure 9b is protonated.

One of the unusual features of the primary event of the rhodopsin photobleaching process is the observation of a deuterium isotope effect [Peters et al., 1977]. It is now generally accepted that the primary event involves an 11-*cis* to 11-*trans* photoisomerization [Akita et al., 1980; Birge, 1981, 1982; Callender and Honig, 1977; Doukas et al., 1984; 1985; Eyring and Mathies, 1979; Hargrave et al., 1984; Honig, 1982; Mao et al., 1980, 1981; Mathies et al., 1982; Ottolenghi, 1980; Yoshizawa, 1972; Yoshizawa and Wald, 1963, 1967]. What remains to be explained is the origin of the deuterium isotope effect since explicit molecular dynamics calculations on the photoisomerization step indicate that this transformation is very insensitive to deuteration [Birge and Hubbard, 1980, 1981]. The most likely explanation is that a proton translocation occurs as a ground-state process following photoisomerization [Doukas et al., 1984, 1985; Honig et al., 1979b]. The binding site model shown in Figure 9b provides a model which can explain the origin of the latter process. An 11-*cis* → 11-*trans* photoisomerization will produce

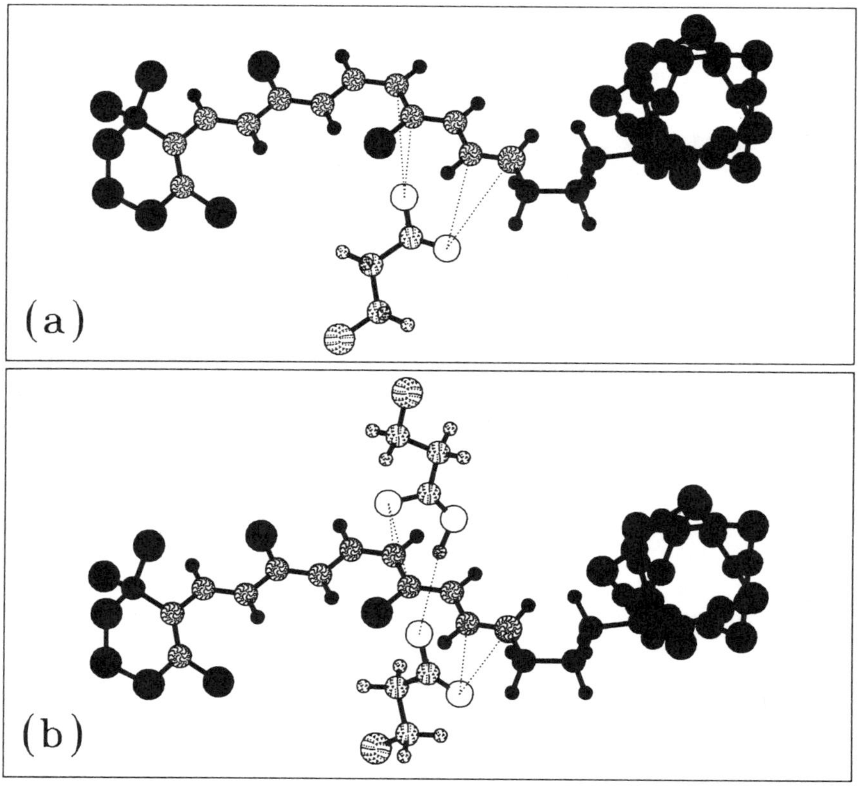


Fig. 9. Two possible binding site geometries which accommodate the available one-photon and two-photon spectroscopic data. The single counterion model (a) accommodates the "two-point charge model" [Honig et al., 1979a] by providing a single glutamic acid residue underneath the plane of the polyene chromophore. The two distinct point charges are provided by the two oxygen atoms (open circles) and the dotted lines indicate nearest atom electrostatic interactions (~3-Å separations). The following transition energies are calculated for the single counterion model: E("$^1B_u^{*+}$" $\leftarrow S_o$) = 22,600 $cm^{-1}$ (442 nm), E("$^1A_g^{*-}$" $\leftarrow S_o$) = 26,160 $cm^{-1}$ (382 nm), [$\Delta$E("$^1A_g^{*-}$"-"$^1B_u^{*+}$") = 3560 $cm^{-1}$]. The two-counterion model (b) is equally capable of accommodating the available spectroscopic data and has the advantage of also rationalizing the observed deuterium isotope data of Peters et al. [1977] (see text). The following transition energies are calculated for the two counterion model: E("$^1B_u^{*+}$" $\leftarrow S_o$) = 21,700 $cm^{-1}$ (461 nm), E("$^1A_g^{*-}$" $\leftarrow S_o$) = 25,800 $cm^{-1}$ (388 nm), [$\Delta$E("$^1A_g^{*-}$"-"$^1B_u^{*+}$") = 4100 $cm^{-1}$]. Although the lysine residue is shown *trans* to the chromophore about the C = N bond, note that the calculations are not sensitive to lysine geometry nor to switching aspartic for glutamic acid residues.

extensive charge separation which will destabilize the counterion environment. The 11-*trans* conformation will be stabilized by translocation of the proton from the upper acid residue to the lower residue so that the upper residue becomes the "primary counterion" in bathorhodopsin. The resulting shift in the absorption maximum of the primary photoproduct following proton translocation may be the spectroscopic phenomenon which is experimentally observed to exhibit a deuterium isotope effect. This possibility is currently under investigation. Molecular dynamics calculations are being used to simulate the primary event based on the model shown in Figure 9b.

## ACKNOWLEDGMENTS

This work was supported in part by grants to R.R.B. from the National Science Foundation (CHE-7916336) and the National Institutes of Health (GM-34548).

## REFERENCES

Akita H, Tanis SP, Adams M, Balogh-Nair V, Nakanishi K (1980): Nonbleachable rhodopsins retaining the full natural chromophore. J Am Chem Soc 102:6370–6372.

Arnaboldi M, Motto MG, Tsujimoto K, Balogh-Nair V, Nakanishi K (1979): Hydroretinals and hydrorhodopsins. J Am Chem Soc 101:7082–7084.

Bagley K, Dolliger G, Eisenstein L, Vittitow J, Zimanyi L, Ebrey TG, Nelson B (1983): Fourier-transform infrared studies of rhodopsin and its photoproducts at low temperatures. Biophys J 41:337a.

Birge RR (1981): Photophysics of light transduction in rhodopsin and bacteriorhodopsin. Ann Rev Biophys Bioeng 10:315–354.

Birge RR (1982): Simulation of the primary event in rhodopsin photochemistry using semiempirical molecular dynamics theory. In Alfano RR (ed): "Biological Events Probed by Ultrafast Laser Spectroscopy." New York: Academic Press, pp 299–317.

Birge RR (1983): One-photon and two-photon excitation spectroscopy. In Kliger DS (ed): "Ultrasensitive Laser Spectroscopy." New York: Academic Press, pp 109–174.

Birge RR (1984): An introduction to two-photon spectroscopy. In Sandorfy C, Theophanides T (eds): "Spectroscopy of Biological Molecules." Boston: Reidel, pp 457–471.

Birge RR, Pierce BM (1979): A theoretical analysis of the two-photon properties of linear polyenes and the visual chromophores. J Chem Phys 70:165–178.

Birge RR, Hubbard LM (1980): Molecular dynamics of *cis-trans* isomerization in rhodopsin. J Am Chem Soc 102:2195–2205.

Birge RR, Hubbard LM (1981): Molecular dynamics of *trans-cis* isomerization in bathorhodopsin. Biophys J 34:517–534.

Birge RR, Schulten K, Karplus M (1975): Possible influence of a low-lying "covalent" excited state on the absorption spectrum and photoisomerization of 11-cis retinal. Chem Phys Lett 31:451–454.

Birge RR, Bennett JA, Pierce BM, Thomas TM (1978): Two-photon spectroscopy of the visual chromophores. Evidence for a lowest excited ${}^1A_g{}^{*-}$-like $\pi\pi^*$ state in all-trans retinol (Vitamin A). J Am Chem Soc 100:1533–1539.

Birge RR, Bennett JA, Hubbard LM, Fang HL, Pierce BM, Kliger DS, Leroi GE (1982a):

Two-photon spectroscopy of all-*trans*-retinal. Nature of the low-lying singlet states. J Am Chem Soc 104:2519–2525.

Birge RR, Bocian DA, Hubbard LM (1982b): Origins of inhomogeneous broadening in the vibronic spectra of visual chromophores and visual pigments. J Am Chem Soc 104:1196–1207.

Birge RR, Pierce BM, Murray LP (1984): Two-photon spectroscopy of biological molecules. In Sandory C, Theophanides T (eds): "Spectroscopy of Biological Molecules." Boston: Reidel, pp 473–486.

Birge RR, Murray LP, Pierce BM, Akita H, Balogh-Nair V, Findsen LA, Nakanishi K (1985): Two-photon spectroscopy of locked-11-*cis*-rhodopsin: Evidence for a protonated Schiff base in a neutral protein binding site. Proc Natl Acad Sci USA 82:4117–4121.

Blatz PE, Lin M, Balasubramaniyan P, Balasubramaniyan V, Dewhurst PB (1969): A new series of synthetic visual pigments from cattle opsin and homologs of retinal. J Am Chem Soc 91:5930–5931.

Buchert J, Stefancic V, Doukas AG, Alfano RR, Callender RH, Pande J, Akita H, Balogh-Nair V, Nakanishi K (1983): Picosecond kinetic absorption and fluorescence studies of bovine rhodopsin with a fixed 11-ene. Biophys J 43:279–283.

Callender RH, Honig B (1977): Resonance raman studies of visual pigments. Ann Rev Biophys Bioeng 6:33–55.

Christensen RL, Kohler BE (1973): Low resolution optical spectroscopy of retinyl polyenes: Low lying electronic levels and spectral broadness. Photochem Photobiol 18:293–301.

Christensen RL, Kohler BE (1974): Excitation spectroscopy of retinal and related polyenes. Photochem Photobiol 19:401–410.

Cooper A (1979a): Energetics of rhodopsin and isorhodopsin. FEBS Lett 100:382–384.

Cooper A (1979b): Energy uptake in the first step of visual excitation. Nature 282:531–533.

Cooper A (1982): Calorimetric measurements of light-induced processes. Methods Enzymol 88:667–673.

Corsetti JP, Kohler BE (1977): Ground and excited state dipole moments of all-*trans* retinal and all-*trans*-retinylidene-*n*-butylamine in solution. J Chem Phys 67:5237–5243.

Das PK, Becker RS, Hannak D, Bayer E (1979a): New model visual pigments. Spectroscopy of Poly(ethylene glycol) peptide Schiff base of retinal. J Am Chem Soc 101:239–240.

Das PK, Kogan G, Becker RS (1979b): Spectroscopy of polyenes III. Absorption and emission spectral investigation of polyene Schiff bases and protonated Schiff bases related to visual pigments. Photochem Photobiol 30:689–695.

Dinur U, Honig B, Schulten K (1980); On the nature of excited electronic states in cyanine dyes: Implications for visual pigment spectra. Chem Phys Lett 72:493–497.

Doukas AG, Junnarkar MR, Alfano RR, Callender RH, Kakitani T, Honig B (1984): Fluorescence quantum yield of visual pigments: Evidence for subpicosecond isomerization rates. Proc Natl Acad Sci USA 81:4790–4794.

Doukas AG, Junnarkar MR, Alfano RR, Callender RH, Balogh-Nair V (1985): The primary event in vision investigated by time-resolved fluoresence spectroscopy. Biophys J 47:795–798.

Eyring G, Mathies R (1979): Resonance raman studies of bathorhodopsin: Evidence for a protonated Schiff base linkage. Proc Natl Acad Sci USA 76:33–37.

Favrot J, LeClercq JM, Roberge R, Sandorfy C, Vocelle D (1979): Intermolecular interactions in visual pigments. The hydrogen bond in vision. Photochem Photobiol 29:99–108.

Granville MF, Holtom GR, Kohler BE, Christensen RL, D'Amico K (1979): *Cis-trans* photoisomerization of 1,3,5,7-octatetraene in *n*-hexane at 4.2 K. J Chem Phys 70:593–594.

Granville MF, Holtom GR, Kohler BE (1980): Experimental conformation of the dipole forbidden character of the lowest excited singlet state in 1,3,5,7-octatetraene. J Chem Phys 72:4671–4675.

Hargrave PA, McDowell JH, Feldman RJ, Atkinson PH, Mohana Rao JK, Argos P (1984): Rhodopsin's protein and carbohydrate structure: Selected aspects. Vision Res 24:1487–1499.

Harosi FI, Favrot J, LeClercq JM, Vocelle D, Sandorfy C (1978): Photochemistry of visual pigments: An interpretation of spectral changes in terms of molecular associations and isomerization. Rev Can Biol 37:257–271.

Honig B (1982): Theoretical aspects of photoisomerization in visual pigments and bacteriorhodopsin. In Alfano RR (ed): "Biological events probed by ultrafast laser spectroscopy." New York: Academic Press, pp 281–297.

Honig B, Dinur U, Nakanishi K, Balogh-Nair V, Gawinowicz MA, Arnaboldi M, Motto MG (1979a): An external point-charge for wavelength regulation in visual pigments. J Am Chem Soc 101:7084–7086.

Honig B, Ebrey T, Callender RH, Dinur U, Ottolenghi M (1979b): Photoisomerization, energy storage, and charge separation: A model for light energy transduction in visual pigments and bacteriorhodopsin. Proc Natl Acad Sci USA 76:2503–2507.

Honig B, Dinur U, Birge RR, Ebrey TG (1980): The isomer dependence of oscillator strengths in retinal and related molecules. Spectroscopic assignments. J Am Chem Soc 102:488–494.

Hudson BS, Kohler BE (1973): Polyene spectroscopy: The lowest energy excited singlet state of diphenyloctatetraene and other linear polyenes. J Chem Phys 59:4984–5002.

Hug G, Becker RS (1976): Solvent and temperature effects on natural radiative lifetimes of some substituted polyenes. J Chem Phys 65:55–63.

Hurley JB, Ebrey TG, Honig B, Ottolenghi M (1977): Temperature and wavelength effects on the photochemistry of rhodopsin, isorhodopsin, bacteriorhodopsin and their photoproducts. Nature 270:540–542.

Kakitani H, Kakitani T, Rodman H, Honig B (1985): On the mechanism of wavelength regulation in visual pigments. Photochem Photobiol 41:471–479.

Kliger DS (1979): Spectroscopic and photochemical studies of model visual pigment chromophores. Int J Quant Chem 16:809–817.

LeClercq J-M, Sandorfy C (1981): On the possibility of protein-chromophore charge transfer in visual pigments. Photochem Photobiol 33:361–365.

LeClercq J-M, Dupuis P, Sandorfy C (1982): Possibility of a double well potential in the proton bridge of visual pigments and bacteriorhodopsin. Croatica Chemica Acta 55:105–119.

Liu RSH, Asato AE (1982): Synthesis and Photochemistry of stereoisomers of retinal. Methods Enzymol 88:506–516.

Liu RSH, Denny M, Grodowski M, Asato AE (1979): Photoisomerization of 10-fluoro-, 14-fluoro-, and 3-dehydroretinal. The polar character of the excited state intermediates. Nouv J Chim 3:503–505.

Liu RSH, Matsumoto H, Asato AE, Denny M, Shichida Y, Yoshizawa T, Dahlquist FW (1981): Synthesis and properties of 12-fluororetinal and 12-fluororhodopsin. A model system for $^{19}F$ NMR studies of visual pigments. J Am Chem Soc 103:7195–7201.

Liu RSH, Asato AE, Denny M (1983): Medium- and substituent-directed stereoselective photoisomerization of polyenes in the vitamin A series. Application of the Dellinger-Kasha model. J Am Chem Soc 105:4829–4830.

Mao B, Ebrey TG, Crouch R (1980): Bathoproducts of rhodopsin, isorhodopsin I, and isorhodopsin II. Biophys J 29:247–256.

Mao B, Tsuda M, Ebrey T, Akita H, Balogh-Nair V, Nakanishi K (1981): Flash photolysis and low temperature photochemistry of bovine rhodopsin with a fixed 11-ene. Biophys J 35:543–546.

Mateescu GD, Muccio DD, Copan WG, Abrahamson EW (1981): Nuclear magnetic resonance studies of rhodopsin and bacteriorhodopsin. J Photochem 17:63.
Mateescu GD, Copan WG, Muccio DD, Waterhous DV, Abrahamson EW (1983): Synthesis and high resolution C-13 and N-15 NMR of visual pigments and related systems. In Duncan WP, Susan AP (eds): "Proceedings of the International Symposium on Synthesis and Applications of Isotopically Labeled Compounds." Amsterdam: Elsevier, pp 123–132.
Mathies R (1982): Resonance raman spectroscopy of rhodopsin and bacteriorhodopsin isotopic analogs. In Packer L (ed): "Methods Enzymol." New York: Academic Press 88:633–643.
Monger TG, Alfano RR, Callender RH (1979): Photochemistry of rhodopsin and isorhodopsin investigated on a picosecond time scale. Biophys J 27:105–116.
Murray LP, Birge RR (1985): Two-photon spectroscopy of the Schiff base of *all-trans*-retinal. Nature of the low-lying $\pi\pi^*$ singlet states. Can J Chem 63:1967–1971.
Muthukumar M, Weimann LJ (1978): On the origin of the bathochromic shift in rhodopsin. Chem Phys Lett 53:436–438.
Oseroff AR, Callender RH (1974): Resonance raman spectroscopy of rhodopsin in retinal disk membranes. Biochemistry 13:4243–4348.
Ottolenghi M (1980): The photochemistry of rhodopsin. Adv Photochem 12:97–200.
Papermaster DS, Dreyer WJ (1974): Rhodopsin content in the outer segment membranes of bovine and frog retinal rods. Biochemistry 13:2438–2458.
Peters K, Applebury ML, Rentzepis PM (1977): Primary photochemical event in vision: Proton translocation. Proc Natl Acad Sci USA 74:3119–3123.
Pierce BM, Birge RR (1982): Lasing properties of several near-IR dyes for a nitrogen laser-pumped dye laser with an optical amplifier. IEEE J Quant Elect QE-18:1164–1170.
Rothschild KJ, Cantore WA, Marrero H (1983): Fourier transform infrared difference spectra of intermediates in rhodopsin bleaching. Science 219:1333–1335.
Schaffer AM, Waddell WH, Becker RS (1974a): Visual pigments. IV. Experimental and theoretical investigations of the absorption spectra of retinal Schiff bases and retinals. J Am Chem Soc 96:2063–2068.
Schaffer AM, Yamaoka T, Becker RS (1974b): Visual pigments—V. Ground and excited-state acid dissociation constants of protonated all-*trans* retinal Schiff base and correlation with theory. Photochem Photobiol 21:297–301.
Shichida Y, Yoshizawa T, Kobayashi T, Ohtani H, Nagakura S (1977): Squid hypsorhodopsin and bathorhodopsin by a picosecond laser photolysis. FEBS Lett 80:214–216.
Shichida Y, Kobayashi T, Ohtani H, Yoshizawa T, Nagakura S (1978): Picosecond laser photolysis of squid rhodopsin at room and low temperatures. photochem photobiol 27:335–341.
Shriver J, Mateescu G, Fager R, Torchia D, Abrahamson EW (1977): Unprotonated chromophore-protein bond in visual pigments from C-13 NMR spectra: Nature 270:271–274.
Siebert F, Mäntele W, Gerwert K (1983): Fourier-transform infrared spectroscopy applied to rhodopsin. The problem of the protonation state of the retinylidene Schiff base reinvestigated. Eur J Biochem 136:119–127.
Takemura T, Das PK, Hug G, Becker RS (1976): Emission spectroscopy and state ordering of retinals. J Am Chem Soc 98:7099–7101.
Takemura T, Das PK, Hug G, Becker RS (1978): Visual pigments. 8. Hydrogen bonding effects on fluorescence properties of retinals. J Am Chem Soc 100:2626–2630.
Waddell WH, West JL (1980): Photochemistry of linear polyenes related to vitamin A. 13-demethylretinal and 14-methylretinal. J Phys Chem 84:134–139.
Waddell WH, Hopkins DL, Uemura M, West JL (1978): Influence of alkyl substitution on the

trans → cis photoisomerization of *all-trans*-retinal and related polyenes. J Am Chem Soc 100:1970–1972.

Warshel A (1976): Bicycle-pedal model for the first step in the vision process. Nature 260:679–683.

Warshel A, Barboy N (1982): Energy storage and reaction pathways in the first step of the vision process. J Am Chem Soc 104:1469–1476.

Weiss RM, Warshel A (1979): A new view of the dynamics of singlet *cis-trans* photoisomerization. J Am Chem Soc 101:6131–6133.

Yoshizawa T (1972): The behaviour of visual pigments at low temperatures. In Dartnell HJA (ed): "Handbook of Sensory Physiology—Photochemistry of Vision." Berlin: Springer-Verlag, chpt. VII, pp 146–179.

Yoshizawa T, Wald G (1963): Pre-lumirhodopsin and the bleaching of visual pigments. Nature 197:1279–1286.

Yoshizawa T, Wald G (1967): Photochemistry of iodopsin. Nature 214:566–571.

*Applications of Fluorescence in the Biomedical Sciences, pages 129–140*


# 6 Single-Molecule Fluorescence Detection: A Feasibility Study Using Phycoerythrin

*Richard A. Mathies and Lubert Stryer*

The high sensitivity of fluorescence detection makes it an ideal method for analyzing biomolecules and cells, as exemplified by fluorescence immunoassays and fluorescence-activated cell sorting [for reviews, see Voller et al., 1981; Steinkamp, 1984; Milby and Zare, 1984; Muirhead et al., 1985]. The current detection limits are about $10^3$ fluorescent molecules in cell sorting and $10^6$ molecules in immunoassays. Can the ultimate in sensitivity, the observation of a single biomolecule, be attained? Inspiration for this quest comes from the work of Hurst et al. [1977]. They succeeded in detecting a single atom of cesium in the presence of $10^{19}$ atoms of a different kind in the gas phase. In another pioneering study, Hirschfeld [1976a] detected single antibody molecules tagged with a polymer containing about 80 fluoresceins. These fluorescent-labeled antibody molecules were spread on a microscope slide and observed with a detection system that viewed only a narrow surface layer. Discrete 2-msec-long light pulses corresponding to the emission of individual molecules were seen when the slide was mechanically swept past a 1 $\mu$m × 5 $\mu$m illuminated aperture.

The detection of single molecules in bulk liquid is a more formidable challenge because of the large background signal. The work of Dovichi et al. [1983, 1984] using a flow cytometry system is an important contribution to

Department of Chemistry, University of California, Berkeley, California 94720 (R.A.M.), Department of Cell Biology, Sherman Fairchild Center, Stanford University School of Medicine, Stanford, California 94305 (L.S.).

enhancing the sensitivity of fluid phase detection. A flowing $10^{-11}$ M solution of rhodamine 6G gave a signal equal to that of a solvent blank. At this concentration, there were 66 rhodamine molecules in the 11-pL probed volume. Their statistical analysis of the dependence of fluorescence intensity on sample concentration gave an extrapolated detection limit of $8.9 \times 10^{-14}$ M, which corresponds to an average of 0.6 rhodamine molecules in the probed volume. A key aspect of this approach is that new fluorescent molecules are flowed into the intensely illuminated volume to replace those photochemically destroyed by light.

It is evident that the attainment of single-molecule sensitivity in analyses of biological systems requires the simultaneous optimization of fluorescent labels and instrumentation. A new class of fluorescent reagents, called phycofluors, has recently been developed [Oi et al., 1982; Glazer and Stryer, 1983, 1984]. Phycofluors are conjugates of intensely fluorescent algal phycobiliproteins with molecules that confer biological specificity. Phycoerythrin, a member of this class of reagents, is an attractive fluorescent tag because it has a very high absorbance coefficient ($2.4 \times 10^6$ $cm^{-1}$ $M^{-1}$ at 562 nm) due to its 34 bilin chromophores and a very high fluorescence quantum yield ($Q = 0.98$). In fact, the fluorescence intensity of a molecule of phycoerythrin is more than 20-fold greater than that of a molecule of fluorescein. The red emission and large Stokes shift of phycoerythrin are also advantageous in discriminating fluorescence from Rayleigh scattering and background fluorescence. Phycoerythrin and other phycobiliproteins can readily be conjugated to biomolecules. Furthermore, they are stable on storage and their emission is constant over a broad range of pH and not quenched by most molecules in cells and body fluids. These ancient light harvesters, more than two billion years old [Glazer, 1984], are now being used as versatile and powerful analytical reagents in medical diagnostics, immunology, and developmental biology [reviewed in Glazer and Stryer, 1984; Kronick and Grossman, 1983; Parks et al., 1983; Shapiro et al., 1983; Zoumbos et al., 1985].

We report here pilot studies exploring the use of phycoerythrin as a fluorescent label for single-molecule detection. This article begins with a presentation of the theoretical basis of single-molecule detection. The photodestruction yield is a limiting factor in attaining ultimate sensitivity. Flow experiments giving the photodestruction quantum yields of phycoerythrin and fluorescein are then described. An emission spectrum arising from three molecules of phycoerythrin in the probed illuminated volume is then shown. The fluorescence signal from three molecules is several times that of the solvent background and clearly resolved from the water Raman band. These results indicate that phycoerythrin is a very promising label for single-molecule fluorescence detection.

# THEORY

## Fluorescence Emission Rates

The rate of absorption of photons $k_a$ (photons/sec) by a chromophore having an absorption cross section $\sigma_a$ ($cm^2$ /molecule) illuminated by an incident beam of intensity $I$ (photons/$cm^2$ sec) is given by

$$k_a = \sigma_a I \tag{1}$$

This equation is valid when the rate of excitation is small compared with the excited state decay rate $1/\tau$, where $\tau$ is the excited state lifetime of the chromophore. The absorption cross section is related to the molar absorbance coefficient $\epsilon$ ($cm^{-1}$ $M^{-1}$) by $\sigma_a = 3.8 \times 10^{-21}\ \epsilon$. For a Gaussian beam with a radius $\omega$ (where the intensity is $e^{-2}$ of its peak value) and power P (photons/sec), the mean intensity $I$ is equal to $P/(\pi\ \omega^2)$, and so

$$k_a = 3.8 \times 10^{-21}\ \epsilon\ P/(\pi\omega^2) \tag{2}$$

The rate of fluorescence emission $k_f$ (photons/sec) is the product of $k_a$ and the fluorescence quantum yield $Q$. Hence,

$$k_f = \sigma_a\ I\ Q = 3.8 \times 10^{-21}\ \epsilon\ Q\ P/(\pi\omega^2) \tag{3}$$

What are the highest attainable fluorescence emission rates? It is instructive to calculate $k_f$ for some fluorescent chromophores under experimentally realizable conditions. For fluorescein, $\epsilon = 8 \times 10^4$ $cm^{-1}$ $M^{-1}$ at 488 nm. Suppose that 65 mW ($1.6 \times 10^{17}$ photons/sec) is focused to a radius of 4 $\mu$m to give an incident intensity of $3.2 \times 10^{23}$ photons/$cm^2$ sec. The absorption rate under these conditions, calculated from eq. (2), is $10^8$ photons/sec. The mean interval between absorptions is 10 nsec. The fluorescence quantum yield is 0.99 and so the emission rate for a single fluorescein molecule illuminated in this way is $10^8$ photons/sec. This calculated emission rate is close to the maximum possible because the excited state lifetime of fluorescein is 4.4 nsec. For phycoerythrin, the same emission rate could be attained with a lower incident light intensity because phycoerythrin has a higher absorbance coefficient than does fluorescein and a similarly high fluorescence quantum yield. Phycoerythrin has an $\epsilon$ of $1.4 \times 10^6$ $cm^{-1}$ $M^{-1}$ at 514.5 nm. Excitation of a molecule of phycoerythrin ($Q = 0.98$) with 5 mW ($1.3 \times 10^{16}$ photons/sec) focused to a radius of 4 $\mu$m to give an incident intensity of $2.6 \times 10^{22}$ photons/$cm^2$ sec would result in $1.3 \times 10^8$ emitted photons/sec. The 2.1-nsec excited state lifetime of phycoerythrin allows it to have a two-fold higher duty cycle than that of fluorescein.

## Maximum Photon Output of a Fluorescent Molecule

The preceding calculations show that a single highly fluorescent molecule can emit $10^8$ photons per second when excited with a suitably intense laser beam. However, the duration of this process is limited by photodestruction of the fluorescent species. The rate of photodestruction $k_d$ ($sec^{-1}$) of a chromophore is given by

$$k_d = \sigma_a I \phi \tag{4}$$

where $\phi$ is the quantum yield of photodestruction. The fluorescence intensity of a stationary sample will decrease exponentially with a decay time of $1/k_d$ because of photodestruction of the fluorescent chromophore.

The total number of photons $n_f$ that can be emitted by a fluorescent molecule is given by

$$n_f = \int_0^\infty k_f \, e^{-k_d t} \, dt = k_f / k_d \tag{5}$$

Substituting for $k_f$ and $k_d$ according to eq. (3) and (4) yields

$$n_f = Q/\phi \tag{6}$$

Thus, the maximum number of photons that can be obtained by intensely exciting a fluorescent chromophore is simply equal to the ratio of its quantum yield of fluorescence to that of photodestruction. It is evident from this treatment that the measurement of photodestruction yields is of fundamental importance in the development of higher-sensitivity fluorescence detection methods.

## Photodestruction Quantum Yields

The quantum yield of photodestruction can be determined by measuring the dependence of fluorescence intensity on the rate of flow of a fluorescent sample through an illuminated volume. The basis of this approach is described here. With a flowing sample, a steady state is achieved in which the change in concentration due to photodestruction is balanced by the flow of new molecules into the illuminated volume. A theory relating the sample composition in the illuminated volume to the flow rate, light intensity, and spectral characteristics of the photolabile species was developed for the purpose of obtaining resonance Raman spectra of visual pigments [Mathies et al., 1976; Callender et al., 1976]. In those studies, the flow rate required to obtain the resonance Raman spectrum of a nearly unphotolyzed sample of rhodopsin was calculated from the known quantum yield of photoisomerization. In this study, we use the same equations but reverse the process. Here we measure the dependence of fluorescence intensity on flow rate to determine the quantum yield of photodestruction. The key variable is the photoalteration parameter $F$, which is given by

$$F = (P\epsilon\phi\ 3.8 \times 10^{-21})/(\sqrt{\pi}\ \omega v) \tag{7}$$

where $v$ is the flow velocity of the sample through the illuminated volume. The experimental arrangement is shown in Figure 1. The normalized fluorescence intensity $S'$ is defined as $S(F)/S(0)$, where $S(F)$ is the observed fluorescence intensity at a particular value of $F$, and $S(0)$ is the fluorescence intensity that would be observed if the molecule were not photolabile (or, equivalently, in the limit of $F$ approaching 0). $S'$ is a function only of $F$ (eq. 12 in Mathies et al. [1976]).

$$S' = \frac{1}{\pi}\int_{-\infty}^{\infty}\int_{-\infty}^{\infty} [\exp - (s^2+t^2)]\exp - \left\{ \sqrt{\frac{2}{\pi}}\, F\int_{-\infty}^{s} \exp - (s'^2+t^2)\, ds' \right\} ds\, dt \tag{8}$$

A plot of $S'$ versus the logarithm of the flow velocity $v$ reveals the photodestruction quantum yield because the other quantities in eq. (7) are known. Applications of this method will now be presented.

## RESULTS

### Photodestruction Quantum Yields of Fluorescein and Phycoerythrin

Photodestruction yields were determined by measuring the dependence of fluorescence intensity on the velocity of flow of fluorescent molecules through an intensely illuminated region. An argon-ion laser beam was focused to a diameter of a few micrometers within a flowing stream of fluorescent sample (Fig. 1). The emission at right angles was imaged on the entrance slit of a double monochromator and a photon-counting system was used for detection. For fluorescein, the intensity of the 488-nm excitation beam focused to a radius of 4 $\mu$m was $4.4 \times 10^{23}$ photons/cm$^2$ sec. The absorption cross section, corresponding to $\epsilon = 8 \times 10^4$ cm$^{-1}$ M$^{-1}$, was $3 \times 10^{-16}$ cm$^2$/

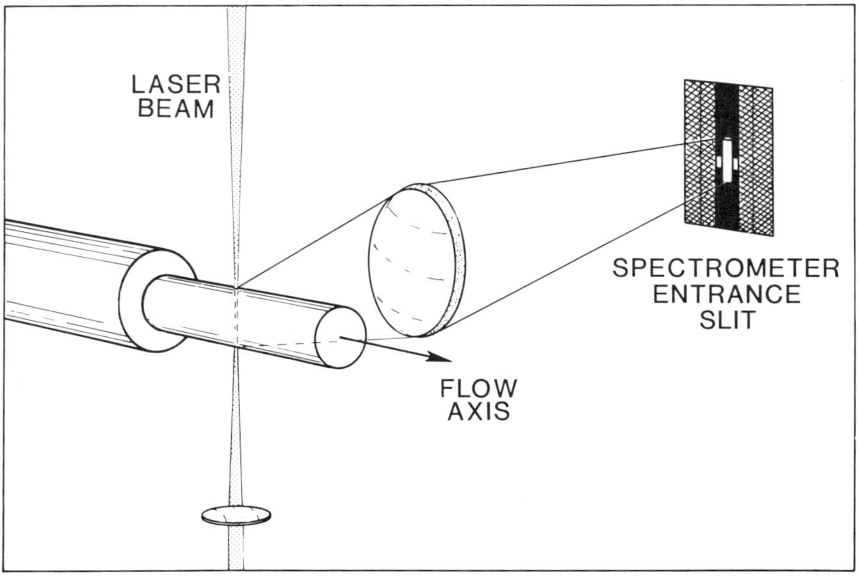


Fig. 1. Experimental arrangement for measuring quantum yields of photodestruction of fluorescent chromophores. An argon-ion laser beam (488 or 514.5 nm, 15–90 mW) was focused on a flowing sample of fluorescent molecules. The beam was focused to a radius between 4 and 20 $\mu$m. The diameter of the pipet was 0.14 cm. The flow rate was adjusted to between 0.009 and 43 cm/sec by a syringe pump. The emission was imaged on the entrance slit of a Spex 1401 monochromator by an f/1 8:1 magnification lens, apertured to a slit height of 1 mm, and detected by an RCA 31034 photomultiplier operated in the photon-counting mode. The band pass of the monochromator was 30 $cm^{-1}$. The excitation, emission, and flow axes were mutually orthogonal.

molecule. Hence, the absorption rate for a single fluorescein molecule was $1.4 \times 10^8$ $sec^{-1}$, and so the mean interval between absorptions was 7 nsec. As shown in Figure 2, the fluorescence intensity decreased from 82% to 16% of its maximal value when the flow was slowed from 10 to 0.1 cm/sec. The time spent by a fluorescein molecule in the illuminated volume increased from 0.08 to 8 msec over this range, which increases its probability of being destroyed. The observed fluorescence intensity as a function of flow velocity fits the curve calculated according to eq. (7) and (8) for a photodestruction quantum yield of $2.7 \times 10^{-5}$ (Fig. 2).

A similar experiment was carried out for B-phycoerythrin by using a 15-mW 514.5-nm excitation beam focused to a radius of 18 $\mu$m to give an

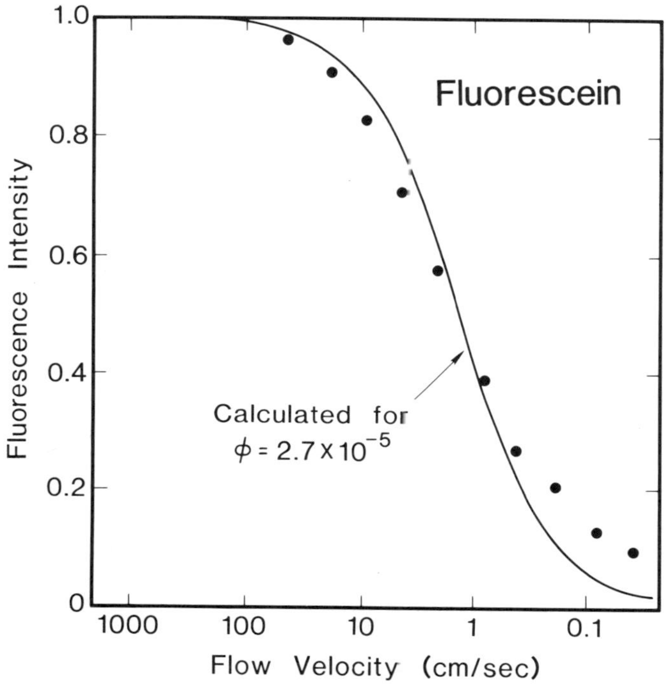


Fig. 2. Determination of the photodestruction quantum yield of fluorescein in basic aqueous solution. The fluorescence intensity was measured as a function of flow velocity. The flowing sample was excited with a 90-mW 488-nm beam focused by a 32-mm focal length lens to a radius of 4 $\mu$m. The incident power was $2.2 \times 10^{17}$ photons/sec. The mean time between photon absorptions was 7 nsec. The best fit to the experimental data (●) is given by a photodestruction quantum yield $\phi$ of $2.7 \times 10^{-5}$ (solid line).

incident intensity of $3.8 \times 10^{21}$ photons/cm$^2$ sec. The absorption cross section at this wavelength, corresponding to $\epsilon = 1.4 \times 10^6$ cm$^{-1}$ M$^{-1}$, is $5.4 \times 10^{-15}$ cm$^2$/molecule. Hence, the absorption rate for a single phycoerythrin molecule is $2.1 \times 10^7$ sec$^{-1}$, and so the mean interval between absorptions is 48 nsec. As shown in Figure 3, the fluorescence intensity decreased from 90% to 28% of its maximal value when the flow was slowed from 10 to 0.1 cm/sec. The time spent by a phycoerythrin molecule in the illuminated volume increased from 0.36 to 36 msec when the flow was slowed over this range. The observed dependence of the fluorescence intensity on flow velocity for B-phycoerythrin is best fit by a curve calculated for a photodestruction quantum yield of $1.1 \times 10^{-5}$ (Fig. 3). The fit for phycoerythrin is not as good as the one for fluorescein, indicating that not all

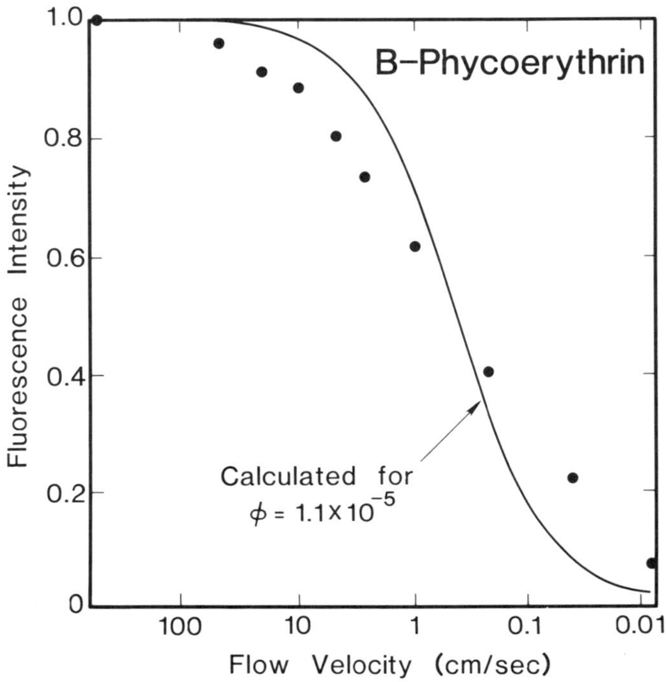


Fig. 3. Determination of the photodestruction quantum yield of B-phycoerythrin in aqueous buffer solution. The fluorescence intensity was measured as a function of flow velocity. The flowing sample was excited with a 15-mW 514.5-nm beam focused by a 133-mm focal length lens to a radius of 18 $\mu$m. The incident power was $3.9 \times 10^{16}$ photons/sec. The mean time between photon absorptions was 48 nsec. The best fit to the experimental data (●) is given by a photodestruction quantum yield $\phi$ of $1.1 \times 10^{-5}$.

molecules of B-phycoerythrin are destroyed by light with the same efficiency. This heterogeneity is most likely due to the presence of multiple bilin chromophores in phycoerythrin.

## Fluorescence Detection of Three Molecules of Phycoerythrin

Knowledge of the photodestruction yield of B-phycoerythrin enabled us to design an experiment to detect just a few molecules of phycoerythrin in the imaged illuminated volume. The strategy is to focus the incident beam and set the flow velocity to obtain a photoalteration parameter $F$ of 0.3, which gives a normalized fluorescence intensity $S'$ of about 0.8. The sample concentration and the imaged illuminated volume are then adjusted to give a small number of phycoerythrin molecules in the detected volume. A 5 ×

$10^{-13}$ M solution of B-phycoerythrin was flowed through the illuminated volume. The incident intensity was $5 \times 10^{22}$ photons/cm$^2$ sec. The interval between photon absorptions was 3.8 nsec, close to the limit imposed by the 2.1-nsec excited state lifetime of phycoerythrin. The average number of molecules of phycoerythrin present in the 10-pL detected volume was 3. As shown in Figure 4, these 3 phycoerythrin molecules gave a fluorescence emission well above the background at its 576-nm emission peak and clearly separate from the water Raman band. The fluorescence peak intensity at 576 nm was 130 counts per second (cps) and the solvent background at this wavelength was 30 cps. The integrated intensity over the fluorescence emission band was about 2,700 cps.

## DISCUSSION

Our measurements of the dependence of fluorescence intensity on flow velocity (Figs. 2, 3) directly give photodestruction quantum yields. For fluorescein in aqueous solution, $\phi = 2.7 \times 10^{-5}$. Hirschfeld [1976b] found that the emission intensity of a fluorescein solution decayed exponentially with a time constant of 3 msec when it was excited with a 488-nm beam having an intensity of $2.9 \times 10^{22}$ photons/cm$^2$ sec. The photodestruction yield calculated by us from this decay time (assuming that diffusion can be neglected) is $3.6 \times 10^{-5}$, in good agreement with our value. For B-phycoerythrin it was $\phi = 1.1 \times 10^{-5}$. Thus, the degree of photolability of phycoerythrin is quite similar to that of fluorescein. The total number of photons $n_f$ that can be emitted by these chromophores can now be calculated from eq. (6). For fluorescein, $n_f = 3.7 \times 10^4$ photons, and for phycoerythrin, $n_f = 8.9 \times 10^4$ photons.

The fluorescence emission spectrum shown in Figure 4 comes from an average of 3 phycoerythrin molecules in the detected volume at any instant. The fluorescence peak intensity at 576 nm is about five times as large as the solvent background at this wavelength. Hence, it should be feasible to detect a single phycoerythrin molecule in the imaged illuminated volume. It is important to emphasize that flow is required in this experiment to bring in new phycoerythrin molecules to replace those destroyed by light. In the experiment shown in Figure 4, the transit time through the beam was 0.12 msec, which means that 26,000 phycoerythrin molecules actually passed through the detected volume per second. Obtaining a discrete signal from a single molecule of phycoerythrin would be feasible if the detection efficiency were sufficiently high. A phycoerythrin molecule emits about $10^5$ photons before it is destroyed by light. A detection efficiency (counts per emitted photon) of $10^{-3}$ would result in 100 counts arriving in a millisecond if the light intensity were adjusted to give an emission rate of $10^8$ sec$^{-1}$. The

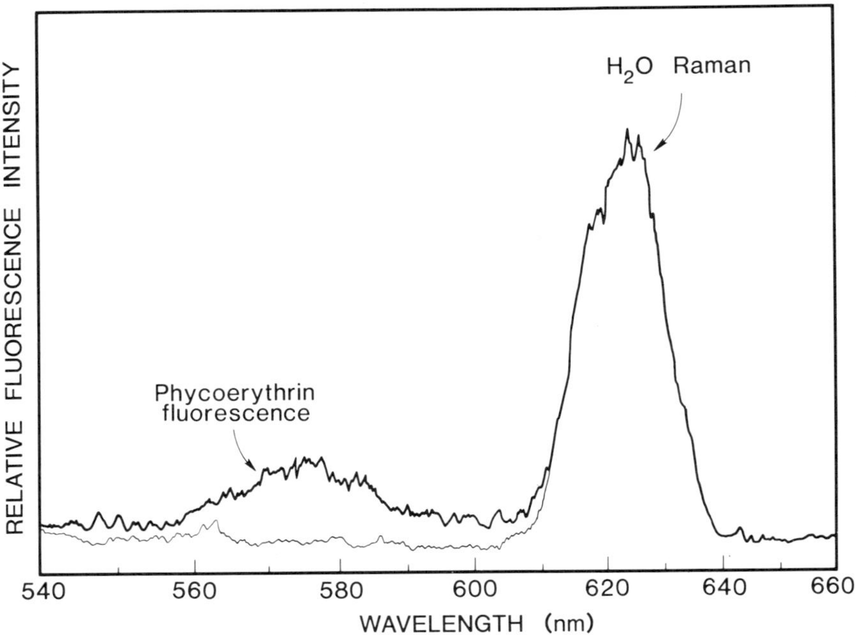


Fig. 4. Observation of three molecules of B-phycoerythrin in the detected volume. A $5 \times 10^{-13}$ M solution of B-phycoerythrin was flowed at a velocity of 8.6 cm/sec. The flowing sample was excited with a 15-mW 514.5-nm beam focused to a radius of 5 $\mu$m. The incident power was $3.9 \times 10^{16}$ photons/sec and the photoalteration parameter F was 0.3. The time between absorptions was 3.8 nsec and the transit time through the beam was 0.12 msec. An average of 3 molecules of phycoerythrin were present in the detected volume of 10 pL. The phycoerythrin emission peak at 576 nm is clearly distinguishable from the background and from the water Raman band near 625 nm. The phycoerythrin fluorescence peak intensity was 100 cps and the integrated intensity of the fluorescence emission was about 2,700 cps. The detection time constant was 1 sec, the monochromator scan rate was 20 $cm^{-1}$/sec, and the detection bandwidth was 30 $cm^{-1}$.

detection efficiency of our apparatus was only $2 \times 10^{-7}$ because it was designed for Raman spectroscopy, which requires a very high spectral resolution. This value is too small to observe a photon burst from a molecule of phycoerythrin. However, by increasing the spectral bandwidth from 30 to 800 $cm^{-1}$ and by optimizing the optical arrangement, it should be feasible to increase the detection efficiency to $10^{-3}$. The challenge then will be to achieve sufficient rejection of Rayleigh and Raman scattering to enable a burst of photons from a single phycoerythrin to be detected.

In general, three criteria must be met to achieve single-molecule detection. First, the product of the emission rate of the fluorescent chromophore and the detection efficiency of the instrument must be greater than the dark count rate of the photomultiplier during the observation interval. This means that the emission rate $k_f$ [as defined in eq. (3)] must be greater that about $10^5$. This condition can be fulfilled only if the excited state lifetime is less than about $10^{-5}$ second. Chromophores with nanosecond excited-state lifetimes are ideal and those with microsecond lifetimes are close to the borderline. Emitters with millisecond lifetimes, such as chelates of europium and terbium, perform admirably in many fluorescence immunoassays [Soini and Kojola, 1983], but they are not suited to single-molecule detection because they cannot emit more than $10^3$ photons per second. Chemiluminescent chromophores also cannot be used for single-molecule detection in the absence of an additional stage of amplification. Second, the fluorescent chromophore employed must be sufficiently photostable. As discussed above, the product of $n_f$ and the detection efficiency must be greater than about 100 to see a photon burst from a single fluorescent molecule. This means that the quantum yield of photodestruction must be less than about $10^{-4}$. Very little is known about the factors governing the photolability of fluorescent chromophores. Research into this facet of fluorescence spectroscopy is likely to be highly rewarding. Third, the fluorescence from a single molecule should be greater than the background fluorescence. Also, Rayleigh and Raman scattering must be rejected to a high degree. The specific requirements imposed by this criterion will depend on the particular system analyzed. It is evident, however, that much is gained by using fluorescent chromophores that have very high absorbance coefficients, high quantum yields, and large Stokes shifts. Phycoerythrin is a very attractive fluorescent tag for single-molecule detection because it combines these favorable characteristics. The fluorescence signal to background ratio observed with phycoerythrin (Fig. 4) is about 100 times greater than that obtained with rhodamine 6G [Dovichi et al., 1983, 1984]. Much of this difference is due to the 40-fold greater absorbance coefficient of phycoerythrin compared with rhodamine 6G at 514.5 nm. The detection of single antibody molecules [Hirschfeld, 1976a] depended on labeling them with a polymer containing between 80 and 100 fluoresceins. It seems likely that the product of fluorescence quantum yield and absorbance coefficient must be greater than $10^6$ $cm^{-1}$ $M^{-1}$ to achieve single-molecule detection.

## ACKNOWLEDGMENTS

This research was supported by grants from the National Institutes of Health and the National Science Foundation. We thank Dr. Alexander Glazer for a gift of phycoerythrin.

## REFERENCES

Callender RH, Doukas A, Crouch R, Nakanishi K (1976): Molecular flow resonance Raman effect from retinal and rhodopsin. Biochemistry 15:1621–1629.

Dovichi NJ, Martin JC, Jett JH, Keller RA (1983): Attogram detection limit for aqueous dye samples by laser-induced fluorescence. Science 219:845–847.

Dovichi NJ, Martin JC, Jett JH, Trkula M, Keller RA (1984): Laser-induced fluorescence of flowing samples as an approach to single-molecule detection in liquids. Anal Chem 56:348–354.

Glazer AN (1984): Phycobilisome: A macromolecular complex optimized for light energy transfer. Biochim Biophys Acta 768:29–51.

Glazer AN, Stryer L (1983): Fluorescent tandem phycobiliprotein conjugates. Biophys J 43:383–386.

Glazer AN, Stryer L (1984): Phycofluor probes. TIBS 10:423–427.

Hirschfeld T (1976a): Optical microscopic observation of single small molecules. Appl Optics 15:2965–2966.

Hirschfeld T (1976b): Quantum efficiency independence of the time integrated emission from a fluorescent molecule. Appl Optics 15:3135–3139.

Hurst GS, Nayfeh MH, Young JP (1977): One-atom detection using resonance ionization spectroscopy. Phys Rev 15:2283–2292.

Kronick MN, Grossman PD (1983): Immunoassay techniques with fluorescent phycobiliprotein conjugates. Clin Chem 29:1582–1586.

Mathies R, Oseroff AR, Stryer L (1976): Rapid-flow resonance Raman spectroscopy of photolabile molecules: rhodopsin and isorhodopsin. Proc Natl Acad Sci USA 73:1–5.

Milby KH, Zare RN (1984): Antibodies, lasers, and chromatography. Am Clin Prod Rev 3:14–19.

Muirhead KA, Horan PK, Poste G (1985): Flow cytometry: Present and future. Bio Tech 3:337–356.

Oi VT, Glazer AN, Stryer L (1982): Fluorescent phycobiliprotein conjugates for analyses of cells and molecules. J Cell Biol 93:981-986.

Parks DR, Hardy RR, Herzenberg LA (1983): Dual immunofluorescence—new frontiers in cell analysis and sorting. Immunol Today 4:145–150.

Shapiro HM, Glazer AN, Christenson L, Williams JM, Strom TB (1983): Immunofluorescence measurement in a flow cytometer using low-power helium-neon laser excitation. Cytometry 4:276–279.

Soini E, Kojola H (1983): Time-resolved fluorometer for lanthanide chelates—a new generation of nonisotopic immunoassays. Clin Chem 29:65–68.

Steinkamp JA (1984): Flow cytometry. Rev Sci Instrum 55:1375–1400.

Voller A, Bartlett A, Bidwell D (1981): "Immunoassays for the 80s." Baltimore: University Park Press.

Zoumbos NC, Gascon P, Djeu JY, Trost SR, Young NS (1985): Circulating activated suppressor T lymphocytes in aplastic anemia. N Engl J Med 312:257–265.

*Applications of Fluorescence in the Biomedical Sciences, pages 141–157*


# 7 Fluorescence Photoactivation and Dissipation (FPD)

*Bennie R. Ware, Laura J. Brvenik, Richard T. Cummings, Ruth H. Furukawa, and Grant A. Krafft*

Attachment of flourescent labels to molecular species has become a powerful approach to the study of the structure and dynamics of complex systems. Fluorescence microscopy provides high-contrast images of the spatial distribution of specifically labeled species. Time resolution of fluorescent images may reveal the dynamics of spatial rearrangements of labeled species. However, once the labeled species have attained a steady state or equilibrium distribution, the fluorescent images become invariant, irrespective of whether individual species are still in motion. The technique of fluorescence photobleaching recovery (FPR) extends the power of time-resolved fluorescent imaging by creating a tracer inequivalence among otherwise equivalent labeled species [Peters et al., 1974; Edidin et al., 1976; Jacobson et al., 1976; Axelrod et al., 1976]. This inequivalence is accomplished by photobleaching a region of the sample by using a brief pulse of high-intensity light in a spatially defined region of the sample. If all labeled species are free to move about in a random fashion, the bleached region will eventually disappear. The time constant for the disappearance of the bleached region will be determined by the translational diffusion coefficient of the labeled species and the square of the characteristic dimension of the bleached region. If some fraction of the bleached species is not free to move, then that fraction of the bleaching will persist, permitting determination of the fraction of immobile species. Since the first introduction of the FPR technique, the primary

Department of Chemistry, Syracuse University, Syracuse, New York 13210.

application has been the study of the diffusion of specifically labeled species in biological membranes. The FPR technique has also been applied successfully to the study of motion in synthetic bilayers, films, solutions, colloids, and cell cytoplasm [for reviews see Cherry, 1979; Peters, 1981; Webb, 1981; Ware, 1984].

Although the FPR technique is simple in concept, considerable experimental refinement has been required to achieve acceptable accuracy for specific applications. One intrinsic disadvantage of the FPR approach has not previously been addressed. FPR inherently involves the measurement of a subtractive signal in the presence of a background of bright fluorescence. This property limits the contrast achievable and the signal-to-noise ratio to be expected for a given system, and it also renders difficult or impossible the long-term monitoring of those species in the region of photobleaching as they diffuse throughout the specimen. We are seeking to address this problem by the introduction of a methodology which is essentially the positive analog of FPR. This new approach, which we call fluorescence photoactivation and dissipation (FPD), potentially has several inherent advantages over conventional FPR methodology. The essential requirement for an FPD measurement is the attachment of a molecular label that is initially non-fluorescent but which will become fluorescent when illuminated by a short pulse of intense light. Such a molecule is called a photoactivable fluorophore (PAF).

The photochemical generation of stable fluorescent molecules from nonfluorescent precursors has been noted in the literature for many years. One of the earliest observations of this type of process was described by Lewis et al. [1940], who studied the photoisomerization of *cis*-stilbene (nonfluorescent) to *trans*-stilbene (blue fluorescence). Since then, examples of photogenerated fluorescence via *cis-trans* isomerization have been reported for several classes of compounds, including indigo dyes [Rogers et al., 1957], β-styryl compounds [Feigl et al., 1955; Wheelock, 1959; Lippert and Luder, 1962], and tetraphenylcyclooctatetraene [White et al., 1969; Sharafy and Musckat, 1971]. These *cis-trans* isomerizations are typically reversible under photochemical conditions, such that an equilibrium mixture of fluorescent and nonfluorescent isomers is generated.

Other photochemical processes also lead to stable fluorescent molecules; *cis*-stilbenes participate in oxidative photocyclization reactions and are thus converted to fluorescent phenanthrene derivatives [Mallory and Mallory, 1984]. This process is the basis of a clinical assay for blood serum levels of the breast cancer drug tamoxifen, and its metabolite 4-hydroxytamoxifen, which form highly fluorescent 10-phenylphenanthrene derivatives [Mendenhall et al., 1978]. Irradiation of aryl azides and subsequent C-H insertion generates secondary aryl amines, which often are fluorescent (e.g., aminonaphthalene). This process was employed in a photoaffinity labeling experiment of nerve cell sodium channels in which photolysis of an aryl azide analog of tetrodotoxin resulted in simultaneous covalent attachment and generation of a fluorescent label [Angelides, 1981]. Additional reactions known to generate fluorescent products include the photo-Fries rearrangement of naphthyl esters to fluorescent hydroxynaphthyl lactones [Bellers, 1971], photo-oxidation of nonfluorescent dihydroaromatic systems to fluorescent aromatic molecules (e.g., dihydrofluorescein → fluorescein) [Uchida, 1963], and the photoisomerization of diarylisoxazoles to fluorescent diaryloxazoles [Ullman and Singh, 1966; Singh and Ullman, 1967]. Huffman et al., [1970] studied several 3-aroyl-2-(2-furyl) chromones which participate in a photo-induced rearrangement to generate fluorescent 1-arylfuro [3,4b]chromone-3-acroleins. This reaction and several others that generate fluorescent products have been reviewed by Zweig [1973], who discussed the potential applications of these reactions in organic imaging processes and as mechanistic probes of photochemical processes.

We present here the first report of the use of photoactivable fluorophores as specific markers for the study of diffusion by tracer transport. Design considerations for new types of PAF molecules and for the FPD apparatus are presented, and the first successful FPD measurements are described.

## FPD APPARATUS

In principle any FPR apparatus could be used for an FPD measurement. However, in practice it is most convenient to employ PAF labels that are not activated by visible light. Photoactivation is best accomplished in the near-ultraviolet (UV) range, followed by illumination with visible light for monitoring the created fluorescence. A schematic diagram of our FPD apparatus is shown in Figure 1. The photoactivating laser light is provided by the UV lines of a Spectra Physics model 2025 $Ar^+$laser. The monitoring beam is usually one of the blue or green lines from a spectra Physics model 164 $Ar^+$ laser. The two laser beams are rendered coincident by a set of optics, but the coincidence criterion is not critical. The incident beams enter a Zeiss Universal Microscope and pass through a pattern (P), which is generally a Ronchi ruling (grating) of known period. A dichroic mirror reflects both the incident beams onto the microscope stage, on which is mounted the specimen of interest. During the collection phase fluorescence passes through the dichroic mirror and through a barrier filter onto a photomultiplier tube.

The sequence of an FPD measurement is as follows: With the grating translated at constant velocity, the monitoring beam illuminates the sample to detect any residual fluorescence due to the inherent fluorescence of the PAF or the fluorescence of any fluorophore that may have formed spontaneously. The grating is then stopped and held stationary while the shutter controlling

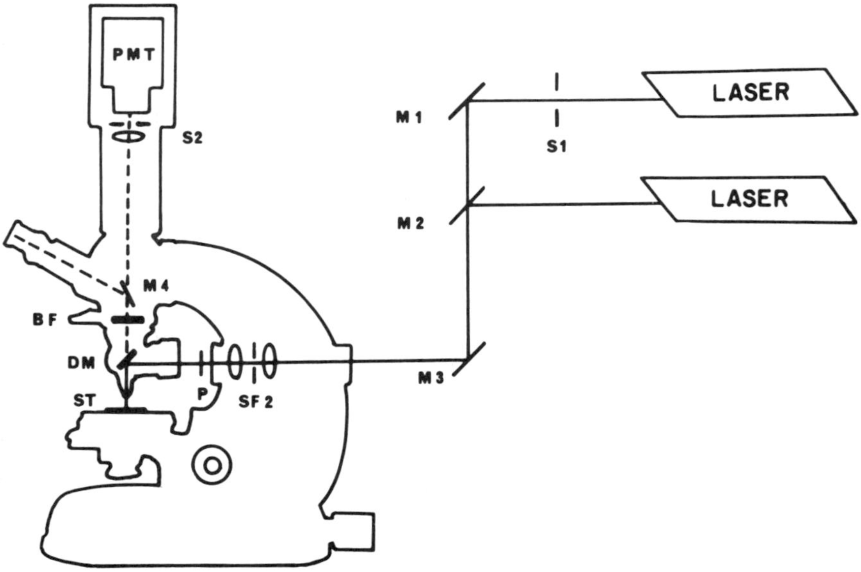


Fig. 1. Schematic diagram of a fluorescence photoactivation dissipation (FPD) instrument. The specimen is on the stage (ST) of a research microscope with epi-fluorescence optics. The activating laser is stopped by a shutter (S1) which is opened for the short period of activation. The monitor and activating beams are rendered coincident and directed to the microscope by a series of mirrors (M1–M3). After passing through a spatial filter (SF2) and a Ronchi ruling pattern (P) the beams are directed by a dichroic mirror (DM) to the stage. Fluorescence passes through the barrier filter (BF) to be detected by the photomultiplier tube (PMT).

the UV photoactivation beam is opened for a brief, measured time. The shutter is then closed, and again the sample is illuminated only with the visible monitoring beam while the grating is translated at constant velocity. The photoactivation in the sample then consists of a set of stripes spaced by the period of the image of the grating on the microscope stage. As the monitoring beam passes through the translated grating, it becomes essentially a striped laser beam, with the stripes translated at a constant velocity. Hence the photoactivated pattern and the stripes of the monitoring beam fall into and out of phase with a frequency that is proportional to the velocity of the translation of the grating. This velocity is controlled by a variable-speed DC motor so that the frequency of modulation of the fluorescence may be tuned to any desired value. The output of the photomultiplier tube is hence a rising photocurrent that is modulated at the frequency determined by the translation of the pattern. Both the DC level and the modulated AC signal are processed separately, as has been described previously in our description of the modulation detection method of FPR [Lanni and Ware, 1982]. Briefly, the AC signal is passed through a tuned amplifier and then is sent to an envelope detector that consists of a peak voltmeter and zero crossing detector in parallel. This device produces discrete values of the envelope function at the modulation frequency. The output of the envelope detector and the DC signal and other timing sequence information are directed to parallel analog-to-digital inputs of a Nova 4C minicomputer for analysis.

In effect, the modulation AC envelope $\Xi(t)$ is a measure of the spatial Fourier component of the fluorescence intensity with wave vector K′ given by

$$K' = 2\pi/L \tag{1}$$

where L is the spacing of the projection of the grating lines in the specimen. For a solution of a single particle undergoing simple translational diffusion, the relaxation of the envelope function will follow the simple diffusion equation

$$E(t) = E(0)\ e^{-DK'^2t} \tag{2}$$

If the solution contains a mixture of small species with high mobility and large species with low mobility, then the AC component decays by diffusion in a biexponential fashion that can be represented by the equation

$$E(t) = E(0)\ (f_{HM}e^{-D_{HM}K'^2t} + f_{LM}e^{-D_{LM}K'^2t}) \tag{3}$$

where $f_{HM}$ and $f_{LM}$ are the respective fractions of high- and low-mobility species, and $D_{HM}$ and $D_{LM}$ are the respective diffusion coefficients. Always $f_{HM} + f_{LM} = 1$. If $D_{LM} \ll D_{HM}$ then the second term will appear as a constant, $f_{LM}$. If the time scales of recovery are separated by substantial magnitudes, the time scale of the measurement may be changed readily by changing the spacing of the grating. In FPD, as in FPR, this modulation detection scheme has the advantages of (1) ease of determination of pattern size, (2) convenient functional form of data, (3) insensitivity to specimen motion, and (4) relative insensitivity to precision of superposition of photoactivation and monitor beams. The separate utilization of the DC signal may also be useful in the FPD measurement, though the form of the change in the DC signal is of course quite different. In the ideal FPD measurement, the DC signal will be raised after photoactivation from an initial value of nearly zero to some new finite value of photocurrent that remains over the time scale of dissipation of the spatial pattern. This condition, however, can only be met if there are many periods of the spatial pattern within the spatial domain of the size of the photoactivating beam. In practice, the DC level will rise over the time scale of the creation of the photoactivable fluorophore out of the laser beam and then will fall due to diffusion of PAF out of the beam and possibly due to undesired photobleaching of the PAF by the monitoring beam. Hence, the DC signal is used primarily as a diagnostic for certain technical aspects of the measurement.

## DESIGN AND SYNTHESIS OF PHOTOACTIVABLE FLUOROPHORES (PAFs)

A photoactivable fluorophore (PAF) is a nonfluorescent molecule that can be converted to a stable fluorescent molecule when irradiated by light. The design of PAFs requires consideration of optimal characteristics of the PAF molecules, of the ultimate fluorophore molecules, and of the photoactivation process which effects the conversion. A number of ideal characteristics of PAF molecules are summarized in Figure 2.

PAF

-- Non-fluorescent
-- Chemically and thermally stable
-- Activated by light not strongly absorbed by ultimate fluorophore
-- Capable of covalent attachment to target molecules

PHOTOACTIVATION PROCESS

-- Efficient (high quantum efficiency)
-- Induced by relatively low energy light
-- Selective (no other processes induced)
-- Clean (no damaging by-products generated)

ULTIMATE FLUOROPHORE

-- Highly fluorescent
-- Excited by non-activating wavelengths
-- Chemically, thermally and photochemically stable

Fig. 2. Ideal characteristics of photoactivable fluorophores (PAFs).

In addition to consideration of these characteristics, the design of PAF molecules also entails consideration of desired spectral and photoactivation parameters, PAF polarity, size limitations, and functional group specificity for covalent attachment, as dictated by particular applications of the PAFs. In general, PAF design should result in a molecule that will be transformed into an ultimate fluorophore similar in structure and spectral characteristics to conventional probes already employed in fluorescence studies.

In order to design a particular organic molecule as an ideal PAF, it is useful to focus on a target fluorophore structure that, by a conceptual retrosynthetic transform, would be converted to a structurally modified nonfluorescent PAF. Two types of retrosynthetic transforms are capable of rendering the ultimate fluorophore target nonfluorescent. One type would involve attachment of a quenching auxiliary to the ultimate fluorophore molecule by a connecting chain, thereby resulting in a nonfluorescent PAF molecule. A second type would involve alteration or removal of the fluorescent chromophore, perhaps by disruption or deconjugation of an extended pi system through bond breakage or conformational change, or by removal or alteration of an integral functional group.

The photoactivation process for a PAF molecule necessarily requires a photochemical reaction which reverses the retrosynthetic transform by which the target fluorophore was rendered nonfluorescent. In the case of a fluorophore linked to a quenching auxiliary, photoactivation would involve cleavage or alteration of the quencher. In other cases, activation processes would involve photochemical bond formation, rearrangement, or functional group manipulation to restore a fluorescing chromophore.

In a schematic sense, three general types of PAFs can be envisioned, as illustrated in Figure 3. A type I PAF consists of a photoremovable quenching auxiliary, Q, attached to a fluorescent molecule F. A type II PAF consists of two parts, A and B, which react with each other (bond formation) or rearrange when irradiated to form the ultimate fluorophore F. A type III PAF might be considered a hybrid of the first two, consisting of X and Y, and a photoremovable protecting group P which prevents X and Y from reacting or rearranging via a thermal/chemical process that otherwise would occur spontaneously.

Figure 4 presents an example of each type of PAF. The type I PAF consists of a substituted nitrobenzyl quenching chromophore coupled via a carbamate linkage to a fluorescent aromatic amine. Irradiation induces an intramolecular redox reaction of the nitrobenzyl group, resulting in decarboxylative cleavage [Patchornick et al., 1970] (vide infra) to generate the free fluorescent amine. The type II PAF is an aroyl furyl chromone which is nonfluorescent by virtue of its cross-conjugated pi system. Irradiation induces a rearrangement (vide infra) to a furochromone which contains a fluorescent chromophore. The

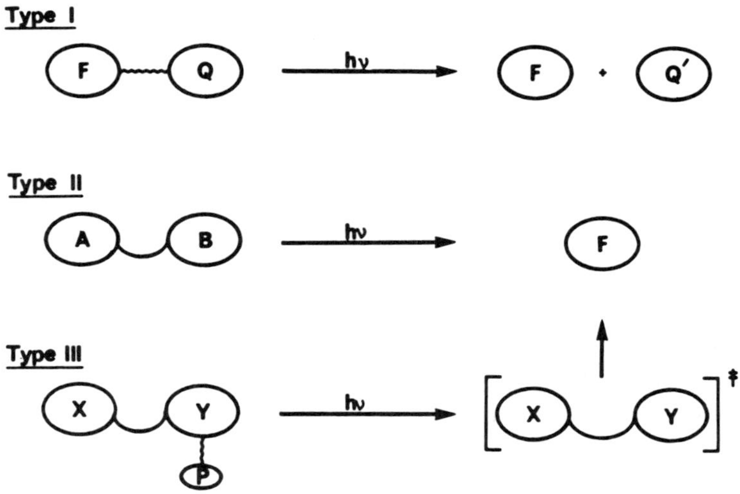


Fig. 3. Schematic representation of three types of PAFs.

type III PAF is an analogue of fluorescamine, a reagent which reacts with primary amino groups to form a fluorescent conjugate. Photo-deprotection of the proximally situated primary amino group of this analogue permits intramolecular cyclization to form the fluorescent conjugate.

We have prepared a number of the type I and type II PAFs illustrated in Figure 4. The N-anthrylnitrobenzyl carbamates (e.g., PAFI-1) were prepared by coupling nitrobenzyl chloroformates with 2-aminoanthracene (Fig. 5). Similarly, N-fluoranthenyl carbamates (e.g., PAFI-2) were prepared from 3-aminofluoranthene. In order to attach the carbamate PAFs to larger molecular species, an aminopropyl linking arm was attached to the carbamate nitrogen via alkylation with the disilyl-protected aminopropyl bromide. After deprotection, the free amine was coupled to carboxyl-functionalized polystyrene spheres (380-Å diameter) by treatment of an aqueous suspension of spheres and amine with 1-cyclohexyl-3-(2-morpholinoethyl)carbodiimide metho-p-toluenesulfonate at room temperature for 36 hr.

The synthesis of the 3-aroyl-2-(2-furyl)chromones (e.g., PAFII-1) is outlined in Figure 6. Aldol condensation of the bis-aryl diketone with furaldehyde derivatives (catalytic piperidine in refluxing ethaenol) provides the dihydrochromone derivatives which were oxidized to the chromones with selenium dioxide in refluxing dioxane. Aldol condensation with 5-phthalimidomethylfuraldehyde, selenium dioxide oxidation, and deprotection via hy-

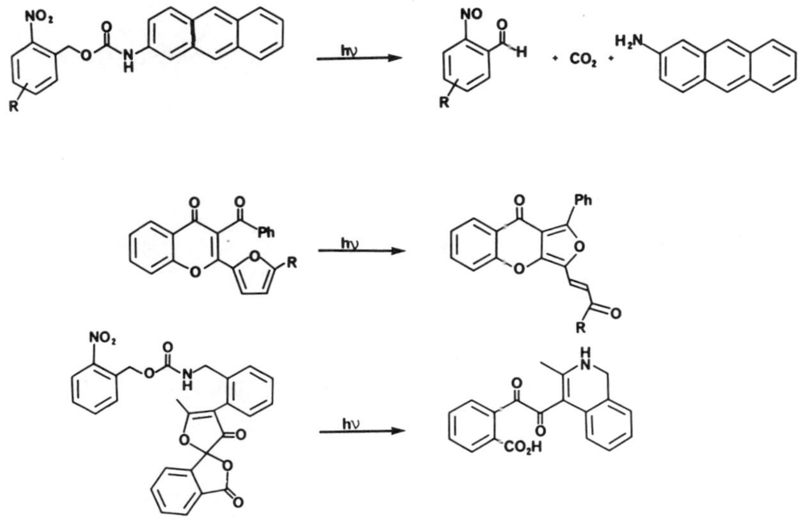


(1)

(2)

(3)

Fig. 4. Examples of three different PAFs and photoactivation to ultimate fluorophores.

drazinolysis provided the aminomethyl derivative PAFII-2 which is capable of covalent attachment to larger molecular species.

The 3,4,5-trimethoxy-2-nitrobenzyl group is an effective quencher of the aminoanthracene fluorescence, eliminating all emission at 510 nm and greater than 98% of the fluorescence emission at 410 nm as shown in figure 7. This quenching presumably occurs by resonant energy transfer from the excited state of the aminoanthracene to the trimethoxynitrobenzyl chromophore. Quenching by less substituted nitrobenzyl chromophores was significantly less effective, due to poorer overlap of the energy-accepting transition of the quencher with the energy-donating excited state of the fluorophore.

Irradiation of PAFII-1 in methanol or water at 350 nm generates 2-aminoanthracene (maximum fluorescence at 514 nm) and 3,4,5-trimethoxy-nitrosobenzaldehye. The quantum efficiency for this activation was $0.22 \pm 0.05$. The mechanism that has been proposed for this reaction [Berson and Brown, 1955] is outlined in Figure 8. The excited aromatic nitro group abstracts a hydrogen atom to form a biradical, which rearranges to form the nitrosobenzaldehyde hemiacetal. Subsequent decarboxylation liberates aminoanthracene and trimethoxynitrosobenzaldehyde. This decarboxylation step is a thermal reaction with kinetics that may be influenced by solvent, pH, and temperature. FPD experiments of rapidly diffusing molecules require that the kinetics of this step be fast, relative to the photoactiva-

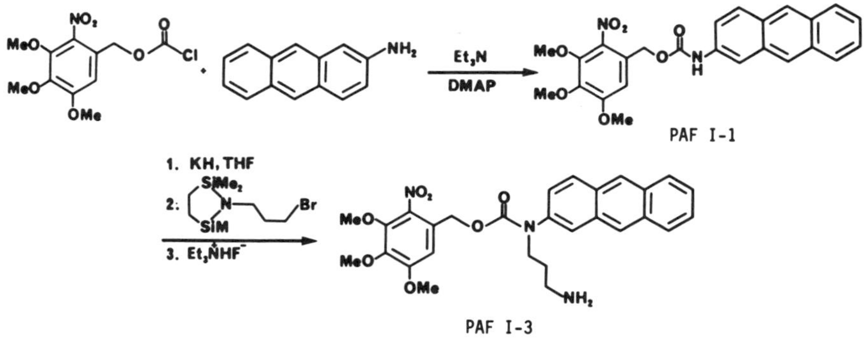


Fig. 5. Synthetic route to N-anthryl nitro-benzyl carbamates.

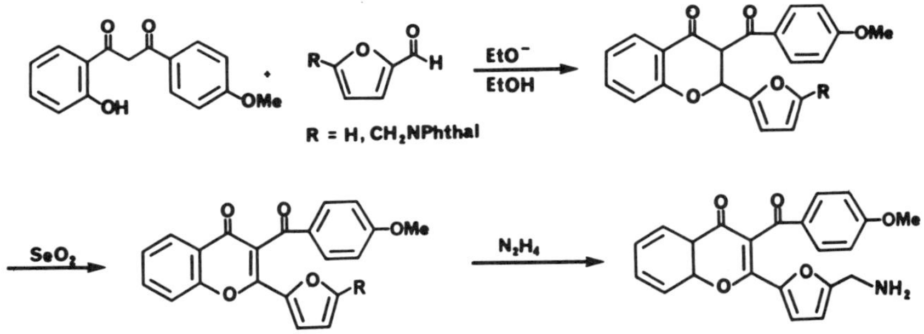


Fig. 6. Synthetic route to 3-aroyl-2-(2-furyl)-chromones.

tion time and diffusion coefficient of the species being studied.

Irradiation of PAFII-1 in dichloromethane at 350 nm produces 1-p-anisyl-furol[3,4b]chromone-3-acrolein, which fluoresces maximally at 527 nm. The quantum efficiency for this process is quite low [3–5%, Huffman et al., 1970]. The proposed mechanism by which this conversion takes place [Huffman et al., 1970] is outlined in Figure 9. Fragmentation of the furan ring leads to a carbene intermediate, which can cyclize to the anisoyl carbonyl to form the fluorescent product, or cyclize to the acrolein carbonyl to regenerate the original chromone. This pathway and other possible reaction pathways of the reactive carbene are undoubtedly responsible for the low reaction quantum efficiency.

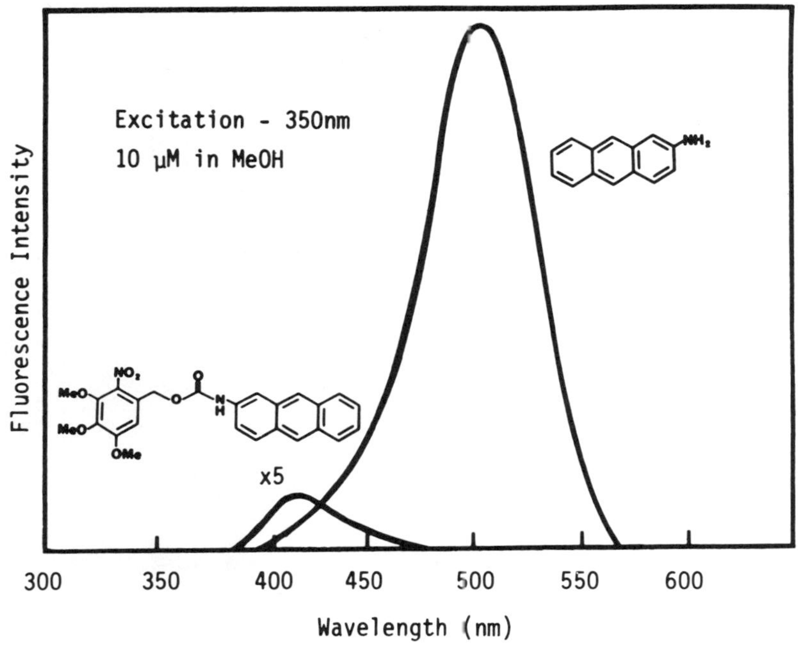


Fig. 7. Fluorescence quenching of 2-amino anthracine by 3,4,5-trimethoxynitrobenzyl group.

## FPD DATA

Although the design and synthesis of PAF molecules with optimized properties will require considerable iterative refinement, the PAFs already in hand are adequate to demonstrate the feasibility of the FPD technique. To date the greatest photoactivation fluorescence yield has been achieved with the furyl chromones (PAF II-1). Representative data for this molecule dissolved in $CH_2Cl_2$ are shown in Figure 10. The upper trace shows the DC photocurrent i(t) (inverted scale), which is a measure of the total fluorescence intensity. The lower trace shows the modulation envelope E(t), which is the measured contrast of the photoactivated pattern, as described previously. Time 0 corresponds to the end of the photoactivation period, which in this measurement had a duration of 100 ms. For all measurements to be described here, the photoactivation intensity was about $10^3 W/cm^2$ at 351.1 nm and 363.8 nm combined, and the monitoring intensity was about 1 $W/cm^2$ at 457 nm. The DC level in Figure 10 is seen to rise after the photoactivation over a time scale of about 200 ms, which is a result of the settling time of

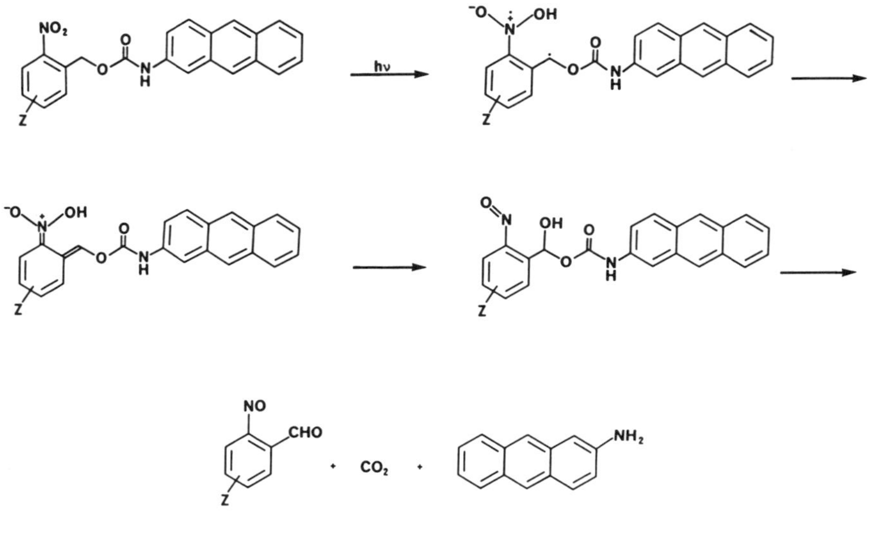


Fig. 8. Mechanism of nitrobenzyl carbamate photocleavage.

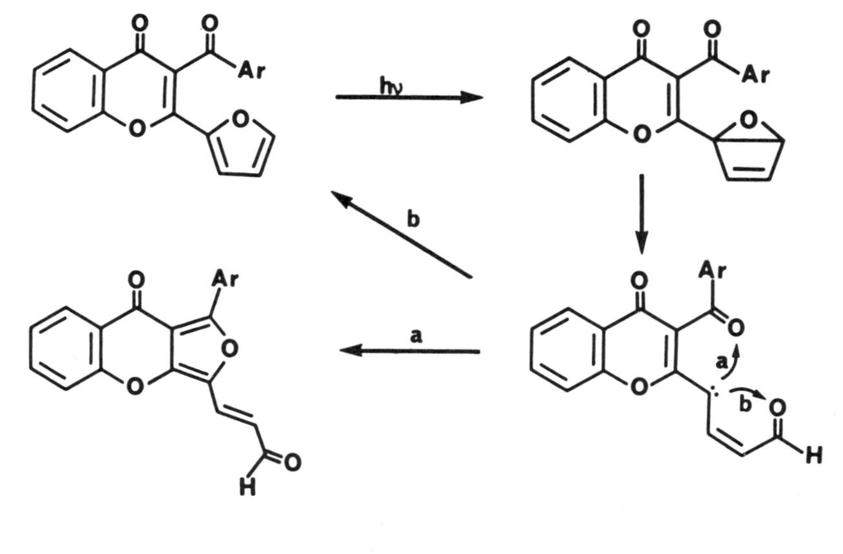


Fig. 9. Mechanism for the conversion of 3-aroyl-2-(2-furoyl) chromones to 1-arylfuro [3,4b] chromone-3-acroleins.

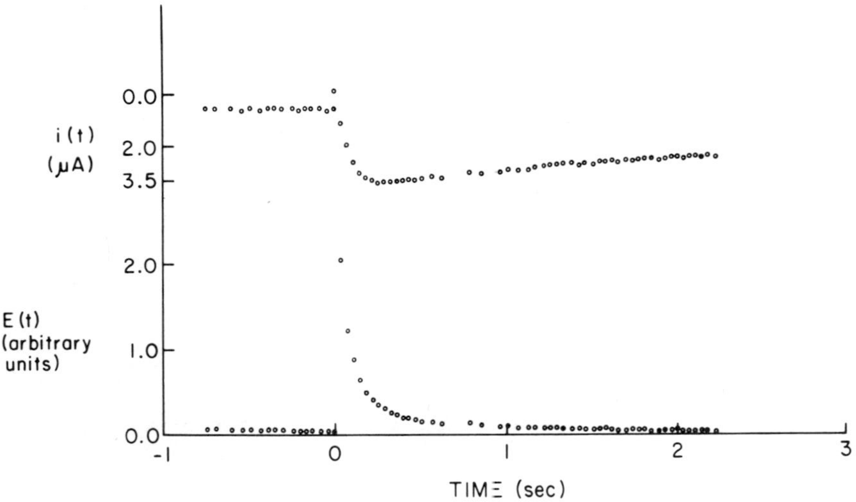


Fig. 10. FPD data for the diffusion of PAF II-1 in $CH_2Cl_2$. The upper trace shows the photocurrent i(t) and the lower trace shows the modulation envelope E(t) as a function of time. The dye concentration was 1% by weight, the temperature was 5.5°C, and the experimental wave vector K′ was 853 $cm^{-1}$.

the current meter employed. This settling time does not influence the response time of the modulation signal. The modulation signal decays exponentially as predicted and with a very high signal-to-noise ratio. The diffusion coefficient determined by computer fit to E(t) is $9.8 \times 10^{-6} cm^2/s$, which is a reasonable value for this temperature (5.5°C). The long-term reduction of the DC photocurrent is caused by the diffusion of PAF out of the monitoring beam. The time constant of this change is much longer because the characteristic dimension of the beam is much greater than that of the grating pattern. Variation of the temperature of the PAF-$CH_{2c}Cl_2$ system produced changes in the measured diffusion coefficient of the anticipated magnitude. Addition of a high-molecular-weight polymer to the system would be expected to reduce the measured diffusion coefficient of the tracer PAF molecule. This expectation was verified by the addition of polystyrene (Polysciences, mol. wt. $6.7 \times 10^5$). The measured diffusion coefficients were reduced as anticipated. Sample FPD data for this experiment are shown in Figure 11. The measured diffusion coefficient from these data was $7.89 \times 10^{-6} cm^2/s$ at 15°C. It was observed in this and other experiments that the addition of polystyrene to the PAF II-1 solution increased the photoactivated fluores-

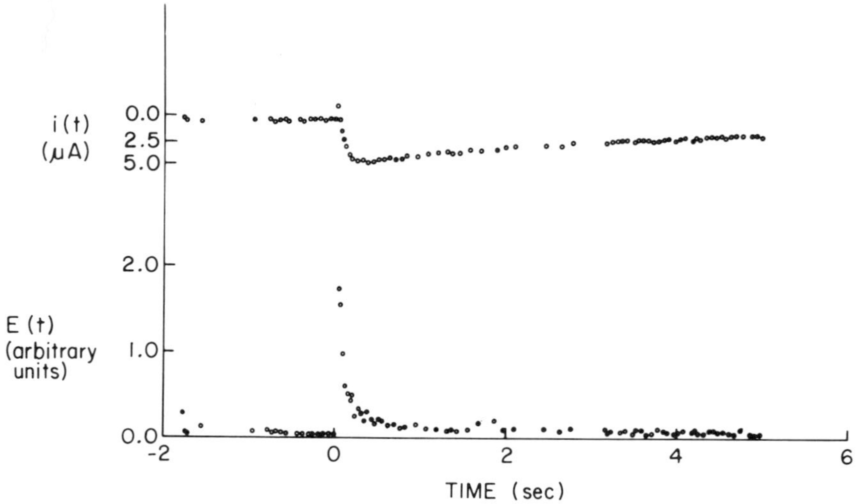


Fig. 11. FPD data for the diffusion of PAF II-2 in $CH_2Cl_2$ containing 5% polystyrene by weight. The dye concentration was 2.6% by weight, the temperature was 15°C, and K′ was 853 $cm^{-1}$.

cence yield by as much as a factor of four. It was also observed that degassing of samples produced up to twice as much photoactivated fluorescence and that many samples produced more photoactivated fluorescence at lower temperature. We have not yet determined whether these increased fluorescence yields are attributable to increased PAF production or to increased fluorescence yield through reduced quenching, but we speculate that the latter is more likely.

The first FPD data using PAF molecules to label larger species have been obtained for polystyrene latex spheres labelled with PAF I-1 (as described previously). Sample data are shown in Figure 12. The rise in fluorescence was only about 20%, even with a 400-ms activation pulse, but both the DC and the modulation signal show the dissipation by diffusion with high signal-to-noise ratio. The diffusion coefficient from the computer fit was $2.3 \times 10^{-7}$ $cm^2/s$ (at 5.5°C) which is substantially larger than the theoretically expected value of $0.72 \times 10^{-7} cm^2/s$. The discrepancy is probably attributable to the presence of adsorbed dye on the spheres which exchanges readily with the solution, yielding a diffusion coefficient that is a weighted average of the values expected for dye and for spheres [Zero et al., 1983]. This expectation is supported by the observation that washing the spheres by pelleting, resuspending, and sonicating led to a marked reduction in signal. Further development of the labeling chemistry to permit adequate levels of labeling without adsorption is in progress.

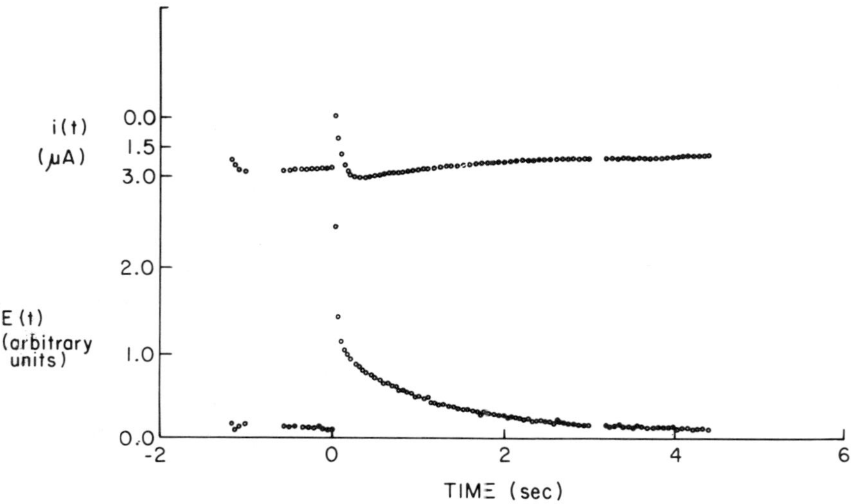


Fig. 12. FPD data for the diffusion of polystyrene latex spheres labelled with PAF I-11. The temperature was 5.5°C and K′ was 2,145 $cm^{-1}$.

## CONCLUSION

This article constitutes the first report of a new technique for the measurement of the motion of specifically labeled particles. Fluorescence photoactivation and dissipation (FPD) is essentially the positive analog of fluorescence photobleaching recovery (FPR). All of the applications of FPR should be amenable to experimentation by FPD and additional applications involving the long-term monitoring of labeled species are made possible by the FPD method. Although the data presented here are clearly preliminary, it is evident that the FPD technique successfully measures translational diffusion of specific molecules with a high signal-to-noise ratio. The key to development of the technique will be the continued design and synthesis of photoactivable fluorophores (PAFs) with optimized photochemical and photophysical properties.

## ACKNOWLEDGMENTS

We are grateful to Dr. Frederick Lanni, formerly a member of our group and now at the Center for Fluorescence Research in Biomedical Sciences at Carnegie-Mellon University, for observations and comments that led us to this line of investigation. This work was supported by grant No. CA 35954 to G.A. Krafft from the National Institutes of Health and grant No. PCM-8306006 to B.R. Ware from the National Science Foundation. G.A. Krafft is a fellow of the American Cancer Society.

## REFERENCES

Angelides K (1981): Fluorescent and photoactivable fluorescent derivatives of tetrodotoxin to probe the sodium channel of excitable membranes. Biochemistry 20:4107–4118.

Axelrod D, Koppel D, Schlesinger E, Elson E, Webb W (1976): Mobility measurement by analysis of fluorescence photobleaching recovery kinetics. Biophys J 16:1005–1069.

Bellers D (1971): Photo-Fries rearrangement and related photochemical [1,8]-shifts (J=3,5,7) of carbonyl and sulfonyl groups. Adv Photochem 8:109–159.

Berson J, Brown E (1955): Studies on Dihydropyridines. II. The Photochemical Disproportionation of 4-(2′-Nitrophenyl)-1,4-dihydropyridines, J Am Chem Soc 77:447–450.

Cherry R (1979): Rotational and lateral diffusion of membrane proteins. Biochim Biophys Acta 559:289–327.

Edidin M, Zagyanski Y, Lardner T (1976): Measurement of membrane protein lateral diffusion in single cells. Science 191:466–468.

Feigl F, Feigl H, Goldstein O, (1955): A sensitive and specific test for coumarin through photocatalysis. J Am Chem Soc 77:4162–4163.

Huffman D, Kuhn C, Zweig A (1970): Photoisomerization of 3-aroyl-2-(2-furyl) chromones. An example of quenching of a photochemical reaction by a product. J Am Chem Soc 92:599–605.

Jacobson K, Wu E, Poste G, (1976): Measurement of the translational mobility of concanavalin A in glycerol-saline solutions, and on the cell surface by fluorescence recovery after photobleaching. Biochim Biophys Acta 433:215–222.

Lanni F, Ware B (1982): Modulation detection of fluorescence photobleaching recovery. Rev Sci Instrum 53:905–908.

Lewis G, Magel T, Lipkin D (1940): The absorption and emission of light by "cis-" and "trans-" stilbenes and the efficiency of their photochemical isomerization, J Am Chem Soc 62:2973–2980.

Lippert E, Luder W, (1962): Photochemical "cis-trans" isomerization of "p"-dimethylaminocinammic acid nitrile. J Phys Chem 66:2430–2433.

Mallory F, Mallory C, (1984): Photocyclization of stilbenes and related molecules. Organic React 30:1–456.

Mendenhall D, Kobayashi H, Shih F, Sternson L, Higuchi T, Fabian C, (1978): Clinical analysis of tamoxifen, an anti-neoplastic agent, in plasma. Clin Chem 24:1518–1524.

Peters R. (1981): Translational diffusion in the plasma membrane of single cells as studied by fluorescence microphotolysis. Cell Biol Int Rep 5:733–760.

Peters R, Peters J, Tews KH, Bahr W, (1974): A microfluorimetric study of diffusion in erythrocyte membranes. Biochim Biophys Acta 367:282–294.

Rogers D, Margerum J, Wyman G, (1957): Spectroscopic studies on dyes. IV The fluorescence spectra of thioindigo dyes. J Am Chem Soc 79:2464–2468.

Sharafy S, Muszkat K, (1971): Viscosity dependence of fluorescence quantum yields. J Am Chem Soc 93:4119–4125.

Singh D, Ullman E, (1967): Photochemical transposition of ring atoms in 3,5-diarylisoxazoles. An unusual example of wavelength control in a photochemical reaction of azirines. J Am Chem Soc 89:6911–6916.

Uchida K, (1963): The photosensitized oxidation of leuco-uranine. III The kinetics of an acridine-sensitized photo-oxidation in an aerated solution. Bull Chem Soc Jpn 36:1097–1106.

Ullman E, Singh B, (1966): Photochemical transposition of ring atoms in five-membered heterocycles. The photoarrangement of 3,5-diphenyl-isoxazole. J. Am Chem Soc 88:1844–1845.

Ware B, (1984): Fluorescence photobleaching recovery. Am Lab 16(4):16–28.

Webb W, (1981): Luminescence measurements of macromolecular mobility. Ann NY Acad Sci 366:300–314.

Wheelock C, (1959): The fluorescence of some coumarins. J Am Chem Soc 81:1348–1352.

White E, Friend C, Stein R, Maskill H, (1969): A probable "trans"-cyclooctatetraene. J Am Chem Soc 91:523–525.

Zero K, Cyr D, Ware B, (1983): Tracer diffusion of counterions through a solution of polyelectrolyte. J Chem Phys 79:3602–3603.

Zweig A, (1973): Photochemical generation of stable fluorescent compounds (photofluorescence) Pure Appl Chem 33:389–410.

*Applications of Fluorescence in the Biomedical Sciences, pages 159–202*


# 8 Fluorescence Probe Studies of Proteins and Membranes

*Bruce S. Hudson, Danni L. Harris, Richard D. Ludescher, Anthony Ruggiero, Alison Cooney-Freed, and Sarina A. Cavalier*

Fluorescence provides a unique method for the investigation of basic physical properties of biopolymers and biological structures. The high sensitivity of fluorescence, combined with the rapid time scale of the process and advances in measurement techniques, permits real-time monitoring of very rapid internal motions and relaxation processes. These rapid processes are becoming of increasing interest because of the recognition of thermal fluctuations as a potentially important aspect of the function of enzymes [Karplus and McCammon, 1981, 1984]. The importance of the dynamic, "fluid" nature of membranes has been recognized for some time [Singer and Nicholson, 1972]. This increased interest has been encouraged by the emergence of dynamical simulation methods that can presumably be used as a basis for "observing" the nature of rapid motions in more detail than can ever be obtained from experiment. Experimental observations at the limit of temporal resolution are needed to test these simulation methods.

Time-resolved fluorescence polarization measurements directly monitor molecular reorientation processes that are both calculable and intuitively understandable. This is also the fluorescence measurement most directly comparable to observations made by other techniques. The ability to measure reorientational motion on a very short and real-time basis is one of the attractive features of the technique. Spectral and total excited-state decay dynamics are more indirectly related to the behavior of the chromophore surroundings and are, in this respect, of less interest. However, it is neces-

Department of Chemistry and Institute of Molecular Biology, University of Oregon, Eugene, Oregon 97403.

sary to have at least a minimal understanding of these other aspects of fluoresence in order to interpret polarization experiments.

The ability to correlate dynamic behavior with function at a detailed molecular level is greatly enhanced by our emerging ability to make specific modifications of biopolymers at sites where structural, functional, or thermodynamic information indicates that significant perturbations will result. Species variation, spontaneous or induced mutations, genetic engineering and chemical modifications all play a role in this capability. In fact, the potential of these modification methods now exceeds our ability to design and execute experiments. Some systematic plan is needed to choose optimum systems for study.

Fluorescence methods depend, ultimately, on the response of some fluorescent component, either intrinsic or extrinsic, to its environment following optical excitation. Detailed evaluation of sophisticated fluorescence experiments requires considerable information concerning the excited-state behavior of the chromophore in question. The direction of the emission and absorption transition dipoles, possible excited-state reactions or relaxation processes, the dependence of spectral and excited-state relaxation properties on the nature of the surrounding environment, and the "perturbation" effects associated with the use of extrinsic probes are all properties of a chromophore that must be understood before a detailed evaluation of fluorescence experiments in terms of motional behavior can be made. These are the realm of molecular spectroscopy and photophysics.

The present article presents the results of recent studies using state-of-the-art subnanosecond laser fluorescence methods aimed at a basic understanding of the behavior of fluorophores in biological systems. The principles of fluorescence experiments, including experimental methodology, are first outlined. This section is followed by a discussion of several membrane and protein systems with emphasis on the detailed interpretation of these experiments. Possible ambiguities in the interpretation of these experiments are then discussed and strategies for the resolution of these ambiguities are outlined. The point of view adopted here is that the lessons learned from years of study of the molecular spectroscopy and photophysical behavior of polyatomic molecules should provide a framework for understanding and designing experiments aimed at biophysical questions.

## BASIC CONCEPTS OF FLUORESCENCE

The basic concepts of molecular spectroscopy as used in fluoresence experiments are, of course, well established [Lakowicz, 1983; Pesce et al., 1971; Becker, 1969; Birks, 1970]. The application of these principles in a particular situation, however, involves many empirical observations and generalities derived from these observations that need to be examined for their validity in each case. Many of the basic observations have been derived from experiments performed on relatively simple chromophores (e.g., naphthalene) in well-defined environments (e.g., single crystals of durene) observed at cryogenic temperatures. The specific behavior of a complex chromophore in a complex environment at physiological temperature may require inclusion of processes not present in these idealized cases. Furthermore, the increased temporal resolution of fluoresence methods and other increasingly sophisticated aspects of these experiments may reveal behavior not previously seen.

## Fluorescence Experiments

Fluorescence measurements may be used to determine several more or less independent aspects of the light emitted following electronic excitation. These distinct experimental observations depend on fundamentally different processes and they must, therefore, be distinguished at all stages of the discussion. Besides being different physical processes, there is also a distinction to be made at the level of interpretability. Put simply, one can ask the question "if one knew the properly statistically averaged time-dependent positions of all of the *atoms* of the system, could one calculate the experimental observable or would one have to know what the *electrons* are doing?" The motion of the atoms is, in principle, a calculable quantity; the behavior of the electrons is presently not realistically calculable for these complex systems.

The spectral distribution of the emitted light can be measured. The peak of the fluorescence emission spectra of many fluorophores is very sensitive to the polarity of the environment. This is especially true for polar species. Chromophores that are centrosymmetric, and thus have no dipole moment in any electronic state, are, to a first approximation, insensitive to polarity but may be sensitive to the polarizability of the environment. The overall shape of the fluorescence spectrum primarily contains information about the difference in the geometry of the chromophore between its ground and excited states. This is generally not of much interest from the biophysical point of view except if one is trying to understand the electronic excited states of intrinsic chromophores that are involved in light absorption in biological processes or where the optical properties of the chromophore have been extensively characterized in terms of structure/spectra correlations.

The intensity of fluorescence emission of a chromophore, ultimately characterized by its quantum yield for emission, is often a very sensitive function of its environment. In this respect this quantity can be used to measure very small structural changes. Similarly, the lifetime of the fluorescence emission often depends critically on the interactions of the excited chromophore with neighboring groups. Both of these quantities are "composite" in the sense that they depend on both radiative and radiationless decay behavior. Even when these are separated by measurement of both lifetime and quantum yield, interpretation is often model dependent.

Collisional quenching processes monitored in terms of the decreased quantum yield or lifetime in the presence of a quencher provide useful information about the degree of exposure of a fluorophore to its solvent environment [for examples see Lehrer, 1976; Lakowicz and Weber, 1973a,b; Ludescher et al., 1985]. This exposure may be either structural or dynamic in nature in the sense that quenching rates less than diffusion controlled may be due to restricted access caused either by static steric constraints or the requirement for rare structural fluctuations. This phenomenon should be interpretable at a microscopic dynamic level so long as the rate-limiting step is diffusion to the neighborhood of the fluorophore. Since the detailed mechanism of quenching is not known, rates that are near diffusion limited are difficult to interpret quantitatively.

Fluorescence energy transfer, as monitored by intensity or lifetime behavior, is a time-honored tool in biophysical chemistry. Here the interpretation of the results at a detailed level depends on an understanding of both the transition dipole orientation and motional behavior of the donor and acceptor species. Qualitative interpretation is possible without knowledge of this information and thus this method is often very useful.

Time-dependent fluorescence anisotropy decay experiments are, in principle, capable of unique interpretation at a molecular (atomic motion) level. By this we mean that a simulation of the motions of the chromophore, given a knowledge of the transition dipole direction in the molecule framework, can be used to calculate the anisotropy decay behavior without any additional assumptions. It is this aspect of fluorescence that is emphasized here.

## Fluorescence Decay Behavior

The conventional kinetic scheme for the behavior of polyatomic molecules in condensed phases following optical excitation is shown in Figure 1. Initial excitation to any vibrational level of an excited singlet state ($S_n$) is followed by very rapid radiationless internal conversion to the lower vibrational levels of the lowest energy singlet state ($S_1$). This radiationless decay process is normally sufficiently rapid that the population yield of the lowest-energy singlet state following excitation to any level is essentially unity. This means that the emission spectral distribution and the fluorescence lifetime are independent of the excitation wavelength. Thermal equilibrium is established within the $S_1$ state vibrational levels. Subsequent decay of this excited singlet state proceeds through radiative fluorescence emission, radiationless decay to the ground state, and radiationless decay to the triplet manifold. All of these proceses are first order in the population of the excited lowest-energy singlet state.

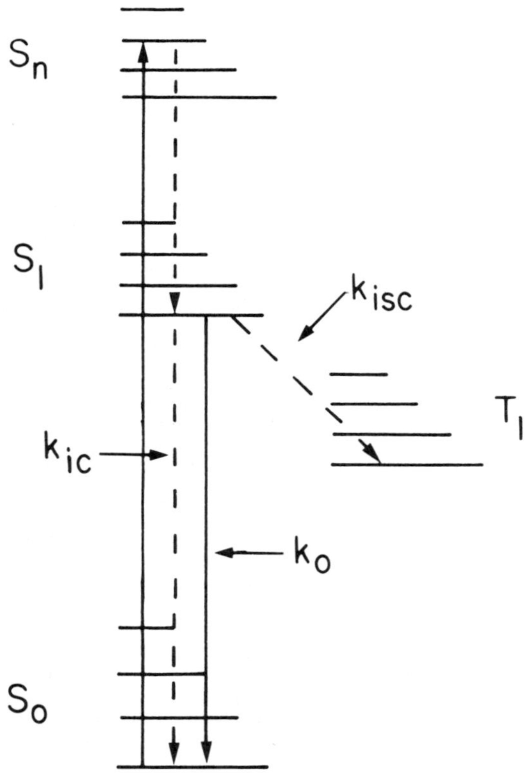


Fig. 1. The conventional kinetic scheme for electronic excitation of a polyatomic molecule in condensed phases. Radiationless transitions are indicated as dashed lines; radiative transitions as solid lines. The labels $S_0$, $S_1$ and $S_n$ indicate singlet states; $T_1$ is the lowest energy triplet state. The first order rate constants $k_o$, $k_{ic}$ and $k_{isc}$ refer to radiative emission, internal conversion and intersystem crossing, respectively.

According to this scheme the time dependence of the fluorescence emission will be described by a single decay constant given by

$$k = k_o + k_{isc} + k_{ic} = k_o + k_{nr} \tag{1}$$

where $k_o$ is the intrinsic fluorescence decay rate, $k_{isc}$ is the rate of intersystem crossing to the triplet states, $k_{ic}$ is the rate of radiationless decay to the ground state, and $k_{nr}$ is the total nonradiative decay rate of the $S_1$ state. The total decay rate, $k$, is $1/\tau$, where $\tau$ is the fluorescence lifetime. The quantum yield of emission as fluorescence in this case is

$$Q = k_o/k = k_o\tau = \tau/\tau_o \tag{2}$$

Equivalently, the more fundamental rate constants $k_o$ and $k_{nr}$ can be extracted from the measurements of $k$ and $Q$ from

$$k_o = k\,Q \tag{3}$$

$$k_{nr} = k - k_o \tag{4}$$

The radiative rate constant is related to the strength of the transition between $S_1$ and $S_0$. Strongly allowed transitions have short radiative lifetimes. The interpretation of radiationless decay rates is more complex but, generally speaking, more fluid environments result in more rapid radiationless decay.

Modifications of this basic scheme include quenching effects that add terms to decay rate dependent on the concentration of quencher, excimer formation where an excited molecule interacts with a ground-state molecule of the same kind to form a stable emissive species, energy transfer processes that depend on the concentration, spatial and orientational distribution of acceptor species, and excited state reaction processes (uni- or bimolecular) that result in transformation of the initially populated species into another species that may be fluorescent. There are, in addition, more complex processes that depend on the illumination intensity at high light levels.

There are well-known exceptions to this standard scheme, even in the absence of the processes listed above. The most famous of these is the fluorescence of azulene where emission occurs from the second excited singlet state with very little emission coming from direct excitation of the lowest singlet state. This is due to the large energy separation between the second and first excited singlets in this species. Another example of $S_2 \rightarrow S_0$ emission has just been reported for the linear conjugated polyene with seven double bonds, tetradeaheptaene [Synder et al., 1985]. Again, a large $S_2$–$S_1$ energy gap occurs for this species.

Recent developments in time-resolved fluorescence techniques have resulted in the demonstration that the simple first-order decay following $S_n \rightarrow S_1$ relaxation may be more the exception than the rule if one looks in more detail. One example is the observation of transient $S_2$ emission occurring in competition with radiationless decay to $S_1$. This has been observed for the diphenylpolyenes including diphenylhexatriene, a commonly used membrane probe [Felder et al., 1982]. In biopolymer applications, deviation from the simple scheme most often takes the form of the observation of nonsingle exponential decay behavior of the fluorescence, or dependence of the lifetime on excitation or emission wavelength. The specific cases of parinaric acid and its parent chromophore, octatetraene, in solution and lipid bilayer membranes and tryptophan fluorescence in proteins will be discussed below. The general situation is that the origin of these deviations from the simple scheme is not easy to determine and, in fact, the mechanism leading to decay complexity may depend on the particular environment even for a specific chromophore.

### "Homogeneous" Versus "Heterogeneous" Behavior

There are two limiting cases leading to complex behavior that are probably worth defining so that specific schemes can be classified and so that the kinds of experiments that distinguish these limits can be examined. In one limiting case the complex decay behavior is related to a variety of forms of the fluorophore in a preexisting ground-state population. These "species" may be chemical forms, conformations, or solvent environments. This is termed the "heterogeneous" case because distinct molecules in the population behave individually according to the simple first-order decay but there is heterogeneity in the population. This behavior depends on the variability of radiative or, more often, radiationless decay rates with changes in the detailed initial state of the chromophore. The magnitude of this effect therefore depends both on the distribution of the initial state and on the sensitivity of the chromophore to its initial state. Fluorescence methods are perhaps unique in that the temporal scale of the measurements is such that configurations that interconvert rapidly on the time scale of other measurements are detected as distinct species in the nanosecond time regime.

The other extreme limit is the case where all chromophores are "kinetically equivalent" in the sense that each molecule is identical in terms of its decay behavior and the "nonexponential" behavior arises from complex kinetic processes subsequent to excitation. Such processes include excited-state reactions leading to emitting products (including exiplexes), relaxation of the solvent or biopolymer to a new configuration, or reorientation of the chromophore so that new quenching mechanisms are possible. These processes, whatever they are, must occur on the time scale of the excited-state emission. There is always a thermal distribution of environments for a chromophore that results in a distribution of spectroscopic and decay properties. Thus, the true homogeneous case can only be approximated by a situation where the chromophore is insensitive to the details of its environment or where energetic constraints restrict the range of possible configurations.

The homogeneous and heterogeneous cases can be distinguished on the basis of their behavior when the excitation wavelength is changed. In the heterogeneous case the individual components of the preexisting distribution of distinct species will usually have distinct absorption spectra. In this case variation of the excitation wavelength will select a particular subset of the distribution and the resulting fluorescence decay behavior will reflect the properties of that particular subset. In the homogeneous case variation of the excitation wavelength will not result in changes in the decay behavior. This separation of these two cases does not occur for the emission wavelength variation of the decay behavior because the subset of the heterogeneous distribution may have distinct emission spectra and, in the homogeneous case, the emission spectrum may change in time as the relaxation process proceeds. There is, however, a specific prediction of the homogeneous mechanism that the fluorescence spectrum will shift to lower energies at longer times as the initially prepared excited state relaxes to its new, lower, energy configuration. In the heterogenous case the spectrum observed at long times may be either red or blue shifted with respect to the early time spectrum depending on the nature of the correlation between the lifetime and the emission spectrum.

Real situations will, naturally, be somewhere between these two limiting cases. In the heterogeneous case the distribution must be established by some equilibrium process that may have components on the fluorescence time scale. In this case the individual subsets of the distribution will interconvert on the same time scale as the emission and the decay kinetics will have amplitudes and rate constants depending on both intrinsic subpopulation and interconversion rates. Where complex decay is observed, the homogeneous case involves some relaxation process between a dominant ground-state configuration and a relaxed configuration that is a higher energy state for the ground state. At finite temperature this "excited" configuration, corresponding to the relaxed state following excitation, will always have some small population. There will therefore be some "heterogeneous" contribution to the population in all cases at finite temperature.

If we think of the "structure" of a solution on a truly instantaneous basis there will be a wide distribution of local environments. Some features of this distribution will interconvert very rapidly and this part of the distribution will be averaged over so far as any finite time measurement is concerned. Other features of the environmental distribution that may be important in determining the excited-state kinetic behavior of a solute may persist for times that are comparable to or longer than the fluorescence time scale. From this point of view, what one expects for a solution is some kind of a continuous distribution of excited-state behavior, in contrast to discrete relaxation times. It is important to note than in analysis of real data, the distinction between such a continuous distribution and several discrete components cannot be made easily. Thus, although we report the decomposition of fluorescence decays in terms of exponential components, this does not necessarily mean that there are actually discrete structures associated with each component. A better way to view the situation may be to say that the fit exponentials give a measure of the mean of the distribution, in the form of the average lifetime, and that the width of the distribution is approximated by the standard deviation of the lifetimes. An analysis of protein tryptophan fluorescence data in this format is given in a recent publication [Ludescher et al., 1985].

## Decay of Fluorescence Anisotropy

Excitation of a randomly oriented sample of chromophores results in photoselection of a subset of the angular distribution with a preference for those molecules with their transition dipoles oriented along the polarization of the incident radiation. The resulting emission is polarized unless reorientational motion occurs much more rapidly than emission. In the limit of no reorientational motion and in the absence of $S_n \rightarrow S_1$ radiationless decay processes, the resulting emission has a preferential intensity parallel to the excitation polarization. In this case the intensity parallel to the excitation direction will have an intensity three times larger than perpendicular to the excitation polarization. The anisotropy defined by

$$r = (I_{\parallel} - I_{\perp})/(I_{\parallel} + 2I_{\perp}) \tag{5}$$

then has a value of two-fifths. If there is radiationless decay from $S_n$ to $S_1$, and if the orientation of the transition dipole for the $S_1$ to $S_0$ transition differs from the photoselected $S_0 \rightarrow S_n$ transition, then the initial anisotropy, designated $r_o$, will be less than two-fifths and will be given by

$$r_o = (2/5)P_2[\cos(\theta)] = (2/5)[3\cos^2(\theta) - 1]/2 \tag{6}$$

where $\theta$ is the angle between the $S_o \rightarrow S_n$ and $S_0 \rightarrow S_1$ transition dipoles. This initial anisotropy can be determined from a measurement of the anisotropy in a rigid matrix. The value of $r_o$ depends on the excitation wavelength because of the variation of the particular photoselected initial state $S_n$. In principle, excitation at the longest absorbing wavelengths should be into the lowest energy vibrational levels of $S_1$ and should yield a value of $r_o$ of 2/5 because the photoselected state is the emitting state.

In order to interpret anisotropy behavior in microscopic terms it is necessary to know the orientation of the emitting transition dipole relative to the molecular axis system. This must be determined from polarized single crystal experiments or from theory. It is the reorientation of this vector that is measured in an anisotropy experiment. If, as is the case for tryptophan, the limiting anisotropy $r_o$ observed with long wavelength radiation does not approach the value of 2/5, then there is an ambiguity as to the orientation of the transition dipole with respect to the molecular axis system. This is discussed by Ichiye and Karplus [1983] and below.

The time dependence of the decay of the anisotropy from this initial value depends only on the motional behavior of the transition dipole of the chromophore. Most experiments are analyzed in terms of an expression of the form

$$r(t) = \sum_i r_i e^{-t/\phi_i} + r_\infty \tag{7}$$

where the $\phi_i$ are reorientational relaxation times, the $r_i$ are the corresponding amplitudes, and $r_\infty$ is the limiting anisotropy. The sum of the $r_i$'s and $r_\infty$ is $r_o$. Agreement between the value of $r_o$ obtained in this way and the value measured in the absence of motion indicates that all of the motional components are being detected. The constant term occurs if the system exercises restricted anisotropic reorientational motion on the time scale of the fluorescence experiment. If this restricted motion is cylindrically symmetric then

$$r_\infty = r_o \, [\langle P_2[\cos(\chi)]\rangle]^2 \tag{8}$$

where $\chi$ is the angle between the average orientation of the transition dipole (the "director") and the instantaneous value and $\langle\,\rangle$ denotes an ensemble average. For unrestricted reorientational motion $r_\infty = 0$. For motion of a spherical object there is only one reorientational correlation time $\phi$. The problem of the general form of the decay of the anisotropy for more general cases has been considered by several authors. Szabo has recently presented a comprehensive treatment [Szabo, 1984].

If there are several types of chromophores in the sample then the anisotropy is a weighted average of the contributions from each component,

$$r(t) = \sum_i f_i(t) r_i(t) \tag{9}$$

where $f_i(t)$ is the time-dependent fraction of the fluorescence contributed by the $i^{th}$ component at time $t$ and $r_i(t)$ is the anisotropy decay of the $i^{th}$ component. The $f_i$'s are normalized to sum to unity.

If the individual environments or fluorophore species have distinct lifetimes, $\tau_i$, then the fractional contribution to the fluorescence from the $i^{th}$ species is given by

$$f_i = x_i e^{-t/\tau_i} \,/\, \sum_i x_i e^{-t/\tau_i} \tag{10}$$

where $x_i$ is the fraction of the fluorophore in the $i^{th}$ environment. At time zero the individual components contribute in proportion to their fractional population but at long times the species with short lifetimes will become relatively less important. At very long times only the longest-lived species contributes to the fluorescence and the observed anisotropy then reduces to the value for this component at long times. A complex time dependence can be observed at intermediate times as discussed elsewhere [Ludescher, 1984a; Ludescher et al., 1986a] and as illustrated below.

## EXPERIMENTAL METHODS

The experimental method used in the studies described below is single-photon counting detection of fluorescence following excitation with either synchrotron radiation [Wolber and Hudson, 1981, 1982] or the doubled output of a synchronously pumped dye laser [Ludescher et al., 1985; Small et al., 1984]. In the laser experiments the time resolution of the instrumentation is determined by the jitter in the transit time of the photomultiplier used to detect single photons. This is about 800 ps for conventional photomultipliers and about 150 ps for microchannel plate detectors. We have used both such devices in our experiments.

The fluorescence decay observed in these experiments is the convolution of the true decay with the instrument response function. Numerical iterative deconvolution [Grinvald and Steinberg, 1974; Cross and Fleming, 1984] or methods-of-moments techniques [Isenberg, 1973a,b; Isenberg et al., 1973; Small and Isenberg, 1976, 1977; Small et al., 1984] can be used to reduce the effect of this instrument broadening so that decays about a factor of five shorter than the instrument response function can be determined. Thus, with current devices the temporal resolution of this method is on the order of 30 ps. This, however, is realistically the case only when the decay is a single exponential. The quantitative evaluation of several closely spaced decays in the short time regime is still problematical.

An alternative method for determining fluorescence and anisotropy decay behavior is the measurement of the phase shift and amplitude decrease of a periodically modulated excitation beam due to the finite "response time" of the fluorescence of the excited sample [Gratton and Limkeman, 1983; Jameson et al., 1984; Gratton et al., 1984]. This technique is also discussed in another article in this volume. Below we present a comparison of the results of these two methods for a particularly complicated fluorescence decay.

## STUDIES OF MEMBRANE DYNAMICS AND LIPID-PROTEIN INTERACTIONS WITH PARINARIC ACID

### General Properties of Parinaric Acid and Its Fluorescence

Parinaric acid, 9,11,13,15-octadecatetraenoic acid, is a naturally occurring 18-carbon conjugated tetraene fatty acid with useful properties as a probe of membrane structure and lipid binding to proteins [Sklar et al., 1975, 1976,

1977a–c; Tecoma et al., 1977; Berde et al., 1978]. This fatty acid is not very widely distributed in nature and is most easily isolated [Sklar et al., 1981] from seeds of the Fiji Island makita tree, *Parinarium glaberrimum*, where it occurs as a triglyceride for food storage. The carboxyl functionality of parinaric acid permits its use as a methyl or cholesterol ester or as one chain of a phospholipid [Tsai et al., 1981] as well as in the form of the free fatty acid. The major feature of interest in terms of the use of parinaric acid as a lipid analog is the close structural similarity it shows to components of biological membranes and commonly studied model membrane lipid bilayers. This structural similarity permits the biosynthetic incorporation of parinaric acid into biological membranes. The form of the fatty acid found in the seed has the *cis* geometry at the 9-10 and 15-16 double bonds. The *all trans* isomer is easily formed by iodine-catalyzed isomerization. The ready availability of two slightly different structural isomers of this membrane probe permits detailed examination of bilayer structure.

The tetraene chromophore of parinaric acid provides it with a strong absorption band in the 325–290-nm region. The extinction coefficient of this band is about 80,000 L/mole cm. The absorption spectrum of parinaric acid is structured due to a vibrational progression in the polyene chain single- and double-bond vibrations. The relatively sharp structure of this absorption band permits the detection of small spectral shifts in the absorption maxima. The fluorescence emission of parinaric acid is very widely spaced from this strong absorption region, having a maximum near 420 nm. This wide separation of the absorption and emission bands is due to the fact that the lowest energy excited state responsible for the emission of tetraenes is not the state responsible for the strong absorption band but is, instead, due to a forbidden transition at lower energy [Hudson and Kohler, 1972, 1973, 1974, 1984; Hudson et al., 1982]. Because this emitting transition is forbidden, the intrinsic radiative lifetime of parinaric acid is relatively long, ca. 100 ns. The intensity of this transition is vibronically borrowed from the nearby strongly allowed transition. The polarization of the emission is at least roughly aligned along the polyene chain and the value of $r_0$ observed in rigid matrices or extrapolated to zero time is very close to 2/5. The fluorescence quantum yield of parinaric acid in aqueous solution is very low but becomes considerable when the probe is in a hydrophobic environment. The fluorescence spectrum of parinaric acid is essentially independent of the polarity or polarizability of the probe environment.

The spectral properties of parinaric acid make it a good acceptor for energy transfer from tryptophan residues of proteins [Sklar et al., 1976]. Furthermore, the fluorescence quantum yield of the tetraene fatty acids is increased dramatically on binding to hydrophobic sites in proteins. These properties have been used to characterize the fatty acid binding site of albumins [Sklar et al., 1977; Berde et al., 1979] and preferential distribution of parinaric acid in bilayers relative to proteins [Kimelman et al., 1979].

Parinaric acid is not very useful as a general fluorescent label. This is due to its rather low fluorescence quantum yield and its photochemical lability, including an irreversible photochemical dimerization reaction [Morgan et al., 1980; Williamson et al., 1983]. The primary utility of this fluorophore is in detailed studies of membrane dynamics where the known location and orientation of the chromophore permits detailed evaluation of the motions involved.

## Steady-state Fluorescence of Parinaric Acid in Model Bilayers

Phospholipid bilayers have a characteristic thermal phase transition between a solidlike gel state with extended acyl chains and a high-temperature fluidlike liquid crystalline phase with rapidly reorienting acyl chains containing numerous *gauche* conformations. The behavior of parinaric acid as a probe of lipid bilayers was established in a series of papers in the middle and late 1970's [Sklar et al., 1975, 1976, 1977a,b]. At the thermal gel-to-liquid-crystalline phase transition of membranes, this probe in any of its forms exhibits a decrease in quantum yield and lifetime and a decreased steady-state polarization. The fluid, liquid crystalline phase is characterized by a low quantum yield and lifetime and low anisotropy due to rapid chain motions. The lifetime and quantum yield of parinaric acid are much more sensitive to environment than are those observed for diphenylhexatriene (DPH).

Another observation from our earlier work that is relevant to the interpretation of current studies presented below is that parinaric acid exhibits a blue shift in its fluorescence excitation and absorption spectra on going from the higher-density gel phase to the lower-density liquid crystalline phase (see Fig. 2). This shift in the spectral maximum is due to the lower polarizability of the high-temperature phase. For a given chemical environment, the polarizability is related to the density. Calibration of the position of the spectral maxima [Sklar et al., 1977a] permits a determination of the density change in the bilayer from this spectral observation. The results are in good agreement with direct density change determinations [Sklar et al., 1977b]. Thus, the environment of the probe can be characterized in terms of its density by determination of the excitation maximum. Conversely, if the probe exists in more than one environment, variation of the excitation wavelength will change the relative contributions of the heterogeneous environmental components.

These observations concerning the behavior of parinaric acid in bilayers are, for the most part, independent of the particular form of the probe, i.e., *cis* or *trans* isomer, free fatty acid or phospholipid. There is, however, one characteristic difference between the *cis* and *trans* isomers that is observed

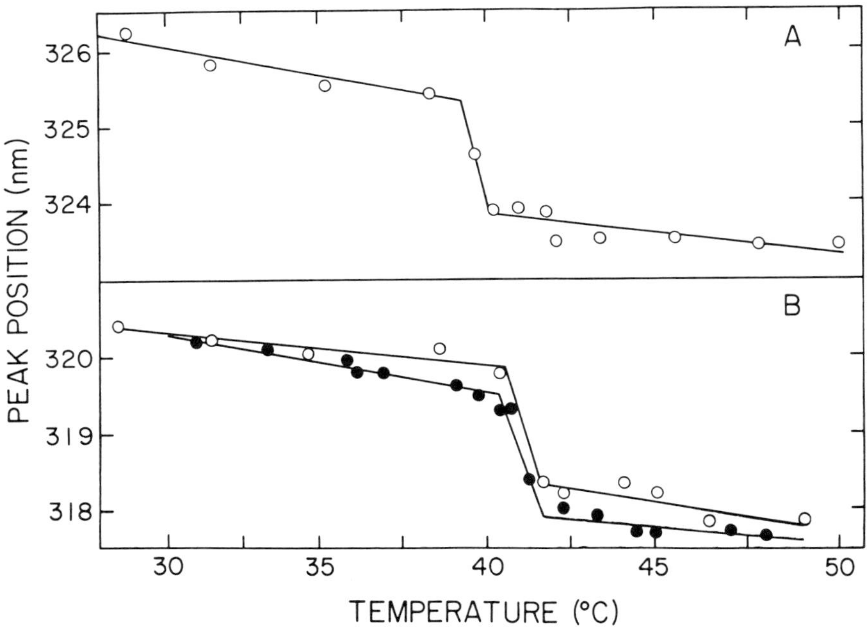


Fig. 2. The shift of the first maximum in the fluorescence excitation spectrum of parinaric acid in dipalmitoylphosphatidylcholine (DPPC) bilayer vesicles as a function of temperature. The upper panel (A) is for *cis*-parinaric acid and the lower panel (B) is for *trans*-parinaric acid. The solid symbols in the lower panel refer to measurements for the absorption spectrum. The thermal phase transition is seen at 41 °C [from Sklar et al., 1977b].

when two phases of phospholipid coexist, either as two distinct vesicle populations of different composition or as laterally separated phases. The *cis* isomer partitions nearly equally between these two phases but the *trans* isomer preferentially accumulates in the solid phase with a partition coefficient of 4–9 [Sklar et al., 1977b, 1979; Sklar, 1980]. This differential partitioning behavior is also observed for phospholipids containing a parinaroyl chain [Pugh et al., 1982; Welti, 1982; Welti and Silbert, 1982]. This is as expected on the basis of the shapes of these two probe species and the dominant structures of the acyl chains in the two phases.

Another distinction between the *cis* and *trans* forms is that the *trans* isomer exhibits a greater sensitivity in its quantum yield and lifetime to changes in its environment. This behavior, in conjunction with the partitioning behavior discussed above, makes *trans*-parinaric acid uniquely sensitive to the presence of small amounts of "solid-phase" regions in lipid bilayers.

**Fluorescence Anisotropy Decay in Bilayers**

Time-resolved fluorescence lifetime and anisotropy decay determinations for parinaric acid in model membrane systems have been reported [Wolber and Hudson, 1981]. The major observation (Fig. 3) is that the time dependence of the anisotropy of parinaric acid is very different in the two bilayer phases. In solidlike gel phase the anisotropy is high and essentially constant in time. In the fluidlike liquid crystalline phase there is a very rapid decay of the anisotropy to a nearly constant value. The magnitude of this residual anisotropy at long times is a measure of the order parameter after this initial rapid reorientational motion. The value of the order parameter obtained in this way is in good agreement with values estimated for these conditions from deuterium nuclear magnetic resonance (NMR) measurements. This has the interesting implication that there are no slow reorientational processes with significant amplitudes occurring on the time scale between nanoseconds and tens of microseconds [Wolber and Hudson, 1981, 1982]. Thus, the fluorescence measurements appear to be of the proper time scale to detect the significant motions in these model membrane systems. These measurements are able to separate the *rate* of reorientational motion from the *amplitude* of this motion. This is not possible with steady-state fluorescence anisotropy determinations or conventional magnetic resonance techniques.

The motion of parinaric acid in these bilayers leading to rapid depolarization of the fluorescence could be of two kinds: the entire acyl chain could move as a unit in a rigid fashion or there could be excitation of *gauche* conformations in the methylene segment of the upper part of the probe. Excitation or migration of "kinks," correlated *gauche* +, *trans*, *gauche* − configurations, does not significantly reorient the tetraene chromophore. The conjugated section of the polyene is rather rigid and internal excitation of this segment is unlikely, at least in the ground electronic state. Rigid reorientation of the entire probe chain is expected to be collective in nature because of the large angle swept out. In this case the motions of the bilayer chains might resemble the behavior of a wheat field. *Trans-gauche* excitation in the upper part of the probe chain might result from statistically uncorrelated motions. This excitation depends on the flexibility of the methylene segment. The amplitude of this kind of motion depends on the number of methylene segments. An experiment is in progress where an analog of parinaric acid is being synthesized which has few methylene segments between the carboxyl group and the tetraene chromophore. If the acyl chain motion is of the rigid body variety then this probe should have the same behavior as that observed for parinaric acid. If *trans-gauche* excitations are required for depolarization this analog should have a decreased amplitude for its depolarization. A more complex interpretation of acyl chain dynamics based on very recent experiments is discussed below.

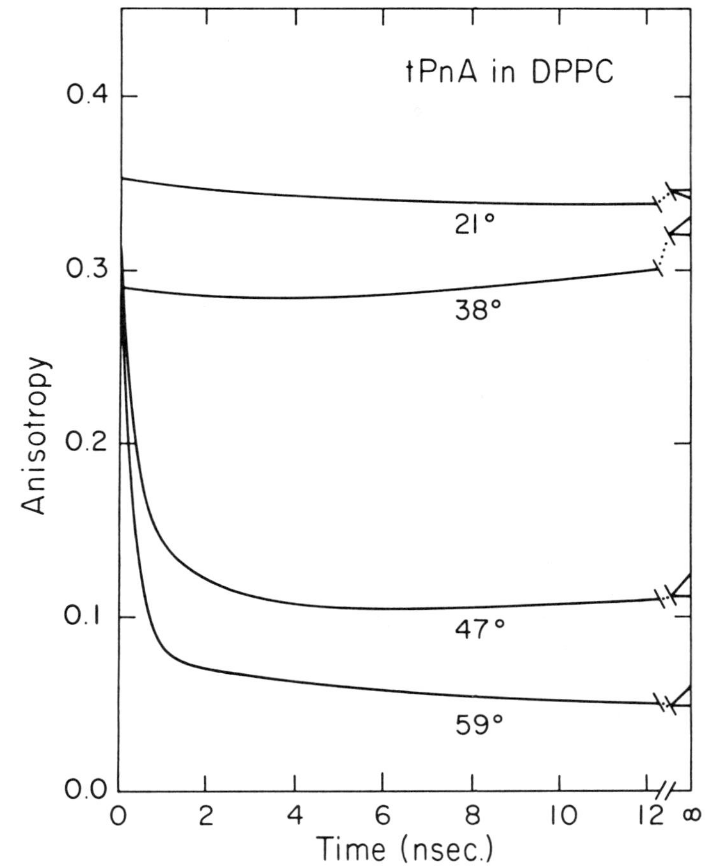


Fig. 3. The time dependence of the anisotropy of the fluorescence of *trans*-parinaric acid (tPnA) in DPPC vesicles of the indicated temperatures. The upper two curves correspond to temperatures below the thermal phase transition and the lower two curves are above the transition at 41°C (see Fig. 2) [from Wolber and Hudson, 1981].

## The Effect of Protein Components on Anisotropy Decay

The effect of incorporated proteins on the chain dynamics of model membranes as measured by parinaric acid has been established in two papers [Kimelman et al., 1979; Wolber and Hudson, 1982]. In each case the incorporated protein was the coat protein of the M13 bacteriophage. In the life cycle of the virus the coat protein exists in the *Escherichia coli* membrane prior to final phage assembly. This protein has a very hydrophobic central sequence and is believed to exist in the bilayer as a simple transmembrane alpha helix. The presence of the protein tends to abolish the thermal transition of one-component phospholipid bilayers. Energy transfer measurements from

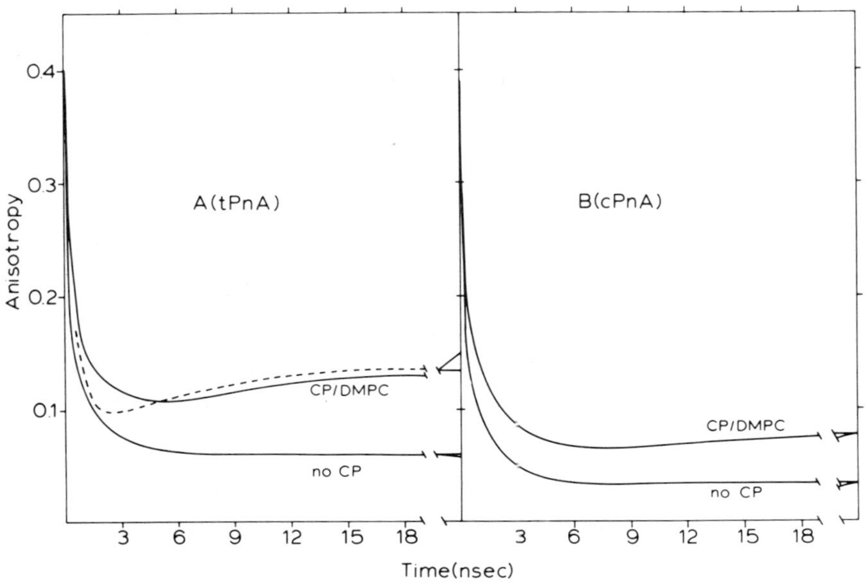


Fig. 4. The time dependence of the anisotropy of *trans* (t) (left, panel A)- and *cis(c)*parinaric acid (right, panel B) in DMPC (dimyristoylphosphatidylcholine) vesicles in the absence and presence of the coat protein (CP) of the M13 virus. The coat protein is present at a molar ratio of 1 CP/30 lipid molecules. The temperature is 30°C. The thermal-phase transition of DMPC occurs at 23°C. The dashed line in panel A is a calculated anisotropy decay based on a heterogeneous environment model [from Wolber and Hudson, 1982].

the single tryptophan of the M13 coat protein to the random distribution [Wolber and Hudson, 1979] of parinaric acid in the bilayer have been used to characterize the nature of the lipid surrounding the protein. The major conclusion is that the protein disorders the bilayer in its vicinity relative to the solid gel phase. This conclusion is based on the avoidance of the vicinity of the protein by the *trans* isomer, consistent with a relative lack of straight-chain structures. On the other hand, the protein orders the bilayer relative to fluidlike liquid crystalline state as revealed by an increase in the limiting anisotropy in the presence of the protein (Fig. 4). This behavior is consistent with the loss of the phase transition upon incorporation of the protein since the difference between the acyl chain structure at low and high temperatures in the vicinity of the protein is diminished.

The upward curvature observed for the anisotropy decay of parinaric acid in the presence of the M13 coat protein is of significance. (Some curvature of this type is also seen in the gel-phase anisotropy decays of Fig. 3.) The interpretation of this rising anisotropy at long times is based on the combination of facts that the protein induces an increase in the fluorescence lifetime relative to that seen for the fluid bilayer and increases the anisotropy (reduces motional freedom) for chains in the vicinity of the protein. At short times all probe species contribute to the fluorescence and therefore to the anisotropy in proportion to their amount. However, at longer times the short-lived species distant from the protein become less important and do not contribute as much to the anisotropy. The longer lifetime for the probe species adjacent to the protein means that at long times the observed anisotropy is that due to these "perturbed" chains. Since the anisotropy is higher for the chains near the protein than that for the bulk lipid, the observed anisotropy rises at long times. The smaller effect for the *cis* isomer is due to the smaller change in its lifetime with change in environment. This behavior can be analyzed to provide an estimate for the number of perturbed chains. This is found to be roughly the number of chains immediately adjacent to the protein [Wolber and Hudson, 1982].

## Time-Resolved Fluorescence Decays: Heterogeneous Environments

In some very recent experiments we have used the superior time resolution of our present laser-based single-photon counting apparatus to reexamine the behavior of parinaric acid in liquid-crystalline phase phospholipids. These studies confirmed previous results [Wolber and Hudson, 1981] that the fluorescence decay even in the nominally one-component high-temperature phase cannot be described with a single exponential. The best analyses seem to require three exponential terms. The shortest lifetime component (ca. 0.8 ns) has a relatively small amplitude (ca. 10%). The amplitude of this short-lifetime component is essentially constant as a function of temperature. Rather more interesting behavior is observed for the intermediate- and long-lifetime components. Results for DPPC (dipalmitoylphosphatidylcholine) are shown in Figure 5. This phospholipid has a thermal phase transition at 41 °C. The long-lifetime component persists at temperatures well above the phase transition and decreases in amplitude rather abruptly at about 48–55 °C. The amplitude of this long lifetime does not depend on the probe concentration, the particular isomer of parinaric acid used, the method of preparation of the bilayer vesicles, or the thermal history of the sample [Wolber and Hudson, 1982].

We associate this residual long-lifetime species with parinaric acid chains in an environment that is characterized as "solidlike." This association is based on the fact that parinaric acid in the low-temperature solid gel phase exhibits a long lifetime. Parinaric acid is able to sense the presence of this "phase" because of its sensitivity to its local environment. We can confirm the nature of the probe environment by measuring the amplitude of this long-lifetime component as a function of the excitation wavelength. The solidlike environment, because of its higher density, should exhibit a relative red shift

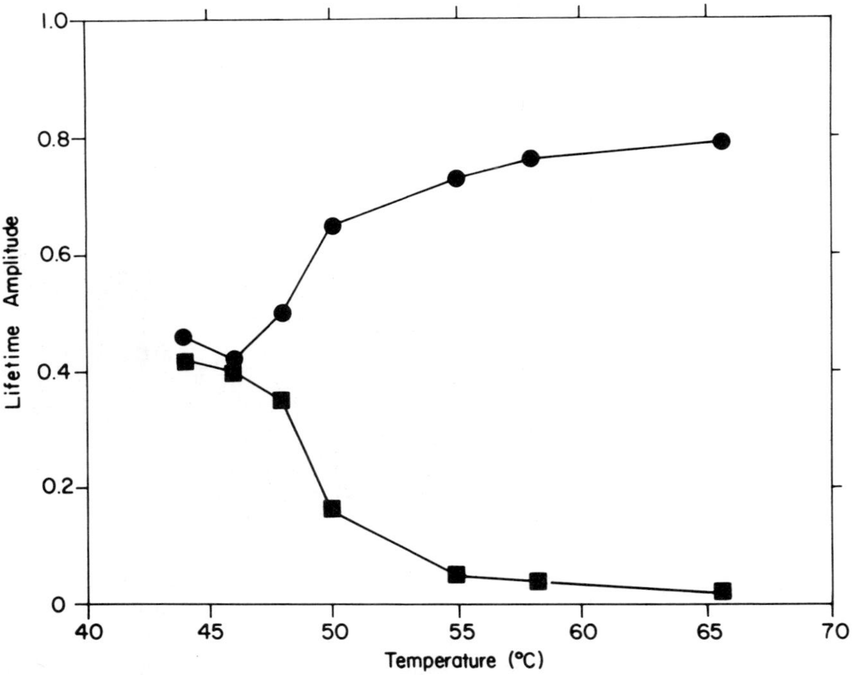


Fig. 5. Variation in the amplitudes of the long- (■) and intermediate-(●) lifetime component amplitudes of *trans*-parinaric acid in DPPC vesicles as a function of temperature. A very-short-lifetime component has an essentially constant amplitude of about 8% [Ruggiero et al., 1984; Ruggiero and Hudson, unpublished].

in its excitation maximum as discussed above. Thus, variation of the excitation wavelength should result in enhancement of the long-lifetime component with a characteristic spectral signature shifted to the red of the overall absorption or fluorescence excitation. This is indeed observed. The fact that differential excitation is observed demonstrates that these solid-like environments have a characteristic lifetime that is long compared to the fluorescence lifetime of several nanoseconds.

We also expect that this long-lived species, if it is indeed reflecting a solidlike environment, should have a high limiting anisotropy compared to the fluid environment with a shorter lifetime. This should result in upward curvature for the anisotropy decay at long times. This is shown in Figure 6 for the case of parinaric acid in DEPC (dielaidoylphosphatidylcholine) vesicles. The limiting anisotropy observed depends on the excitation wavelength in the fashion expected. Thus the environment sensed by parinaric acid giving rise to a long lifetime has a red-shifted absorption and a high asymptotic anisotropy, both characteristic of an ordered, high-density, solidlike phase.

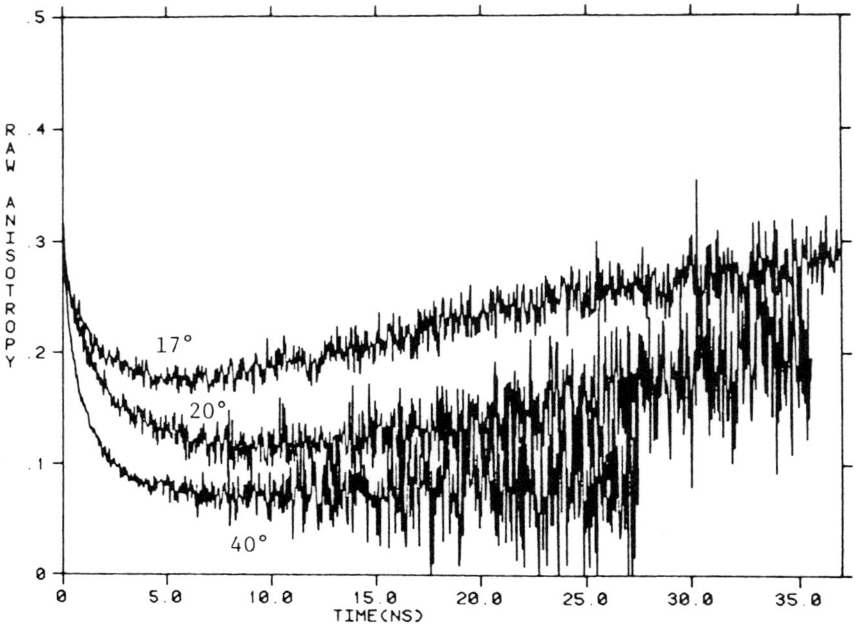


Fig. 6. The time dependence of the anisotropy of *trans*-parinaric acid in dielaidoylphosphatidylcholine (DEPC) vesicles at the indicated temperatures. This lipid has a thermal-phase transition at about 10°C. Excitation is at 302 nm and emission is at 420 nm [Ruggiero and Hudson, unpublished].

Addition of cholesterol to phospholipid bilayers increases the amplitude of this long-lifetime component indicating that this sterol somehow stabilizes the solidlike environment (Fig. 7). The effect on the anisotropy behavior is particularly dramatic (Fig. 8). At low cholesterol levels a large increase in the anisotropy is observed at long times. This is interpreted as being due to an environment with a long lifetime and a very high asymptotic anisotropy. At high cholesterol levels all of the bilayer is converted into this solidlike phase.

Our present interpretation of all of these results is that phospholipid bilayers exhibit behavior similar to that seen for critical systems where long-range correlations give rise to transient fluctuations in the order parameter. This has been predicted on theoretical grounds [Jahnig, 1981a–c] and is consistent with the results of acoustic absorption measurements [Sakanishi et al., 1979; Mitaku and Okano, 1981; Mitaku and Date, 1982; Mitaku et al., 1983]. The lifetime of these critical density fluctuations has been estimated from the resonance behavior of the acoustic absorption to be on the order of 30 ns for temperatures slightly above the main chain melting transition. This is consistent with the fluorescence results. In particular, a lifetime for these

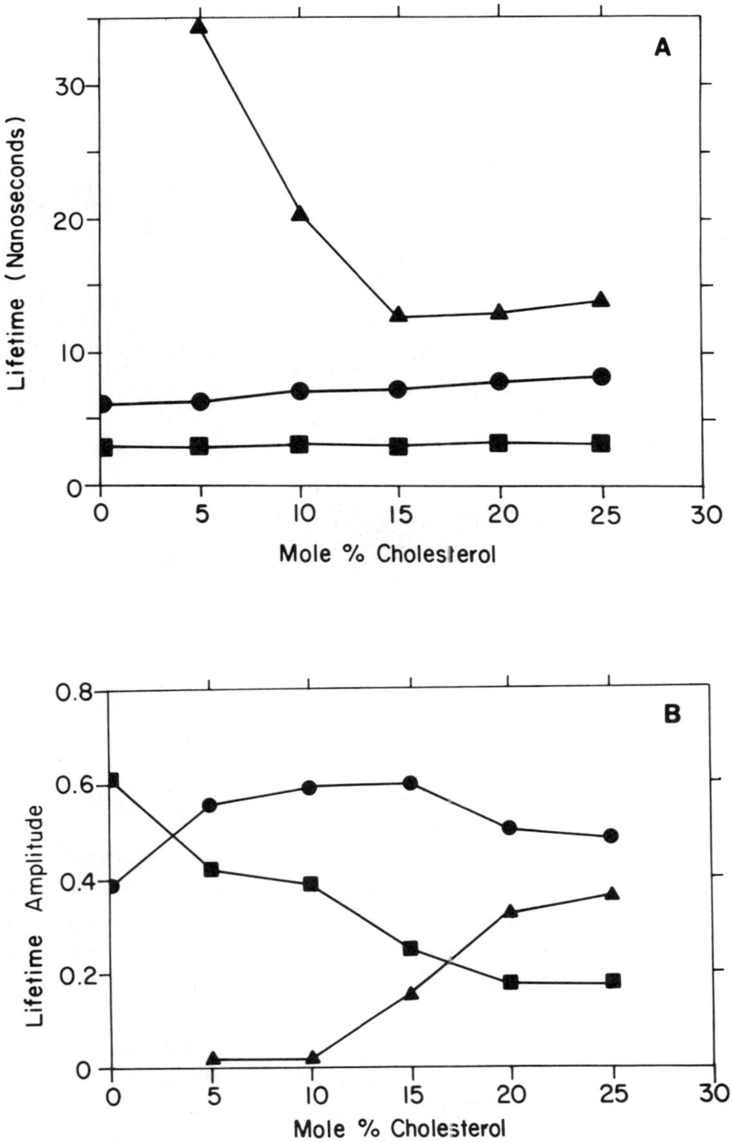


Fig. 7. Variation in the lifetimes (A) and amplitudes (B) of the components of the fluorescence decay of *trans*-parinaric acid as a function of cholesterol content in egg phosphatidylcholine vesicles. The symbols refer to the long- (▲), intermediate- (●), and short- (■) lifetime or component amplitude in each case. The temperature is 23°C. The long-lifetime component is not observed in the absence of cholesterol [Ludescher, 1984b; Ruggiero et al., 1984; Ludescher and Hudson, unpublished].

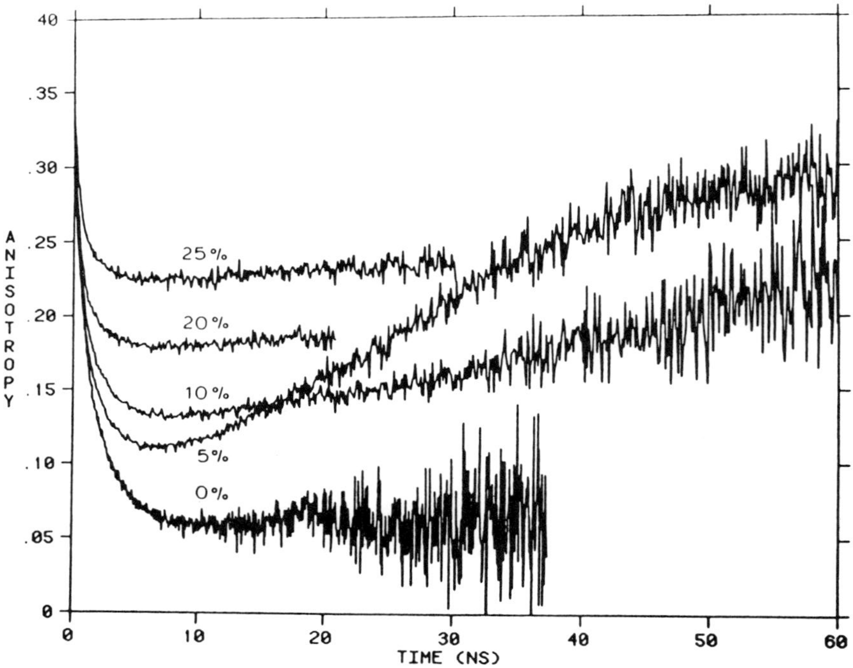


Fig. 8. The anisotropy decay of the fluorescence of *trans*-parinaric acid in egg PC vesicles at several cholesterol mole fractions. The sample temperature is 23°C [from Ludescher, 1984b; Ruggiero et al., 1984; Ludescher and Hudson, unpublished].

domains in this time regime would not permit lateral equilibration of the probe species between the fluid and solidlike regions. This is consistent with the lack of preferential partitioning of *cis* and *trans* parinaric acid between the two regions. We view the bilayer as a flickering structure with transient alignment and packing of the chains trapping the probe species in that environment and then dissipating to reform the fluid domain on a time scale somewhat longer than the fluorescence lifetime but shorter than the time required to establish lateral spatial equilibrium of the probe. Cholesterol must stabilize these high-density regions. A simple model for this stabilization consistent with very recent monolayer results [Weis and McConnell, 1984, 1985; Fischer et al., 1984] consists of cholesterol binding to the interface between the two structural types, orienting its planar surface [Bloch, 1983] toward the solidlike domains and its methylated face toward the fluidlike domains.

The fluorescence decay observed for parinaric acid in solution is also not describable by a single exponential decay [Parasassi et al., 1984; Ruggiero and Hudson, unpublished]. Two, or more likely, three components are needed to describe the emission. As an example, the decay observed at 20°C in hexane solution consists of 10% of a 0.8-ns component, 30% at 2.5 ns, and 60% at 5.5 ns. Similar behavior is observed for 1,3,5,7-octatetraene. It must be emphasized that this apparent decomposition of the decay into discrete components may not reflect the true description, which may, instead, correspond to a continuum of decay rates. In recent studies we have shown that the mechanism of this nonexponential behavior in solution is quite distinct from that observed in bilayers. In particular, variation of the excitation wavelength does not result in changes in the observed decay. This indicates that, in contrast to the bilayer case, the mechanism leading to the complex decay in solution is "homogeneous" and probably does not represent distinct transient environments. We believe this behavior in solution is due to a photochemical process leading to several emitting excited-state conformers that may thermally revert to the *all trans* form on returning to the ground state. The major point here is that distinctly different behavior can be observed for the same chromophore under different conditions. This does not necessarily mean that distinct physical processes are involved in different cases but only that their intrinsic rates are influenced by specific environments. In this example, it may be the case that the fluid environment of bilayers exhibits a complex homogeneous behavior similar to that of solutions but that the rate of the process leading to nonexponential decay behavior is slowed down in the solidlike environments *and* that these environments preexist with different excitation behavior due to a ground-state distribution that is sensitive to its environment.

## STUDIES OF PROTEIN DYNAMICS VIA TRYPTOPHAN ANISOTROPY DECAY

### General Properties of Tryptophan Fluorescence

Tryptophan is the dominant intrinsic fluorescent chromophore of most proteins. For this reason the emission of the indole side chain of this residue has been extensively studied [for recent reviews see Creed, 1984; Lumry and Hershberger, 1978; Lakowicz, 1983]. Proteins that contain single tryptophan residues exhibit fluorescence behavior that cannot be described by single exponential decays. The two lifetimes reported by Grinvald and Steinberg [1976] for proteins containing a single tryptophan fall in the range from 0.3 to 2.5 ns for the short lifetime and from 2.4 to 6.1 ns for the long lifetime. In a specific case discussed below it appears that three or even four exponential terms are needed to adequately describe the fluorescence decay. The origin of this complexity is not clear and the solution to this problem may be important in terms of the interpretation of the time dependence of the anisotropy of this residue. In particular, it is important to know if this behavior represents a heterogeneous or a homogeneous situation.

Indole and its derivatives exhibit complex photophysical behavior. The near-UV transition is due to two overlapping transitions, $^1L_a$ and $^1L_b$. The stronger $^1L_a$ transition is diffuse and broad and overlies the weaker $^1L_b$ transition [Konev, 1967; Yamamoto and Tanaka, 1972; Valeur and Weber, 1977]. Solvent polarity differentially shifts the $^1L_a$ energy due to its larger excited-state dipole moment [Strickland et al., 1970]. Dual fluorescence from both states occurs in some indole derivatives but apparently not in tryptophan [Andrews and Forster, 1974]. Stable complexes (exciplexes), which may or may not be fluorescent, can form between the excited state of the indole chromophore and many polar molecules [Walker et al., 1967; Lasser et al., 1977; Lumry and Hershberger, 1978]. There is also appreciable Rydberg character to these near-UV transitions [Lami, 1977] due to the mixing of higher-energy ionizable states with, probably, the $^1L_a$ state. This is related to the electron ejection that can occur when indole derivatives are excited to high vibrational levels of $S_1$ [Joschek and Grossweiner, 1966; Bent and Hayon, 1975]. Electron loss from vibrationally relaxed, electronically excited molecules can also occur due to either intra- or intermolecular charge transfer interactions with nearby electronegative groups [Feitelson, 1971]. Electron loss may be the predominate temperature-dependent nonradiative decay channel for tryptophan [Kirby and Steiner, 1970]. Charge transfer reactions in conjunction with ground-state rotamers [Szabo and Rayner, 1980] have recently been invoked to explain the double exponential decay of tryptophan and its derivatives [Chang et al., 1983; Petrich et al., 1983]. Solvent reorientation around the larger dipole moment of the $^1L_a$ excited state has a pronounced effect on the Stokes shift of indole as well as tryptophan fluorescence and may also influence the nonradiative decay rates of these fluorophores [Eisinger and Navon, 1969]. Ground-state heterogeneity, as well as mobility on the fluorescent time scale that changes the orientation and proximity of nearby chemical groups, may influence the relative contributions of all of these nonradiative pathways.

The spectral distribution, lifetime, and quantum yield of a tryptophan residue of a protein are fairly characteristic of its environment and may be used to classify the residue as buried, surface, or exposed [Burstein and Vedenkina, 1973]. In addition to the complex photophysics of indole, there are two other factors that must be considered when discussing the multiexponential decay of tryptophan in proteins [see Ross et al., 1981]. First, each lifetime species may represent a different ground-state conformer of the tryptophan. These conformers include different rotamers about the $C_\beta$-$C_\gamma$ bond of the tryptophan as well as different conformations of the rest of the protein. The second factor to consider is the influence of what may be loosely called excited-state reactions of the tryptophan. These include solvent or protein reorientation around the excited-state fluorophore, the formation of exciplexes, and reorientations of the chromophore during the excited-state lifetime that significantly change the environment of the indole ring.

The complexities of tryptophan fluorescence make the interpretation of fluorescence anisotropy decays somewhat ambiguous. If the decay is homogeneous then the weighting functions of equation 9 are the same for all proteins. In this case, the complex lifetime dependence, whatever its origin, can be simply factored and and observed anisotropy decay is the proper average for the sample. Alternatively, it may be the case that the anisotropy decay behavior for all proteins is the same despite differences in fluorescence lifetime. Again, the expression for the average anisotropy factors and the observed anisotropy is interpretable. Conceptual difficulty is only incurred when both the lifetime and the anisotropy decay behavior are distinct for subpopulations of the sample.

The $^1L_a$ and $^1L_b$ transitions of tryptophan are nearly orthogonal in polarization [Valeur and Weber, 1977] (see Fig. 9). The overlap of these two transitions leads to a strong variation in the value of $r_0$ with excitation wavelength. This, by itself, does not complicate the interpretation of fluorescence anisotropy experiments. However, there is a complication due to the presumably related fact that the limiting anisotropy observed even with the longest excitation wavelength is only about 0.3, rather than the theoretical value of 2/5. This means that the emission dipole is not colinear with the absorption dipole and leads to the need to average over both orientations. This becomes important when a molecular dynamics simulation is being compared to the results of experiment [Ichiye and Karplus, 1983].

### Strategies for Evaluation and Elimination of the Complexities of Tryptophan Fluorescence: Alternate Chromophores—Azulene and Benzo-b-thiophene

We have been investigating a chemical approach to eliminating the complexity of fluorescence decay behavior of the indole chromophore. The idea is to replace the indole chromophore by another similar ring system with structural and chemical properties that are as similar as possible to indole but whose spectroscopic properties are simpler, or at least different, from those of indole. Two such candidates have been identified: azulene and benzo-b-thiophene. The structures of these species are compared to that of indole in Figure 9. Preliminary characterization of the spectral properties of the chromophores of these species have been completed.

The choice of these two ring systems is based on the following rationale. The series indole/benzo-b-thiophene/azulene represents a progression in the separation between the $^1L_a$ and $^1L_b$ excited electronic states. In indole these

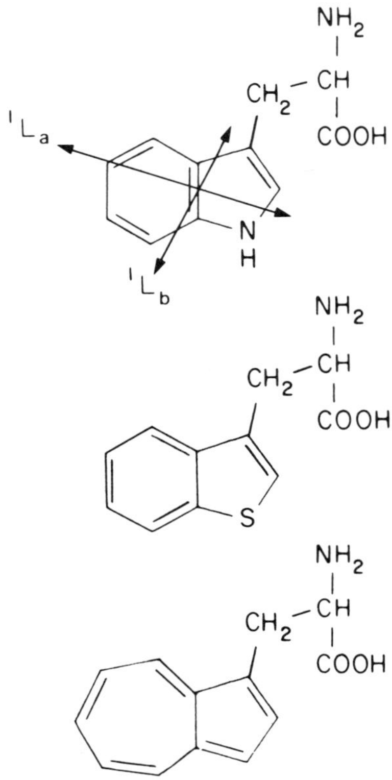


Fig. 9. The structures of tryptophan and two tryptophan analogs. These are chromophoric derivatives of alanine with the chromophore being (top to bottom) indole, benzo-b-thiophene and azulene. The transition dipole directions for the indole chromophore of tryptophan are indicated.

two states strongly overlap as discussed above. This may be one of the reasons for the complexity of tryptophan fluorescence decay and certainly leads to the dependence of $r_o$ on excitation wavelength. In benzo-b-thiophene these two states are clearly separated with the $^1L_b$ state being of lower energy. The value of $r_o$ determined for this chromophore in a rigid organic glass is close to the value of 2/5 as expected and decreases only at rather short excitation wavelengths because of the high energy of the $^1L_a$ transition. For azulene, these two states are widely separated, being, in fact, at opposite ends of the UV-visible spectrum. Here the $^1L_b$ state is again the lowest (near 700 nm) but the emitting state is the UV $^1L_a$ state ($S_2 \rightarrow S_0$ emission). This separation of these two states in these two chromophores, besides simplifying the polarization behavior, should eliminate any dual-emission contributions to the complexity of the fluorescence decay. Furthermore, in contrast to indole (or tryptophan), we have shown that the emission spectra of these chromophores are essentially invariant to changes in the polarity of the solvent. This should eliminate any contributions from dipolar relaxation to complex decay kinetics. Another rationale for the choice of these chromophores is that the ionization potential in each case is significantly higher than that for indole (presumably due to lack of low-energy Rydberg-like excitations of the ring nitrogen group of indole). This should eliminate, or at least reduce, the contributions to kinetic complexity originating from ionization and neighboring group quenching effects.

There remains the difficulty of effecting the replacement of tryptophan by one of these chromophores. Attempts at biosynthetic incorporation of the azulene analog into an *E. coli* tryptophan auxotroph have been unsuccessful thus far. It is possible that selective mutational procedures can be devised that will result in success along these lines. Also, it may be possible to find conditions where the provision of both tryptophan and one of the analogs in the growth medium will result in partial substitution of some specific protein of interest. Such methods have been used for the incorporation of fluorotryptophan derivatives. The fluorescence experiments can make use of selective excitation to examine only the modified proteins. For some proteins, including phospholipase $A_2$ discussed below, it is possible to perform semisynthetic modification to incorporate either of these species [Slotboom and de Haas, 1975; Slotboom et al., 1978; Jansen, 1979; van Scharrenberg et al., 1982].

There still remains the issue of the effect of such ring modification on the dynamics of the protein. In particular, these indole analogs do not contain the N-H group and therefore cannot act as hydrogen bond donors. The influence of N-H hydrogen bonding on the ring dynamics can be examined, as discussed below, by comparison of tryptophan dynamics for proteins that differ in the presence and absence of indole N-H hydrogen bond acceptor groups. Hydrogen bonding of the indole N-H group may not have a large effect on the anisotropy decay of tryptophan. The reason for this is that a major reorientation motion for the emission dipole of tryptophan is motion about the $C_\beta$, $C_\gamma$ bond. This motion is important because the emitting transition dipole is roughly perpendicular to this bond (Fig. 9). This motion preserves the orientation of the N-H bond vector and therefore the hydrogen bond can be maintained.

If the anisotropy decay behavior for one of these indole analogs in a protein differs significantly from that observed for tryptophan, this could be due to structural differences or to the fact that the indole measurements are complicated by the photophysical behavior discussed above. One of the objectives of a detailed interpretation of anisotropy decay behavior for tryptophan residues in proteins is a testing of molecular dynamics calculations. Therefore, if a significant difference is observed for different chromophores, a dynamics calculation should be able to simulate this effect if it is structural in origin. The idea here is that these simulation methods should be applicable not just to the native protein, but to any "perturbed" system as well.

The azulene analog of tryptophan has several other interesting properties for fluorescence studies. For example, to the extent that it is a good analog of tryptophan, it can be used to measure tryptophan binding equilibria in cases where tryptophan itself does provide a discernible spectroscopic signal. We have used this tryptophan analog to measure binding to serum albumin where the intrinsic tryptophan fluorescence of the protein obscures that from the amino acid itself. Another intriguing feature of this analog stems from its $S_2 \rightarrow S_0$ emission and the fact that its $S_0 \rightarrow S_1$ absorption is at low energy. This permits energy transfer experiments where the azulene ring acts as a donor to some other fluorescent species that, in turn, can act as an energy donor back to azulene. This double transfer experiment for a particular pair of chromophores decreases, but does not entirely eliminate, the ambiguity of energy transfer distance measurements due to the orientation dependent term in the transfer expression.

## Phospholipase $A_2$, Its Zymogen, and Its Complexes With Lipids

We now discuss two specific protein systems, phospholipase $A_2$ and T4 phage lysozyme, in terms of the fluorescence and anisotropy decay behavior of their tryptophan residues. Pancreatic phospholipase $A_2$ (phosphatide acyl-hydrolase EC 3.1.1.4) catalyzes the stereospecific hydrolysis of the 2-acyl ester bond of 3-sn-phosphoglycerides. The pancreatic enzyme is synthesized in a precursor form, prophospholipase $A_2$, of molecular weight 14,600 daltons. The zymogen is converted to the fully active 13,800-dalton form by proteolytic cleavage of a seven amino acid fragment at the N-terminus. Prophospholipase and the processed form both hydrolyze short-chain monomeric phospholipids with comparable efficiency. When phospholipids are present in an aggregated micellar form, however, the activity of the enzyme is increased several hundred-fold. The proenzyme does not exhibit this dramatic increase in activity toward "organized" substrates. This enhancement of activity for the enzyme is believed to be due to a conformational change of the protein upon binding to the micelle due to a hydrophobic "interfacial recognition site" distinct from the active site [Verger et al., 1973; Brockerhoff, 1973; Pieterson et al., 1974].

Phospholipase $A_2$ has a single tryptophan residue at position 3 in the processed enzyme (position 10 in the zymogen). This tryptophan residue is believed to be part of the interfacial recognition site. The basic evidence for this conclusion is that semisynthetic modification of the N-terminal region with replacement of the tryptophan by an alanine removes the enhancement of the reaction rate observed at the critical micelle concentration [Slotboom and de Haas, 1975; Slotboom et al., 1978; Jansen, 1979]. There are considerable changes in the absorption and fluorescence properties of the tryptophan residue upon binding to micellar substrates and analogs [van Dam-Mieras et al., 1975; Hille et al., 1981; Ludescher et al., 1985]. These spectral effects are consistent with transfer of the tryptophan from an aqueous to a hydrophobic environment [Burstein and Vendenkina, 1973]. The micelle formed by the enzyme with the nonhydrolyzable substrate analog n-hexadecylphosphocholine has a well-defined structure with two proteins and 80 lipid monomers [van Dam-Mieras et al., 1975]. The pro-enzyme does not bind to these micelles.

Some other interesting features of this enzyme are the presence of seven disulfide bonds, making it a very thermally stable protein, a "charge relay" histidine/glutamate structure in the active site, the availability of a variety of different forms due to species variation, and the ability to chemically modify the region near the N-terminus. The crystal structure of the bovine form has been refined to 1.7-Å resolution [Dijkstra et al., 1978, 1981a,b]. This crystal structure shows that the N-terminal alanine residue of the enzyme is anchored by hydrogen bonding. The crystal structure of the zymogen of the porcine enzyme show that the N-terminal region is disordered [Dijkstra et al., 1982]. These features, plus the involvement of the tryptophan in an interesting lipid-protein interaction and the availability of three forms (proenzyme, enzyme, and enzyme/micelle complex), make this a particularly interesting candidate for the study of tryptophan fluorescence and dynamics. The relatively small size of the enzyme and its probable rigidity due to the disulfide bond network make it an attractive candidate for comparison of the results of molecular dynamics simulations with experiment.

The tryptophan residue at position 3 is partially exposed to the aqueous environment and is hydrogen bonded in a bifurcated bond to an asparagine and a water molecule. The tryptophan is also stacked on an adjacent glutamine residue. Replacement of this invariant glutamine [van Scharrenburg et al., 1982] results in a considerable decrease in the fluorescence quantum yield of the protein. This is due to a conformational change because of the involvement of this residue in anchoring the N-terminal region. Model-building studies of this tryptophan site discussed below show that there is considerable conformational freedom for rotation about the tryptophan beta-gamma bond even without displacement of the protein structure from its equilibrium position and without breaking the hydrogen bond formed by the indole N-H group.

We have recently completed studies of the decay of the fluorescence and the anisotropy of porcine pancreatic phospholipase $A_2$, its proenzyme, and a micelle complex [Ludescher et al., 1985, 1986a]. An analysis of a fluores-

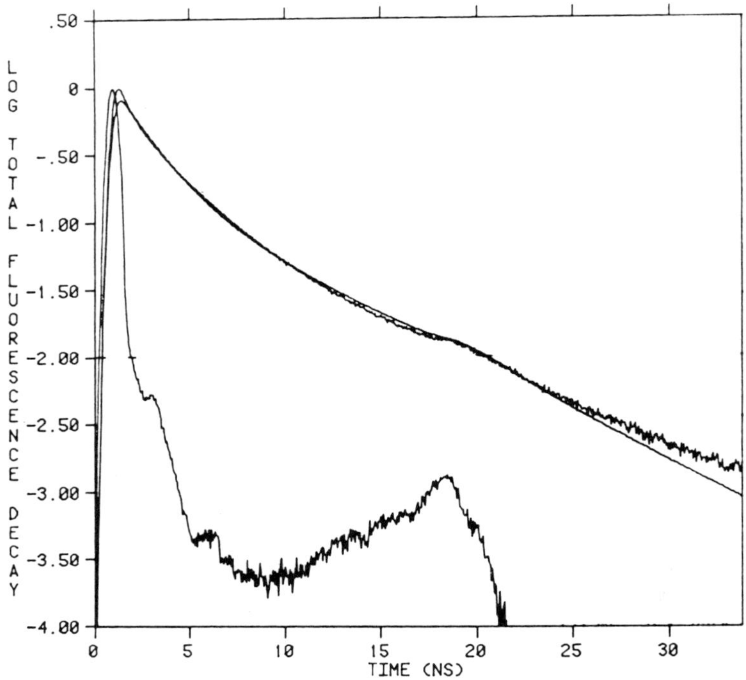


Fig. 10. A logarithmic plot (base 10) of the decay of the fluorescence intensity of prophospholipase $A_2$ at 20°C. The instrument response function is shown at the bottom. The smooth curve drawn through the data is the best two-exponential decay that can be constructed. It includes a scatter subtraction and a shift in the instrument response due to wavelength variation as variables. The residual deviation demonstrates that this decay requires at least three exponential components [from Ludescher et al., 1985].

cence decay using an optimized two-exponential fit is shown in Figure 10. This figure is intended to demonstrate that a two-exponential fit to the data is inadequate. Three or perhaps four components are needed for an adequate fit. The corresponding extracted decay parameters are given in Tables 1 and 2. These fluorescence decays are clearly complex.

Table 1 is a test of the independence of the results of these multicomponent decays to changes in the numerical analysis method. Duplicate experiments are also presented. This is intended primarily for experts in this field. Also included in Table 1 is a comparison of the results of our pulsed fluorescence measurements with recent multifrequency phase-modulation methods presented at this meeting. The agreement is excellent and demonstrates the ability of both techniques to extract multiexponential decay behavior. This result is particularly significant because the artifactual errors of these two methods are certainly quite different.

**TABLE 1. Comparison of the Results of Three- and Four-Component Analyses for Phospholipase $A_2$ at 20° Using Different Analysis[1] Methods (All lifetimes in ns)**

| | 3-component analysis | | | | | | |
|---|---|---|---|---|---|---|---|
| Method | $a_1$ | $\tau_1$ | $a_2$ | $\tau_2$ | $a_3$ | $\tau_3$ | $\chi^2$ |
| All data[2] | 0.071 | 6.93 | 0.382 | 2.49 | 0.546 | 0.512 | 11.5 |
| From peak[2] | 0.075 | 6.70 | 0.377 | 2.34 | 0.549 | 0.422 | 4.8 |
| From peak[2,4] | 0.077 | 6.71 | 0.394 | 2.56 | 0.529 | 0.606 | 25.0 |
| From peak[3] | 0.072 | 7.13 | 0.416 | 2.86 | 0.522 | 0.623 | 2.6 |
| 10% peak[2] | 0.073 | 7.20 | 0.411 | 2.67 | 0.516 | 0.684 | 1.7 |
| 10% peak[2,4] | 0.079 | 6.84 | 0.410 | 2.66 | 0.511 | 0.712 | 2.7 |
| 10% peak[3] | 0.064 | 7.56 | 0.416 | 2.86 | 0.520 | 0.774 | 1.4 |
| MOM[5] | 0.074 | 7.29 | 0.412 | 2.82 | 0.514 | 0.834 | |
| Phase[6] | 0.089 | 6.95 | 0.395 | 2.79 | 0.515 | 0.704 | |

| | 4-component analysis | | | | | | | | |
|---|---|---|---|---|---|---|---|---|---|
| Method | $a_1$ | $\tau_1$ | $a_2$ | $\tau_2$ | $a_3$ | $\tau_3$ | $a_4$ | $\tau_4$ | $\chi^2$ |
| From peak[2] | 0.039 | 7.83 | 0.250 | 3.09 | 0.286 | 1.07 | 0.425 | 0.241 | 2.1 |
| From peak[2,4] | 0.039 | 7.86 | 0.197 | 3.68 | 0.292 | 1.86 | 0.472 | 0.542 | 24.1 |
| 10% peak[2] | 0.043 | 7.98 | 0.275 | 3.22 | 0.304 | 1.26 | 0.378 | 0.378 | 1.2 |
| 10% peak[2,4] | 0.056 | 7.10 | 0.308 | 2.86 | 0.351 | 0.90 | 0.285 | 0.231 | 2.0 |

[1]"All data" means that the entire data set (1,024 channels minus an initial background region) were used in the calculation of $\chi^2$ in a least-squares iterative reconvolution calculation. "From peak" means that only the data from the peak of the excitation pulse were used in the calculation of $\chi^2$; "10% peak" means that only the data from the point where the excitation pulse had fallen to 10% of its original value were used in the calculation of $\chi^2$. (All data points are used in the calculation of the convolution integral in all cases.)

[2]No scatter subtraction was included.

[3]A variable scatter subtraction factor was included in the analysis.

[4]A duplicate experiment.

[5]Analysis of the original data set (all points) by using the method-of-moments (MOM) with moment index displacement MD 2 and lambda supression of 0.36 [see Isenberg 1973a,b; Isenberg et al., 1973; Small and Isenberg, 1976, 1977; and Small et al., 1984].

[6]A third independent experiment performed by using phase modulation methods [E. Gratton, private communication].

Table 2 presents the temperature dependence of the fluorescence decays and compares the enzyme and the zymogen. The mean lifetime and the standard deviation of the lifetime distribution do not depend significantly on whether a three- or four-component analysis is used. These characteristics of the distribution decrease smoothly with increasing temperature and the values for the zymogen are clearly different from those of the enzyme. The zymogen with its disordered tail and more exposed tryptophan has a longer lifetime and a greater width to its lifetime distribution than does the enzyme with its anchored N-terminal structure.

The decays of the fluorescence anisotropy for phospholipase $A_2$, its zymogen, and its complex with a nonhydrolyzable substrate analog are shown in Figures 11 and 12; the extracted parameters are given in Table 3 for the enzyme and its zymogen. Table 4 presents the results for the complex. This complex has also been studied with parinaric acid. In each case the anisotropy decays require two or three temporal components for adequate fits. We associate the longest correlation time component with overall particle rotation in each case. This is confirmed by the dependence of this correlation time on the solution viscosity (Fig. 12) and by the expected larger value observed for the micelle complex. The values of these rotational correlation times for the motions of the proteins are somewhat longer (by about a factor of two) than expected on the basis of Stoke's law for an unhydrated, spherical protein but this can be explained on the basis of either hydration or aspherical geometry. On the other hand, the observed rotational correlation times for the micelle complex are considerably smaller than expected on the basis of the known size of the particle. This is also observed for the longest rotational correlation time for micelles labeled with parinaric acid. A possible mechanism for this enhanced rotational diffusion for the lipidic particles is that there is a rapid motion due to translational diffusion that appears to be isotropic because of the small size and small radius of curvature of these particles.

**TABLE 2. Decay Components, Average Lifetimes, and Lifetime Distribution Standard Deviations for $proPLA_2$ and $PLA_2$ (All lifetimes in ns)[1]**

| Protein | Temperature (°C) | $a_1$ | $\tau_1$ | $a_2$ | $\tau_2$ | $a_3$ | $\tau_3$ | $a_4$ | $\tau_4$ | $\langle\tau\rangle$ | $S^2$ |
|---|---|---|---|---|---|---|---|---|---|---|---|
| 3-component analysis | | | | | | | | | | | |
| $proPLA_2$ | 8 | 0.302 | 9.45 | 0.328 | 4.52 | 0.370 | 1.44 | | | 4.87 | 3.3 |
| $proPLA_2$ | 20 | 0.311 | 7.78 | 0.357 | 3.14 | 0.331 | 0.92 | | | 3.85 | 2.8 |
| $proPLA_2$ | 40 | 0.054 | 5.61 | 0.419 | 2.03 | 0.527 | 0.60 | | | 1.47 | 1.2 |
| $PLA_2$ | 8 | 0.065 | 8.44 | 0.396 | 3.54 | 0.539 | 0.91 | | | 2.44 | 2.0 |
| $PLA_2$ | 20 | 0.073 | 7.20 | 0.411 | 2.68 | 0.516 | 0.68 | | | 1.98 | 1.7 |
| $PLA_2$ | 40 | 0.055 | 5.64 | 0.422 | 2.06 | 0.524 | 0.64 | | | 1.51 | 1.2 |
| 4-component analysis | | | | | | | | | | | |
| $proPLA_2$ | 8 | 0.220 | 9.90 | 0.258 | 6.10 | 0.320 | 2.42 | 0.202 | 0.85 | 4.70 | 3.3 |
| $proPLA_2$ | 20 | 0.239 | 8.12 | 0.264 | 4.11 | 0.303 | 1.52 | 0.195 | 0.40 | 3.56 | 2.9 |
| $proPLA_2$ | 40 | 0.029 | 6.31 | 0.201 | 2.69 | 0.360 | 1.24 | 0.411 | 0.36 | 1.31 | 1.2 |
| $PLA_2$ | 8 | 0.022 | 10.37 | 0.214 | 4.99 | 0.345 | 2.16 | 0.420 | 0.63 | 2.30 | 2.0 |
| $PLA_2$ | 20 | 0.043 | 7.98 | 0.275 | 3.22 | 0.304 | 1.26 | 0.378 | 0.38 | 1.76 | 1.8 |
| $PLA_2$ | 40 | 0.029 | 6.33 | 0.200 | 2.73 | 0.373 | 1.28 | 0.397 | 0.39 | 1.37 | 1.2 |

[1] $proPLA_2$, prophospholipase; $PLA_2$, phospholipase $A_2$.
[2] The standard deviation of the lifetime distribution.

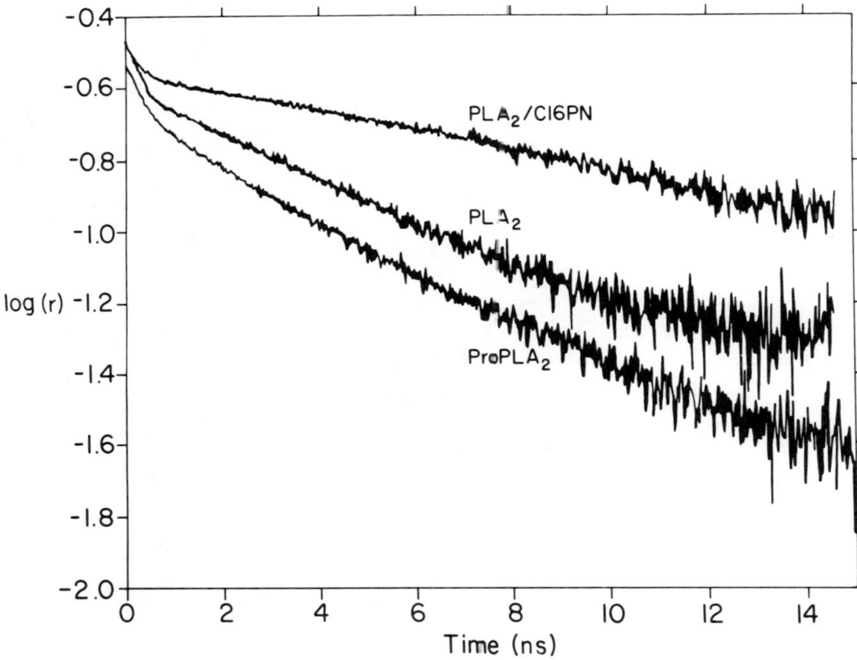


Fig. 11. A logarithmic plot (base 10) of the undeconvoluted anisotropy decays of the fluorescence of pro-phospholipase (pro-$PLA_2$), phospholipase $A_2$ ($PLA_2$), and the $PLA_2$ complex with hexadecylphosphocholine ($PLA_2/C_{16}PN$) at 20°C [from Ludescher, 1984b; Ludescher et al., 1986a].

**TABLE 3. Fit Anisotropy Decay Parameters for Trp-3 in pro$PLA_2$ and $PLA_2$ at 8, 20, and 40°C[1] (All relaxation times in ns)**

| Sample | Temperature (°C) | $r_1$ | $\phi_1$ | $r_2$ | $\phi_2$ | $r_3$ | $\phi_3$ | $r_o$ |
|---|---|---|---|---|---|---|---|---|
| pro$PLA_2$ | 8 | [0.06][2] | 0.04 (7) | 0.030 (3) | 1.7 (3) | 0.208 (4) | 8.4 | — |
| | 20 | 0.06 (12) | 0.2 (2) | 0 08 | 2.7 (4) | 0.16 | 8.2 (4) | 0.30 |
| | 40 | 0.08 (16) | 0.1 (3) | 0.03 (3) | 0.6 (4) | 0.19 | 3.4 | 0.30 |
| $PLA_2$ | 8 | 0.1 (4) | 0.1 (2) | | | 0.247 | 9.95 (8) | 0.35 |
| | 20 | [0.057][3] | 0.18 (4) | | | 0.243 | 6.97 (4) | — |
| | 40 | 0.07 (11) | 0.1 | | | 0.222 | 3.32 (2) | 0.29 |

[1]Excitation at 300 nm, emission collected through a broad-band filter (310–390-nm transmission, maximum at 350 nm). The values in parentheses are standard deviations in the last digits(s), i.e., 0.030(3) is equivalent to 0.030 ± 0.003. If no figures are indicated the error is one unit in the last digit. These errors, which are calculated by the fitting program, are statistical measures of the correlation between the parameters and are therefore only approximate. The values in brackets are obtained by subtracting the sum of the more accurately determined values of $r_1$ from 0.3, the anisotropy of tryptophan with 300-nm excitation in the limit of no motion. The actual fit values of $r_1$ for these cases are clearly too large: [2]$r_1$ = 0.5(9); [3]$r_1$ = 0.5(1).

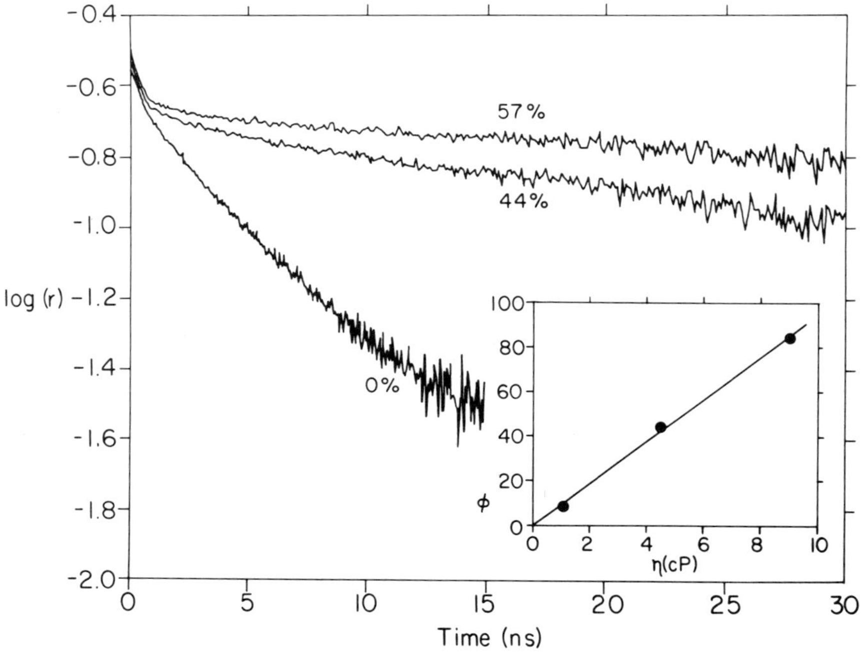


Fig. 12. The effect of glycerol on the anisotropy decay of the fluorescence of pro-phospholipase $A_2$. The inset shows a plot of the long-time rotational correlation time as a function of viscosity [from Ludescher, 1984b; Ludescher et al., 1986a].

The limiting anisotropy at zero time obtained in these experiments is of interest. For the tryptophan anisotropies this value is quite close to the value of about 0.3 expected on the basis of measurements for tryptophan in rigid media. This indicates that no rapid motions with significant amplitudes are being missed in the data collection or analysis. Besides indicating full inclusion of rapid motional components, this agreement between the extrapolated zero time and irrotational anisotropy values indicates the validity of the data analysis methods.

We now turn to a discussion of the rapid internal motions revealed by these measurements. The rotational correlation times for these internal motions appear to be on the order of 100 ps but the particular values are very tentative. The amplitudes of these motions are of greater interest at the present time. The amplitude of the shortest time-scale motion is roughly 0.06 for both phospholipase and its zymogen. By using the observed values of the zero time anisotropy, this corresponds to a root mean square angular motion of about $\pm$ 15 degrees. This value is consistent with examination of molecular models and rotation of the indole ring about the beta-gamma bond until

**TABLE 4. Fit Anisotropy Decay Parameters for Trp-3 in $PLA_2/C_{16}PN$ and tPnA in $C_{16}PN$ Micelles at 8, 20, and 40°C[1] (All relaxation times in ns)**

| Sample | Temperature (°C) | $r_1$ | $\phi_1$ | $r_2$ | $\phi_2$ | $r_3$ | $\phi_3$ | $r_o$ |
|---|---|---|---|---|---|---|---|---|
| $PLA_2/C_{16}PN$ | 8 | 0.02 (4) | 0.2 (2) | | | 0.286 | 27.6 (3) | 0.31 |
| | 20 | [0.024][2] | 0.07 (7) | | | 0.276 | 16.1 | — |
| | 40 | 0.02 (56) | 0.04 (90) | | | 0.268 | [6.8][3] | 0.29 |
| $tPnA/C_{16}PN$ | 8 | 0.133 | 0.24 | 0.06 | 1.25 | 0.204 | 29.6 | 0.40[4] |
| | 20 | 0.151 | 0.15 | 0.08 | 0.74 | 0.171 | 15.4 | 0.40 |
| | 40 | | | 0.18[5] | 0.29 | 0.151 | 6.7 | 0.33 |

[1]Tryptophan excitation at 300 nm, emission collected through a broad-band filter (310–390 nm, maximum at 350 nm). Parinaric acid excitation at 300 nm, emission collected through a Corning O-52 long-pass filter (transmission cutoff at 350 nm). The values in parentheses are standard deviations in the last digits(s). If no figures are indicated the error is one unit in the last digit. These errors, which are calculated by the fitting program, are statistical measures of the correlation between the parameters and are therefore only approximate. $PLA_2$ = phospholipase $A_2$; $C_{16}PN$ = N-hexadecylphosphocholine; tPnA = *trans*-parinaric acid.

[2]This value is an estimate of the amplitude of the rapid component determined by subtracting the more accurately determined value of $r_3$ from 0.3. The fit value of $r_1 = 0.1(2)$.

[3]This value is an underestimate because of downward curvature in the anisotropy decay.

[4]The $r_o$ of tPnA is 0.4.

[5]Least-squares iterative reconvolution analysis only finds two exponentials. The low value of $r_o$ indicates that the fast decay is not being resolved.

positions are reached where van der Waals radii distances are reached. This particular reorientation has a large effect on the anisotropy because the beta-gamma bond is roughly perpendicular to the emission transition dipole. As noted above, this bond is nearly colinear with the indole N-H bond so that hydrogen bonding is maintained throughout the motion. The lack of apparent temperature dependence of the amplitude of this motion is consistent with motion within a rigid protein and this is, in turn, consistent with the disulfide and salt bridge structure of the protein.

The rapid internal motion of the tryptophan residue of the zymogen appears to be different from that of the enzyme. The analysis of Table 3 is in terms of an additional motional component with an amplitude on the order of 0.05 and a time scale from 0.5 to 3 ns. An alternative two-component analysis combines the two rapid components resulting in an overall amplitude of 0.09–0.14 with an average time scale of about 1 ns. This corresponds to a root mean square angular excursion of $\pm$ 19–25°. In either analysis it is clear that there is an increased motional freedom in the proenzyme with a somewhat longer scale than that seen for the enzyme. This result is consistent with the structural observation that the amino-terminal segment of the proenzyme is disordered in the crystal structure [Dijkstra et al., 1982].

The rapid internal motion of the tryptophan residue of the enzyme is greatly restricted upon formation of the micelle. Here the amplitude of the fast motion is only 0.02 corresponding to an angle of ± 9°. This restriction of the tryptophan motion is consistent with location of this residue in the micelle structure in close contact with the lipid components.

In conclusion, the anisotropy decay behavior observed for this enzyme, the zymogen, and the lipid complex makes sense on structural grounds, indicating that the measurements are substantially correct and are not severely distorted by the photophysical complexity of tryptophan discussed above.

## T4 Phage Lysozyme: A Protein Engineered for Fluorescence Studies

The lysozyme of T4 phage is an extensively studied enzyme with a number of properties that make it an ideal candidate for the detailed study of the internal motions of proteins. The crystal structure of this 18,700-dalton protein has been determined to 1.7-Å resolution [Matthews and Remington, 1974; Remington et al., 1978; Weaver and Matthews, 1985]. The overall structure of wild-type T4 lysozyme is roughly ellipsoidal with a diameter of about 3 nm and a length of about 5 nm. An active site cleft separates the ellipse into two distinct lobes. Numerous mutants have been isolated or designed and characterized in terms of their thermal stability [Ellwell and Schellman, 1975, 1977, 1979; Schellman et al., 1981]. Five of these mutants have had their crystal structures determined [Alber et al., 1985; Grutter et al., 1979, 1983]. The wild-type protein contains three tryptophan residues at positions 126, 138, and 158. Two of these residues (126 and 158) are at the protein surface and one, Trp 138, is a buried residue near the active site of the protein. This multiplicity of fluorescent groups adds considerable complexity to the interpretation of fluorescence and anisotropy decay behavior because of the overlap of the signals from each group. However, all of the modified forms where any or all of the Trp residues have been replaced by Tyr groups have been prepared and are being studied by using fluorescence methods. These same species and others containing additional substitutions are also being characterized in terms of thermal stability, isotope exchange rate at the indole N-H group, and crystallography.

We propose to use the fluorescence anisotropy behavior of these T4 lysozyme forms to test molecular mechanics simulation calculations of protein dynamics. The anisotropy decay can be calculated from these simulations [Ichiye and Karplus, 1983]. Direct comparison can be made with experiment by convoluting the calculated anisotropy with the instrument response function. This avoids numerical deconvolution of the experimental anisotropy data. The Trp 138 residue is the most interesting from this point of view because its movement requires significant fluctuations of the entire protein structure.

**TABLE 5. Fluorescence Decay of T4 Lysozyme (W126Y/W158Y) Containing Only Trp 138: Excitation at 300 nm, Emission at 340 nm (All lifetimes are in ns)**

| Temperature (°C) | $a_1$ | $\tau_1$ | $a_2$ | $\tau_2$ | $\langle \tau \rangle$ | S |
|---|---|---|---|---|---|---|
| pH 6.1 (native form) | | | | | | |
| 10 | 0.86 | 1.49 | 0.14 | 3.76 | 1.81 | 1.07 |
| 20 | 0.85 | 1.34 | 0.15 | 3.56 | 1.67 | 0.79 |
| 25 | 0.85 | 1.23 | 0.15 | 3.39 | 1.55 | 0.77 |
| 30[1] | 0.80 | 0.93 | 0.19 | 2.69 | 1.25 | 0.69 |
| 40[1] | 0.74 | 0.79 | 0.25 | 2.38 | 1.18 | 0.69 |
| 50[1] | 0.69 | 0.65 | 0.31 | 2.05 | 1.08 | 0.64 |

| Temperature (°C) | $a_1$ | $\tau_1$ | $a_2$ | $\tau_2$ | $a_3$ | $\tau_3$ | $\langle \tau \rangle$ | S |
|---|---|---|---|---|---|---|---|---|
| pH 2, 3 M guanidinium HCl (denatured form) | | | | | | | | |
| 20 | 0.35 | 0.55 | 0.50 | 2.2 | 0.17 | 4.36 | 2.04 | 1.31 |
| 30 | 0.32 | 0.49 | 0.55 | 1.9 | 0.13 | 3.62 | 1.64 | 0.98 |

[1]A very small component with a lifetime of 7 ns is also observed at these higher temperatures.

Fluorescence experiments have been performed on the modified forms of T4 phage lysozyme (W126Y/W158Y), containing Trp 138 as the only tryptophan, and (W138Y/W158Y), containing Trp 126 as the only tryptophan. In each case the other two Trp residues have been replaced by tyrosines. The results of these experiments for the Trp 138 protein are presented in Tables 5 and 6. The fluorescence decays for this Trp 138 form exhibit clear double exponential behavior with the major fraction being a short-lifetime component. The amplitudes, lifetime values, and the average lifetime all show a smooth variation with temperature. The decays seem to be well represented by two exponential terms but at the higher temperatures there may be a very small (0.3%) component at a longer lifetime. Denaturation at low pH and guanidine HCl results in a slight increase in the lifetime and a large increase in the width of the overall lifetime distribution indicating a wide range of environments.

The protein containing only Trp 126 has a very different fluorescence decay behavior from that of the Trp-138 residue. At 20°C the decay is given by 24% of a 0.84-ns component, 75% of a 2.4-ns component, and 1% of an 8-ns component. The major difference relative to the native forms at the same temperature is the very much smaller value of the very-short-lifetime component.

The anisotropy decay data show the overall particle tumbling in the 5–15-ns time range and a fast internal motion with a time constant on the order of 150–350 ps. The time constant for the overall motion is consistent with the

**TABLE 6. Decay of the Anisotropy of the Fluorescence of T4 Lysozyme (W126Y/W158Y): Excitation at 300 nm, Emission at 340 nm, pH 6.1 (All relaxation times are in ns)**

| Temperature (°C) | $r_1$ | $\phi_1$ | $r_2$ | $\phi_2$ | $r_o$ |
|---|---|---|---|---|---|
| 10 | | | 0.30 | 13.3 | 0.30 |
| 20 | 0.025 | 0.350 | 0.27 | 11.0 | 0.30 |
| 35 | 0.035 | 0.345 | 0.24 | 7.4 | 0.28 |
| 40 | 0.041 | 0.255 | 0.25 | 7.0 | 0.29 |
| 45 | 0.031 | 0.167 | 0.24 | 6.7 | 0.27 |
| 50 | 0.022 | 0.275 | 0.24 | 5.8 | 0.26 |

size and shape of the protein given a reasonable value of 10% hydration. The value of the anisotropy extrapolated to time zero, $r_o$, is close to the value of 0.30 expected for an immobilized tryptophan residue with this excitation frequency. At 10°C the Trp-138 ring appears to be immobilized so that only overall particle tumbling is observed. Increasing the temperature results in a significant amplitude for internal motion. The drop of the value of $r_o$ below the value of 0.3 at higher temperatures indicates that there is also a motional component that is too fast to be resolved. Thus, the complex variation of the amplitude and rate of the resolved fast motion probably represents a change-over from the situation at 20°C (where all of the internal motion is resolved) to one where, at intermediate temperatures, two fast components are averaged in the analysis. At the highest temperatures the very fast component is missed completely. Some information on the rate of this very fast component can probably be obtained by collecting data sets at higher resolution, i.e., by using a shorter time range, and by collecting many more counts. A three-exponential fit for the anisotropy may then be justified.

The most interesting aspect of this anisotropy behavior so far as understanding protein dynamics is concerned is that the temperature variation of the rates and amplitudes are fairly large in a reasonable temperature range. Furthermore, the rate of the internal motion, at least near 20°C, is sufficiently slow such that there is no difficulty in determining both the rate and amplitude to an accuracy of ca. 10%. On the other hand, at higher temperatures the simulation calculations will only have to be extended to a few hundred picoseconds before all of the significant internal motion should be reproduced.

This single Trp-138 form of the protein denatures at about 55°C at pH 6.1. The denatured form obtained at pH 2 in GuHCl shows an anisotropy behavior at 20 and 30°C characterized by an $r_o$ of about 0.2, two internal components in the 0.5- and 1.7-ns range, and a small (0.02–0.05) limiting value at long times. The low value of $r_o$ indicates that some very fast motion is being missed; the finite asymptotic value at long times may indicate that the mobility of the Trp side chain is somewhat hindered even in the unfolded form and that overall tumbling of the denatured form is slow.

The next step in this project will be the introduction of other modifications of the protein structure and a determination of the effects of that perturbation on the tryptophan dynamics. Some specific modifications of particular interest are replacement of Gln 105, to which Trp 138 is hydrogen bonded, and replacement of Ala 146 by Thr, a substitution that is known to result in movement of the Trp-138 residue. Changes in the structure more distant from the tryptophan residue will be used to determine the degree of propagation of structural effects as they influence protein dynamics.

## CONCLUSIONS

Current pulsed laser fluorescence methods are capable of resolving very rapid internal motions of components of biopolymers. The illustrations presented above utilize fluorescent groups, parinaric acid and tryptophan, with very different photophysical behavior. Much of this difference probably stems from the large dipole moment change upon excitation of tryptophan and the effective lack of any permanent dipole moment in the ground or excited states of parinaric acid. The very strong absorption transition of parinaric acid makes its excitation sensitive to changes in local density which seem to be present in lipid bilayers. A heterogeneous model appears to be required to interpret the fluorescence data. The large dipole moment change and quenching effects observed for the indole residue of tryptophan make it very sensitive to neighboring groups and local polarity. Exposure of this group to a wide range of environments appears to increase the spread of its lifetime distribution. Elevated temperatures result in a decrease in this spread, probably because of very rapid averaging between environments. A homogeneous model is used to interpret the data.

Despite these complexities, the anisotropy decay behavior obtained from these experiments seems to make sense in terms of available structural information and the trends expected as the temperature is increased. The temporal resolution appears to be adequate to resolve all of the motion in many cases and the experimental data are internally consistent. A detailed comparison with simulation calculations seems feasible and warranted at this point.

## REFERENCES

Alber T, Grutter MG, Gray TM, Wozniak JA, Weaver LH, Chen B-L, Baker EN, Matthews BW (in press): Structure and stability of mutant lysozymes from bacteriophage T4. In: UCLA Symposia on Molecular and Cellular Biology. "Protein Structure, Folding and Dynamics." New York: Alan R. Liss, Inc.

Andrews LJ, Forster LS (1974): Fluorescence characteristics of indoles in non-polar solvents: Lifetimes, quantum yields and polarization spectra. Photochem Photobiol 19:353–60.

Becker RS (1969): "Theory and Interpretation of Fluorescence and Phosphorescence." New York: Wiley-Interscience.

Bent DV, Hayon E (1975): Excited state chemistry of aromatic amino acids and related peptides III. J Am Chem Soc 97:2612–19.

Berde CB, Hudson B, Simoni RD, Sklar LA (1979): Human serum albumin: Spectroscopic studies of binding and proximity relationships for fatty acids and bilirubin. J Biol Chem 254:391–400.

Birks JB (1970): "Photophysics of Aromatic Molecules." London: Wiley-Interscience.

Bloch KR (1983): Sterol structure and membrane function. CRC Critical Rev Biochem 14:47–92.

Brockerhoff H (1968): Substrate specificity of pancreatic lipase. Biochim Biophys Acta 159:296–303.

Burstein EA, Vedenkina NS, Ivkova MN (1973): Fluorescence and the location of tryptophan residues in protein molecules. Photochem Photobiol 18:263–279.

Chang MC, Petrich JW, McDonald DP, Fleming GR (1983): Nonexponential fluorescence decay of tryptophan, tryptophylglycine and glycyltryptophan. J Am Chem Soc 105:3819–3824.

Creed D (1984): The photophysics and photochemistry of the near-UV absorbing amino acids. I. Tryptophan and its simple derivatives. Photochem Photobiol 39:537–562.

Cross AJ, Fleming GR (1984): Analysis of time-resolved fluorescence anisotropy decays. Biophys J 46:45–56.

Dijkstra BW, Kalk KH, Drenth J, Vandermaelen PJ (1978): Three dimensional structure and disulfide bond connections in pancreatic phospholipase $A_2$. J Mol Biol 124:53–60.

Dijkstra BW, Kalk KH, Hol WGJ, Drenth J (1981a): Structure of bovine pancreatic phospholipase $A_2$ at 1.7 Å resolution. J Mol Biol 147:97–123.

Dijkstra BW, Drenth J, Kalk KH (1981b): Active site and catalytic mechanism of phospholipase $A_2$. Nature 289:604–6.

Dijkstra BW, van Nes GJH, Kalk KH, Brandenburg NP, Hol WG, Drenth J (1982): The structure of bovine pancreatic prophospholipase $A_2$ at 3.0 Å resolution. Acta Cryst B38:793–9.

Eisinger J, Navon G (1969): Fluorescence quenching and isotope effect in tryptophan. J Chem Phys 50:2069–71.

Ellwell M, Schellman J (1975): Phage T4 lysozyme. Physical properties and reversible unfolding. Biochim Biophys Acta 386:309–23.

Ellwell M, Schellman J (1977): Stability of phage T4 lysozymes I. Native properties and thermal stability of wild type and two mutant lysozymes. Biochim Biophys Acta 494:367–83.

Ellwell M, Schellman J (1979): Stability of T4 phage lysozymes II. Unfolding with guanidinium chloride. Biochim Biophys Acta 580:327–38.

Feitelson J (1971): The formation of hydrated electrons from the excited state of indole derivatives. Photochem Photobiol 13:2612–87.

Felder TC, Choi KJ, Topp MR (1982): Picosecond observation of fluorescence from $B_u$ states of diphenylpolyenes. Chem Physiol 64:175–82.

Fischer A, Losche M, Mohwald H, Sackmann E (1984): On the nature of the lipid monolayer phase transition. J Phys Lett 45L:785–91.

Gratton E, Limkeman M (1983): A continuously variable frequency cross-correlation phase fluorimeter with picosecond resolution. Biophys J 44:315–32.

Gratton E, Jameson DM, Hall RD (1984): Multifrequency phase and modulation fluorometry. Ann Rev Biophys Bioengineering 13:105–124.

Grinvald A, Steinberg IZ (1974): Analysis of fluorescence decay kinetics by the method of least squares. Anal Biochem 59:583–98.
Grinvald A, Steinberg IZ (1976): The fluorescence decay of tryptophan residues in native and denatured proteins. Biochim Biophys Acta 427:663–78.
Grutter MG, Hawkes RB, Matthews BW (1979): Molecular basis of thermostability in the lysozyme from bacteriophage T4. Nature 277:667–9.
Grutter MG, Weaver LH, Gray TM, Matthews BW (1983): Structure, function and evolution of the lysozyme from bacteriophage T4. In Matthews CK, Kutter E, Mosig G, Berget P (eds): "Bacteriophage T4." Washington: American Society for Microbiology.
Hille JDR, Donne-Op den Kelder GM, Sauve P, Haas GH de, Egmond MR (1981): Physico-chemical studies of the interaction of pancreatic phospholipase $A_2$ with a micellar substrate analog. Biochemistry 20:4068–73.
Hudson B, Kohler BE (1972): A low-lying weak transition in the polyene $\alpha,\omega$-diphenylocta-tetraene. Chem Phys Lett 14:299–304.
Hudson B, Kohler BE (1973): Polyene spectroscopy: The lowest energy excited singlet state of diphenyloctatraene and other linear polyenes. J Chem Phys 59:4984–5002.
Hudson B, Kohler BE (1974): Linear polyene electronic structure and spectroscopy. Ann Rev Phys Chem 25:437–460.
Hudson B, Kohler BE (1984): Electronic structure and spectra of finite linear polyenes. Synth Metals 9:241–252.
Hudson B, Kohler BE, Schulten BE (1982): Linear polyene electronic structure and potential surfaces. In Lim EC (ed): "Excited States." New York: Academic Press, Vol. 6, pp 1–95.
Ichiye T, Karplus M (1983): Fluorescence depolarization of tryptophan residues in proteins: a molecular dynamics study. Biochemistry 22:2884–93.
Isenberg I (1973a): On the theory of fluorescence decay experiments I. Non-random distortions. J Chem Phys 59:5696–707.
Isenberg I (1973b): On the theory of fluorescence decay experiments II. Statistics. J Chem Phys 59:5708–13.
Isenberg I, Dyson RD, Hanson R (1973): Studies on the analysis of fluorescence decay data by the method of moments. Biophys J 13:1090–115
Jahnig F (1981a): Critical effects from lipid-protein interaction in membranes I. Theoretical description. Biophys J 36:329–45.
Jahnig F (1981b): Critical effects from lipid-protein interaction in membranes II. Interpretation of experimental results. Biophys J 36:347–57.
Jahnig F (1981c): The ordered-fluid transition in lipid bilayers. Mol Cryst Liq Cryst 63:157–70.
Jameson DM, Gratton E, Hall RD (1984): The measurement and analysis of heterogeneous emissions by multifrequency phase and modulation fluorimetry. Appl Spect Rev 20:55–106.
Jansen EHJM (1979): NMR studies on phospholipases $A_2$. Ph.D. Dissertation. State University at Utrecht, Netherlands.
Joschek H-I, Grossweiner LI (1966): Optical generation of hydrated electrons from aromatic compounds. II. J Am Chem Soc 88:3261–68.
Karplus M, McCammon JA (1981): The internal dynamics of globular proteins. CRC Crit Rev Biochem 9:293–349.
Karplus M, McCammon JA (1984): Dynamics of proteins: Elements and function. Ann Rev Biochem 53:263–300.
Kimelman D, Tecoma ES, Wolber PK, Hudson B, Wickner WT, Simoni RD (1979): Protein-lipid interactions: Studies of the M13 coat protein in dimyristoylphosphatidylcholine vesicles using parinaric acid. Biochemistry 18:5874–5880.
Kirby EP, Steiner RF (1970): The influence of solvent and temperature upon the fluorescence of indole derivatives. J Phys Chem 74:4480–90.

Konev SV (1967): "Fluorescence and Phosphorescence of Proteins and Nucleic Acids." New York/London: Plenum Press.
Lakowicz JR, Weber G (1973a): Quenching of fluorescence by oxygen. A probe for structural fluctuations in proteins on the nanosecond time scale. Biochemistry 12:4161–71.
Lakowicz JR, Weber G (1973b): Quenching of protein fluorescence by oxygen. Detection of structural fluctuations on the nanosecond time scale. Biochemistry 12:4171–9.
Lakowicz JR (1983): "Principles of Fluorescence Spectroscopy." New York-London: Plenum Press.
Lami H (1977): On the possible role of a mixed valence-Rydberg state in the fluorescence of indoles. J Chem Physiol 67:3274–81.
Lasser N, Feitelson J, Lumry R (1977): Exciplex formation between indole derivatives polar solutes. Israel J Chem 16:330–4.
Lehrer SS (1976): Perturbation of intrinsic protein fluorescence. In Chen RF, Edelhoch H (eds): "Biochemical Fluorescence: Concepts." New York-Basel: Marcel Dekker, Vol. 2, pp 515–544.
Ludescher RD (1984a): Artifacts in the decay of the fluorescence anisotropy: the effect of multiple lifetimes. Biophys J 45:379a.
Ludescher RD (1984b): "Nanosecond Dynamics in Proteins and Lipids: A Time-Resolved Fluorescence Anisotropy Study." PhD Dissertation. Eugene, OR: University of Oregon.
Ludescher RD, Volwerk JJ, de Haas GH, Hudson B (1985): Complex photophysics of the single tryptophan of porcine pancreatic phospholipase A2, its zymogen and an enzyme/micelle complex. Biochemistry 24:7240–7249.
Ludescher RD, Volwerk JJ, de Haas GH, Jost PC, Hudson B (1986a): Rotational dynamics of the single tryptophan of procine pancreatic phospholipase A2, its zymogen and an enzyme/micelle complex. A steady-state and time-resolved anisotropy study. Biochemistry (submitted).
Ludescher RD, Hudson S, Hudson B (1986b): The effect of lifetime heterogeneity on fluorescence anisotropy decay. Biophys J (submitted).
Lumry R, Hershberger M (1978): Status of indole photochemistry with special reference to biological applications. Photochem Photobiol 27:819–840.
Matthews BW, Remington SJ (1974): Three dimensional structure of the lysozyme from bacteriophage T4. Proc Natl Acad Sci USA 71:4178–4182.
Mitaku S, Okano K (1981): Ultrasonic measurements of two-component lipid bilayer suspensions. Biophys Chem 14:147–58.
Mitaku S, Date T (1982): Anomalies of nanosecond ultrasonic relaxation in the lipid bilayer transition. Biochim Biophys Acta 688:411–21.
Mitaku S, Jippo T, Kataoka R (1983): Thermodynamic properties of the lipid bilayer transition: Pseudocritical phenomena. Biophys J 42:137–44.
Morgan CG, Hudson B, Wolber PK (1980): The photochemical dimerization of parinaric acid in lipid bilayers. Proc Natl Acad Sci USA 77:26–30.
Parasassi T, Conti F, Gratton E (1984): Study of heterogeneous emission of parinaric acid isomers using multifrequency phase fluorometry. Biochemistry 23:5660–5664.
Pesce AJ, Rosen C-G, Pasby TL (1971): "Fluorescence Spectroscopy: An Introduction for Biology and Medicine." New York: Marcel Dekker.
Petrich JW, Chang MC, McDonald DB, Fleming GR (1983): On the origin of non-exponential fluorescence decay in tryptophan and its derivatives. J Am Chem Soc 105:3824–32.
Pieterson WA, Vidal JC, Volwerk JJ, Haas GH de (1974): Zymogen catalyzed hydrolysis of monomeric substrates and the presence of a recognition site for lipid-water interfaces in phospholipase $A_2$. Biochemistry 13:1455–60.
Pugh EL, Kates M, Szabo AG (1982): Studies on fluorescence polarization of 1-acyl-2-*cis*- or

*trans*-parinaroyl *sn*-3-glycerophosphorylcholines in model systems and microsomal membranes. Chem Phys Lipids 30:55–69.
Remington SJ, Anderson WF, Owen J, Ten Eyck LF, Grainger CT, Matthews BW (1978): Structure of the lysozyme from bacteriophage T4: An electron density map at 2.4 A resolution. J Mol Biol 118:81–91.
Ross JBA, Rousslang KW, Brand L (1981): Time resolved fluorescence and anisotropy decay of the tryptophan in adrenocorticotropin-(1-24). Biochemistry 20:4361–69.
Ruggiero A, Ludescher R, Hudson B (1984): Multiple environments in lipid bilayers: Effects on fluorescence lifetimes and polarization anisotropy decays. Biophys J 45:336a.
Sakanishi A, Mitaku S, Ikegami A (1979): Stabilizing effect of cholesterol on phosphatidylcholine vesicles observed by ultrasonic velocity measurement. Biochemistry 79:2636–42.
Schellman JA, Lindorfer M, Hawkes R, Grutter M (1981): Mutations and protein stability. Biopolymers 20:1989–99.
Singer SJ, Nicholson GL (1972): The fluid mosaic model of the structure of cell membranes. Science 175:720–31.
Sklar LA, Hudson B, Simoni RD (1975): Conjugated polyene fatty acids as membrane probes: Preliminary characterization. Proc Natl Acad Sci USA 72:1649–1653.
Sklar LA, Hudson BS, Simoni RD (1976): Conjugated polyene fatty acids as fluorescent membrane probes: Model system studies. J Supramol Struct 4:449–465.
Sklar LA, Hudson B, Petersen M, Diamond J (1977a): Conjugated polyene fatty acids as fluorescent probes: Spectroscopic characterization. Biochemistry 16:813–819.
Sklar LA, Hudson B, Simoni RD (1977b): Conjugated polyene fatty acids as fluorescent probes: Synthetic phospholipid membrane studies. Biochemistry 16:819–828.
Sklar LA, Hudson B, Simoni RD (1977c): Conjugated polyene fatty acids as fluorescent probes: Binding to bovine serum albumin. Biochemistry 16:5100–5108.
Sklar LA, Miljanich GP, Dratz AD (1979): Phospholipid lateral phase separation and the partition of *cis*-parinaric acid and *trans*-parinaric acid among aqueous, solid lipid, and fluid lipid phases. Biochemistry 18:1707–1716.
Sklar LA (1980): The partition of *cis*-parinaric acid and *trans*-parinaric acids among aqueous, fluid lipid and solid lipid phases. Mol Cell Biochem 32:169–177.
Sklar LA, Hudson B, Simoni RD (1981): Parinaric acid from parinarium glaberrimum. Methods Enzymol Lipids (Part D) 72:479–482.
Slotboom AJ, de Haas GH (1975): Specific transformations at the N-terminal region of phospholipase $A_2$. Biochemistry 14:5394–99.
Slotboom AJ, Jansen EHJM, Pattus F, Haas GH de (1978): Semisynthetic studies on phospholipase $A_2$. In Offord RD, Dibello C (eds): "Semisynthetic Peptides and Proteins." New York, London, San Francisco: Academic Press, pp 315–48.
Small EW, Isenberg I (1976): Index displacement in analyzing fluorescence time-decay data. Biopolymers 15:1093–1100.
Small EW, Isenberg I (1977): On moment index displacement. J Chem Phys 66:3347–51.
Small EW, Libertini L, Isenberg I (1984): Construction and tuning of a monophoton decay fluorometer with high resolution capabilities. Rev Sci Instrum 55:879–85.
Snyder R, Arvidson E, Foote C, Harrigan L, Christensen RL (1985): Electronic energy levels in long polyenes: $S_2 \rightarrow S_0$ emission in all-trans-1,3,5,7,9,11,13-tetradecaheptaene. J Am Chem Soc 107:4112–4117.
Strickland EH, Horwitz J, Billups C (1970): Near ultraviolet absorption bands of tryptophan using indole and 3-methylindole as models. Biochemistry 9:4914–21.
Szabo A (1984): Theory of fluorescence depolarization in macromolecules and membranes. J Chem Phys 81:150–167.
Szabo AG, Rayner DM (1980): Fluorescence decay of tryptophan conformers in aqueous

solution. J Am Chem Soc 102:554–63.
Tecoma E, Sklar LA, Simoni RD, Hudson B (1977): Conjugated polyene fatty acids as fluorescent probes: Biosynthetic incorporation of parinaric acids by *Escherichia coli* and studies of phase transitions. Biochemistry 16:829–835.
Tsai A, Hudson BS, Simoni RD (1981): Preparation of parinaric acid derivatives. Methods Enzymol Lipids (Part D) 72:483–485.
Valeur B, Weber G (1977): Resolution of the fluorescence excitation spectrum of indole into its $^1L_a$ and $^1L_b$ excitation bands. Photochem Photobiol 25:441–4.
van Scharrenburg GJM, Puijk WC, Egmond MR, van der Schaft PH, de Haas GH, Slotboom AJ (1982): Effects of substitution of the absolutely invariant glutamine-4 and phenylalanine-5 in bovine pancreatic phospholipase $A_2$ on enzymatic activity and substrate binding. Biochemistry 21:1345–1352.
van Dam-Mieras MCE, Slotboom AJ, Pietersen WA, de Haas, GH (1975): The interaction of phospholipase $A_2$ with micellar interfaces. The role of the N-terminal region. Biochemistry 14:5387–94.
Verger R, Mieras MCE, de Haas GH (1973): Action of phospholipase A at interfaces. J Biol Chem 248:4023–34.
Walker MS, Bednar TW, Lumry R (1967): Exciplex studies. II. Indole derivatives. J Chem Phys 47:1020–8.
Weaver LH, Matthews BM (1985): Structure of bacteriophage T4 lysozyme refined at 1.7 Å resolution. J Mol Biol (in press).
Weis RW, McConnell HM (1984): Two-dimensional chiral crystals of phospholipid. Nature 310:47–49.
Weis RW, McConnell HM (1985): Cholesterol stabilizes the liquid crystalline interface in phospholipid monolayers. Biophys J 47:44a.
Welti R (1982): Partition of parinaroyl phospholipids in mixed head group systems. Biochemistry 21:5690–5693.
Welti R, Silbert DF (1982): Partition of parinaroyl phospholipid probes between solid and fluid phosphatidylcholine phases. Biochemistry 21:5685–5689.
Williamson HW, Morgan CG, Fuller S, Hudson B (1983): Melittin induces fusion of phospholipid vesicles. Biochim Biophys Acta 732:668–674.
Wolber PK, Hudson B (1979): An analytic solution to the forster energy transfer problem in two dimensions. Biophys J 28:197–210.
Wolber PK, Hudson B (1981): Fluorescence lifetime and time resolved polarization anisotropy studies of acyl chain order and dynamics in lipid bilayers. Biochemistry 20:2800–2810.
Wolber PK, Hudson B (1982): Bilayer acyl chain dynamics and lipid protein interactions: The effect of the M13 bacteriophage coat protein on the decay of the fluorescence anisotropy of parinaric acid. Biophys J 37:253–262.
Yamamoto Y, Tanaka J (1972): Polarized absorption spectra of crystals of indole and its related compounds. Bull Chem Soc Jpn 45:1362–66.

*Applications of Fluorescence in the Biomedical Sciences, pages 203–223*


# 9 Determination of the Coordination Geometry of the Calcium (II) Binding Sites in the Insulin Hexamer Using Europium (III) Fluorescence Fine Structure as a Probe

*George K. Alameda, G. Alan Schick, Jeffrey L. Evelhoch, and Robert R. Birge*

It is well known that calcium is present in high concentrations in insulin granules [Herman et al., 1973; Schäfer and Klöppel, 1974; Howell et al., 1975; Ravuzzolu et al., 1976], and it has been suggested that calcium plays an important role in the function and control of insulin [Hellman et al., 1976; Howell et al., 1978; Anderson and Berggen, 1979; Kohnert et al., 1979; Molaisse et al., 1975; Hellman, 1977; Wollheim et al., 1978]. In order to ascertain better its biological importance, characterization of the calcium presence in insulin, both chemically and structurally, is necessary. X-ray crystallographic studies of insulin show that the insulin hexamer is torus shaped (~50 Å in diameter, ~35 Å high; see Fig. 1) and contains two Zn(II) ions located 8.5 Å from the center of the hexamer on a threefold axis of symmetry which traverses the central cavity [Blundell et al., 1972]. Ca(II) binding sites on the surface of crystalline insulin hexamer [Pitts et al., 1980] and in the center of the hexamer in solution [Sudmeier et al., 1981] have been reported. In a previous effort to characterize the central binding sites of

Departments of Biological Sciences and Chemistry, Center for Fluorescence Research in Biomedical Sciences, Carnegie-Mellon University, Pittsburgh, Pennsylvania 15213.

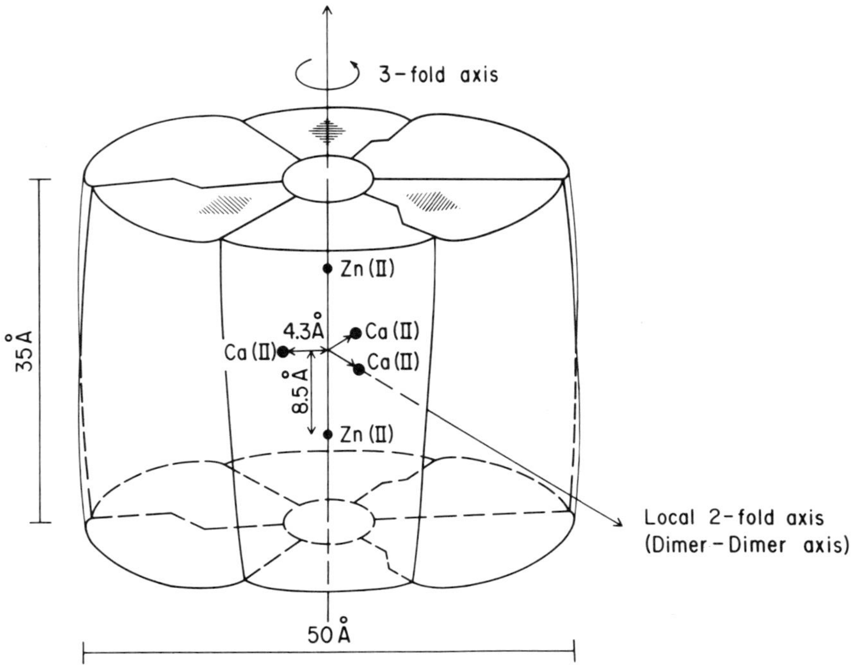


Fig. 1. Three-dimensional structure representation of the two-Zn(II) insulin hexamer showing the Ca(II) binding sites. The jagged lines between monomers on the top and bottom of the hexamer indicate a dimer-dimer interface. The straight lines between monomers on the top and bottom of the hexamer indicate a monomer-monomer interface within a dimer.

Ca(II) in the two-Zn(II) insulin hexamer, we reported the laser-induced luminescence of insulin-bound Eu(III) in solution [Alameda et al., 1985]. The study provided evidence for the presence of three equivalent Ca(II) binding sites at the center of the hexamer as illustrated in Figure 2. The Ca(II) binding sites are symmetrically oriented and lie in a plane which passes through the middle of the central cavity of the hexamer.

The purpose of the present work is to characterize further the binding sites of Ca(II) in the central cavity of the two-Zn(II) insulin hexamer. We present laser-induced fluorescence data for samples in $D_2O$ of $EuZn_2In_6$ (In = insulin) at 77K and $Eu_3Zn_2In_6$ at 77K and 293K. The fine structure of the $^5D_0 \rightarrow {}^7F_2$ transition of Eu(III) bound in the Ca(II) binding sites is analyzed in terms of a ligand field perturbation potential. Although Eu(III) does not

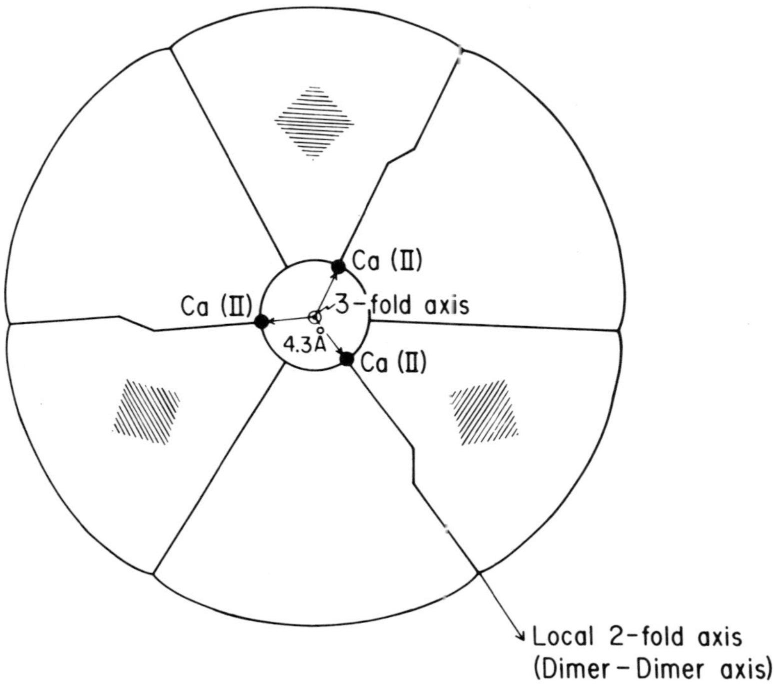


Fig. 2. Same as Figure 1, except viewed from above.

occur naturally in biological systems, it is a useful probe for Ca(II) binding sites in macromolecules [Neiboer, 1975; Ruben, 1977); and studies indicate that such substitutions are isomorphic and generally maintain biological activity [Matthews and Weaver, 1974; Moews and Kretsinger, 1975; Ruben, 1975]. On this basis we are able to extract information regarding the local coordination sphere of Ca(II) bound in insulin.

## EXPERIMENTAL PROCEDURES

### Instrumentation

Laser-induced fluorescence of Eu(III) in insulin was collected by using a (PDP 11/03) computer-controlled Princeton Applied Research (PAR) model 1205A optical multichannel analyzer (OMA) (see Fig. 3). Details of the apparatus are described elsewhere [Alameda et al., 1981]. Ultraviolet radiation from an argon ion laser (363.8 and 351.1 nm) was used for excitation. At 90° to the laser beam axis, a Schott glass bandpass filter (type G6455, 500–1,000 nm) allowed only the red sample fluorescence to fall on the slits

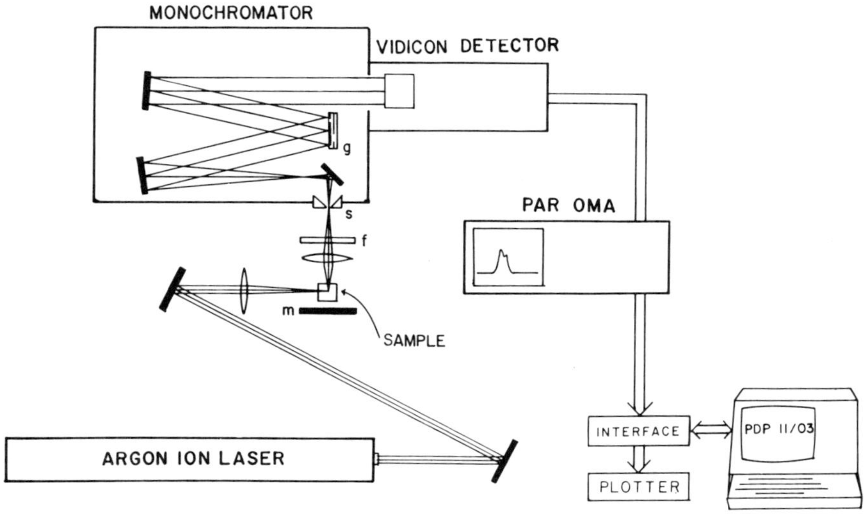


Fig. 3. Experimental setup for the optical multichannel analyzer. An argon ion laser is focused into a quartz cuvette. Mirror (m) reflects part of the resultant sample fluorescence through a lens focused on the polychromator slits. In the path is a Schott glass, type G6455, 500–1000-nm bandpass filter (f) to prevent scattered laser light from entering the polychromator. Light dispersed from the grating falls on a vidicon detector where a linear array of 500 light-sensitive elements record the fluorescence spectrum. The Princeton Applied Research optical multichannel analyzer (PAR OMA) electronics read each of the elements of the array and average the recorded value in memory. Later, a PDP 11/03 computer scans the OMA memory and plots the spectrum on a digital plotter.

of an Instruments SA, Inc., model HR-320 polychromator. Attached to the output of the polychromator is a vidicon detector consisting of a linear array of 500 light-sensitive elements. The OMA allows the signal on the detector to be integrated and then averaged into digital memory. The linear array is then automatically cleared and another scan is taken. Through the use of successive integrations, low-level optical signals can be recorded with a high degree of resolution (resolution for this system is 0.4 Å at 546 nm). The PAR OMA is equipped with two independent digital memories to allow collection of and correction for any spectral background. This feature proved useful for removal of the broad tyrosin fluorescence from the insulin signal.

## Materials

Insulin (bovine, crystalline) was purchased from Sigma Chemical Co. (lot 49C-0197). Chelex 100 (100–200 mesh, Na form) was purchased from Bio-Rad. Europium nitrate (99.9 atom % Eu) was purchased from Research

Organic/Inorganic Chemical Corp. Deuterium oxide (99.8 atom % D) was purchased from Matheson, Coleman, and Bell.

### Methods

Insulin stock solutions (5 mg/ml) were prepared by mixing the crystalline protein with $H_2O$ and then adjusting the pH to 10.5 with dilute NaOH to completely dissolve the protein. The solution was passed over a Chelex 100 column (30 × 1 cm, titrated to pH 10.5 with dilute $HNO_3$) to remove all traces of zinc and other contaminating metal ions. Atomic absorption spectroscopy confirmed the absence of metals. The metal-free insulin was lyophilized and stored at 4°C.

Stock solutions of $Eu(NO_3)_3$ and $ZnSO_4$ were made approximately $1 \times 10^{-2}$ M in $D_2O$. Samples of metal-free insulin were dissolved in $D_2O$ followed by adjustment of the pH to 8.0 with dilute metal-free $H_2SO_4$. The metal ions were then added from the stock solutions. To insure complete removal of $H_2O$ (which very effectively quenches Eu(III) emission [see Alameda et al., 1985]), the samples were lyophilized and redissolved in $D_2O$. Samples used for the low-temperature spectra were mixed with glycerol to aid in the formation of a clear solid sample at 77K.

## THEORETICAL

The observed optical transitions of lanthanide ions in crystals and solutions correspond to transitions among states of the $4f^n$ configuration. These transitions can be either magnetic- or electric-dipole in nature. Typically, magnetic-dipole transitions are quite weak yet can exhibit comparable intensities to those of electric-dipole transitions in cases in which the latter are "forbidden." For Eu(III) these two types of transitions can exhibit similar intensities as exemplified by the comparable magnitudes of the $^5D_0 \rightarrow {}^7F_1$, (magnetic-dipole allowed) and $^5D_0 \rightarrow {}^7F_2$ (first-order electric-dipole allowed) emissions observed from $D_2O$ solutions [Alameda et al., 1985]. However, in the present case of insulin-bound Eu(III), the $^5D_0 \rightarrow {}^7F_1$ emission is substantially weaker than the $^5D_0 \rightarrow {}^7F_2$ emission [Alameda et al., 1985] indicating that the magnetic-dipole character is relatively weak. We will therefore confine our theoretical treatment to electric-dipole transitions.

Although the electric-dipole transitions of the free lanthanide ion are parity forbidden, the noncentrosymmetric interactions of binding ligands lead to mixing states with opposite parity. Theoretical treatment of this mixing has been presented [Van Vleck, 1937; Broer et al., 1945] and is based on coupling of states of opposite parity by means of odd terms in the ligand field potential. A detailed treatment would involve modeling both energy splittings and intensities of the fine structure measurements; however, simple approaches which model only transition probabilities (intensities) have been quite useful in interpreting lanthanide luminescence in terms of environmental effects [Judd, 1962; Ofelt, 1962; Korolkov et al., 1967]. Through the use of multipole operator and the Wigner-Eckart theorem, the observed lanthanide fluoresecence fine structure can be related to the lanthanide coordination geometry.

The Hamiltonian $\mathcal{H}$ for a lanthanide ion in a ligand field may be written as

$$\mathcal{H} = \mathcal{H}_F + V \tag{1}$$

where $\mathcal{H}_F$ is the free ion Hamiltonian and V is the potential provided by the ligand field [Condon and Shortly, 1935]. This potential can be expressed by expanding $V$ in terms of spherical-harmonics:

$$V = \sum_{t,p} A_{tp} D_p^{(t)} \tag{2}$$

where $D_p^{(t)}$ is the tensor representation of the multipole operator described by Racah [1943] and $A_{tp}$ are parameters which describe the particular symmetries of the ligand field. This result is based on the fact that the interactions between each pair of electrons can be written as an expansion in Legendre polynomials

$$V_{ij} = e^2 \sum_t \frac{r_j{}^t}{r_i^{t+1}} P_t (\cos \omega_{ij}) \tag{3}$$

where $\omega_{ij}$ is the angle subtended by the radial vectors of electron $i$ and $j$. Finally, by using the spherical-harmonic addition theorem [Condon and Shortly, 1935] $V$ can be written as

$$V = e^2 \sum_{i,j,t} \frac{\mathrm{r}_j^t}{r_i^{t+1}} \cdot \frac{4\pi}{2t+1} \sum_p Y_{tp}^*(\theta_i,\phi_i)\ Y_{tp}(\theta_j,\phi_j) \tag{4}$$

It is customary to write eq. (4) as the product of two terms [Wybourne, 1965]

$$A_{tp} = \sum_i \frac{e^2}{r_i^{t+1}} \left[\frac{4\pi}{2t+1}\right]^{1/2} Y_{tp}^* (\theta_i,\phi_i) = \sum_i \frac{e^2}{r_i^{t+1}} C_p^{*(t)} (\theta_i,\phi_i) \tag{5A}$$

and

$$D_p^{(t)} = \sum_j r_j^t \left[\frac{4_\pi}{2t+1}\right]^{1/2} Y_{tp}(\theta_j,\phi_j) = \sum_j r_j^t\, C_p^{(t)}(\theta_j,\phi_j) \tag{5B}$$

Placement of the lanthanide ion in a ligand field causes even states of the $4f^n$ configuration $|\gamma JM\rangle$ and $|\gamma' J' M'\rangle$ to become slightly admixed with odd states, $|K\rangle$, of some excited state configuration $4f^5 n\ell$. This yields new states, A and B, such that

$$\langle A| = \langle \gamma JM| - \sum_K \frac{\langle \gamma JM|V|K\rangle\langle K|}{E(\gamma JM) - E(K)} \tag{6A}$$

and

$$|B\rangle = |\gamma' J' M'\rangle - \sum_K \frac{|K\rangle\langle K|V|\gamma' J' M'\rangle}{E(\gamma' J' M') - E(K)} \tag{6B}$$

where $K \equiv \gamma'' J'' M''$; $\gamma_1 J_1 M_1$ are the quantum numbers for the intermediate states of the $4f^5 n\ell$ configuration. It should be noted that only odd terms in $V$ ($t = 1, 3, 5, \ldots$) produce admixing and, hence, electric-dipole-allowed transitions.

The probability for an electric-dipole transition can be written as

$$\langle A|D_q^{(1)}|B\rangle = -\sum_K \frac{\langle \gamma JM|V|K\rangle\langle K|D_q^{(1)}|\gamma' J' M'\rangle}{E(\gamma JM) - E(K)} - \sum_K \frac{\langle \gamma JM|D_q^{(1)}|K\rangle\langle K|V|\gamma' J' M'\rangle}{E(\gamma' J' M') - E(K)} \tag{7}$$

Calculation of $\langle A|D_q^{(1)}|B\rangle$ cannot be made without knowledge of the intermediate state of $K$. However, by following the procedures of Judd [1962] and Ofelt [1962], this quantity can be approximated by first expanding $|\gamma JM\rangle$ in linear combinations of perfect Russell-Saunders (RS) states, $|\gamma SLJM\rangle$, and then replacing the energy denominators by a single average denominator, $\Delta E_{AV}$. The use of $\Delta E_{AV}$ is based on the assumption that the intermediate states are degenerate, or rather that the mixing occurs only from addition of terms from one excited-state configuration [Judd, 1962; Ofelt, 1962].

For the present work Eu(III) is used exclusively. The observed fluorescence of Eu(III) is composed primarily of transitions from the $^5D_0$ level to the $^7F_{J'}$ manifold. This results in a simplified form of the general expression for $\langle A|D_q^{(1)}|B\rangle$ [Koralkov et al., 1966].

$$
\begin{aligned}
\langle A|D_q^{(1)}|B\rangle &= \langle {}^5D_{00}|D_q^{(1)}|{}^7F_{J'M'}\rangle \\
&= \sum_{t,p,\ell,\lambda} A_{tp} \begin{pmatrix} 1 & t & J' \\ q & p & M' \end{pmatrix} \\
&\times [(2J'+1)(2L'+1)]^{1/2}
\end{aligned}
$$

$$
\times\ 42\langle\psi_i|\psi''\rangle\langle\psi''|\psi'_j\rangle \sum_{L_1} (-1)^{L_1}\cdot(2L_1+1)
$$

$$
\times \begin{Bmatrix} 1 & J' & t \\ L' & L_1 & S' \end{Bmatrix} \begin{Bmatrix} S' & L_1 & 1 \\ \ell & 3 & L'' \end{Bmatrix} \begin{Bmatrix} L_1 & L' & t \\ 3 & \ell & L'' \end{Bmatrix} \cdot (2\ell+1)
$$

$$
\times \begin{pmatrix} 3 & 1 & \ell \\ 0 & 0 & 0 \end{pmatrix} \begin{pmatrix} \ell & t & 3 \\ 0 & 0 & 0 \end{pmatrix} \left[\Delta E_{AV}^{-1} + (-1)^{J'}\cdot\Delta E_{AV}^{-1}\right]
$$

$$
\times\ \langle r\,(f,\ell)\rangle\langle r^t(f,\ell)\rangle \tag{8}
$$

where $\lambda$ indicates summation over $i$, $j$ and the quantum numbers $\gamma SLJM$ and $\gamma' S'L'J'M'$, and quantities in ( ) and {} represent 3-$j$ and 6-$j$ symbols, respectively. The integrals $\langle r^t(f,\ell)\rangle \equiv \langle 4f^6|r^t|4f^5n\ell\rangle$ and $\langle\psi_a|\psi_b\rangle$ are overlap integrals involving pure RS wavefunctions.

Since Eu(III) is in a $4f^6$ free ion configuration, the chelated states must be mixed with the terms from $4f^5nd$ and or $4f^5ng$ excited-state configurations (the dominating contributions would involve states for which $n = 5$). The assumption that the energy denominators in eq. (7) can be replaced by $\Delta E_{AV}$ therefore amounts to presuming that the energy differences between the $^5D_0$ and $^7F_j$ levels are small compared to those between the $4f^6$ and $4f^55d$ or $4f^55g$ configurations. The former differences are on the order of $10^4$ cm$^{-1}$, whereas the latter are approximately $10^5$ cm$^{-1}$ [Judd, 1962]. Although these values indicate the validity of this approximation to be marginal at best, differences in $\Delta E$ on the order of 10% will not significantly alter the salient results of the present treatment.

One further simplification of eq. (8) is to reduce the sum over $L_1$ by using the Biedenharn-Eliott sum rule [Judd, 1963],

$$
\sum_{L_1} (-1)^{L_1}(2L_1+1) \begin{Bmatrix} 1 & J' & t \\ L' & L_1 & S' \end{Bmatrix} \begin{Bmatrix} S' & L_1 & 1 \\ \ell & 3 & L'' \end{Bmatrix} \begin{Bmatrix} L_1 & L' & t \\ 3 & \ell & L'' \end{Bmatrix}
$$

$$
= (-1)^{J'+L''+S'+L'+t+\ell+1} \begin{Bmatrix} J' & 3 & 3 \\ \ell & t & 1 \end{Bmatrix} \begin{Bmatrix} 3 & 3 & J' \\ L' & S' & L'' \end{Bmatrix} \tag{9}
$$

Eq. (8) can now be written as

$$\langle D^5_{00}|D_q^{(1)}|F_{J'M'}\rangle = \sum_{t,p,\lambda} A_{tp}\begin{pmatrix} 1 & t & J' \\ q & p & M' \end{pmatrix} \sum_{\ell} [(2J'+1)\,(2L'+1)]^{1/2}$$

$$\times \begin{Bmatrix} J' & 3 & 3 \\ \ell & t & 1 \end{Bmatrix} Z(J')\,[1+(-1)^{J'}]\,\Delta E_{AV}^{-1} \cdot Z(\ell,t) \tag{10}$$

where

$$Z(J') = \sum_{i,j,L''} \langle \psi_i|\psi''\rangle\langle\psi''|\psi'_j\rangle \begin{Bmatrix} 3 & 3 & J' \\ L' & S' & L'' \end{Bmatrix} (-1)^{L''}$$

and

$$Z(\ell,t) = 42(2\ell + 1)\begin{pmatrix} 3 & 1 & \ell \\ 0 & 0 & 0 \end{pmatrix}\begin{pmatrix} \ell & t & 3 \\ 0 & 0 & 0 \end{pmatrix} \langle r(f,\ell)\rangle\langle r^t(f,\ell)\rangle$$

Korolkov et al. [1966] have shown that $Z(J)$ remains almost constant for $J = 1 - 6$ for the $D_0 \rightarrow {}^7F_J$ transitions. This was based on evaluation of $\Sigma_\psi''$ $\langle\psi|\psi''\rangle\langle\psi''|\psi'\rangle$ by means of fractional parentage coefficients and implies that the fluorescing state $\psi$ and the ground state $\psi'$ are built up from the products of one-electron functions of the $4f^55\ell$ configuration. In the same paper these workers analyzed $Z(\ell,t)$ for two cases where the intermediate state consists of either the $4f^55d$ or the $4f^55g$ configuration. Intensity ratios for transitions of the type ${}^5D_0 \rightarrow {}^7F_{J+2}/{}^5D_0 \rightarrow {}^7F_J$ can now be calculated based upon the previous observation that $Z(J)$ remains fairly constant and that the determining factor for the relative intensity is

$$\approx \sum_{\ell} (2J'+1)^{1/2} \begin{Bmatrix} J' & 3 & 3 \\ \ell & t & 1 \end{Bmatrix} Z(\ell,t) \tag{11}$$

The results of the calculations for odd-order terms in the potential are given in Table 1.

There are several important features to be discussed. First, only even values of $J'$ are allowed due to the presence of the $[1 + (-1)^{J'}]$ term in eq. (10). Second, odd terms of order $t = 1$ induce allowedness only in the ${}^5D_0 \rightarrow {}^7F_0$ and ${}^5D_0 \rightarrow {}^7F_2$ transitions. Likewise, third-order terms allow

**TABLE 1. Relative Intensities of $^5D_0 \rightarrow {}^7F_J$ Transitions**

| | | I ($^5D_0 \rightarrow {}^7F_{J+2}$) : ($^5D_0 \rightarrow {}^7F_J$)[3] | |
|---|---|---|---|
| $t$[1] | $J$[2] | $\ell = 2$[4] | $\ell = 4$[4] |
| 1 | 0 | 0.013 | 0.0022 |
| 3 | 2 | 6.1 | 0.074 |
| 5 | 4 | 59 | 0.012 |

[1]Order of term in the ligand field potential expansion.
[2]Quantum number for a level in the $^7F_J$ manifold.
[3]Relative Intensity of the $^5D_0 \rightarrow {}^7F_{J+2}$ transition to that of $^5D_0 \rightarrow {}^7F_J$ transition.
[4]Quantum number indicating the excited state configuration $4f^5n\ell$.

mixing for only the $^5D_0 \rightarrow {}^7F_2$ and $^5D_0 \rightarrow {}^7F_4$ transitions, and so forth. This restriction results from symmetries imposed by the trials defined from the 6-$j$ symbol in eq. (11). For example, the triads of the 6-$j$ in eq. (11) are: ($J'$, 1, $t$), (3, $\ell$, $t$), ($J'$, 3, 3), (3, $\ell$, 1). Evaluation of a triad requires consideration of the triangular conditions; i.e., for general triad ($J_1$, $J_2$, $J_3$) the triangular conditions are $J_1 + J_2 - J_3 \geqslant 0$, $J_1 - J_2 + J_3 \geqslant 0$ and $-J_1 + J_2 + J_3 \geqslant 0$ otherwise, the 6-$j$ symbol is zero. Table 2 summarizes the triangular conditions for the triad ($J'$, 1, $t$) of eq. (11) for $J' = 0, 2$, or 4 and $t = 1, 3$.

Next, we see from Table 1 that the contribution to the intensity of the $^5D_0 \rightarrow {}^7F_2$ line from first-order terms of $D_q^{(t)}$ is approximately one one-hundredth of the intensity of the $^5D_0 \rightarrow {}^7F_0$ transition and is true for both excited state models. This result is inconsistent with all observations made for Eu(III)

**TABLE 2. 6-$j$ Symbol Transition Restrictions for the ($J'$,1,$t$) Triad of Eq. (11)**

| Triangular conditions | $t = 1, J' = 0$ | $t = 1, J' = 2$ | $t = 1, J' = 4$ |
|---|---|---|---|
| $J' + 1 - t \geqslant 0$ | 0 | 2 | 4 |
| $J' - 1 + t \geqslant 0$ | 0 | 2 | 4 |
| $-J' + 1 + t \geqslant 0$ | 2 | 0 | −2 |
| | (allowed) | (allowed) | (not allowed) |
| Triangular conditions | $t = 3, J' = 0$ | $t = 3, J' = 2$ | $t = 3, J' = 4$ |
| $J' + 1 - t \geqslant 0$ | −2 | 0 | 2 |
| $J' - 1 + t \geqslant 0$ | 2 | 4 | 6 |
| $-J' + 1 + t \geqslant 0$ | 4 | 2 | 0 |
| | (not allowed) | (allowed) | (allowed) |

chelate compounds—that the $^5D_0 \rightarrow {}^7F_2$ transition is the most intense of the $^5D_0 \rightarrow {}^7F_J$ emission manifold [Axe, 1963; Korolkov et al., 1967; Alameda et al., 1985]. Accordingly, we are left with the primary contribution to the intensity of the $^5D_0 \rightarrow {}^7F_2$ transition being from the third-order term. On the other hand, the third-order term indicates that for the $4f^55d$ configuration the $^5D_0 \rightarrow {}^7F_4$:$^5D_0 \rightarrow {}^7F_2$ ratio is approximately 6:1. Again, this is inconsistent with experiment, so we are left with the conclusion that the $4f^55d$ configuration is most likely not an important intermediate state. The $4f^55g$ configuration, however, predicts that the $^5D_0 \rightarrow {}^7F_4$:$^5D_0 \rightarrow {}^7F_2$ ratio is 0.074, a value which agrees with the experimental observations of the relative fluorescence intensities [Alameda et al., 1985]. This finding is consistent with that of Axe [1963] for Eu(III) ethylsulfate.

## RESULTS AND DISCUSSION

In the previous section it was shown that for $^5D_0 \rightarrow {}^7F_{J'}$ emission the line intensities could be derived from eq. (10). Included in that equation is a term which [see eq. (12)] can be used to describe the fine structure of a given transition based upon the geometry of the ligand field.

$$I_{rel} \propto \left[ \sum_{t,p} A_{tp} \begin{pmatrix} 1 & t & J' \\ q & p & M' \end{pmatrix} \right]^2 \qquad (12)$$

For the $^7F_2$ state, five sublevels are possible based upon the $M'$ quantum number; i.e., $|J'M'\rangle = |20\rangle, |2\pm1\rangle, |2\pm2\rangle$. We recall that the major contribution to the intensity of the $^5D_0 \rightarrow {}^7F_2$ transition is from the third-order terms in the ligand field perturbation potential. Thus, having specified $t = 3$, $J' = 2$, and $M' = 0, \pm1, \pm2$, the 3-$j$ symbol of eq. (12) can be used to determine which multipole terms are associated with each of the five sublevels.

The result of that evaluation is shown in Table 3 and summarized below.

$|20\rangle$; $A_{30}, A_{31}, A_{3-1}$,
$|2\pm1\rangle$; $A_{30}, A_{31}, A_{3-1}, A_{32}, A_{3-2}$,
$|2\pm2\rangle$; $A_{31}, A_{3-1}, A_{32}, A_{3-2}, A_{33}, A_{3-3}$

Since these terms are based upon third-order associated Legendre polynomials [see eq. (5A)], they have the same angular dependence as the f wave-functions. Through consideration of the maximum overlap between

**TABLE 3. Allowed Coefficients for the Various States of $^5D_0 \rightarrow {}^7F_2$ Transitions**

$$\begin{bmatrix} 1 & 3 & 2 \\ q & p & M' \end{bmatrix}^1 \quad q + p + M' = 0^2$$

| $M'^3$ | $p^4$ | $q^5$ |
|---|---|---|
| 0 | 0 | 0 |
| 0 | 1 | −1 |
| 0 | −1 | 1 |
| 1 | 0 | −1 |
| 1 | −1 | 0 |
| 1 | −2 | 1 |
| −1 | 0 | 0 |
| −1 | 1 | 0 |
| −1 | 2 | −1 |
| 2 | −2 | 0 |
| 2 | −1 | −1 |
| 2 | −3 | 1 |
| −2 | 2 | 0 |
| −2 | 1 | 1 |
| −2 | 3 | −1 |

[1] 3–*j* symbol of eq. (12).
[2] Conditions for nonzero value of 3–*j* symbol in eq. (12).
[3] Quantum number for $|2M\rangle$ state of Eu(III).
[4] Integer from associated Legendre polynomial making $A_{tp}$ coefficient in eq. (12).
[5] Integer from associated Legendre polynomial representing the dipole operator for the electric-dipole transition in eq. (7).

Eu(III) and the bonding ligands, the structure of the sublevels $|J'M'\rangle$ can be determined under different coordination geometries. A six-coordinate octahedral Eu(III) complex will have the largest overlap with orbitals having $f_{z(x^2-y^2)}$ and $f_{z(2z^2-3x^2-3y^2)}$ symmetry. Therefore, the structure of the $|J'M'\rangle$ sublevels must consist of linear combinations of $A_{30}$, $A_{32}$, and $A_{3\text{-}2}$ terms. Likewise, an eight-coordinate cubic complex will have maximum overlap with orbitals of $f_{x(y^2-z^2)}$ symmetry and the $|J'M'\rangle$ sublevels will consist of $A_{31}$, $A_{3\text{-}1}$, $A_{33}$, $A_{3\text{-}3}$ terms. Table 4 gives the relative sublevel intensities based upon full evaluation of the 3–*j* symbol of eq. (12) [Korolkov et al., 1967]. The observed fine structure for an eight-coordinate system would be made up of the central $|20\rangle$ transition with the doublet $|2\pm1\rangle$ one-third greater in intensity (Table 4). The $|2\pm2\rangle$ doublet will be somewhat less intense depending upon the magnitude of the $A_{33}$, $A_{3\text{-}3}$ coefficients. However, for a six-coordinate complex the fine structure can be either a pair of doublets or a pair of doublets and a singlet, again depending on the magnitudes of $A_{30}$, $A_{32}$, $A_{3\text{-}2}$.

**TABLE 4. Relative Intensities of $^5D_0 \rightarrow {}^7F_2$ as a Function of $A_{3J}$[1]**

| $\mid J',M' \rangle$ | 8 coordinate | 6 coordinate |
|---|---|---|
| $\mid 20\rangle$ | $1B$ | $3A$ |
| $\mid 2{\pm}1\rangle$ | $4/3 \cdot B$ | $2A + 2C$ |
| $\mid 2{\pm}2\rangle$ | $1/6B+D$ | $1C$ |

[1] $A = 1/35\ (A_{30})^2$, $B = 3/25\ [(A_{3-1})^2 + (A_{31})^2]$,
$C = 1/35\ [(A_{3-2})^2 + (A_{32})^2]$, $D = 1/7\ [(A_{3-3})^2 + (A_{33})^2]$.

The fine structure spectra of $EuZn_2In_6$ at 77K and $Eu_3Zn_2In_6$ at 77K and 293K were analyzed by using a multiple nonlinear least-squares regression for three, four, and five bands (Gaussian and Lorentzian) of equal full width at half maximum (FWHM). The results of the fits are tabulated in Table 5. Line shapes of the individual components were assumed to be either Gaussian or Lorentzian. A recent study of the $^5D_0 \rightarrow {}^7F_0$ transition of Eu(III) in borate glass indicates that the line width primarily results from inhomogeneous broadening and is on the order of 60 $cm^{-1}$ [Morgan et al., 1981]. Results for the fit of the $EuZn_2In_6$ emission at 77K to 3, 4, and 5 Gaussians are shown in Figure 4. Little improvement was observed in the standard deviation, $S^2$, for the fit in going from four to five bands. The same was true for the fit of the $EuZn_2In_6$ emission to 4 and 5 Lorentzians (see Fig. 5, Table 5), but with a larger standard deviation due to the poor fit along the tails of the peak. Figures 6 and 7 show the fit of 3 and 4 Gaussians and Lorentzians for $Eu_3Zn_2In_6$ emission at 77K. In each case results were similar to those obtained for $EuZn_2In_6$ at 77K.

The room temperature spectra of $Eu_3Zn_2In_6$ fit to 4 Gaussians and 4 Lorentzians are shown in Figure 8. This spectrum is not as well resolved as those of the two previous low-temperature samples, but the fit to four bands yields an acceptable result (see Table 5). Figure 9 presents the individual bands for the fit of $EuZn_2In_6$ at 77K and $Eu_3Zn_2In_6$ at 77K and 293K to 4 Gaussians and to 4 Lorentzians.

The above analysis indicates that the fine structure of the $^5D_0 \rightarrow {}^7F_2$ transition of insulin-bound Eu(III) consists of a pair of doublets. Since the $\mid 20\rangle$ transition appears to be very small or completely absent, the fine structure is best rationalized in terms of a distorted octahedron. As described above, the $A_{3J}$ coefficients have the same symmetry as third-order associated Legendre polynomials. Thus, the relative peak heights can be simulated on the basis of the equations in Table 4. Figure 10 shows the peak heights for the octahedral complex as a function of the angle, $\theta$, between the twofold axis and an oxygen ligand from the B-13 glutamate (see Fig. 11). Figure 10 indicates that the $\mid 20\rangle$ transition is quite sensitive to the position of the B-13 glutamate carboxyl oxygens and reaches a minimum value near 40°. The

**TABLE 5. Summary of Numerical Fitting of 3, 4, and 5 Gaussians and Lorentzians to $EuZn_2In_6$ at 77K and $Eu_3Zn_2In_6$ at 77K and 293K**

| $EuZn_2In_6$ at 77K | | | | | |
|---|---|---|---|---|---|
| 3 Gaussians $S^2=0.0378$ FWHM=112.4 $cm^{-1}$ | | 4 Gaussians $S^2=0.0268$ FWHM=86.3 $cm^{-1}$ | | 5 Gaussians $S^2=0.0264$ FWHM=85.9 $cm^{-1}$ | |
| Peak location ($cm^{-1}$) | Intensity | Peak location ($cm^{-1}$) | Intensity | Peak location ($cm^{-1}$) | Intensity |
| 16,126.2 | 0.3509 | 16,126.5 | 0.4744 | 16,126.5 | 0.4761 |
| 16,211.2 | 0.7727 | 16,209.1 | 0.8730 | 16,208.9 | 0.8738 |
| 16,314.5 | 0.6741 | 16,297.1 | 0.7373 | 16,296.3 | 0.7333 |
| | | 16,365.6 | 0.2443 | 16,363.6 | 0.2543 |
| | | | | 16,491.2 | 0.0250 |
| $EuZn_2In_6$ at 77K | | | | | |
| 3 Lorentzians $S^2=0.0383$ FWHM-89.9 $cm^{-1}$ | | 4 Lorentzians $S^2=0.0336$ FWHM=80.8 $cm^{-1}$ | | 5 Lorentzians $S^2=0.0335$ FWHM=81.1 $cm^{-1}$ | |
| 16,142.8 | 0.3730 | 16,142.0 | 0.4174 | 16,141.8 | 0.4162 |
| 16,214.1 | 0.7388 | 16,213.2 | 0.7594 | 16,213.3 | 0.7615 |
| 16,305.5 | 0.6850 | 16,296.4 | 0.6179 | 16,296.6 | 0.6236 |
| | | 16,344.4 | 0.1808 | 16,346.1 | 0.1823 |
| | | | | 16,522.8 | 0.0001 |

| $Eu_3Zn_2In_6$ at 77K | | $Eu_3Zn_2In_6$ at 293K |
|---|---|---|
| 3 Gaussians $S^2 = 0.0408$ | 4 Gaussians $S^2 = 0.0242$ | 4 Gaussians $S^2 = 0.0135$ |
| 3 Lorentzians $S^2 = 0.0287$ | 4 Lorentzians $S^2 = 0.0179$ | 4 Lorentzians $S^2 = 0.0372$ |

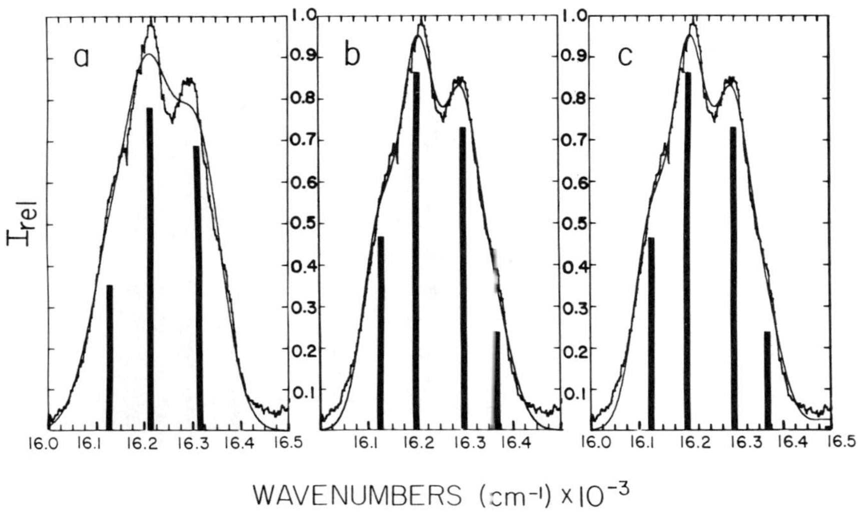


Fig. 4. Emission spectrum of $EuZn_2In_6$ at 77K showing the results of multiple linear least-squares regressions for (a) 3 Gaussians, (b) 4 Gaussians, and (c) 5 Gaussians. Solid bars indicate individual peak heights and locations.

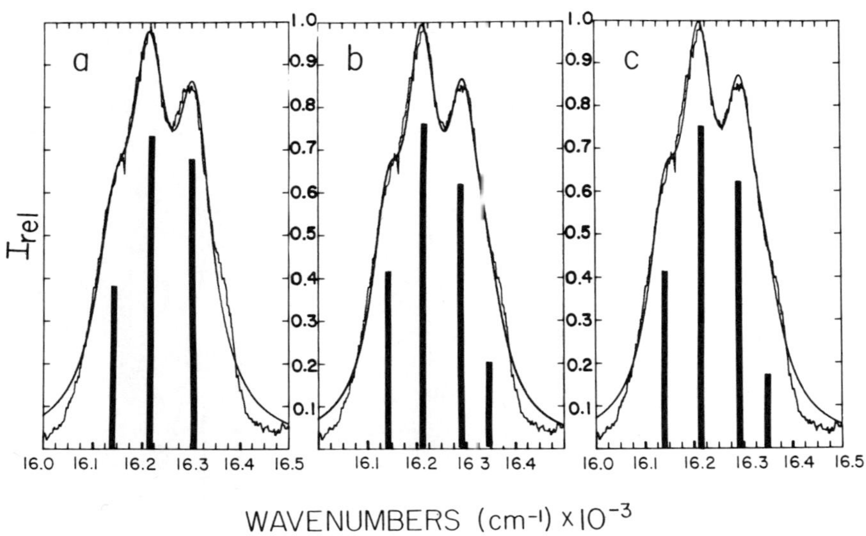


Fig. 5. Emission spectrum of $EuZn_2In_6$ at 77K showing the results of multiple linear least-squares regressions for (a) 3 Lorentzians, (b) 4 Lorentzians, and (c) 5 Lorentzians. Solid bars indicate peak heights and locations.

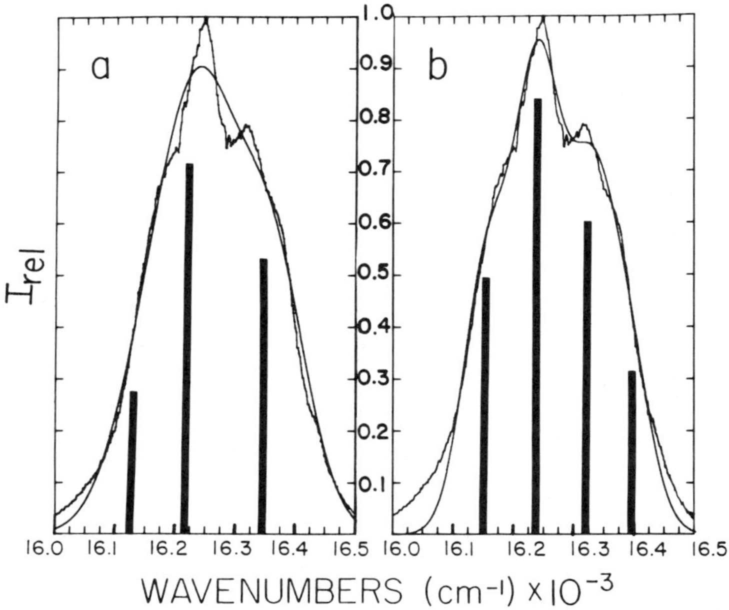


Fig. 6. Emission spectrum of $Eu_3Zn_2In_6$ at 77K showing the results of multiple linear least-squares regressions for (a) 3 Gaussians and (b) 4 Gaussians. Solid bars indicate individual peak heights and locations.

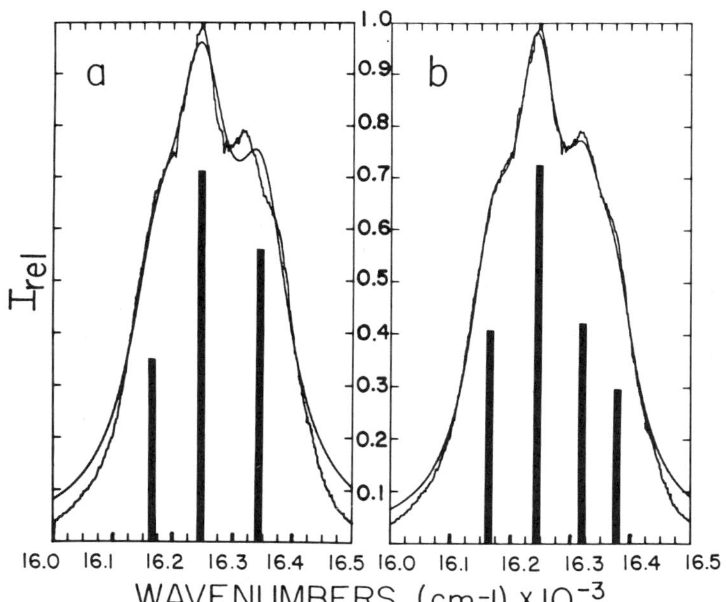


Fig. 7. Emission spectrum of $Eu_3Zn_2In_6$ at 77K showing the results of multiple linear least-squares regressions for (a) 3 Lorentzians and (b) 4 Lorentzians. Solid bars indicate individual peak heights and peak locations.

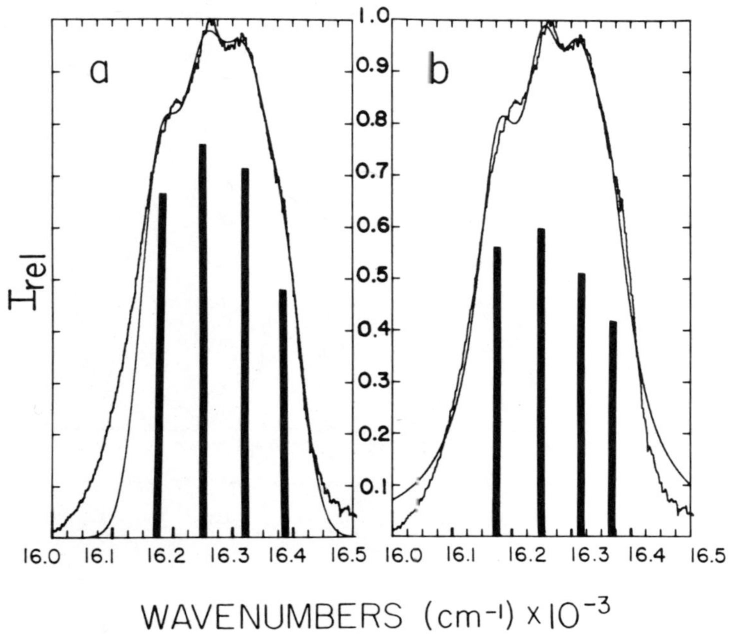


Fig. 8. Emission spectrum of $Eu_3Zn_2In_6$ at 293K showing the results of multiple linear least-squares regressions for (a) 4 Gaussians and (b) 4 Lorentzians. Solid bars indicate peak heights and peak locations.

equations presented in Table 4 are not able to account for any asymmetric distortion of the octahedral complex which is most likely responsible for the skewing of each pair of doublets. Figure 10b gives the average value of the peak heights for the pair of doublets obtained from $EuZn_2In_6$ at 77K (see Figs. 9c and 10a).

By using the angle which minimizes the $|20\rangle$ transition (39°) and assuming the Eu(III) lies in the plane of the carboxyl groups, the Eu(III)/oxygen distances can be estimated to be $1.80 \pm 0.2$ Å (see Fig. 11). This error is based on the inability of the spectrometer to detect the intensity of the $|20\rangle$ transition below a relative value of 0.05 (Fig. 10).

## CONCLUSIONS

A relative intensity analysis of the laser-induced fluorescence fine structure of Eu(III) bound in the two-zinc insulin hexamer is most consistent with a model in which the Eu(III) atoms are situated in a distorted octahedral

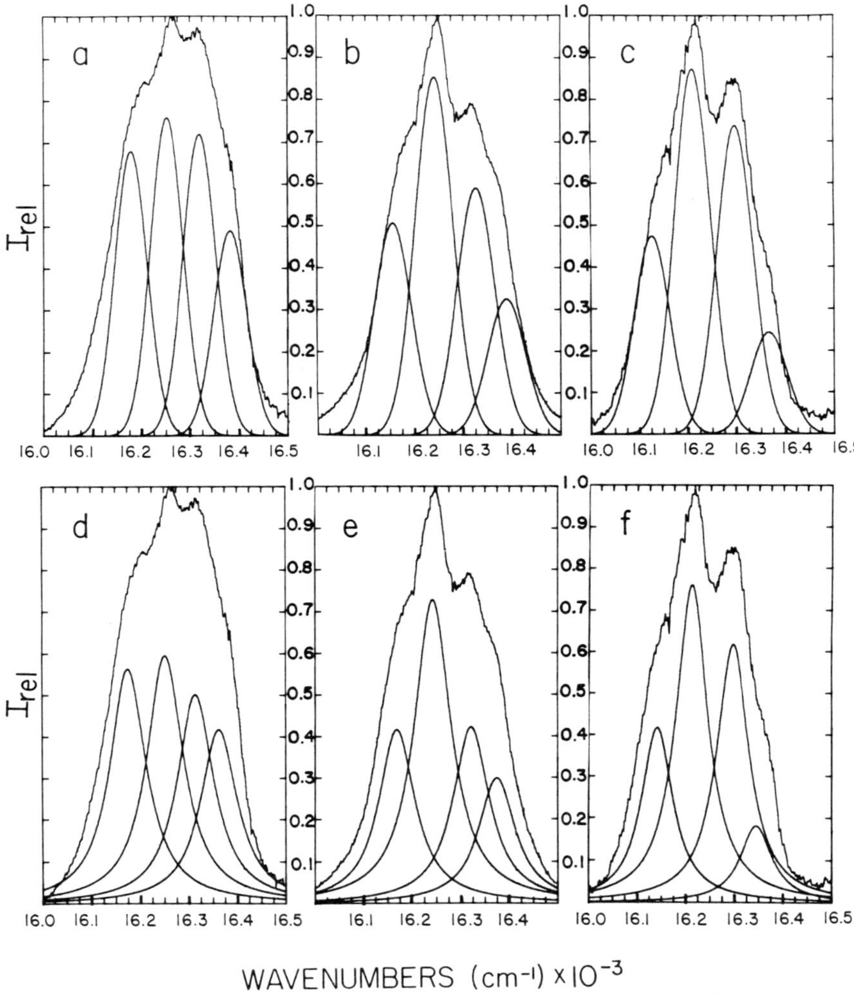


Fig. 9. Emission spectra showing the results of deconvolution by multiple linear least-squares regression for (a) $Eu_3Zn_2In_6$ at 293K and 4 Gaussians, (b) $Eu_3Zn_2In_6$ at 77K and 4 Gaussians, (c) $EuZn_2In_6$ at 77K and 4 Gaussians, (d)$Eu_3Zn_2In_6$ at 293K and 4 Lorentzians, (e) $Eu_3Zn_2In_6$ at 77K and 4 Lorentzians, and (f) $EuZn_2In_6$ at 77K and 4 Lorentzians.

environment. This result is consistent with previous fluorescence studies of insulin-bound Eu(III). Since Eu(III) is known to isomorphically replace Ca(II) in proteins, we suggest that the same coordination number exists for insulin-bound Ca(II). We suggest further that a coordination number of only six for Eu(II) [or Ca(II)] bound in insulin is due to the lack of available solvent molecules at the center of the insulin hexamer.

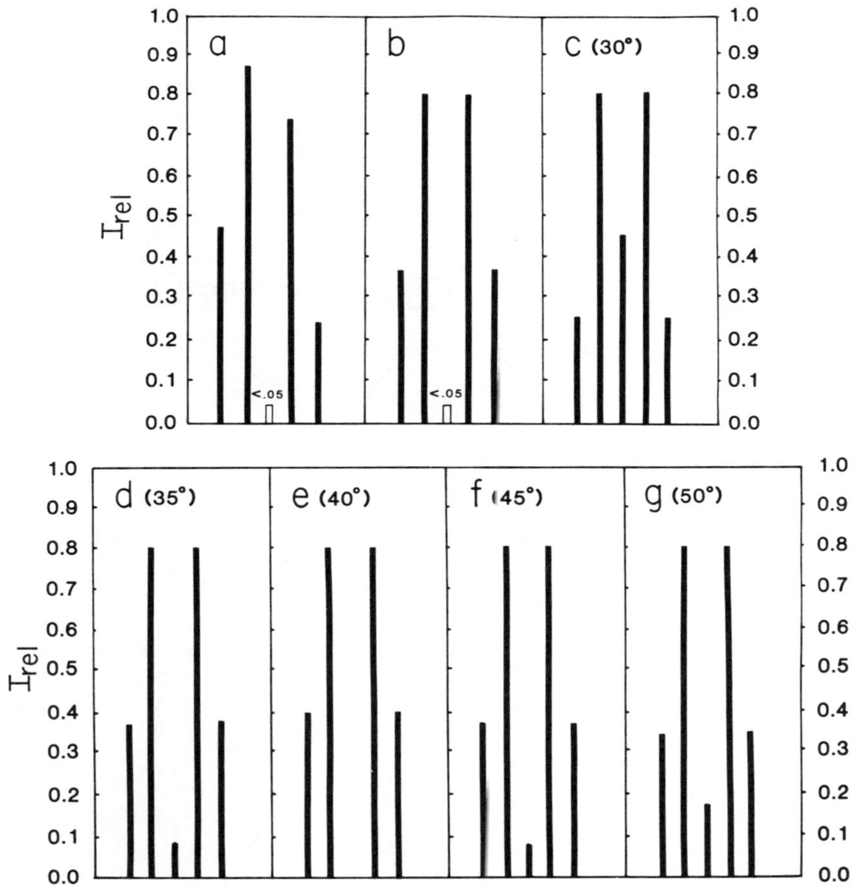


Fig. 10. Results of numerical simulation of the peak heights for the octahedral complex as a function of the angle $\theta$, the angle between the twofold axis and an oxygen ligand from the B-13 glutamate (see Fig. 11). Figure 10b gives the average value of the peak heights for each pair of doublets from the deconvolution of the $EuZn_2In_6$ spectrum at 77K (Figs. 9c and 10a). It can be seen that the best fit to the experimental data occurs at an angle of distortion near 40°.

## ACKNOWLEDGMENTS

This work was supported in part by grants to R.R.B. from the National Institutes of Health (EY-02202) and the National Science Foundation (CHE-7916336). The authors gratefully acknowledge helpful discussions and exchange of unpublished data with G.G. Dodson. We would also like to acknowledge D.F. Bocian, M.F. Dunn, M.C. Storm, and J.L. Sudmeier for their helpful comments throughout this investigation.

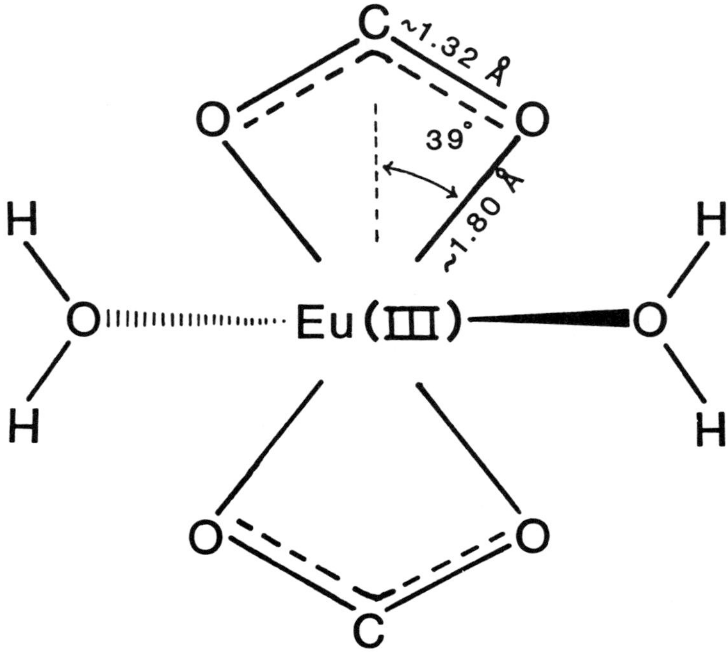


Fig. 11. Three-dimensional structure representation of Eu(III) coordination geometry in the Ca(II) binding site of the two-Zn(II) insulin hexamer. Based upon the fluorescence fine structure, the Eu(III) oxygen bond angle should be 39°. Assuming a carbon-oxygen bond distance of 1.32 Å, the Eu(III)-oxygen bond length should be approximately 1.80 Å.

## REFERENCES

Alameda GK, Bennett J, Birge RR (1981): General purpose microcomputer based data acquisition and laser control circuitry for use in one-photon laser spectroscopy. Rev Sci Instrum 52:1664.

Alameda GK, Sudmeier JL, Evelhoch JL, Birge RR (1985): Characterization of the calcium(II) binding sites in the insulin hexamer using europium(III) fluorescence as a probe. Biochemistry 24:1757.

Anderson T, Berggren P-O (1979): Characterization of exchangeable and nonexchangeable calcium in β-cell granules. Acta Endocrinol S277:8.

Axe J (1963): Radiative transition probabilities within $4f^n$ configurations: The fluorescence spectrum of europium ethylsulfate. J Chem Phys 39:1154.

Blundell T, Dodson G, Hodgkin P, Mercola D (1972): Insulin: The structure in the crystal and its reflection in the chemistry and biology. Adv Prot Chem 26:279.

Broer LJF, Gorter CJ, Hoogschagen J (1945): On the intensities and the multipole character in the spectra of the rare earth ions. Physica 11:231.

Condon, EU, Shortly GH (1935) "The Theory of Atomic Spectra." Cambridge University Press.

Hellman B, Sehlin J, Täljedal I-B (1976): Calcium and secretions: Distinction between two pools of glucose-sensitive calcium in pancreatic islets. Science 194:1421.

Hellman B (1977): Mechanism of insulin secretion. In Esmann V (ed): "Regulatory Mechanisms of Carbohydrate Metabolism." London: Pergamon Press, p 41.
Herman L, Sato T, Hales CH (1973): The electron microscope localization of cations of pancreatic islets of Langerhans and their possible role in insulin secretion. J Ultrastruct Res 42:298.
Howell SL, Montaque W, Tyhurst M (1975): Calcium distribution in islets of Langerhans: A study of calcium concentrations and of calcium accumulations in $\beta$-cell organelles. J Cell Sci 19:395.
Howell SL, Tyhurst M, Duvefelt H, Anderson A, Hellerström C (1978): Role of zinc and calcium in the formation and storage of insulin in the pancreatic $\beta$-cell. Cell Tissue Res 188:107.
Judd BR (1962): Optical absorption intensities of rare-earth ions. Phys Rev 127:750.
Judd BR (1963): "Operator Techniques in Atomic Spectroscopy." New York: McGraw-Hill.
Kohnert KD, Hahn HJ, Gylfe E, Borg H, Hellman B (1979): Calcium and pancreatic $\beta$-cell functions. Mol Cell Endocrinol 16:205.
Korolkov VS, Makhanek AG, Kuznetsova VV (1966): The intensity distribution of the luminescence spectra of Eu coordination compounds. Opt Spectry 21:246.
Korolkov VS, Makhanek AG, Kuznetsova VV (1967): On the effects of the symmetry of environment on the luminescence spectrum of the Eu(III) ion. Opt Spectry 23:499.
Malaisse WJ, Herchuelz A, Levy J, Somers G, Devis G, Ravazzola M, Malaisse-Lagae F, Orci L (1975): Insulin release and the movements of calcium in pancreatic islets. In Carifoli E (ed): "Calcium Transport in Contraction and Secretion. Amsterdam: North-Holland, p. 211.
Matthews BW, Weaver LH (1974): Binding of lanthanide ions to thermolysin. Biochemistry 13:1719.
Moews PC, Kretsinger RH (1975): Terbium replacement of calcium in carp muscle calcium-binding parvalbumin: An x-ray crystallographic study. J Mol Biol 91:229.
Morgan JR, Chock EP, Hopewell WD, El-Sayed MA, Orbach R (1981): Origins of homogeneous and inhomogeneous line widths of the $^5D_0$–$^7F_0$ transition of Eu(III) in amorphous solids. J Phys Chem 85:747.
Neiboer J (1975): The lanthanide ions as structural probes in biological and model systems. Struct Bond 22:1.
Ofelt GS (1962): Intensities of crystal spectra of rare-earth ions. J Chem Phys 37:511.
Pitts JE, Wood SP, Horuk S, Bedarkar S, Blundell TL (1980): Pancreatic hormone storage granules. The role of metal ions and polypeptide oligomers. In: "Insulin Chemistry, Structure and Function of Insulin and Related Hormones." New York: de Gruyter, p. 63.
Racah G (1943): Theory of complex spectra. Phys Rev 63:367.
Ravazzola M, Malaisse-Langae F, Amherdt M, Perrelet A, McManse WJ, Orci L (1976): Patterns of calcium localization in pancreatic endocrine cells. J Cell Sci 27:107.
Ruben J (1975): The lanthanide as spectroscopic and magnetic resonance probes in biological system. NaturwissenSchaften 62:172.
Ruben J (1977): Lanthanides as spectroscopic and magnetic resonance probes. In Wassermann RH (ed.): "Calcium Binding Proteins." New York: North-Holland, p. 21.
Schäfer H-J, Klöppel G (1974): The significance of calcium in insulin secretion. Virchows Arch A 362:231.
Sudmeier JL, Bell SJ, Storm MC, Dunn MC (1981): Cadmium-113 studies of bovine insulin; the two-zinc insulin hexamer specifically binds one calcium ion. Science 212:560.
Van Vleck JH (1937): The puzzle of rare-earth spectra in solids. J Phys Chem 41:67.
Wollheim CB, Kikuchi M, Renold AE, Sharp GW (1978): The roles of intercellular and extracellular Ca(II) in glucose-stimulated biphasic insulin release by rat islets. J. Clin Invest 801:451.
Wybourne BC (1965): "Spectroscopic Properties of Rare Earths." New York: Wiley and Sons.

*Applications of Fluorescence in the Biomedical Sciences, pages 225–244*


# 10 Biochemical Applications of Frequency-Domain Fluorometry

*Joseph R. Lakowicz*

Fluorescence spectroscopy is often used in biochemical research because of the suitable time scale of the phenomenon. Emission typically occurs within nanoseconds of light absorption, which is also the time scale of dynamic processes in biochemical samples. These processes include decay of the fluorescence intensity, rotational diffusion of proteins and membrane-bound fluorophores, and spectral relaxation of biopolymer-bound fluorophores. Because of the complexity of biological macromolecules it is often necessary to resolve multiexponential decays of fluorescence intensity or anisotropy. For example, the decay times of fluorophores are sensitive to their local environment, and biological macromolecules often contain fluorophores in several distinct locations. It is desirable to resolve the emitting species, which may be accomplished by resolution of the decay law of the fluorescence emission. In this chapter we describe the use of a relatively new method, variable frequency phase-modulation fluorometry, to resolve complex fluorescence decay processes.

Considerable information is also contained in the time-resolved decays of fluorescence anisotropy. For instance, the anisotropy decay depends upon the size, shape, and rigidity of the rotating unit. Segmental motions of aromatic amino acids in proteins are revealed by apparent correlation times which are too rapid to be explained by overall rotational diffusion [Lakowicz

Department of Biological Chemistry, University of Maryland School of Medicine, Baltimore, Maryland 21201.

et al., 1983a; Lakowicz and Maliwal, 1983]. Hence the hydrodynamic and dynamic properties of the macromolecule can be revealed by the time-resolved decays of fluorescence anisotropy.

Time-resolved emission spectra are also sensitive to the dynamics of macromolecules. In this case the macromolecule responds to the altered electronic charge distribution of the excited state. This response probably consists of small changes in distance and/or orientation between polar groups on the fluorophore and on the macromolecule. These geometric changes are seen as time-dependent shifts in the emission spectra to lower energies or longer wavelengths. At present, the molecular interpretation of these shifts is difficult and ambiguous, primarily because most theories of environmental effects on fluorescence spectra regard the environment as a continuum. In reality, the macromolecule provides a specific environment for the fluorophore, and theories are needed to account for such effects. One approach is to divide the effects into specific and general solvent effects, as reviewed previously [Lakowicz, 1983]. In this way one distinguishes the pseudocontinuous dielectric properties of the solvent from its potential to form specific hydrogen bonds with the fluorophore. A more detailed and preferable approach is to specifically treat the fluorophore-solvent interactions, as suggested by MacGregor and Weber [1981]. In the anticipation that such theories will be developed we described time-resolved emission spectra for several limiting cases, such as a stepwise relaxation process and a continuous relaxation process.

In another chapter (Lakowicz, this volume) we reviewed the two dominant methods for resolution of fluorescence decay kinetics, these being time-correlated photon counting and frequency-domain fluorometry. In this chapter we describe in detail the use of frequency-domain fluorometry for determination of anisotropy decays and time-resolved emission spectra. Phase-modulation fluorometers which operate over a range of frequencies (1–220 MHz) have been recently described [Gratton and Limkemann, 1983; Lakowicz and Maliwal, 1985]. The operational principles of these instruments are described in the above references.

## ANALYSIS OF FREQUENCY-DOMAIN DATA

### Decay of Fluorescence Anisotropy

Frequency-domain fluorometers appear to be particularly well suited for the measurement of anisotropy decays. This is probably because the signal due to rotational diffusion is measured directly and because this phase angle does not increase monotonically with frequency, but, rather, is nearly a Lorentzian. Additionally, frequency-domain fluorometers provide two measurements of the anisotropy decay, the phase shift (Δ) and the ratio of polarized modulated amplitudes (Λ).

Assume a sample is excited with amplitude-modulated light at a circular modulation frequency $\omega$, polarized vertically (Fig. 1). The emission is observed through a polarizer which may be oriented parallel (||) or perpendicular (⊥) to the excitation polarization. A second PMT is sometimes used as a phase reference, but this is often unnecessary with modern instruments. The phase angle difference (Δ) between the polarized components is determined by the details of the anisotropy decay. The origin of the phase shift Δ may be understood as follows. The phase angle of the perpendicular component (· · ·) is longer than the rotation-free phase angle (——) because this polarizer selects for those molecules which have rotated prior to emission. Conversely, the phase angle of the parallel component (---) is shorter because the fluorophores which emit with this polarization are decaying by both emission and rotation away from the vertical orientation. In addition to these phase angles we also use the ratio of the polarized modulated emission (Λ). This ratio varies from that expected for the steady-state anisotropy (low

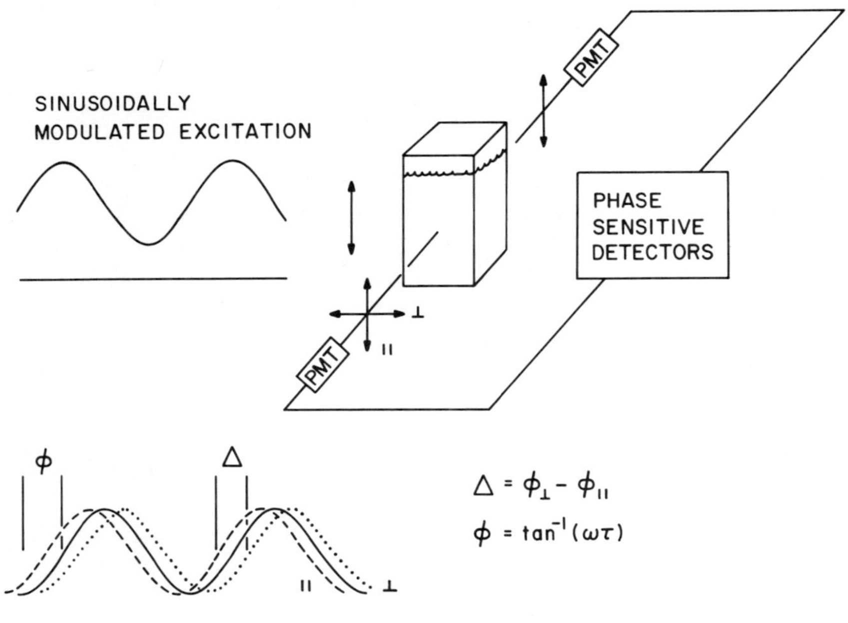


Fig. 1. Schematic description of differential polarized phase fluorometry [from Lakowicz, 1983].

frequency) to that expected in the absence of rotational diffusion (high frequency). Additional details are presented elsewhere [Lakowicz et al., 1985; Lakowicz and Maliwal, in preparation].

The time-resolved anisotropies are determined by the method of nonlinear least squares. One measures the frequency-dependent phase difference ($\Delta_\omega$) between the perpendicular ($\phi_\perp$) and the parallel ($\phi_\parallel$) components of the modulated emission ($\Delta_\omega = \phi_\perp - \phi_\parallel$) and the amplitude ratio ($\Lambda_\omega$) of the parallel ($m_\parallel$) and the perpendicular ($m_\perp$) components of the modulated emission ($\Lambda = m_\parallel / m_\perp$). These values are compared with those expected on the basis of an assumed anisotropy decay law.

Following $\delta$-function excitation the decays of the parallel and perpendicular components of the emission are given by

$$I_\parallel(t) = \frac{1}{3} I(t)\ [1 + 2r(t)] \tag{1}$$

$$I_\perp(t) = \frac{1}{3} I(t)\ [1 - 2r(t)] \tag{2}$$

where $r(t)$ is the time-resolved decay of anisotropy. Generally, $r(t)$ can be described as a multiexponential decay,

$$r(t) = r_o \sum_i g_i\, e^{-t/\theta_i} \tag{3}$$

where $r_o$ is the limiting anisotropy in the absence of rotational diffusion, $\theta_i$ the individual correlation times, and $g_i$ the associated amplitudes. The expected values of $\Delta_\omega$ ($\Delta_{c\omega}$) and $\Lambda_\omega$ ($\Lambda_{c\omega}$) can be calculated from the sine and cosine transforms of the individual polarized decays:

$$N_i = \int_0^\infty I_i(t) \sin \omega t\, dt \tag{4}$$

$$D_i = \int_0^\infty I_i(t) \cos \omega t\, dt. \tag{5}$$

The frequency-dependent values of $\Delta_\omega$ and $\Lambda_\omega$ are given by

$$\Delta_{c\omega} = \arctan \left( \frac{D_\parallel N_\perp - N_\parallel D_\perp}{N_\parallel N_\perp + D_\parallel D_\perp} \right) \tag{6}$$

$$\Lambda_{c\omega} = \left( \frac{N_\parallel{}^2 + D_\parallel{}^2}{N_\perp{}^2 + D_\perp{}^2} \right)^{1/2} \tag{7}$$

where the $N_i$ and $D_i$ are calculated at each frequency. The parameters describing the anisotropy decay are obtained by minimizing the squared deviations between measured and calculated values,

$$\chi^2 = \sum_{\omega} \frac{1}{\sigma_{\Delta\omega}^{2}} (\Delta_\omega - \Delta_{c\omega})^2 + \sum_{\omega} \frac{1}{\sigma_{\Lambda\omega}^{2}} (\Lambda_\omega - \Lambda_{c\omega})^2 \tag{8}$$

In this expression $\sigma_{\Delta_\omega}$ and $\sigma_{\Lambda_\omega}$ are the estimated experimental uncertainties in the measured quantities ($\Delta_\omega$ and $\Lambda_\omega$). The goodness-of-fit is usually judged by the value of reduced $\chi^2$,

$$\chi_R^2 = \frac{\chi^2}{\nu} = \frac{\chi^2}{2N - p} \tag{9}$$

where the number of degrees of freedom is given by $\nu = 2N - p$. $N$ is the number of modulation frequencies and $p$ is the number of floating parameters. If the assumed model is appropriate for the sample, and if the errors are random and properly estimated by $\sigma_{\Delta\omega}$ and $\sigma_{\Lambda\omega}$, then $\chi^2_R$ is expected to be near unity.

We used three models to describe the decays of anisotropy. For an isotropic rotator we used

$$r(t) = r_o\, e^{-t/\theta} \tag{10}$$

For the anisotropic rotators perylene and 9-aminoacridine we used an anisotropy decay law with two or three correlation times,

$$r(t) = r_o \sum_i g_i\, e^{-t/\theta_i} \tag{11}$$

with $\Sigma_i\, g_i = 1$, $g_i$ as floating parameters and $r_o$ fixed. For perylene in lipid bilayers we used the hindered model

$$r(t) = (r_o - r_\infty)e^{-t/\theta} + r_\infty \tag{12}$$

where $r_\infty$ is the limiting anisotropy observed at times longer than the rotational correlation time of the fluorophore. Equation 11 with two and three correlation times was also used for perylene.

**Calculation of Time-Resolved Emission Spectra**

Construction of time-resolved emission spectra from the frequency-domain data is a straightforward extension of the analysis of multiexponential decays [Lakowicz et al., 1984a,b]. Time-resolved emission spectra can be calculated from the impulse response functions, if these are known at each emission wavelength. The impulse response functions are determined by using measurements of the phase $[\phi_\omega(\lambda)]$ and modulation $[m_\omega(\lambda)]$ of the emission over a range of emission wavelengths ($\lambda$) and modulation frequencies ($\omega$). Typically, we use about ten emission wavelengths and ten modulation frequencies for each wavelength. These data were fit, by using the method of nonlinear least squares, to the function

$$I(\lambda,t) = \sum_{i=1}^{3} \alpha_i(\lambda)\, e^{-t/\tau_i} \tag{13}$$

In the presence of solvent relaxation the decay at each emission wavelength $\lambda$ is probably more complex than a sum of exponentials. However, a three-exponential decay is usually adequate to represent the measured quantities, and hence adequate for the construction of the time-resolved spectra.

Calculation of the time-resolved emission spectra requires that the impulse response functions be normalized [Easter et al., 1976]. The total intensity at each wavelength must be normalized to the steady-state intensity at this same wavelength $[F(\lambda)]$. Let

$$H(\lambda) = \frac{F(\lambda)}{\int_0^\infty I(\lambda,t)\, dt} \tag{14}$$

$$= \frac{F(\lambda)}{\alpha_1(\lambda)\tau_1 + \alpha_2(\lambda)\tau_1 + \alpha_3(\lambda)\tau_3} \tag{15}$$

Then, the appropriately normalized functions are given by

$$I'(\lambda,t) = H(\lambda)\, I(\lambda,t) \tag{16}$$

$$= \sum_i \alpha'_i(\lambda)\, e^{-t/\tau_i} \tag{17}$$

where $\alpha_i'(\lambda) = H(\lambda)\, \alpha_i(\lambda)$. For any desired time ($t_i$), the time-resolved emission spectrum is obtained by plotting the values of $I'(\lambda,t_i)$ for all the measured wavelengths.

## ANISOTROPY DECAYS FROM FREQUENCY-DOMAIN FLUOROMETRY

At present there is considerable interest in measuring nanosecond or picosecond anisotropy decays of protein fluorescence or of labeled macromolecules. The impetus for such measurements is the result of quenching measurements which indicate that proteins possess significant segmental flexibility [Lakowicz and Weber, 1973a,b; Eftink and Ghiron, 1981] and of molecular dynamic calculations which predict fluctuations on the picosecond time scale [Karplus and McCammon, 1981; McCammon and Karplus, 1980; Gurd and Rothgeb, 1979].

Frequency-domain fluorometry can provide reliable measurements of anisotrophy decays on the picosecond timescale. This is illustrated by measurements of the rotational diffusion of POPOP in ethanol at 45°C and in hexane at 40°C (Fig. 2). The phase angle differences between the perpendicular and parallel components of the emission are shown in Figure 2. The differential polarized phase angle ($\Delta_\omega$) is dependent upon the rotational rate. For a single correlation time a plot of $\Delta_\omega$ versus log $\omega$ is nearly Lorentzian in shape. This shape is not evident in Figure 2 because the rotational rate of POPOP is fast in these solvents, and the maximum value of $\Delta_\omega$ is displaced beyond our maximum frequency of 200 MHz. The lines represent the calculated values of $\Delta_\omega$ for the correlation times shown on Figure 2. We could easily determine correlation times as short as 0.11 and 0.047 nsec for POPOP in ethanol and hexane at 45°C, respectively. For such rapid diffusive motions it is difficult to know if the decay of anisotropy is single or double exponential. In our measurements there was no substantial improvement in $\chi_R{}^2$ upon inclusion

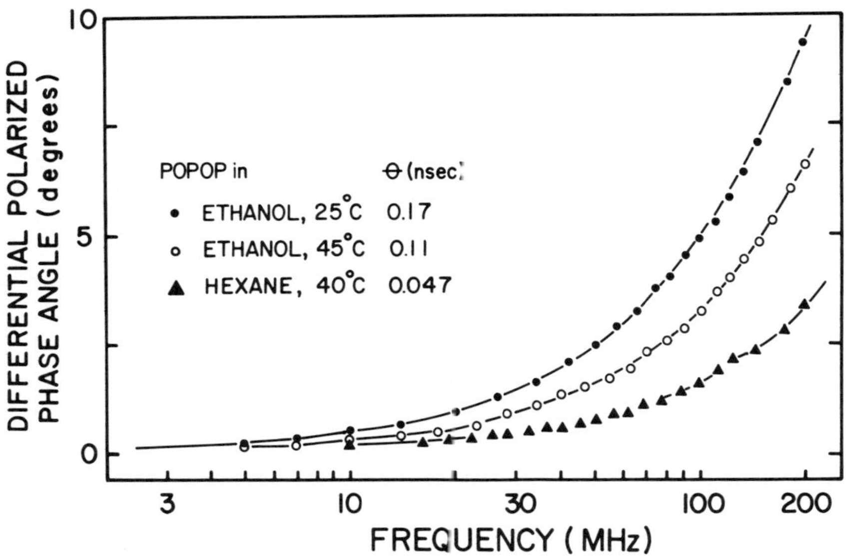


Fig. 2. Determination of picosecond anisotropy decays by using frequency-domain fluorometry. Differential polarized phase angles are shown for p-bis[2-(5-phenyloxazolyl)] benzene (POPOP) in fluid solvents [from Lakowicz and Maliwal, 1985].

of an additional correlation time. It seems probable that the frequency range of our instrument can be extended to 400 MHz. Hence, the measurements may soon be directly comparable to molecular dynamics calculations on proteins.

**Resolution of Multiexponential Decays of Fluorescence Anisotropy**

There are many instances in which the anisotropy decay is expected to be more complex than a single exponential. Two important cases are due to the asymmetrical shapes of proteins and to the complex diffusive motions of probes in lipid bilayers. In these instances one attempts to recover the anisotropy decay parameters in order to characterize shape of the protein or the microenvironment in the membrane.

Preliminary data indicate that the frequency-domain fluorometry provides adequate resolution to determine complex anisotropy decays, even when these occur on the subnanosecond time scale. This is illustrated in Figure 3 for perylene in propylene glycol at 28°C. Perylene is a platelike molecule, and the apparent correlation times are known to be in the ratio of about 7 to 1 [Barkley et al., 1978]. It was not possible to account for the measured data (-•-, -△-) by using a single correlation time (---). This is evident from the mismatch seen in Figure 3, the systematically varying deviations between the measured and calculated values (Fig. 4), and the unacceptable value of $\chi_R^2 = 102$. Inclusion of a second correlation time in the fitting algorithm (equation 11) results in a decrease in $\chi_R^2$ to the expected value and random deviations (Fig. 4). (The low value of $\chi_R^2$ is because our errors are smaller than estimated.) It is remarkable that the value of $\chi_R^2$ decreases 340-fold upon inclusion of the second correlation time. This result indicates that we could resolve still more complex anisotropy decays, or a two-component decay for which the correlation time are shorter than the values of 0.2 and 1.3 nsec in Figure 3. This capability is likely to be valuable for interpreting the details of anisotropy decays of tryptophan residues in proteins in terms of known molecular features of their structures.

A still more difficult resolution is shown in Figure 5, which shows data for 9-aminoacridine in glycerol. Visually, the distinction between the one- and two-correlation time fits to these data was modest (Fig. 6). Statistically, the differences were substantial. At each temperature the value of $\chi_R^2$ decreased about threefold upon inclusion of the second correlation time. Additionally, at each temperature the ratio of the two correlation times was about 1.6. This result is significant in several ways. First, the correlation times are in good agreement with those obtained by Barkley et al. [1978] using time-correlated photon counting. However, these workers did not observe significant decreases in $\chi_R^2$, whereas our two correlation time fits are strongly supported by the analysis. For example, with 60 degrees of freedom there is

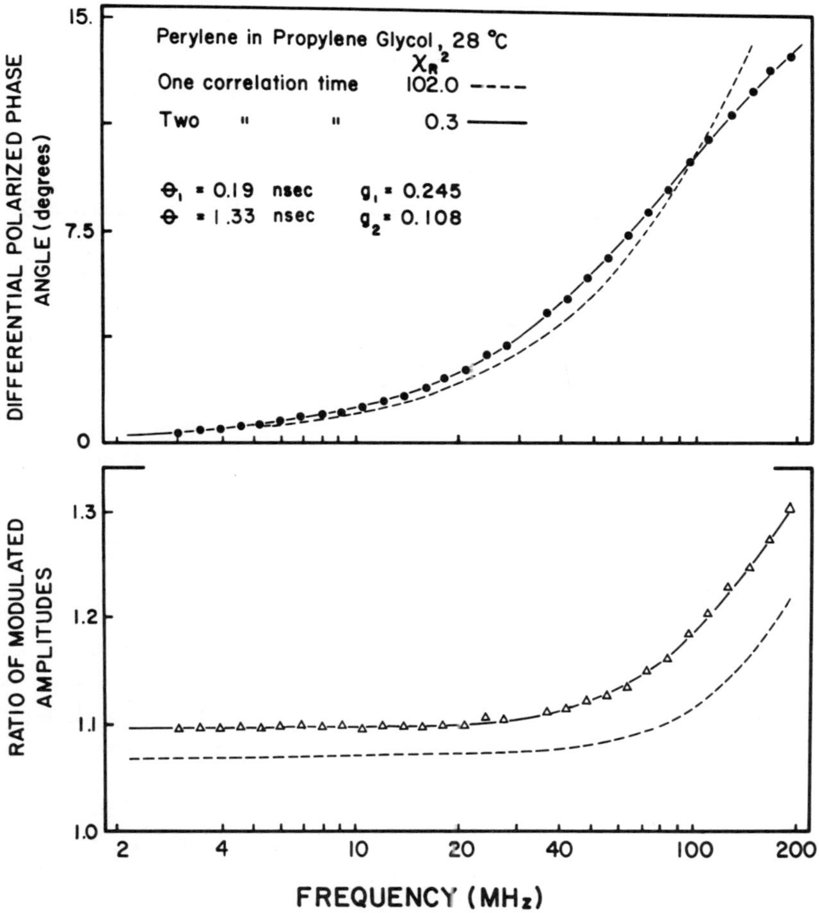


Fig. 3. Resolution of a two-component anisotropy decay on the subnanosecond time scale.

an insignificant probability (< 0.001) that the one-correlation time model is appropriate [Bevington, 1969]. Secondly, many proteins have axial ratios in the range of 1.5 and above. Consequently, we expect the frequency-domain method to provide useful measurements of the axial ratios of globular proteins.

Another example of a complex anisotropy decay is provided by the anisotropic molecule perylene in dimyristoyl-L-$\alpha$-phosphatidylcholine (DMPC) bilayers. Previous studies from this laboratory [Lakowicz and Knutson, 1980] indicated that below the transition temperature (< 25°C) the rotational motions of perylene were hindered by the bilayer. Since perylene is itself asymmetric and displays two correlation times in isotropic solvents, we

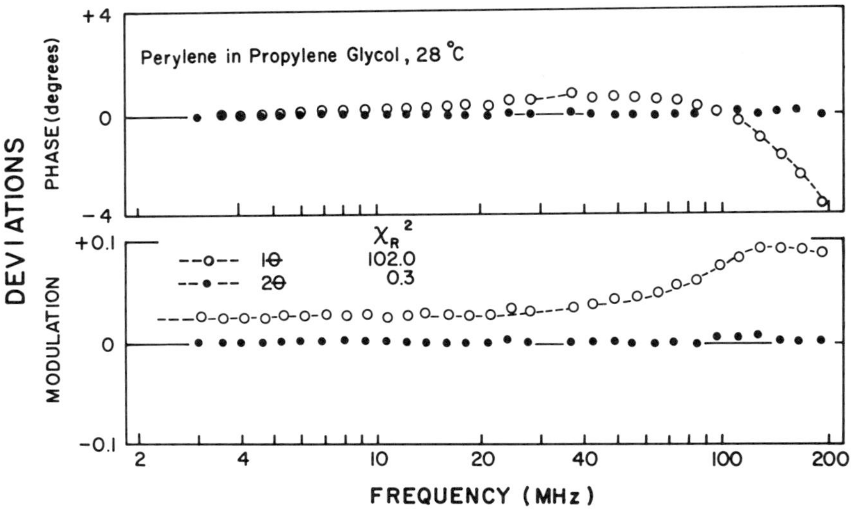


Fig. 4. Deviations between the measured and calculated data for perylene in propylene glycol at 28°C. Deviations are shown for the fit using one (--o--) and two (—•—) correlation times.

expected that three correlation times could be present in the membrane. These data and fits are shown in Figure 7. The hindered model (---; equation 12) was completely inadequate to account for the data ($\chi_R^2 = 106$). Even the fit with two correlation times (— —) is easily judged to be inadequate ($\chi_R^2 = 14.4$). The three-correlation time fit provides a reasonable value of $\chi_R^2 = 0.5$, and reasonable correlation times 0.2, 1.4, and 47 nsec). We attribute the two shorter correlation times to rapid motions of perylene within the nonhindered environment, and the 47-nsec correlation time to the hindered motion. It is remarkable that one could obtain a 28-fold decrease in $\chi_R^2$ for distinguishing two versus three rotational correlation times. To the best of our knowledge this is the first published resolution of three rotational rates.

As a final example of anisotropy decays from frequency-domain measurements data are shown for ethidium bromide intercalated into double-helical DNA (Fig. 8). It was unambiguously necessary to use three correlation times to account for the data as $\chi_R^2$ decreased tenfold relative to the two-correlation time fit. In this case the molecular interpretation of the three correlation times is not clear. These data probably require analysis with more complex anisotropy decay laws, as predicted by DNA by Barkley and Zimm [1979]. Such an analysis is not expected to be difficult, irrespective of the complexity of the anisotropy decay law. We now know it is possible and practical to use numerical transforms for equations 3 and 4 in place of the analytical transforms used until this time.

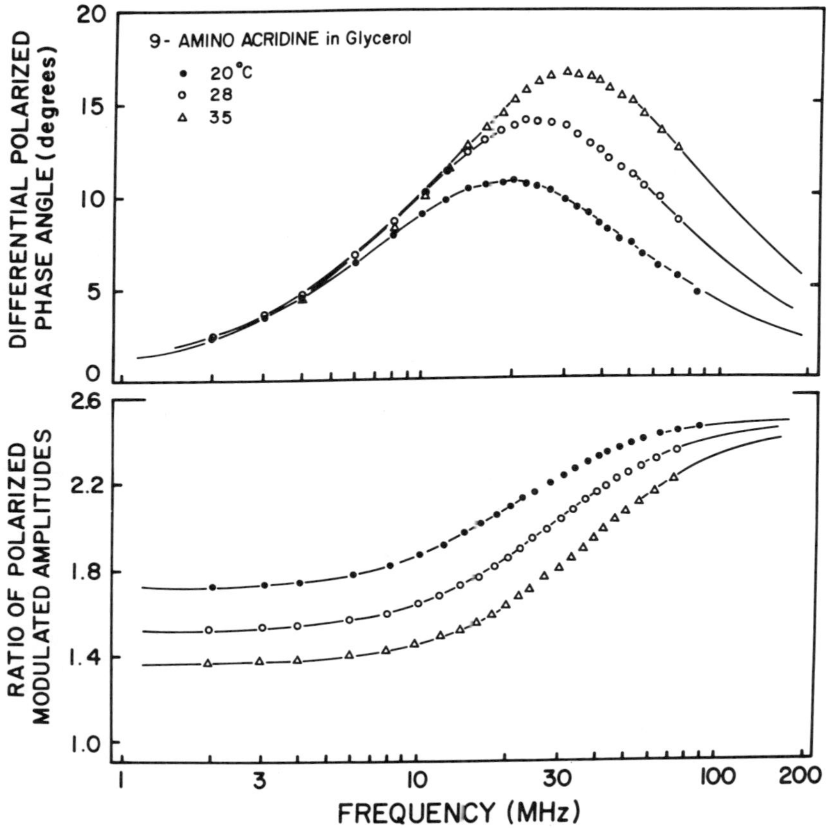


Fig. 5. Resolution of two closely spaced correlation times for 9-aminoacridine in glycerol. At each temperature the ratio of the two correlation times is near 1.6 to 1. The values of $\chi_R^2$ are approximately three and one for the fits using one and two correlation times, respectively [from Lakowicz and Maliwal, 1985].

In summary, the results presented in Figures 2–8 demonstrate that the frequency-domain method provides remarkable resolution of complex decays of fluorescence anisotropy.

## Time-Resolved Emission Spectra

Fluorophores in the excited state often undergo time-dependent interactions with their surroundings. For solvent-sensitive fluorophores such interactions result in time-dependent spectral shifts to lower energies or longer wavelengths. A generally held view is that only pulse fluorescence methods are capable of yielding time-resolved emission spectra. We now know that

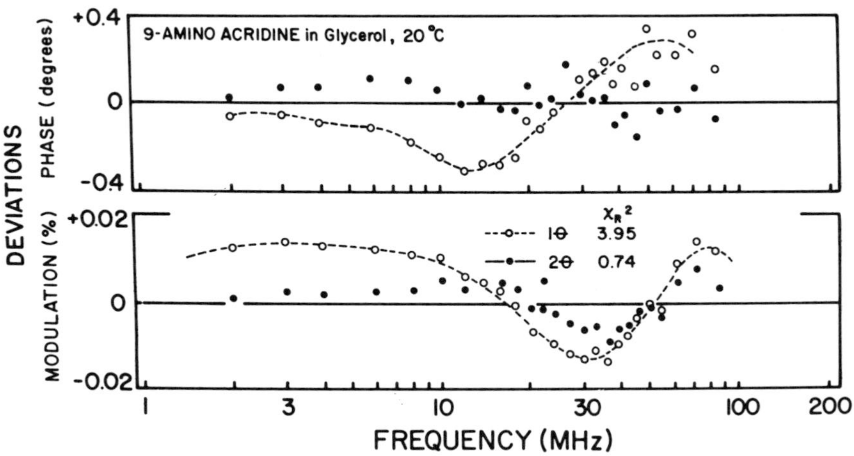


Fig. 6. Deviations between one and two correlation time fits for 9-aminoacridine in glycerol [from Lakowicz and Maliwal, 1985].

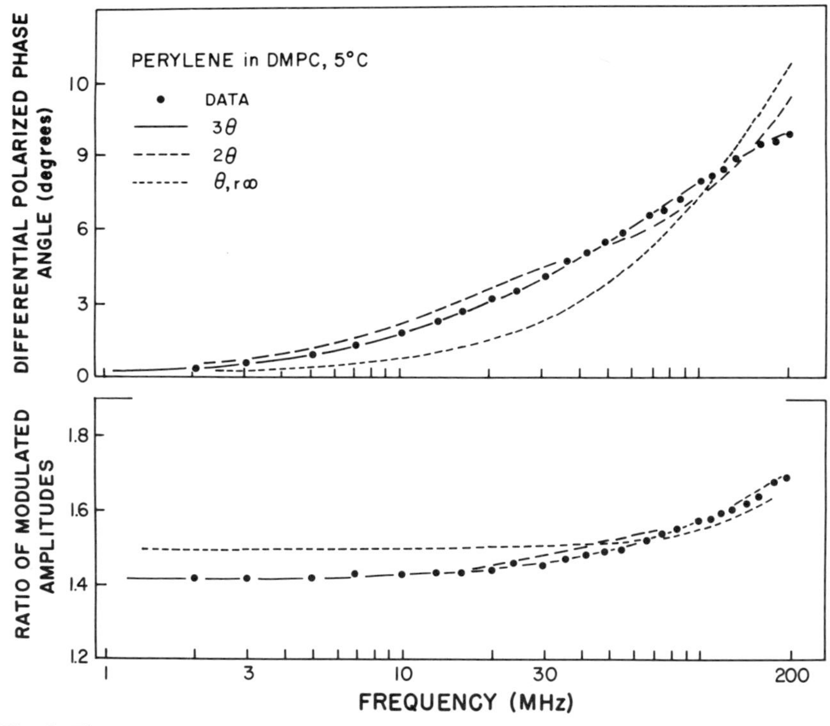


Fig. 7. Three-component anisotropy decay of perylene in DMPC bilayers. The values of $\chi_R^2$ for the three, two correlation time and hindered model fits are 0.5, 14.4, and 106, respectively [from Lakowicz et al., 1985].

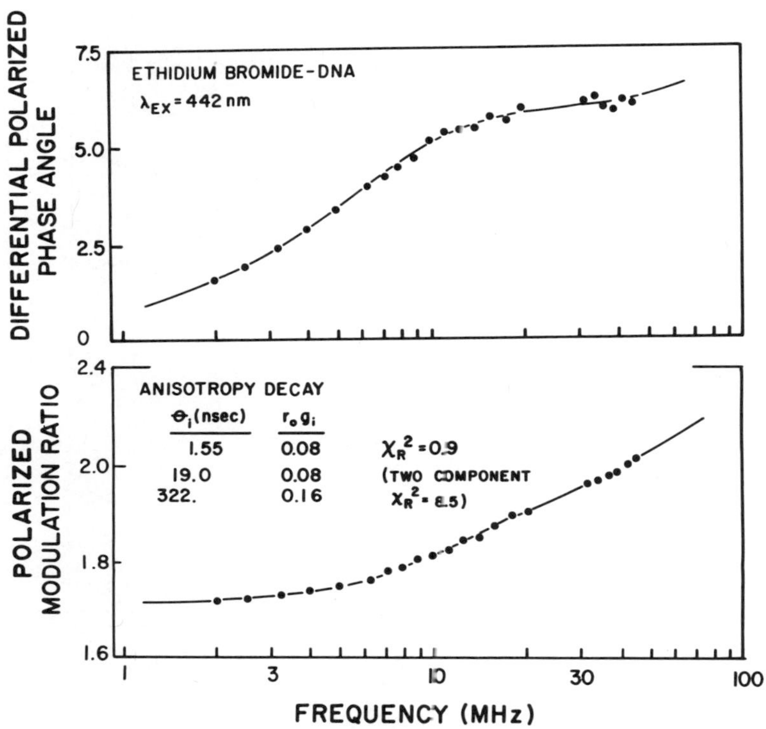


Fig. 8. Anisotropy decay of ethidium bromide bound to DNA.

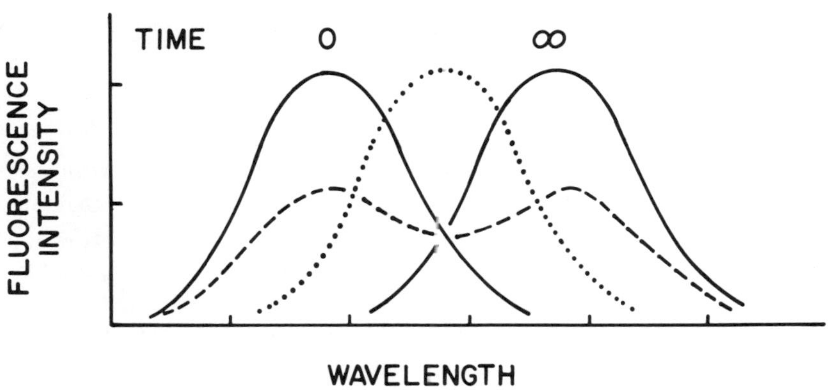


Fig. 9. Schematic description of continuous and stepwise models of solvent relaxation. At intermediate times a stepwise process displays two spectral components (---) whereas a continuous process displays only a single emission spectrum (...) [from Lakowicz et al., 1984].

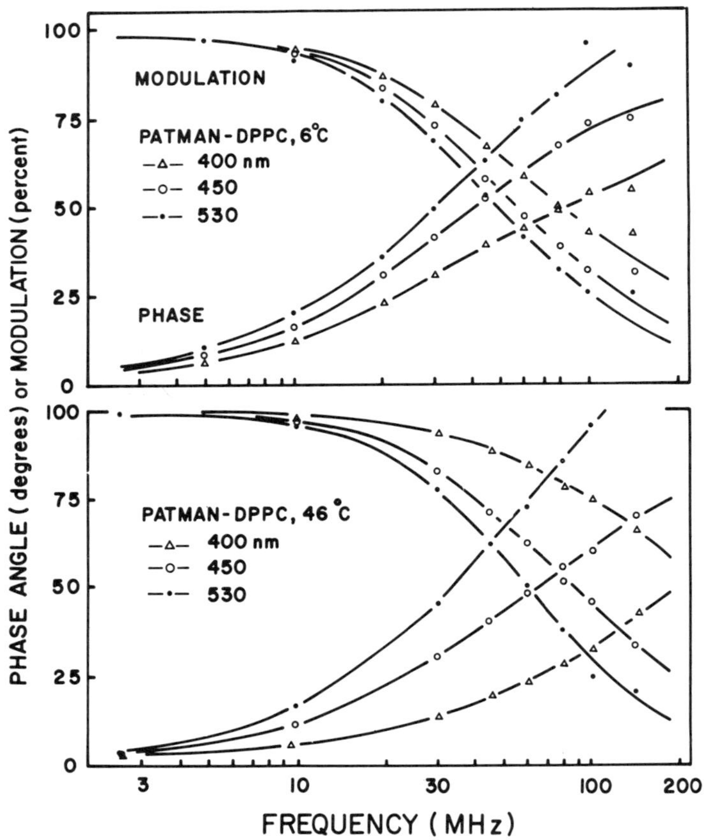


Fig. 10. Frequency-domain data for 6-palmitoyl-2-{[-(trimethylamino)ethyl] methylamino} naphthalene (PATMAN) labeled vesicles [from Lakowicz et al., 1984].

frequency-domain fluorometry can reveal considerable detail about time-dependent spectral shifts [Lakowicz et al., 1984a,b]. To illustrate the ability of frequency-domain fluorometry to determine time-resolved emission spectra we present data for spectral relaxation of the problem TNS (2-p-toluidinyl-6-naphthalenesulfonic acid) and PATMAN (6-palmitoyl-2-{[-(trimethylamino)ethyl] methylamino} naphthalene) bound to lipid vesicles. The comparative time-dependent properties of these probes illustrate continuous and stepwise relaxation, respectively.

The time-resolved emission spectra can be interpreted in terms of two phenomenological models, one being a continuous model and the second a two-state model. One can imagine these two models as being the result of specific (stepwise) and general (continuous) relaxation. A stepwise process could be the result of a specific hydrogen bonding reaction which shifts the spectrum, or some other discrete process such as complex formation or a molecular rearrangement. The continuous process could be the collective result of the small movements of a large number of molecules surrounding the fluorophore. The spectral properties of these extreme cases are illustrated in Figure 9. In both models, it is assumed that following a $\delta$-pulse excitation, the emission spectra shift from an initially excited state (0) to an equilibrium state ($\infty$). The centers of gravity of the emission spectra of these limiting states are given by $\bar{\nu}_0$ and $\bar{\nu}_\infty$, respectively.

Assume that the emission spectra could be observed at any desired instant. At longer times, the spectra will be shifted to longer wavelengths. For continuous relaxation, the normalized emission spectrum is assumed to slide to longer wavelengths, without any change in the emission spectral distribution. Hence, irrespective of the time chosen to observe the emission, a constant spectral shape is expected (...) and the time-dependent spectral widths are expected to be independent of time. Overtly different results are expected for the two-state model. In this case, the initially excited and the equilibrium states are assumed to be distinct species, each with a characteristic emission spectrum. As the reaction proceeds, the emission from the initially excited state decays, followed by the appearance of a new emission at longer wavelengths (Fig. 9). At intermediate times, one would observe emission from both states (---) and the spectral half-width is expected to be larger.

These two models provide a useful framework within which to consider spectral relaxation. Consider the instantaneous emission spectra at times comparable to the spectral decay rate. For continuous relaxation, one expects these time-resolved spectra to have the same spectral shape. Hence, if one measures the time-resolved spectral width, this should be independent of time. Contrasting results are expected for the two-state model. In this case, the spectral width is expected to be larger at intermediate times because of the contribution of both the initially excited and the equilibrium excited states.

Time-resolved emission spectra can be constructed from the frequency-domain data. In fact, such determinations are quite easy, in spite of claims to the contrary [Teale, 1983]. These spectra are reconstructed from the wavelength-dependent data. Frequency-dependent phase and modulation data for PATMAN-labeled dipalmitoyl-L-$\alpha$-phosphatidylcholine (DPPC) vesicles are shown in Figure 10. It is not practical to display the data at all emission wavelengths. Instead, we chose to show the data obtained at three emission wavelengths, located on the blue, center, and red sides of the emission. These data are characteristic of an excited-state reaction [Lakowicz and Balter,

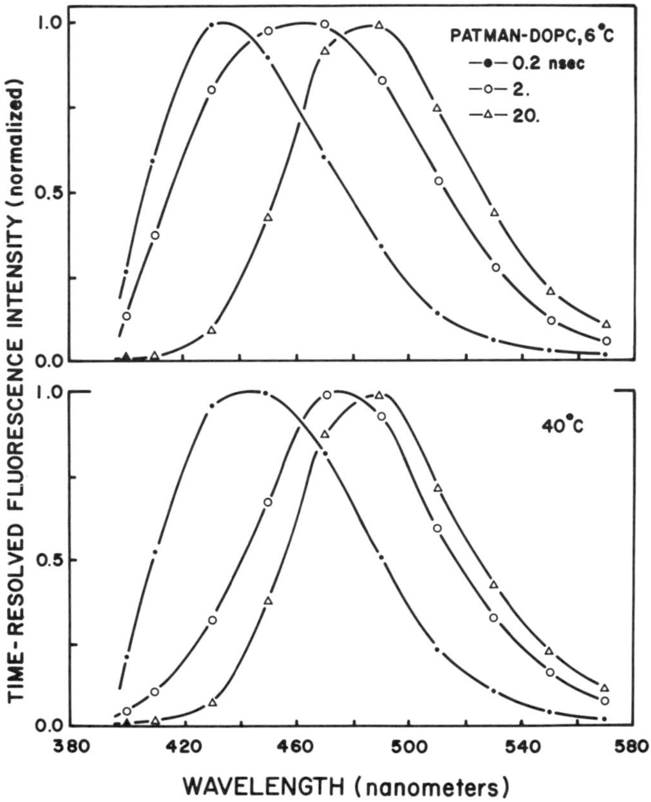


Fig. 11. Time-resolved emission spectra of PATMAN-DOPC and 2-p-toluidinyl-6-naphthalenesulfonic acid (TNS) DOPC [from Lakowicz et al., 1984].

1982]. That is, the phase angles increase and the modulation decreases with increasing emission wavelength. Furthermore, the data are unambiguous in demonstrating the occurrence of a time-dependent process which occurs subsequent to excitation. The apparent phase and modulation lifetimes at each frequency increase with modulation frequency (data not shown), and the measured phase angles exceed 90°. This is not possible for a mixture of directly excited fluorophores.

Time-resolved emission spectra of the PATMAN-labeled vesicles are shown in Figure 11. To facilitate the comparison of these spectra we chose to display, in each case, time-resolved spectra at 0.2, 2, and 20 nsec. A dramatically broadened spectrum is seen at 2 nsec at 6°C. The spectral widths are obviously more narrow at 0.2 and 20 nsec, indicating that spectral relaxation is incomplete at 0.2 nsec and that it is mostly complete at 20 nsec. At higher

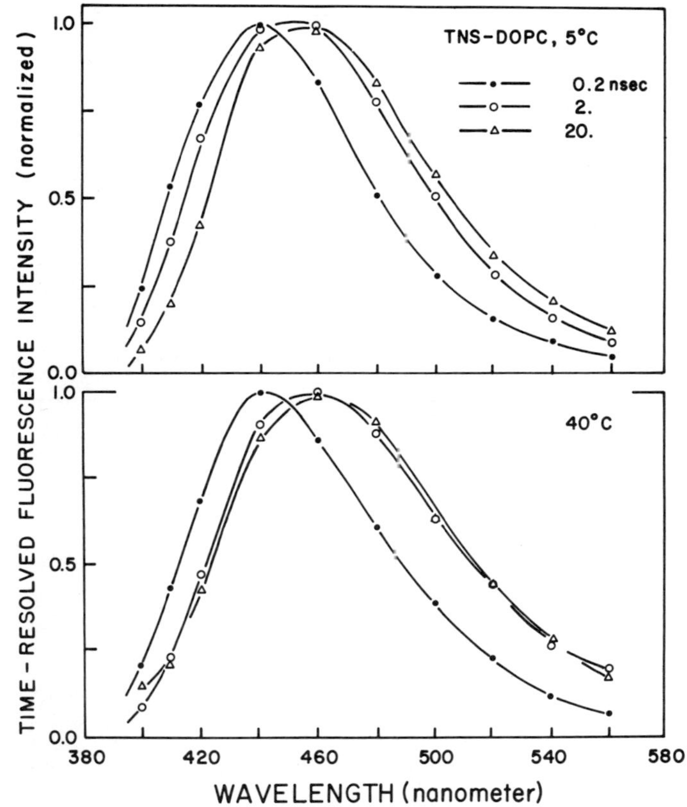


Fig. 11 (continued).

temperature, the spectral relaxation appears to be more rapid. This is evident from the nearly relaxed spectrum seen at 2 nsec at 40°C (Fig. 11).

Also shown in Figure 11 are time-resolved emission spectra for TNS-labeled vesicles of dioleoyl-L-$\alpha$-phosphatidylcholine (DOPC). Once again, faster relaxation is seen at the higher temperature. However, broadening of the spectral distribution is not evident at intermediate times, as was the case for PATMAN-labeled DOPC. This indicates that spectral relaxation of TNS is better described by the continuous model, whereas relaxation of PATMAN is better described by the stepwise model. The spectral decays are more easily visualized by examining the time-dependent center of gravity. This parameter is proportional to the average energy of the emission. The decay is more rapid at higher temperatures (Fig. 12). A more detailed analysis of these decays is necessary before comments can be made concerning the number of exponential components.

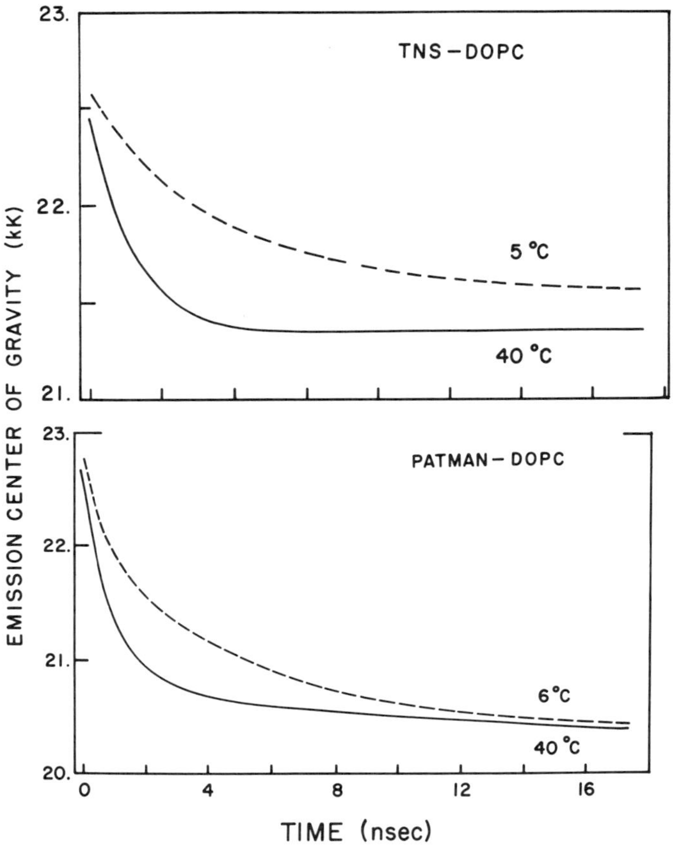


Fig. 12. Time resolved emission center of gravity of PATMAN-DOPC and TNS-DOPC [from Lakowicz et al., 1984].

The distinction between the stepwise and continuous models is most evident upon comparing the time-resolved spectral half-widths. These values are shown for PATMAN and TNS-labeled vesicles in Figure 13. The PATMAN-labeled vesicles display a dramatic and rapid increase in spectral width, which decreases at longer times. In contrast, the spectral widths of TNS are essentially constant over the course of the emission. These results indicate that spectral relaxation of membrane-bound PATMAN is a stepwise process, whereas membrane-bound TNS appears to be better described as a continuous process. We do not wish to claim that either fluorophore relaxes precisely by either mechanism. Rather, we wish to emphasize that considerable detail about spectral relaxation can be obtained from the frequency-domain data.

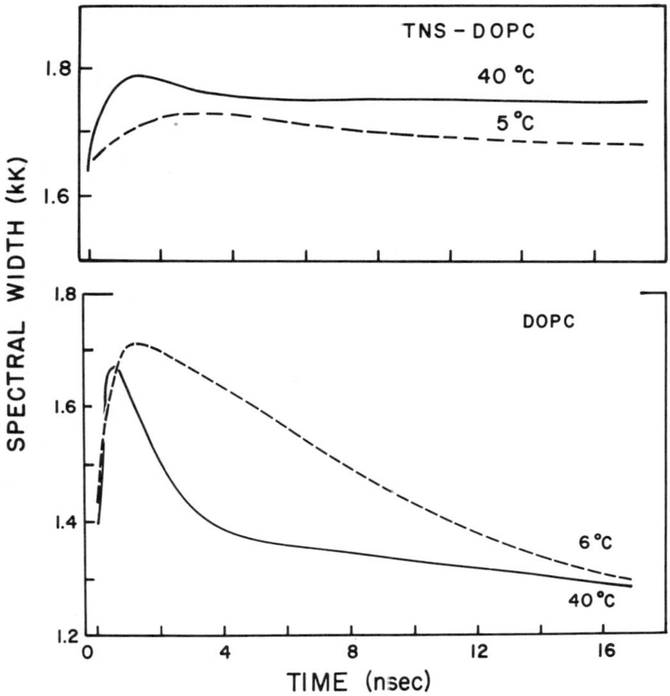


Fig. 13. Time-resolved spectral half-width of PATMAN-DOPC and TNS-DOPC [from Lakowicz et al., 1984].

## CONCLUSION

Frequency-domain fluorometry now provides impressive resolution of complex anisotropy decays and of time-resolved emission spectra. These capabilities will result in its widespread use in chemical and biochemical research.

## ACKNOWLEDGMENTS

The author thanks Dr. Badri P. Maliwal for measurements on DNA and perylene. This work was supported by grants from the National Science Foundation (PCM 82-10878, 81-06910, and DBM-8502853), from the NIH (GM 29318), and from the Department of Defense (DAAG 29-85-G-0017).

## REFERENCES


Barkley MD, Zimm B (1979): Theory of twisting and bending of chain macromolecules: Analysis of the fluorescence depolarization of DNA. J Chem Physiol 70:2991–3007.

Barkley MD, Kowalczyk A, Brand L (1978): Fluorescence decay studies of anisotropic rotation of small molecules. J Chem Physics 75:3581–3591.

Bevington PR (1969): "Data Reduction and Data Analysis for the Physical Sciences." NY: McGraw-Hill.

Easter JH, DeToma RP, Brand L (1976): Nanosecond time-resolved emission spectroscopy of a fluorescent probe adsorbed to L-$\alpha$-egg lecithin vesicles. Biophys J 16:571–583.

Eftink MR, Ghiron C (1981): Fluorescence quenching studies with proteins. Anal Biochem 114:199–227.

Gratton E, Limkeman M (1983): A continuously variable frequency phase fluorometer with picosecond resolution. Biophys J 44:315–324.

Gurd FRN, Rothgeb TM (1979): Motions in proteins. Adv Protein Chem 33:73–165.

Karplus M, McCammon JA (1981): The internal dynamics of globular proteins. CRC Crit Biochem 9:293–349.

Lakowicz JR (1983): "Principles of Fluorescence Spectroscopy." NY: Plenum Press.

Lakowicz JR, Weber G (1973a): Quenching of fluorescence by oxygen—A probe for structural fluctuations in macromolecules. Biochemistry 12:4161–4170.

Lakowicz JR, Weber G (1973b): Quenching of protein fluorescence by oxygen-detection of structural fluctuations in proteins on the nanosecond timescale. Biochemistry 12:4171–4179.

Lakowicz JR, Knutson JR (1980): Hindered depolarizing rotations of perylene in lipid bilayers: Detection by lifetime-resolved anisotropy measurements. Biochemistry 19:905–911.

Lakowicz JR, Balter A (1982): Theory of phase-modulation fluorescence spectroscopy for excited state processes. Biophys Chem 16:99–115.

Lakowicz JR, Maliwal BP (1983): Oxygen quenching and fluorescence depolarization of tyrosine residues in proteins. J Biol Chem 258:4794–4801.

Lakowicz JR, Maliwal BP (1985): Construction and performance of a variable-frequency phase-modulation fluorometer. Biophys Chem 21:61–78.

Lakowicz JR, Maliwal BP, Cherek H, Balter A (1983): Rotational freedom of tryptophan residues in proteins. Biochemistry 22:1741–1752.

Lakowicz JR, Cherek H, Laczko G, Gratton E (1984a): Time-resolved fluorescence emission spectra of labeled phospholipid vesicles, as observed using multi-frequency phase-modulation fluorometry. Biochim Biophys Acta 777:183–193.

Lakowicz JR, Gratton ER, Cherek H, Maliwal BP, Laczko G (1984b): Determination of time-resolved emission spectra and anisotropies of a fluorophore-protein complex using frequency-domain phase-modulation fluorometry. J Biol Chem 259:10967–10972.

Lakowicz JR, Cherek HC, Maliwal BP, Gratton E (1985): Time-resolved fluorescence anisotropies of diphenylhexatriene and perylene in solvents and lipid bilayers obtained from multi-frequency phase-modulation fluorometry. Biochemistry 24:376–383.

MacGregor RB, Weber G (1981): Fluorophores in polar media. Spectral effects of the Langevin distribution of electrostatic interactions. Proc Natl Acad Sci USA 366:140–154.

McCammon JA, Karplus M (1980): Simulation of protein dynamics. Ann Rev Phys Chem 31:29–45.

Teale FJW (1983): Phase Fluorometry. In Cundall RB, Dale RE (eds): "Time-Resolved Fluorescence Spectroscopy in Biochemistry and Biology." NY: Plenum, p 77.

*Applications of Fluorescence in the Biomedical Sciences, pages 245–254*


# 11 Triplet Probes and the Rotational Diffusion of Membrane Proteins

*Peter B. Garland and John J Birmingham*

The application of triplet probes and optical methods for the study of Brownian rotational diffusion of membrane proteins has been extensively reviewed in the past few years [Cherry, 1979; Jovin et al., 1981; Garland and Johnson, 1985; Garland, 1984]. It is not our intention in this article to cover that ground again. Instead we wish to emphasize the way in which studies of the rotational diffusion of membrane proteins have, in several aspects, lagged behind the usage of lateral diffusion measurements. Nevertheless, measurements of lateral diffusion and rotational diffusion of membrane proteins are complementary to each other in their information content, and we shall describe initiatives that may enable the optical determination of rotational diffusion of membrane proteins to approach the performance of lateral diffusion measurements both in sensitivity and applications.

## OPTICAL METHODS COMPARED

Table 1 summarizes and compares the principles, performance, and requirements for optical measurements of either rotational or lateral diffusion of membrane proteins when extrinsic probes are used. The only method presently available for measuring rotational diffusion of membrane proteins at a sensitivity comparable to that available for lateral diffusion is that using fluorescence depletion [Johnson and Garland, 1981, 1982]. This has a sensitivity of about $10^4$ molecules, whereas lateral diffusion measured by fluorescence recovery after photobleaching [Axelrod et al., 1976; Koppel et al., 1976; Peters, 1981] is sensitive to about $10^3$ of a probe such as rhodamine.

However, the time resolution of the fluorescence depletion method is poor because of spurious signals arising in the first few $\mu$sec following the photo-

Department of Biochemistry, Dundee University, Dundee DD1 4HN, Scotland, U.K.

**TABLE 1. A Comparison of Measurement Methods for Rotational and Lateral Diffusion of Membrane Proteins, Using Extrinsic Probes**

| Feature | Rotational diffusion | Lateral diffusion |
|---|---|---|
| Photochemistry | Triplet state formation | Photobleaching |
| Photoactivation | $< 1$ μsec | Few hundred msec |
| Probe molecule | Forms triplet state | Fluorescent |
| Measure | Decay of photoinduced anisotropy (two channels) | Recovery of fluorescence photo-bleaching (one channel) |
| Polarization optics | Yes | No |
| Time domain | μsec–msec | sec–min |
| Cuvette detection (sensitivity, mole) | Extinction ($10^{-9}$–$10^{-10}$), delayed emission ($10^{-10}$–$10^{-12}$) | Not applicable |
| Microscopic detection (sensitivity, molecules) | Delayed emission ($10^8$–$10^7$), fluorescence depletion, ($10^4$–$10^5$) | Fluorescence, $10^3$ or less |
| Sample conditions | Strictly anaerobic | Aerobic |
| Equipment | Complex | Simple |
| Data collection and processing | Complex, flash interference; multiple lifetimes | Relatively simple |
| Single cell use | Difficult (fluorescent depletion) | Routine |

selection flash. Further loss of time resolution also arises from the inability of available pulsed laser power from a modulated continuous wave (CW) argon ion laser to achieve sufficient ground-state depletion of a fluorescent triplet probe in less than a few μsec, which is too long in at least some cases of membrane protein rotation. And even if much higher laser powers were achieved, as with a two-laser version of the apparatus developed by Johnson and Garland [1981], unacceptable temperature rises would occur at the membrane target, if enough laser energy was used to cause significant ground-state depletion of a probe, such as rhodamine, which has a low quantum yield for triplet formation [Axelrod, 1977; Johnson and Garland, 1982].

## PHASE-MODULATION METHODS

In order to avoid the limitations of time resolution described above, we decided to explore the application of phase-modulation methods in place of pulse methods for the measurement of slow rotations in the μsec-msec time domain. The essential differences between the two methods are illustrated in

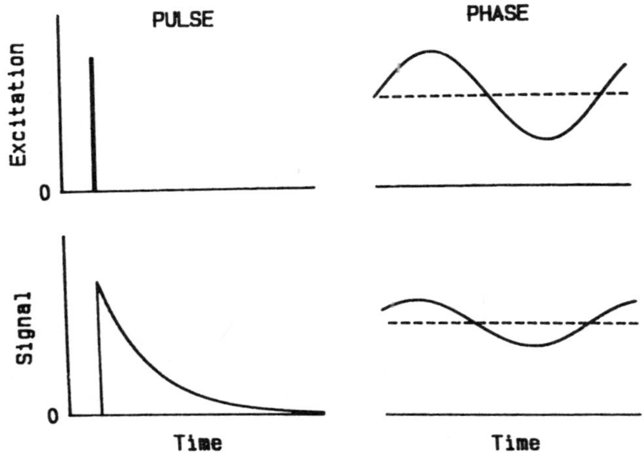


Fig. 1. Comparison of excitation and emission waveforms in pulse and phase methods for measuring fluorescence or phosphorescence lifetimes. The upper panels are the excitation waveforms, the lower panels the emission signal, with the pulse method on the left and phase method on the right.

Figure 1. In the probe method, an intense but brief flash of light causes photoactivation of the molecules under study to some excited state. The subsequent decay of the excited-state population is then detected usually (but not necessarily) by the associated emission of light as excited molecules relax to the ground state. For fluorescence the excited-state lifetime (first singlet) is typically a few nsec, whereas for phosphorescence (triplet) it is many $\mu$sec or even msec. With the phase method the excitation is continuous but with its intensity sinusoidally modulated with time. The modulation frequency is in the region of the reciprocal of the lifetime under study: e.g., 10–200 MHz for fluorescence, 0.1–50 kHz for phosphorescence. The emitted light is also intensity modulated, but delayed in phase and demodulated with respect to the excitation (Fig. 1). For excitation modulated at an angular frequency $\omega$ the relationship between the phase shift $\phi$ and lifetime $\tau$ is given by

$$\phi = \tan^{-1}(\omega\tau) \tag{1}$$

The demodulation M is the ratio of the modulations (the ratio of the AC amplitude to the average signal) of the excitation and signal waveforms, and is given by

$$M = [1 + (\omega\tau)^2]^{-0.5} \tag{2}$$

If the system of a single exponential decay (in pulse terminology) is replaced by several noninteracting components, each giving rise to a single exponential decay, then the phase and modulation of the resultant sinusoidal signal waveform in response to sinusoidally modulated excitation is given by

$$\phi = \tan^{-1}(S/G), \quad M = (S^2 + G^2)^{-0.5} \tag{3}$$

$$S = \sum_i f_i M_i \sin \phi_i \tag{4}$$

$$G = \sum_i f_i M_i \cos \phi_i \tag{5}$$

The values of $\phi_i$ and $M_i$ for each component are given by equations 1 and 2, where $f_i$ is the fractional intensity of the ith component ($\Sigma_i f_i = 1$). It is important to note that the fractional intensity, or amplitude, refers to the zero time amplitude of the component in a classical pulse method. The fractional intensity $f_i$ does not by itself indicate the relative contribution of a component to the total emission obtained with continuous excitation, but the product $f_i\tau$ does.

If the excitation light is linearly polarized and the resulting signal (fluorescence, phosphorescence, triplet absorption, ground-state depletion) is detected through polarizers oriented either parallel (‖) or perpendicular (⊥) to the excitation phase, then the differential signals of interest in obtaining rotational rates are

$$\Delta = \phi_\perp - \phi_\parallel, \quad Y = M_\parallel / M_\perp \tag{6}$$

The differential polarized phase angle Δ is the phase angle difference between the signal waveforms detected through either perpendicular or parallel polarizers. The modulation ratio Y is the ratio of the modulation amplitudes of the two signal waveforms.

Phase-modulation methods have been used exclusively for fluorescence lifetime work, in the MHz frequency region. Recent advances in instrumentation using electro-optically modulated CW lasers have enabled data analysis to be carried out in the frequency domain, with the consequence that the previous superiority of time-resolved pulse methods has now been largely removed or even reversed. For fuller accounts of multifrequency phase-modulation fluorescence methods the reader is referred to Gratton et al. [1984a,b], Lakowicz et al. [1985], and Beecham et al. [1983].

## A LOW-FREQUENCY PHASE-MODULATION APPARATUS

### Construction

Due to the low-frequency range (0.1–50 KHz) it was not necessary to adopt the design features of high-frequency machines [Lakowicz and Maliwal, 1985]. First, we used an acousto-optic rather than an electro-optic modulator to control the excitation beam energy. Second, because our interest is in probes excited in the visible spectral range, we used an argon ion laser with a choice of lines in the blue-green region. An air-cooled model with 75-

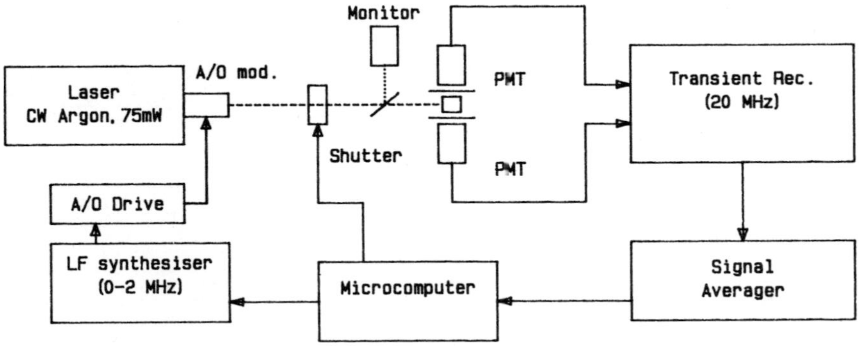


Fig. 2. Components of a multifrequency phase-modulation apparatus for measurement of emission lifetimes in the μsec to msec time domain. Proceeding clockwise from the top left corner, the components are: an argon ion laser, 75mW all-line maximum power (Lexel model 65.07); an acousto-optic modulator operating in an analogue mode (Coherent model 340); a mechanical shutter to block the laser beam between measurements; a beam splitter and monitor, a cuvette or sample position flanked by photomultipliers (PMT) guarded with appropriate filters; a 20-MHz sample rate transient recorder, operating at 10-MHz sample rate for dual channel (Datalab model 920); a signal averager (Nicolet model 1170); a microcomputer (Hewlett Packard model 86A); a low-frequency (LF) waveform synthesizer (Philips model 5190); and the drive unit for the acousto-optic (A/O) modulator. The whole system is controlled by the microcomputer.

mW power is more than adequate. Third, the photomultiplier signals were captured directly with a digital transient recorder and signal averager, and then transferred to computer disk for subsequent calculation of phase angle and modulation. Cross-correlation circuits [Spencer and Weber, 1969] with a phase meter and modulation meter were not required, nor was an additional phase-locked frequency synthesizer. Figure 2 gives a block diagram of our apparatus. The choice of components was largely dictated by their availability in our laboratory, where they had been acquired for time-resolved work [Garland, 1981; Johnson and Garland, 1981]. The only new requirement was a digital low-frequency waveform synthesizer. The present transient recorder and signal averager combination will shortly be replaced by a totally programmable signal averager. It will then be possible to collect and store the signal waveform for 20–30 modulation frequencies in 1–2 minutes. We are also exploring the use of simultaneous multiple-modulation frequencies followed by Fourier transform analysis. If successful this would reduce the data collection time to a few seconds at most.

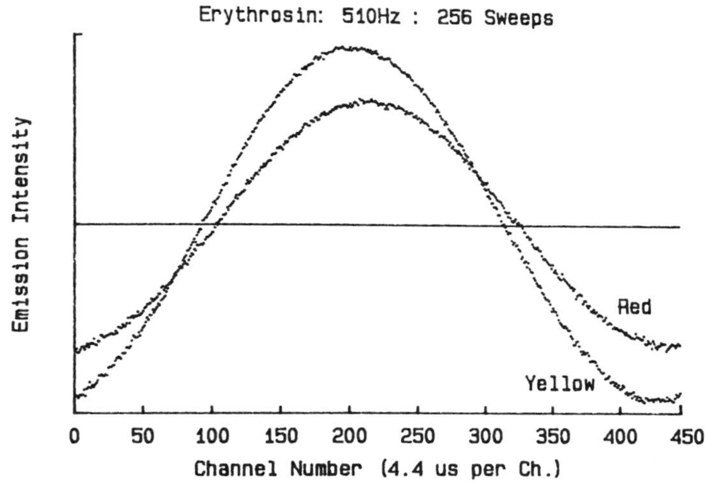


Fig. 3. Emission waveforms for fluorescence and phosphorescence from sinusoidally excited erythrosin (1 $\mu$M in solid Poly(methyl methacrylate). The apparatus described in Figure 2 was used. Phosphorescence was detected with a red-sensitive photomultiplier guarded with an RG 650 filter (red channel in the figure). Fluorescence was measured with a red-insensitive photomultiplier guarded with a KV550 filter (yellow channel). Delayed fluorescence makes only a small contribution [Garland and Moore, 1979]. The mean values of the emissions in the red and yellow channels were normalized to a common value. The phosphorescence signal (red) is delayed and demodulated with respect to the fluorescence signal (yellow).

## Data Analysis

The modulation frequency set by the digitizal frequency synthesizer is accurate to $\pm 1 \times 10^{-6}$ of the frequency setting. Thus it remained to determine the phase and amplitude of the signals. For this purpose we have used the phosphorescence of erythrosin as an emission signal because such emission is, in principle, fully described by existing theories and data analysis for fluorescence. The much more sensitive fluorescence depletion method requires a different theory and analysis, and these will be described elsewhere.

We have analyzed phosphorescence emission waveforms by fitting them by a Fourier least-squares method with a sine wave. The phase and amplitude of the best-fit sine wave were then used in all subsequent calculations using phase angle and modulation. Such calculations were made over 20–30 frequency values, and the resulting plots of the frequency dependence of phase angle or modulation were curve-fitted by a nonlinear least-squares method [Marquardt, 1963; Bevington, 1969] to yield excited-state lifetimes and

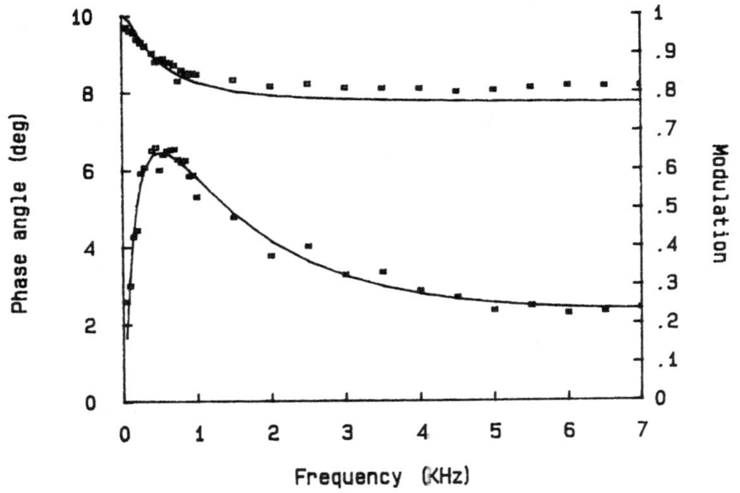


Fig. 4. Frequency-dependence of phase and modulation of emission from sinusoidally excited erythrosin [1 $\mu$M in poly (methyl methacrylate)]. Emission was detected through a RG 650 filter, and includes both prompt fluorescence and phosphorescence. The continuous lines were computed for best fit by a nonlinear least-squares method, and gave three components of amplitudes 0.999, 0.0003, and 0.00013, and lifetimes of 0.44, 181, and 572 $\mu$S, respectively. These amplitudes refer to the initial amplitudes in a conventional pulse experiment.

amplitudes [Lakowicz et al., 1984]. We should again emphasize that the amplitudes or fractional intensities correspond to those of a conventional pulse experiment, i.e., the pre-exponential terms for a multicomponent system.

## Performance

Figure 3 shows signals arising from the modulated excitation of erythrosin, a triplet probe which has the following properties: prompt fluorescence lifetime of 0.5 nsec, triplet lifetime(s) of many $\mu$sec, quantum yield for fluorescence of about 0.01, and quantum yield for phosphorescence of about 0.003 [Garland and Moore, 1979; Porter et al., 1974]. The lifetime for fluorescence is so fast that excitation modulation in the sub-MHz range causes neither phase delay nor demodulation with respect to the excitation waveform. Fluorescence and phosphorescence can be measured via yellow- and red-sensitive channels, respectively, but the red-sensitive channel also carries a contribution from the red end of the fluorescence emission spectrum [Garland and Moore, 1979]. Thus it was to be expected that analysis of the

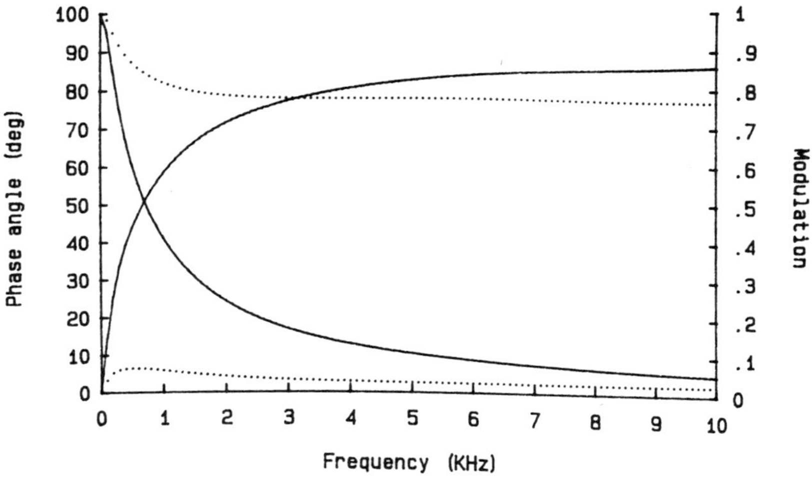


Fig. 5. Computer-simulated frequency-dependence of the modulation and phase angle of sinusoidally excited emission arising from either two components (continuous lines, amplitudes 0.69 and 0.31, lifetimes 180 and 590 μsec) or three components (interrupted lines, amplitudes 0.999, 0.003, and 0.00013, and lifetimes 0.44, 180, and 590 μsec. At zero frequency the modulation is 1.0 and the phase angle zero degrees.

frequency-dependence of phase angle or modulation would yield a very fast lifetime component in addition to the much slower triplet lifetimes. This was indeed the case (Fig. 4). The best fit to the data gave the following lifetimes and fractional intensities: 0.44, 181, and 572 μsec, 0.999, 0.0003, and 0.00013, fractional intensity respectively.

It is neither necessary nor possible in this low-frequency application to determine accurately the lifetime and fractional intensity of fluorescence mixed with phosphorescence. The admixture of short-lived fluorescence with phosphorescence lowers the phase-shift and demodulation that would otherwise arise from phosphorescence alone. This point is emphasized by the computer simulations shown in Figure 5. Nevertheless, operation in the frequency domain has enabled us to clearly resolve at least two phosphorescence lifetimes from contaminating prompt fluorescence. The coefficient of variation of the phase measurements was approximately ± 1%, corresponding to ± 0.1% at 10° phase shift. We have compared time-resolved pulse methods with frequency-resolved phase methods for analyzing phosphorescence lifetimes from the same sample, and our preliminary conclusion is that the phase method gave us superior resolution.

## CONCLUSION

As a move toward putting measurements of the rotational diffusion of membrane proteins on the same basis of sensitivity and convenience as holds for lateral diffusion measurements, we have successfully developed a prototype apparatus for multiple-frequency phase-modulation measurements of phosphorescence emission from the triplet state of erythrosin, in the 0.1–50 kHz frequency range. The same approach but with suitably modified apparatus and data analysis will be developed for use of the fluorescence depletion method.

## ACKNOWLEDGMENTS

This work was supported by the Medical Research Council and the Royal Society. We thank Unilever Research for the loan of a digital frequency synthesizer.

## REFERENCES

Axelrod D (1977): Cell surface heating during fluorescence photobleaching recovery experiments. Biophys J 18:129–131.

Axelrod D, Koppel DE, Schlessinger J, Elson E, Webb WW (1976): Mobility measurement by analysis of fluorescence photobleaching recovery kinetics. Biophys J 16:1055–1069.

Beecham JM, Knutson JR, Ross JBA, Turner BW, Brand L (1983): Global resolution of heterogeneous decay by phase/modulation fluorometry: Mixtures and proteins. Biochemistry 22:6054–6058.

Bevington PR (1969): "Data Reduction and Error Analysis for the Physical Sciences." New York: McGraw-Hill Book Co., pp 204–206.

Cherry RJ (1979): Rotational and lateral diffusion of membrane proteins. Biochim Biophys Act 559:289–327.

Garland PB (1981): Fluorescence photobleaching recovery: Control of laser intensities with an acousto-optic modulator. Biophys J 33:481–482.

Garland PB (1984): Optical probes and the detection of conformational changes in membrane proteins. In Chapman D (ed): "Biological Membranes." London: Academic Press, Vol 5, pp 279–288.

Garland PB, Moore CH (1979): Phosphorescence of protein-bound eosin and erythrosin. A possible probe for measurments of slow rotational mobility. Biochem J 183:561–572.

Garland PB, Johnson P (1985): Rotational diffusion of membrane proteins. In Martonosi AN (ed): "The Enzymes of Biological Membranes." Plenum Publishing Corp, Vol 1, pp 421–439.

Gratton E, Jameson DM, Hall RD (1984a): Multifrequency phase and modulation fluorometry. Ann Rev Biophys Bioeng 13:105–124.

Gratton E, Limkeman M, Lakowicz JR, Maliwal BP, Cherek H, Lazko G (1984b): Resolution of mixtures of fluorophores using variable-frequency phase and modulation data. Biophys J 46:479–486.

Johnson P, Garland PB (1981): Depolarization of fluorescence depletion: A microscopic method for measuring rotational diffusion of membrane proteins on the surface of a single cell. FEBS Lett 132:252–256.

Johnson P, Garland PB (1982): Fluorescent triplet probes for measuring the rotational diffusion of membrane proteins. Biochem J 203:313–321.

Jovin TM, Bartholdi M, Vas WLC, Austin RG (1981): Rotational diffusion of biological macromolecules by time-resolved delayed luminescence (Phosphorescence, fluorescence) anisotropy. Ann NY Acad Sci 366:176–196.

Koppel DE, Axelrod D, Sclessinger J, Elson EL, Webb WW (1976): Dynamics of fluorescence marker concentration as a probe of mobility. Biophys J 16:1315–1329.

Lakowicz JR, Maliwal BP (1985): Construction and performance of a variable-frequency phase-modulation fluorimeter. Biophys Chem 21:61–78.

Lakowicz JR, Laczko G, Cherek H, Gratton E, Limkeman M (1984): Analysis of fluorescence decay kinetics from variable frequency phase shift and modulation data. Biophys J 46:463–477.

Lakowicz JR, Cherek H, Maliwal BP (1985): Time-resolved fluorescence anisotropies of diphenylhexatriene and perylene in solvents and bilayers obtained from multifrequency phase-modulation fluorimetry. Biochemistry 24:376–383.

Marquardt DW (1963): An algorithm for least-squares estimation of nonlinear parameters. J Soc Ind Appl Math 11:431–441.

Peters R (1981): Translation diffusion in the plasma membrane of single cells as studied by fluorescence microphotolysis. Cell Biol Int Rep 5:733–760.

Porter G, Reid ES, Tredwell CJ (1974): Time resolved fluorescence in the picosecond region. Chem Phys Lett 29:469–472.

Spencer RD, Weber G (1969): Measurements of subnanosecond fluorescence lifetimes with a cross-correlation phase fluorometer. Ann NY Acad Sci 158:361–376.

# PART II

# Imaging and Quantitative Fluorescence Microscopy

*Applications of Fluorescence in the Biomedical Sciences, pages 257–288*


# 12 Imaging at Low Light Level in Fluorescence Microscopy

*Gary R. Bright and D. Lansing Taylor*

Quantitative fluorescence microscopy is a powerful tool for studying the spatial and temporal dynamics of macromolecules, molecules, and ions in living cells for at least five important reasons: (1) Sensitivity—the detection of fluorescence is more sensitive than methods based on light absorption, scattering, or interference. With existing technology, as few as ca. 50 fluorescent molecules can be detected as a cluster on surfaces or inside cells [Barak and Webb, 1981]. This number should be improved with advances in probe chemistry, as well as in instrumentation. (2) Specificity—single classes of molecules can be studied even in a mixture of many other molecules by selecting the proper fluorescent probes and filter combinations [Haugland, 1983; Waggoner, this volume]. Multiple compartments can be visualized in concert with each other to study the interrelationships of various components and compartments. The fluorescent probes can be directed to specific molecules either by covalent attachment to purified molecules or by designing the physical-chemical properties of the probes to complement the target molecule (see Waggoner, this volume). (3) Spectroscopy—fluorescence spectroscopic measurements such as fluorescence anisotropy and resonance energy transfer can yield molecular information about the molecules or the immediate environment (see Lakowicz, this volume). (4) Spatial resolution—the use of the fluorescence microscope permits the analysis of fluorescence parameters at

Department of Biological Sciences, Center for Fluorescence Research in Biomedical Sciences, Carnegie-Mellon University, Pittsburgh, Pennsylvania 15213.

the limit of resolution of the light microscope. Therefore, two- and three-dimensional [Agard 1984; Lanni, this volume] maps of molecular distributions, as well as molecular activity, are possible in living cells. (5) Temporal resolution—fluorescence techniques are also valuable since they can detect and quantify those events which occur with a frequency equal to or greater than the inverse of the time between absorption and emission of light. By using many common fluorescent probes, those biological processes occurring with a rate on the order of about $10^8$ $sec^{-1}$ or slower can be detected and measured. These processes include membrane potential, microviscosity, mobility, orientation of molecules, ligand binding, the formation of supramolecular structures, and some chemical reactions [Lakowicz, 1983; Taylor et al., 1984].

The use of quantitative imaging techniques in conjunction with fluorescence microscopy has the potential of optimizing the acquisition of information at the molecular level in time and space, even within living cells. The purpose of this chapter is to discuss some of the parameters that must be considered when a low-light-level video microscope system is used for quantitative imaging of fluorescent molecules. The parameters discussed are fundamental both for characterizing the limits of capability of the system as well as for providing the necessary data for partially rectifying some of these limitations by using image processing techniques. In practical terms the establishment of these parameters is an indication of the amount of information the system is capable of recording. Emphasis is placed on video cameras used as the photodetector since most of the limitations are imposed on the system by this less-than-ideal detector.

The historical roots of the use of imaging techniques with quantitative microscopy in cell and molecular biology can be traced back to many fields and to investigators. The following references are not meant to be comprehensive, but indicate the diversity of some of the fundamental studies: cytochemistry [Ornstein, 1957; Chance and Legallais, 1959; Thaer, 1966; Goldman, 1968; Udenfriend, 1969; Ruck, 1970; Thaer and Sernetz, 1973], cell biology [Chance et al., 1962; Taylor and Wang, 1978; Taylor and Wang, 1980; Kohen et al., 1981], chemical kinetics [Rotman, 1961; Chance, 1962], luminescence [Reynolds, 1968; Gilkey et al., 1978; Reynolds and Taylor, 1980], TV-based cytometry [West, 1969; Willingham and Pastan, 1978; Ploem et al., 1979; Reynolds and Taylor, 1980], and laser-based microscopy [Siemens et al., 1982; Jacobson et al., 1983; McGregor et al., 1984]. A brief review of recent applications has also appeared [Arndt-Jovin et al., 1985]. The continued developments in instrumentation, fluorescence probe chemistry, and biological probes should make this approach even more powerful in the future.

## IMAGING TECHNOLOGIES

The fundamental purpose of the imaging system is to visually duplicate the object. With a microscope this entails visually duplicating a magnified object. Since the optics are not perfect and the electronics are not free from noise, the imaging system necessarily provides an approximation rather than an exact replica of the object. There are some basic parameters that define all imaging systems regardless of whether they are based on a camera [e.g., vidicon or charge-coupled device (CCD)] or a scanning system (laser-scanning microscope or scanning confocal microscope). There is not yet enough information to make a definitive statement about the relative merits of an imaging system based on a camera or a photomultiplier with a scanning mechanism. The present discussion will focus on the use of video cameras. A general discussion of video microscopy has appeared recently [Inoué, 1986].

**Conventional Vidicon Television Cameras** [McGinty, 1984; Biberman and Nudelman, 1971a,b]

Vidicon is a generic term for a photoconductive camera tube. The type of vidicon is determined by the target material. A chalnicon camera tube, for example, is defined as having a cadmium selenide target, whereas a silicon vidicon has a silicon target. Most targets are complex multilayer semiconductors. In a vidicon the object is imaged onto the target resulting in a localized electron depletion that is prortional to the local light intensity. Thus, a charged pattern appears on the target. The charged pattern (image) is read off the target by a scanning electron beam that neutralizes the charged pattern. The video signal is generated from the current generated by this charge neutralization. For an interlaced camera the scanning beam does not scan every line of the frame in succession. Every other line is scanned to form one field, followed by the alternate lines to form the second field. The two fields make up one video frame. Less common vidicon cameras exist that scan in a non-interlaced pattern.

**Silicon-Intensified Target Cameras** [Nudelman, 1979]

The silicon-intensified target camera (SIT) is similar to the conventional vidicon camera tube. It includes an added gain factor over the standard vidicon by incorporating an intensification stage within the camera structure. In this case, the cathode is a photoemitter rather than a photoconductor. The photoelectrons emitted from the cathode impinge upon a silicon diode array target causing a charged pattern to form. This pattern is scanned by the electron beam as in the vidicon. An ISIT is also available which is a SIT camera with an additional conventional image intensifier as a first stage.

**Conventional Image Intensifiers** [Schnitzler and Morton, 1971; Coleman and Boksenberg, 1976; Nudelman, 1979]

An image intensifier exploits the photoelectric effect to convert the stream of photons from the object into a stream of electrons. Conventional image intensifiers consist of a vacuum tube with two electrodes, a photocathode and an anode. An image is focused onto the photocathode and each area of the photocathode emits electrons in proportion to the light intensity of the image in the area. The emitted electrons are accelerated by the bias potential applied between the electrodes, thus increasing their kinetic energy. A visible image is formed by focusing the electrons on a phosphorescent screen which emits light proportional to the flux rate. High-gain intensifiers require magnetic or electric fields to focus the electrons.

**Microchannel Plate Intensifier** [Lampton, 1981; Timothy, 1981; Talmi, 1983]

A new type of image intensifier that uses microchannel plates (MCP) has recently become available. An MCP is effectively an array of hollow glass channels with the inner surfaces coated with an electron-emissive material. Each channel acts as a channel electron multiplier, amplifying the photoelectrons as would a standard photomultiplier but without separate dynodes. Each channel electron multiplier corresponds to 1 pixel [see Lampton, 1981]. For most practical purposes the MCP can be viewed as an array of photomultipliers. Since the photoelectrons are constrained within the channel, large degrees of amplification can occur without extensive degradation of resolution.

**Photon Counting Imaging** [Boksenberg and Burgess, 1972; Firmani et al., 1982]

A relatively new approach to imaging by using conventional devices is photon counting imaging [Boksenberg and Burgess, 1972]. Photon counting imaging is ideally suited for very-low-light-level imaging and relatively stable objects. The photon flux per unit area must be low enough such that the arrival of each photon is visible. The basic approach to image photon counting is to amplify the photoelectron image to a level where all the scintillations on the output screen of an intensifier (see two sections earlier) can be clearly discerned above the readout noise of a continuously scanning low-lag vidicon camera used to view the intensifier output. After suitable processing, the image information is stored event by event in a computer memory. A system based on these principles has recently become commercially available (Hamamatsu Photonics [Wick, 1985]).

An alternate approach involves placing a resistive anode readout device behind two or more microchannel plates. The resistive anode provides the coordinates of a photoelectron when it strikes it. Thus as photoelectrons hit, the pixel value at the appropriate coordinates is incremented. Resolutions as high as 500 × 500 pixels have been reported [Firmani et al., 1982; Timothy, 1981; Lampton, 1981]. Devices based on this technology are also now commercially available (Surface Science Laboratories, Instrument Technology, LTD).

**Charge-Coupled Device Imagers** [Howes and Morgan, 1979; Collet, 1985]

The charge-coupled device (CCD) imager is a solid-state sensor. It is a two-dimensional discrete array of fixed-position detectors. The discrete nature of the device results in stable pixels with well-defined geometries. CCDs are capable of 14-bit gray-scale resolution although at a relatively slow readout rate. This large dynamic range allows a wider range of intensities within a given scene to be imaged compared to vidicon-based systems. CCDs of adequate resolution for microscopy (500 × 500 to 1,024 × 1,024 pixels) have only recently become available. CCDs attached to image intensifiers are also available. As an integrating detector, CCD's provide the same advantages as photography. The CCD can integrate a signal over a desired time interval and the readout noise is introduced only once. The full applicability to quantitative fluorescence microscopy has not yet been assessed.

**Confocal Scanning Microscope** [Brackenhoff, 1979; Ash, 1980; Cox, 1984; Wilson and Sheppard, 1984]

The confocal scanning microscope (CSM) is based on the object-scan principle. The specimen is mounted on a very-high-resolution mechanical scanning stage. The stage moves the specimen point by point past a fixed optical system. The light reflected, transmitted, or emitted is recorded by a photomultiplier. A pinhole is placed in front of the photomultiplier. The confocal configuration places the light source, illuminated specimen point, and detector pinhole conjugate to each other. The image is assembled point by point as the object is scanned. This system provides a very shallow depth of field. The dynamic range is very large due to the use of a photomultiplier.

**Laser Scanning Microscope** [Shoemaker et al., 1982; Hansen, 1985; see Ploem, this volume]

The laser scanning microscope (LSM) is based on an incident light scan principle. The object is scanned point by point using a finely focused laser beam. The laser beam can be scanned by different mechanisms, such as rotating polygon mirrors, acoustooptic deflectors, or servocontrolled galvanometers. Maneuverability of the laser beam under computer control allows the fluorescence of a given location or a series of locations to be measured quickly without exposing the whole specimen. The reflected, absorbed, or emitted light is recorded by a photomultiplier. The LSM can provide a base for other biophysical instruments such as fluorescence recovery after photobleaching (FRAP) and fluorescence lifetime measurements. A laser scanning microscope has recently become commercially available (Carl Zeiss, NY; Newport Electro-optics Systems, Melbourne, FL; Meridian Instruments, Okemos, MI).

## CONSIDERATIONS OF FLUORESCENCE IN MICROSCOPY

An understanding of basic fluorescence spectroscopy is essential to successful applications of fluorescence microscopy. Luminescence is the emission of photons from electronically excited states. Luminescence requires the input of energy in order to raise the target molecules to an excited state. This energy may come from chemical reactions (chemiluminescence), enzyme-catalyzed reactions or interactions with biological macromolecules (bioluminescence), or absorption of light energy (fluorescence, phosphorescence). Most applications in the biomedical sciences have emphasized fluorescence, although other types of luminescence can be imaged.

Emission of light as fluorescence is usually not the major pathway for deexcitation of the excited state. The amount of emitted energy compared to the amount of excitation energy (quantum yield of fluorescence) is generally rather low in fluorescently labeled macromolecules. The dissipation of this energy involves many more processes than just light emission. Radiationless dissipation pathways include collisional quenching by solvent and other molecules, intersystem crossing, internal conversion, and chemical reaction with either the ground state or the excited state of the fluorophore, as well as loss as heat (vibrational energy) [Birks, 1970; Turro, 1978; Lakowicz, 1983].

A fluorescence microscope is a highly efficient fluorometer. It is able to collect a higher percentage of the emitted fluorescence from the sample than a standard fluorometer. The fluorescence emitted from the sample is collected over a wider angle because the numerical aperture (N.A.) of the collection optics of the microscope is much higher than in a standard fluorometer. For this same reason much higher excitation intensities are reached at the specimen plane. Excitation intensity measurements should be made at the specimen plane with the objective to be used in the specific experiment. This can conveniently be measured by focusing the objective onto a wide field-of-view photodiode.

The single most important factor limiting the use of fluorescence for imaging is the bleaching of the fluorescent probe. Bleaching is a term applied to indicate the irreversible loss of fluorescence emission resulting from prolonged excitation. It is thought to involve reaction of the excited-state molecule with singlet oxygen resulting in an irreversible loss of the fluorophore. Deoxygenation of the solution will dramatically reduce the level of photobleaching. An alternative is to create a reducing environment which will reduce the singlet oxygen [Bock et al., 1985]. Both of these methods of reducing photobleaching are obviously impractical for living biological systems, but are practical for fixed specimens such as those used for immunofluorescence. Fluorescence bleaching presents the ultimate limit to both the level of excitation used and the amount of time the sample is illuminated. It is bleaching of the fluorescent probe that has encourged the use of very low levels of excitation light and the use of multiple-stage-intensified video cameras for imaging fluorescence. These limitations unfortunately determine the types of experiments that can be performed and also define the limits of measurement within a given experiment. There is a trade-off of excitation energy vs. having enough emission for a signal to be recorded.

The level of photobleaching limits the number of images that can be recorded and over what time period. We do not discuss any decrease in fluorescence due to biological processing of the probe by the cell or other cellular-based artifacts in this review (see Waggoner, this volume). The purpose of this review is to define some of the factors and limitations of the video cameras commonly used in low-light-level imaging fluorescence microscopy. We also do not consider nonimaging applications such as measurement of fluorescence spectra (see Wampler, this volume) or other quantitative fluorescence measurements (see Lanni, this volume).

## COMPONENTS OF A MODERN QUANTITATIVE FLUORESCENCE MICROSCOPE IMAGING SYSTEM

An imaging system is a continuum of components involved in forming, recording, processing, and analyzing images. Table 1 lists the basic components of an imaging system. Selected references are included to guide the reader. Figure 1 and Table 2 list some of the specific components of a modern quantitative fluorescence microscope system. Here, we will focus our attention on the fundamental parameters of video cameras when used as quantitative photodetectors.

## PRACTICAL CONSIDERATIONS IN USING PHOTODETECTORS AS QUANTITATIVE IMAGING DEVICES

### Radiometry Vs. Photometry

The nomenclature for light measurement can be confusing and is frequently misused. "Photometric" is often used when "radiometric" is correct. Photometric units denote a unit of measure that is weighted to the wavelength

**TABLE 1. Basic Components of a Microscope Imaging System**

| Component | Role in system |
|---|---|
| Microscope/optics | Formation of an image[1] |
| Photodetector | Record light intensity and location[2] |
| A/D converter | Digitize signal[3] |
| Computer/image processor | Store location and relative intensity of signal[4] |
| D/A converter | Convert digitized signal back into analog signal for presentation to a display device[3] |
| Display | Visual display of raw image, processed image[4] |
| Image analysis/hardware/ software | Extraction of quantitative information from image[4] |

[1]Slayter [1970], Hecht and Zajac [1974].
[2]Biberman and Nudelman [1971a,b].
[3]Castleman [1979], Baxes [1984].
[4]Pratt [1978], Castleman [1979], Green [1983], Baxes [1984].

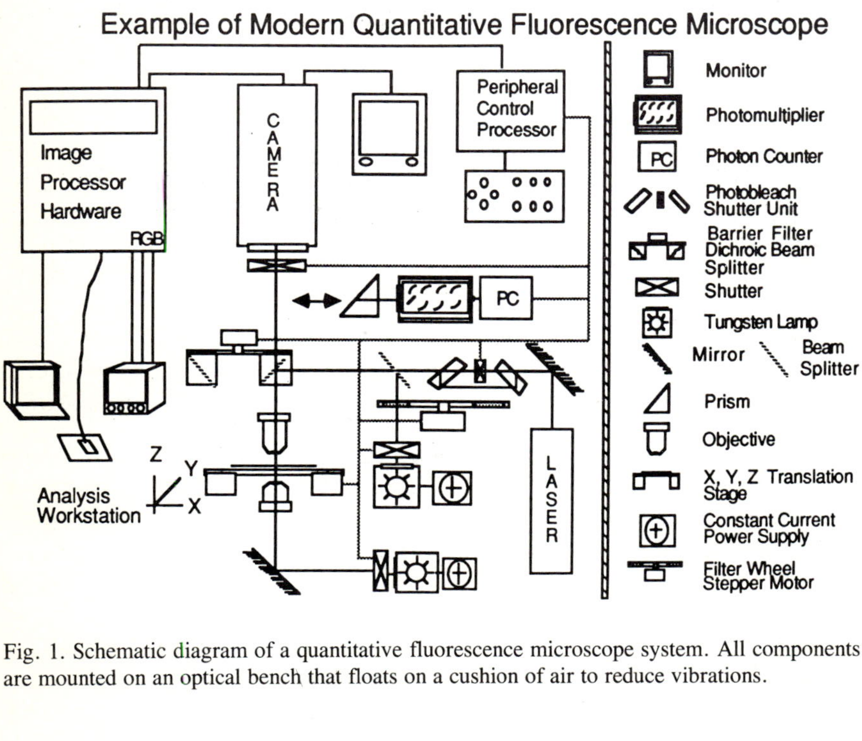


Fig. 1. Schematic diagram of a quantitative fluorescence microscope system. All components are mounted on an optical bench that floats on a cushion of air to reduce vibrations.

**TABLE 2. Major Components of a Modern Quantitative Fluorescence Microscope**

| | |
|---|---|
| Scanning stage | Used for automatically locating cells and allows coverage of large areas using a high-magnification objective; Z-axis control provides automatic focus adjustments |
| Temperature-controlled stage | For maintenance of constant temperature and is programmable from microprocessor for temperature shift experiments |
| Shutter | For minimizing excitation light exposure during nonacquisition time |
| Stepper motor | For automatic changing of excitation and emission filters |
| Photomultiplier | A large dynamic range detector for quantitation of localized spots and comparison with camera |
| Prism | Provides means of switching between camera and photomultiplier as a recording device |
| Barrier filter/dichroic beam splitter unit | Provides a means of automatically changing sets of dichroic mirrors and barrier filters. This allows automatic recording of multiple fluorescent probes in the sample |
| Peripheral control processor | Provides separate microprocessor for real-time control of peripherals either manually from keypad or via communication with image processor |
| Laser | Provides subfield illumination and high-intensity bleaching beam, and a source of high spectral purity |
| Photobleach shutter unit | Provides shutter-beam splitter mechanism for fluorescence recovery after photobleaching (FRAP) and general control of laser beam |
| Camera | Provides 2-dimensional detector providing high-speed recording of low-light-level signals; a second high-resolution camera may be added to provide imaging for conventional transmitted light microscopy |
| Image processor | Contains specialized hardware for processing images at near real-time rates; provides color analysis workstation for interactive manipulation and display of image data |

response of the human eye [Biberman, 1971] (see Fig. 2). Photometric units are clearly not appropriate, since the eye is not the measuring device in quantitative microscopic imaging and in particular there is interest in fluorescent probes that emit light energy outside the spectral response of the eye. Also, there is usually an interest in a narrow range of wavelengths. Physical units are more appropriate for quantifying the radiant energy emitted from fluorescent molecules. Table 3 compares the physical process with both the physical units and the psychophysical units.

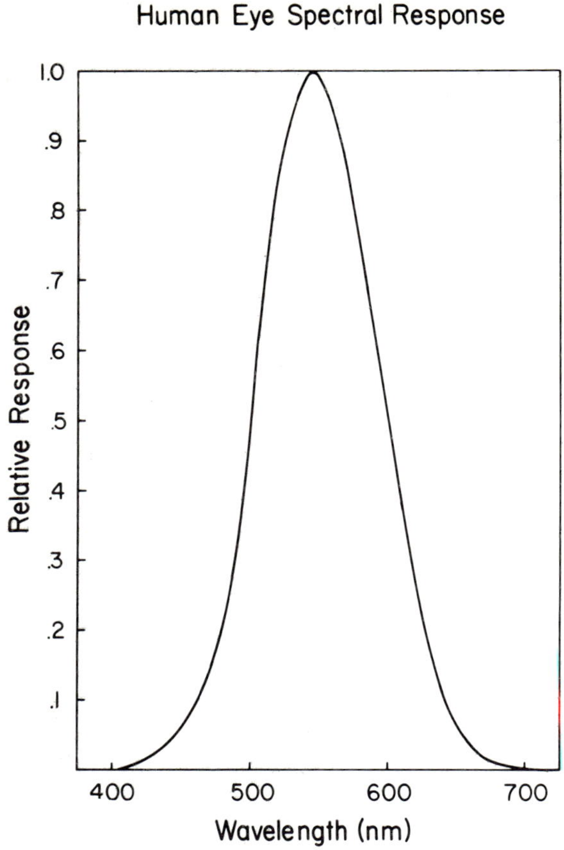


Fig. 2. Photopic spectral response of the human eye. Data from Fogiel [1981].

## Noise

In an imaging system, noise ultimately determines the ability to distinguish a structure from the background. Noise can be defined very generally as any intensity variation in the image that is not a part of the object. There are two basic types of noise, signal-independent and signal-dependent. Signal-independent noise exists both in the presence and absence of the signal and therefore is present in a fixed amount. It is primarily due to thermal noise of the intensifier and camera targets and the video amplifier. Signal-dependent noise exists only in the presence of the signal and varies with the signal. It is exemplified by photoelectron noise, shot noise, and phosphor grain noise

**TABLE 3. Light Measurement Nomenclature and Relationships[1]**

| Physical process | Radiometric term | Radiometric unit | Photometric term | Photometric unit |
|---|---|---|---|---|
| Total amount of radiant energy | Radiant energy | Joule | Luminous energy | Lumen-sec |
| Radiant energy per area | Radiant density | Joule/$m^2$ | Luminous density | Lumen-sec/$m^2$ |
| Radiant energy per second | Radiant flux | Watt (joule/sec) | Luminous flux | Lumen |
| Radiant flux per area | Irradiance | Watt/$m^2$ | Illuminance | Lux (lumen/$m^2$) Footcandle (lumen/$ft^2$) |
| Radiant flux received at a point from a direction | Radiant intensity | Watt/sr | Luminous intensity | Candela (lumen/sr) |
| Radiant intensity per area | Radiance | Watt/sr-$m^2$ | Luminance | Candela/$m^2$ |

[1]Adapted from Biberman [1971; see: Biberman, 1971; Fogiel, 1981; Sears et al., 1982].
sr, stearadian.

[see Lavin, 1971; Biberman and Nudelman, 1971a,b; Rose, 1973]. Examples of some sources of noise in quantitative microscope imaging systems are listed in Table 4.

The parameter used in discussing noise is the signal-to-noise ratio (SNR). The SNR is a measure of the relative magnitude of the signal with respect to the noise present. An ideal imaging system would present an output SNR equivalent to the input SNR. However, each component of the imaging system adds some noise to the image. The noise added by the various components (see Table 4) will result from different physical processes. A discussion of the physical basis of various types of noise is beyond the scope of this review [see DeHaan, 1960; Lavin, 1971; Biberman and Nudelman, 1971a,b; Rose, 1973]. In quantitative imaging it is useful to determine the sources and amount of various noise components, since some types of noise can be significantly reduced.

In video systems the SNR is defined as the signal-to-root-mean-square (rms)-noise ratio and is measured as:

$$SNR = \frac{\text{mean signal } (V_S) - \text{background } (V_B)}{\text{signal standard deviation } (V_{Nrms})}$$

**TABLE 4. Examples of Sources of Noise in Imaging Systems[1]**

| | |
|---|---|
| Electronic noise | Shot noise |
| | Thermal noise |
| | Readout noise |
| | Fixed pattern noise |
| Optical noise | Photon statistical noise |
| | Contribution from out of focus object |
| | Light scatter from specimen (in focus) |
| | Light scatter from optics |
| Digital imaging noise | Quantization noise |
| | Round-off error in calculations |

[1]See: DeHaan [1960], Nudelman [1969], Lavin [1971], Oppenheim and Schafer [1975], Castleman [1979], Green [1983].

Figure 3 illustrates the basic definitions in determining the SNR [Hall, 1971b; Miller, 1971]. The SNR is frequently discussed using the units of decibels (dB). The following equation shows the conversion of SNR to the decibel scale.

$$SNR(db) = 20 * \log (SNR)$$

For a thorough discussion of SNR in imaging see Rose [1973] and Dainty and Shaw [1974].

Noise in an imaging system limits the amount of information that can ultimately be extracted from an image. It defines the lower limit of the dynamic range (see following section) of the camera and therefore determines the sensitivity. In quantitative fluorescence microscopy the sensitivity of the imaging device determines the amount of excitation energy that must be used to obtain a usable image. The sensitivity of a camera is determined by the photocathode quantum gain (the number of photoelectrons released with the absorption of each photon), the intensifier gain (the amount of amplification of the photoelectron image), and the noise level. Figure 4 presents a relative comparison of the light levels necessary to produce a "usable picture" for several intensified and nonintensified video cameras. Figure 5 shows an image recorded by three different intensified cameras under exactly the same conditions. A less-sensitive camera requires a greater excitation light intensity with the concomitant greater photobleaching rates. This ultimately limits the number of images that can be recorded during the experiment. The presence of noise will also affect the performance of many image-processing algorithms. Many functions, such as edge detection algorithms, enhance the noise to a greater extent than the signal.

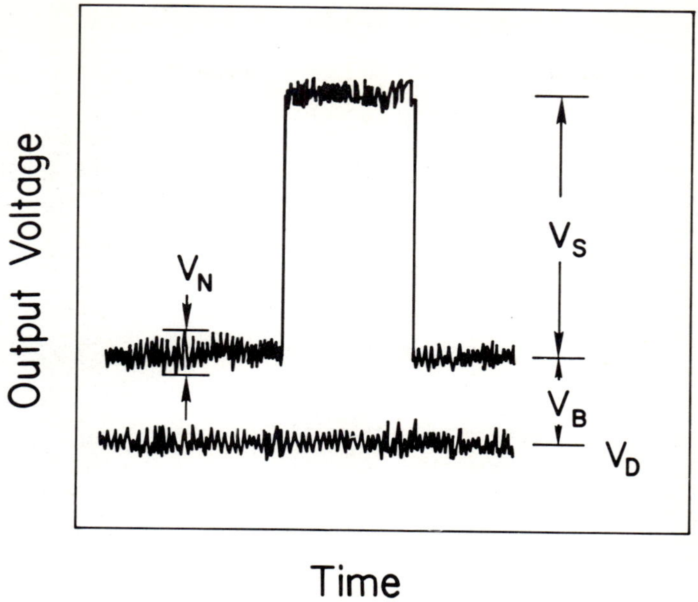


Fig. 3. Schematic oscilloscope-like view of a single video line through an image illustrating the parameters used to define the signal-to-noise ratio. $V_S$—signal voltage (image of object), $V_B$—background voltage (image without object present), $V_D$—dark voltage (camera shutter closed), $V_N$—envelope noise voltage, $V_{Nrms}$ (rms noise) = 0.66 $V_N$. Adapted from Hall [1971b].

Since most noise is due to random processes, the noise in consecutive video frames is uncorrelated. Thus the SNR can be effectively improved by averaging the signal over several frames. Since the SNR improvement is proportional to the square root of the number of frames [Oppenheim and Schafer, 1975], the upper practical limit on averageing is about 256 frames (compare Fig. 6A,B). It is important to point out that the image of human observer sees is an integrated image. The eye and brain integrate the field of view for approximately 0.2 sec [Rose, 1973]. This is equivalent to about six video frames (@ 30 frames per second). Averaging fewer than six frames will result in apparent degradation in the image over what the viewer sees in real time on the video monitor.

**Radiometric Response**

An ideal camera will respond to light within a particular range of intensities and will produce voltage levels that are linearly proportional to the light

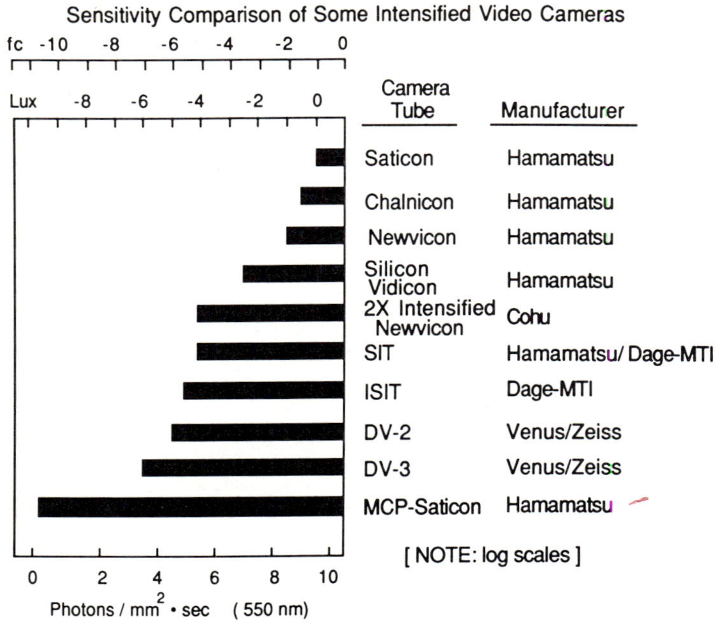


Fig. 4. A comparison of the relative sensitivity of some video cameras. The data were taken from manufacturers' literature. Values are only for nonquantitative comparison since the assessment of "minimum usable picture" is a subjective judgment. Some nonintensified cameras are included for comparison.

intensity in the object. The intensity response transfer function describes the response of the system to changes in the input signal. Although in many cameras some portion of their response curves is linear there is usually a nonlinear portion at each end (see Fig. 7). Figure 8 illustrates the intensity response curve measured on a Dage-MTI ISIT operated at maximum gain. It's clear that this camera, when used at its maximum sensitivity, is unsuitable for quantitative analysis; the nonlinearity and dynamic range are severe problems. When operated at lower gain, the ISIT performs significantly better.

As can be seen in Figure 7, once the transfer function is defined several parameters can be derived. The dynamic range of the camera is primarily determined by the minimum detectable signal and by the maximum signal that can be stored on the target during one frame time. The minimum detectable signal, as discussed previously is determined by the noise level.

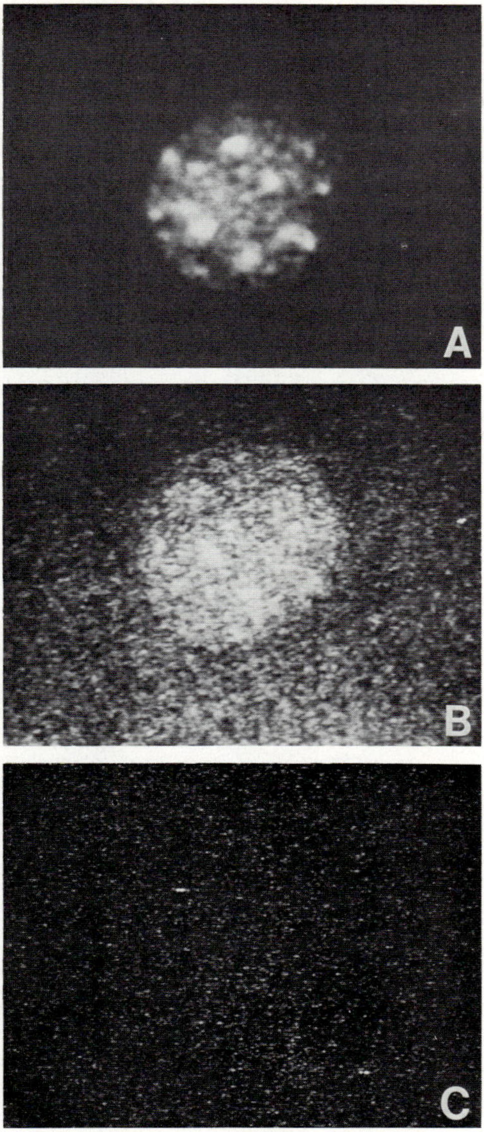


Fig. 5. Images of a Hoechst-stained nucleus recorded with constant excitation light level with three different cameras. A. Hamamatsu MCP-Saticon (VIM) operated in single photon counting mode. B. Dage-MTI ISIT operated at maximum gain. C. Hamamatsu SIT operated at maximum gain.

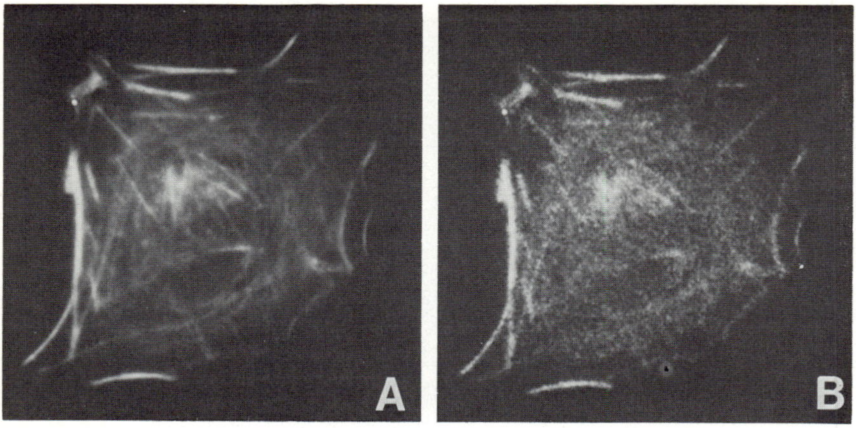


Fig. 6. Integrated and nonintegrated images of a 3T3 cell with actin stained with rhodamine-phalloidin. A. A single frame. B. 256 integrated (averaged) frames. Recorded with a Dage-MTI ISIT.

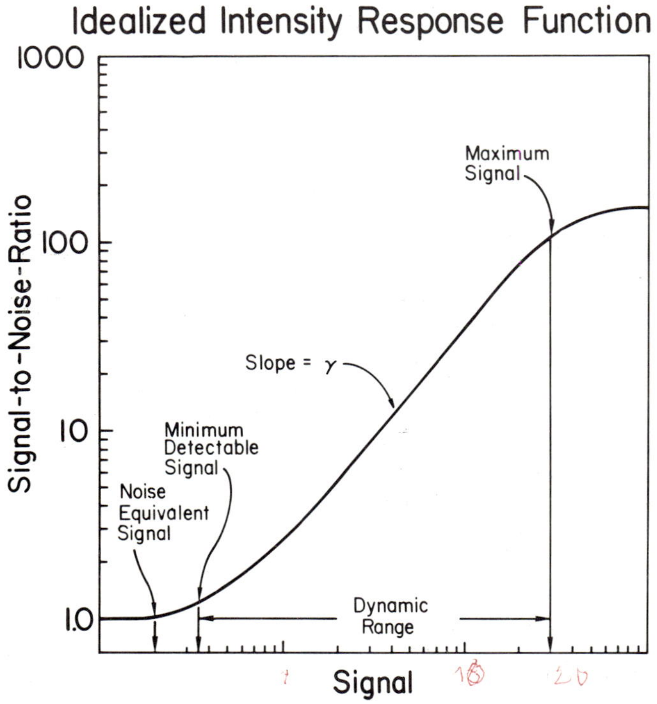


Fig. 7. An idealized intensity response transfer function.

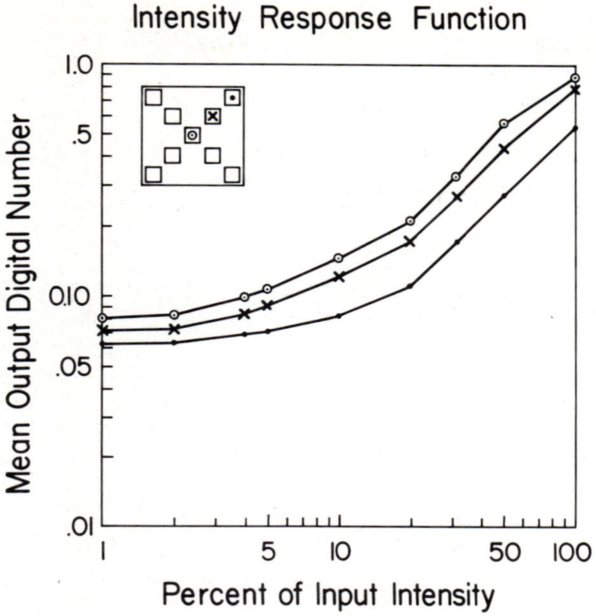


Fig. 8. An intensity response function of a Dage-MTI ISIT operated at very high gain (video amplifier at 50% full scale, intensifier high voltage at 90% full scale). The insert identifies the areas of the image where each of the curves was derived.

The maximum signal is determined by the ability of the scanning beam to fully recharge the target. The slope of the linear portion is called gamma or the tonal transfer function. The tonal transfer function is related to the number of shades of gray that represent the range of the signal and is determined by the characteristics of the photocathode material.

The second basic radiometric response function, shading, relates the intensity response function to the spatial position in the image. Different regions of the camera target can respond differently to light. The use of intensifiers magnifies the effect considerably due to the electrostatic focusing mechanism [Schnitzler and Morton, 1971]. Commonly in intensified cameras the center of the target will have a different response function than the outer edges (see Fig. 8). Imaging a uniformly illuminated field usually results in the center being brighter than the edges. The slopes of the response curves may also be different. This shading effect can be as large as 30% from center to edge (see Fig. 9).

In a fluorescence imaging experiment the typical goal is to make as many measurements as possible over time, and therefore very low doses of excita-

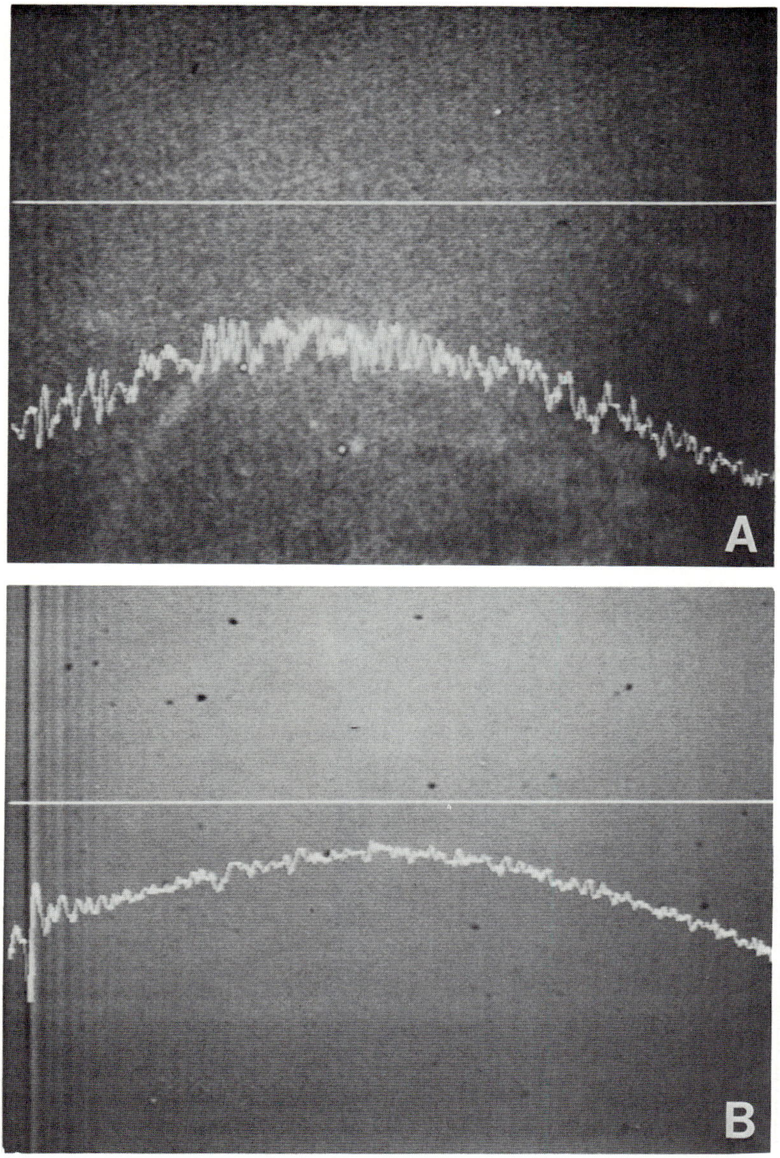


Fig. 9. Illustration of shading effect in two intensified video cameras. A. Dage-MTI ISIT operated at moderate gain. B. Hamamatsu SIT operated at moderate gain. Both represent integrated images of approximately 256 frames. The straight white line identifies the video line plotted by the lower line.

tion light are required. Dose is defined as the intensity of excitation (I) multiplied by the total time (t) of illumination. The same dose can be obtained with very short duration at higher intensity or by longer duration at lower light intensities. It is yet to be determined whether a short duration of higher intensity is better than a longer duration of lower intensity in experiments using living cells. This issue involves the biological responses to illumination, as well as issues relating to the selection of an appropriate camera. Minimizing the excitation light intensity results in a lower fluorescence signal. Therefore, the low-light-level camera is often used near the nonlinear portion of the curve (see Fig. 8). In a quantitative experiment, a visible image does not necessarily indicate a usable image. The degree of nonlinearity will determine the minimum usable image. The minimum "usable" image can be defined as an image that has a signal-to-noise ratio near the lower end of the linear portion of the intensity response transfer curve. In addition, it may be necessary to image two compartments of widely different intensities. This would require an appropriate dynamic range (the ratio of maximum signal to minimum detectable signal), the limits of which are usually the nonlinear portions of the response function.

To measure the intensity transfer curve, a source of constant light is needed. Two basic approaches include the use of a light-emitting diode as a specimen [Reynolds, 1980; Young, 1983] or a quantum counter [see Demas, 1982; Lakowicz, 1983] in a flat capillary tube [Sernetz and Thaer, 1973]. The lamp current is kept constant and the intensity is adjusted to various intensities with a series of neutral density filters. The average value of a small area is plotted vs. the input signal on a log-log scale. The intensity response transfer function is a plot of log input signal vs. log output signal. In practice, values for both shading and intensity response functions can be determined with the same data.

Corrections for a nonlinear intensity response function can be performed. The corrections involve developing a function that maps the recorded response curve to a linear curve representing the input signal. Once the mapping function is determined it can be applied to experimental images. Shading corrections involve applying different mapping functions to different spatial positions. Usually the image is divided into small areas rather than performing the correction on a pixel-by-pixel basis [see Pratt, 1978; Castleman, 1979; Green, 1983; Baxes, 1984].

## Spectral Response

Optical and opto-electronic devices respond differently to light of different wavelengths. In the optical components this is due to the transmission characteristics of the optical material itself (e.g., optical cement) and to the

**TABLE 5. Camera Tube Comparison**

| Manufacturer | Image tube | Spectral response[1] | Lag (%) | Gamma |
|---|---|---|---|---|
| Hamamatsu[2] | Vidicon | | 20 | 0.65 |
| Hamamatsu | Chalnicon | △ | 20 | 0.95 |
| Hamamatsu | Saticon | | 2.5 | 0.95 |
| Hamamatsu | Newvicon | + | 10 | 1.0 |
| Cohu[3] | Intensified Newvicon | + | 20 | 1.0 |
| Hamamatsu | Silicon Vidicon | ○ | 12 | 1.0 |
| Hamamatsu | SIT | □ | 7 | 1.0 |
| Dage MTI[4] | SIT | □ | | 8 |
| Dage MTI | ISIT | □ | | 8 |
| Venus/Zeiss[5] | DV-2 | Extended red-multialkali | | 0.6 |
| Venus/Zeiss | DV-3 | Extended red-multialkali | | 0.6 |
| Hamamatsu | MCP-Saticon | Various | | 1.0 |

[1]See figure 10 for spectral response.
Suppliers of information: [2]Robert Wick, Photonic Microscopy Inc., Hamamatsu Photonics, Oak Brook, IL. [3]Cohu Inc., San Diego, CA. [4]Jim Jones, Dage-MTI, Michigan City, IN. [5]Carl Zeiss Inc., NY.

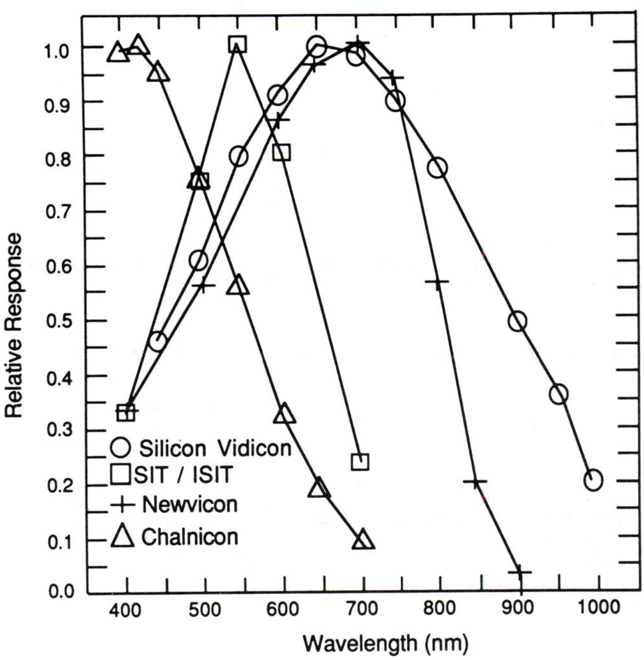


Fig. 10. Representative spectral response curves for several intensified and nonintensified video cameras.

antireflective coatings. The spectral response of the electro-optic devices (e.g., intensifiers and cameras) is defined by the photocathode material. These photocathodes are in most cases the same ones typically found in photomultipliers. The wavelength limit at the low end of the spectral curve is due to the glass window. If the goal is to monitor multiple fluorescent probes, the imaging system must be capable of detecting all the desired wavelengths. With the development of fluorescent probes for use in the near infrared (Waggoner, this volume), there is intense interest in cameras that are sensitive to wavelengths from the visible (~400 nm) to the near infrared (~900 nm). Table 5 and Figure 10 illustrate the relative spectral response of several cameras. If the camera is preceded by an intensifier then it is the spectral response of the first intensifier stage that is important.

Representative spectral response curves are generally available from camera manufacturers. For assessing a camera for a particular application these will often be adequate. For quantitative applications, however, it will be necessary to characterize the spectral response of a particular camera. The spectral response of an imaging device can be determined either in absolute terms (amps/watt) or in relative terms. In practice, the relative response is usually what is important. The relative response can be determined by using a series of narrow-bandwidth filters or a monochrometer [Hall, 1971a,b; Miller, 1971].

The photocathode of the imaging device should be selected to match the emission wavelengths of the fluorescent probes of interest. For a broad range of probes a broad-range general-purpose photocathode is most suitable. Some experiments, however, may dictate specialized cameras. In some cases a specially selected photocathode with a high quantum yield at a particular wavelength can make the difference in being able to image a weak signal.

## Geometric Response

Most vidicon cameras introduce a certain amount of geometric distortion into the image. This is due to the inaccuracies in alignment of the scanning electron beam. In most nonintensified vidicons the distortion is usually so minor as not to be visible. Intensified cameras, on the other hand, generally have pronounced pin cushion distortion. This is primarily due to the electrostatic focusing mechanism of the intensifier. A magnetically focused intensifier will introduce less distortion [Reynolds, 1968; Schnitzler and Morton, 1971].

Figure 11 illustrates the pin cushion distortion in an ISIT adjusted to high gain. The edges of the image are significantly curved. This effect is easy to see in this test target image because the image is known to be made up of straight vertical lines. It is important that the two-dimensional integrity of the object be preserved throughout the imaging process. This dictates that very-

Fig. 11. A standard video resolution test target illustrates the pin cushion geometric distortion common in intensified cameras. A Dage-MTI ISIT was mounted on an Optoliner (Optical Instruments Corp., Buena Park, CA) for projection of the target.

high-quality optical components be used. In practice the geometric distortions of high-quality optical components tend to contribute minimally to the overall distortions of the system.

Corrective geometric operations (e.g., spatial warping, rotation, translation) in image processing provide for the spatial reorientation of pixel data within an image. Correction for these distortions is routinely used in spacecraft-based imaging and medical imaging [Pratt, 1978; Castleman, 1979; Green, 1983; Baxes, 1984].

### Dynamic Response

The ability of an imaging system to respond on an appropriate time scale with an acceptable SNR is defined as the dynamic response of the system. The ability of the system to respond in time to the changing signal defines the adequacy of the system for the particular experiment. The camera defines the limits of the dynamic response, since the optical system has an instantaneous response.

Lag is the undesirable image retention caused by the incomplete erasure of the charged pattern by a single scan of the electron beam (see vidicon description). Some of the information accumulated in the exposure interval for one frame appears in the succeeding frame. Lag is defined as the percentage of the signal remaining in the "third field" [Redington, 1971; Hall, 1971b]. The third field is effectively the first field of the next frame. Lag is characterized by the decay in the output current following a sudden cutoff of light to the camera. In general, lag will be greater at lower light levels for a given camera. Table 5 lists values of lag for several commonly used cameras.

Lag limits the rate of change that may take place in the sample and still be able to record a nonblurred image. This is particularly important in dynamic biological systems. Signal averaging is commonly used in low-light-level imaging to attain an acceptable SNR. However, signal averaging of a dynamic system can lead to significantly lower resolution if the time scale of the changes in the sample are on the same order as the averaging time of the system. Therefore, a system that gives an adequate SNR after 3 sec of integration will lead to significant degradation in resolution if the sample changes on the time scale of seconds. A more useful imaging system for such an experiment will be one with higher gain capable of attaining an adequate SNR in less than a second. Thus the dynamic response must be taken into account in selecting a camera for a given experiment.

**Resolution**

Evaluation of the resolution of a detector is a complex issue. The simplest definition of resolution is "limiting resolution." Classically this is determined by forming the image of a diminishing bar (see Fig. 11) or wedge pattern on the photocathode and determining the finest spacing that is still recognized as a pattern rather than a uniform gray field. Limiting resolution commonly referred to in TV lines/picture height means that if this number of lines, half black and half white (100% contrast), were arranged alternately over a distance equal to the image height they would barely be visible in the image (2–5% contrast). For most purposes it is more useful to express resolution in line pairs/mm at the photocathode.

Resolution defines the ability to resolve two closely spaced structures. A simple measure of resolution is the sampling density. This can be measured using a stage micrometer [Young, 1983]. If a system has a sampling density of 2 pixels per $\mu$m and the structure of interest is 0.5 $\mu$m then the structure is only 1 pixel wide, resulting in significant sampling error (aliasing). No aliasing will occur if the sampling density is greater than or equal to twice the highest frequency in the object. This value is known as the Nyquist frequency [Oppenheim and Schafer, 1975]. Micrometer slides with up to 2-

$\mu$m spacings are available commercially (Ealing Corp., South Natick, MA).

The figure of merit usually quoted for evaluating the performance of image-forming devices is limiting resolution. A more accurate system for analyzing image quality uses the concept of the optical transfer function (OTF). This function is made up of two parts—a real part called the modulation transfer function (MTF) and an imaginary part called the phase transfer function (PTF). The MTF is the ratio of the modulation at each spatial frequency present in the final image to the modulation at the same spatial frequency in the object. The PTF is a measure of distortion and is defined as the difference in phase between the image formed by the optical system under test and the image that would be formed by an equivalent error-free system. For a conceptual discussion of the OTF see Cornsweet [1970; Scott and Frauenhofer, 1971; Dainty and Shaw, 1974; Inoué, 1986].

The MTF may be thought of as a frequency response function. An MTF describes the ability of the imaging system to record a given frequency. The ability to image sharp edges requires the imaging system be able to record higher frequency. The higher the frequency response the sharper the edge. Having sharp edges means two closely spaced edges are more easily resolved and therefore a system capable of imaging higher frequencies will necessarily have greater resolution. Measuring the MTF of a system requires the use of two-dimensional sine wave targets of various frequencies. These targets are very difficult to make and are not available for high magnification microscopes. In practice the contrast trasnfer function (CTF) or edge spread function is measured and converted mathematically to the MTF [Nudelman, 1969; Hall, 1971a; Young, 1983].

The CTF is an approximation of the MTF. The CTF is determined by using a square wave target (Fig. 11). The normalized contrast at each unit of spacing is plotted against the known frequency (Fig. 12). Square wave targets are easier to construct. An alternative approach for determining the MTF is to evaluate the edge response function. As stated above the image of a sharp edge will contain the highest frequencies the system is capable of recording. The MTF is the Fourier transform of the edge response function [Lavin, 1971].

One of the more useful aspects of the MTF is that the MTF of the whole system is the product of the MTFs of the component parts, although this is only true when all the elements have linear radiometric responses. There is no single MTF that is sufficient for a system evaluation. There is rather an an MTF for each spacial patch of the two-dimensional target, one for an object area on the optical axis, one for each of several areas at varying distances from the optical axis, and one each for different directions of the spatial frequency lines. Within each of these patches, there is also an MTF for each wavelength region [Hall, 1971a; Lavin, 1971].

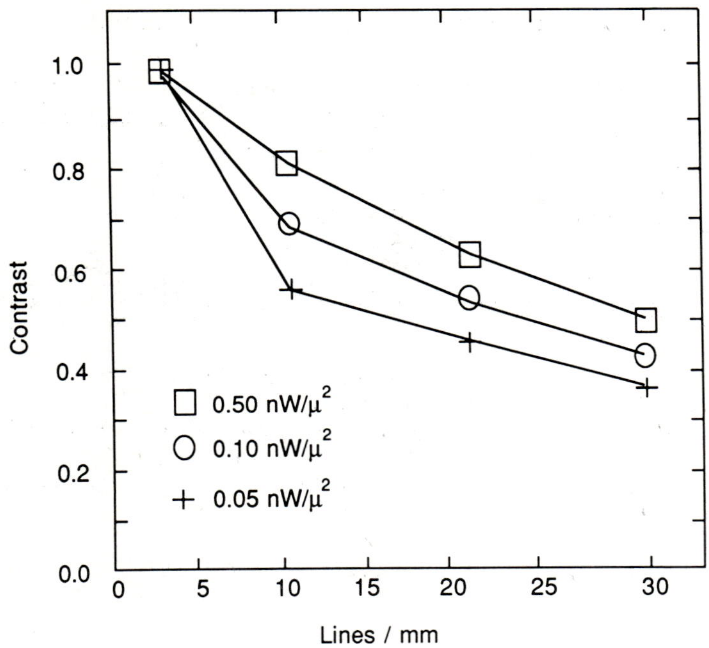


Fig. 12. The contrast transfer function of a Dage-MTI ISIT at several irradiances. Contrast [% contrast = $(I_{max} - I_{min})/(I_{max} + I_{min})$] is plotted against the known spacing of the bar pattern on the camera faceplate.

## SOME APPLICATIONS OF FLUORESCENCE IMAGING TECHNIQUES IN CELL AND MOLECULAR BIOLOGY

Fluorescence microscopy has been used for many years. Table 6 lists some recent applications of fluorescence microscopy. The trend has been to take a more quantitative approach in the application of fluorescence microscopy. One of the recent exciting applications of quantitative fluorescence microscopy is ratio imaging. Ratio imaging can yield a spatial map of a variety of cellular parameters normalized for pathlength including: 1) A probe could be excited at two wavelengths to produce a ratio value related to the concentration. 2) Steady state fluorescence polarization measurements could be made on whole cells by generating a ratio image from sequential images obtained with orthogonal states of polarization. 3) A map of local concentration of a fluorescent protein analogue could be made relative to a contrasting fluorescent molecule using a combination of fluorescent analogue

**TABLE 6. Some Applications of Imaging Techniques in Cell and Molecular Biology**

| | |
|---|---|
| Distribution | |
| Proteins | [See Taylor et al., this volume; Webb and Gross, this volume; Plant et al., 1985; Benson et al. 1985] |
| Lipids | [See Pagano and Sleight, 1985] |
| Organelles | [Johnson et al., 1981; see Waggoner, this volume] |
| Fluorescence recovery after photobleaching (FRAP) | [McGregor et al., 1984; Salmon and Wadsworth, this volume; Taylor et al., this volume] |
| Total internal reflection fluorescence (TIRF) | [Axelrod et al., 1983; Lanni et al., 1985] |
| Membrane potential | [Grinvald et al., 1981; see London et al., this volume] |
| pH | [Tanasugarn et al., 1984] |
| Calcium | [Gilkey et al., 1978; Taylor et al., 1980a; Rose and Lowenstein, 1975, 1976] |
| Ratio imaging | [Tanasugarn et al., 1984] |
| 3D-reconstruction | [Agard and Sedat, 1983; Mathog et al., 1985; Coggins et al., 1985; Fay et al., 1986; see Lanni, this volume] |
| Fluorescence polarization | [Axelrod, 1979] |
| Molecular imaging | |
| Actin | [Yanagida et al., 1984] |
| DNA | [See Yanagida et al., this volume] |

cytochemistry [Wang et al., 1982] and ratio imaging. 4) Resonance energy transfer between suitable labeled molecules could be detected by obtaining a ratio of emission intensities of the acceptor and the donor [Tanasugarn et al., 1984].

An example of the application of ratio imaging is the measurement of the pH of the cytoplasm of single living cells [Tanasugarn et al., 1984]. The fluorescence emission of fluorescein is very sensitive to the pH of the solvent when excited with 489nm light. The emission is largely insensitive to pH when excited with 452nm light. Thus the experiment involves recording an image of the cell using 489nm excitation quickly followed by recording an image using 452nm excitation light. Corresponding background images are recorded with no cell in the field. The following equation defines the ratio.

$$\text{RATIO} = \frac{\text{489\_image} - \text{489\_bkg}}{\text{452\_image} - \text{452\_bkg}}$$

The mathematical manipulations are carried out with a digital image processor.

Ratio imaging has also recently been applied to the measurement of cytoplasmic calcium concentration in single living cells [Williams et al., 1985; Sawyer et al., 1985; Keith et al., 1985]. For a more general discussion of quantitative microscopy, including nonimaging applications, see Lanni (this volume).

## CONCLUSIONS AND PROSPECTUS

The video camera is only one part of the overall system. We have limited our discussion to the camera because it is the limiting factor in most systems. The common intensified video cameras used in fluorescence microscopy are less than ideal detectors. The recorded image is distorted in a variety of ways. However, since the distortions can be well characterized the image can be corrected. In general these distortions limit but do not prevent the use of the video camera as a quantitative detector.

The application of image processing to fluorescence microscopy goes beyond its application to correction for detector deficiencies. There are several catagories of image-processing techniques. *Image enhancement* provides subjective improvement of particular features by manipulating the intensity distributions or by using techniques such as pseudocolor mapping. *Image detection and characterization* considers the detection of a particular object or class or objects and the estimation of various parameters about those objects such as shape, area, or orientation. Edge enhancement algorithms are common detection procedures. *Image restoration* involves the corrections applied to an image to compensate for limitations in the image formation process. Common algorithms include radiometric response nonlinearities and geometric distortions, as discussed in earlier sections. *Image reconstruction* can be used in microscopy to generate a three-dimensional image from serial optical sections of the object. *Image analysis* is an assessment of the image, in the context of the experiment, for comparison with other images. Analysis involves segmentation (dividing the image into parts) and classification. How an image is analyzed depends upon the goal of the particular experiment. Algorithms include thresholding, segmentation schemes, edge enhancement, and texture analysis.

The quantitative fluorescence microscope with automated components (e.g., filter changers) integrated with a sophisticated image processor provide all the elements of a multiparameter microscope system. By choosing fluorescent probes with excitation and emission wavelengths in the range of 400–900 nm it becomes possible to correlate, for example, the relationships among calcium ion concentration, cytoplasmic pH, mitochondrial membrane potential, and actin distribution changes within the same cell upon some external stimulus. The system provides a very favorable environment for studying the interrelationships of various compartments both temporally and spatially in a dynamic living cell.

## ACKNOWLEDGMENTS

We would like to thank Gregory Fisher, Michelle Skrzypczak, and Steve Remis for their helpful comments on this manuscript and Lerson Tanasugarn, Fred Lanni, Alan Waggoner, and Jadwiga Rogowska for many helpful discussions. We would also like to thank the manufacturers who supplied technical data. We particularly would like to thank Hamamatsu Photonics, K.K. for their continued support and assistance. This work has been supported by grants from the NIH (AM-18111), NIH Program Project (GM-34639), NSF (DCB-8416855), NSF (DMB-A414772), and the Council for Tobacco Research, USA, Inc. #1412 to DLT. GB has been supported by a Leukemia Society Postdoctoral Fellowship.

## REFERENCES

Agard DA, Sedat JW (1983): Three-dimensional architecture of a polytene nucleus. Nature 302:676–681.

Agard DA (1984): Optical sectioning microscopy, cellular architecture in three dimensions. Ann Rev Biophys Bioeng 13:191–219.

Arndt-Jovin DJ, Robert-Nicoud M, Kaufman SJ, Jovin TM (1985): Fluorescence digital imaging microscopy (DIM) in cell biology. Science 230:247–256.

Ash EA (1980): "Scanned Image Microscopy." NY: Academic Press.

Axelrod D (1979): Carbocyanine dye orientation in red cell membrane studied by microscopic fluorescence polarization. Biophys J 26:557–573.

Axelrod D, Thompson NL, Burghart T (1983): Total internal reflection fluorescence microscopy. J Microsc 129:19–28.

Barak LS, Webb WW (1981): Observation of dynamics of individual receptor complexes on cultured human fibroblasts. J Cell Biol 90:595–604.

Baxes GA (1984): "Digital Image Processing: A Practical Primer." Englewood Cliffs, NJ: Prentice-Hall.

Benson DM, Bryan J, Plant AL, Gotto Jr AM, Smith LC (1985): Digital imaging fluorescence microscopy: Spatial heterogeneity of photobleaching rate constants in individual cells. J Cell Biol 100:1309–1323.

Biberman LM (1971): Luminance, radiance, and temperature. In "Photoelectronic Imaging Devices. Volume 1." Biberman LM, Nudelman S (eds): NY: Plenum Press, pp 9–38.

Biberman LM, Nudelman S (1971a): "Photoelectronic Imaging Devices. Volume 1. Physical Processes and Methods of Analysis." NY: Plenum Press.

Biberman LM, Nudelman S (1971b): "Photoelectronic Imaging Devices. Volume 2. Devices and Their Evaluation." NY: Plenum Press.

Birks JB (1970): "Photophysics of Aromatic Molecules." NY: Wiley-Interscience.

Bock G, Hilchenbach M, Schauenstein K, Wick G (1985): Photometric analysis of antifading reagents for immunofluorescence with laser and conventional illumination sources. J Histochem Cytochem 33:699–705.

Boksenberg A, Burgess DE (1972): An image photon counting system for optical astronomy. Adv Electron Electron Phys 33B:835–849.

Brackenhoff GJ (1979): Imaging modes in confocal scanning light microscopy. J Microsc 117:233–242.

Castleman KR (1979): "Digital Image Processing." Englewood Cliffs, NJ: Prentice-Hall.

Chance B (1962): Kinetics of enzyme reactions within single cells. Ann NY Acad Sci 97:431–448.

Chance B, Legallais V (1959): Differential microfluorometer for the localization of reduced pyridine nucleotide in living cells. Rev Sci Instrum 30:732–735.

Chance B, Legallais V, Schoener B (1962): Metabolically linked changes in fluorescence emission spectra of cortex of rat brain, kidney and adrenal gland. Nature 195:1073–1075.

Coggins J, Fogarty K, Fay F (1985): Development and application of a 3-dimensional artificial visual system. Proc 9th Symp Comp Appl Med Care (in press).

Coleman CI, Boksenberg A (1976): Image intensifiers. Contemp Phys 17:209–236.

Collet M (1985): Make way for solid-state imagers. Photonics Spectra Sept:103–113.

Cornsweet TN (1970): "Visual Perception." NY: Academic Press.

Cox IJ (1984): Scanning optical fluorescence microscopy. J Microsc 133:149–154.

Dainty JC, Shaw R (1974): "Image Science." NY: Academic Press.

DeHaan EF (1960): Signal-to-noise ratio of image devices. Adv Electron Electron Phys 12:291–306.

Demas JN (1982): Measurement of photon yields. Optic Radiat Measurements 3:195–248.

Fay F, Fogarty K, Coggins J (1986): The analysis of molecular distribution in single cells using a digital imaging microscope. In de Weer P, Salzburg B (eds): "Optical Methods in Cell Physiology." J. Wiley, in press.

Firman C, Ruiz E, Carlson CW, Lampton M, Paresce F (1982): High resolution imaging with a two dimensional resistive anode photon counter. Rev Sci Instrum 53:570–574.

Fogiel M (1981): "The Optics Problem Solver." NY: Research & Education Assoc.

Gilkey JC, Jaffee LF, Ridgeway EB, Reynolds GT (1978): A free calcium wave traverses the activating egg of the madaka *Oryzias latipes*. J Cell Biol 76:448–466.

Goldman M (1968): "Fluorescent Antibody Methods." NY: Academic Press.

Green WB (1983): "Digital Image Processing: A Systems Approach." NY: Van Nostrand Reinhold.

Grinvald A, Ross WN, Farber I (1981): Simultaneous optical measurements of electrical activity from multiple sites on processes of cultured neurons. Proc Natl Acad Sci USA 78:3245–3249.

Hall JA (1971a): Evaluation of direct-view imaging devices. In Biberman LM, Nudelman S (eds): "Photoelectronic Imaging Devices. Volume 2." NY: Plenum Press, pp 53–76.

Hall JA (1971b): Evaluation of signal-generating image tubes. In Biberman LM, Nudelman S (eds): "Photoelectronic Imaging Devices. Volume 2." NY: Plenum Press, pp 77–115.

Hansen EW, Allen RD, Strohbehn JW, Chaffee MA, Farrington DL, Murray WF, Pillsbury TA, Riley MF (1985): Laser scanning phase modulation microscope. J Microsc (in press).

Haugland RP (1983): Covalent fluorescent probes. In Steiner RF (ed): "Excited States of Biopolymers." NY: Plenum Press.

Hecht E, Zajac A (1974): "Optics." Reading, MA: Addison-Wesley.

Howes MJ, Morgan DV (1979): "Charge-Coupled Devices and Systems." NY: John Wiley & Sons.

Inoué S (1986): "Video Microscopy." NY: Plenum Press.

Jacobson K, Elson E, Koppel D, Webb W (1983): International workshop on the application of fluorescence photobleaching techniques to problems in cell biology. Fed Proc 42:72–79.

Johnson LV, Walsh ML, Bockus BJ, Chen LB (1981): Monitoring of relative mitochondrial membrane potential in living cells by fluorescence microscopy. J Cell Biol 88:526–535.

Keith CH, Ratan B, Maxfield FR, Bajer A, Shelanski ML (1985): Local cytoplasmic calcium gradients in living mitotic cells. Nature 316:848–850.

Kohen E, Thorell B, Hirschberg J, Woiters A, Kohen C, Bartick P, Salmon J-M, Viallet P, Schahtschabel D, Rabinovitch A, Minty D, Meda P, Westerhoff H, Nestor J, Ploem J (1981): Microspectrofluorometric procedures and their applications in biological systems. In Wehry EL (ed): "Modern Fluorescence Spectroscopy, Vol. 3." NY: Plenum Press, pp 295–346.

Lakowicz J (1983): "Principles of Fluorescence Spectroscopy." NY: Plenum Press.

Lampton M (1981): The microchannel image intensifier. Sci Am 245:62–71.

Lanni F, Waggoner AS, Taylor DI (1985) Structural organization on interphase 3T3 fibroblasts studied by total internal reflection fluorescence microscopy. J Cell Biol 100:1091–1102.

Lavin HP (1971): System analysis. In Biberman LM, Nudelman S (eds): "Photoelectronic Imaging Devices. Volume 1." NY: Plenum Press, pp 333–374.

Mathog D, Hochstrasser M, Sedat JW (1985): Light microscope based analysis of three-dimensional structure: applications to the study of *Drosophila* salvary gland nuclei. I. Data collection and analysis. J Microsc 137:241–242.

McGinty GP (1984): "Video Cameras: Operation and Servicing." Indianapolis: Howard W. Sams & Co.

McGregor GN, Kapitza H-G, Jacobson KA (1984): Laser-based fluorescence microscopy of living cells. Laser Focus Nov:85–93.

Miller LD (1971): Transfer characteristics and spectral response of television camera tubes. In Biberman LM, Nudelman S (eds): "Photoelectronic Imaging Devices. Volume 1." NY: Plenum Press, pp 267–290.

Nudelman S (1969): Intensifiers: detective quantum efficiency, efficiency contrast transfer function and the signal-to-noise ratio. Adv Electron Electron Phys 28B:577–587.

Nudelman S (1979): Image intensification devices and applications. In McLean TP, Schagen P (eds): "Electronic Imaging." NY: Academic Press, pp 263–307.

Oppenheim AV, Schafer RW (1975): "Digital Signal Processing." Englewood Cliffs, NJ: Prentice-Hall.

Ornstein L (1957): New horizons in fluorescence microscopy. J Mt Sinai Hosp 24:6.

Pagano RE, Sleight RG (1985): Defining lipid transport pathways in animal cells. Science 229:1051–1057.

Plant AL, Benson DM, Smith LC (1985): Cellular uptake and intracellular localization of benzo(a)pyrene by digital fluorescence imaging microscopy. J Cell Biol 100:1295–1308.

Ploem J, Verwoerd N, Bonnet J, Koper G (1979): An automated microscope for quantitative cytology combining television image analysis and stage scanning microphotometry. J Histochem Cytochem 27:136–143.

Pratt WK (1978): "Digital Image Processing." Englewood Cliffs, NJ: Wiley.

Redington RW (1971): Introduction to the vidicon family of tubes. In Biberman LM, Nudelman S (eds): "Photoelectronic Imaging Devices. Volume 2." NY:Plenum Press, pp 263–274.

Reynolds GT (1968): Image intensification applied to microscope systems. Adv Opt Electron Microsc 2:1–40.

Reynolds GT (1972): Image intensification applied to biological problems. Q Rev Biophys 5:295–347.

Reynolds GT (1980): Criteria for evaluation of low light level intensifier-vidicon detectors for microscopy and spectroscopy. Technical Report No. 24. Princeton NJ: Dept. Physics, Princeton University.

Reynolds GT, Taylor DL (1980): Image intensification applied to light microscopy. Bioscience 30:586–591.

Rose A (1973): "Vision: Human and Electronic." NY: Plenum Press.

Rose B, Loewenstein W (1975): Calcium ion distribution in cytoplasm visualized by aequorin:

Diffusion in cytosol restricted by energized sequestering. Science 190:1204–1206.
Rose B, Loewenstein W (1976): Permeability of a cell junction and the local cytoplasmic free ionized calcium concentration: a study with aequorin. J Membr Biol 28:87–119.
Rotman B (1961): Measurement of activity of single molecules of $\beta$-D-galactosidase. Proc Natl Acad Sci USA 47:1981–1991.
Ruck F (1970): Principles and some applications of cytofluorometry. In Wied GL, Bohr GF (eds): "Introduction to Quantitative Cytochemistry II." NY: Academic Press, pp 431–450.
Sawyer DW, Sullivan JA, Mandel GL (1985): Intracellular free calcium localization in neutrophils during phagocytosis. Science 230:663–666.
Schnitzler AD, Morton GA (1971): Cascade image intensifiers. In Biberman LM, Nudelman S (eds): "Photoelectronic Imaging Devices. Volume 2." NY: Plenum Press, pp 119–131.
Scott F, Frauenhofer D (1971): The modulation transfer function and methods of measurement. In Biberman LM, Nudelman S (eds): "Photoelectronic Imaging Devices. Volume 1." NY: Plenum Press, pp 291–306.
Sears FW, Zemansky MW, Young HD (1982): "University Physics." Reading, MA: Addison-Wesley Publ.
Sernetz M, Thaer A (1973): Microscapillary fluorometry and standardization for microscope fluorometry. In Thaer AA, Sernetz M (eds): "Fluorescence Techniques in Cell Biology." NY: Springer-Verlag, pp 41–50.
Shoemaker RL, Bartels PH, Hillman DW, Jonas J, Kessler D, Shack RV, Vukobratovich D (1982): An ultrafast laser scanner microscope for digital image analysis. IEEE Trans Biomed Eng BME-29:82–91.
Siemens A, Walter R, Liaw L-H, Berns M (1982): Laser-stimulated fluorescence of submicrometer regions within single mitochondria of rhodamine-treated myocardial cells in culture. Proc Natl Acad Sci USA 79:466–470.
Slayer EM (1970): "Optical Methods in Biology." NY: Wiley-Interscience.
Talmi Y (1983): "Multichannel Image Detectors. Volume 2." Washington: Am. Chem. Soc.
Tanasugarn L, McNeil P, Reynolds G, Taylor DL (1984): Microspectrofluorometry by digital image processing: Measurement of cytoplasmic pH. J Cell Biol 98:717–724.
Taylor DL, Wang Y-L (1978): Molecular cytochemistry: Incorporation of fluorescently labeled actin into living cells. Proc Natl Acad Sci USA 75:857–861.
Taylor DL, Wang Y-L (1980): Fluorescently labeled molecules as probes of the structure of living cells. Nature 284:405–410.
Taylor DL, Blinks JR, Reynolds GT (1980a): Contractile basis of ameboid movement. VIII. Aequorin luminescence during ameboid movement, endocytosis, and capping. J Cell Biol 86:599–607.
Taylor DL, Wang Y-L, Heiple J (1980b): Contractile basis of ameboid movement. VII. Distribution of fluorescently labeled actin in living amoebae. J Cell Biol 86:590–598.
Taylor DL, Amato PA, Luby-Phelps K, McNeil P (1984): Fluorescent analog cytochemistry. Trends Biochem Sci 9:88–91.
Thaer A (1966): Instrumentation for microfluorometry. In Wied GL, Bohr GF (eds): "Introduction to Quantitative Cytochemistry II." NY: Academic Press, pp 409–426.
Thaer A, Sernetz M (1973): "Fluorescence Techniques in Cell Biology." NY: Springer-Verlag.
Timothy JG (1981): Curved-channel microchannel array plates. Rev Sci Instrum 52:1131–1142.
Turro NJ (1978): "Modern Molecular Photochemistry." Reading, MA: Benjamin/Cummings Publ.
Udenfriend S (1969): "Fluorescence Assay in Biology and Medicine. Volume 11." NY: Academic Press.

Wang Y-L, Heiple J, Taylor DL (1982): Fluorescent analog cytochemistry of contractile proteins. Meth Cell Biol 25(Pt.B):1–11.

West S.S. (1969): Fluorescence microspectrophotometry of supravitally stained cells. In Pollister AW (ed): "Physical Techniques in Biological Research. Volume 3, Part C, Cells and Tissues." pp 253–321.

Wick RA (1985): Grabbing images at very low light levels. Photonics Spectra May:133–136.

Williams DA, Fogarty KE, Tsien RY, Fay FS (1985): Calcium gradients in single smooth muscle cells revealed by the digital imaging microscope using Fura-2. Nature 318:558–561.

Willingham M, Pastan I (1978): The visualization of fluorescent proteins in living cells by video intensification microscopy. Cell 13:501–505.

Wilson T, Sheppard C (1984): "Theory and Practice of Scanning Optical Microscopy." NY: Academic Press.

Yanagida T, Nakane M, Nishiyama K, Oosawa F (1984): Direct observation of motion of single F-actin filaments in the presence of myosin. Nature 307:58–60.

Young IT (1983): The use of digital image processing techniques for the calibration of quantitative microscopes. Proc Soc Photo-optical Inst Eng 397:326–335.

*Applications of Fluorescence in the Biomedical Sciences, pages 289–300*


# 13 New Instrumentation for Sensitive Image Analysis of Fluorescence in Cells and Tissues

*J.S. Ploem*

Fluorescence methods are increasingly applied in clinical cytometry. New instrumental developments have made this possible. A first example consists of the laser scanning microscope (LSM), which enables digital storage of an image of a scanned area in computer memory. In this instrument entire microscope fields are scanned by mirrors vibrating in opposite directions or by a rotating polygon mirror, deflecting the laser light onto the specimen. In an instrument produced by Zeiss-Oberkochen, West-Germany [Wilke, 1982], light from an argon laser fills the entire entrance pupil of the objective for each microscope step. All the light is concentrated on only one small spot in the microscope preparation.

Advantages of this laser scanning microscope are the low level of excitation light on optical parts and the very low microscope objective magnification (e.g., 2.5× dry objective) that can be used. This enables the examination of fluorescence in large areas of microscope specimens, such as, e.g., in brain tissue.

In the laser scan microscope the fluorescence of each spot illuminated by the laser is measured with a photomultiplier (PM). The system therefore results in a good signal-to-noise ratio with selected PM tubes. The digitized signals are directly stored in computer memory by using a high-resolution analog-to-digital converter. The Zeiss type of image scanner is however, much slower than TV-based systems, such as the new low-magnification field microscope from E. Leitz, Wetzlar, West Germany [Wasmund and Ploem, 1983].

Department of Histochemistry and Cytochemistry, Leiden University, Wassenaarseweg 72, AL Leiden 2333 , The Netherlands.

The development of low-magnification superwide field microscopes, especially designed for image analysis, has contributed significantly to the analysis of weakly fluorescent images. These microscopes can combine fluorescence detection with conventional modes of microscopy, such as phase-contrast, absorbance microscopy, and reflectance microscopy. These new instruments make use of immersion objectives with maximal numerical aperture (NA) and superwide fields, and permit a relatively low magnification. This is obtained by direct projection of the image of the objective on a TV tube without using a conventional projective eyepiece.

With the new Leitz multiparameter microscope bright fluorescence images can be obtained. Their brightness is about 25 times higher than those of a normal fluorescence microscope. By using conventional plumbicon tubes such an instrument can perform image analysis on most immunofluorescence applications.

A third instrumental development combines scanning electron microscopy (SEM) with fluorescence light microscopy (LM) [Ploem and Thaer, 1980, 1981; Wouters and Koerten, 1982]. This has been achieved by incorporating an epi-fluorescence microscope into the vacuum chamber of a scanning electron microscope. Fluorescence exciting light can be guided onto the specimen in the vacuum chamber of the SEM and the fluorescence can be observed through the light microscope objective in the vacuum chamber. The image is transferred to the outside by a light pipe connection to an epi-fluorescence microscope.

Specimens have to be mounted on thin coverglass slips and treated in the usual way with gold coating to make them visible in the SEM. Cytochemical staining methods have been developed which allow measurement of a fluorescent Feulgen DNA stain while maintaining a good surface cytomorphology, based on the gold coating in the scanning electron microscope [Wouters et al., 1984, 1985]. This instrument opens new possibilities in clinical cytometry, since cancer cells, detected on the basis of strongly increased anaploid DNA content, can be studied for abnormal surface structures with SEM.

All three instrumental developments of the type described above permit a much wider use of quantitative methods in fluorescence cytochemistry and histochemistry. In contrast to the older microfluorometry equipment, the newer instruments permit an effective pattern recognition. Strongly fluorescent cells or cellular organelles can often be automatically isolated from the background by thresholding of the image.

In specimens showing an irregular staining, gradient-type algorithms applied to the stored digitized image can isolate cells and cellular organelles from the background. The isolated image structures can thereafter be analyzed for a number of parameters.

With conventional fluorescence microscopy it was not feasible to use 2.5× dry objectives with low NA, whereas in the laser scan microscope, large tissue areas can be examined with low magnification for cytochemical fluorescent stains. This permits, e.g., an integrated measurement of the total neurotransmitter fluorescence in a nervous network or in a group of nerve cells.

The LM/SEM instrument allows the application of a powerful combination of parameters measured both in scanning electron and light microscopy.

## LITERATURE

### Laser Scan Microscopes

Laser scan microscopes have been developed by a number of workers [Bartels et al., 1981; Deinet et al., 1983, Shoemaker et al., 1982; Wilke, 1982; Wilson and Sheppard 1984]. Some of the original versions of the instruments were not directed to fluorescence applications. However, fluorescent scanning applications have been included in the newer designs.

The very powerful laser scan used with some of these instruments can in principle concentrate an enormous amount of energy on each pixel of the microscope field [Bartels et al., 1981; Shoemaker et al., 1982]. With the more rapid scanning systems, the dwelling time of the illuminating beam is relatively short. Consequently problems of destruction of the fluorophore or the biological carrier can be mastered.

With this instrumentation very high scanning rates can be achieved [Bartels et al., 1981]. Real speed limits are determined by the mechanical requirements of vibrating mirrors or rotating polygons and the energy of the laser available. In other words, the signal-to-noise ratio is dependent on the number of photons that are available per time unit. It can be expected that the instrument using very rapidly rotating polygon mirrors [Bartels et al., 1981] will provide similar and in some cases even greater speed than the flow cytometry systems. In many cases the data acquisition rates are so high that present equipment based on single microprocessors cannot process the data. Future developments in image processing architecture, such as parallel processing, are expected to solve the type of data rate problem of the high-speed laser scanning microscopes.

The two main types of laser scanning microscopes, the ultrafast [Bartels et al., 1981] and the slower types [Wilke, 1982], aim at different applications. The first is designed for rapid automation of cytology and hematology specimens and the latter for investigations which require a combination of microscopic methods (phase-contrast, reflection, fluorescence, interference contrast).

In addition it is possible to use more than one fluorescence excitation wavelength with the latter instrument, using mercury or xenon lamps, so that combination fluorescence is feasible in conventional microscopy. The fluorescence signal can be combined with phase-contrast, which is very important for living cell investigations. Wilson and Sheppard [1984] have mainly focussed on the resolution problem. Laser scanners using confocal microscopy achieve higher resolving power than is feasible with conventional light microscopy. In addition, valuable information can be obtained with this type of instrument regarding the three-dimensional structure of cells as revealed by fluorescence probes.

Since fluorescence image cytometry provides morphological information, even about small organelles, the potential of the fluorescence image analysis for the recognition of cells and cell organelles after fluorescence cytochemical staining is considerable.

**Instrumentation for TV-Based Image Analysis**

TV-based systems have been used by various workers for the analysis of fluorescent stains [Wachsmuth et al., 1984]. Due to the required sensitivity, mostly silicon intensified target (SIT) and intensified SIT (ISIT) cameras have been used.

An early application of image analysis techniques in fluorescence is described by Hillman et al. [1972]. He used a TV camera connected to a video digitizer to read photographic negatives of fluorescently labeled tissues. In this way he converted qualitative pictures into quantitative data.

Satisfactory results have been obtained by using normal TV cameras such as plumbicons [Al et al., 1978]. Their applications are limited to cases where relatively strong fluorescence staining is available. For weaker fluorescence most TV cameras need an additional procedure for building up sufficiently strong fluorescent signals to achieve good detectability. To this purpose special instruments have been made commercially available. A particularly sensitive instrument has been developed by Hamamatsu [Reynolds and Taylor, 1980; Allen and Allen, 1983]. These instruments combine video-enhanced microscopy with ultrahigh-sensitivity TV cameras. The advantage of using these extremely sensitive cameras lies specifically in the application to the study of living cells labeled with fluorescent markers, which can support only very weak excitation light.

Very few of these instruments have been used in a way similar to flow cytometry for the analysis of large numbers of cells or the detection of rarely occurring cell types. For this type of application, the adding of fluorescence signals during many TV cycle periods would slow down the instrument too much. Most likely, superwide-field low-magnification microscopy using high

NA objectives will (in combination with adding the fluorescence signal over a few TV cycles) provide a good compromise of sensitivity and speed.

Fluorescence microscopy in combination with image analysis has also been used in intravital methods [Gahm et al., 1985]. Data acquisition is made by a highly sensitive TV camera.

### Light Microscopy/Scanning Electron Microscopy (LM/SEM)

These two microscope techniques complement each other. The SEM provides an image, which gives three-dimensional information of the surface with a relatively great depth of focus and high resolution. The secondary emission mode usually used in the SEM to generate the surface image reveals no information concerning the internal cell structures. Although it is sometimes possible to identify cells with an SEM only, a better understanding of the cell structure and function will be obtained by a correlation between morphological, physical, and cytochemical properties. With light microscope methods the internal structure of the cell can be studied, especially after appropriate cytochemical staining.

In the past SEM and LM have frequently been combined, but in different instruments. The preparation then has to be transported from one instrument to the other, inducing relocation problems. Furthermore, cell loss or damage may occur when the cover glass used in LM investigation is removed for the SEM investigation. The latter problem may be avoided when the cover glass is, e.g., placed on a nail-polish rim. Using two different instruments for LM and SEM investigation is time-consuming, provides limited possibilities for alternating LM and SEM for reinvestigations, and limits the numbers of investigated cells for reasons of efficiency.

To circumvent the problems of using two separate microscopes, an instrument has been developed for the combined use of LM and SEM [Hartmann et al., 1978; Ploem, 1981; Ploem and Thaer, 1980, 1981]. For investigation with the combined microscope, the cells are put on cover glasses, whereafter they are cytochemically stained for LM and critical point dried and gold coated for SEM [Wouters et al., 1984].

## INSTRUMENTATION

### Laser Scan Microscope (LSM)

The commercial design of the Zeiss (Oberkochen, West Germany) laser scan microscope provides a large variety of optical methods, such as phase-contrast, reflection contrast, interference contrast, and absorption microscopy. This is achieved by careful alignment of the illumination of the entrance pupil of all objectives which are likely to be used with this instrument. The lack of speed in this type of laser scanner is compensated by the availability of multiparameter microscope data, obtainable by this variety of microscope methods.

With the LSM a 2.5× dry objective can be used, enabling examination and measurement of fluorescence in large tissue areas.

For each microscope step, light from the laser fills the entire entrance pupil of the objective, whereafter all the light is concentrated on only one small spot in the preparation. An advantage of this procedure is that the load of exciting light on the optical parts, such as the objective and the immersion oil, is relatively low. Roughly estimated the energy load on the entrance pupil of the objective amounts to only about 1 mW, and this results in extremely low autofluorescence of optical parts and immersion oil. With normal fluorescence microscopes the amount of exciting light reaching the entrance pupil of an objective can have an energy in the range of 1 W. The low level of autofluorescence from optical parts in the Zeiss LSM significantly contributes to the extreme sensitivity of this type of microscope.

The laser scan microscope (LSM) can be used with a conventional light source as well as with a laser. When a laser is used as light source, the laser beam is scanned over the probe by means of two vibrating mirrors. The fluorescence signal is then digitized, stored in a digital memory, and visualized on a cathode ray tube (CRT) display

With the laser light source, transmitted light microscopy as well as incident light microscopy are possible. For transmitted light microscopy, a standard condenser is used to obtain absorbance, phase, and differential interference contrast images. Reflection and fluorescence contrast images can be obtained with incident light. Because of the implemented laser, it is not possible to choose another wavelength for excitation as the laser and the accompanying optical parts are used in a monochromatic way (488 nm from an air-cooled argon laser).

As a conventional light source, the instrument only permits the use of a tungsten lamp guided by a glass fiber. With this setup however, fluorescence microscopy is not possible.

The fluorescence measurements are performed with a photomultiplier. The digitized signals are stored in computer memory.

**Modifications of the LSM Instrument.** As already mentioned, it is not possible to observe a complete fluorescence image in the eyepieces. It is also not possible to observe a double-stained probe when other wavelengths for excitation are to be used. Furthermore the focussing is difficult, since the observer has to wait 2 full seconds before the image is seen on the CRT screen. Preparations consisting of small samples may be difficult to locate when cell structures can only be recognized on the basis of their fluorescence. For these reasons several components have been added to make normal epi-illumination fluorescence microscopy possible.

The changes consist of (1) a connection for a mercury lamp, like a HBO-50 or HBO-100, without interfering with laser scan microscopy; (2) a mirror to switch the mercury lamp illumination into the light path; (3) a field diaphragm to alter the illuminated area; (4) filters for fluorescence microscopy with UV, blue, and green excitation; (5) and a filter compartment for the separate analysis of different wavelengths of emitted light [Ploem, 1984]. Thus double-stained probes can be analyzed.

**Theoretical Calculation of LSM Sensitivity.** Since the total energy of the laser is focussed into one single small spot in the microscope specimen, it was expected that sensitivity would be high. To estimate the sensitivity more quantitatively, experiments were performed with test beads from Becton and Dickinson. These particles have a known amount of FITC molecules (diameter in microns: 4.3, number of FITC molecules: 2,900). In this way it is possible to extrapolate the sensitivity of the microscope in terms of detectable amount of FITC molecules. Experimentally determined, the fluorescence sensitivity of the laser scan microscope amounted to about ten FITC molecules per square micrometer. In comparison, the "normal" microscope would need more than a hundred times more FITC molecules to obtain an image on a highly sensitive photo film. For microscopes using image intensifiers and SIT-TV cameras the sensitivity will of course be higher than in the normal microscope.

### Modular Image Analysis Computer (MIAC) in Combination With Automated Multiparameter Microscopy

The new Leitz multiparameter microscope (Wetzlar, West Germany) uses a 40× objective with an NA of 1.30, enabling the analysis of microscope fields of 500 × 500 $\mu$m with a pixel separation of about 1 $\mu$m. Due to demagnification, the total microscope magnification reaching the TV tube is only 20×. This results in bright fluorescence images. Their brightness is about 25 times higher than in a normal fluorescence microscope.

The multiparameter microscope permits simultaneous observation—using one wide-field immersion oil objective (NA 1.30)—of an image with low magnification (20×) and high magnification (80×). The instrument has the facilities to provide the low-magnification channel with a plumbicon tube for measurement purposes and the high-magnification channel with an SIT-TV camera for the analysis of the highly magnified field. It can be expected that such instruments will meet most requirements in fluorescence cytochemical work, including speed and high sensitivity.

The image-analyzing computer is constructed around a Versa Module Europe (VME) bus and a high-speed TV bus. The instrument can adapt 17 triple slot eurocards and two double slot cards for interfacing a microscope, a terminal, and a printer. As main processor the MIAC contains a Motorola 68,000 microprocessor. The basic idea of this instrument is to combine rapid

**TABLE 1. Basic Configuration of the Modular Image Analysis Computer (MIAC)**

| |
|---|
| Analog to digital converter |
| Memory address processor |
| Memory board for storage of images (1 megabyte) |
| Memory board for storage of images (1 megabyte) |
| Memory address processor |
| Memory board for storage of images (1 megabyte) |
| Memory board for storage of images (1 megabyte) |
| Gray value processor for mathematical morphology |
| Motorola microcomputer board + 1 Megabyte memory |
| Binary processor (mathematical morphology) |
| Binary processor (mathematical morphology) |
| Second motorola microcomputer board for a slave processor + 1 Megabyte memory |
| DNA measuring card |
| Individual analysis card |

image analysis based on mathematical morphology [Meyer, 1979; Serra, 1982] with conventional pattern recognition analysis run in Motorola 68,000 microprocessors. A basic configuration of the instrument is depicted in Table 1.

The binary processor and gray-value processor, using mathematical morphology algorithms, take care of rapid scene segmentation and analysis of particles in the microscope field. The basic time cycle for this type of operation of the instrument is 20 msec.

With the MIAC 512 × 512 pixels, as detected by a plumbicon tube, are stored directly in the computer memory by using a high-speed analog-to-digital conversion. Although these systems suffer from the limited signal-to-noise resolution of TV-based systems, they are extremely fast for biomedical work, since an image of 512 × 512 pixels can be treated in only 20 msec.

## LM/SEM

The LM/SEM instrument consists of a Leitz 1200 B scanning electron microscope with special adaptations, connected to a Leitz light microscope. The adaptation in the SEM consists of a special stage in the vacuum chamber. Onto this stage an exchangeable oil-immersion objective and a medium-power objective with a long free-working distance, serving as a condenser, are mounted. The centering and focussing devices for the objective and the handle for swinging the condenser in and out the light path are outside the vacuum chamber. The condenser can be swung out of the light path of the optical axis to permit undisturbed electron-beam passage.

For LM investigations with transmitted light a tungsten light source is used. The light is guided by a glass fiber into the vacuum chamber. Here the light is led by mirrors and a prism to the condenser. After traversing the condenser, the light passes through the specimen and is collected by the oil-immersion objective. The light leaves the chamber via an optical bridge and then enters the LM part of the instruments. The resulting image can be studied through eyepieces or a micrograph can be taken.

For LM fluorescence microscopy light is obtained from a high-pressure mercury arc attached to the LM part of the instruments. The light reaches the vacuum chamber via the optical bridge, whereafter it is concentrated onto the specimen by the oil-immersion objective (epi-illumination). The emitted fluorescence light returns to the LM via the same pathway. The fluorescence image can be observed through the eyepieces or photographed. A microfluorometer attached to the LM enables fluorescence intensity measurements.

## APPLICATIONS

### Laser Scanning Microscope

A very low magnification (2.5× dry objective) can be used for the detection of immunofluorescence in specimens treated for histochemical evaluation of large parts of the brain. With this low magnification, the energy of the laser is simply concentrated in a larger spot of the specimen, resulting in a sufficiently strong fluorescence. The area under investigation using a 2.5× objective is 3.75 × 3.75 mm.

The low magnification in laser scan microscopes means that connectivity of elements in peripheral nerve and brain tissue as detected with specific monoclonal antibodies against neurotransmitters or their precursors can be studied with simultaneous storage of all the fluorescence signals in computer memory. In the stored digitized fluorescence image, segmentation based on contrasting algorithms can be used to isolate nerve fibers from the background. This again permits an integrated measurement of the total neurotransmitter-based fluorescence in a network or in a group of nerve cells.

Immunofluorescently stained structures in the floor of the fourth ventricle and axons in the iris of the rat were studied as examples. The iris of the rat was studied after immunofluorescence (IF) antibody incubation against noradrenaline was well as after paraformaldehyde vapor treatment yielding catecholamine fluorescence. The catecholamine fluorescence has an excitation peak at 410 nm, but due to the very low background fluorescence in the LSM, the reaction product can even be excited at 488 nm. The electronic magnification still shows the same detail up to a factor of 80× due to the overresolution of the scanned area.

The very dark background of the LSM results in a high image contrast. This characteristic permits the observation of immunofluorescently stained DNA adducts, which requires a very sensitive fluorescence microscope. It can be applied, e.g., to environmental research, where it is important to demonstrate very low concentrations of drugs or chemical noxae that may attach to DNA. With the LSM cells could be observed which could not be identified with a normal fluorescence microscope.

Another example of low-level fluorescence work is hybridocytochemistry. Especially at low magnification the LSM provides bright fluorescence images.

Due to the combination of phase-contrast and fluorescence at low magnification, the LSM is also very useful in cytotoxicity tests. Countings can be obtained in phase-contrast, whereas the lymphocytes positive for a certain antibody can be observed in the fluorescence mode.

## Modular Image Analysis Computer

Microscope fields of 500 × 500 μm contain, e.g., several thousands of human red blood cells. If monoclonal antibodies are used to detect variant erythrocytes in such large fields, 2,000 erythrocytes can be analyzed for the presence of one single fluorescent cell in 20 msec at a pixel separation of 1 μm. In reality, however, the speed of such instruments depends on the time needed for focussing and elimination of fluorescence artefacts. On the other hand, only a few seconds are basically needed to determine a percentage of a subset of T-lymphocytes by using a panel of monoclonal antibodies.

Investigations [van Driel-Kulker et al., 1980, 1982; Meyer and van Driel-Kulker, 1980; Ploem, 1982] in our laboratory with a less advanced image analysis system (LEYTCAS) have led to the programming of the MIAC for the automated screening of cervical smears. Preliminary investigations have shown that the MIAC is very suitable for automated screening and measuring purposes. The analysis of the large microscope fields is very efficient and the resolution is still high.

## LM/SEM

Application model studies [Wouters et al., 1985] have shown that cytochemical staining of the cells does not damage surface morphology as observed with the SEM. For instance, the surface morphology of quantitatively stained lymphocytes was completely comparable to lymphocytes that were only treated for SEM morphology. Also the DNA measurements were not significantly influenced by SEM treatment of the cells. The coefficient of variation in the measurement of the DNA content in 50 lymphocytes was only 6.1%. This experiment shows that correlation studies between cell cycle and surface morphology are feasible.

Cells identified by their aneuploid DNA content can be studied with SEM. Other studies have shown that such cells have a very high correlation to premalignant and malignant changes [van Driel-Kulker and Ploem, 1982; Ploem, 1982] in cervical cancer. Surface morphology changes concurring with malignancy can with this method be detected by using a parameter independent from the morphology, namely a strongly increased anaploid DNA content. Reliable correlations between malignancy-associated changes, detected with SEM, can be observed.

## REFERENCES


Al I, Ploem JS, Tanke HJ (1978): Fluorescence microscopy for biomedical image analysis using a television based system (LEYTAS).Proc R Microsc Soc 13:58–59.

Allen RD, Allen NS (1983): Video-enhanced microscopy with a computer frame memory. J Microsc 129:3–17.

Bartels PH, Buchroeder RA, Hillman DW, Jonas JA, Kessler D, Shoemaker RM, Shack RV, Towner D, Vukobratovich D (1985): Ultrafast laser scanner microscope. Anal Quant Cytol 3:55–60.

Deinet W, Linke M, Mueller R, Sander I (1983): Laser-scanning-mikroskop mit automatischer Fokussierung. Microsc Acta 87:129–138.

van Driel-Kulker AMJ, Meyer F, Ploem JS (1980): Automated analysis of cervical specimens using the TAS Microsc Acta (Suppl) 4:73–81.

van Driel-Kulker AMJ, Ploem JS (1982): The use of LEYTAS in analytical and quantitative cytology. IEEE Trans Biomed Eng BME-29 (no 2):92–100.

Gahm T, Reinhardt E, Witte S, Bloss WH (1985): Fluoreszenz-mikroskopische fernsehtechnische Untersuchung der Mikrozirkulation im Rattenmesenterium. Zeiss Inform 27:49–51.

Hartmann H, Hund A, Moll SH, Thaer A (1978): Attachment for combined scanning electron and light microscopical examinations. Beitr Elektronenmikroskop Direksabb Oberfl 11:381.

Hillman GR, Johnston D, Kwan S-W, Carney D, Childa G (1972): Histochemical applications of image analysis techniques. Proc IEEE Int Symp Med Imag Image Interp.

Meyer F (1979): Iterative image transformations for automatic screening of cervical smears. J Histochem Cytochem 27:128–135.

Meyer F, von Driel-Kulker AMJ (1980): Automatic screening of Papanicolaou stained cervical smears with the TAS. Microsc Acta (suppl) 4:82–95.

Ploem JS (1981): A new instrument permitting simultaneous scanning electron microscopy and fluorescence microscopy of the same specimen by integrating a LM in a SEM vacuum chamber. Cytometry 2:121.

Ploem JS (1982): Automated methods in immunofluorescence studies. In Wick et al. (eds) "Immunofluorescence technology." Selected theoretical and clinical aspects. Amsterdam: Elsevier Biomedical Press, pp 73–94.

Ploem JS (1984): Organic and biological surfaces: Fluorescence microscopy. In Echlin P. (ed): "Analysis of Organic and Biological Surfaces. John Wiley and Sons Inc, pp 609–628.

Ploem JS, Thaer A (1980): Luminescence studies with an integrated instrument permitting SEM and fluorescence microscopy of the same specimen. Proc R Microsc Soc 15:9–10.

Ploem JS, Thaer A (1981): A light microscope integrated in the vacuum chamber of a SEM. Proc R Microsc Soc 16:253.

Reynolds GT, Taylor DL (1980): Image intensification applied to light microscopy. Bioscience 30:586–592.

Serra J (ed.) (1982): "Image Analysis and Mathematical Morphology." London: Academic Press Inc.

Shoemaker RL, Bartels PH, Hillman DW, Jonas J, Kessler D, Shack RV, Vukobratovich D (1982): An ultrafast laser scanner microscope for digital image analysis. IEEE Trans Biomed Engi, Vol BME-29, no. 2:82–91.

Wachsmuth, ED, Becker M, Grieve AP, Maurer-Schultze B (1984): Fully automated and fast image analysis of autoradiographs with a TAS-Leitz. Determination of size, Feulgen fluorescence and grain counts of individual nuclei and their evaluation by a simplified cluster analysis. Histochemistry 81:227–236.

Wasmund HF, Ploem JS (1983): Ein computergesteuertes Mikroskop als Baustein eines automatisierten Prescreening Stystems. Microsc Acta (Suppl) 6:135–144.

Wilke V (1982): Laser scanning in microscopy. Proc R Microsc Soc 17/4:21.

Wilson T, Sheppard C (1984): Theory and practice of scanning optical microscopy. London: Academic Press.

Wouters CH, Koerten HK (1982): Combined light microscope and scanning electron microscope, a new instrument for cell biology. Cell Biol. Rep 6:955–959.

Wouters CH, Hesseling SC, Daems WTh, Ploem JS (1984): A centrifugation method for standardized sedimentation of mononuclear human blood cells on glass for scanning electron microscopy. J Micros 136:315.

Wouters CH, Koerten HK, Bonnet J, Daems WTh, Ploem JS (1985): Quantitative DNA measurements in an instrument combining scanning electron microscopy and light microscopy. J Microsc (in press).

*Applications of Fluorescence in the Biomedical Sciences, pages 301–319*


# 14 Microspectrofluorometry With an Intensified Vidicon Detector and Whole Image Spectral Scanning

*John E. Wampler*

Since the first applications to biological systems by George Reynolds [1964, 1968, 1972], the contribution of image-intensified video microscopy has been growing. The advantages of the videotape record, extreme light sensitivity, and, more recently, image processing often outweigh the disadvantages of relatively low spatial resolution and loss of color discrimination. Recent advances in instrumentation design and lowered costs due to growing commercial uses have continued to expand these applications. Several recent reviews discuss the growth in hardware and applications [Reynolds and Taylor, 1980; Wampler, 1985a,b].

Detector systems formed by coupling video cameras to high-gain image intensifiers have found use in a variety of disciplines with perhaps the earliest and most extensive development efforts occurring in astronomy [Meaburn, 1976]. In their most sophisticated form, these detector systems can "see" single photons. The first such system for biological applications was developed by Reynolds and co-workers [Reynolds et al., 1978]. It employed a very-high-gain ($\sim 4 \times 10^6$), low-noise four-stage intensifier, coupled to either an SIT (silicon-intensified target) or SEC (secondary electron conductor) vidicon. The only major limitation of their system was the physical size of the intensifier itself and its supporting electronics. The IDG low-light video system [Rich and Wampler, 1981] was designed to be similar in performance, but more compact. It employs a small two-stage MCP (microchannel plate) intensifier requiring less sophisticated support electronics. It

Department of Biochemistry, Instrument Design Group, University of Georgia, Athens, Georgia 30602.

also achieves single-photon sensitivity, but with considerably higher noise. However, image-processing techniques can be used to significantly reduce the effect of noise [Wampler, 1985c] and the detector system is small enough to easily mount in either a conventional or an inverted microscope configuration.

Since these two systems were developed, commercial low-light video cameras have been manufactured with sensitivities which approach single-photon detection [see Wampler 1985b, for review]. Some of these cameras are even more compact. More recently, a photon-counting camera system (model C-1966-20) has been marketed by Photonic Microscopy, Inc. (Oak Brook, IL). The system utilizes an unusually high-gain image intensifier coupled to a vidicon camera and includes digital-processing electronics.

All of these systems, commercial and custom designed, have one very important characteristic when compared to conventional video cameras alone. They have a very broad dynamic operating range. Most are able to form an image with illumination levels varying over a millionfold and, with a variable aperture to limit exposure of the detector, even wider operating ranges can be obtained. Broad dynamic range is an advantage of any system where the first stage detector is an image intensifier.

While most microspectrofluorometers employ photomultiplier tubes for quantitative analysis, we have been investigating the potential of these very sensitive camera systems as detectors in this application. When coupled to digitizing circuitry and computer processing, they offer the advantage of whole image, spatially resolved analysis. The major disadvantage is that even though they have single-photon sensitivity, they are much noisier than photomultiplier tubes. This problem can be partially overcome, however, by repetitive measurements (i.e., signal averaging). This paper analyzes the potential for whole-image spectral data acquisition when one uses such low-light video detectors in combination with a simple variable spectrum interference filter. This type of monochromator has a relatively broad bandwidth, but maintains good image quality. Thus it allows for spatial as well as spectral resolution. While spectral resolution is somewhat limited, useful analyses can be made including spectral shape analysis via calculation of the moments, as recently described [Mulkerrin and Wampler, 1982].

In the limit of the ideal microspectrofluorometers, there are five desired features:

1. Spatial Resolution—sub-$\mu$m
2. Spectral Resolution—$\sim$ 10 nm
3. Temporal Resolution—$\mu$- to msec
4. Low excitation irradiance— $<50$ $\mu W/cm^2$
5. Sensitivity—$\mu$molar concentrations of fluorophores

Of course, this ideal is not reached with a video-based detector. For instance, standard video frame rates (US standard = 30 frames/sec) limit temporal resolution and the low signal-to-noise of such detectors imposes further time limits if multisample averaging is used as described above. Sufficient spectral and spatial resolution is demonstrated by the system described here, but only by trading sensitivity and temporal resolution. In applications where such spectral resolution is not needed, some of the sensitivity can be retained by substituting broad-band filters for the monochromator.

An obvious question which might arise in response to this discussion is whether a single-photon-sensitive detector is necessary for this application. Sample irradiance levels of 50 $\mu W/cm^2$ of green light represent $1.3 \times 10^{14}$ photons per sec/$cm^2$. This is, of course, a large number. However, a typical 100-$\mu$m square area would receive only $1.3 \times 10^{10}$ hv $sec^{-1}$. In one video frame time (1/30th sec), this is $4 \times 10^8$ hv $frame^{-1}$. If the sample is 10 $\mu$m thick and the fluorophore has an extinction coefficient of $5 \times 10^4$, a 10-$\mu$M concentration would absorb about 0.1% of this light. With a quantum yield of 50%, the entire fluorescence band in a frame time would be represented by only $2 \times 10^5$ photons. For spectral scanning, even with relatively wide band widths, only a small portion of these photons will be detected at each wavelength position ($<$10% at the maximum) and, even with a high collection-angle objective lens, the collection efficiency and transmittance of the optical system will further degrade the signal. Thus the detector receives only a few thousand photons and converts only 10% of these to an electron signal. The bright macroscopic irradiance is represented by a quantum-limited fluorescence image on the microscopic scale. Not only must the detector be single-photon sensitive but some form of averaging or counting is essential.

## INSTRUMENTATION

### Hardware

The major components of the low-light video system have been previously described [Rich et al., 1981; Rich and Wampler, 1981; Wampler, 1985c]. Two cameras are employed, a one-inch Chalnicon and a one-inch SIT vidicon. Both are manufactured by Hamamatsu TV Co. (sold in the US by Hamamatsu Systems, Waltham, MA). The Chalnicon is a vidicon employing a $CdSe/As_2S_3$ target with performance characteristics similar to a Plumbicon (lead oxide target) or Saticon (Se, As, $Te/Sb_2S_3$ target), but with higher sensitivity. The vidicon tube used in the SIT camera is manufactured by Thomson-CSF and has a silicon diode array target intensified by electron optics and high potential acceleration of photoelectrons from an S-20 photocathode. This camera is significantly more sensitive than a conventional vidicon (see below), but still does not approach single-photon sensitivity.

The MCP image intensifier is a very compact device (19 × 25 × 52 mm) manufactured by ITT's Electro-Optical Products Division (Fort Wayne, IN). It incorporates two chevron-configuration microchannel plates, proximity coupled to an input photocathode (S-20) and an output phosphor screen (P-20). It is coupled to the camera by means of an Olympus OM-1, 50 mm, f/1.4 camera lens.

The camera, intensifier, and optional eyepiece magnifier are fitted with complementary bayonet mounts. The microscope head also has a mounting flange. Thus any combination of the component parts can be assembled as a system. A typical combination is diagrammed in Figure 1. The microscope nose piece is mounted permanently onto the monochromator/filter wheel body. A removable beam splitter/combiner in the nose piece allows for epifluorescence excitation. Excitation lamps, objective lenses, and the nose piece were obtained from Zeiss (Carl Zeiss, Inc., New York, NY). The excitation and emission monochromator/filter wheel assemblies are identical. They have stepper motor drives geared so that one step is approximately 1 nm of spectral scan. Each revolution of these wheels requires 800 steps. The wheels themselves (Fig. 1 insert) contain a permanently mounted graded spectrum interference filter (Optical Coating Laboratory, Inc.), three holes for 1-inch-diameter filters, and movable shutter vanes. The input to each monochromator is controlled by Uniblitz electronic shutters (A.W. Vincent Assoc. Inc., Rochester, NY).

The entire microscope/camera system assembly is mounted on a rigid, half-inch aluminum plate which in turn can be moved by a hydraulic piston for fine movement and a stepping motor for coarse movement. It moves along precision 1-inch steel rods carried on close-tolerance roller bearings. The mounting, optical table, focusing, and stage positioning system is as previously described [Rich and Wampler, 1981] with modifications to convert the system to an inverted microscope configuration.

The cameras, motors, and other mechanical hardware are interfaced to a Nova 820 minicomputer (Data General Corporation, Westboro, MA) as previously described [Rich and Wampler, 1981]. Data acquisition and camera control are via a Hamamatsu C-1000 camera controller as previously discussed.

**Software**

The software for this sytem is written in Nova assembler language with the software design of SPECOS (the Spectroscopy Operating System [Wampler, unpublished]) with extensions for handling image data. Many of the features of this software have been discussed previously [Wampler et al.,

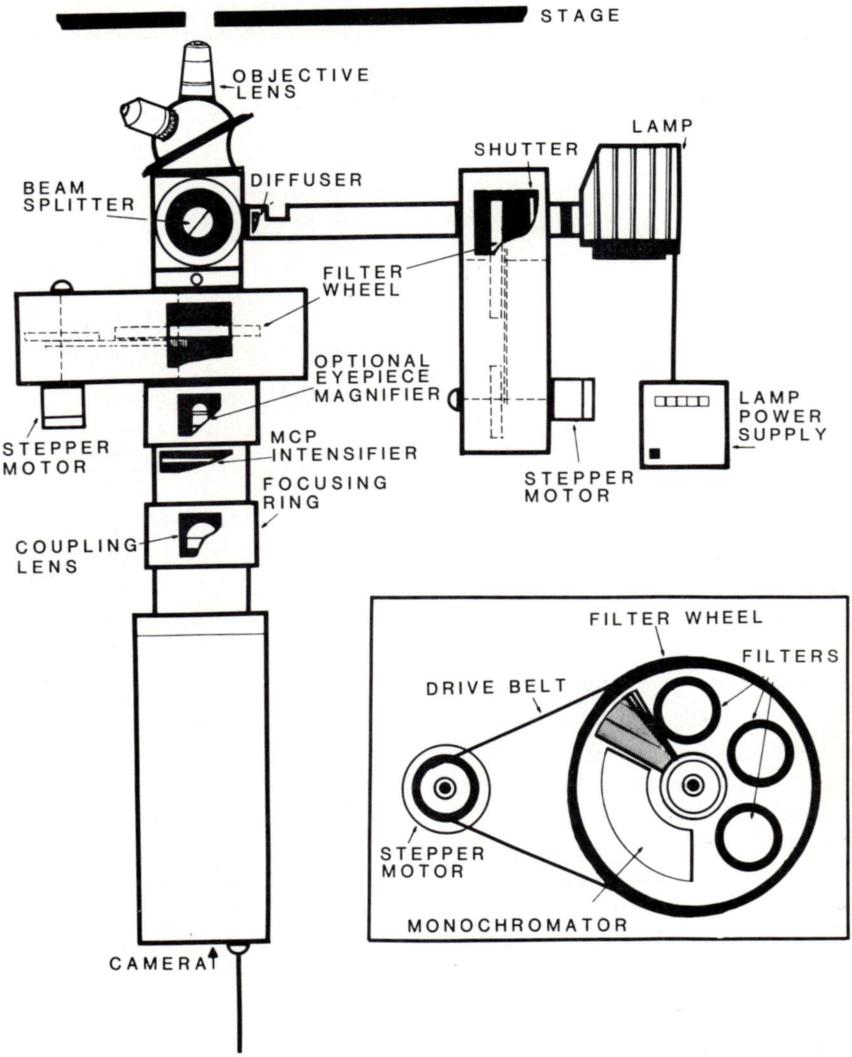


Fig. 1. The microscope head showing the relative positions of the various components. Stepper-motor-driven filter wheels (Insert) allow computer selection of both emission and excitation wavelengths. MCP, microchannel plate.

1979]. This software is menu driven. Selections for the spectral scanning menu are outlined in Table 1. Other menus offer selections of the SPECOS package with functions previously outlined [Wampler et al., 1979].

Data acquisition is controlled by the location of three cursors on the TV screen. Cursor S marks the data collection point. Its position, as well as those of the other two cursors, are recorded on the terminal screen display.

**TABLE 1. WAVESCAN Menu**

| Selection | Function |
|---|---|
| R | Set reference point at M cursor position |
| S | Set sample point at M cursor position |
| C | Enter lens calibration data |
| I | Initialize the monochromators |
| − | Selects difference mode data collection |
| / | Selects ratio mode data collection |
| X | Selects for scanning of excitation |
| E | Selects for scanning of emission |
| M | Enter position and move active monochromator |
| W | Enter two wavelengths for ratio measurements |
| = | Take ratio at selected wavelengths |
| D | Take dark current at S and R cursor points |
| 〉 | Scan spectrum with current scan parameters |
| G | Evaluate response ratio between S and R |

Cursor R is used for two purposes. In measurements of spatial dimensions, this cursor is the origin. Measurements can be made in any desired units of distance after proper lens calibration. Typically, calibration is in microns with the aid of a stage micrometer. For spectral imaging, the cursor R marks a reference point which can be used either for a ratio or a difference measurement. These two modes are selected via the keyboard (see menu selections, Table 1).

The moving cursor, M, is controlled by two interface potentiometers. It is used to locate the position of the other two by moving it to a desired location and then selecting either the keyboard letter "R" for the reference cursor or "S" for the spectral sampling point.

Spectral data can be taken as either a ratio or a difference. Since these cameras are not perfect detectors, the actual measurement equations are complicated by two necessary corrections. Ir and Is are corrected for dark current signals, Dr and Ds respectively, and for the response difference of the detector system at the two locations, Rs. For difference spectroscopy the spectral value, $S_D$, is given by equation 1.

$$S_D = (Is - Ds) - (Ir - Dr) \times Rs \qquad (1)$$

For ratio measurement the equation is as follows:

$$S_R = (Is - Ds)/[(Ir - Dr) \times Rs] \qquad (2)$$

The menu allows for assignment of dark currents and the gain correction by the user. Scan range, scan averaging, repetitive data cycles, and the number of data points per scan are set under the main SPECOS menu. Two wavelength positions for either excitation or emission ratio measurements and mode selection (difference versus ratio; excitation versus emission) choices are also made from the WAVSCAN menu. The current excitation or emission ratio value is displayed along with the other descriptive data with the on-screen display of menu selections.

Both spectral and image data can be averaged automatically. The program allows assignment of a number of automatic repetitive cycles, $C$, and a number of repetitive scans, $S$. Each point in each image or spectral scan is evaluated $C$ times. In addition the scanning process is repeated $S$ times. Thus, each datum in an image or spectrum represents the average of $C \times S$ samples of both the reference and the sample cursor positions.

Under this menu, selection "I" initializes the system. Following this selection the program requests the wavelength settings of both monochromators. Other selections from this menu (Table 1) set the lens calibration factor and move the active monochromator to a new location.

The data manipulation features of SPECOS programs have been outlined previously [Wampler et al., 1979]. For this work, the moments calculation was employed where the mean wavelength position of a spectrum is determined by equation 3 and the dispersion by equation 4.

$$\text{Mean} = \frac{\Sigma\, X_i Y_i}{\Sigma\, Y_i} \tag{3}$$

$$\text{Dispersion} = \sqrt{\frac{\Sigma\,(\text{Mean} - X_i)^2 Y_i}{\Sigma\, Y_i}} \tag{4}$$

where $X_i$ and $Y_i$ are the wavelength and spectral datum values, respectively, at each wavelength position, $i$.

## METHODS AND MATERIALS

Macroscopic fluorescence measurements were performed on the IDG on-line spectrofluorimeter system [Wampler, 1976]. Microcapillaries were used for standardization following the recommendations of Sernetz and Thaer [1970]. Capillaries were drawn from borosilicate glass tubing to an average internal diameter of 200 $\mu$m.

Microdroplets were prepared according to the procedure of Mendoza [1984] by using a 1% v/v concentration of the surfactant Span 85 in n-pentane. The advantage of this solvent is that its refractive index is similar to that of water (1.36 versus 1.33) and that the microdroplets settle to the coverslip in the inverted configuration. Luminol chemiluminescence reactions were carried out as previously described [Rich et al., 1981].

## RESULTS AND DISCUSSION

### Sensitivity

The absolute sensitivity of one of our detector systems (SIT plus intensifier) was determined previously [Rich and Wampler, 1981] with the result that approximately one out of every six green photons is detected. There is excess gain so that these single photon events can be visualized as fairly bright spots. The SIT/intensifier combination, while extremely sensitive, is noisy, has lower resolution than some other low-light cameras, and since it tends to be used with lower total image flux, gives a picutre which is less appealing visually. We might initially question whether this combination is a good detector system for fluorescence. To test this question, we have compared its response characteristics to the Chalnicon, both with and without the intensifier. Table 2 gives the results of our test of relative sensitivity. These tests involve measurement of a small aperature in the slide plane illuminated by a green light-emitting diode with an alkaline battery supply source. The Chalnicon is more sensitive ($\sim 10\times$) than the typical Plumbicon or Saticon camera tube, but has similar linearity and resolution. The SIT alone is even more sensitive (as shown in the table). The nominal gain of the intensifier at maximum accelerating voltage is $\sim$500,000. Since it is optically coupled to the camera, its contribution to the overall gain of the combination is substantially less, as shown.

Other characteristics of the three detectors are given in Table 3. Besides the added sensitivity, incorporation of such an intensifier into a camera system gives the net system broad spectral response and a very wide dynamic range. Since the detector area of these two cameras is about one-half that of the intensifier, a 2:1 demagnification is employed which also improves the spatial resolution of the combined system.

For fluorescence studies where photobleaching is not a problem, the Chalnicon/intensifier combination gives a visually appealing, moderate-resolution image comparable to that of several of the commercial low-light cameras. Neither camera by itself has enough sensitivity for practical application. The SIT/intensifier combination gives a somewhat noisier, lower-resolution image than the Chalnicon/intensifier, but has the advantage that it can be used with much lower excitation levels. Typically, the excitation intensity needed to adequately visualize a sample with the Chalnicon/intensifier combination is at least 100-fold higher than for the SIT/intensifier combination.

**TABLE 2. Relative Sensitivities of the Camera Systems**

| Camera[1] | Intensifier[1] | Relative sensitivity[2] |
|---|---|---|
| Chalnicon | − | 10 |
| Chalnicon | + | $1.2 \times 10^5$ |
| SIT | − | 3,930 |
| SIT | + | $5.7 \times 10^6$ |

[1]Both camera and intensifier operated at optimum settings and maximum gain.
[2]Sensitivity to green photons as compared to a typical lead oxide vidicon (Plumbicon).

**TABLE 3. Characteristics of the Three Detectors**

| Parameter | Chalnicon | SIT | Intensifier |
|---|---|---|---|
| Resolution | 800 TV lines | 500 TV lines | 550 line pairs |
| Linearity | 0.95 | 1.0 | 1.0 |
| Sensitivity | 10 | 3,930 | $5 \times 10^5$ |
| Lag (%) | 20 | 10 | <1 |
| Shading (%) | 9.9 | 23 | <10 |
| Gain adjust | 1:1 | 100:1 | 10,000:1 |

SIT, silicon-intensified target.

## Spectral Performance

As pointed out previously [Wampler, 1978], spectra and spectrally resolved data from continuous graded interference filters have relatively low wavelength resolution. In addition their "slit function" differs markedly from the ideal triangular shape. However, with most fluorophores common to biological studies where sharp lines are not encountered, spectral shape and position can be defined adequately with this device. Figure 2 shows the uncorrected spectra of two typical fluorophores (acridine yellow and coriphosphene O) taken first on the on-line fluorometer system and then, by using microcapillary tubes, on the microscope system. These data are uncorrected for wavelength responsiveness, but illustrate that the system can easily resolve such spectra and that the slit function and bandwidth difference (10 nm for the fluorometer, 12 nm for the filter wheel) does not greatly broaden these spectra.

Wavelength positioning with the stepping motor drive is accurate to less than a nanometer with one step equal to 0.92 nm. Spectral positioning at the viewing port is linear over the full range from 370 to 750 nm. Half-bandwidth is nearly constant at 12 nm.

Since the filter is continuously variable, the analyzing wavelength will vary somewhat over the field of view. This is a minor problem, however,

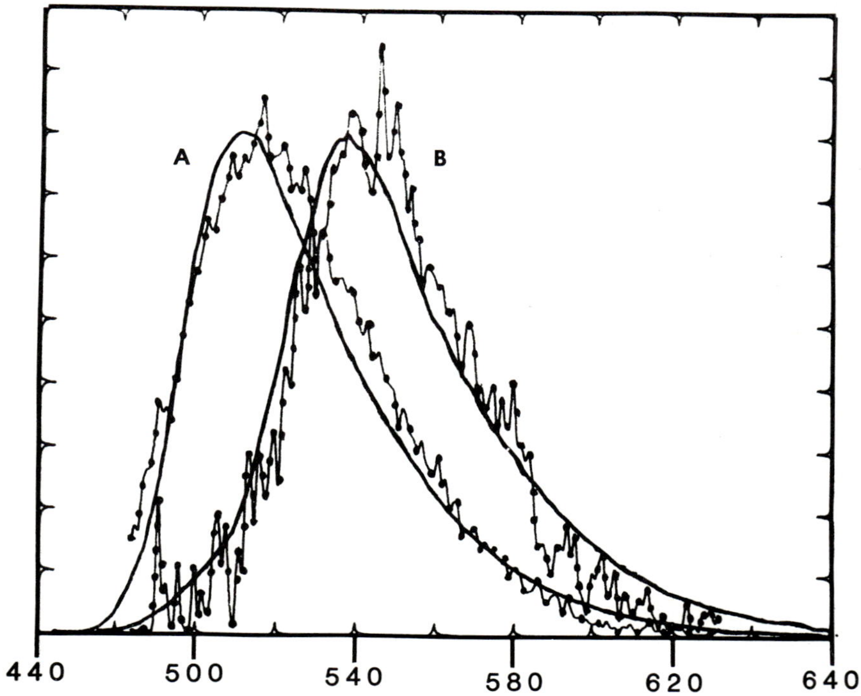


Fig. 2. Comparison of the fluorescence spectra of acridine yellow (A) and coriphosphene O (B). Solid lines taken with the on-line fluorimeter system [Wampler, 1976]. Lines with dots taken by using the video microscope system. Spectra are uncorrected. Each dye in ethanol at concentrations to produce peak absorbance less than 0.1.

with less than a nanometer variation radially (horizontally) in the worst case and only 8 nm from the bottom to the top of the field. Excitation bandwidth is slightly broader (~20 nm) but homogeneous over the field due to use of a diffuser plate.

## Linearity

If these cameras are to be used for quantitative measurements, their response must be linear and predictable. Linearity was determined by three different procedures. In experiments similar to those described by Rich et al. [1981], the kinetics of light decay for luminol chemiluminescence was monitored. This decay is exponential. Both cameras with and without the intensifier record true exponential decays over one decade and then significant deviations occur.

Fluorescent microdroplets containing fluorescein were also used to test the system's linearity [Mendoza, 1984]. By using the measuring cursor to

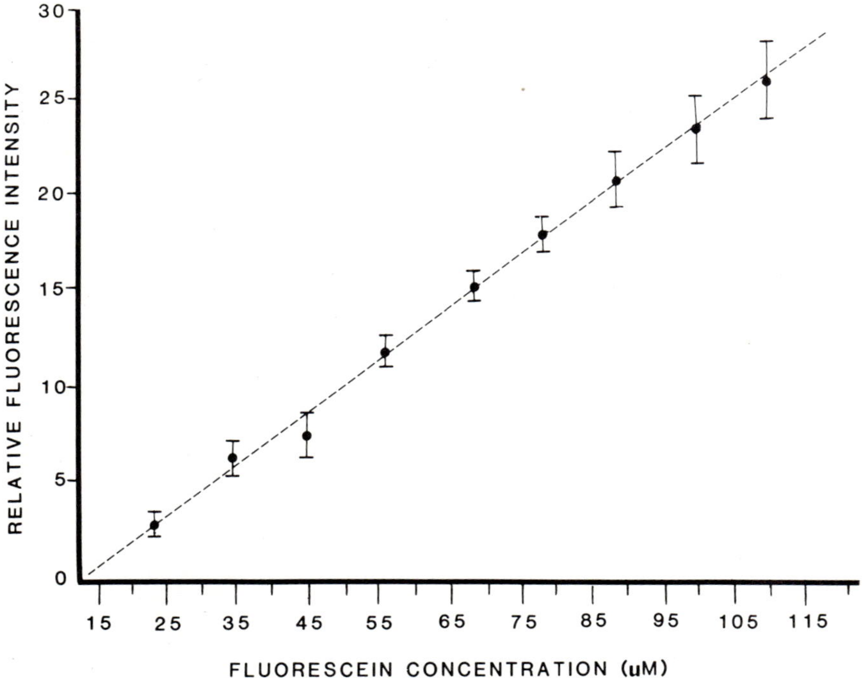


Fig. 3. Fluorescence of fluorescein in aqueous buffer (pH 6.8) with the excitation monochromator set at 489 nm and emission viewed through a #16 Wratten filter. For additional details, see text. [from Mendoza, 1984].

select only droplets of one size (42 $\mu$m), droplets loaded with different fluorescein concentrations were examined. Figure 3 shows the linear response to fluorescein concentration where the measurement used is the integrated intensity from a 10-$\mu$m square area at the center of the fluorescent droplet. For improved sensitivity, the fluorescence was viewed through a cut-on filter.

Since decay profiles indicated some nonlinearity in performance when signal levels drop below ~10% of full scale, it is important to evaluate the impact of this problem on quantitative spectral measurements and identify its source. Red-and-green fluorescent paper was used for this test since it exhibited no photobleaching over the full period of study. Spectra were measured with a wide variety of lighting and gain variations.

Spectral moments are extremely sensitive indicators of shape changes and distortion [Mulkerrin and Wampler, 1982]. In Figure 4, the moments of various spectra are plotted as a function of peak signal. As the total signal decreases, there is a definite trend of change in the moments indicating subtle

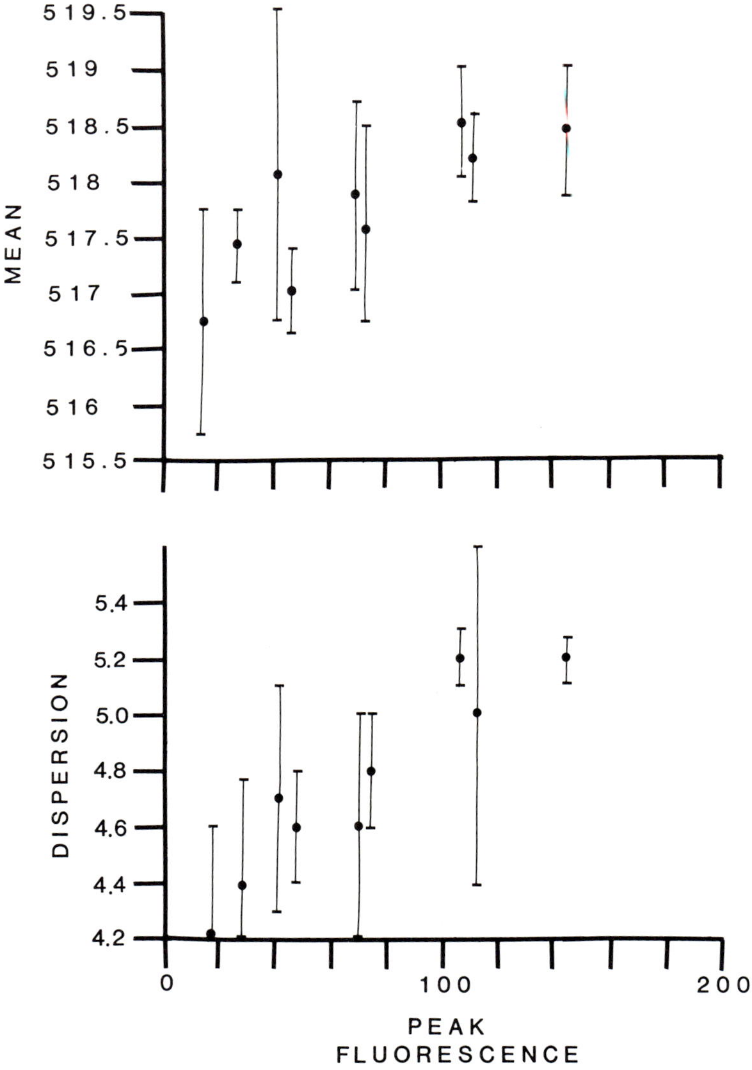


Fig. 4. Moment parameters versus peak fluorescence signal. The average values of three or four determinations of the mean and dispersion for spectra of green fluorescent paper. Error bars represent ± 1 standard deviation.

shape changes in the measured spectra. These changes are very small, but consistent. The entire range of change of the mean position, for example, is approximately the same as the variation that would be caused by a 2-nm wavelength positioning error.

Is this nonlinearity due to the camera or the intensifier? To test this question, spectra were run over widely varying lighting conditions, but peak

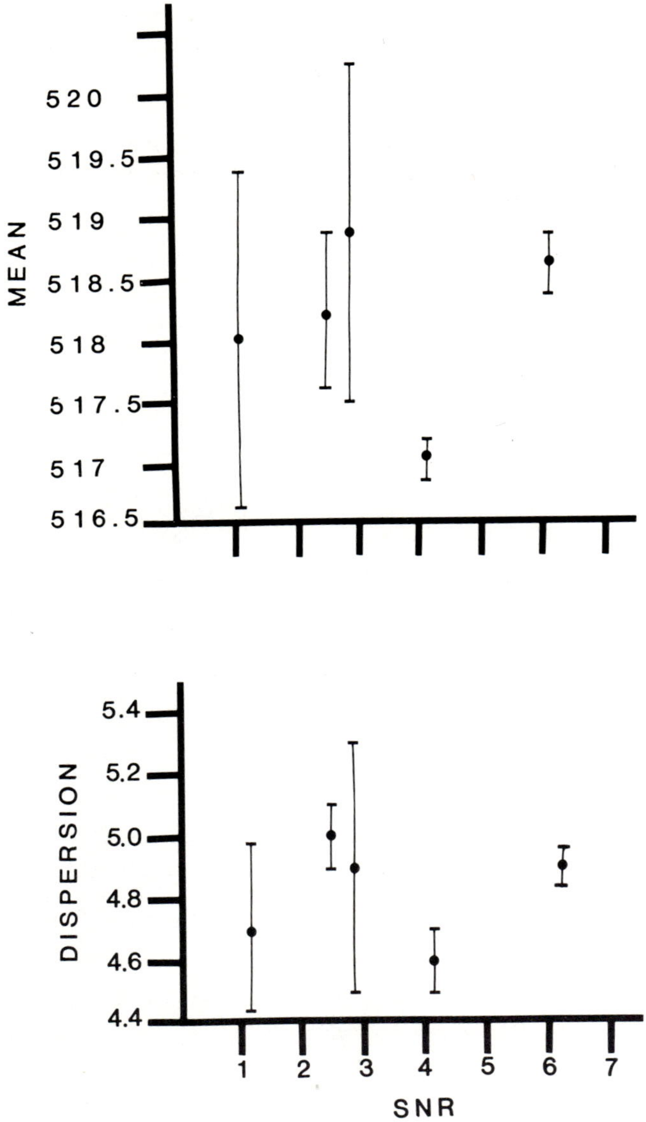


Fig. 5. Moment parameters versus signal-to-noise ratio for a constant peak signal adjusted via intensifier gain. Same conditions as Figure 4.

signal was maintained high and constant by adjusting the intensifier's gain over several orders of magnitude. Figure 5 shows the moments in this case. No trend was observed. Thus the intensifier's linearity over a broad dynamic range is verified and a solution to the problem is illustrated by the difference between Figures 4 and 5. Providing that intensifier gain is adjusted to produce a similar peak signal within fairly broad limits, spectra are reproducible and

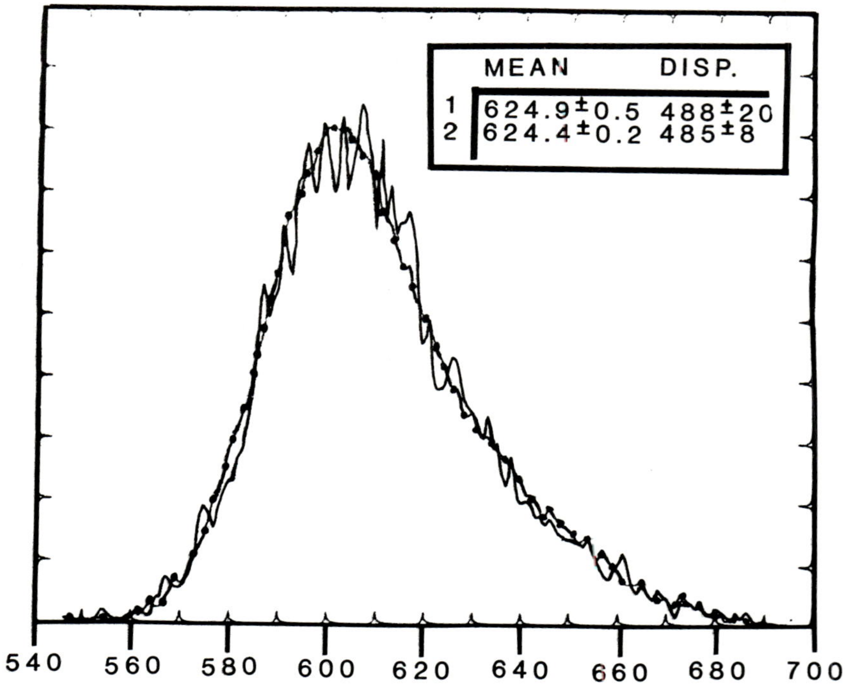


Fig. 6. Red fluorescence from red-dyed paper at high excitation level (line with dots) and at a 100 fold lower level (noisy line). Intensifier gain was adjusted so that the same average peak signal was obtained in both cases.

their moments accurately reflect their shape and position. This is illustrated by the spectra of Figure 6. These two spectra represent data collected with over a 100× gain difference.

## Applications

A variety of fluorochromes are now used to monitor changes within living cells, as discussed in this volume. Many of these reagents respond as detectors with changes not only in quantum yield but with changes in spectral shape. Such changes offer real power for in vivo analysis since they allow accurate quantitative measurements by using spectral data obtained on a relative, not an absolute, scale. As an example, one of the most studied and used probes is fluorescein either unmodified, as the carboxymethyl analog, or conjugated to carrier macromolecules [Sernetz and Thaer, 1972; Thomas et al., 1976; Heiple and Taylor, 1980; Babcock, 1983].

To investigate the potential of this instrument system for in vivo analysis, the precision of pH measurements using fluorescein was investigated. Software was developed to implement excitation (or emission) ratio measure-

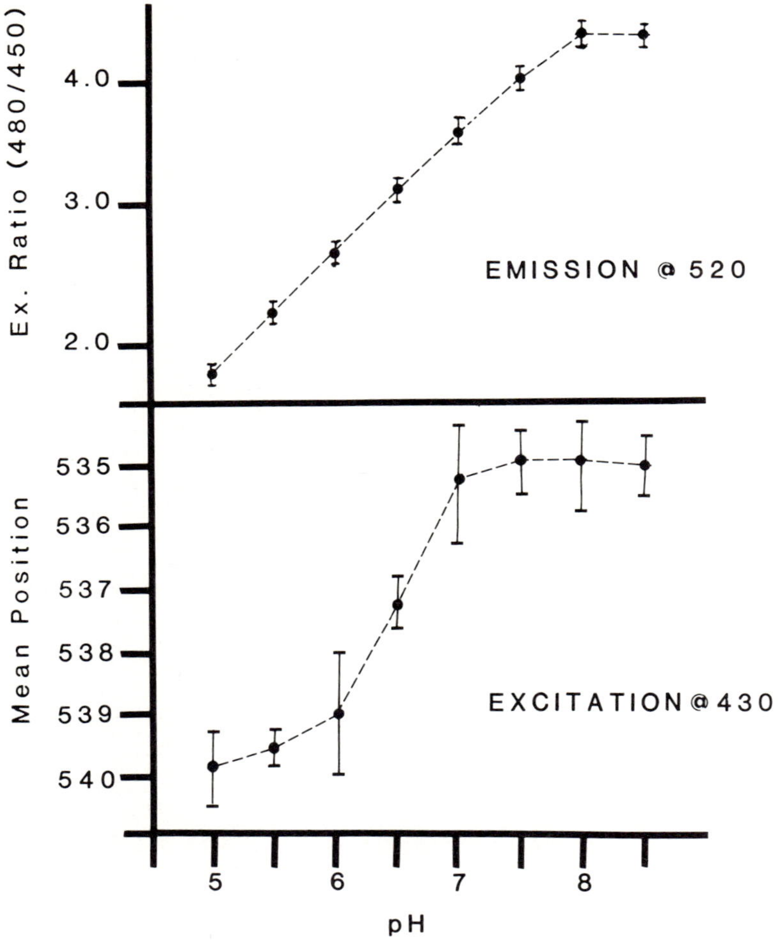


Fig. 7. Standard curves for 500 μM fluorescein in physiological buffers (120 mM KCl, 20 mM NaCl, 10 mM buffer) in capillary tubes (average 200 μm internal diameter). Average values are plotted with error bars representing ± 1 standard deviation. Buffers used were MES (2-[N-Morpholino]ethanesulfonic acid), MOPS (3-[N-Morpholino]propanesulfonic acid), and HEPES (N-2-Hydroxyethylpiperazine-N′-2-ethanesulfonicacid). Wavelength selection for both excitation and emission via graded spectrum interference filters as discussed in the text. The excitation ratios presented are corrected for scattered light, but are uncorrected for refractive effects in the capillary tubes. The mean wavelength position for uncorrected fluorescein emission spectra were calculated for the wavelength range from 490 to 640 nm.

ments and this approach was compared with a new method [Wampler and Mendoze, unpublished] where the first moment is used as the analytical parameter. First, microcapillaries were used to obtain a standard curve with the results shown in Figure 7. In both cases, the data collection cycle was

**TABLE 4. Comparison of Methods of Standardization and the Effect of Microdroplet Size[1]**

| Method | Size | Value | CV (%) |
|---|---|---|---|
| Microcapillary | 209 | 2.15 | 3 |
| Microdroplet | 44 | 2.24 | 18 |
| | 26 | 2.06 | 15 |
| | 22 | 2.22 | 18 |
| | 17 | 2.16 | 37 |

[1]Standard solution: 0.5 mM fluorescein at pH 5.5. Average reading and coefficient of variation (CV) from 1,200 measurements of the excitation ratio 480/450 with emission at 525 nm.

adjusted so that the number of values averaged per reading gave the same total measurement time (~100 sec/final average). With excitation ratios, each final datum value represents the average of 1,200 separate reference and sample measurements. Each spectral mean position is based on 1,800 such measurements.

Microcapillaries allow fairly high excitation intensities without significant photobleaching over the measurement time. This approach is compared to use of microdroplets in Table 4 where much lower excitation intensity is required to avoid photobleaching. For these samples, the lamp output was reduced 100-fold and the intensifier's gain increased to compensate. The result is a lower single-to-noise ratio (SNR), but equivalent values (Table 4). Droplet size does not appear to effect the measurement.

By using monochromators for both excitation and emission wavelength selection, pH measurements should be possible even in the presence of other fluorophores. Figure 8 shows a cell of the protonema of the moss, *Physcomitrella patens*, after loading with fluorescein diacetate for 2 min (10-μM concentration, pH 7). The fluorescence seen with the monochromator set at 525 is entirely due to fluorescein and is localized in the chloroplasts! Controls show complete absence of any chloroplast fluorescence. Even with organelles this size (~6 μm), an excitation ratio of 3.1 can be measured with a coefficient of variation of 6% [McCurdy and Wampler, unpublished]. Further work is required before this value can be translated into pH, but both the precision and specificity of the measurement are encouraging.

The combination of excitation or emission ratio measurements with spectral moments may also allow us to more clearly define the intracellular environment of a probe. Mendoza [1984] has shown that even when the pH of the cytoplasm of the aquatic mold *Rhizidiomyces* is equilibrated with the external medium, the moments of fluorescein spectra reveal a environment that is different from that of free fluorescein or the fluorescein bovine serum

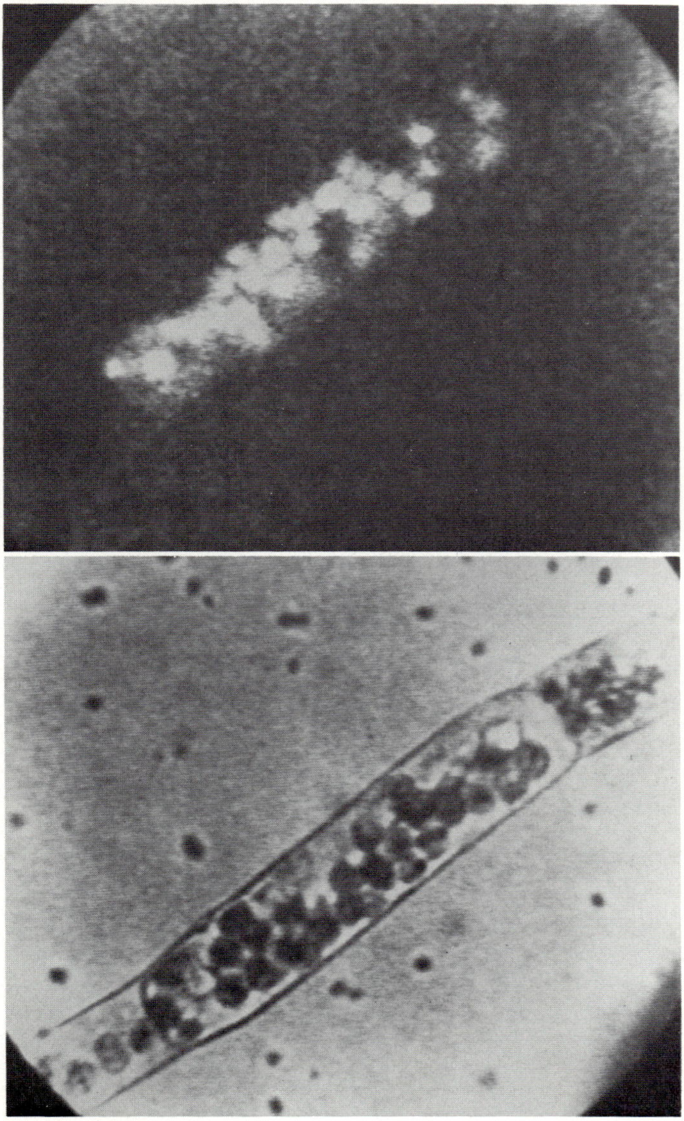

Fig. 8. Protonema cell of the moss *Physcomitrella patens*. Upper—video micrograph of a single cell, transmitted light image at 525 nm. Lower—fluorescence of the same cell. Cells were loaded with fluorescein diacetate (10 mM in pH 7 HEPES) for 2 min at room temperature.

albumin complex in the equilibration buffers (Fig. 9). Thus pH measurements using only one parameter may be very misleading. This also supports the recommendation for in situ calibration [see Heiple and Taylor, 1980].

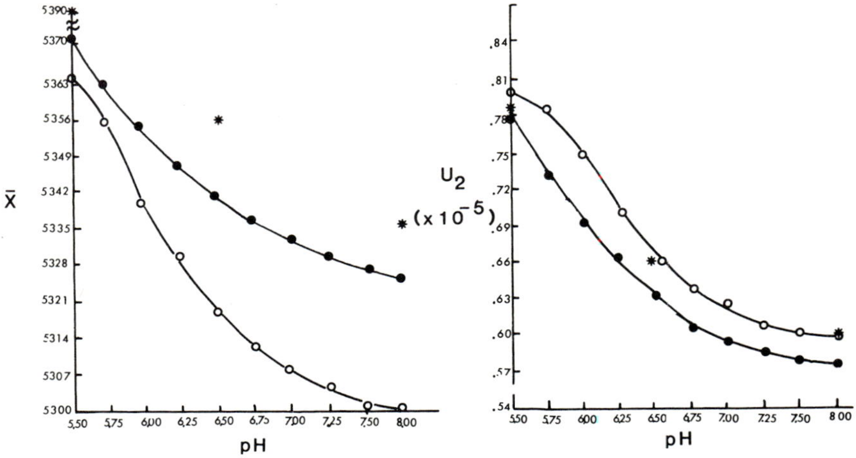


Fig. 9. Variations in the mean position and the variance, $U_2$ (dispersion squared) for fluorescein as a function of pH. Open circles— 50 $\mu$M fluorescein in physiological buffer (120 mM $K^+$, 20 mM $Na^+$, 50 mM buffer, 10 mM $Mg^{++}$, 190 mM $Cl^-$). Closed circles—50 $\mu$M fluorescein in physiological buffer with 100 $\mu$M bovine serum albumin. Asterisks—*Rhizidiomyces* cell suspension loaded with fluorescein diacetate (10 $\mu$M FDA in pH 7.0 HEPES buffer) with pH equilibration buffers (pH 5.5 and pH 6.5 using potassium acetate (100 mM) with 20 mM NaCl; pH 8.0 using 5 mM methylamine with 10 mM HEPES and 20 mM NaCl). These experiments were performed in standard fluorescence cuvettes with 2-ml volumes by using the on-line fluorimeter system [data from Mendoza, 1984].

## ACKNOWLEDGMENTS

The work upon which this paper is based was supported in part by the NIH (grant No. GM 28209), the National Science Foundation (PCM 79-12316), and the University of Georgia Research Foundation. It includes or is based on work by graduate students Michael Mulkerrin, Leo Mendoza, and Joan Chen, and postdoctoral student Dr. David McCurdy, and on engineering, design, and fabrication work by Mr. Ned Rich, Mr. John Gilbert, and the University Instrument Shop under the direction of Mr. Ralph Morton. The author gratefully acknowledges these important contributions.

## REFERENCES

Babcock DF (1983): Examination of the intracellular ionic environment and of ionophore action by null point measurements employing the fluorescein chromophore. J Biol Chem 258:6380–6389.

Heiple JM, Taylor DL (1980): Intracellular pH changes in single motile cells. J Cell Biol 86:885–890.

Meaburn J (1976): "Detection and Spectrometry of Faint Light." Boston: D. Reidel Publishing Co.

Mendoza LG (1984): "Quantitation of an Intracellularly Trapped Fluorescent Probe in Single Living Cells Using Image Intensified Video Fluorescence Microscopy." University of Georgia: M.S. Thesis.

Mulkerrin MG, Wampler JE (1982): Empirical evaluation of the use of moments in describing fluorescence spectra. Anal Chem 54:1778–1782.

Reynolds GT (1964): Evaluation of an image intensifier system for microscopic observations. IEEE Trans Nucl Sci NS 11(3):147–151.

Reynolds GT (1968): Image intensification applied to microscope systems. Adv Opt Elect Microsc 2:1–40.

Reynolds GT (1972): Image intensification applied to biological problems. Q Rev Biophys 5:295–347.

Reynolds GT, Milch J, Gruner S (1978): High sensitivity image intensifier—TV Detector for X-ray diffraction studies. Rev Sci Instrum 49:1241–1249.

Reynolds GT, Taylor DL (1980): Image intensification applied to light microscopy. Bioscience 9:586–592.

Rich ES, Wampler JE (1981): A flexible, computer controlled video microscope capable of quantitative spatial, temporal and spectral measurements. Clin Chem 27:1558–1568.

Rich ES, Grover C, Wampler JE (1981): The spatial distribution of light emission from liquid phase Bio- and chemiluminescence. Photochem Photobiol 33:727–736.

Sernetz M, Thaer A (1970): A capillary fluorescence standard for microfluorometry. J Microsc 91:43–52.

Sernetz M, Thaer A (1972): Microfluorimetric binding studies of fluorescein-albumin conjugates in single fibroblasts. Anal Biochem 50:98–109.

Thomas JA, Cole RE, Langworthy TA (1976): Intracellular pH measurement with spectroscopic probe generated *in situ*. Fed Proc 35:1455.

Wampler JE (1976): Fluorescence spectroscopy with on-line computers. In Wehry EL (ed): "Modern Fluorescence Spectroscopy." New York: Plenum Pub. Co., pp 1–43.

Wampler JE (1978): Measurements and physical characteristics of luminescence. In Herring PJ (ed): "Bioluminescence in Action." London: Academic Press, pp 1–48.

Wampler JE (1985a): Instrumentation. Seeing the light and measuring it. In Burr J (ed): "Chemiluminescence and Bioluminescence Today." New York: Marcel Dekker Inc., pp 1–44.

Wampler JE (1985b): Low-light video systems. In Van Dyke K (ed): "Bioluminescence and Chemiluminescence: Instruments and Applications, Vol II." Boca Raton, Florida: CRS Press, Inc., pp 123–146.

Wampler JE (1985c): Enhancing real-time perception of quantum limited images from a doubly intensified SIT camera system. Computer Vision, Graphics and Image Processing, in press.

Wampler JE, Mulkerrin MG, Rich ES (1979): Instrumentation and techniques for analysis of hydrogen peroxide. Clin Chem 25:1628–1634.

*Applications of Fluorescence in the Biomedical Sciences, pages 321–345*


# 15 Video-Connected Fluorescence Microscopy of Large DNA Molecules, Chromatin, and Chromosomes

*Mitsuhiro Yanagida, Kosuke Morikawa, Yasushi Hiraoka, Seiji Matsumoto, Tadashi Uemura, and Shun Okada*

The mobile images of individual DNA molecules in solution were successfully visualized by light microscopy by the fluorescence staining method [Morikawa and Yanagida, 1981; Matsumoto et al., 1981; Yanagida et al., 1983]. The method is a combination of fluorescent probes, which specifically bind to DNA, with a video-connected fluorescence microscope. In this paper, we report further developments in the method, especially in regard to various probes applicable for visualization of DNA molecules, quantitative estimation of fluorescence intensity, agents to minimize the damage of specimens, and effects of the probes on DNA superhelicity. Experimental results obtained on the dynamic structures of single DNA molecules in solution are summarized.

The method can also be applied to observe chromatin fibers, mitotic chromosomes, and the compartmentalized distribution of the interphase nuclear DNA in yeast [Toda et al., 1981; Umesono et al., 1983; Hiraoka et al., 1984; Hiraoka and Yanagida, unpublished results]. Flexible chromatin filaments devoid of histones H2A and H2B were released from isolated nuclei treated with high salt concentrations, and their behavior was indistinguishable from the naked DNA filaments. Three-dimensional features of the mitotic chromosomes and the interphase nuclear chromatin region were deduced by analyzing a series of video micrographs.

Specimens under physiological solution conditions were thus directly observed from the molecular level of a single DNA filament to the subcellular level of nuclear DNA. The "intact" images of DNA molecules obtained are strikingly different from those taken by electron microscopy, which are static in a high vacuum and often extensively deformed during fixation and dehy-

Department of Biophysics, Faculty of Science, Kyoto University, Kyoto 606, Japan.

dration. Hence the fluorescent probes have given us a novel and continuous view of DNA, chromatin, and chromosomes, filling a part of the large gap between largely different orders of DNA structures.

## METHOD AND APPARATUS

Light microscopy cannot distinguish two objects within 200 nm but this does not necessarily mean that objects smaller than 200 nm are invisible. If the object emits or scatters a light intense enough to be distinguished from the surrounding area, such an object can be visualized. For example, bacterial flagella or microtubules with diameters of 20–30 nm can be seen by dark-field microscopy.

In fluorescence microscopy, the specimen bound with a fluorescence probe emits fluorescent light by absorption and excitation of the incident light. If the emitted fluorescent light is intense enough, an object smaller than 200 nm can be recognized. The double helical DNA is, however, very thin, having a diameter of 2 nm, which is only one one-hundredth of the resolution limit of light microscopy. Therefore, the probes used should emit strong fluorescence upon binding to DNA and the image-recording system should be highly sensitive to weak light. The probes applicable for visualization of DNA molecules are listed in Table 1.

The apparatus used was described previously [Yanagida et al., 1983]. A Hamamatsu C-1000-12, silicon intensified target camera attached to an epi-fluorescence microscope was employed for taking the mobile video images. Appropriate filters and dichroic mirrors were selected for each probe. The static images were taken by a conventional camera system with an exposure time of 1–10 sec [Matsumoto et al., 1981]. As shown in Figure 1, the static image (a) is better in resolution than the video image (b).

A video image processor performed the analogue-digital conversion of the video signals derived directly from the television camera or from the video recorder. The intensity of the video signals is quantified into 16 grades (Fig. 2). The digitized gray scale is proportional to the intensity of fluorescent light. This was proved by using combinations of neutral density filters (data not shown). The video signals could be sliced at an appropriate level (Fig. 2b,d) and be integrated in a given area.

## FLUORESCENT PROBES APPLICABLE FOR VISUALIZATION OF DNA MOLECULES

The 11 fluorescent probes tested were DAPI (4′,6-diamidino-2-phenylindole dihydrochloride) [Dann et al., 1971; Williamson and Fennell, 1975], DIPI [4′,6-bis(2′-imidazolinyl-4′,5′-H)-2-phenylindole dihydrochloride]

**TABLE 1. Fluorescence Probes Applicable to Visualization of DNA Molecules[1]**

| Probes | Concentration (μg/ml) | Excitation max (nm) | Emission max (nm) | Enhancement | Fluorescence microscopy | | | | |
|---|---|---|---|---|---|---|---|---|---|
| | | | | | | Fading | | Cutting | |
| | | | | | Intensity | − | + | − | + |
| DAPI | 1 | 365 (UV) | 450 | 20 | ++++ | R | S | F | VIF |
| DIPI | 1 | 372 (UV) | 448 | 13 | ++++ | R | S | F | VIF |
| H33258 | 5 | 374 (UV) | 472 | >100 | +++ | * | S | * | IF |
| Acridine orange | 2 | 503 (blue) | 523 | 6 | ++ | * | R | * | F |
| Acridine yellow | 2 | 461 (blue) | 491 | 1.3 | ++ | * | R | * | F |
| Ethidium bromide | 20 | 524 (green) | 584 | 18 | ++ | S | VS | VIF | VIF |
| 288/45 | 1 | 396 (UV) | 448 | 4.4 | ++ | R | S | F | VIF |
| 186/134 | 1 | 365 (UV) | 436 | 1.1 | + | ND | ND | ND | ND |

[1] +, −, the presence and the absence of 2% 2-mercaptoethanol; UV, blue, green, filters for fluorescence microscopy; R, rapid; S, slow; VS, very slow; F, frequent: IF, infrequent; VIF, very infrequent; ND, not determined; *, invisible; DAPI, 4′, 6-diamidino-2-phenylindole dihydrochloride; DIPI, 4′, 6-bis (2′-imidazolinyl-4′, 5′-H-)2-phenylindole dihydrochloride; H33258, Hoechst 33258; 288/45, 2′, 6-bis (2′imidazolinyl-4′, 5′-H)-2′, 6-bisindole; 186/134, 2-[4′(p-amidinophenoxy)phenyl]-6-amindoindole.

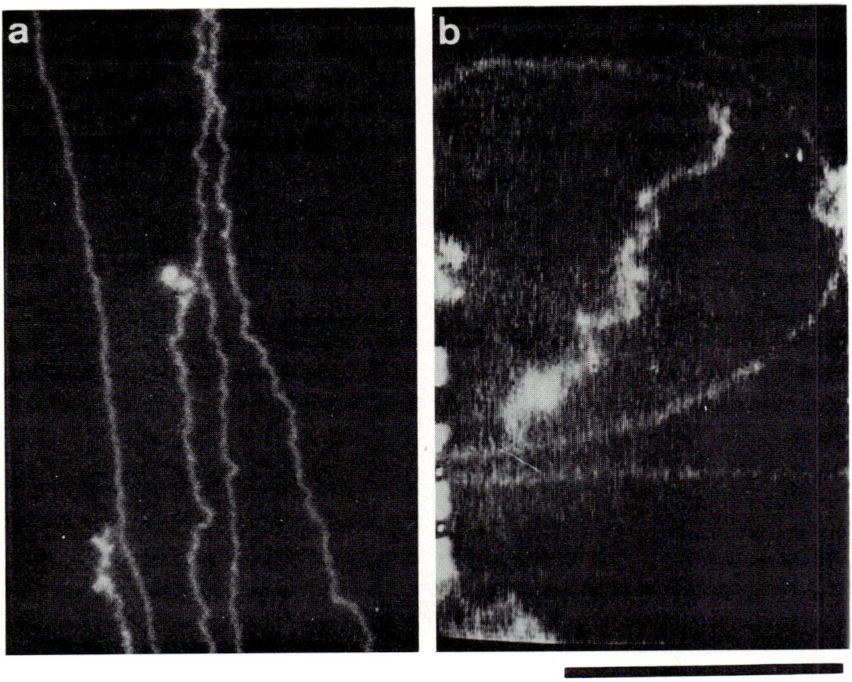


Fig. 1. Fluorescence micrographs of purified DNA molecules of phage T4 in 1 mM sodium cacodylate at pH 6.6 containing 0.2 M NaCl and DAPI (abbreviations in legends follow explanation in Table 1 footnote) (1.0 $\mu$g/ml). a. A static image taken by film. b. A video image. The bar indicates 10 $\mu$m.

[Dann et al., 1971; Schnedl et al., 1977], H33258 (Hoechst 33258) [Hilwig and Gropp, 1972], acridine yellow, acridine orange, ethidium bromide, 288/45 [2′,6-bis(2′imidazolinyl-4′,5′-H)-2′, 6-bisindole], 186/134 {2-[4′(p-amidinophenoxy)phenyl]-6-amidinoindole}, quinacrine mustard [Casperson et al., 1968], mithramycin [Schweizer, 1976], and ellipticine [5,11-dimethylpyrido-(4,3-b)carbazole] [Paoletti et al., 1979]. The first eight dyes were found to produce the images of individual DNA molecules (Fig. 3, Table 1). Stained DNA molecules showed different fluorescence colors according to the probes employed; blue fluorescence was emitted by DNA associated with DAPI, DIPI, or H33258, while yellow-green and orange-red fluorescence were produced with acridine yellow and ethidium bromide, respectively. No clear filamentous images of DNA could be obtained, however, by quinacrine mustard or by mithramycin even in very high concentrations of the probes (200 $\mu$g/ml). Ellipticine produced aggregates of DNA molecules; thick bundles of DNA strands were often seen. THis dye broke the DNA filaments into shorter pieces.

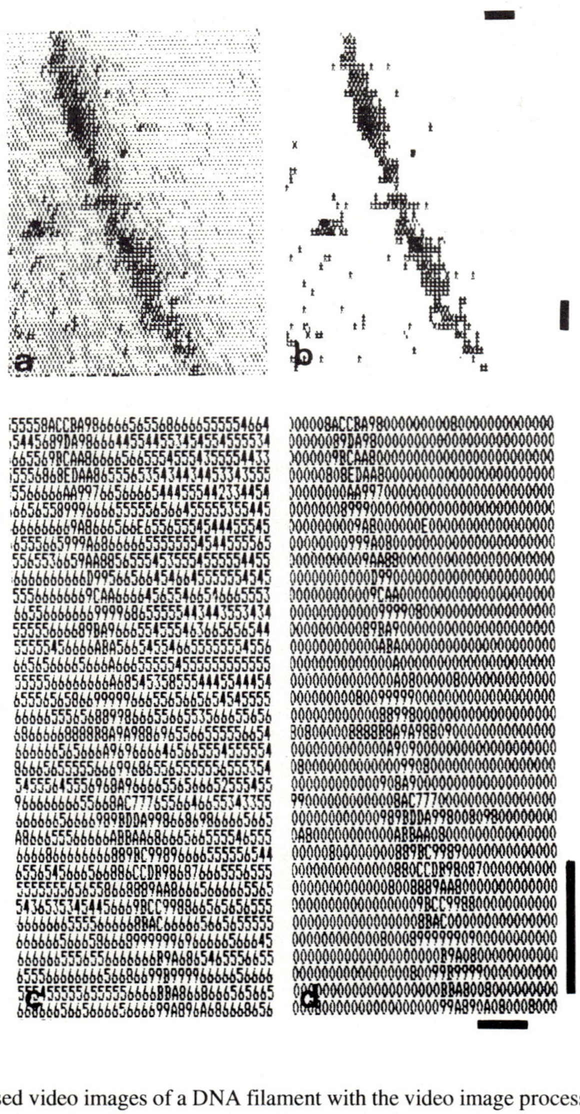

Fig. 2. Processed video images of a DNA filament with the video image processor (constructed by Sun Contact Lens Co., Kyoto); (a) dotted image; (b) the same image with the background noise removed; (c) digital image. The fluorescence intensity was expressed with 16 steps (the order of increasing intensity; from 0–9 and from A–F). The area marked by a rectangle in b is represented here; (d) the same image but the intensity below 6 was sliced. The digitized images have different magnification ratios between x and y directions. The bars indicate 1 μm.

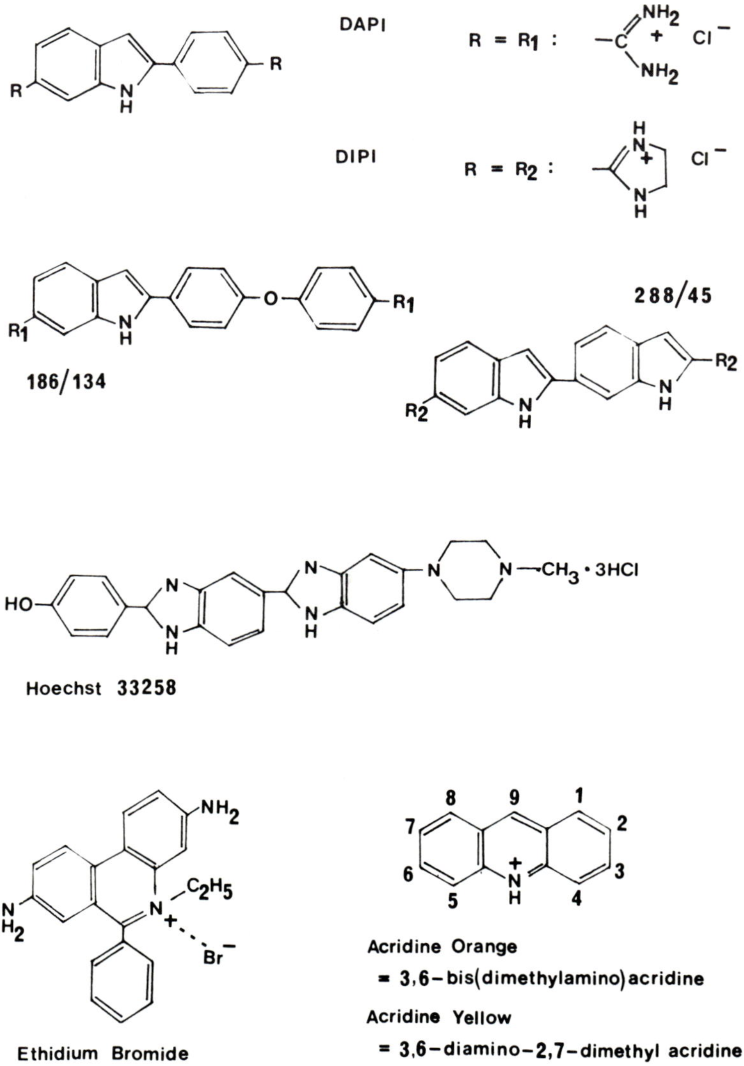


Fig. 3. Fluorescent probes applicable to visualization of DNA molecules.

The optimal concentrations of the probes were 1–2 μg/ml except ethidium bromide and H33258 (the concentration of DNA was fixed at 1 μg/ml). Although DNA molecules could be visualized by much lower probe concentrations (e.g., less than 0.1 μg/ml for DAPI), the quality of the images was not satisfactory and the signal-to-noise ratios were low.

Excitation and emission maxima upon binding to DNA were measured for each probe with a fluorescence spectrophotometer (Table 1). Interference filters used for fluorescence microscopy are also given in parentheses. It should be mentioned that ultraviolet illumination failed to produce images of ethidium-bound DNA molecules, but excitation by green light gave clear images of the filaments. Excitation spectra of ethidium bromide in association with DNA have two maxima at 300–330 nm and at 520 nm, and fluorescence intensity excited at 300–330 nm is higher than that at 520 nm [LePecq and Paoletti, 1967]. The reason for this failure in observing ethidium-bound DNA under UV illumination is not understood.

Fluorescence enhancement was defined by the ratio of fluorescence intensity of a probe with DNA at an emission maximum to that without DNA (Table 1). The microscopic fluorescence intensity was semiquantitatively estimated by the light intensity on a video monitor screen in the presence of 1% ME (2-mercaptoethanol; see below for the reason for the addition of ME). These two parameters were not always correlated. For example, H33258 had the highest fluorescence enhancement but its microscopic intensity upon binding to DNA was lower than those of DAPI or DIPI, which gave the highest microscopic intensity. The microscopic fluorescence intensities must be influenced by various factors.

## FLUORESCENCE FADING AND DNA STRAND BREAKAGE

One percent ME strikingly reduced the light-induced damages of specimens. Fluorescence fading and DNA strand breakage were major technical difficulties at an early stage of our study (Fig. 4a,b). These were found to be greatly diminished in the presence of 1% ME (Fig. 4c); 0.5% ME was less effective. Dithiothreitol also reduced the damage if similar concentrations were employed. The speed of fading and the frequency of DNA strand breakage with or without ME were variable depending upon the probes employed (Table 1). In the case of H33258, the effect of ME was most drastic. DNA was invisible in the absence of ME, but became visible in its presence; the fluorescent images were intense, the fluorescence decay was relatively slow, and the DNA strand cutting became infrequent. In acridine dyes, DNA also became visible in 1% ME, but the images were less intense.

The fading speed was measured with a video image processor. The fluorescence intensity of a single DNA molecule was integrated in a fixed area for a time-lapse series of video images. Ratios of the integrated values for time $t$ to that for time zero ($I_t/I_0$) were plotted versus time on a semilogarithmic scale (Fig. 5).

$I_t/I_0$ of DAPI-stained DNA without ME (empty circles in Fig. 5) rapidly decreased to 0.01 at 10 sec, but in the presence of ME (filled circles), the fading was negligible even after 60 sec. In contrast, the fluorescence decay of ethidium-bound DNA (empty squares) was slow in the absence of ME and became even slower in the presence of ME (filled squares). Even after 10

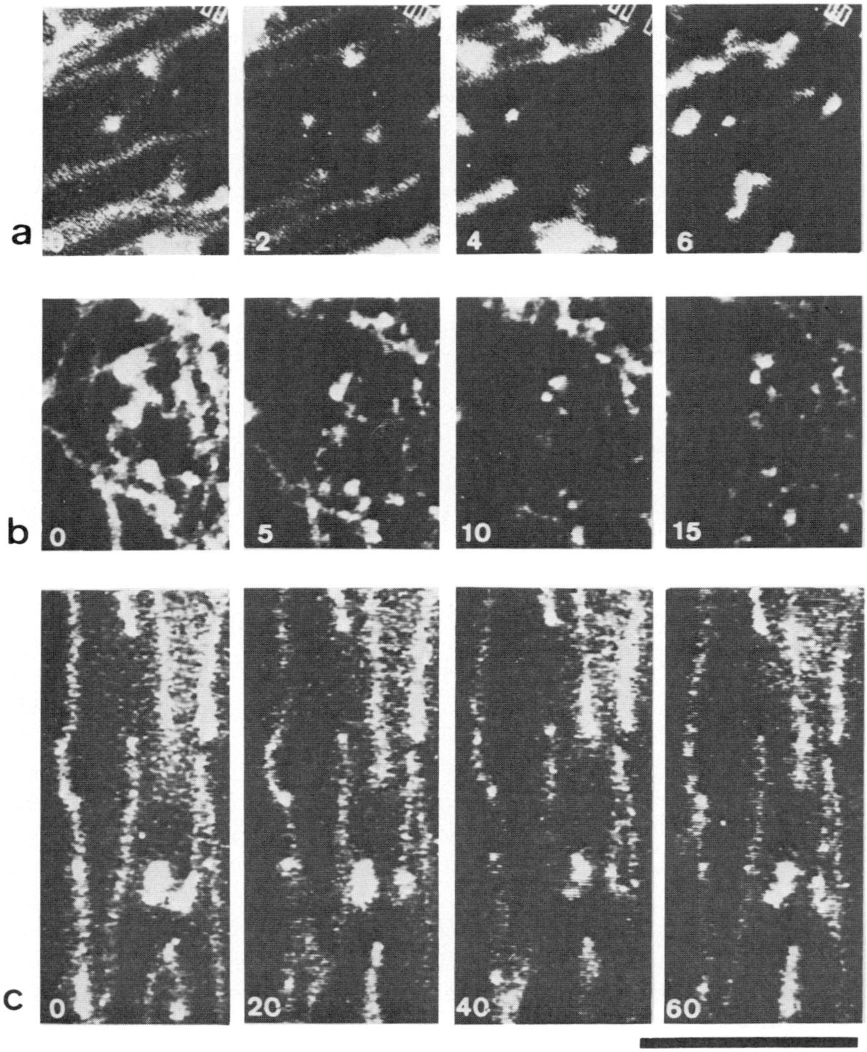


Fig. 4. One percent 2-mercaptoethanol (ME) preserves the fluorescently labeled DNA molecules under UV illumination. DAPI-stained T4 phage DNA filaments in the cacodylate buffer were quickly bleached and broken into pieces in the absence of ME (a,b) but were well preserved in the presence of ME (c). The number in each micrograph indicates the irradiation time in seconds. The bar = 10 $\mu$m.

min, the beam damage was still not significant. The fluorescence decay seemed to be a single hit mechanism because *log* $I_t/I_0$ obtained was proportional to time. ME reduced the decay rate by 130-fold for DAPI and 40-fold for ethidium bromide (Table 2). The decay rates for ethidium-, DAPI- and

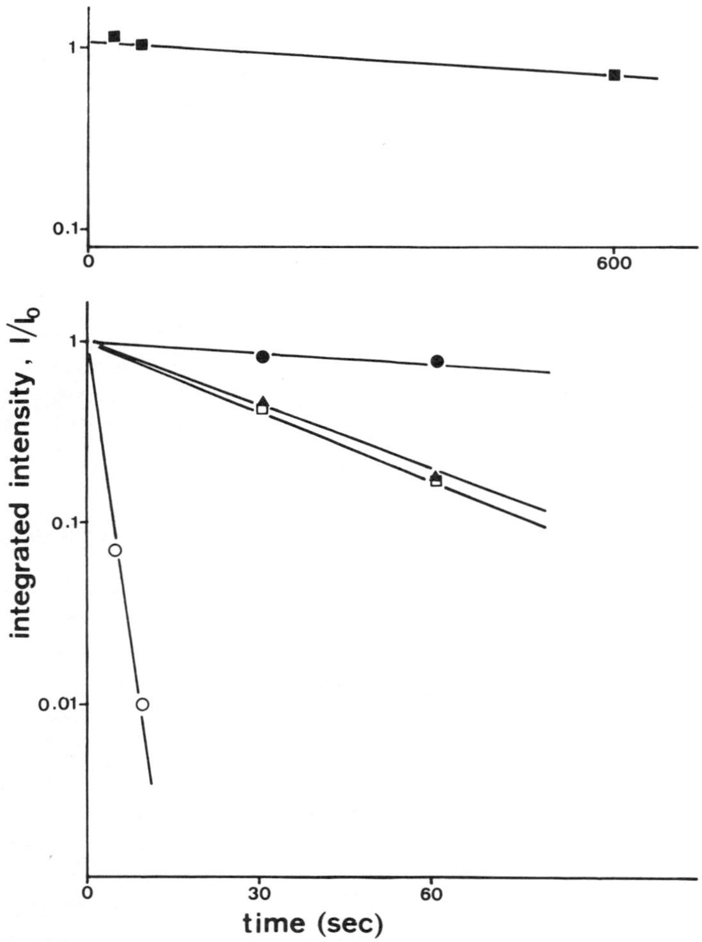


Fig. 5. The rates of fluorescence decay measured with the video processor. The intensity of I of a DNA filament was integrated in a given area (see Fig. 2) for a series of video images taken at appropriate time intervals. The values $I_t/I_0$ ($I_0$ being the value for time zero) are plotted in *log* scale versus time. The filled and empty circles indicate DAPI-stained DNA with and without ME, respectively; the filled and empty squares indicate ethidium bromide with and without ME; the filled triangles indicate H33258 with ME.

H33258-DNA in the presence of ME differed significantly and were about 1:5:36.

The microscopic fluorescence intensity of DNA bound to ethidium bromide was low, whereas the fluorescence decay was the slowest. The slow decay did not seem to result from a high concentration of the dye (20 $\mu$g/ml; Table 1); even at a lower concentration of the dye (5 $\mu$g/ml), the same slow decay was observed. Therefore, ethidium-bound DNA appeared to be most

**TABLE 2. The Rates of Fluorescence Decay in the Absence or the Presence of 1% 2-Mercaptoethanol**

| Probes | 2-mercaptoethanol | Rates of decay ($sec^{-1}$) |
|---|---|---|
| DAPI | − | 0.48 |
| | + | 0.0037 |
| Ethidium | − | 0.029 |
| bromide | + | 0.00075 |
| H33258 | − | * |
| | + | 0.027 |

* invisible.

resistant to irradiation. The decay speed of fluorescence was rather independent of the concentration of probes; the rates observed in DAPI-DNA did not differ over a wide range of concentrations (1–100 μg/ml).

## EFFECTS OF FLUORESCENT PROBES ON DNA SUPERHELICITY

Superhelical changes in closed-circular duplex DNA upon association with different fluorescent probes were determined by agarose gel electrophoresis [Keller, 1975]. The supercoiled pBR322 DNA was first relaxed with the fission yeast type I DNA topoisomerase [Uemura and Yanagida, 1984] in the presence of increasing concentrations of a probe, and then the bound molecules were removed by isoamyl alcohol so that the plasmid DNA could acquire negative superhelicity, depending upon the degree of unwinding by the bound probe molecules. In the case of well-known intercalating ethidium bromide, the bound molecule reduces the average rotation angle in the double helix by 26° [Wang, 1974; Keller, 1975].

The distance between a migrating electrophoretic band (or the midpoint of ladders) and the position of the relaxed plasmid DNA was measured for increasing probe concentrations. The ratios of the obtained distances to that between the relaxed and the supercoiled bands, which represent the changes in superhelical density, were plotted against the concentrations (μg/ml) of each fluorescent probe (Fig. 6a,b). In Table 3, approximate concentrations (μM) of the probes at the ratio of 0.5 are shown.

Expectedly, intercalating dyes such as ethidium bromide, acridine yellow, and acridine orange acquired a negative superhelicity equal to the original supercoil with low concentrations of the probes (Fig. 6a). However, much higher (30–50 times) concentrations were needed for H33258, DAPI, and DIPI (Fig. 6b). H33258 and DAPI had the least effect on superhelicity (Table 3). The curve for 288/45 was similar to that for acridine orange, indicating its intercalating property. The curve for 186/134 showed two distinct steps for untwisting the closed circular DNA at lower and higher concentrations.

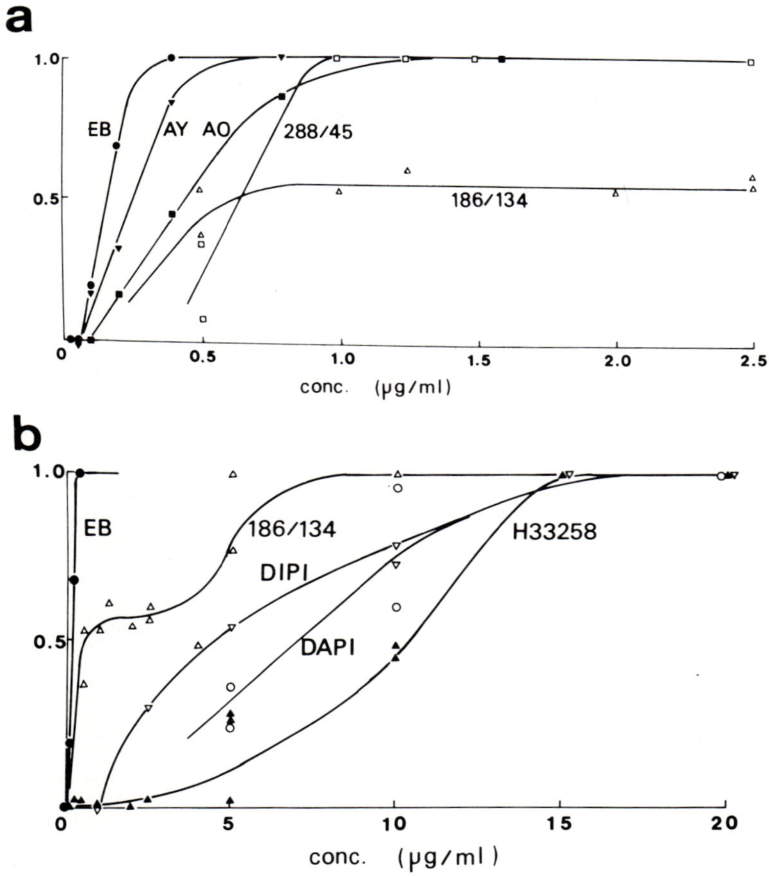


Fig. 6. Changes of electrophoretic mobilities of closed circular pBR322 DNA versus concentrations of different fluorescent probes. Supercoiled pBR322 plasmid DNA (1 μg/ml) was mixed with different concentrations of a probe (0–20 μg/ml) and incubated with type I DNA topoisomerase at 30°C for 60 min. The reaction was stopped by 0.5% sodium dodecyl sulfate (SDS) and 15 mM EDTA. DNA was treated by isoamylalcohol to remove the bound probe molecules, precipitated with ethanol, and run in 0.5% agarose gel electrophoresis. The distances from the open circle to the migrated band (or the midpoint of ladders of the bands) and from the open circle position to the closed circle were measured and their ratios were plotted versus the probe concentrations. a. Lower concentrations (0–2.5 μg/ml). b. Higher concentrations (0–20 μg/ml). EB, ethidium bromide; AY, acridine yellow; AO, acridine orange.

Ellipticine caused the aggregation of the plasmid DNA: the sample DNA did not enter into agarose gel but remained on the top of the gel. Thus, DAPI appeared to be most appropriate for observing DNA molecules in regard to the fluorescence intensity, the degree of specimen preservation by ME, and the minimal effect on DNA superhelicity.

**TABLE 3. Concentrations of the Probes Required for Untwisting the Supercoils Present in Closed Circular pBR322 DNA to the Ratio of 0.5**

| Probes | Concentration ($\mu$M)[1] | Ratio |
|---|---|---|
| Ethidium bromide | 0.38 | 1.0 |
| Acridine yellow | 1.05 | 2.8 |
| Acridine orange | 1.49 | 3.9 |
| 288/45 | 1.56 | 4.1 |
| 186/134 | 1.47–6.78 | 3.9–17.8 |
| DIPI | 11.18 | 29.4 |
| H33258 | 19.62 | 51.6 |
| DAPI | 20.00 | 52.6 |

[1]Concentrations of the probes. The concentration of pBR322 DNA was fixed at 1.0 $\mu$g/ml.

## VISUALIZATION OF SINGLE DNA MOLECULES

The images of DNA molecules stained with fluorescent probes can be recorded on film [Matsumoto et al., 1981] as well as on videotape [Morikawa and Yanagida, 1981]. DNA in the presence of 10 mM $MgCl_2$ adheres to the surface of the glass slide and can be filmed as still images. The film images, better in resolution than the videotape image, showed details of the morphology of DNA filaments under different solution conditions (Fig. 1) [Matsumoto et al., 1981].

A possibility that the materials observed by DAPI staining might not be DNA was rejected by enzyme digestion experiments [Matsumoto et al., 1981]; the fluorescent objects were made to disappear by DNase but not by proteinase or RNase.

Evidence for observing individual DNA molecules instead of the bundled DNA was obtained as follows [Matsumoto et al., 1981]. The lengths of the thin filaments seen in viral DNA correlated well with those determined by electron microscopy. Measurements of the contour length of the extended DNA filaments showed an average of 40 $\mu$m for bacteriophage T4 DNA and 10 $\mu$m for phage T3 DNA. The maximal lengths of T4 and T3 filaments were about 50 and 12 $\mu$m, respectively, which are in good agreement with their full lengths as determined by electron microscopy. The number of particles observed in a unit area corresponded to the number of DNA molecules calculated from the concentration and the molecular weight of DNA.

The maximal length of linear phage $\lambda$ DNA was about 15 $\mu$m, consistent with its 47-kb length. For open circular $\lambda$ DNA prepared by sealing the linear ends with T4 DNA ligase, circle-like filaments were occasionally seen (Fig. 7), but most of the particles observed were the folded form. The

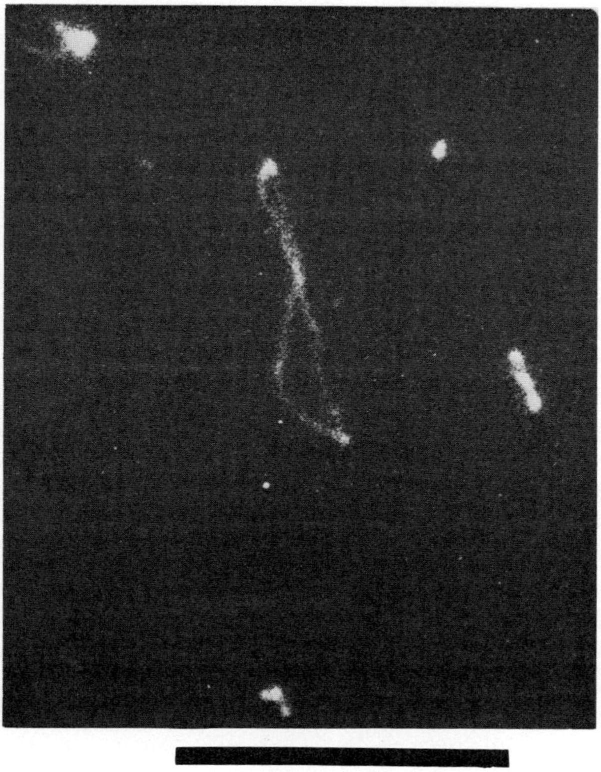

Fig. 7. A fluorescence micrograph of circular λ phage DNA. Circles were occasionally seen but most of the particles were folded. The bar indicates 10 μm.

circular DNA did not stick to the surface of the glass slide even in the presence of 10 mM $MgCl_2$ and therefore, the glass slide was treated beforehand with polylysine.

## STRUCTURES OF DNA MOLECULES IN SOLUTION

Dynamic conformations of DNA molecules in solution observed by fluorescence microscopy were previously reported in detail [Yanagida et al., 1983] and are briefly described here. Folded particles represent the basic conformation of DNA molecules in solution under the minimal shear stress. The particles moved translationally (in average, 1.2 μm/sec for a 50-μm length linear DNA), and at the same time their particle shape fluctuated rapidly among the three forms classified—sphere, ellipsoid, and rod (in average 3.5 μm/sec and maximally 14 μm/sec). The time average-form is ellipsoidal with an axial ratio of about 2.1–2.7 for large DNA molecules. Thus, fluorescence staining offers visual information on Brownian and micro-Brownian motions of large DNA molecules.

DNA conformation in solution has been modeled on the theory of wormlike chains of flexible rods [Kuhn and Kuhn, 1948]. Our measurements showed that the persistence length was in the range of 47–70 nm, which is in good agreement with 50–70 nm from measurments made by other methods such as light scattering [Record et al., 1981]. Other structural parameters showed that the DNA we observed has properties similar to those of the DNA studied by hydrodynamic and light-scattering measurements. We also observed new profiles of DNA molecules in solution [Yanagida et al., 1983], as described below.

Short, thick filaments were observed. They existed in the metastable form, were present in large numbers under reduced shear force, and were shorter and more fluorescent than the thin, extended filaments. These thick filaments were flexible, swaying and exhibiting waves of irregular length; they also extended and shortened (the average rate of movement was 3.6 $\mu$m/sec). The ends of the filaments had a tendency to form folded structures. These thick filaments could elongate into thin filaments by solvent flow with a two to threefold increase in length. After the stream was arrested, the filaments shortened again to become thick filaments. In the still images, the thick filaments showed substructures, coils, and domains, and appeared to consist of waves and zigzags. Fairly regular repetitive segments were seen, suggesting a natural superhelicity in the linear DNA molecules.

The short, thin filaments were shortened very rapidly upon the breakage caused by UV irradiation. Immediately before the breakage, the filaments bent slightly, forming a wave of small amplitude. Shortening of the filament took place quickly to roughly one-third of the initial length; the shortening velocity ranged between 11 $\mu$m/sec and 18 $\mu$m/sec. This rapid and dynamic behavior of DNA molecules in solution clearly demonstrates the elasticity and flexibility of DNA molecules.

## NUCLEAR CHROMATIN REGION AND CHROMOSOMES OF FISSION YEAST

Video-enhanced fluorescence microscopy was initially introduced in our laboratory to study the yeast nuclear chromatin region (defined as the nuclear area where DAPI specifically binds and fluoresces) of the fission yeast *Schizosaccharomyces pombe* [Toda et al., 1981]. *S. pombe* has a small nucleus (2-$\mu$m diameter) containing three chromosomes of total 20,000-kb DNA. The shape of the chromatin region is a hemisphere with two short protrusions in the interphase nucleus (Fig. 8), as determined by three-dimensional reconstructions of the video images viewed at different angles

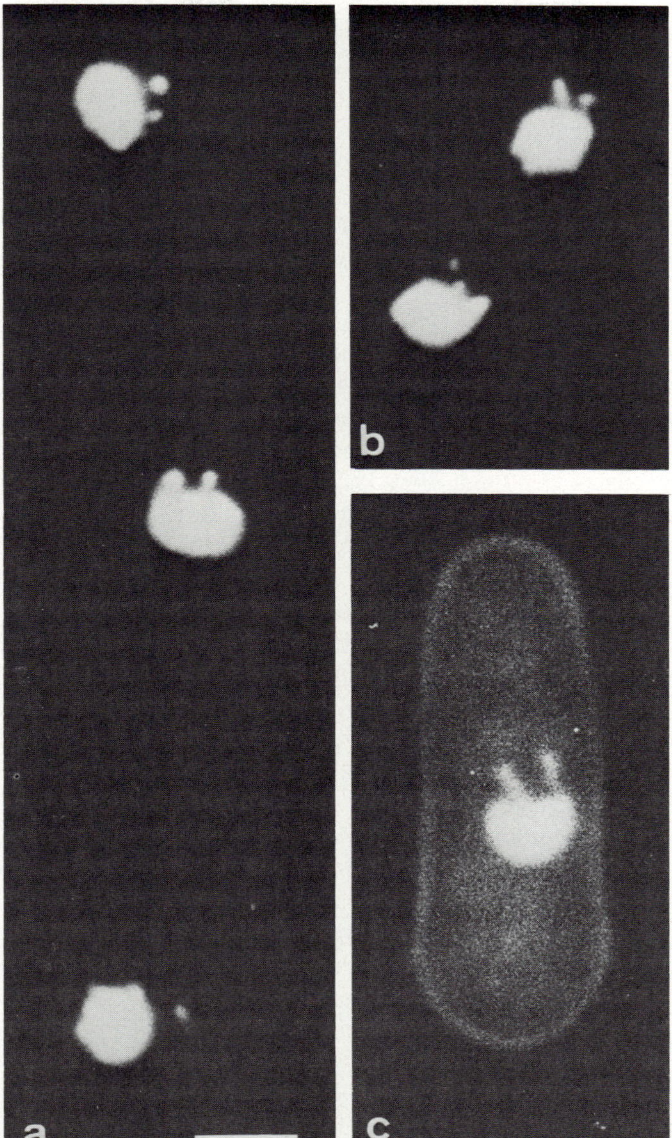


Fig. 8. Fluorescence micrographs of the nuclei and the cell of fission yeast stained with DAPI. The isolated nuclei (a,b) preserved the characteristic hemispherical chromatin region (with two short protrusions) of the interphase cell (c). The bar indicates 2 $\mu$m.

[Toda et al., 1981]. Such compartmentalized distribution of the nuclear DNA was further supported by enzymatic digestions of the isolated nuclei (Fig. 9). DNase digestion (a) destroyed the chromatin region while RNase digestion (b) preserved it. Protease digestion (c) caused the rapid expansion of the chromatin region. Double staining by DAPI and ethidium bromide (d) showed a blue chromatin region and an orange-red nucleolar region [Toda et al., 1981]. RNase digestion of the nucleus stained with ethidium (e), which revealed RNA as well as DNA, removed one hemisphere, leaving the other half-domain closely resembling the nuclear chromatin region (f). The two short protrusions seemed to be nucleolar DNA [Umesono et al., 1983]. Thus, the spherical interphase nucleus consists of the DNA- and RNA-rich half-domains [Toda et al., 1981].

Organization of the fission yeast nuclear chromatin region changed strikingly during mitosis. By investigating the synchronous culture of the wild type, we found sequential structural alteration of the chromatin region from the regular spherical form, compaction into an ellipsoidal form, formation of a U-shaped intermediate, and the rapid segregation into two daughter hemispheres [Toda et al., 1981; Hiraoka et al., 1984]. The mitotic spindle appeared immediately before compaction of the chromatin region, elongated along the cell axis, and quickly disappeared after the completion of nuclear division [Hiraoka et al., 1984]. A number of conditional-lethal mutants defective in the nuclear division [Toda et al., 1983] were isolated by cytological screening procedures with fluorescence microscopy. Among the genes required for nuclear division, the *NDA2* and *NDA3* genes were identified as encoding $\alpha$- and $\beta$-tubulin, respectively [Toda et al., 1984; Hiraoka et al., 1984]. The three-dimensional feature of the U-shaped mitotic chromatin region could be deduced by analyzing the successive video images of a cell viewed at different angles (Fig. 10).

Individual chromosomes of *S. pombe* can now be observed in the cells of a $\beta$-tubulin mutant under a nonpermissive condition [Umesono et al., 1983; Hiraoka et al., 1984]. The three-dimensional features of the condensed chromosomes were obtained by superposing the serial video images taken by steps of through-focusing (Fig. 11). Because the depth of focus is about 0.2–0.3 $\mu$m for maximal resolving power, ten to 15 images taken at different focal planes gave three-dimensional information on the chromosomes located in the nucleus.

## FLEXIBLE CHROMATIN FIBERS DEVOID OF HISTONES H2A AND H2B

Nuclei isolated from exponentially growing cells of the fission yeast *S. pombe* in the stabilizing buffer [phosphate buffer at pH 7.0 containing 1 M sorbitol, 0.1 mM $CaCl_2$, and protease inhibitor phenylmethylsulfonylfluoride

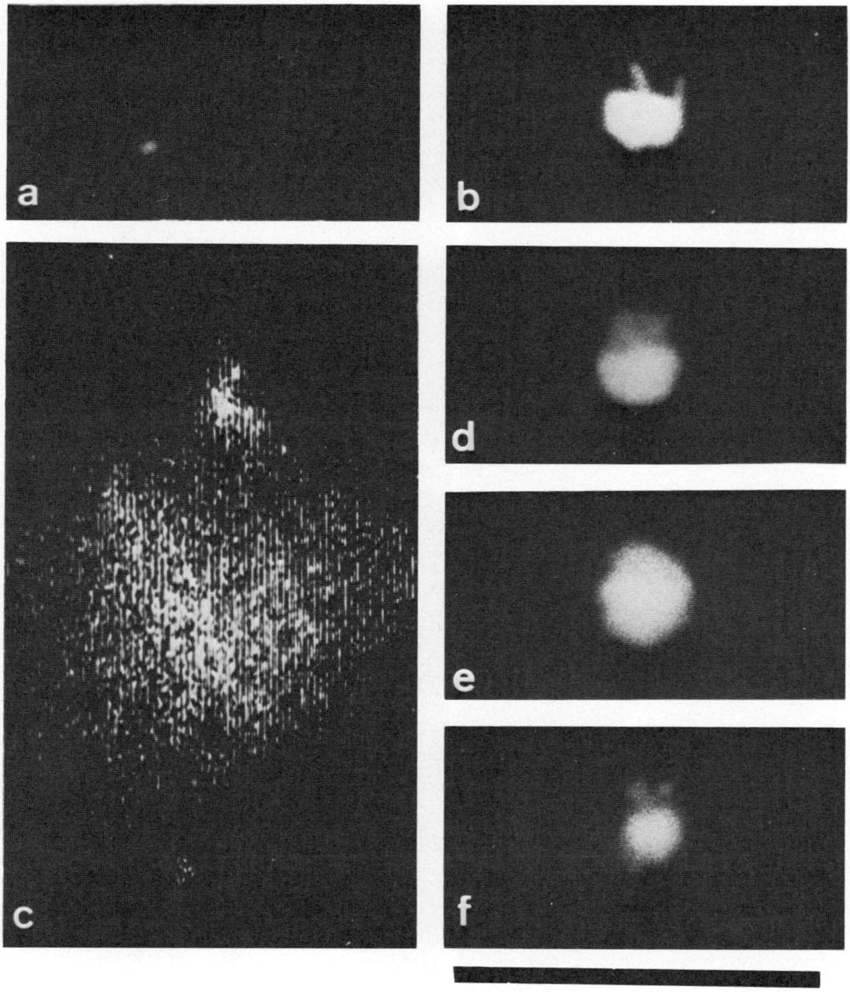


Fig. 9. Enzymatic digestions of the isolated nuclei of *S. pombe*. a. Stained with DAPI and digested with pancreatic DNase I. b. Stained with DAPI and digested with RNase. c. Stained with DAPI and digested with proteinase K. This micrograph is a video image. d. Doubly stained with DAPI and ethidium bromide and no digestion (taken with color film). e. Stained with ethidium bromide and no digestion. f. Stained with ethidium and digested with RNase. The bar indicates 10 $\mu$m.

(PMSF)] were suspended in 2 mM NaCl or in 0.2 mM ethylenediamine tetraaceticacid (EDTA) at pH 7.0 and stained with DAPI (0.1 $\mu$g/ml). The nuclear membranes that were recognized by their slight fluorescence with DAPI were broken and the nuclear chromatin region was loosened and expanded two to three times in diameter (Fig. 12c). A beaded fiber was often

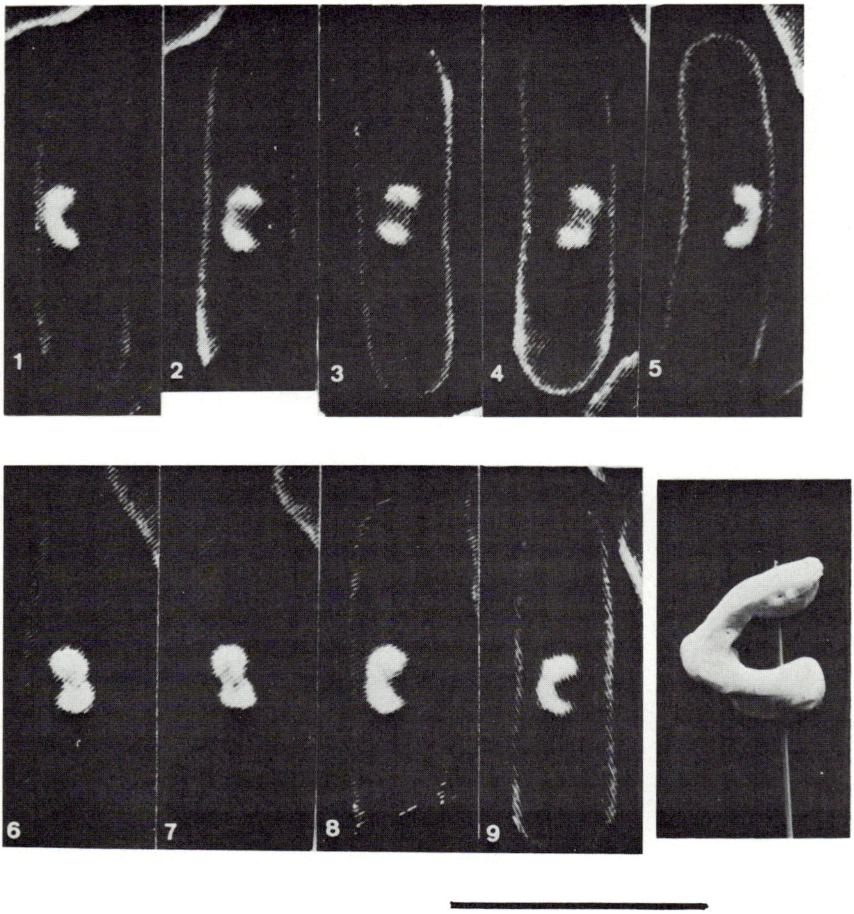


Fig. 10. A series of video images of a mitotic cell showing the U-shaped nuclear chromatin region. The chosen cell slowly rotated around the long cell axis due to a weak water stream, and the video images (1–9) for a single rotation were taken. A clay model is shown. The bar indicates 10 $\mu$m.

released from the nucleolar area of the nuclei (Figs. 13, 15a–d). The nucleolar chromatin could be unfolded at low ionic strength [Umesono et al., 1983].

If 1 mM $MgCl_2$ or 0.1 mM $CaCl_2$ was added, the nuclear chromatin region became compact (Fig. 12b), and such compaction was investigated under different salt concentrations. The divalent cations were effective for compaction in 0.1 mM–0.3 M NaCl, but not in more than 0.4 M NaCl. The chromatin region was swelled in 0.4–0.7 M NaCl regardless of the addition of divalent cations.

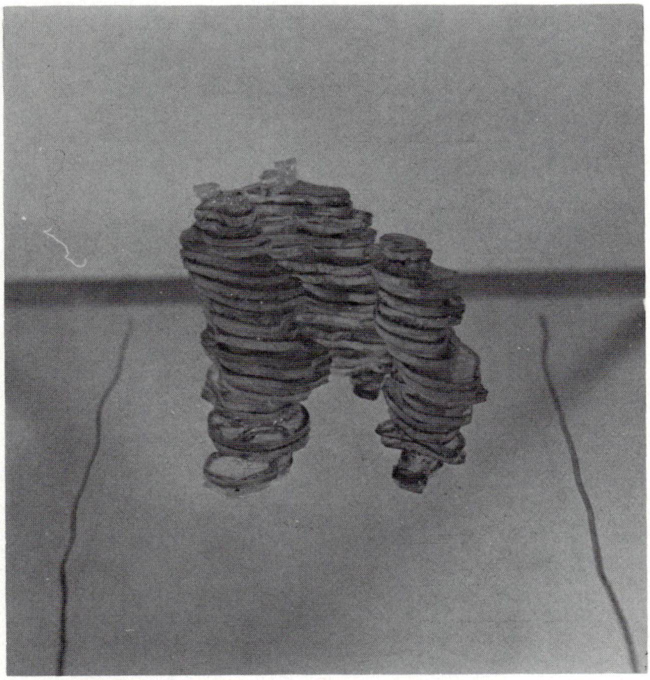

Fig. 11. A three-dimensional reconstruction of condensed chromosomes seen in the β-tubulin mutant of fission yeast [Hiraoka et al., 1984]. Video images taken by through-focusing at 0.1-μm intervals were used.

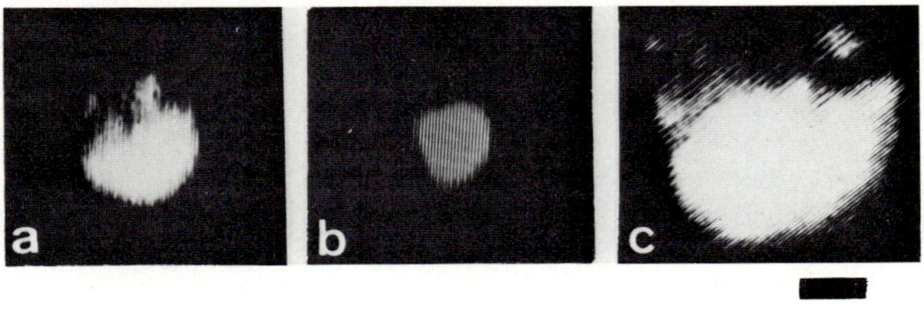


Fig. 12. Video images showing compaction and expansion of the isolated chromatin region. a. In stabilizing phosphate buffer containing 1 M sorbitol and 0.1 mM $CaCl_2$. b. In the same buffer containing 2 mM $MgCl_2$. c. In 2 mM EDTA at pH 7.0. The bar indicates 1 μm.

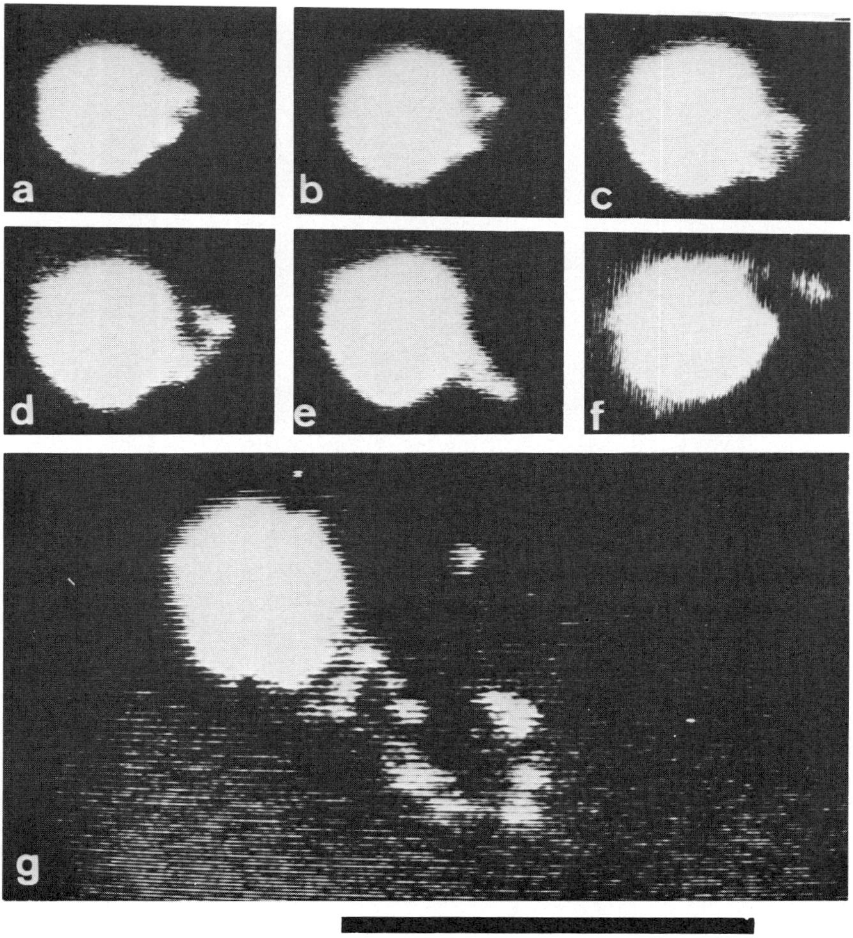


Fig. 13. A time-lapse series of video images of the swelled nucleus in 0.2 mM EDTA at pH 7.0. The short protrusions (the nucleolar DNA) transformed into a beaded filament at low ionic strength. The bar indicates 10 $\mu$m.

A striking structural change occurred at higher salt concentrations. When the nuclei were suspended in 0.8 M NaCl, the long, thin filaments were forced out (Fig. 14). The transition from the swelled chromatin region to the explosion of chromatin fibers was very sharp, and occurred suddenly at more than 0.8 M NaCl. The filaments were very long (more than 300 $\mu$m) and often formed a vast network (Fig. 15e).

These results suggested that the fission yeast nuclear chromatin is organized into a $Mg^{++}$-dependent structure (compaction by addition of $MgCl_2$

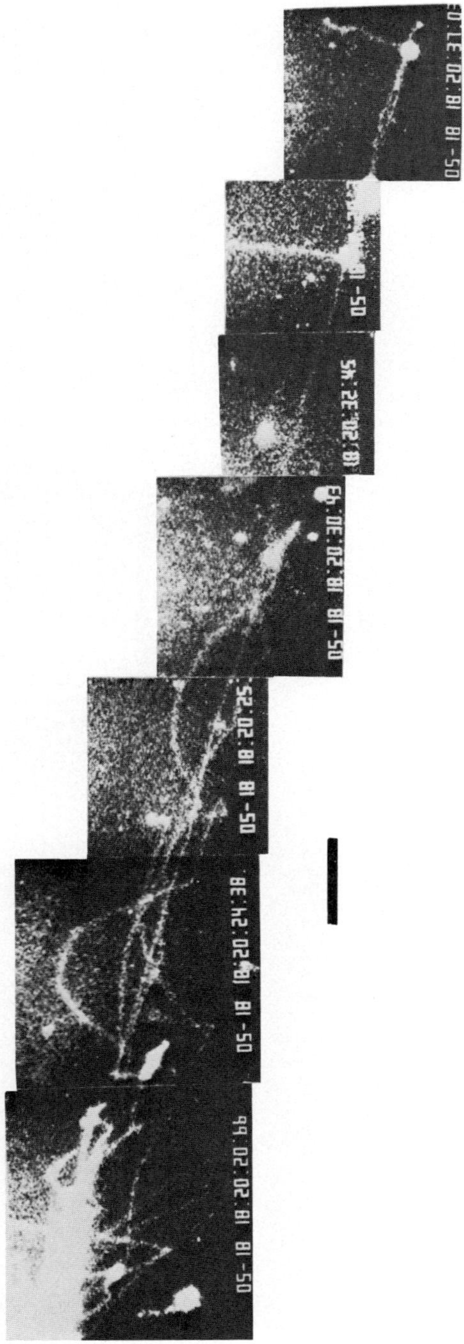


Fig. 14. Thin chromatin filaments released from a fission yeast nucleus in 0.2 mM EDTA and 0.8 M NaCl. This micrograph was constructed from seven frames of video images. The disrupted nucleus is at the upper left corner. The bar indicates 10 $\mu$m.

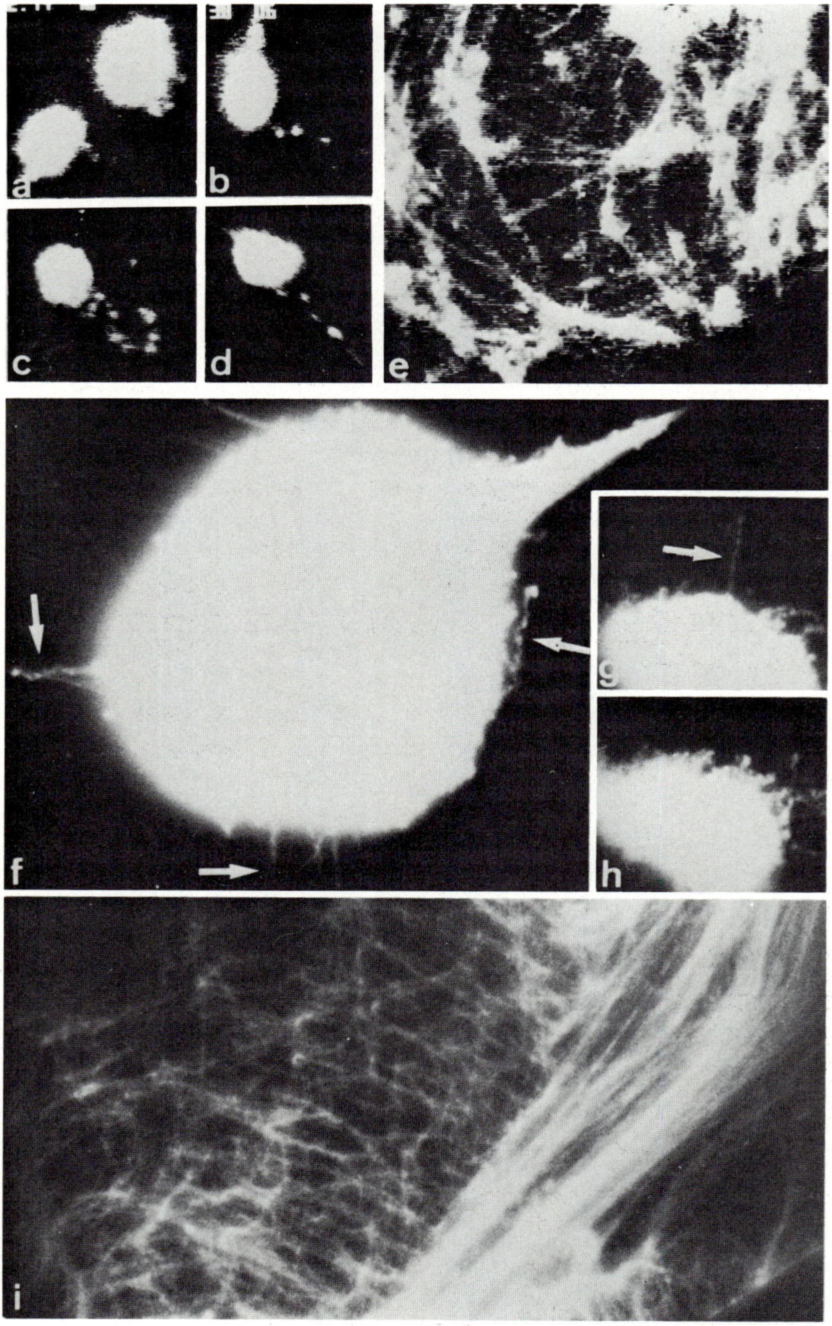
a
b
c
d
e
f
g
h
i

or $CaCl_2$), which is lost at higher salt concentration (above 0.4 M NaCl). The chromatin region (the nuclear membranes previously removed by 2 mM NaCl) can preserve a hemispherical form in 0.1 mM–0.7 M NaCl although its size is varied significantly (Fig. 12c). Such structural integrity is finally broken at 0.8 M NaCl with the explosion into long, thin chromatin filaments. The behavior of these thin filaments in solution was apparently indistinguishable from the naked DNA filaments.

We supposed that this drastic change of chromatin observed at 0.8 M NaCl resulted from the dissociation of certain proteins. The isolated nuclear chromatin was suspended in different salt concentrations and ultracentrifuged and polypeptides of the pellets and the supernatants were analyzed. Results showed that histones H3 and H4 remained with the chromatin fibers (e.g., the pellet fraction) but H2A and H2B were in the supernatant at 0.8 M NaCl (H1 histone was not identified in this organism). Therefore, the filaments seemed to contain histones H3 and H4.

We carried out similar experiments for chicken erythrocytes. At low ionic strength (0.2 mM EDTA pH 7.0), the nuclear chromatin was greatly swelled and segmented fibers were occasionally released from its periphery (Fig. 15f–h). The burst of the thin filaments took place at 0.9–1.0 M NaCl instead of 0.8 M NaCl (Fig. 15i). Neither H2A and H2B nor H1 and H5 were found in the pellet fractions whereas most of the H3 and H4 remained, in association with DNA [Stockley and Thomas, 1979]. Therefore, these thin, flexible filaments are not naked DNA but are associated with histones H3 and H4, probably in the form of an H3-H4 tetramer.

If this interpretation is correct, chromatin filaments containing only H3 and H4 histones are highly flexible, and indistinguishable from the naked DNA in fluorescence microscopy. The depletion of H2A and H2B from native chromatin might cause enormous structural changes, such as the decrease of side-by-side interactions of the nearby chromatin fibers or the increase of chromatin mobility. It should be noted that in the chicken erythrocyte, the removal of H1 and H5 caused only a small expansion of the chromatin region.

---

Fig. 15. Fluorescence micrographs of the nuclei isolated from the fission yeast *S. pombe* (a–e) and the chicken erythrocyte (f–i) are stained with DAPI in the presence of ME under various solution conditions. a–d. In 0.2 mM EDTA at pH 7.0, the fission yeast nuclei were swelled and a beaded fiber was released from the nucleolar region. e. In 0.8 M NaCl containing 0.2 mM EDTA, the nuclei were disrupted and numerous filaments were exploded out. f–h. In 0.2 mM EDTA (pH 7.0), the chicken nuclei were expanded and released segmental filaments at the periphery (indicated by the arrows). i. In 1.0 M NaCl containing 0.2 mM EDTA, the thin chromatin filaments devoid of histones H2A and H2B (also H1 and H5) were forced out by this treatment. The bar indicates 10 $\mu$m.

## ACKNOWLEDGMENTS

We are greatly indebted to Drs. O. Dann and S. Murakami for gifts of fluorescent probes. This work ws supported by grants from the Yamada Science Foundation, the Takeda Science Foundation, and the Ministry of Education, Science, and Culture of Japan.

## REFERENCES

Casperson T, Farber S, Foley GE, Kudynowski J, Modert EJ, Simonsson E, Wagh U, Zech L (1968): Chemical differentiation along metaphase chromosomes. Exp Cell Res 49:218–222.

Dann O, Bergen G, Demont E, Voltz G (1971): Trypanocide diamine des 2-phenyl-bendofuran, 2-phenyl-inden und 2-phenyl-indols. Liebigs Ann Chem 794:68–89.

Hiraoka Y, Toda T, Yanagida M (1984): The *NDA3* gene of fission yeast encodes $\beta$-tubulin; A cold sensitive *nda3* mutation blocks spindle formation and chromosome movement in mitosis. Cell 39:349–358.

Hilwig I, Gropp A (1972): A staining of constitutive heterochromatin in mammalian chromosomes with a new fluorochrome. Exp Cell Res 75:122–126.

Keller W (1975): Determination of the number of superhelical turns in simian virus 40 DNA by gel electrophoresis. Proc Natl Acad Sci USA 72:4876–4880.

Kuhn W, Kuhn H (1948): Rigidity of chain molecules and its determination from viscosity and flow birefringence in dilute solution. J Colloid Sci 3:11–32.

LePecq JB, Paoletti C (1967): A fluorescent complex between ethidium bromide and nucleic acids. J Mol Biol 27:87–106.

Matsumoto S, Morikawa K, Yanagida M (1981): Light microscopic structure of DNA in solution studied by the 4′,6-diaminidino-2-phenylindole staining method. J Mol Biol 152:501–516.

Morikawa K, Yanagida M (1981): Visualization of individual DNA molecules in solution by light microscopy: DAPI staining method. J Biochem (Tokyo) 89:693–696.

Paoletti C, Lesca C, Cros S, Malvy C, Auclair C (1979): Ellipticine and derivatives induce breakage of DNA *in vitro*. Biochem Pharmacol 28:345–350.

Record MT, Mazur SJ, Melancon P, Roe J-H, Shaner SL, Unger L (1981): Double helical DNA: conformations, physical properties, and intercalations with ligands. Ann Rev Biochem 50:996–1024.

Schnedl W, Mikelsaar A-V, Breitenbach M, Dann O (1977): DIPI and DAPI: Fluorescence banding with only negligible fading. Hum Genet 36:167–172.

Schweizer D (1976): DAPI fluorescence of plant chromosomes prestained with actinomycin D. Exp Cell Res 102:408–412.

Stockly PG, Thomas JO (1979): A nucleosome-like particle containing an octamer of the arginine-rich histones H3 and H4. FEBS Lett 99:129–135.

Toda T, Yamamoto M, Yanagida M (1981): Sequential alterations in the nuclear chromatin region during mitosis of the fission yeast *Schizosaccharomyces pombe*. J Cell Sci 52:271–287.

Toda T, Umesono K, Hirata A, Yanagida M (1983): Cold-sensitive nuclear division arrest mutants of the fission yeast *Schizosaccharomyces pombe*. J Mol Biol 168:251–270.

Toda T, Adachi Y, Hiraoka Y, Yanagida Y (1984): Identification of the pleiotropic cell division cycle gene NDA2 as one of two different $\alpha$-tubulin genes in *Schizosaccharomyces pombe*. Cell 37:233–242.

Uemura T, Yanagida M (1984): Isolation of type I and II DNA topoisomerase mutants from fission yeast: Single and double mutants show different phenotypes. EMBO J 3:1737–1744.

Umesono K, Hiraoka Y, Toda T, Yanagida M (1983): Visualization of chromosomes in mitotically arrested cells of the fission yeast *Schizosaccharomyces pombe.* Curr Genet 7:123–128.

Wang JC (1974): The degree of unwinding of the DNA helix by ethidium. J Mol Biol 89:783–801.

Williamson DH, Fennell D Jr (1975): The use of fluorescent DNA binding agent for detecting and separating yeast mitochondrial DNA. In Prescott DM (ed): "Methods in Cell Biology." New York: Academic Press, Vol. 12, pp 335–351.

Yanagida M, Hiraoka Y, Katsura I (1983): Dynamic behaviors of DNA molecules in solution studied by fluorescence microscopy. Cold Spring Harbor Symp Quant Biol 47:177–187.

*Applications of Fluorescence in the Biomedical Sciences, pages 347–376*


# 16 Spatial and Temporal Dynamics of Specific Molecules and Ions in Living Cells

*D. Lansing Taylor, Philip A. Amato, Paul L. McNeil, Katherine Luby-Phelps, and Lerson Tanasugarn*

Understanding the complex chemical processes occurring within living cells is a major challenge in modern cell and molecular biology. The ultimate goal of this laboratory is to map in time and space in living cells the chemical and physical dynamics of the macromolecules, metabolites, and ions involved in cytoplasmic structure and contractility during such events as chemotaxis, phagocytosis, and cell growth. We have adopted this approach because of the gap presently separating our knowledge of biochemical processes studied in vitro from our knowledge of what occurs in the far more complex milieu of living cytoplasm. The task of understanding biochemistry in vivo is complicated by the fact that cytoplasm can vary in its properties from one region to another during cell functions [see Taylor and Condeelis, 1979]. Basic chemical principles that apply to dilute solutions may have little relevance to the highly concentrated and structured environment of cytoplasm. Therefore, it is essential to study the molecular properties of cytoplasm in time and space within living cells.

Our research has been focused on the actin- and myosin-based contractile-cytoskeleton in nonmuscle cells [Taylor et al., 1981a; Taylor and Fechheimer, 1982]. Figure 1 is a simple diagram of a generalized mammalian cell where the spatial variation of the structure of the actin-based cytoskeleton is emphasized. Changes in the local shape and movement of cells are believed to be produced by interactions between actin and a variety of proteins. In vitro biochemical studies have identified a large number of actin-associated

Department of Biological Sciences, Center for Fluorescence Research in Biomedical Sciences, Carnegie-Mellon University, Pittsburgh, Pennsylvania 15213.

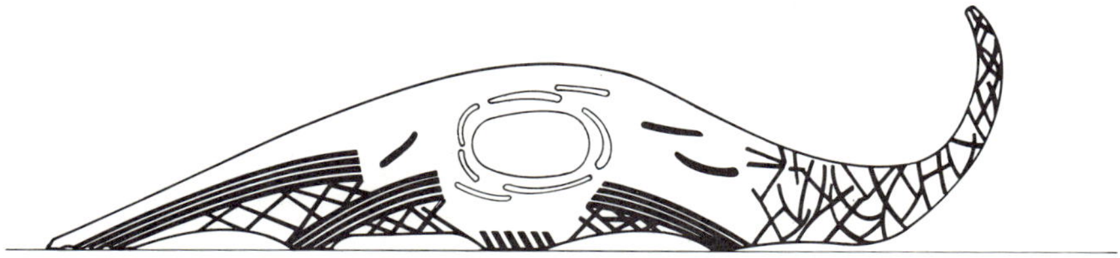

Fig. 1. Diagram of an idealized mammalian cell viewed from the side showing the spatial variation in the structure and organization of the actin-based, contractile cytoskeleton. Actin exists in the cell in three major forms: (1) unassembled subunits, (2) actin filaments crosslinked in an optically isotropic gelled matrix, and (3) actin filaments crosslinked into bundles called stress fibers. These structures are believed to change in time and space to regulate cytoplasmic structure and to cause motility.

proteins that regulate filament assembly-disassembly, crosslinking of filaments into bundles or isotropic meshworks, and force generation with myosin [reviewed in Taylor and Condeelis, 1979; Schliwa, 1981; and Craig and Pollard, 1982]. However, complete understanding of the molecular basis of the function of actin and associated proteins in cellular processes will require temporal-spatial information acquired from living cells.

We have chosen fluorescence spectroscopy in conjunction with fluorescence microscopy as a tool for mapping the physical and chemical dynamics of selected molecules in living cells [see Taylor and Wang, 1980; Wang et al., 1982a; and Taylor et al., 1984, for reviews]. Fluorescence microscopic techniques are potentially powerful in cell and molecular biology for the five major reasons outlined in the previous chapter: sensitivity, specificity, spectroscopy, spatial resolution, and temporal resolution (Bright and Taylor, this volume). Furthermore, the interpretation of the more complicated studies performed in vivo can be clarified by direct comparison with simpler experiments performed in dilute solutions under controlled conditions in vitro.

The purpose of this chapter is threefold: (1) to discuss some of the important practical considerations involved in studying the dynamics of fluorescently labeled molecules in living cells (fluorescent analog cytochemistry); (2) to describe two recent applications of fluorescent analog cytochemistry in cell biology; and (3) to summarize our recent measurements of pCa and pH in living cells and to discuss the relationships between changes in pCa and pH and the structure of the cytoskeleton.

## SOME PRACTICAL CONSIDERATIONS OF FLUORESCENT ANALOG CYTOCHEMISTRY

Fluorescent analog cytochemistry is a recently developed method that allows the study of the distribution and molecular activity of specific cellular

**TABLE 1. Some Fluorescent Analogs of Cytoskeletal Proteins Used in Living Cells**

| Target molecule | Fluorophore | Reference |
|---|---|---|
| 1) Actin | a) 5′ iodoacetamidofluorescein | a) Taylor and Wang [1978] Wang and Taylor [1980] |
| | b) Tetramethylrhodamine isothiocyanate | b) Kreis et al. [1979] |
| | c) Rhodamine B sulfonyl chloride | c) Kreis et al. [1982] |
| 2) Tubulin | a) Dichlorotriazinylamino-fluorescein (DTAF) | a) Keith et al. [1981] Leslie et al. [1984] Wadsworth and Sloboda [1983] |
| 3) Calmodulin | a) Tetramethylrhodamine isothiocyanate | a) Zavortink et al. [1983] Luby-Phelps et al. [1985b] |
| | b) N-(7-dimethylamino-4-methylcoumarinyl)-maleimide | b) Hamaguchi and Iwasa [1980] |
| 4) α-actinin | a) Tetramethylrhodamine isothiocyanate | a) Feramisco [1979] |
| 5) Vinculin | a) DTAF | a) Geiger [1981] |
| 6) Tropomyosin | a) 5-iodoacetamidofluorescein | a) Wehland and Weber [1980] |

components in living cells [see Taylor et al., 1984; Wang et al., 1982a; Taylor and Wang, 1980; Wang et al., 1982b; Kreis and Birchmeier, 1982, for review]. A fluorescent analog consists of a native cellular component (e.g., a purified protein) plus one or more covalently attached fluorescent probes. We originally named the method molecular cytochemistry [Taylor and Wang, 1978] but changed the name to more accurately define the technique [Wang et al., 1982a].

The technique involves five sequential steps: (1) isolating the molecules or organelles to be studied in living cells; (2) labeling such molecules or organelles covalently with a fluorophore; (3) assaying the biochemical activity and solution spectroscopic properties of the analogs in vitro; (4) incorporating the fluorescent analog into cells; and (5) analyzing the fluorescence parameters of the analogs in the living cells. Good analogs of a large number of proteins and lipids have now been prepared [Table 1; see Taylor et al., 1984; Wang et al., 1982a; and Pagano and Longmuir, 1983].

Thus far, we have focused most of our attention on characterizing one fluorescent analog of actin in detail both in solution and in living cells [Taylor et al., 1981b, 1984; Wang and Taylor, 1982; Wang et al., 1982a, 1983]. Actin is important in the determination of the shape and the rigidity as well as the motility of a large number of cell types [Taylor and Condeelis, 1979].

The biochemistry of actin is complex, since it can undergo reversible self-assembly into filaments and also interacts with a diverse collection of actin-binding proteins [Korn, 1982; Pollard and Craig, 1982; Craig and Pollard, 1982; and Schliwa, 1981]. Many details of the basic mechanisms involved in these processes remain to be worked out, both in vitro and in vivo [Bonder et al., 1983; Pollard and Craig, 1982; Wang and Taylor, 1982; Pardee et al., 1982]. We chose to study muscle actin since no dramatic functional differences have been demonstrated between muscle and nonmuscle actins [Korn, 1982]. In the future, it will be valuable to compare the distribution and activity of fluorescent analogs of different isoforms of actin [McKenna et al., 1985].

The sulfhydryl reactive probe, 5′-iodoacetamidofluorescein (IAF) was chosen as our first probe since it has a relatively large quantum yield (Φ), a large extinction coefficient (ε), an excitation maximum in the visible spectrum, and it has been shown that actin has a highly reactive sulfhydryl group at cystein 374 [Taylor and Wang, 1978; Wang and Taylor, 1980]. These properties ensured a good signal-to-noise ratio and a site-specific label.

Some of the interesting fluorescent analogs of actin used in early solution studies were not as useful for living cells since the wavelengths of excitation were in the ultraviolet or near ultraviolet [see Wang and Taylor, 1980], and mammalian cells exhibit autofluorescence over a broad excitation spectrum with a maxima at ca. 380 nm and 460 nm, respectively.

Our laboratory has been investigating a wide range of practical considerations in the use of fluorescent analog cytochemistry over the past 7 years. Some of these considerations have already been discussed in detail [Taylor and Wang, 1978, 1980; Wang and Taylor, 1979, 1980; Wang et al., 1982a; Taylor et al., 1984]. Four of the most important practical considerations are discussed below by using our experience with AF-actin as an example.

**Functional Capability of Analogs In Vitro**

The physiological significance of fluorescence measurements depends on understanding the functinal capabilities of the fluorescent analog. The analog must therefore be studied extensively in vitro before employing it in serious experiments in vivo. The IAF-labeled muscle actin (AF-actin) has now been studied in considerable detail (Table 2). However, there are more parameters that need to be analyzed. For example, it is now possible to measure the critical concentration of actin at the barbed and pointed ends of the filaments [Bonder et al., 1983]. In addition the activities of ADP- and ATP-actin appear to differ [Pollard, 1984]. A complete characterization of an actin analog would include comparison to unlabeled actin in these additional properties. A detailed characterization of fluorescent analogs in vitro is critically important to ensure the physiological significance of experiments performed in vivo.

**TABLE 2. Properties of Iodoacetamidofluorescein-Labeled Actin (AF-Actin)**

| | |
|---|---|
| Site of modification | Cyst. 374 |
| Dye protein | 0.8–1.0 |
| $\Phi$ (G-actin) | 0.23 |
| $\epsilon$ | $6.0 \times 10^4$ $cm^{-1}$ $M^{-1}$ |
| Excitation max. (G-actin) | 498 nm |
| Emission max. (G-actin) | 519 nm |
| $\tau$ (F-actin) | 3.5 nsec |
| G-F self-quench | 15% |
| High shear viscosity | 100% of control |
| Activation of myosin ATPase | 90–100% of control |
| Inhibition of actin assembly by spleen profilin[1] | < 10% |
| Formation of $Ni^{++}$ and $Mg^{++}$ paracrystals | Yes |

[1]Iodoacetamide modified actin [Malm, 1984]. However, there is evidence for recombination of acanthamoeba actin labeled at cyst. 374 with acanthamoeba profilin [Tobacman et al., 1983].

For example, it is apparent from Table 2 that spleen profilin does not interact as well with cyst. 374–modified actin as with control actin. This lowered interaction may or may not be an important parameter in the living cell, but it points out a fundamental rule for fluorescent analog cytochemistry: No single site of modification or single fluorophore should be used in the final analysis of the molecular properties of the target molecule or organelle. At least two different, well-characterized analogs should be employed to increase the physiological significance of the results. We are currently studying several actin analogs that differ in the fluorophore and/or the site of modification.

## Incorporating Analogs Into Cells

The fluorescent analog of a selected molecular species is intended to function as a "reporter" of the molecular activity of the endogeneous population of molecules. In order to do so, fluorescent analogs must be introduced into the cell with a minimum of cell perturbation and so that the final intracellular concentration can be estimated and controlled. Excessively high concentrations of the analog could actually perturb cellular functions, while excessively low concentrations might not yield a high enough signal/noise for analysis.

There are two major classes of methods for incorporating macromolecules into living cells with an acceptable level of stress to the cell: (1) Injection of cells by using microneedles and (2) bulk loading of cell populations [see Celis, 1984, for a review]. The physical microinjection methods have been used for over 50 years [see Chambers and Chambers, 1961]. In contrast, a large number of clever bulk-loading procedures have been developed over the last 5 years. One new method is particularly simple [McNeil et al., 1984].

A major concern in either approach is controlling both the volume of the injection and the final concentration of the fluorescent analog in the cells. This problem is reasonably straightforward with physical microinjection. The injection volumes can either be calibrated by measuring the volumes of injected oil droplets [Kiehardt, 1982; Diacumakos, 1973], or by measuring the radioactivity of cells microinjected with a solution of a radioactive tracer [Stacey and Allfrey, 1976], or by measuring the movement of the miniscus in the injection needle [Graessmann et al., 1980]. We have used the oil droplet method in our studies on amoebae [Taylor et al., 1980a,b], sea urchin eggs [Wang and Taylor, 1979], and mammalian cells [Amato et al., 1983]. In our experience, we can control the volume of injection within a factor of two from cell to cell. Given the rather wide range of cell sizes in some standard cell lines, we have accepted this level of precision for our early studies. Reasonable dose-responses studies can be performed under these conditons by simply varying the concentration of the fluorescent analog in the stock solution. Taking into account the approximate (approx.) twofold variability in volume, we have been able to inject Swiss 3T3 cells with final cytoplasmic actin analog concentrations ranging from ca. 0.02 mg/ml to >0.2 mg/ml. This represents a ratio of injected actin analog to endogenous actin of less than 1% calculated as follows:

Below is a calculation of the number of endogenous actin molecules in "average" mammalian fibroblasts.

1. Approx. 0.75 ng protein/cell.
2. Approx. $10^{-12}$ liter cell volume.
3. Actin makes up approx. 10% w/w of protein. $\therefore$ $7.5 \times 10^{-11}$ g actin/cell.
4. MW of actin $4.5 \times 10^4$ daltons, $\therefore$ $7.5 \times 10^{-11}$ g/$4.5 \times 10^4$ = $1.67 \times 10^{-15}$ moles actin/cell.
5. $6.023 \times 10^{23}$ molecules g/mol, $\therefore$ $6.023 \times 10^{23}$ molecules g/mol $\times$ $1.67 \times 10^{-15}$ = approx. $10^9$ molecules actin/cell.

Below is a calculation of the number of fluorescent analogs of actin injected into an "average mammalian fibroblast.

1. $10^{-13}$ liter injection volume.
2. Actin injection stock conc. 1.0–5.0 mg/ml $\therefore$ $2.2 \times 10^{-5}$ to $1.1 \times 10^{-4}$ moles of actin.
3. Moles actin $\times$ $6.023 \times 10^{23}$ g/mol $\times$ $10^{-13}$ liter = $1.3 \times 10^6$ to $6.0 \times 10^6$ molecules of actin injected/cell. $\therefore$ the actin analog to endogenous actin ratio ranges from ca. $10^{-3}$ to $10^{-2}$.

Bulk-loading methods will have increasing importance in studies where it is valuable to have large populations of cells (> $10^5$) to analyze. Bulk loading

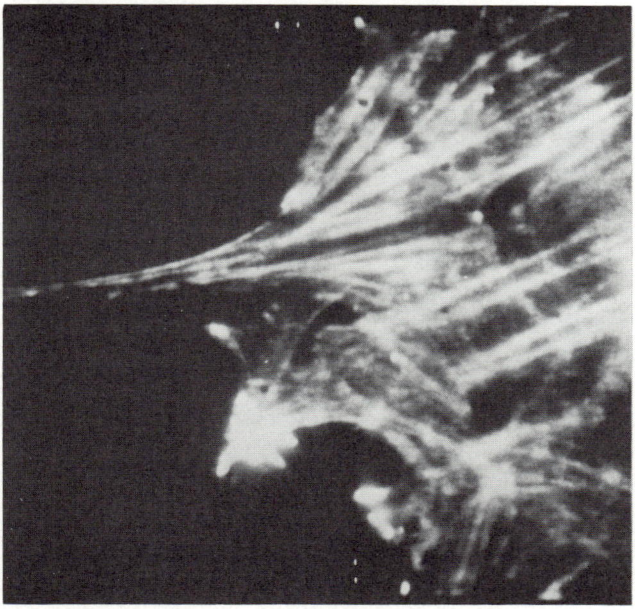

Fig. 2. Image of the leading edge of a living Swiss 3T3 cell containing AF-actin. The image was obtained by using a Zeiss 63.0×, 1.2 N.A. water immersion objective and a Photonic Microscopy Incorporated (PMI) two-stage multichannel plate detector coupled to a PMI digital image processing system. The light intensity at the specimen plane was 35 μW. Image processing consisted of integrating 600 frames and subtracting a background image. Stress fibers and less organized actin- containing structures can be seen, 1.0 cm equals 10 μm.

provides the large sample size required for accurate statistical analysis (see Murphy, this volume) or low final magnification scanning microscopic methods (Ploem, this volume). While it is difficult to control the volume and concentration loaded, a method for determining the concentration of fluorescent macromolecules in cells by flow cytometry has been described, and there is even the potential for sorting subpopulations based on the concentration of the fluorescent analog [McNeil et al., 1984].

Our working rule at this time is to keep the final concentration of the fluorescent analog below 10% of the endogenous concentration of the molecule to minimize cell perturbations. This protocol is easy to follow for proteins like actin which make up a large fraction of the total protein. However, fluorescent analogs of some cellular constituents may be difficult to use if the total cellular concentration of the molecule or organelle is very low. A typical image of a cell injected with AF-actin is shown in Figure 2.

### Do Fluorescent Analogs Incorporate Into Endogenous Structures at the Molecular Level?

Fluorescent analog cytochemistry can yield valuable quantitative information on the molecular properties of specific molecules in living cells only if the fluorescent analogs exchange freely with the endogenous pool of molecules.

We have investigated the exchange of actin analogs with actin subunits within stress fibers of fibroblasts. Stress fibers are complex structures containing not only tens of actin filaments, but also a large number of actin binding proteins. The structural organization of stress fibers has been described by electron microscopy [Buckley and Porter, 1967; Goldman et al., 1979; Langanger et al., 1984; Sanger and Sanger, 1980] and by immunofluorescence [Lazarides and Weber, 1974; Fujiwara and Pollard, 1976].

Immunofluorescence microscopy has demonstrated that myosin and tropomyosin alternate periodically along some stress fibers with $\alpha$-actinin [Lazarides, 1975; Lazarides and Burridge, 1975; Sanger et al., 1983; Weber and Groeschel-Stewart, 1974]. A definite sarcomeric-like distribution of contractile proteins has been shown in some stress fibers [Sanger et al., 1983, 1984], although the functional significance of this organization has not yet been determined. However, it is known that stress fibers are capable of contracting [Isenberg et al., 1976; Kreis and Birchmeier, 1980].

The complex structure of the stress fiber made it a challenging object for testing the ability of fluorescent analogs of actin to exchange with endogenous actin subunits. We developed a cytochemical assay using anti-fluorescein antibodies as a "probe" of our fluorescent analog [Luby-Phelps et al., 1984; Amato and Taylor, 1986]. This approach evolved from our use of antifluorescein antibodies to prepare an actin affinity column [Luna et al., 1982]. In fact, antifluorescein antibodies have been used extensively both in research in immunology [Dandliker and Feigen, 1961; Lopatin and Voss, 1971; Levinson et al., 1975] and in cell biology [Sklar et al., 1981; Soltys and Borisy, 1985].

The preparation and characterization of our antifluorescein antibody has been discussed elsewhere in detail. Briefly, solution spectroscopic, ultrastructural, and cytochemical studies demonstrated that antifluorescein antibodies could recognize and bind to the fluorescein moiety of AF-actin either in the monomeric or the filamentous state [Luby-Phelps et al., 1984]. The antifluorescein method was extended to the ultrastructural level in order to approach the question of molecular incorporation [Luby-Phelps et al., 1984; Amato and Taylor, 1986]. Swiss 3T3 fibroblasts were microinjected with monomeric AF-actin and the time of incorporation was studied by immunoelectron microscopy with horseradish peroxidase, coupled to the secondary antibody targeted to the antifluorescein IgG [Amato and Taylor, 1986].

Fig. 3. A 3T3 cell fixed at 5 min postinjection with AF-actin and labeled with antifluorescein and peroxidase-coupled IgG. A periodic localization of the reaction product is seen along the length of some fibers (arrows). Mag. ×29,700. Reprinted with permission, J Cell Biol.

The injected cells were identified at low magnification in the electron microscope. The adjacent uninjected cells served as controls. The images we saw in cells fixed during the first 5 min after injection fell into two classes: (1) some stress fibers showed a periodic localization of the reaction product that suggested a microheterogeneity in the incorporation of AF-actin along the length of these fibers (Fig. 3); (2) other fibers exhibited a less pronounced periodicity of incorporation along the length of the stress fibers, but showed

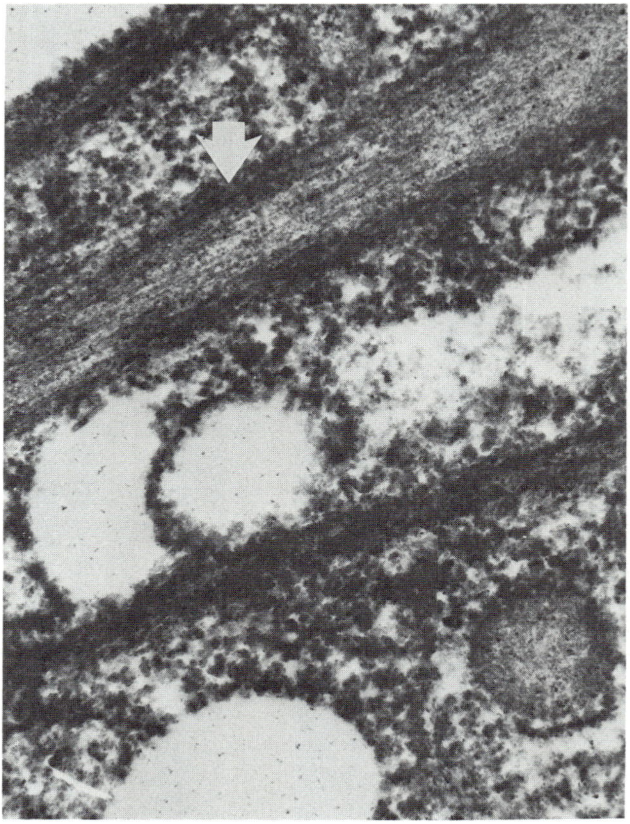

Fig. 4. A 3T3 cell prepared as described in Figure 3. The reaction product is sometimes observed first at the outer surface of stress fibers (arrows). Reaction product can also be seen associated with less structured actin in the background. Mag. ×37,000.

more extensive reaction product at the outer surface of the fibers than in the core of the stress fiber (Fig. 4). By 20 min postinjection, most stress fibers exhibited a uniform distribution of reaction product throughout (Fig. 5). From this, we concluded that AF-actin injected into Swiss 3T3 cells in the monomeric state could incorporate into existing stress fibers within 20 min. At this level of resolution it appeared that the fluorescent analog of actin was probably exchanging with endogenous actin subunits at the molecular level. To more conclusively rule out the possibility that AF-actin simply decorated stress fibers by adventitious binding to existing filaments (templating), we prepared a "templated" cell model by adding AF-actin to fixed and extracted

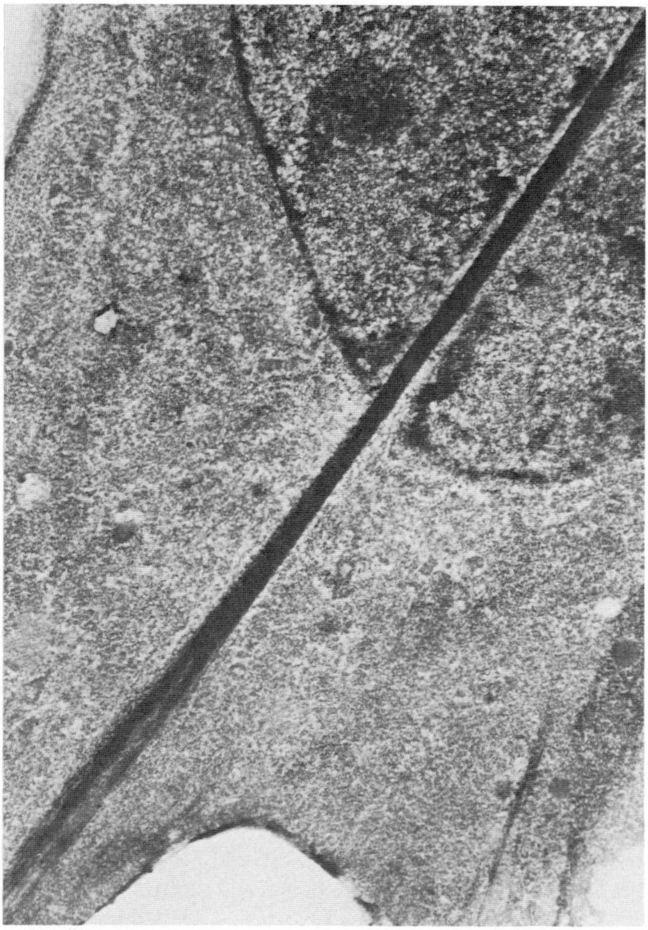

Fig. 5. A 3T3 cell prepared as described in Figure 3 but fixed at 20 min postinjection. Most actin-containing stess fibers show reaction product uniformly throughout by this time period. Mag. ×20,000.

cells in which no incorporation into filaments could occur [Amato and Taylor, 1986]. In these templated models, reaction product was restricted exclusively to the exterior of the fiber bundle. Further study will be required to prove that AF-actin actually incorporates into endogenous filaments in a physiological manner. However, we are satisfied that the available evidence warrants the cautious interpretation of the distribution and activity of actin analogs in living cells.

The apparent periodicity of incorporation seen in some stress fibers could be significant, perhaps reflecting a differential rate of actin subunit exchange related to the local concentration of specific actin binding proteins or the location of filament ends. The appearance of reaction product first at the surface of stress fibers could be due to steric hindrance of the penetration of actin subunits into stress fibers.

**Path Length and Accessible Volume in Cells**

One simple goal of both immunofluorescence techniques and fluorescent analog cytochemistry is to define the local concentration of selected molecules. However, fluorescence images from three-dimensional objects like cells are not simple to interpret. In most fluorescence microscopic studies, an image of the three-dimensional object is projected onto a two-dimensional detector such as a camera. Therefore, distribution of fluorescence intensity is a function of the plane of focus, the environmental sensitivity of the fluorophore, the accessibility of either an antibody or a fluorescent analog to a particular cellular domain, and the total thickness of the cell, in addition to local concentration of the labeled molecule. We have discussed this problem before [Wang and Taylor, 1979; Taylor et al., 1980b, 1984; Wang et al., 1982a], but it continues to be a source of serious artefacts, especially in the immunofluorescence literature. Ultimately, three-dimensional reconstruction of fluorescence in cells will become simpler to perform (see Lanni, this volume). This approach will solve some, but not all, of the issues. The following discussion presents one possible solution that is immediately applicable to the problem.

We originally included a soluble molecule labeled with a heterologous fluorescent probe in experiments using a fluorescent analog. Our early studies utilized proteins such as ovalbumin and BSA which were injected together with the fluorescent analogs labeled with a different probe. Theoretically, the control proteins should diffuse into any cellular domain limited only by the relative molecular dimensions of the cellular domain and the control protein. In contrast, the fluorescent analog of actin could become concentrated in specific domains within the cell as a result of specific molecular interactions.

These early studies were performed by visually comparing the fluorescence of a rhodamine-labeled control protein with the fluorescein-labeled actin. However, we have proposed that ratio image analysis can be used to quantify the distribution of the fluorescent analogs relative to soluble controls [Wang et al., 1982a; Taylor et al., 1984; Tanasugarn et al., 1984]. We are currently working toward routine implementation of this method.

At the same time that we were exploring the use of control proteins as relative indicators of concentration, accessible volume, and functional activ-

**TABLE 3. Dextrans and Proteins in Aqueous Solution ($D_{aq}$) and in 3T3 Cell Cytoplasm ($D_{cyto}$)**

| Molecule[4] | MW | $D_{aq} \times 10^6$ | $D_{cyto} \times 10^7$ [1] | % Mobile (cyto) |
|---|---|---|---|---|
| FITC-dextran | 20,000 | 1.01 + 0.2 | 1.65 + 0.44 | 100 |
| FITC-dextran | 39,000 | 0.7 + 0.026 | 1.21 + 0.41 | 100 |
| FITC-dextran | 65,500 | 0.524 + 0.048 | 0.87 + 0.41 | 100 |
| FITC-lactalbumin | 16,700 | 1.02 + 0.112 | 0.646 + 0.34 | 42 + 4.9[2] |
| FITC-ovalbumin | 43,000 | 0.69 + 0.265 | 0.55 + 0.23 | 78 + 4.6[2] |
| FITC-BSA | 66,000 | 0.665 + 0.028 | 0.64 + 0.535 | 77 + 15.6[2] |
| FITC-IgG | 150,000 | 0.455 + 0.0548 | 0.63 + 0.1732 | 54 + 24.4[2] |
| LRB-calmodulin | 16,700 | 1.02 + 0.126 | 0.38 + 0.14 (fast) | 81 + 5.9 |
| | | | 0.02 + 0.007 (slow) | |
| AF-actin | 42,000 | 0.715 + 0.35[3] | 0.315 + 0.46 | 54 + 8.6 |

[1]Corrected to 22°C. Mobilities calculated from data obtained at 37°C are 1.5 times higher.
[2]FRAP of these proteins yields visible persistent spots, indicating significant interaction with cellular components.
[3]Lanni and Ware [1984].
[4]BSA, bovine serum albumin; FITC, fluorescein isothiocyanate; LRB, lissamine rhodamine B.

ity, we began to test the mobility of proteins in living cells by fluorescence recovery after photobleaching (FRAP) [Wang et al., 1982b; see also Jacobson and Wojcieszyn, 1984, and Kreis et al., 1982]. It became clear that proteins had a different mobility in cytoplasm than in water. Furthermore, the cytoplasmic tracer diffusion coefficient of a single protein varied widely with different cell types.

We applied FRAP to compare the tracer diffusion coefficients and the percent immobile fractions of some selected proteins with those of dextrans [Luby-Phelps et al., 1985] (Table 3). The results demonstrated that in cytoplasm the dextran tracer diffusion coefficients scaled inversely with molecular weight over the range studied and that the dextrans exhibited 100% recovery. This suggests that dextrans do not bind with high affinity to cytoplasmic components. In contrast, all of the proteins exhibited very similar tracer diffusion coefficients [Luby-Phelps et al., 1985; see also Jacobson and Wojcieszyn, 1984], and they all showed sizable immobile fractions indicating severely restricted mobility and/or binding to cytoplasmic components. These results demonstrated that nonspecific soluble proteins were not the best control molecules for fluorescent analog cytochemistry, and that as expected, the physical chemistry of cytoplasm was complex [Jacobson and Wojcieszyn, 1984; Porter, 1984]. We have started using fluorescently labeled dextrans at selected molecular weights as the soluble control mole-

**TABLE 4. Quantitative Fluorescence Microscopy**

| Method | Possible information |
|---|---|
| Microspectrofluorometry | Ligand binding, assembly of subunits, environmental changes near fluorophore |
| Resonance energy transfer | Ligand binding, assembly of subunits |
| Anisotropy of fluorescence | Ligand binding, assembly of subunits, orientation of molecules, rotational diffusion coefficient |
| Fluorescence recovery after photobleaching | Ligand binding, assembly of subunits, translational diffusion coefficient |
| Total internal reflection fluorescence | Distance of selected molecules to substrate, chemical exchange of selected molecules near substrate contact points |
| 3-dimensional reconstruction | Distribution of selected molecules in 3 dimensions |

cules. In addition, the cytoplasmic mobility of dextrans with molecular weights varying over a wide range has been investigated to probe the structure of cytoplasm [Luby-Phelps et al., 1986].

These are just four of the practical considerations that need to be addressed before fluorescent analog cytochemistry can be used to its fullest advantage. However, we feel that the control experiments described above verify the value of the technique as a tool for defining the temporal and spatial dynamics of molecules in living cells and justify its application to research in cell biology.

## TWO RECENT APPLICATIONS OF FLUORESCENT ANALOG CYTOCHEMISTRY IN CELL BIOLOGY

The combined use of fluorescent analog cytochemistry with one or more quantitative fluorescence microscopic techniques should yield important new information about the molecular dynamics in living cells (Table 4). The use of imaging methods permits the acquisition of spatial as well as temporal information. The two examples given below demonstrate the kinds of spatial and temporal information possible with this approach.

## Reorganization of Actin During Stimulus-Response Coupling

Until recently, the reorganization of cytoskeletal proteins during cellular processes such as chemotaxis, phagocytosis, and growth-factor-induced movement could only be studied by fixing cell populations at different time points and describing the average changes in the population of cells by immunofluorescence or electron microscopy. Unfortunately, most cell populations exhibit a wide range of morphologies and a heterogeneity of response to stimuli. Fluorescent analog cytochemistry in conjunction with ultralow-light-level-imaging methods permits the analysis of the molecular dynamics of selected molecules during an entire cell process in the same living cell. Any heterogeneity of response can be detected by studying a large number of cells.

Our first investigation of the dynamics of a fluorescent analog in a single cell was a study on the reorganization of the actin in serum-starved cells in response to 10% serum. We used the Hamamatsu two-stage multichannel plate detector (MCP) to record images of cells over a period of time before and after serum stimulation. The excitation light intensity was kept very low at the specimen plane to reduce the rate of photobleaching to an undetectable level during the course of the experiment. Images were recorded by integrating anywhere between 60 and 1,000 video frames to improve the signal/noise ratio. An image of the background was also subtracted from each image.

Swiss 3T3 cells were place in low serum media (0.1% serum) for 24 hr before injecting AF-actin. The AF-actin incorporated into existing stress fibers and Figure 6 depicts an enhanced image of a cell 1.5 hr postinjection. Serum-starved cells are quiescent and do not move or change shape extensively (Fig. 6) The actin in stress fibers of serum-starved cells became disorganized within a few minutes of stimulation with 10% serum (Fig. 6d) [Herman and Pledger, 1985]. We do not now whether this disorganization involves a reorganization of filamentous actin, disassembly of actin filaments, or both. However, future application of FRAP and fluorescence anisotropy measurements should help us to define the molecular dynamics involved in this transition.

## Rate of Actin Incorporation Into Existing Stress Fibers

The ability of mammalian cells to change shape and to move requires mechanisms for assembling and disassembling actin networks and stress fibers. Since actin filaments are the basis of the structure of stress fibers, we investigated the rate of actin subunit exchange in stress fibers by image analysis and FRAP. We have also attempted to relate the kinetics of exchange to some values from studies in vitro.

Two different types of experiments were performed to measure the rate of actin subunit exchange: (1) cells were injected with monomeric AF-actin and

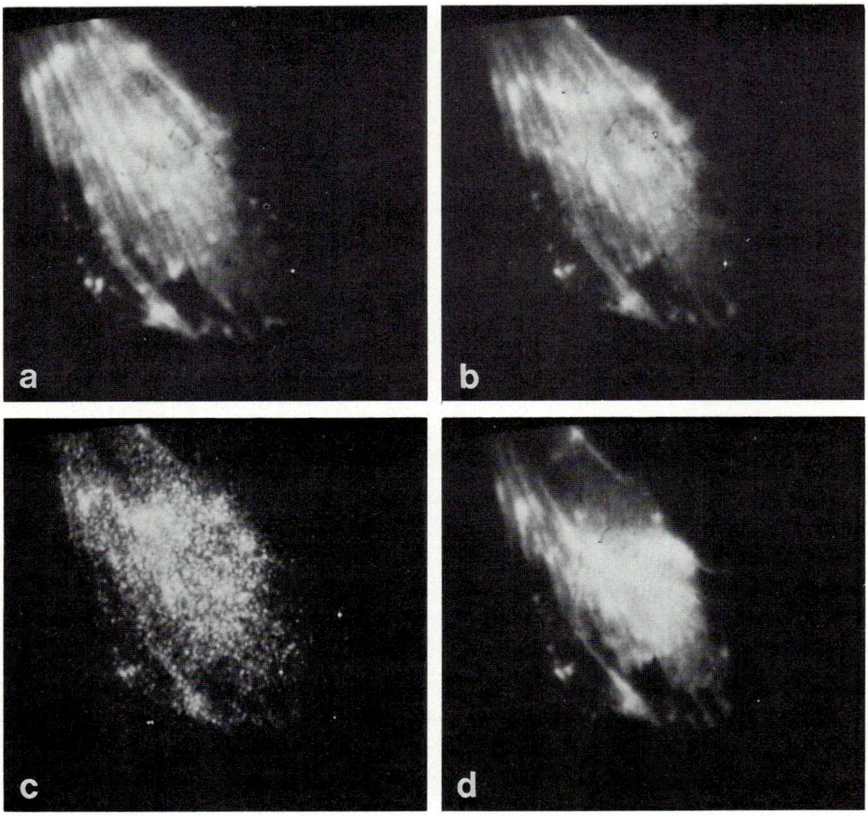


Fig. 6. Living serum-starved Swiss 3T3 cell containing AF-actin. Image processing consisted of integrating 256 frames and subtracting a background. Image obtained 1 hr 30 min postinjection. b. Image obtained 1 hr 50 min postinjection showing no change in this quiescent cell. c. A raw image to demonstrate value of image processing. d. Image of the same cell, 5 min after serum stimulation. Note loss of some stress fibers. Reprinted with permission, Tokyo University Press.

the rate of fluorescence increase (incorporation) over selected stress fibers was measured with the Hamamatsu two-stage MCP camera; and (2) AF-actin was injected into cells, allowed to equilibrate for 2 hr, and then studied by FRAP [Wang et al., 1982b; Salmon et al., 1985; McGregor et al., 1984; Kreis et al., 1982]. These two experimental approaches involve two very different types of potential artefacts. In the incorporation experiment a transient perturbation of the monomeric actin concentration may affect the exchange rate, while in the FRAP experiment the production of highly reactive species during bleaching may cause an as-yet-unknown amount of perturba-

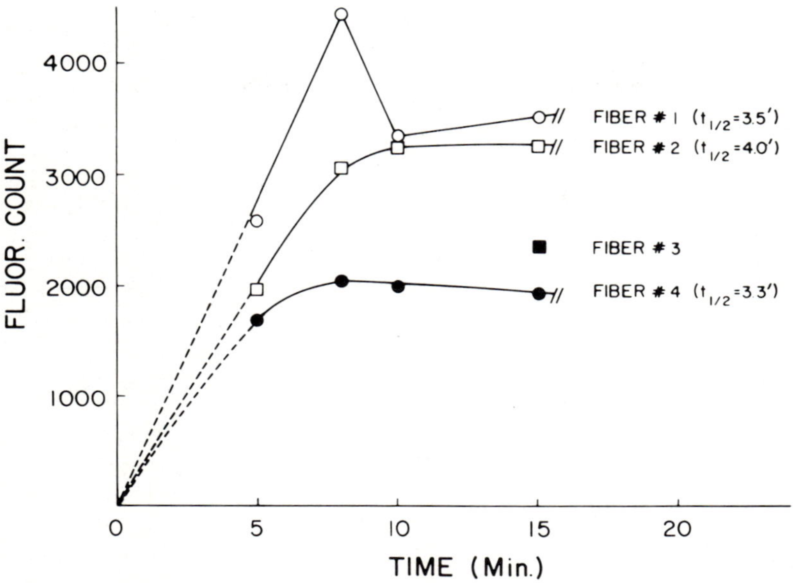


Fig. 7. The rate of increase of fluorescence associated with three separate, preexisting stress fibers following the injection of monomeric AF-actin into 3T3 cells. Fiber #3 formed spontaneously during the experiment. Reprinted with permission, J Cell Biol.

tion. Both experiments involved AF-actin which has a reduced affinity for spleen profilin [Malm, 1984]. It is not clear how this might affect the function of AF-actin in situ. However, the fact that AF-actin exhibits a highly mobile component in cytoplasm indicates that cells do regulate the extent of actin assembly [Wang et al., 1982b; Kreis et al., 1982], perhaps by a mechanism involving profilin. These considerations must be kept in mind in interpreting the results of our study.

We determined the rate of AF-actin incorporation into stress fibers by analyzing the fluorescence intensity over ca. 10-$\mu$m lengths of stress fibers after injecting the monomeric AF-actin. Images were acquired by integrating 900 frames with an excitation intensity of 35 $\mu W/cm^2$ at the specimen plane. At least four time points were obtained for each fiber. Figure 7 depicts the results from one cell. Comparable results were obtained from the other four cells analyzed. Half-times of incorporation ranged from 3.3 to 4.0 min. In addition, at least one fluorescent stress fiber (fiber #3 in Fig. 7) appeared de novo during this time period [Wang, 1985]. We detected no obvious cellular polarity of incorporation of AF-actin into existing stress fibers [Amato and Taylor, 1986].

The rate of exchange of AF-actin already incorporated into stress fibers was determined by fluorescence recovery after photobleaching (FRAP). Cells

**TABLE 5. FRAP[a] of Actin in Living Mammalian Cells**

| Region of cell | Mobility $\times 10^9$ cm$^2$/sec | (sec) $\tau$ | Spot radius ($\mu$m) | % immobile | Cell type |
|---|---|---|---|---|---|
| Fibers and surround | 4.0 (+ 1.4) (video) | 37.00 | 6.0 | 20.0 | SW3T3 |
| | 2.9[b] (+ 0.8) (photometry) | 1.88 | 1.5 | 82.0 | Embryonic chicken gizzard |
| | 48.4 (+ 24) (photometry) | 1.86 | 6.0 | 24.0 | SW3T3 |
| Fibers only | Video[b] | 300–600 | — | — | Embryonic chicken gizzard |
| | Incorporation data | 240 | — | — | SW3T3 |

[a]FRAP, fluorescence recovery after photobleaching.
[b]Kreis et al. [1982].

injected with AF-actin were incubated for 2 hr to ensure complete equilibration of the exogenous and endogenous actin subunits. Selected stress fibers were bleached with a quasi-Gaussian 6.0-$\mu$m radius spot. Recovery was monitored both with a photometer and by a full-field-illumination video method [Amato and Taylor, 1986]. The photometric method allowed the detection of rapidly recovering species, while the slower video method permitted us to observe the spatial recovery in the stress fibers relative to the immediate surrounding area.

Table 5 summarizes our results in Swiss 3T3 cells and compares them to results from chicken embryo cells [Kreis et al., 1982]. Our photometric data from Swiss 3T3 cells demonstrated a high-mobility fraction with 10× greater mobility than the highest mobility observed in chicken embryo cells [Kreis et al., 1982]. In addition, we found that 24% of actin in Swiss 3T3 cells appears to be immobile compared to 82% in chicken embryo cells. The differences measured in these two experiments could reflect differences in cell type, fluorescent analog employed, or the spot size selected for FRAP. We used a rather large spot in our experiments to optimize our ability to detect rapidly recovering species (Table 5).

Comparison of the video FRAP results on exchange of actin into fibers in chicken embryo cells [Kreis et al., 1982] with the rate of incorporation results that we obtained on Swiss 3T3 cells indicated that actin subunits can exchange into filaments very rapidly, even in stress fibers. Semiquantitative analysis of the recovery of fluorescence in the stress fibers of chicken embryo cells indicated a half-time of recovery of 5–10 min [Kreis et al., 1982]. Our results on incorporation of injected AF-actin demonstrated an average half-time of incorporation of ca. 4 min (Table 5). The physiological significance of the high rate of apparent subunit exchange and the site of subunit exchange cannot be determined from these experiments since the stress fibers possess actin filaments that are oriented in both directions along the stress fiber. As would be expected in these bipolar fibers, we found no obvious polarity of recovery in the stress fibers. However, we could compare the rate of subunit exchange to rates of exchange calculated from published critical concentrations at the pointed and barbed ends of purified actin filaments studied in vitro, making the following assumptions:

1. 360 actin subunits/$\mu$m filament.
2. Critical concentration of actin is 0.1 $\mu$M at the barbed end of actin and is 0.5 $\mu$M at the pointed end of actin in the presence of KCl/EGTA [Bonder et al., 1983]. Therefore, at steady state a flux of approx. 1 molecule/sec would occur by treadmilling. Therefore, an actin filament 1.0 $\mu$m long would exchange in 360 sec. (6 min).
3. Actin filaments are 1.0 $\mu$m long in stress fibers.

**Predicted Results Assuming Treadmilling**

Each actin filament in a stress fiber would exchange all subunits by treadmilling in 6 min. This very simple calculation makes assumptions that may not be true in living cells. However, the time for complete exchange of a 1.0-$\mu$m actin filament in vitro does fall into the time frame that we have measured in cells. It must be emphasized that other possible mechanisms of subunit exchange could also exhibit the rates measured. The most important point at this time is that stress fibers are very dynamic and exchange of subunits into stress fibers may occur at rates comparable to the fastest rate calculated in vitro.

The possible role of treadmilling in situ could be approached in living cells by photobleaching regions of the cells where unipolar actin filaments have been demonstrated, such as at the cytoplasmic surface of the membrane. Video analysis of the recovery would determine the involvement of treadmilling. The bleached spot should translate in the direction of treadmilling if it occurs (Salmon, this volume). Wang [1985] has recently performed this experiment and demonstrated reasonable preliminary evidence for actin treadmilling in vivo.

Whether treadmilling occurs or not, it is quite possible that cells regulate the stability of actin filaments by controlling the access to actin at the barbed and pointed ends [Kirschner, 1980]. A combination of studies in vitro and studies in vivo utilizing fluorescent analog cytochemistry should help to define the mechanism(s) of actin subunit exchange. It will be particularly important to investigate different regions of cells undergoing specific cellular processes such as chemotaxis and growth-factor-stimulated movement since the cell mechanism or the rate of exchange may vary in time and space.

In summary, fluorescent analog cytochemistry can be used as a quantitative tool in studies on the spatial and temporal dynamics of selected molecules in living cells. However, the control experiments are very demanding and even the best-characterized analogs such as actin have been incompletely studied at this time. The practical considerations discussed in the previous section are not inclusive, but they demonstrate the type of issues that must be addressed for the successful application of the methods. Future studies will include more detailed characterization of fluorescent analogs in vitro and in situ, as well as more quantitative studies on the dynamics of selected analogs using one or more of the quantitative methods outlined in Table 4. We hope to be able to map the spatial and temporal dynamics of such processes as actin assembly-disassembly; the binding of specific actin binding proteins; and the correlation of these molecular events with changes in presumptive secondary messengers [Tanasugarn et al., 1984; Taylor et al., 1984, 1985].

## Fluorescent and Luminescent Assays of pH and pCa

As part of our study of cytoplasmic structure and contractility, we are attempting to identify and to locate over time and cellular space the cytoplasmic signals regulating the dynamics of the actin-based cytoskeleton. We are presently investigating the cations $H^+$ and $Ca^{++}$ as potential cytoplasmic signals for growth-factor-induced changes in the cytoskeletal architecture described above. Fluorescent and luminescent indicators were selected for these studies according to the following criteria.

1. Indicator should be irreversibly trapped in the cytoplasm and should not easily enter organelle compartments.
2. Indicator should exhibit a large change in signal over the full range of physiological response.
3. Indicator should exhibit a good signal-to-noise ratio in situ.
4. Indicator should be sensitive to small changes in the messenger concentration.
5. Indicator should be specific for the particular messenger.
6. Indicator should be nontoxic and nonperturbing to cell functions.

### pH

Our laboratory has applied the fluorescein fluorescence excitation ratio method originally described by Ohkuma and Poole [1978] to measure pH in single living cells [Heiple and Taylor, 1980, 1982; McNeil et al., 1983; Tanasugarn et al., 1984]. The method depends on the response of the excitation spectrum of fluorescein to changes in pH. The fluorescence emission intensity of fluorescein at 520 nm is measured at two wavelengths of excitation and the ratio of these values is calibrated in terms of pH. We used the technique of microspectrofluorometry by digital image processing, based on ratio imaging, previously introduced as a method for measuring the temporal and spatial changes in cytoplasmic pH, as well as other spectroscopic parameters in living cells [Tanasugarn et al., 1984].

Our digital image ratio microfluorometer has undergone several modifications over the last few years. Most significantly, we have replaced the original silicon intensified target (SIT) camera with the Hamamatsu 2-stage multichannel plate (MCP) camera. The system has been defined in terms of linearity of sensitivity, spatial linearity, signal-to-noise ratio, contrast transfer function, and instrument stability over time [Tanasugarn et al., 1984; Tanasugarn, 1985; Taylor et al., 1985].

We used fluorescein-isothiocyanate-labeled dextran (FITC-dextran) (MW 70,000) as the probe of cytoplasmic pH in Swiss 3T3 cells. This probe was trapped in the cytoplasm by microinjection or by the scrape-loading method [Tanasugarn et al., 1984]. The excitation ratio 489/452 of FITC-dextran exhibits almost a fivefold change over the range from pH 4.5 to pH 7.4. The signal/noise ratio was optimized with the MCP camera by integrating 128 frames, while maintaining the dynamic range of the A/D converter. Excitation wavelengths of 452 and 489 nm with broad-band emission at 520 nm resulted in a good signal/noise relative to the cellular autofluorescence. The FITC-labeled dextran in the ratio method is very sensitive and linear from ca. pH 4.5 to ca. pH 7.3. However, the pKa of fluorescein (6.45) causes the excitation ratio to begin flattening at ca. pH 7.3. Therefore, the probe loses sensitivity above pH 7.3 [Tanasugarn et al., 1984]. Recently, Cassal et al., [1983], reported the use of 4′,5′-dimethylfluorescein coupled to dextran which extended the sensitivity of the ratio method beyond pH 7.3. The change in fluorescence of FITC-labeled dextran is a reasonably specific indicator for changes in cytoplasmic pH. The fluorescein molecules are covalently attached to dextran so that they cannot enter membrane compartments, and the possibility of the formation of dye aggregates is minimal. The ratio signal is relatively insensitive to ionic strength, divalent cation concentration, and probe concentration over a wide range [Heiple and Taylor, 1982]. The FITC-labeled dextran used to measure pH had a fluorescein to glucose residue ratio of 0.004. The intracellular concentration of the probe was ca. 0.3 mg/ml dextran. Therefore, each cell contained ca. $4 \times 10^7$ fluorescein molecules. No cellular perturbations have been observed under these conditions.

The average cytoplasmic pH of interphase Swiss 3T3 cells in complete serum was 6.83 (+ 0.38). No spatial variation of cytoplasmic pH was detected within individual interphase cells, but a large variability of cyto-

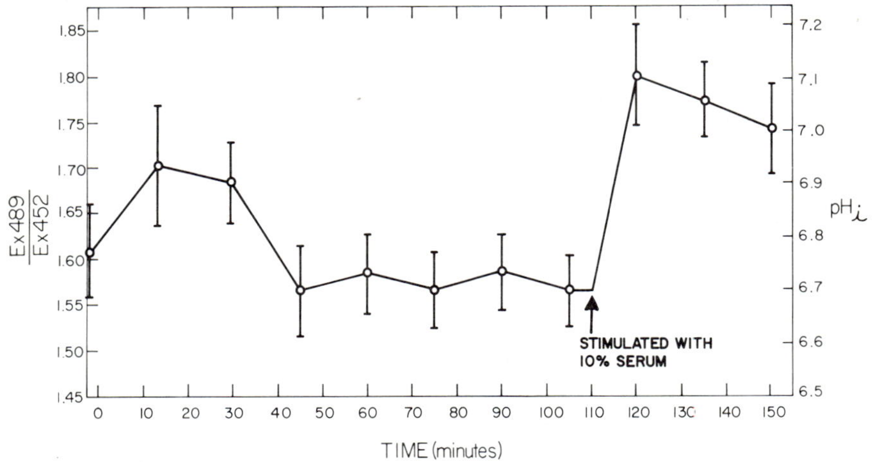


Fig. 8. The average cytoplasmic pH in a single serum-starved 3T3 cell determined by the fluorescein fluorescence ratio imaging method. Alkalinization occurs upon stimulation with 10% serum.

plasmic pH was observed between cells [Tanasugarn et al., 1984]. We have also studied the alkalinization of cytoplasm in serum-starved cells in response to serum stimulation in media not containing bicarbonate. Serum-starved, quiescent Swiss 3T3 cells were stimulated by increasing the serum concentration from 0.1% to 10%. In many cells, the average cytoplasmic pH increased by 0.2 to 0.4 pH units within 2 min of stimulation (Fig. 8). The cytoplasmic pH in the serum-starved state showed no subcellular spatial variation, but the pH ranged from ca. pH 6.6 to pH 7.1 between cells [Tanasugarn, 1985; Taylor et al., 1985]. We are currently refining the ratio imaging method to improve our ability to detect spatial, as well as temporal, changes in cytoplasmic pH in response to serum and growth factor stimulation. We are also investigating the effects of the extracellular ionic content on cytoplasmic pH.

There are now many reports that indicate that cytoplasmic pH increases in response to serum, growth factors, chemotactic factors, mitogens, and hormones [Busa and Nuccitelli, 1984; Cassal et al., 1983]. Many investigations have relied on the use of fluorescein diacetate, which can diffuse across membranes in the unhydrolyzed state. Unfortunately, this probe can penetrate into compartments other than the cytoplasm before it is hydrolyzed and it can also slowly diffuse out of cells. It is more accurate to refer to cellular pH measurements, rather than cytoplasmic pH measurements, when discussing most of these studies. The most significant result of our studies so far is that individual cells in an unstimulated population exhibit a wide range of cytoplasmic pH. Most of the published studies have investigated cell populations which yield an average pH of the population. Heterogeneity in cell populations and changes in individual cells require single-cell methods like those described here.

### pCa

Our laboratory uses the bioluminescent protein aequorin as an intracellular calcium indicator [Taylor et al., 1980a; McNeil et al., 1985, 1986; McNeil and Taylor, 1985]. In our opinion, aequorin is preferable to the present fluorescent indicators of calcium, such as Quin-2 [Tsien et al., 1982], because of its greater sensitivity to changes in $[Ca^{++}]$ between $10^{-7}$ and $10^{-4}$M, because aequorin is far less likely to buffer $[Ca^{++}]$ than Quin-2, and because intense excitation with UV light is not required for aequorin measurements. However, a recent report indicates that a new fluorescent calcium probe, Fura2, may not be subject to some of the limitations of other fluorescent indicators [Poene et al., 1985]. If so, ratio imaging methods could be applied to measure the intracellular free calcium ion concentration with this improved probe [Tanasugarn et al., 1984; Kruskal et al., 1984].

Recent work has focused on measuring transients of free calcium during growth factor stimulation of serum-starved Swiss 3T3 cells [McNeil et al., 1985] and during phagocytosis in murine peritoneal macrophages [McNeil et al., 1986]. Most of our work has been on small cell populations attached to glass substrates (ca. $10^3$–$10^4$ cells), although we have initiated aequorin imaging studies on single cells. The characteristics of aequorin as a calcium indicator [Blinks et al., 1978] and the description of our instrumentation [McNeil and Taylor, 1985] have been described in detail elsewhere, but we will summarize the important points here. Aequorin is a bioluminescent protein with a molecular weight of ca. 20,000 daltons. It consists of an apoprotein and an imidazolopyrazinone chromophore, and emits blue light with an emission maximum at 469 nm when calcium is bound. The light intensity is a product of the aequorin concentration and the free calcium ion concentration [Blinks et al., 1978].

Aequorin can be trapped in the cytoplasm by a variety of injection methods [McNeil and Taylor, 1985]. The dynamic range of sensitivity is from ca. $3.0 \times 10^{-8}$ to ca. $10^{-4}$ M free calcium ion concentration. The signal-to-noise ratio is excellent, since the method is based on bioluminescence, where no excitation light is required. The major source of noise is a small calcium-independent luminescence that limits the lower limit of detection to ca. $3.0 \times 10^{-8}$ M [Blinks et al., 1978]. The rate of luminescence increases ca. $10^4$-fold over the range of free calcium from ca. $3.0 \times 10^{-8}$ M to $10^{-4}$ M. Aequorin luminescence is very specific for calcium. Neither $Mg^{++}$ nor $Mn^{++}$ stimulates luminescence. However, some other minor divalent cations will stimulate luminescence, but their levels are not believed to be modulated

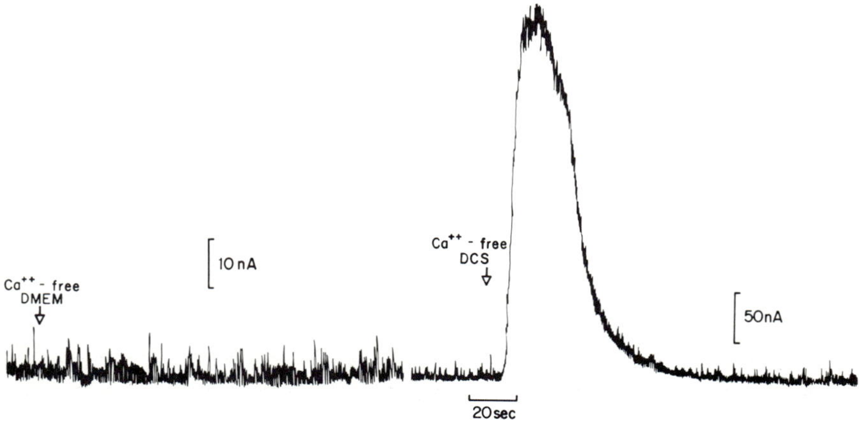


Fig. 9. Aequorin luminescence of serum-starved 3T3 cells (ca. $10^4$ cells) stimulated with 10% serum in the absence of exogenous [$Ca^{++}$]. Cells were first rinsed with 6.0 ml of [$Ca^{++}$]-free DME containing 0.5 mM EGTA (first arrow), and a very slight drop in aequorin luminescence resulted. [$Ca^{++}$]-free 10% dialyzed calf serum DCS (containing 0.5 mM EGTA), presented at the second arrow, resulted in a near normal rise in luminescence.

in cells [Blinks et al., 1978]. Aequorin does not buffer calcium in cells significantly when the intracellular concentration is kept below ca. 0.1 $\mu$M [Blinks et al., 1978]. We have estimated the intracellular concentration of aequorin in scrape-loaded cells to be below 0.5 $\mu$M. Aequorin has also been shown to be nontoxic in all of the cells investigated [Blinks et al., 1978].

The average free calcium ion concentration in a population of ca. $10^4$ serum-starved, subconfluent Swiss 3T3 cells was 89 ( + 20) nM, which was almost twofold higher than that of a population of growing cells. The addition of 10% serum caused a rise in the [$Ca^{++}$] to a maximal value of 5.3 $\mu$M at 20 sec poststimulation, followed by a decrease in the [$Ca^{++}$] to prestimulatory levels within 3 min, where it remained for up to 12 hr [McNeil et al., 1985] (Fig. 9). Platelet-derived growth factor (PDGF) produced a peak elevation in [$Ca^{++}$] to ca. 1.4 $\mu$M at 115 sec poststimulation. The major source of calcium appeared to be intracellular since 70% of the rise was elicited in the absence of extracellular calcium. Phorbol myristate acetate (PMA) inhibited the growth-factor-induced transients of [$Ca^{++}$] when cells were pretreated with this activator of protein kinase C, but such PMA treatment of cells did not inhibit mitogenesis in response to growth factors. These latter results suggested that calcium transients themselves are not required for mitogenesis, and that protein kinase C may be involved in regulating $Ca^{++}$ mobilization in fibroblasts [McNeil et al., 1985].

There is increasing evidence for the role of both cytoplasmic alkalinization and increases in the [$Ca^{++}$] during a variety of stimulus-response events in cells [Rasmussen et al., 1984]. Our initial studies with indicators of pH and pCa are encouraging and we are now making measurements on single cells to determine if any spatial changes in these parameters occur during stimulation. It is interesting that we measure increases in [$Ca^{++}$] within seconds of serum stimulation, while the pH elevation proceeds over minutes. Furthermore, the actin reorganization seems to lag behind both the change in calcium and the change in pH. It is possible that pH- and pCa-sensitive regulatory and structural proteins are involved in the early phase of growth factor stimulation of cells including the stimulation of movement [Brockus and Stiles, 1984; Schlessinger and Geiger, 1981].

In contrast to growth factor stimulation of cells, phagocytic stimulation of macrophages with IgG-coated particles results in ligation of cellular receptors only in localized segments of plasma membrane and, correspondingly, it causes highly localized changes in the actin-based cytoskeleton rather than a generalized alteration in cytoskeletal structure and/or rate of cell motility as for growth factors. These localized changes in the cytoskeleton during phagocytosis are proposed to power and to shape particle engulfment by pseudopods. It is interesting therefore that we have failed, using aequorin, to measure any change in average [$Ca^{++}$] in phagocytosing macrophages. This result is consistent with the hypothesis that cytoplasmic signals regulating phagocytosis are localized only to that segment of cytoplasm involved in phagocytic engulfment. In other words, if [$Ca^{++}$] is a signal for phagocytosis, then large changes in average cytoplasmic free [$Ca^{++}$] might not occur in the phagocytosing cell. We estimate that localized changes in [$Ca^{++}$], occupying 1.4% of the macrophage cytoplasmic volume and not exceeding 0.5 $\mu$M rise in [$Ca^{++}$], would not have been detected in phagocytosing macrophages by our methods [McNeil et al., 1986]. We emphasize therefore that quantitative microscopic fluorescence and luminescence microscopic methods are essential for mapping the spatial, as well as temporal, dynamics of ion fluxes and other cellular parameters.

## Prospectus

The potential of using multiple parameter analyses in living cells to correlate changes in several metabolic and molecular parameters is enormous. By selecting or producing fluorescent probes with excitation and emission maxima ranging from ca. 400 to 900 nm (see Waggoner, this volume), it should be possible to monitor at least five or six parameters at a time. For example, measurements of membrane potential, pCa, pH, actin assembly, and myosin binding could conceivably be correlated in the same cells. Therefore, temporal and spatial maps of molecular activity could be defined in living cells during a wide variety of cellular functions. The use of several quantitative fluorescence spectroscopic methods applied to the microscope (see specific chapters in this volume) should optimize the acquisition of molecular information. Finally, it is now possible to use the fluorescence microscope to make either high-resolution measurements at high magnification, or to make statistically significant measurements on large numbers of cells at very low magnification (Ploem, this volume). The next few years should witness tremendous advances in cell and molecular biology with the aid of these techniques.

## ACKNOWLEDGMENTS

The authors are pleased to acknowledge the suggestions of many colleagues over the years. Dr. Ed Bonder kindly reviewed this paper and made many helpful suggestions.

We would like to thank the present and past members of our laboratory for their contributions to this project. The work reported here has been supported by grants from the NIH (AM-18111), the National Science Foundation (DCB-8416855), and the Council for Tobacco Research, USA, Inc. #1412.

## REFERENCES

Allen D, Blinks J (1979): The interpretation of light signals from aequorin-injected skeletal and cardiac muscle: A new method of calibration. In Ashley CC, Campbell AK, (eds): Detection and Measurement of Free $Ca^{++}$ in Cells. Amsterdam: Elsevier-North Holland, pp 159–174.

Amato P, Taylor DL (1986): Probing the mechanism of incorporation of fluorescently labeled actin into stress fibers. J Cell Biol (in press).

Amato P., Unanue E, Taylor D (1983): Distribution of actin in spreading macrophages: A comparative study on living and fixed cells. J Cell Biol 96:750–761.

Aubin J (1979): Autofluorescence of viable cultured mammalian cells. J Histochem Cytochem 27:36–43.

Blanks J, Mattingly P, Jewell B, Van Leeuwen U, Harper G, Allen D (1978): Practical aspects of the use of aequorin as a calcium indicator: Assay, preparation, microinjection, and interpretation of signals. Methods Enzymol 57:292–328.

Bonder EM, Fishkind DJ, Mooseker MS (1983): Direct measurement of critical concentrations and assembly rate constants at two ends of an actin filament. Cell 34:491–501.

Brockus B, Stiles C (1984): Regulation of cytoskeletal architecture by platelet-derived growth factor, insulin and epidermal growth factor. Exp Cell Res 153:186–197.

Buckley IK, Porter KR (1967): Cytoplasmic fibrils in living cultured cells: A light and electron microscope study. Protoplasma 64:349–380.

Busa W, Nuccitelli R (1984): Metabolic regulation via intracellular pH. Am J Physiol 246:R409–R438.

Cassel D, Rothenberg P, Zhuang Y-X, Deuel T, Glaser L (1983) Platelet-derived growth factor stimulates $Na^{+}/H^{+}$ exchange and induces cytoplasmic alkalinization in NRG cells. PNAS 80:6224–6228.

Celis J (1984): Microinjection of somatic cells with micropipettes: Comparison with other transfer techniques. Biochem J 223:281–291.

Chambers R, Chambers EL (1961): "Explorations into the Nature of the Living Cell." Cambridge: Harvard University Press.

Craig SW, Pollard TD (1982): Actin-binding proteins. Trends Biochem Sci 7:88–92.

Dandliker WB, Feigen GA (1961): Detection of the antigen- antibody reaction by fluorescence polarization. Fed Proc 20:11.

Diacumakos EG (1973): Methods for micromanipulation of human somatic cells in culture. Methods Cell Biol 7:287–311.

Feramisco J (1979): Microinjection of fluorescently labeled $\alpha$-actinin into living fibroblasts. PNAS 76:3967–3971.

Fujiwara K, Pollard TD (1976): Fluorescent antibody localization of myosin in the cytoplasm, cleavage furrow, and mitotic spindle of human cells. J Cell Biol 71:848–875.

Geiger B (1981): Involvement of vinculin in contact-induced cytoskeletal interactions. Cold Spring Harbor Symp. Exp Biol XLVI:671–682.

Graessman A, Graessman M, Mueller C (1980): Microinjection of early SV40 DNA fragments and T-antigen. Methods Enzymol 65:816–825.

Goldman RD, Chojnacki B, Yerna MJ (1979): Ultrastructure of microfilament bundles in baby hamster kidney (BHK-21) cells. J Cell Biol 80:759–766.

Hamaguchi Y, Iwasa F (1980): Localization of fluorescently labeled calmodulin in living sea urchin eggs during early development. Biomed Res 1:502–509.

Heiple J, Taylor DL (1980): Intracellular pH in single motile cells. J Cell Biol 86:885–890.

Heiple J, Taylor DL (1982): An optical technique for measurement of intracellular pH in single living cells. In Nuccitelli R, Beamer DW (eds): "Intracellular pH: Its Measurements, Regulation, and Utilization in Cellular Functions." New York: Alan R. Liss, pp 22–54.

Herman B, Pledger W (1985): Platelet-derived growth factor-induced alterations in vinculin and actin distribution in BALB/C-3T3 cells. J Cell Biol 100:1031–1040.

Isenberg G, Rathke PC, Halsmann N, Franke WW, Wohlfarth-Bottermann KE (1976): Cytoplasmic actomyosin fibrils in tissue culture cells: Direct proof of contraction by laser microbeam dissection. Cell Tissue Res 166:427–443.

Jacobson K, Wojcieszyn J (1984): The translational mobility of substances within the cytoplasmic matrix. PNAS 81:6747–6751.

Kiehardt DP (1982): Microinjection of echinoderm eggs: Apparatus and procedures. Methods Cell Biol 25 (Pt. B):13–31.

Keith C, Feramisco J, Shelanski M (1981): Direct visualization of fluorescein-labeled microtubules in vitro and in microinjected fibroblasts. J Cell Biol 88:234–240.

Kirschner M (1980): Implications of treadmilling for the stability and polarity of actin and tubublin polymers *in vivo*. J Cell Biol 86:330–334.

Korn ED (1982): Actin polymerization and its regulation by proteins from nonmuscle cells. Physiol Rev 62:672–736.

Kreis TE, Birchmeier W (1980): Stress fiber sarcomeres of fibroblasts are contractile. Cell 22:555–561.

Kreis T, Birchmeier W (1982): Microinjection of fluorescently labeled proteins into living cells with emphasis on cytoskeletal proteins. Int Rev Cytol 75:209–228.

Kreis T, Winterhalter K, Birchmeier W (1979): In vivo distribution and turnover of fluorescently labeled actin microinjected into human fibroblasts. PNAS 76:3814–3818.

Kreis T, Geiger B, Schlessinger J (1982): Mobility of microinjected rhodamine actin within living chicken gizzard cells determined by fluorescence photobleaching recover. Cell 29:835–845.

Kruskal B, Keith C, Maxfield F (1984): Thyrotropin-releasing hormone-induced changes in intracellular [$Ca^{++}$ measured by microspectrofluorometry on individual quin2-loaded cells.

J Cell Biol 99:1167–1172.
Langanger G, DeMey J, Moeremans M, Daneels G, DeBrabander M, Small JV (1984): Ultrastructural localization of $\alpha$-actinin and filamin in cultured cells with the immunogold staining (IGS) method. J Cell Biol 99:1324–1334.
Lanni F, and Ware B, (1984): Detection and characterization of actin monomers, oligomers, and filaments in solution by measurement of fluorescence photobleaching recovery. Biophys J 46:97–110.
Lazarides E (1975): Tropmyosin antibody: The specific localization of tropomyosin in non-muscle cells. J Cell Biol 65:549–561.
Lazarides E, Weber K (1974): Actin antibody: The specific visualization of actin filaments in non-muscle cells. Proc Natl Acad Sci USA 71:2268–2272.
Lazarides E, Burridge K (1975): Alpha-actinin: Immunofluorescent localization of a muscle structural protein in non-muscle cells. Cell 6:289–298.
Leslie R, Saxton W, Mitchison TJ, Neighbors B, Salmon E, McIntosh JR (1984): Assembly properties of fluorescein-labeled tubulin in vitro before and after fluorescence bleaching. J Cell Biol 99:2146–2156.
Levison SA, Hicks AN, Portmann AJ, Dandliker WB (1975): Fluorescence polarization and intensity kinetic studies of antifluorescein antibody obtained at different stages of the immune response. Biochemistry 14:3778–3786.
Lopatin DE, Voss EW, Jr (1971): Fluorescein hapten and antibody active-site probe. Biochemistry 10:208–213.
Luby-Phelps K, Amato P, Taylor DL (1984): Selective immunocytochemical detection of fluorescent analogs with antibodies specific for the fluorophore. Cell Motil 4:137–149.
Luby-Phelps K, Lanni F, Taylor DL (1985): The behavior of a fluorescent analog of calmodulin in living Swiss 3T3 cells. J Cell Biol 101:1245–1256.
Luby-Phelps K, Taylor DL, Lanni F (1986): Probing the structure of cytoplasm in living cells. J Cell Biol, in press.
Luna E, Wang Y-L, Voss EW, Branton D, Taylor DL (1982): A stable, high capacity, F-actin affinity column. J Biol Chem 257:13095–13100.
Malm B (1984): Chemical modification of Cys-374 of actin interferes with the formation of the profilactin complex. FEBS 173:399–402.
Mastro A, Keith AD (1984): Diffusion in the aqueous compartment. J Cell Biol 99:1805–1875.
McGregor G, Kapitza H, Jacobson K (1984): Laser-based fluorescence microscopy of living cells. Laser Focus/Electro-Optics November: 84–93.
McKenna N, Meigs JB, Wang Y-L (1985): Identical distribution of fluorescently labeled brain and muscle actins in living cardiac fibroblasts and myocytes. J Cell Biol 100:292–296.
McNeil P, Taylor DL (1985): Aequorin entrapment in mammalian cells. Cell Calcium 6:83–93.
McNeil P, Tanasugarn L, Meigs J, Taylor DL (1983): Acidification of phagosomes is initiated before lysosomal enzyme activity is detected. J Cell Biol 97:692–702.
McNeil P, Murphy RF, Lanni F, Taylor DL (1984): A method for incorporating macromolecules into adherent cells. J Cell Biol 98:1556–1564.
McNeil P, McKenna M, Taylor DL (1985): A transient rise in cytosolic calcium follows stimulation of quiescent cells with growth factors and is inhibitable with phorbal myristate acetate. J Cell Biol 101:372–379.
McNeil P, Swanson J, Wright S, Silverstein S, Taylor DL (1986): Fc-receptor mediated phagocytosis occurs in macrophages without an increase in the average intracellular free calcium ion concentration. J Cell Biol, in press.
Ohkuma S, Poole B (1978): Fluorescence probe measurement of the intralysomal pH in living cells and the perturbation of pH by various agents. PNAS 75:3327–3331.
Pagano RE, Longmuir KJ (1983): Intracellular translocation and metabolism of fluorescent

lipid analogues in cultured mammalian cells. Trends Biochem Sci 8:157–161.
Pardee JD, Simpson PA, Stryer L, Spudich JA (1982): Actin filaments undergo limited subunit exchange in physiological salt conditions. J Cell Biol 94:316–324.
Poene M, Aldarton J, Tsien R, Steinhardt RA (1985): Changes of free calcium levels with stages of the cell division cycle. Nature 315:147–149.
Pollard TD (1984): Polymerization of ADP-actin. J Cell Biol 99:769–777.
Pollard TD, Craig SW (1982): Mechanism of actin polymerization. Trends Biochem Sci 7:55–58.
Porter KR (1984): The cytomatrix: A short history of its study. J Cell Biol 99(1, pt. 2):3s–12s.
Rasmussen H, Kojima I, Kojima K, Zawalich W, Apfeldorf W (1984): Calcium as intracellular messenger: Sensitivity modulation, C-kinase pathway, and sustained cellular response. Adv. Cyclic Nucleotide Res Phosph Res 18:159–193.
Salmon ED, Saxton WM, Leslie RJ, Karow ML, McIntosh JR (1985): Diffusion coefficient of fluorescein labeled tubulin in the cytoplasm of embryonic cells of a sea urchin: Video image analysis of fluorescence redistribution after photobleaching. J Cell Biol 99:2157–2164.
Sanger JW, Sanger JM (1980): Banding and polarity of actin filaments in interphase and cleaving cells. J Cell Biol 86:568–575.
Sanger JW, Sanger JM, Jockusch BM (1983): Differences in the stress fibers between fibroblasts and epithelial cells. J Cell Biol 96:961–969.
Sanger JW, Mittal B, Sanger JM (1984): Interaction of fluorescently-labeled contractile proteins with the cytoskeleton in cell models. J Cell Biol 99:918–928.
Schlessinger J, Geiger B (1981): Epidermal growth factor induces redistribution of actin and actinin in human epidermal carcinoma cells. Exp Cell Res 134:273–279.
Schliwa M (1981): Proteins associated with cytoplasmic actin. Cell 25:587–590.
Sklar L, Oades Z, Jesaitis A, Painter R, Cochrane C (1981): Fluoresceinated chemotactic peptide and high affinity anti-fluorescein antibody as a probe of the temporal characteristics of neutrophil stimulation. PNAS 78:7540–7544.
Soltys B, Borisy G (1985): Polymerization of tubulin in vivo: Direct evidence for assembly onto microtubule ends and from centrosomes. J Cell Biol 100:1682–1689.
Stacey DW, Allfrey VG (1976): Microinjection studies of duck globin messenger RNA translation in human and avian cells. Cell 9:725–732.
Tanasugarn L (1985): Cytoplasmic and Endosomal pH Measurements by Digital Ratio Microfluorometry. Ph.D. Thesis, Harvard University.
Tanasugarn L, McNeil P, Reynolds G, Taylor D (1984): Microspectrofluorometry by digital image processing: Measurement of cytoplasmic pH. J Cell Biol 98:717–724.
Taylor DL, Wang Y-L (1978): Molecular cytochemistry incorporation of fluorescently labeled actin into living cells. PNAS 75:857–861.
Taylor D, Condeelis J (1979): Cytoplasmic structure and contractility in amoeboid cells. Int Rev Cytol 56:57–144.
Taylor DL, Wang Y-L (1980): Fluorescently labeled molecules as probes of the structure and function of living cells. Nature 284:405–410.
Taylor D, Fechheimer M (1982): Cytoplasmic structure and contractility. The solation-contraction coupling hypothesis. Philos Trans R Soc Lond 299:183–197.
Taylor DL, Blinks J, Reynolds GT (1980a): Contractile basis of ameboid movement. VIII. Aequorin luminescence during ameboid movement, endocytosis, and capping. J Cell Biol 86:599–607.
Taylor DL, Wang Y-L, Heiple J (1980b): The contractile basis of amoeboid movement. VII. The distribution of fluorescently labeled actin in living amoebas. J Cell Biol 86:590–598.
Taylor DL, Heiple J, Wang Y-L, Luna EJ, Tanasugarn L, Brier J, Swanson J, Fechheimer M, Amato P, Rockwell M, Daley G (1981a): Cellular and molecular aspects of amoeboid movement. Cold Spring Harbor Symp Quant Biol XLVI:101–111.

Taylor DL, Reidler Spudich JA, Stryer L (1981b): Detection of actin assembly by fluorescence energy transfer. J Cell Biol 89:362–367.
Taylor DL, Amato PA, Luby-Phelps K, McNeil P (1984): Fluorescent analog cytochemistry. Trends Biochem Sci 9:88–91.
Taylor DL, Amato P, Lanni F, Luby-Phelps K, McNeil P, Tanasugarn L (1985): The dynamics of specific molecules and ions in living cells. In: "Yamada Conference on Cell Motility." Tokyo Press, in press.
Tobacman L, Brenner S, Korn ED (1983): J Biol Chem 258:8806–8812.
Tsien R, Pozzan T, Rink T (1982): Calcium homeostasis in intact lymphocytes; Cytoplasmic free calcium monitored with a new, intracellularly trapped fluorescent indicator. J Cell Biol 94:325–334.
Wadsworth P, Sloboda R (1983): Microinjection of fluorescent tubulin into dividing sea urchin cells. J Cell Biol 97:1249–1254.
Wang Y-L (1985) Exchange of actin subunits at the leading edge of living fibroblasts: Possible role of treadmilling. J Cell Biol 101:597–602.
Wang Y-L, Taylor DL (1979): Distribution of fluorescently labeled actin in living sea urchin eggs during early development. J Cell Biol 82:672–679.
Wang Y-L, Taylor DL (1980): Preparation and characterization of a new molecular cytochemical probe: 5-IAF-labeled actin. J Histochem Cytochem 28:1198–1206.
Wang Y-L, Taylor DL (1982): Probing the dynamic equilibrium of actin polymerization by fluorescence energy transfer. Cell 27:429–436.
Wang Y-L, Heiple J, Taylor DL (1982a): Fluorescent analog cytochemistry of contractile proteins. Methods Cell Biol 25 (Pt. B):1–11.
Wang Y-L, Lanni F, McNeil P, Ware B, Taylor D (1982b): Mobility of cytoplasmic and membrane-associated actin in living cells. PNAS 79:4660–4664.
Wang K, Feramisco J, Ash J (1982c): Fluorescent localization of contractile proteins in tissue culture cells. Methods Enzymol 85:514–562.
Wang Y-L, Bonder EM, Mooseker MS, Taylor DL, (1983): Effects of villin on the polymerization and subunit exchange of actin. Cell Motil 3:151–165.
Weber K, Groeschel-Stewart U (1974): Antibody to myosin: The specific visualization of myosin containing filaments in non-muscle cells. PNAS 71:4561–4564.
Wehland J, Weber K (1980): Distribution of fluorescently labeled actin and tropomyosin after microinjection in living tissue culture cells as observed with TV image intensification. Exp Cell Res 127:397–408.
Wojcieszyn J, Schlegel R, Wu E-S, Jacobson K (1981): Diffusion of injected macromolecules within the cytoplasm of living cells. PNAS 78:4407–4410.
Zavortink M, Welsh M, McIntosh J (1983): The distribution of calmodulin in living mitotic cells. Exp Cell Res 149:375–385.

*Applications of Fluorescence in the Biomedical Sciences, pages 377–403*


# 17 Fluorescence Studies of Tubulin and Microtubule Dynamics in Living Cells

*E.D. Salmon and Pat Wadsworth*

Fluorescent analog cytochemistry (FAC) [ Taylor et al., 1984] and fluorescence redistribution after photobleaching (FRAP) [Jacobson et al., 1983] are providing new insights about the polymerization and behavior of spindle and cytoplasmic microtubules within living cells [Keith et al., 1981; Leslie et al., 1984; McIntosh et al., 1985; Salmon, 1984; Salmon et al., 1984b,c; Saxton et al., 1984; Scherson et al., 1984; Wadsworth and Salmon, 1985a,b, 1986; Wadsworth and Sloboda, 1983, 1984]. FAC enables us to "tag" tubulin dimers, the subunits from which microtubules are polymerized, making them visible with fluorescence light microscopy under conditions that are not harmful to the cells, and FRAP techniques allow us to observe the mobility of tubulin in the cell and within cell structures that are composed of microtubules.

Microtubules are dynamic components of the cell cytoskeleton which participate in many processes including ciliary motility, vesicle transport, cell morphogenesis, and mitosis [Dustin, 1984]. As a cell progresses through the cell cycle, the spatial organization and the extent of assembly of microtubules change [Vandre et al., 1984] (Fig. 1). During interphase most microtubules originate at the centrosome region near the nucleus, and they extend toward the cell periphery in a stellate, astral array, the "cytoplasmic microtubule complex" (CMTC). The centrosome is a "microtubule organizing center" (MTOC) in the cell and an efficient nucleation site for microtubule polymerization in vitro. When the cell enters mitosis, the CMTC disassembles, and during mitosis all microtubule assembly is associated with the

Department of Biology, University of North Carolina at Chapel Hill, Chapel Hill, North Carolina 27514.

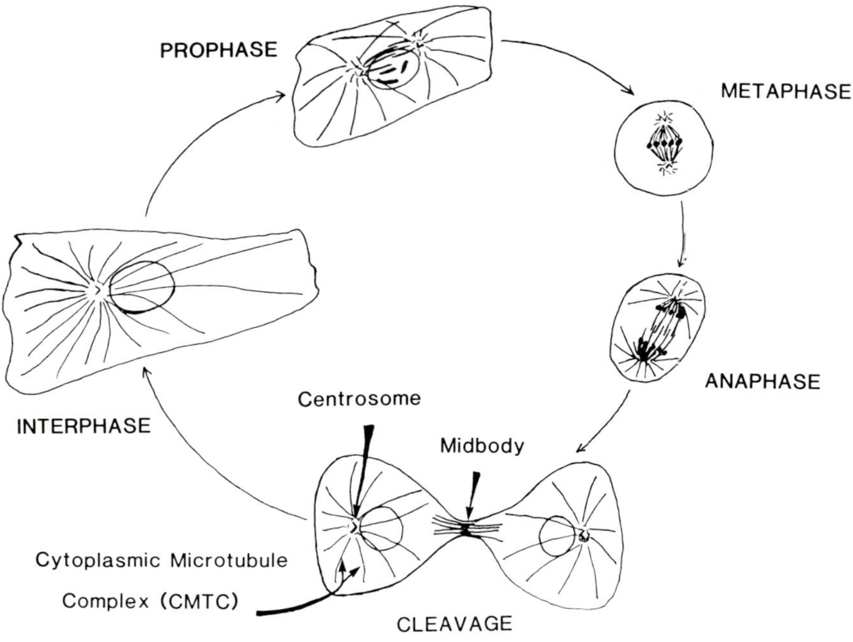


Fig. 1. Rearrangements of microtubule arrays during the cell cycle. The sketch shows major features of the typical distribution of microtubules inferred from immunofluorescent microtubule patterns and electron micrographs. It does not provide a comprehensive picture of the distribution of all microtubules in the cell.

central spindle and asters of the mitotic apparatus. The spindle is formed from a bipolar, overlapping array of microtubules (Fig. 2A). Microtubules attached to a kinetochore are termed "kinetochore microtubules," as opposed to the other "nonkinetochore microtubules" of the half-spindle and asters. During anaphase, chromosomes move poleward as the kinetochore microtubules shorten. Concurrently, the distance between the poles increases, and microtubules grow outward from the centrosomes toward the cell surface to form a new CMTC. A parallel bundle of overlapping microtubules, "the midbody," extends between the dividing daughter cells.

The basic subunit of microtubules is the tubulin dimer [Dustin, 1984]. Each 100-kD dimer is composed of one molecule of $\alpha$ tubulin and one of $\beta$ tubulin, both polypeptides having nearly equal molecular weight. Tubulin dimers arrange themselves into 13 protofilaments that form the cylindrical wall of a microtubule. All microtubules have an intrinsic structural polarity which is specified by the tubulin lattice [McIntosh and Euteneuer, 1984; Telzer and Haimo, 1981]. This polarity is reflected in different polymeriza-

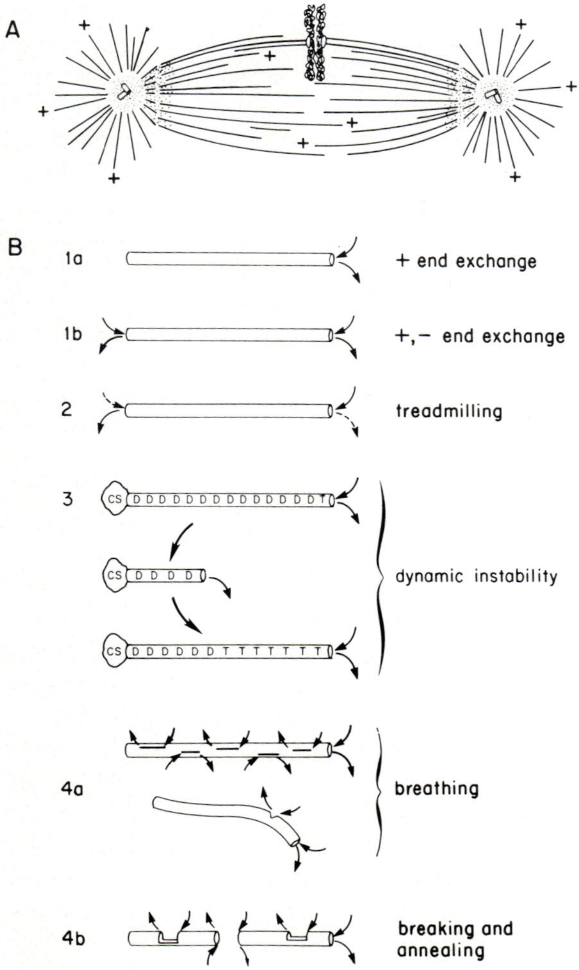


Fig. 2. A. Sketch of the current concept of microtubule arrangements in a metaphase first-division sea urchin spindle. Such a cartoon is based on detailed structural studies but must still be considered preliminary and merely indicative of the true state of affairs. The structure of the spindle poles, in particular, is poorly understood. Only one chromosome pair is drawn for clarity. Kinetochore-microtubules are defined as microtubules attached to kinetochores. B. Four possible pathways of tubulin exchange with spindle microtubules: (1) simple equilibrium subunit exchange by reversible association-dissociation events at one (a) or both (b) ends of a microtubule; (2) treadmilling or head-to-tail polymerization as a consequence of GTP hydrolysis; (3) dynamic instability as a consequence of the loss of tubulin-GTP "caps" at the plus ends; (4) dynamic equilibrium subunit exchange all along the apparent length of a microtubule either by insertion or deletion of tubulin within the microtubule wall (a), or by rapid breaking and reannealing (b). Reprinted with permission, from Salmon et al. [1984c], Salmon [1984].

tion rates at the two ends of a microtubule: polymerization is favored at the (+) end, depolymerization at the (−) end [Purich and Kristofferson, 1984]. Microtubules in the mitotic apparatus and CMTC have the same polarity, (+) distal to the centrosomes.

In vitro purified tubulin self-assembles into microtubules in low-calcium buffers near neutral pH at physiological temperature [Purich and Kristofferson, 1984]. When purified centrosomes are used to nucleate assembly in vitro, the resulting microtubules have the same structure and polarity as microtubules in vivo [Evans et al., 1985]. The mechanism by which microtubules assemble in vitro has been identified as a nucleated, linear condensation polymerization reaction [Purich and Kristofferson, 1984]. The binding of guanosine triphosphate (GTP) to the tubulin dimer greatly promotes polymerization. Hydrolysis of the bound GTP follows incorporation of the dimer into the microtubule. The free energy of GTP hydrolysis frees microtubule polymerization from the thermodynamic constraints of simple helical polymerization and provides the opportunity for more dynamic pathways of assembly (Fig. 2B) [Margolis et al., 1978; Margolis and Wilson, 1981; Mitchison and Kirschner, 1984a–c].

Our principal interest has been the mechanism by which microtubules assemble and disassemble within the mitotic spindle. Observations of spindle birefringence with polarization microscopy have shown that the spindle microtubules can be rapidly disassembled (within 10–30 sec) by cold, increased hydrostatic pressure, and drugs, such as colchicine, which bind to tubulin and prevent assembly [Inoué and Sato, 1967; Inoué, 1981; Salmon, 1975; Salmon et al., 1984a]. The spindle reassembles rapidly after these conditions are returned to normal, which demonstrates that most of the spindle and astral microtubules are in a rapid, labile equilibrium with the cellular pool of tubulin subunits [Inoué, 1981]. The kinetochore microtubules, usually a small fraction (<30%) of the microtubules in a half-spindle, are differentially stable to such treatments [Rieder, 1982], perhaps because both ends of these microtubules are engaged [Salmon et al., 1976; Mitchison and Kirschner, 1984c]. The lability of most microtubules in the spindle asters and the interphase CMTC is similar to that of the nonkinetochore spindle microtubules. The midbody microtubules, however, are even more stable than kinetochore microtubules [Salmon et al., 1976].

For analysis of microtubule dynamics in living cells, we have used tubulin labeled with a fluorescein derivative, 5-(4,6-dichlorotriazin-2-yl) amino fluorescein (DTAF)-tubulin, as a tracer of the endogenous tubulin pool (normal intracellular concentration about 20 $\mu$M). In living cells endogenous tubulin and injected DTAF-tubulin appear to intermingle randomly when microtubules form. A laser microbeam (488-nm wavelength) can then be used to photobleach the fluorophores in selected regions of microtubule arrays at steady-state assembly without producing structural damage to the cells. Polymerization characteristics of microtubules can be inferred from the pattern of fluorescence recovery and redistribution after photobleaching. This work was initiated in collaboration with J.R. McIntosh's laboratory (University of Colorado, Boulder, CO) and our initial findings have been published [Leslie et al., 1984; Salmon et al., 1984b,c; Saxton et al., 1984].

As will be discussed more fully below, there are several proposed models for microtubule assembly (Fig. 2B) which make specific predictions about the rate and pattern of FRAP for the incorporation of tubulin in microtubule arrays at steady-state assembly. For example, very slow FRAP is expected for "simple equilibrium" tubulin subunit exchange solely at the free ends of microtubules [Hill and Kirschner, 1982; Zeeberg et al., 1980]. The "treadmilling" model proposes that GTP hydrolysis drives the preferential assembly and disassembly of tubulin subunits at opposite ends of microtubules, which results in uniform flow of tubulin along the spindle microtubules toward the poles [Margolis et al., 1978; Margolis and Wilson, 1981]. Consequently, FRAP should be detected first near the spindle equator, bleached spots should move poleward during FRAP, and the rate of FRAP will depend inversely on the diameter of the bleached spot [Wadsworth and Salmon, 1986a,1986d]. Both the "dynamic equilibrium" and "dynamic instability" models predict that FRAP will occur rapidly and uniformly [Salmon et al., 1984c].

## DTAF-TUBULIN

Criteria for useful fluorescent analogs of any cellular molecule have been described in detail by Taylor et al., [1984]. The major criterion for a tubulin analog, however, is native assembly ability both in vivo and in vitro.

Only a few fluorophores and labeling procedures have been found that will label tubulin without denaturing the protein. Although there are various methods for preparing fluorescent analogs to tubulin [Wadsworth and Salmon, 1986b], we routinely use a procedure developed in J.R. McIntosh's laboratory by Leslie et al. [1984] to label purified porcine brain tubulin with DTAF, a fluorescein derivative that reacts with $\epsilon$ amino groups and N-terminal amino acids near neutral pH.

DTAF-tubulin is the most extensively characterized fluorescent analog of tubulin [Leslie et al., 1984; McIntosh et al., 1986; Wadsworth and Salmon, 1985b]. After purification of labeled tubulin from denatured protein and microtubule-associated proteins, the stoichiometry of the DTAF/tubulin dimer has varied for different preparations from 0.5 to 1.8. Both $\alpha$ and $\beta$ tubulin are labeled, with more DTAF on $\beta$ tubulin. DTAF-tubulin has been shown to have nearly normal characteristics of microtubule assembly and disassembly in vitro. DTAF-tubulin polymerizes onto microtubule ends and does not bind adventitiously to the walls of microtubules in standard in vitro microtubule reassembly buffers.

We have microinjected DTAF-tubulin into tissue culture cells and sea urchin embryos using glass microneedles [Graessmann et al., 1980; Salmon et al., 1984b,c; Saxton et al., 1984]. Typically, we injected an amount of 15–55 $\mu$M DTAF-tubulin (concentration in the needle), equal to 1–10% of the cell volume. DTAF-tubulin incorporates into microtubules at all phases of the cell cycle, and the cells develop normally when the intracellular concentration of DTAF-tubulin is less than 2–5 $\mu$M [Salmon et al., 1985c; Saxton et al., 1984] (Fig. 3).

## FLUORESCENCE MICROSCOPY, PHOTOBLEACHING, AND RECORDING

The optical system we have recently used in our fluorescence studies is diagrammed in Figure 4. There are three illumination and three imaging optical pathways through a custom-made inverted optical bench microscope [Inoué, 1961; Wadsworth and Salmon, 1986a, 1986d]. The rationale for the

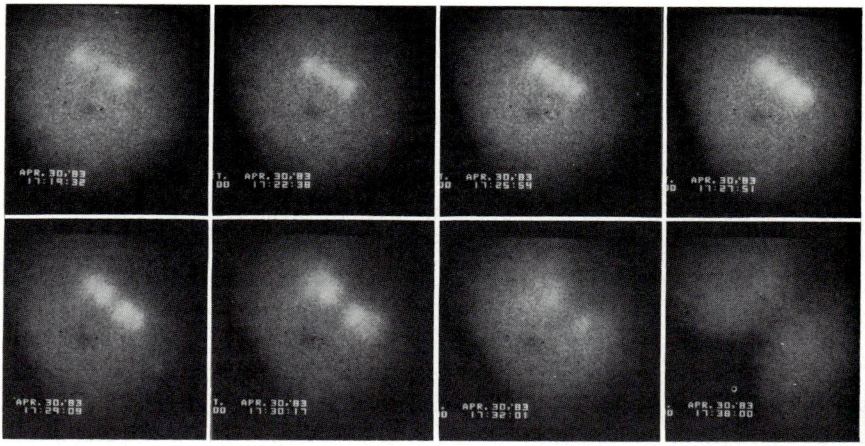

Fig. 3. Time-lapse photographs from a video monitor showing the DTAF-tubulin distribution from prometaphase through telophase and cell cleavage in a first-division embryo of the sea urchin *Lytechinus variegatus*. Full metaphase occurs near 17:25:54, while anaphase begins near 17:27:30. Time in hours:minutes:seconds is given on each frame. Bar, 20 $\mu$m ×275. Reprinted with permission from Salmon et al. [1984c].

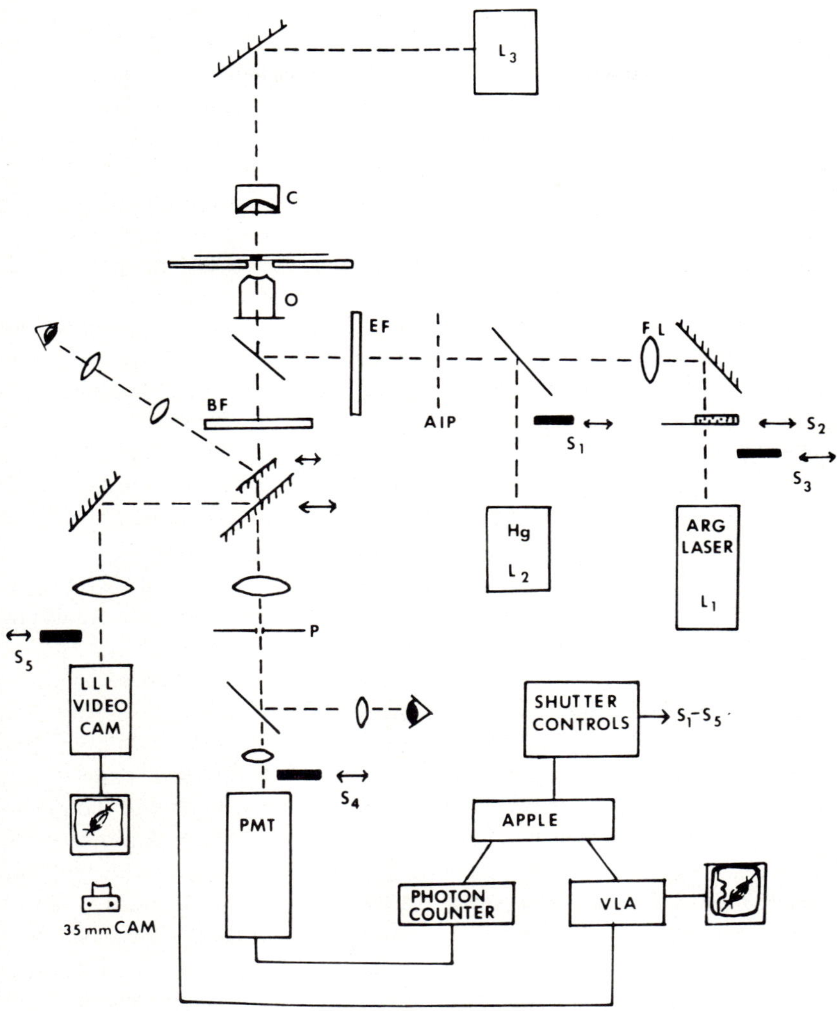


Fig. 4. Optical pathways for fluorescence microscopy, laser photobleaching, and video and photomultiplier recording. Major components are described in the text, in Salmon et al. [1984b] and in Wadsworth and Salmon [1985a, 1986]. Symbols: $L_1$, argon-ion laser; $L_2$, 100-W Hg illuminator; $L_3$, 100-W Hg or 60-W tungsten illuminator; C, condenser; O, objective; EF, fluorescence excitation filters; DM, dichroic mirror; BF, fluorescence blocking filters; AIP, aerial image plane; FL, lens that focuses laser beam at AIP; PMT, photomultiplier; $S_2$, low-transmission shutter for PMT-FRAP; $S_1$, $S_3$, $S_4$, shutters, LLL, low light level; CAM, camera; Apple, Apple II computer; VLA, video line analyzer. For polarization microscopy, a polarizer and a compensator are inserted above the condenser, an analyzer is inserted below the objective, and DM and BF are removed.

design is described in detail elsewhere [Salmon et al., 1984b]. Transillumination is used for polarization and phase contrast microscopy. A DC-regulated 100-W Hg burner provides uniform epi-illumination of large regions of the specimen. Local regions of the specimen are illuminated or photobleached by using a 488-nm laser microbeam (Spectra-physics model 164). Zeiss fluorescein excitation and barrier filters (#487709) are used for DTAF analogs. Fluorescent images are projected to the eye, to a low-light-level video camera (Venus DV2 or DAGE-MTI model 68 SIT), or through a pinhole in a conjugate image plane to a photomultiplier (EMI 9863A). Images from the low-light-level camera are photographed using 0.5-sec exposures [Salmon et al., 1984b]. Electronic shutters (Uniblitz, Vincent Associates) limit whole field epi-illumination during photographic exposures to 0.6 sec to prevent photobleaching and to block illumination of the video camera and photomultiplier tubes during laser photobleaching.

An Apple computer records video voltages along lines scanned through the image to provide line profiles of fluorescence and FRAP (Video-FRAP) [Wadsworth and Salmon, 1986a]. Line profiles have also been obtained from calibrated densitometry of photographic negatives recorded from the video monitor [Salmon et al., 1984b,c; Saxton et al., 1984]. The Apple computer is also used to sample, store, and plot data obtained from the photomultiplier through a photon counter (EMI model C-10).

We have performed fluorescence excitation and photobleaching using two patterns. In most of our recent experiments, we have analyzed FRAP for circular photobleached regions. For photomultiplier measurements (PMT-FRAP), a spherical focusing lens (17.5-mm focal length in most experiments) was used to focus the laser beam on the field diaphragm plane of the Zeiss 4 FL epi-illuminator. The characteristic "1/e" beam radius [Axelrod et al., 1976; Schneider and Webb, 1981] at the point of focus at the specimen depends on the objective and is 5.8 $\mu$m for a Zeiss 10× Acromat, 2.25 $\mu$m for the Zeiss 25× Plan Neofluor, 1.4 $\mu$m for the Zeiss 40× Plan Neofluor, and 0.76 $\mu$m for the Zeiss 63× Plan Apochromat objective. For bleaching, the beam power into the objective was about 100 mW, attenuated by 1–1.8 OD depending on the objective. During the measurement phase of photometric recordings of FRAP, the beam was further attenuated by a factor of 20,000 compared to the bleaching beam intensity with an electronic shutter ($S_2$ in Fig. 4) [Yguerabide et al., 1982]. Spindle fluorescence was usually bleached by 40–60% by a 0.1-sec exposure to the unattenuated laser beam. We have also used a bar pattern of photobleaching in our studies of spindle microtubule dynamics. An 18-mm focal length cylindrical focusing lens and a Zeiss 63×/1.4 NA objective produced the bar pattern [Wadsworth and Salmon, 1986a, 1986d].

## TUBULIN MOBILITY IN THE CYTOPLASM

The tracer diffusion coefficient (D) for tubulin has been measured in two types of cells: sea urchin embryos and BSC-1 tissue culture cells. Initially, the diffusion coefficient was measured in sea urchin embryos by using video measurements of FRAP (Video-FRAP) [Salmon et al., 1984b]. Profiles of fluorescence through circular bleached regions were obtained by calibrated densitometry of video images before and after photobleaching. The diffusion constant was obtained by analysis of fluorescence recovery profiles by using Fick's second law of diffusion, assuming that only radial diffusion occurred within the cylindrical bleached region through the cell. More recently we have used conventional photomultiplier spot measurements (PMT-FRAP) [Axelrod et al., 1976; Jacobson et al., 1983] to obtain values for D in both sea urchin and BSC-1 mammalian tissue culture cells from half-times of fluorescence recovery (Fig. 5). Both techniques yielded similar values for D in the sea urchin cytoplasm, about 4–6 $\times$ $10^{-8}$ $cm^2$/sec at 25°C. Tubulin diffusion in the cytoplasm of BSC-1 cells is 1–2 $\times$ $10^{-8}$ $cm^2$/sec at 32°C, two to six times slower. As with other proteins, the diffusion coefficient of tubulin strongly depends on cell type [Salmon et al., 1984b]. Jacobson and Wojcieszyn [1984] provide an interesting analysis and discussion of the structural and chemical factors which may determine the rate of diffusion of proteins through the cytoplasm.

The major limitation on tubulin mobility in the cell is incorporation into microtubules. As seen in Figure 5, about 65% of the DTAF-tubulin in a region of an interphase BSC-1 cell near the centrosome (see Fig. 1) is relatively immobile compared to the diffusable fraction of DTAF-tubulin in the cell. Depolymerization of the great majority of the CMTC by 10 $\mu$M nocodazole [Debrabander et al., 1981] correspondingly reduces the relatively immobile fraction to less than 15%. The rate of diffusion of nonpolymerized tubulin does not appear to be greatly altered after microtubule depolymerization. Other components of the cytoplasm or cytomatrix affect the mobility of soluble tubulin more significantly [Jacobson and Wojcieszyn, 1984].

## SPINDLE FRAP

When a region between the chromosomes and a pole of a metaphase spindle is photobleached, the majority of bleached fluorescence recovers rapidly, as illustrated by the video images in Figures 6 and 7. In Figure 6, a substantial amount of the initial fluorescence in a large region of the upper half-spindle and aster have been photobleached. Sixty seconds later spindle fluorescence and morphology have returned to near normal. Fluorescence recovery appears to occur uniformly throughout bleached regions of spindles

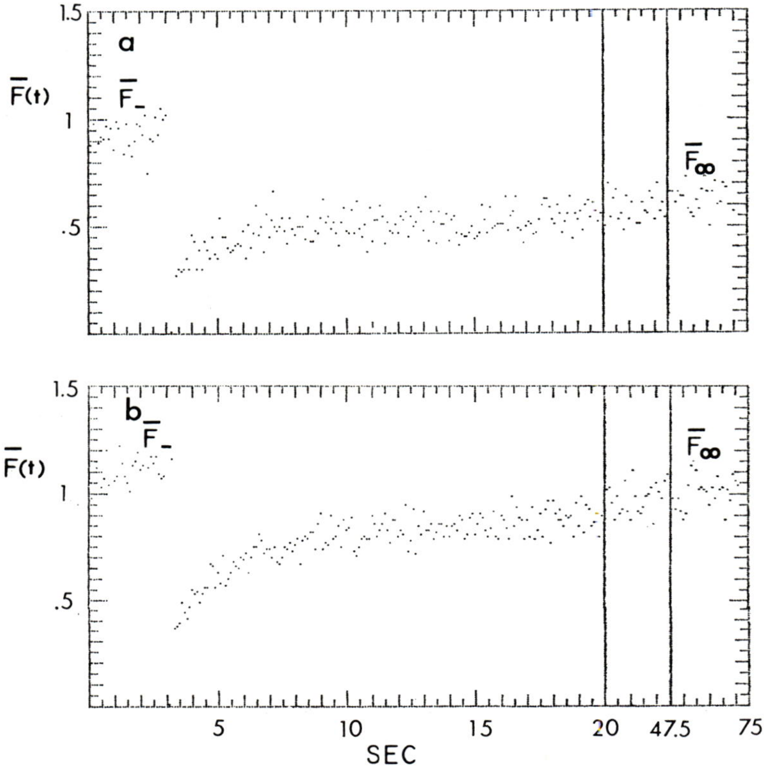


Fig. 5. The lateral diffusion of DTAF-tubulin in the cytoplasm of interphase BSC-1 tissue culture cells measured by PMT-FRAP of (a) a normal cell and (b) a cell pretreated for 30 min with 10 $\mu$M nocodazole to induce the depolymerization of labile microtubules in the cytoplasmic microtubule complex (CMTC). A 50-cm spherical focusing lens (FL in Fig. 3) plus 40×/0.75 NA Zeiss Neofluor phase objective produced a 6.6-$\mu$m-diameter Gaussian laser beam at the specimen. Measurements were made in the cytoplasm about 10 $\mu$m from the centrosome region and 20 $\mu$m from the cell periphery. Photon counts were sampled over 0.1-sec intervals. The first five samples were averaged by the computer and used to normalize the remaining samples which are plotted. Samples are plotted for the time periods 0–20 sec, 45–47.5 sec (between the vertical bars), and 72–75 sec. The percentage of bleached fluorescence recovered in 75 sec in a is 38% and in b is 85%.

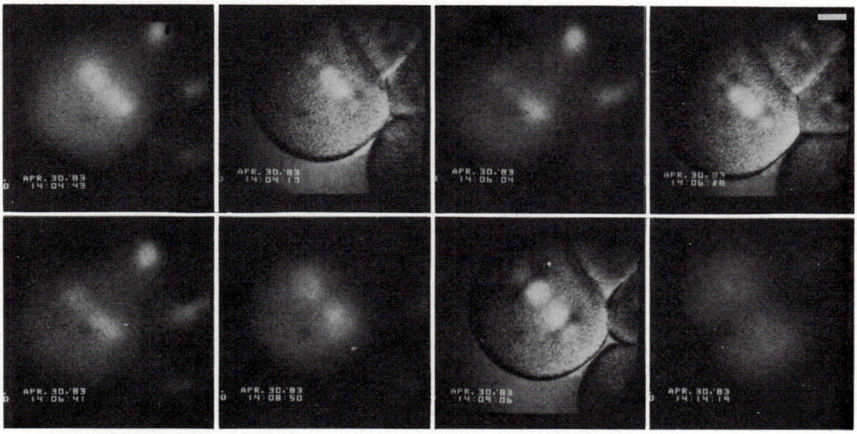

Fig. 6. FRAP in a first-division metaphase spindle in an embryo of the sea urchin *Lytechinus variegatus*. Bleaching for 8 sec with a 12-$\mu$m diameter laser beam, focused on the specimen using a Zeiss 25×/0.6 NA Neofluor phase objective, ended at 14:06:03. Fluorescence in the region of the upper half-spindle and aster, almost completely bleached by laser irradiation, recovered substantially within 60 sec. The fluorescence (a,c,e,f,h) and polarization (b,d,g) micrographs demonstrate that photobleaching does not significantly alter the normal amount and distribution of spindle and astral birefringence, morphological changes in assembly during mitosis, or the timing of mitotic events. Bar, 20 $\mu$m. ×275. Reprinted with permission from Salmon et al. [1984c].

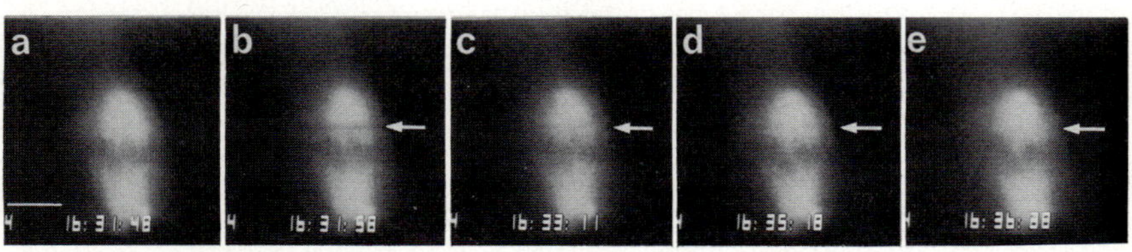


Fig. 7. FRAP of a metaphase BSC-1 cell microinjected with DTAF-tubulin. The lower half-spindle was photobleached midway between the chromosomes and the pole for 0.1 sec with a bar bleach pattern (approx. 1.6 $\mu$m wide; see text for details). Cell before bleaching (a); immediately after bleaching (b); and after 73 sec (c), 200 sec (e), and 271 sec (e) of recovery. Arrows mark the position of the bleached region. Bar, 10 $\mu$m. ×700. Reprinted with permission from Wadsworth and Salmon [1986d].

and asters [Salmon, 1984; Salmon et al., 1984c, Saxton et al., 1984]. A similar result is seen when a 1.6-$\mu$m-wide bar pattern is bleached across the half-spindle of the BSC-1 spindle in Figure 7 [Wadsworth and Salmon, 1985a, 1986d].

We have quantitated the kinetics of FRAP using standard PMT-FRAP methods [Axelrod et al., 1976; Wadsworth and Salmon, 1986d], since translation of the bleached regions during fluorescence recovery was not apparent from video FRAP images. Figures 8 and 9 illustrate FRAP records for metaphase sea urchin and BSC-1 spindles. For each record, there are three major phases of fluorescence recovery. There is an initial, rapid phase which corresponds to the rate of diffusion of tubulin subunits in the cytosol within the spindle and in the cytoplasm above and below the spindle along the optical path of the laser beam. The sea urchin embryo is much thicker along the optical path of the laser beam than is the BSC-1 cell (about 6 times) and the magnitude of the rapid diffusion phase in the FRAP kinetics is greater. Both the sea urchin and BSC-1 central spindles are about the same size, about 24 $\mu$m long and 6 $\mu$m wide at the equator.

The second phase of fluorescence recovery we term the "fast incorporation phase" because it appears to correspond with the rate of tubulin incorporation into the majority of spindle fiber microtubules at steady state. Fluorescence recovery during the incorporation phase can be fit closely with an exponential function [Salmon et al., 1984c; Saxton et al., 1984; Wadsworth and Salmon, 1986a]:

$$\overline{F}_{\infty} - \overline{F}(t) = [\overline{F}_{\infty} - \overline{F}(0)]\, e^{-kt} \qquad (1)$$

where $\overline{F}_{\infty}$ is the normalized fluorescence in the bleached region 365 sec after bleaching, $\overline{F}(t)$ is the normalized fluorescence after bleaching where time $t = 0$ immediately after bleaching, and $k$ is a first-order rate constant (Fig. 8b). The half-time of fluorescence recovery is given by

$$t_{1/2} = \ln 2 \,/\, k \qquad (2)$$

where $k$ is determined from the shape of a best-fit straight line through plots of ln $[\overline{F}_{\infty} - \overline{F}(t)]$ vs. time [Wadsworth and Salmon, 1986a, 1986d]. Table 1 lists half-times measured for sea urchin and BSC-1 spindles. At physiological temperature the half-time of fluorescence recovery for the sea urchin (16 sec) is similar to the value for BCS-1 cells (estimated to be 24 sec at 37°C from an Arrhenius plot of the half-time data in Table 1 [Wadsworth and Salmon, 1986a]). Similar half-times have been determined by video FRAP for BSC-1 and $PtK_2$ metaphase spindles.

The third phase of FRAP is a very slow incorporation phase. About 12% of the initial bleached fluorescence in the sea urchin spindle (about 25% for

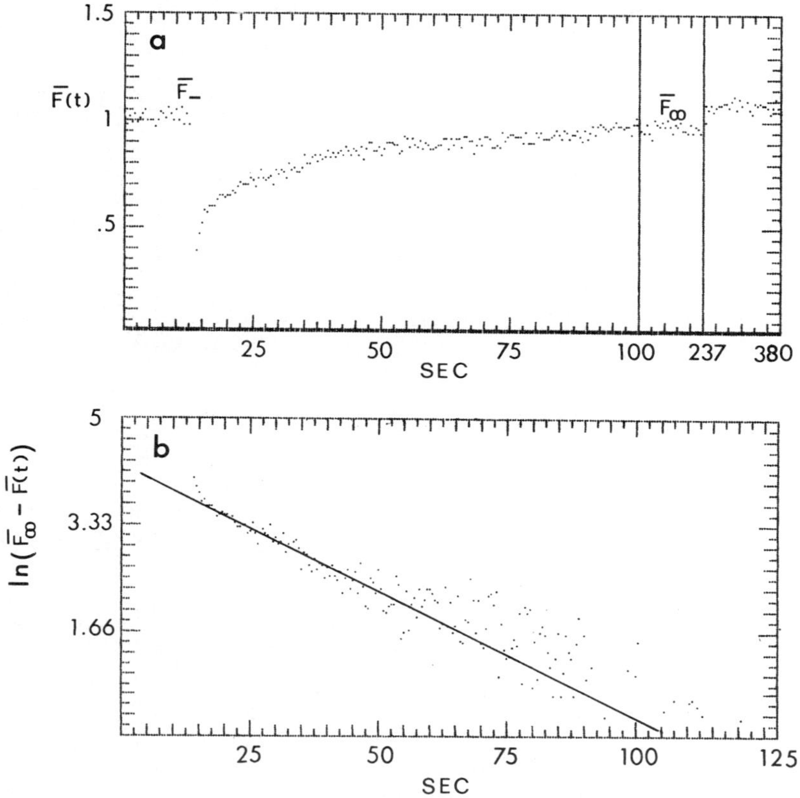


Fig. 8. Computer-generated records of FRAP for a metaphase sea urchin spindle at 25°C obtained by photometric measurements. A Zeiss 25×/0.6 NA Neofluor phase objective was used to produce a 4.2-$\mu$m-diameter Gaussian laser beam through a central half-spindle region. The number of photon counts as recorded every 0.5 sec. Fluorescence measurements, $F$(t), were normalized by dividing by the average number of photon counts for five samples preceding the recording data. A total of 7680 samples were taken in the experiment. The first 200 samples were plotted, then samples #450–475 (between the first and second vertical bars), and then samples #720–750 (second vertical bar to the end). At the end of the record, the spindle has gone into metaphase and fluorescence increases because microtubule assembly increases. Fluorescence was about 5,000 pulses per second (pps) in the spindle, compared with 2,500 pps in the adjacent cytoplasm. $\bar{F}_-$ is the average prebleach fluorescence and $\bar{F}_\infty$ is the average recovered fluorescence after bleaching for 0.2 sec at 17 sec. B. The rapid incorporation phase of fluorescence recovery was analyzed according to Eq. 1 by plotting $\ln[\bar{F}_\infty \bar{F}(t)]$ vs. time and fitting a straight line to the data as described in the text.

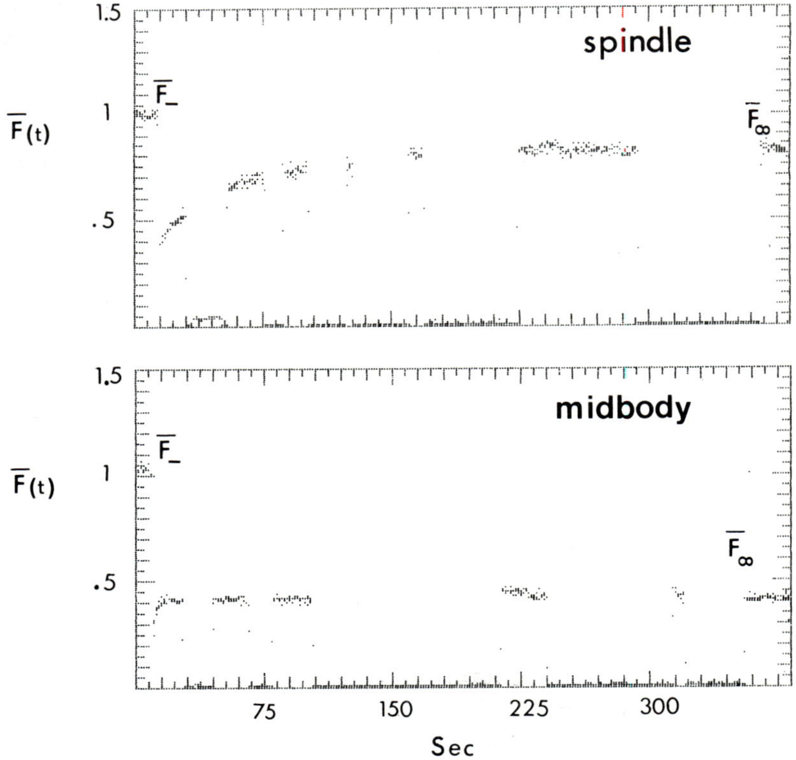


Fig. 9. Comparison of FRAP for a metaphase spindle and a midbody in BSC-1 tissue culture cells at 32°C. A Zeiss 40×/0.75 NA Neofluor phase objective produced a 2.8-μm-diameter Gaussian laser beam. Parameters and sampling protocol were as described in Figure 8. Spindle fluorescence was about 6,000 pps compared to 1,000 pps in the adjacent cytoplasm.

BSC-1 spindles at 32°C) does not recover within 365 sec. Part of this unrecovered, bleached fluorescence, $R\% = [\bar{F}_{\infty} - \bar{F}(0)]/[\bar{F}_{-} - \bar{F}(0)]$, may be the result of bleaching a significant fraction of the total cell tubulin, particularly when the half-spindle fluorescence is bleached greater than 50% in the BSC-1 cells [Saxton et al., 1984]. A large fraction of the unrecovered fluorescence is probably also due to the slow exchange of tubulin within the differentially stable kinetochore fiber microtubules. We have not yet tried to measure the rate of fluorescence recovery from the very slow phase.

Both $t_{1/2}$ and $R\%$ are sensitive functions of temperature [Wadsworth and Salmon, 1986a] (Table 1). This suggests that the microtubules which persist at lower temperatures are more stable than the majority of microtubules at

**TABLE 1. Comparison of FRAP[1] for Metaphase Spindles in Sea Urchin Embryos and BSC-1 Culture Cells**

| Cell type | Temperature (°C) | Beam Diameter ($\mu$m) | $t_{1/2}$ (sec)[2] | R%[2] | n |
|---|---|---|---|---|---|
| Sea urchin[3] | 23–25 | 4.5 | 16 ± 4.5 | 88 ± 11 | 12 |
| BSC-1[3,4] | 32 | 4.5 | 41 ± 10 | 69 ± 15 | 9 |
| | 32 | 2.8 | 43 ± 13 | 74 ± 22 | 8 |
| | 26 | 2.8 | 75 ± 33 | 53 ± 25 | 7 |
| | 23 | 2.8 | 107 ± 9 | 39 ± 20 | 4 |
| | 20 | 2.8 | 144 ± 18 | 47 ± 26 | 5 |

[1]Fluorescence redistribution after photobleaching.
[2]Calculated as described in the text.
[3]Wadsworth and Salmon [1986a].
[4]Wadsworth and Salmon [1986d].

physiological temperature. We are currently exploring further the energetics of spindle microtubule dynamics.

Our spindle FRAP results thus far have shown that at steady state, tubulin throughout a great majority of spindle fiber microtubules (probably the labile nonkinetochore microtubules) exchanges with the tubulin pool with a half-time of about 20 sec at physiological temperature, an extremely rapid rate.

## FRAP OF OTHER MICROTUBULE ARRAYS

The rate and percentage of FRAP was found initially by Saxton et al. [1984] to depend on the type of microtubule in living cells, and they are different for microtubules in vivo and in vitro [Leslie et al., 1984]. Figure 9 compares the FRAP for a BSC-1 metaphase spindle and a midbody complex, and Figure 10 compares the FRAP for the interphase CMTC of BSC-1 cells and for a preparation of reassembled microtubules in vitro. The $t_{1/2}$ we have measured using PMT-FRAP for the CMTC is 200 ± 61 sec at 32°C, and $R\% = 77 \pm 17$ ($N = 15$). This value, similar to the half-time obtained by Saxton et al. [1984] using Video-FRAP, is about ten times greater than the half-time for spindle FRAP. The incomplete recovery of bleached fluorescence suggests that about 25% of the microtubules in the CMTC are much more stable than the majority of microtubules.

Steady-state incorporation of tubulin into either midbody microtubules or reassembled microtubules in vitro is extremely slow (Figs. 9, 10). Since microtubules of the midbody are highly resistant to microtubule depolymerization conditions [Salmon et al., 1976], little FRAP is expected. Although in vitro microtubules in reassembly buffer are rapidly depolymerized by cool-

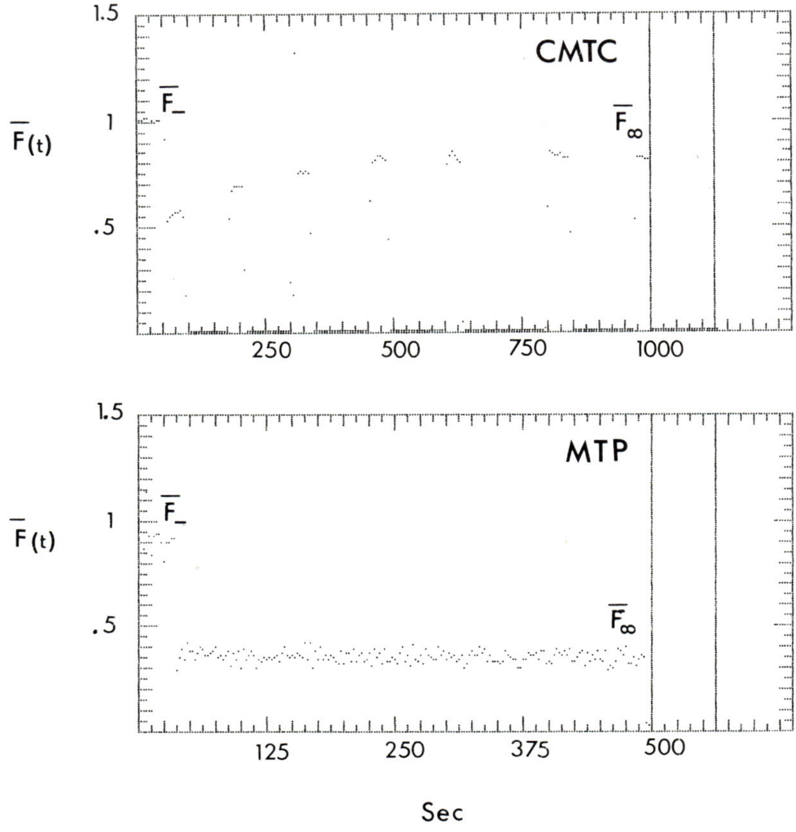


Fig. 10. Comparison of FRAP at 32°C of the interphase cytoplasmic microtubule complex (CMTC) in a living cell and microtubules reassembled in vitro from purified brain microtubule protein (MTP). A Zeiss 25×/0.6 NA Neofluor phase objective produced a 4.5-$\mu$m-diameter Gaussian laser beam. Photon counts were sampled at 5-sec intervals for the CMTC experiment and 2.5 sec intervals for the MTP experiment. Other parameters and procedures were as described in Figure 8. In the CMTC experiment, the laser microbeam was positioned about 10 $\mu$m away from the centrosome region. Nonfluorescent MTP was purified by three cycles of temperature-dependent assembly-disassembly as described by Leslie et al. [1984]. About 0.3 mg/ml of DTAF-tubulin was added to 3 mg/ml of MTP and assembly was initiated by warming to 37°C in reassembly buffer that contained 0.1 M PIPES (pH = 6.8), 0.5 mM $MgSO_4$, 1 mM EGTA, and 0.5 mM GTP. After 1 hr, a sample was sealed between a slide and coverslip, spaced about 30 $\mu$m apart and held at 32°C, for analysis of FRAP. Similar preparations made within 30 min after initiation of assembly yielded complete recovery of bleached fluorescence, as described in the text.

ing, as are labile microtubules in the living cell, FRAP is extremely slow compared to the majority of microtubules in the spindle and CMTC.

Leslie et al. [1984] found a similar result for microtubules assembled in vitro by nucleated growth from purified centrosomes. The microtubules were grown from purified brain microtubule protein and immobilized at their (−) ends by the attachment of the centrosome to the coverslip [Evans et al., 1985]. The whole aster was photobleached, and less than 10% of the initial bleached fluorescence was recovered after 60 min.

For self-assembled microtubules in suspension, the rate of FRAP decreases with time, after the initiation of assembly. At early stages of assembly, FRAP that is not associated with subunit diffusion appears due to the diffusion of whole microtubules, which becomes slower as the average length of a microtubule in the preparation becomes longer (apparently because of end-to-end microtubule annealing) at steady state and approaches zero when gelation occurs [Pryer, Wadsworth, and Salmon, unpublished observations]. FRAP has been used successfully to analyze actin polymerization dynamics in vitro [Lanni et al., 1981], and we are currently exploring further this use of FRAP to characterize the dynamics of in vitro microtubule assembly. Qualitatively, FRAP of microtubules reassembled in standard in vitro buffers appears to result from the mobility of whole microtubules, and not from the rapid turnover of tubulin throughout the microtubules as appears to occur for labile microtubules in vivo.

## CONTROLS AND CORRELATIONS

The major assumption in our experimental approach is that FRAP measures the steady-state dynamics of tubulin within microtubules in the living cells. This assumption appears justified for several reasons. The rate and extent of FRAP are similar when a small region of the spindle is bleached, when a bar is bleached across the half-spindle, and when the whole half-spindle is bleached. Thus the movement of whole microtubules is not responsible for the FRAP. Photobleaching does not alter spindle birefringent retardation (Fig. 6) and does not change microtubule assembly in interphase CMTC as viewed by high-voltage electron microscopy or antitubulin immunofluorescence methods [Saxton et al., 1984]. This evidence supports our assumption that FRAP measures steady-state tubulin incorporation into microtubules, not a reassembly response to irradiation damage such as that produced by UV microbeam irradiation [Forer, 1965; Leslie and Pickett-Heaps, 1983]. Consistent with this conclusion is the observation that photobleaching half-spindle fluorescence several times during mitosis does not noticeably alter the rate of mitosis or cytokinesis [Salmon et al., 1984c, Saxton et al., 1984].

Leslie et al. [1984] have extensively studied the effects of photobleaching on microtubule assembly in vitro. Extensive photobleaching of DTAF-tubulin that has been polymerized into microtubules in vitro did not cause microtubules to depolymerize or fragment. No effects on microtubule polymerization and little photoactivated crosslinking of proteins were observed under conditions typical for laser photobleaching experiments with living cells: 0.1–1-sec bleach periods, less than 70% bleach, in a reducing environment containing 2 mM glutathione, and a ratio of DTAF-tubulin to unlabeled tubulin in the microtubules of 0.1 or less. However, photobleaching was found to kill polymerization activity and induce crosslinking of tubulin and microtubule-associated proteins for longer bleaching periods in nonreduced buffers at higher ratios of DTAF-tubulin to unlabeled tubulin.

The in vivo FRAP results correspond nicely to other data on spindle microtubule dynamics. First, the incorporation of DTAF-tubulin in microtubules to steady-state levels should occur at rates and extents corresponding to the FRAP characteristics following microinjection of DTAF-tubulin into cells. There is a direct correspondence between the rate of FRAP and the rate of "forward" incorporation [Hamaguchi et al., 1985; McIntosh et al., 1986; Saxton et al., 1984; Wadsworth and Salmon, 1986a]. Visually, the magnitude of fluorescence in the spindle appears to achieve steady-state levels within 1–2 min following injection of DTAF-tubulin into metaphase BSC-1 cells. Within 10 min following injection, the magnitude of fluorescence is nearly identical to the fluorescence obtained after spindles are initially disassembled by cooling and then reassembled by rewarming to 37°C (Fig. 11). Given available evidence, there is a quantitative correlation between the rate of FRAP and the rate of uptake of tubulin into the spindle and CMTC, and a qualitative correlation in the extent.

The second correlation is that the non-steady-state rate and extent of microtubule disassembly, which result when steady-state assembly is abruptly blocked, should occur at rates and extents equal to or greater than the kinetics of FRAP. We have used colchicine or other drugs, such as nocodazole, which bind to the same site on tubulin as does colchicine, to abruptly block microtubule assembly in living cells [Salmon et al., 1984a; Cassimeris et al., 1985], and the approximate half-times for depolymerization are given in Table 2 for spindles, CMTC, and mid-bodies in BSC-1 cells. There is a direct correlation with the rate of FRAP for each type of microtubule array. For both spindles and the CMTC there is a minor fraction of microtubules which are differentially stable as predicted by the incomplete recovery of fluorescence after photobleaching, but the percentage of the total initial population of microtubules is not known yet.

## ANALYSIS OF MICROTUBULE ASSEMBLY MODELS

The FRAP results have demonstrated that a minor population of microtubules in the mitotic spindle (probably the kinetochore microtubules) and in the CMTC is differentially stable, similar to the midbody microtubules. This

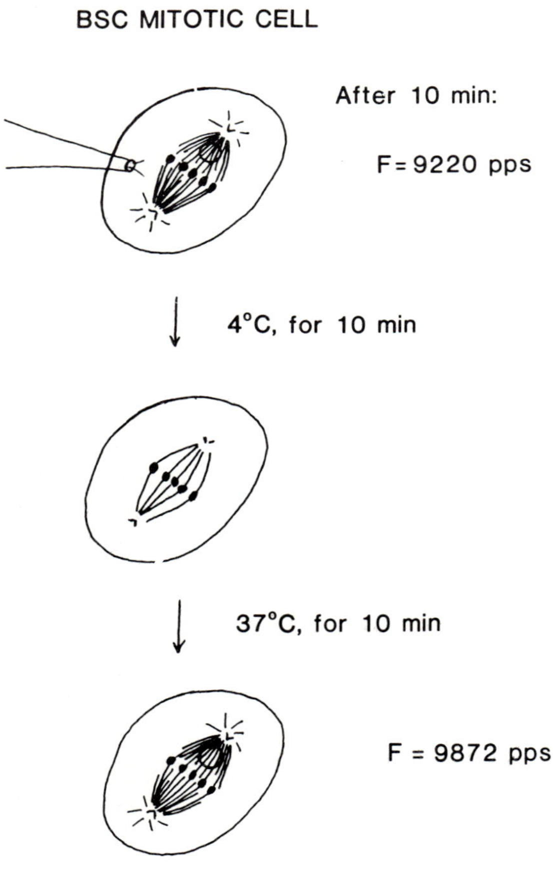


Fig. 11. A demonstration that tubulin turnover is rapid throughout the nonkinetochore spindle and astral microtubules. A prometaphase BSC-1 cell was injected with DTAF-tubulin. After 10 min at 37°C the fluorescence was 9220 pps, measured by using an attenuated 2.5-$\mu$m-diameter Gaussian laser beam passing through the central region of a half-spindle. The cell was cooled to 4°C for 10 min to depolymerize completely the nonkinetochore and astral microtubules and then rewarmed to 37°C to induce reassembly. When fluorescence in the half-spindle was measured 10 min later, it was slightly higher than it had been before cooling.

**TABLE 2. Comparison for Different Microtubule Arrays of the Rate of FRAP and the Rate of Depolymerization Induced by Nocodazole or Cooling**

| | Rate or Half-time (sec) | | |
|---|---|---|---|
| Microtubule array | FRAP (32°C)[1] | 10μM nocodazole (37°C) | 4°C |
| BSC-1 spindle (non-k) | < 40 | 10[2,3] | Very fast |
| BSC-1 CMTC | 200 | 160[3] | Fast |
| BSC-1 Midbody | Very slow | Very slow[3] | Very slow |
| In vitro centrosome aster | Very slow[4] | — | — |
| In vitro gelled microtubules | Very slow[1] | Very slow[1] | Fast[1] |

[1]This report; Saxton et al. [1984].
[2]Salmon et al. [1984a].
[3]Cassimeris et al. [1985].
[4](37°C), Leslie et al. [1984].

differential stability was predictable by their stability to other microtubule depolymerization agents. Why these microtubules are stable is of considerable interest, but the underlying mechanism is unknown. There is evidence that microtubules can be stabilized by "capping" both ends with a complex that blocks tubulin association-dissociation [Salmon et al., 1976; Rieder, 1982; Leslie and Pickett-Heaps, 1983; Mitchison and Kirschner, 1984c]. The binding of microtubule-associated proteins along the length of microtubules may promote stability, and stabilization may also depend on ATP-dependent reactions [Debrabander et al., 1981; Bershadsky and Gelfand, 1981]. Preliminary evidence shows that the percentage of FRAP of both the spindle and the CMTC is substantially reduced when BSC-1 cells are treated with inhibitors of ATP production [Wadsworth and Salmon, 1986c].

What is the dominant steady-state pathway of microtubule assembly for the dynamic majority of microtubules in the spindle and CMTC? We will consider the spindle first and then the CMTC, because the conclusions are similar but the details are slightly different. The FRAP results support two major conclusions about the steady-state assembly of the labile nonkinetochore spindle microtubules: (1) tubulin throughout these microtubules is exchanged with tubulin in the cytoplasmic pool with a half-time of about 20 sec at physiological temperature, and (2) tubulin incorporates into microtubules at sites throughout the half-spindle.

The rapid and extensive FRAP of the nonkinetochore spindle microtubules clearly demonstrates that simple equilibrium exchange of tubulin subunits at microtubule ends is not the dominant pathway of steady-state microtubule assembly [Salmon, 1984; Salmon et al., 1984c]. Figure 12 diagrams the incorporation of new tubulin subunits at the (+) end of a microtubule after photobleaching. The mean length of incorporation, $\bar{x}$, is given by

$$\bar{x} = 2(k_d / \pi) / 1{,}634 \quad \mu m \tag{3}$$

where $k_d$ is the tubulin dissociation, $t$ is time after photobleaching, and 1,634 is the approximate number of tubulin dimers/$\mu$m of microtubule [Hill and Kirschner, 1982; Zeeburg et al., 1980]. For the sea urchin spindle, the $k_d$ has been estimated to be about 1,000 dimers/sec for an average steady-state microtubule length of 5.5 $\mu$m [Salmon et al., 1984a]. From Eq. 3, the time required to exchange all tubulin dimers in a microtubule that is 1 $\mu$m long is 1,051 sec, and for a microtubule 5.5 $\mu$m long it is 31,833 sec. The exact distribution of microtubule lengths over the 10–12-$\mu$m-long half-spindle is not known, but these times are at least two orders of magnitude too long compared to the 16-sec half-time of spindle FRAP observed. The same argument can be made for the BSC-1 spindle.

Poleward treadmilling of microtubules as proposed by Margolis and Wilson (Fig. 2B) predicts initial recovery of fluorescence proximal to the metaphase plate region, translocation of the bleached region toward the pole during fluorescence recovery, and continuous poleward flow of tubulin poleward due to the "head to tail" steady-state assembly of the microtubules. In actuality, when the whole half-spindle is photobleached, fluorescence recovery appears uniform throughout the half-spindle (Fig. 6) [Salmon et al., 1984c; Saxton et al., 1984], and no poleward translation of a narrow bleached bar pattern occurs during fluorescence recovery (Fig. 7) [Wadsworth and Salmon, 1986a,c, 1986d]. We have tested for poleward flow of tubulin by comparing the rate of FRAP for circular beams of different diameters. If FRAP is due to flow, then the half-time of fluorescence recovery should be proportional to the beam diameter. We found that the rate and extent of FRAP are independent of beam diameter (Table 1) [Wadsworth and Salmon, 1986a, 1986d]. The FRAP of nonkinetochore spindle microtubules is not

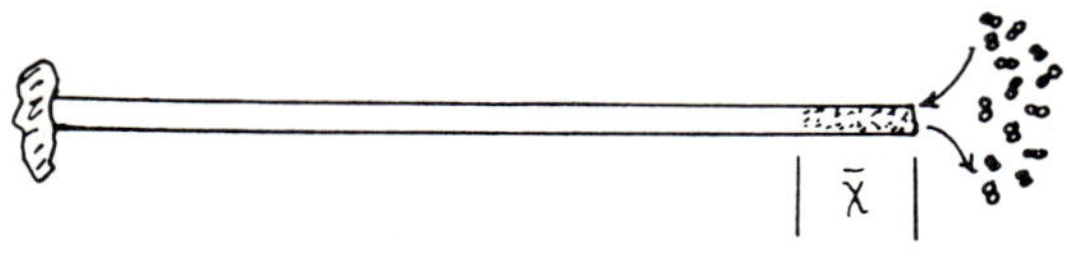


Fig. 12. Sketch of the incorporation of tubulin subunits into a free end of microtubule at steady-state by simple equilibrium association-dissociation events. $\bar{x}$ is the mean distance from the end of a microtubule in which tubulin dimers are replaced as a function of time.

accounted for by the treadmilling model, which means that treadmilling is not likely to be the dominant pathway of steady-state microtubule assembly in the spindle.

Either the dynamic equilibrium model, initially proposed by Inoué [Inoué and Sato, 1967; Inoué and Ritter, 1975] or the dynamic instability model proposed by Mitchison and Kirschner [1984a–c] (Fig. 2b) could account for the rapid, uniform, and extensive FRAP of spindles [Salmon 1984, Salmon et al., 1984c; Wadsworth and Salmon, 1986a, 1986d]. The dynamic equilibrium model proposes that tubulin enters and leaves the microtubule polymer at sites all along the length of the microtubule, for which there is some evidence from cryo-electron microscopy [Mandlekow and Mandelkow, 1985]. Microtubules could also be continuously breaking and reannealing along their length [Salmon et al., 1984c]. The dynamic instability model proposes that tubulin exchanges only at the end of a microtubule, the (+) end of microtubules grown from the centrosomes and spindle poles. Rapid and extensive exchange of tubulin within the nonkinetochore spindle microtubles could occur if these microtubules were continuously, but asynchronously, going through alternating, relatively persistant phases of rapid growth and rapid shortening as Mitchison and Kirschner propose. They suggest that the transition between growth and shortening is regulated by the hydrolysis of GTP ligated to tubulin after incorporation into the microtubule lattice. When GTP hydrolysis "catches up" with elongation (Fig. 2B), subunit association is inhibited, and rapid depolymerization ensues [Carlier et al., 1984; Mitchison and Kirschner, 1984a–c]. Experiments to determine sites of tubulin incorporation at the resolution of single microtubules are required to distinguish between the dynamic equilibrium and dynamic instability models.

The average length of microtubules in the interphase CMTC of BSC-1 cells appears, from immunofluorescence micrographs of cells stained with antibodies to tubulin, to be greater than 20 $\mu$m [Vandre et al., 1984]. The same arguments can be made for the pathways of tubulin incorporation into the microtubules of the interphase CMTC as have been made above for the nonkinetochore spindle microtubules. FRAP is too fast to be the result of simple equilibrium exchange of tubulin subunits at microtuble ends, and there is no evidence for translation of the bleached region during FRAP [Saxton et al., 1984]. The same conclusion was reached from FRAP studies of the CMTC with a fluorescent analog to a microtubule-associated protein, MAP2, purified from mammalian brain [Scherson et al., 1984].

Recently, Soltys and Borisy [1985] have provided evidence that DTAF-tubulin does not incorporate at sites along the length of microtubules of the CMTC. Interphase tissue culture cells were injected with DTAF-tubulin and then fixed and extracted at various times after injection. To locate regions of DTAF-tubulin incorporation into microtubules they compared photographs of the distribution of a rhodamine-labeled antibody to tubulin, which stains all microtubules, to corresponding photographs of the distribution of insoluble DTAF-tubulin. Only DTAF-tubulin incorporation at the ends of microtubules distal from the centrosome and growth of new microtubules from the centrosome were reported to occur over several minutes following injection of the DTAF-tubulin into the cell. Because injection of DTAF-tubulin into the cell raises the concentration of total tubulin within the cell, these results do not strictly represent incorporation at steady state. Nevertheless, these initial findings do indicate that tubulin incorporation does not occur at sites along the length of cytoplasmic microtubules at steady state. Similar studies for microtubules within the mitotic spindle are not yet available.

We have recently analyzed the dynamics of cytoplasmic microtuble depolymerization in human monocytes when assembly is abruptly blocked with the tubulin-binding drug nocodazole [Cassimeris et al., 1985]. Cells on coverslips were fixed at various times after treatment with nocodazole to block assembly abruptly and then extracted and stained for immunofluorescence by using tubulin antibodies. All microtubules extend from the centrosome within a cell and they were traced from images on a video monitor generated by using an SIT video camera and focusing up and down in the microscope. Individual microtubles were found to depolymerize rapidly once disassembly was initiated, but initiation occurred very asynchronously among the population of microtubules. This behavior is directly predictable from the dynamic instability model.

An interesting correlation occurs between the differences in the rate of FRAP of cold-labile microtubule arrays and their sensitivity to colchicine or colchicinelike drugs (Table 2). Nonkinetochore spindle and astral microtubules and the CMTC are rapidly depolymerized by cooling, as are reassembled microtubules in vitro [Purich and Kristofferson, 1984]. However, these microtubles have significantly different responses to colchicine which correspond to the large differences in FRAP in vivo and in vitro (Table 2). Colchicine binds to the tubulin dimer at a site not exposed when the dimer is assembled into the wall of a microtubule [reviewed in Salmon et al., 1984a]. The colchicine-tubulin dimer complex (TC) is the active agent which promotes microtubule depolymerization in vivo [Wadsworth and Salmon, 1986a; McIntosh et al., 1986], apparently by interfering with microtubule elongation. In vitro, TC blocks microtubule elongation [Dustin, 1984], but rapid and extensive depolymerization of microtubules does not occur as it does in vivo. An interesting explanation for the difference in response to colchicine in vitro and in vivo is that in the living cell conditions favor strongly the dynamic instability mode of steady-state microtubule assembly, whereas in the in vitro reassembly buffers the simple equilibrium mode is strongly favored. In vivo, TC complex blocks microtubule elongation, but normal, rapid depolymerization continues, while TC also blocks new microtubule growth. Consequently, there is a net loss of microtubule polymer. In vitro, the conditions which produce the rapid, catastrophic depolymerization of microtubules in vivo are absent, and simple equilibrium exchange of tubulin subunits dominates the dynamics of the microtubules.

## FUTURE CONSIDERATIONS

Fluorescent analog cytochemistry and FRAP have shown that most microtubules in the spindle and CMTC are much more dynamic than indicated from previous studies of microtubule assembly in vitro. Attention is now focused on the dynamic instability model of microtubule assembly, which does provide simple explanations for the complex behavior of microtubules within the living cell [Mitchison and Kirschner, 1984c]. Approaches similar to the recent studies of Soltys and Borisy [1985] should provide clear information on sites of tubulin incorporation into the various types of spindle microtubules at the different stages of mitosis by using secondary labels for DTAF-tubulin, or other tubulin analogs, for electron microscopy. Of particular interest are the sites of tubulin incorporation and dissociation for the differentially stable kinetochore fiber microtubules during chromosome movements. If the dynamic instability model correctly explains the dynamics of the spindle microtubules, then microtubule growth is much faster than expected from microtubule growth rates measured in vitro for tubulin concentrations comparable to the concentration of tubulin in cells [Salmon 1984; Salmon et al., 1984c]. It would be useful to know what oligomer is the polymerizing subunit of tubulin in vivo. Once the dynamics of microtubules are established, then valuable information on the dynamics of microtubule-associated proteins can be achieved by using FRAP approaches [McIntosh et al., 1986; Scherson et al., 1984]. These approaches will be valuable, in particular, for determining the molecular and physiological basis of regulation of microtubule assembly in living cells.

## ACKNOWLEDGMENTS

We thank Nancy Pryer, David Battigelli, John Brockington, Lynne Cassimeris, Dick McIntosh, Bill Saxton, Roger Leslie, and Ken Jacobson for their helpful discussions. Thanks to Nancy Salmon for editing and to Lans Taylor for waiting forever for this paper. This work is supported by NIH GM-24364.

## REFERENCES

Axelrod D, Koppel DE, Schlesinger J, Elson E, Webb WW (1976): Mobility measurement by analysis of fluorescence photobleaching recovery kinetics. Biophys J 16:1055–1069.

Bershadsky AD, Gelfand VI (1981): ATP-dependent regulation of cytoplasmic microtubule disassembly. Proc Natl Acad Sci USA 78:3610–3613.

Carlier M-F, Hill TL, Chen Y (1984): Interference of GTP hydrolysis in the mechanism of microtubule assembly: An experimental study. Proc Natl Acad Sci USA 81:771–775.

Cassimeris LU, Wadsworth P, Tioran M, Salmon ED (1985): Dynamic instability of cytoplasmic microtubules in vivo. J Cell Biol 101: 24a.

Debrabander M, Geuen G, Nuydens R, Willebrords R, DeMey J (1981): Microtubule assembly in living cells after release from nocodazole block: The effects of metabolic inhibitors taxol and pH. Cell Biol Int Rep 5:913–927.

Dustin P (1984): "Microtubules." New York: Springer-Verlag.

Evans L, Mitchison T, Kirschner M (1985): Influence of the centrosome on the structure and nucleation of microtubules. J Cell Biol 100:1185–1191.

Forer A (1965): Local reduction of spindle fiber birefringence in living *Nephrotoma sutarales* (Loew) spermatocytes induced by ultraviolet microbeam irradiation. J Cell Biol 25:96–117.

Graessmann A, Graessmann M, Mueller C (1980): Microinjection of early SV40 DNA fragments and T antigen. Methods Enzymol 65:816–825.

Hamaguchi Y, Toriyama M, Sakai H, Hiramoto Y (1985): Distribution of fluorescently labeled tubulin injected into sand dollar eggs from fertilization through cleavage. J Cell Biol 100:1262–1272.

Hill T, Kirschner M (1982): Bioenergetics and kinetics of microtubule and actin filament assembly-disassembly. Int Rev Cytol 78:1–125.

Inoué S (1961): Polarizing microscope: Design for maximum sensitivity. In Clarke GL (ed): "The Encyclopedia of Microscopy." New York: Reinnold, pp 480–485.

Inoué S (1981): Cell division and the mitotic spindle. J Cell Biol 91(pt2):131s–147s.

Inoué S, Sato H (1967): Cell motility by labile association of molecules. J Gen Physiol 50:259–292.

Inoué S, Ritter H (1975): Dynamics of microtubule organization and function. In Inoue S, Stephens R (eds): "Molecules and Cell Movement." New York: Raven Press, pp 3–30.

Jacobson K, Wojcieszyn J (1984): The translational mobility of substances within the cytoplasmic matrix. Proc Natl Acad Sci USA 81:6747–6751.

Jacobson K, Elson E, Koppel D, Webb W (1983): International workshop on the application of fluorescence photobleaching to problems in cell biology. Fed Proc 42:72–79.

Keith CH, Feramisco JR, Shelanski M (1981): Direct visualization of fluorescein-labeled microtubules in vitro and in micro-injected fibroblasts. J Cell Biol 88:234–240.

Lanni F, Taylor DL, Ware BR (1981): Fluorescence photobleaching recovery in solutions of labeled actin. Biophys J 35:351–364.

Leslie RJ, Pickett-Heaps JD (1983): Ultraviolet microbeam irradiation of mitotic diatoms: Investigation of spindle elongation. J Cell Biol 96:548–561.

Leslie RJ, Saxton WM, Mitehison TJ, Neighbors B, Salmon ED, McIntosh JR (1984): Assembly properties of fluorescein—labeled tubulin *in vitro* before and after fluorescence bleaching. J Cell Biol 99:2146–2156.

Mandelkow, M-E, Mandeldow E (1985): Unstained microtubules studied by cryo-electron microscopy. Substructure, supertwist and disassembly. J Mol Biol 181:123–135.

Margolis RL, Wilson L (1981): Microtubule treadmills—possible molecular machinery. Nature 293:705–711.

Margolis RL, Wilson L, Keifer BI (1978): Mitotic mechanism based on intrinsic microtubule behavior. Nature 272:450–452.

McIntosh JR, Euteneuer U (1984): Tubulin hooks as probes for microtubule polarity: An analysis of the method and an evaluation of data on microtubule polarity in the mitotic spindle. J Cell Biol 98:525–536.

McIntosh JR, Saxton WM, Stemple DL, Leslie RJ, Welsh MJ (1986): Dynamics of tubulin and calmodulin in the mammalian mitotic spindle. Ann NY Acad Sci (in press).

Mitchison T, Kirschner M (1984a): Microtubule assembly nucleated by isolated centromeres. Nature 312:232–237.

Mitchison T, Kirschner M (1984b): Dynamic instability of microtubule growth. Nature 312:237–242.

Mitchison T, Kirschner M (1984c): Microtubule dynamics and cellular morphogenesis. In Borisy GG, Cleveland DW, Murphy DB (eds): "Molecular Biology of the Cytoskeleton." Cold Spring Harbor, NY: Cold Spring Harbor Laboratory, pp 27–44.

Purich DL, Kristofferson D (1984): Microtubule assembly: A review of progress, principles and perspectives. Adv Protein Chem 36:133–211.

Rieder CL (1982): The formation, structure, and composition of the mammalian kinetochore and kinetochore fiber. Int Rev Cytol 79:1–57.

Salmon ED (1975): Spindle microtubules: Thermodynamics of in vivo assembly and role in chromosome movement. Ann NY Acad Sci 253:383–406.

Salmon ED (1984): Tubulin dynamics in microtubules of the mitotic spindle. In Borisy GG, Cleveland DW, Murphy DB (eds): "Molecular Biology of the Cytoskeleton." Cold Spring Harbor, NY: Cold Spring Harbor Laboratory, pp 99–110.

Salmon ED, Goode D, Maugel TK, Bonar DB (1976): Pressure-induced depolymerization of spindle microtubules. III Differential stability in HeLa cells. J Cell Biol 69:443–454.

Salmon ED, McKeel M, Hays T (1984a): Rapid rate of tubulin dissociation from microtubules in the mitotic spindle in vivo measured by blocking polymerization with colchicine. J Cell Biol 99:1066–1075.

Salmon ED, Saxton WM, Leslie RJ, Karow ML, McIntosh JR (1984b): Diffusion coefficient of fluorescein-labeled tubulin in the cytoplasm of embryonic cells of the sea urchin. J Cell Biol 99:2157–2164.

Salmon ED, Leslie RJ, Saxton WM, Karow ML, McIntosh JR (1984c): Spindle microtubule dynamics in sea urchin embryos: Analysis using a fluorescein-labeled tubulin and measurements of fluorescence redistribution after laser photobleaching. J Cell Biol 99:2165–2174.

Saxton WM, Stemple DL, Leslie RJ, Salmon ED, Zavortink M, McIntosh JR (1984): Tubulin dynamics in cultured mammalian cells. J Cell Biol 99:2175–2186.

Scherson T, Kreis TE, Schlessinger J, Littauer UZ, Borisy GG, Geiger B (1984): Dynamic interactions of fluorescently labeled microtubule associated proteins in living cells. J Cell Biol 99:425–434.

Schneider MB, Webb W (1981): Measurement of submicron laser beam radii. Appl Optics 20:1382–1388.

Soltys B, Borisy GG (1985): Polymerization of tubulin in vivo: Direct evidence for assembly onto microtubule ends and from centrosomes. J Cell Biol 100:1682–1689.

Taylor DL, Amato PA, Luby-Phelps K, McNeil P (1984): Fluorescence analogue cytochemistry. Trends Biochem Sci 9:88–91.

Telzer BR, Haimo LT (1981): Decoration of spindle microtubules with dynein: Evidence for uniform polarity. J Cell Biol 89:373–378.

Vandre DD, Kronebusch P, Borisy GG (1984): Interphase-mitosis transition: Microtubule rearrangements in cultured cells and sea urchin eggs. In Borisy GG, Cleveland DW, Murphy DB (eds): "Molecular Biology of the Cytoskeleton." Cold Spring Harbor, NY: Cold Spring Harbor Laboratory, pp 1–16.

Wadsworth P, Sloboda R (1983): Microinjection of fluorescent tubulin into dividing sea urchin eggs. J Cell Biol 97:1249–1254.

Wadsworth P, Sloboda RD (1984): Interaction of bimane-labeled fluorescent tubulin with the isolated mitotic apparatus. Cell Motil 4:183–196.

Wadsworth P, Salmon ED (1986a): Microtubule dynamics in mitotic spindles in living cells. Ann NY Acad Sci (in press).

Wadsworth P, Salmon ED (1986b): Fluorescent analogs to tubulin. Methods Enzymol (in press).

Wadsworth P, Salmon ED (1986c): Analysis of microtubule dynamics in mitotic spindles using fluorescence redistribution after photobleaching. J Cell Biol (in press).

Wadsworth P, Salmon ED (1986d): Analysis of the treadmilling model for mitosis at metaphase using measurements of fluorescence redistribution after photobleaching. J Cell Biol (in press).

Yguerabide J, Schmidt JA, Yguerabide EE (1982): Lateral mobility in membranes detected by fluorescence recovery after photobleaching. Biophys J 39:69–75.

Zeeburg B, Reid R, Caplow M (1980): Incorporation of radioactive tubulin into microtubules at steady state: Experimental and theoretical analysis of diffusional and directional flux. J Biol Chem 255:9891–9899.

*Applications of Fluorescence in the Biomedical Sciences, pages 405–422*


# 18 Patterns of Individual Molecular Motions Deduced From Fluorescent Image Analysis

*Watt W. Webb and David Gross*

Detection of individual cell-surface receptor molecules and the microscopic imaging of their changing distributions on living cells in culture is made possible by the availability of sufficiently bright fluorescent labels and sensitive detection systems. Our development of a brightly fluorescent, biologically active derivative of low-density lipoprotein (LDL) has enabled observations of the distribution and movements of the individual LDL receptor molecules that are specific for LDL on the surface of human fibroblasts.

We have investigated the mobility of LDL receptors on the surface of the mutant line GM2408A which lacks the ability to internalize LDL at coated pits. Cluster size distribution, tracking of individual molecules on the cell surface during Brownian motion and concerted drive motion, and internalization have been analyzed by using digital-intensified video images. Electrophoretic migration, "equilibrium" segregation, and derivative responses have been analyzed by statistical analysis of sets of fluorescence photomicrographs. This research showing various aspects of the dynamics of individual receptors on living cell surfaces has been carried out by Larry Barak, Eugene Nothnagel, David Tank, William Fredericks, and David Gross in our laboratories at Cornell during intermittent studies over a period of several years. Although still far from complete, this research illustrates the experimental potential of the ability to track and analyze the behavior of individual cell-surface receptors.

Department of Applied and Engineering Physics, Cornell University, Ithaca, New York 14853.

The potential of these methods will become available for other receptors as suitable bright labels are developed. One promising approach depends on binding a single antibody to brightly fluorescent phycobiliprotein.

## THE RECEPTOR MOBILITY PROBLEM

The characteristically small and restricted diffusibility and modulatable immobilization of membrane proteins on the surfaces of living eucaryotic cells present a problem that resists understanding after more than a decade of systematic measurements with fluorescent markers of proteins and lipids and careful structural and biochemical studies [Webb et al., 1981]. Fluorescence images readily provide a measure of the distribution, at about 1-$\mu$m resolution, of cell-surface proteins specifically labeled with fluorescent antibodies or toxin or nonspecifically labeled directly or with chemical ligands as haptens. This technology ordinarily allows detection of receptor concentrations as low as a few hundred per square micron. Below this concentration, autofluorescence of the cell usually becomes obtrusive. Image-intensified video cameras and digital image recording with background subtraction allow convenient quantitation of these data at shot-noise-limited rates of, say, one image per second.

The dynamics of diffusibility of many membrane components have been studied extensively on many cell types in culture by using fluorescence photobleaching recovery (FPR, alias FRAP) techniques [Axelrod et al., 1976; Jacobson et al., 1982, 1983]. In these experiments the concentration of fluorescent label is locally perturbed by irreversibly photobleaching the chromophore locally within a bright laser beam focused to a spot or pattern. Then the kinetics of recovery of the fluorescence intensity within the photobleached pattern are measured as fresh fluorescent-labeled membrane components diffuse into the bleached areas. Quantitative analysis provides the diffusion coefficient of the labeled components. They can reveal systematic cooperative molecular transport that might be generated by cytoskeletal contractions, membrane flows, and aggregation, provided that they occur on a macroscopic scale.

Local fluorescence intensity fluctuations also provide measures of molecular diffusion, chemical reactions, flow, and transport in an applied force field [Magde et al., 1972, 1974, 1978; Elson and Magde, 1974; Elson and Webb, 1975; Webb, 1976]. However, these fluorescence correlation spectroscopy (FCS) methods have not been particularly successful in studies on living cells due to interference by extraneous fluctuations.

FPR measurements during the last 15 years have established distinctive characteristics of protein diffusion on the surfaces of most vertebrate cells in culture [Webb et al., 1981]. Generally protein diffusion coefficients D are several orders of magnitude smaller than the lipid fluidity would allow and usually a significant fraction, sometimes all, of each protein is virtually immobile. Anchorage of membrane proteins to the cytoskeleton, to an insoluble matrix or to the glycocalix provides a conceivable source of the extra restraints on protein diffusion in cell membranes, but more complex processes seem to be implicated. Their identity remains to be established. There is one clear feature: the diffusion restraints can be eradicated by treatment of cells to form blebs or bulbous cell forms where the plasma membrane is separated from the bulk of the cytoskeletal cortex. On blebs the membrane proteins diffuse nearly as fast, $D \approx 3 \times 10^{-9}$ $cm^2/sec$, as expected for free protein molecules in the viscous lipid membrane. Pure proteins reconstituted into model lipid membranes also diffuse as fast, $D \approx 5 \times 10^{-9}$, as expected from lipid viscous drag. The rhodopsin in disc membranes of visual rod cells which have little associated cytoskeleton also diffuses at the viscous limit [Poo and Cone, 1974]. Electrophoresis of proteins on the cell surface sometimes seems to increase their post-electrophoresis diffusibility. If clustering is induced by incubation after electrophoresis, it may also reduce diffusibility [Poo, 1981].

Because FPR requires very strong illumination during photobleaching, it is reasonable to suspect that photo-induced cross linking generates protein diffusion restraints. Although protein denaturation or cross linking certainly can be induced by protracted exposure to visible light, no evidence of this effect has been found in rather extensive controls on FPR [Webb et al., 1981]. Some favorite controls are based on repeated bleaches of the same spot with the same or different fluorophores. Also, increased mobility on blebs argues against photo-induced immobilization. Most convincing, however, is the elementary general observation that cell physiology requires restraints on membrane protein mobility as necessary and essential properties of the cell surface; otherwise, no cell-surface organelle could be preserved for significant times. Thus modulatable diffusion restraints are necessary properties of the cell surface. For example, the acetylcholine receptor is nearly immobile in patches and in neuromuscular junctions while it diffuses at typically slow rates elsewhere on the cell surface.

Many properties of cell-surface receptors depend critically on their mobility and effects of ligand binding on mobility and diffusibility. Here we take the term mobility to imply possible molecular movements that may be driven in ways not ordinarily encompassed by the definition of diffusion from a physical point of view as a random dissipative process.

To understand these somewhat mysterious dynamic processes, it is necessary to find more effective probes than FPR or fluorescence photographs. Recall that FPR is based on the time course of measurements at one point on the cell surface corresponding to one resolution-limited pixel in an image.

Fluorescence photography provides topographical concentration data but suffers from limited sensitivity and lack of useful time dependence. What one would like is the means to observe the dynamics of individual receptors on the living cell surface with higher spatial resolution. Realization of this capability at the resolution of the light microscope is the objective of the work summarized in this paper. The molecule with which we have begun is the low-density lipoprotein receptor (LDL-R).

## LOW-DENSITY LIPOPROTEIN AS A SPECIFIC MARKER OF ITS RECEPTOR MOLECULES

The LDL-R binds LDL specifically and serves to convey it across cell membranes via interaction with clathrin-coated pits, surface organelles that provide the pathway for the receptor-mediated internalization of many cell-surface ligands. LDL carries cholesterol ester from the plasma to lysosomes within fibroblastic cells where it is metabolized in a key step of its regulation.

The feature of LDL that has made it particularly useful for the study of receptor dynamics is the development of a fluorescent analog that is so bright that individual biologically active fluorescence-labeled particles are readily visible bound to their receptors on the living cell surface. This derivative of diI-LDL was developed in our laboratory by L.S. Barak several years ago and has provided the foundation for the research that is reported here [Barak and Webb, 1981].

Of particular interest in the behavior of the LDL receptor is a sustained human cell line, J.D. or GM2408A, obtained from a victim of a familial hypercholesterolemia that occurs due to a mutation which apparently prevents binding of the LDL-R to coated pits and drastically reduces its internalization rate [Brown and Goldstein, 1976]. DiI-LDL retains the biological properties of native LDL and can be readily synthesized by inserting about 30–50 copies of a fluorescent lipid analog 3,3′ indocarbocyanine into the LDL particle where we presume that it replaces some of the phosphatidyl choline in the outer layer [Barak and Webb, 1981].

Barak and Webb [1982] measured the diffusion of the LDL receptor on J.D. cells and found that although it was very slow it still diffused fast enough to supply coated pits with LDL-R at a sufficient rate to accommodate normal internalization [Barak and Webb, 1982; Goldstein et al., 1984]. In spite of the exceptionally slow diffusion on intact cells, we found the diffusion of the LDL-R on blebs to increase by more than two orders of magnitude, approaching the value limited by lipid viscosity.

Individual particle trajectories were recorded on videotape through a fluorescence microscope with a silicon-intensified video camera; they were plotted by hand to derive the incremental mean square displacements as a function of diffusion time. From the slopes of the plots of the displacements in Brownian random motion the diffusion coefficents were estimated. This crude experiment was the first dynamic experiment using the detectability of individual LDL particles and is described here to illustrate the simple idea that we have subsequently pursued with more elegant instrumentation.

## APPLICATION OF DIGITAL IMAGE-INTENSIFIED VIDEO FLUORESCENCE MICROSCOPY TO INDIVIDUAL LDL-R DYNAMICS

The first point to consider in appraising video technologies is the remarkable effectiveness of ordinary photography. Figure 1 shows a typical distribution of LDL-R labeled with dil-LDL on a J.D. cell as recorded on standard ASA 400 film that was push processed a bit. Each bright dot is a single LDL-R or a small unresolved cluster of less than five liganded receptors. There are only a few thousand particles per cell, and each individual or cluster is clearly resolved. The size of the sharpest bright dots is determined by the resolution of the microscope and of course some are blurred by their position relative to the plane of focus. The photographic process is cheap, has not degraded the resolution, and shows the desired particles. A comparable video image of dil-LDL-R on a J.D. cell is shown in Figure 2. So why invest in elaborate image-intensified video cameras and digital image storage and processing? The reasons become clear below.

The exposure to enough excitation radiation required to record one photograph photobleaches much of the fluorophore because the quantum efficiency of the photographic processes is very low. Thus around 99% of fluorescence photons are lost. Photobleaching limits the number of recordable images and thus the length of the sequence available for recording dynamic processes. Strong whole cell light exposure damages many cells. Image-intensified quantum detectors provide many orders of magnitude of improved sensitivity. Precise quantitative fluorescence emission power measurements are required to determine whether our light spots correspond to a single receptor or a cluster. The fluorescence power of these sources is not given by the size of their saturated photographic image spots. To obtain sufficient quantitative dynamic range in photography a series of exposures would be required. In contrast the video system provides precise intensity data which is easily analyzed via digital processing. Thus the sacrifice of some of the spatial resolution of photography may be traded for the quantitative intensity analysis and the improved sensitivity due to higher quantum efficiency of intensified video detection and digital recording and analysis. The spatial resolution of video methods can, of course, be improved to match the state of the art in photography but only at great expense at the present time.

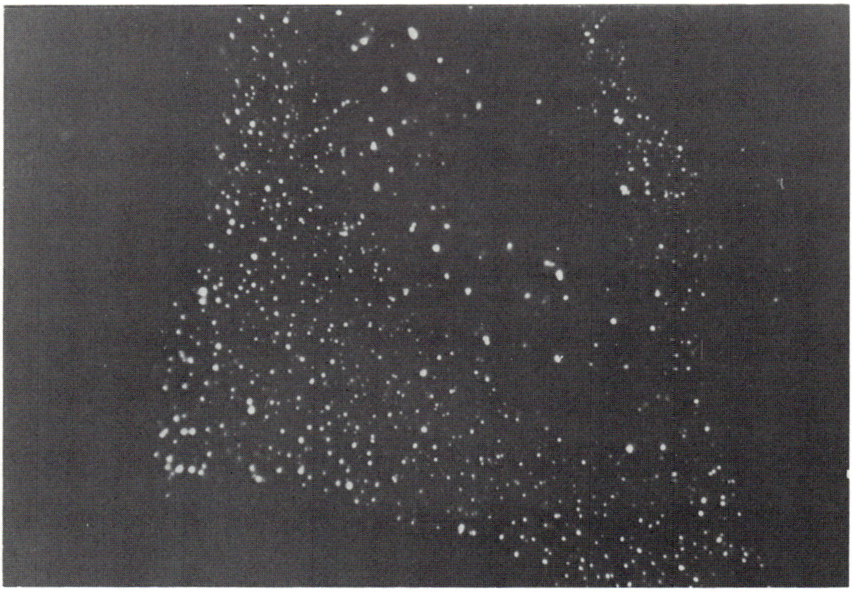

Fig. 1. Fluorescence photomicrograph of diI-low-density lipoprotein (LDL) receptor complexes on one J.D. cell recorded on film. Dimension of cell about 30 μm.

The video image acquisition and recording system and the software for digital control and analysis used for this research evolved in our laboratory just slightly too early to benefit from commercial systems and software development that now provide comparable capability off the shelf.

## DIGITAL VIDEO MICROSCOPY

All measurements described here were carried out on a Zeiss Universal Microscope (Carl Zeiss, Inc., 1 Zeiss Drive, Thornwood, NY 10594) equipped with phase-contrast optics, mercury arc epi-illumination for fluorescence excitation, and a 100× 1.25 N.A. phase-contrast objective. The microscope was interfaced to a Venus TV2M image intensifier vidicon camera (Venus Scientific, Inc., 339 Smith Street, Farmingdale, NY 11735) for these experiments although the Venus TV3M three-stage intensifier we currently use is superior for dimmer sources. The video output of the camera is fed into the digitizing input of a Grinnell model 274 image-processing system (Grinnell Systems, 2159 Bering Drive, San Jose, CA 95131) which is under the interactive control of a PDP 11/23 computer (Digital Equipment Corporation, Maynard, MA 01754). Digitized video images are stored in

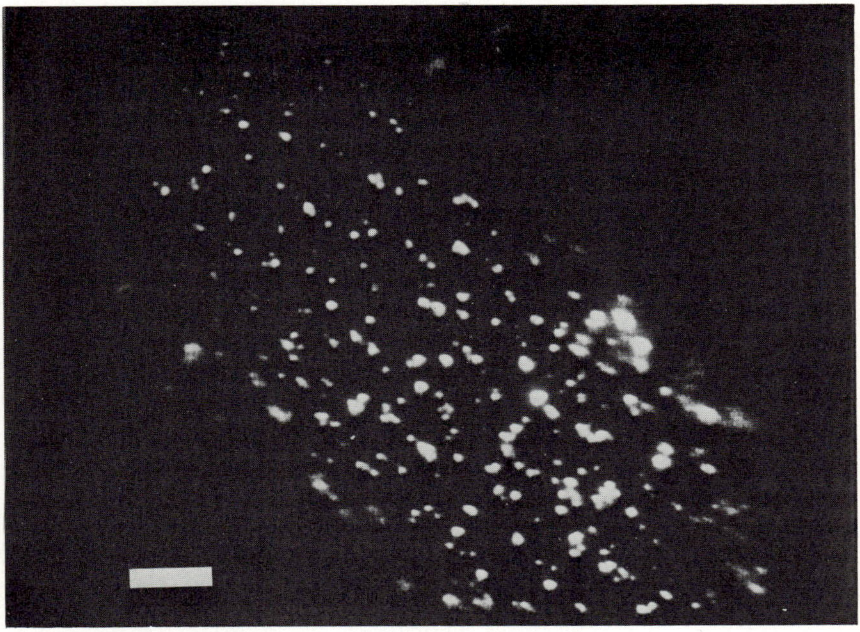

Fig. 2. Video image of diI-LDL receptor complexes on a J.D. cell recorded by using intensified video camera system for comparison with photographic image. Bar is 5 $\mu$m.

real time in the three 512 × 480 × 8 bit volatile memory planes of the image-processing system and are stored by computer data transfer on an 80 Mbyte Winchester hard disk (Model 5380, Kennedy Company, 1600 Shamrock Avenue, Monrovia, CA 91016). Figure 3 presents a schematic diagram of one version of the complete system.

The video signal from the camera is digitized to eight-bit resolution by the image-processing system. The digitized image is routed through video lookup tables, a 16-bit arithmetic and logic unit, a zoom and pan subsystem, and through a digital-to-analog converter which generates red-green-blue (RGB) video signals which drive a high-resolution color monitor (model C-3910, Mitsubishi Electric Company, Tokyo, Japan). All signal routing is under interactive computer control. Digital images stored in any of the three memory planes can be rapidly analyzed by image-processing hardware for the gray-level intensity content of a full frame, the maximum and minimum gray levels in a frame, and for the gray-level values in any of the 245,760 pixels in the frame. Specialized interactive software has been developed to measure the intensity content and spatial distribution of the digital images

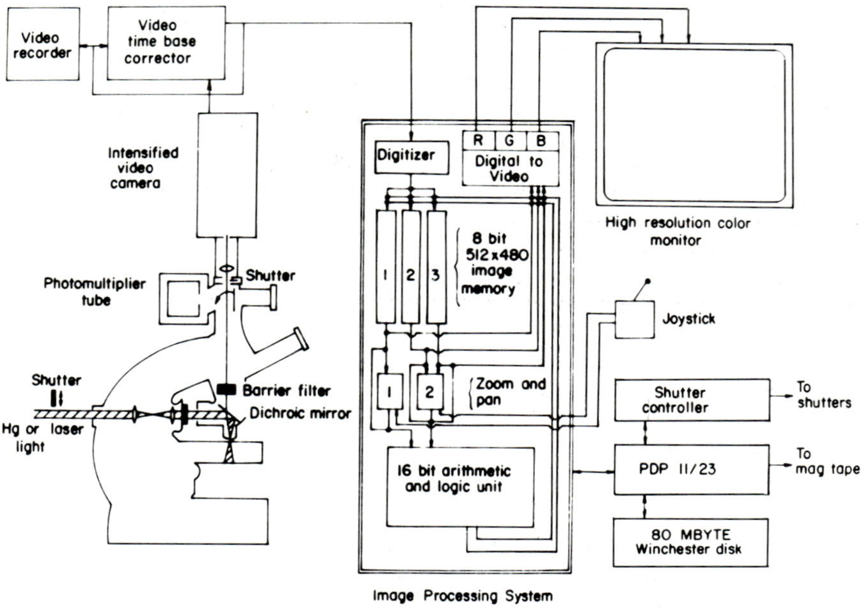


Fig. 3. Schematic diagram of video microscopy system with digital recording and analysis.

contained in the memory planes. In particular, software to aid in the collection of the fluorescence power from a large number of fluorescent spots in an image was developed. The software allows the user to store a local background intensity which is subtracted from the integrated intensity of a bright spot, also selected by the user.

Because the diI-LDL particles are so bright, only modest readily available image intensification is needed to record them, but the demands for quantitative intensity measurements with background subtraction, time and area integration, and pattern recording and analysis call for digital video analysis.

To illustrate the use of this system we describe two experiments, first, measurement of the clustering of LDL-R, and second, tracking of the movement of individual molecules.

## FLUORESCENCE EMISSION POWER AS AN ASSAY FOR LDL-CLUSTER SIZE

On normal cells LDL-R bind to coated pits ordinarily forming clusters of, say, 5 to 15 LDL-R per coated pit. On J.D. cells smaller clusters and individuals are found presumably because binding to coated pits is lost. How strongly LDL-R bind to each other remains an interesting issue. Since the size of the clusters of LDL-R expected on human fibroblasts is too small for spatial resolution of the individual receptors, the only measure of cluster size that is available is their fluorescence emission power [Gross and Webb, 1984]. Fortunately, the number of diI molecules absorbed per diI-LDL in our labeling process is distributed approximately in accord with Poisson statistics. Thus with a (measured) average number of diI per LDL of about $\bar{n} = 36$ the variance is $\sqrt{\bar{n}}$ or six. Since self-quenching of the fluorescence is negligible, the fluorescence power of an LDL particle will vary by roughly $\sigma_{\bar{n}} \approx 1/\sqrt{\bar{n}}$ or 16.7%. Thus cluster sizes of up to about five particles should be resolvable by the quantization of their fluorescence powers. Such an experiment is described below.

Consider the diI-LDL distribution on the J.D. cell shown in Figure 2. To measure the fluorescence power associated with each particle or cluster, the digital record is summed over the cluster image and the neighborhood background is subtracted. A bar graph of the distribution of these fluorescence power values is shown in Figure 4 with the statistical fit calculated by using the appropriate Poisson distribution of $\bar{n}$. Clearly it fits, with the only adjustable parameters being the numbers of particles per cluster size. The value of $\bar{n}$ was checked by absolute-intensity measurements which agreed with that determined by video microscopy. Figure 5 shows comparable data on two other cell types including normal human fibroblasts where larger clusters occur. The distribution of the numbers $N$ of LDL-R in the clusters is readily measured with integer quantization up to about six. In this case about one-fourth of the clusters on J.D. cells are singlets, one-fourth doublets, one-fourth triplets, and the remainder are primarily four or five diI-LDL-R complexes in the cluster.

The precision of the technique is illustrated by the data in Figure 5 which compares clustering of LDL-R on J.D. mutants, normal human fibroblast (prepared at low temperature to limit internalization), and an epidermoid carcinoma line of A-431 cells. The figure shows for each case the percentage of cluster sizes up to seven.

To determine whether there is a systematic spatial distribution of large or small cluster sizes, the clusters can be labeled in pseudocolor by cluster size for display. Figures 6a and 6b show such generated images, but created for black-and-white display by placing the counted number of liganded LDL-Rs at their respective locations. The case illustrated is a part of normal fibroblasts briefly incubated at room temperature to allow some cluster growth and at least incipient internalization to form large clusters. The large clusters $N > 6$, which form the brightest spots in this reproduction, are slightly grouped especially along an edge of one of the cells.

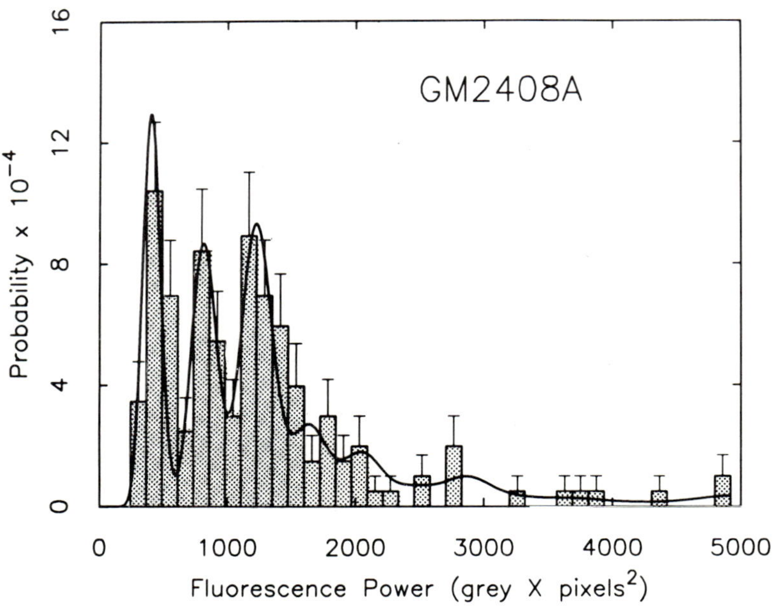


Fig. 4. Fluorescence power probability distribution for diI-LDL bound to J.D. cells. The bar graph is the raw (binned) data while the smooth curve is the theoretical multiple Poisson distribution.

## TRACKING OF THE MIGRATION OF INDIVIDUAL LDL-R ON J.D. CELLS

With the objective of gaining insight into the nature of the restraints on protein diffusion on cell surfaces, we have developed a procedure for tracking individual molecules with the help of the video system [Gross and Webb, 1983]. An early illustrative experiment in which about 20 molecules were tracked on a J.D. cell for about 20 minutes at room temperature shows a diversity of behavior amongst the individuals in the sample. Figure 7a is the initial fluorescence image at the start of a series of time-lapse frames collected at 20-second intervals. Figure 7b is near the end of the series. The bright object at one edge is a fluorescent contaminant on the cell. The dissimilarity in the distribution of diI-LDL on the cell at different times demonstrates the motion of some of the particles. More revealing, however, is the map of the trajectories of some of these same objects in Figure 7c. The outline of the cell is indicated by a continuous line. In this early experiment brighter spots were selected for convenience so the cases illustrated are probably clusters.

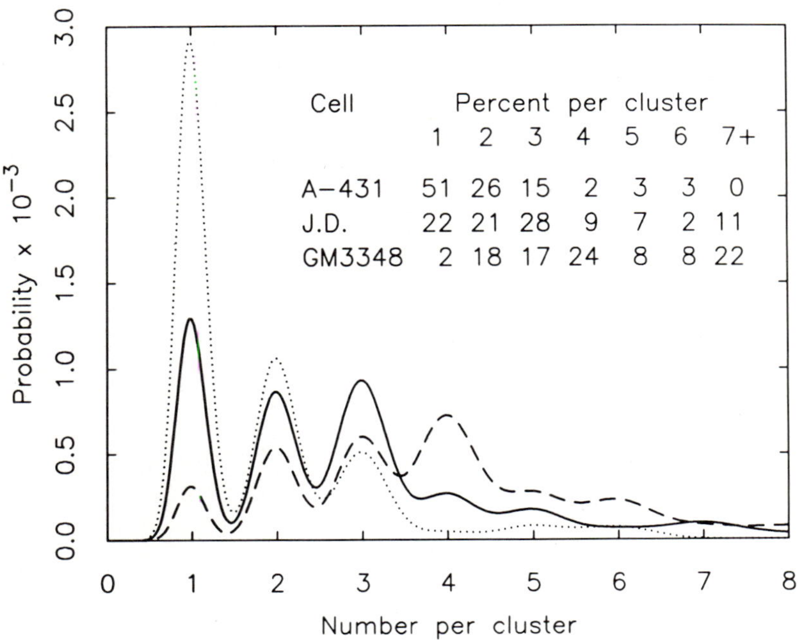


Fig. 5. Cluster size distributions compared for A-431 (dotted), J.D. (solid), and normal GM3348 (dashed) cells. The table section indicates the relative population of each cluster size for cells as indicated.

The process of disintegration or formation of individual clusters is not illustrated here but that is not to imply that cluster sizes are fixed. The particle motion can be analyzed to determine effective diffusion coefficients *D* from the statistics of the Brownian motion and any concerted velocity can be recognized in the features of plotted trajectories.

Note in Figure 7c the variety of patterns of motion illustrated. Some particles are virtually immobile with $D < 10^{-12}$ cm$^2$/sec; others display slow quasirandom motion in all directions with D $\approx 10^{-11}$–$10^{-12}$ cm$^2$/sec. The values of *D* obtained by FPR experiments several years ago under similar conditions were around $10^{-11}$ cm$^2$/sec. A few particles near the edges of the cell show preferential velocities of drift toward or away from the edges. The motion of the group of three particles near the bottom right corner of the figure is clearly correlated at concerted velocity toward the edge of the cell of around 0.2 $\mu$m/min. The minor transverse random component of this motion approaches the jitter of the recording technique and is representable by a fictive $D < 10^{-12}$ cm$^2$/sec. These experiments are just reaching a stage

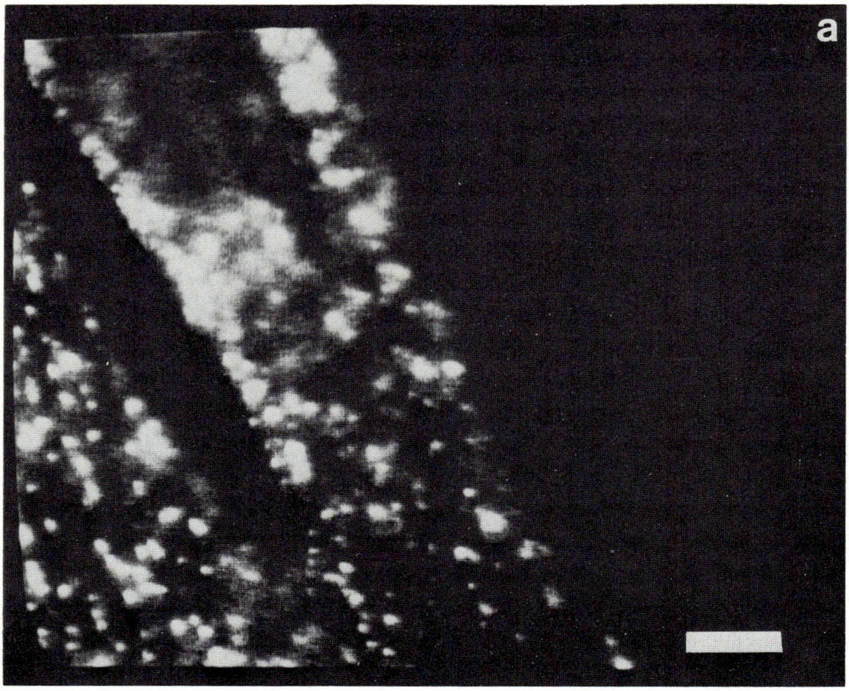


Fig. 6. The digitized image of two normal GM3348 fibroblast stained with diI-LDL at 22°C for 15 minutes (a), a condition which allows some endocytotic processing of LDL to proceed. The cluster distribution for this cell was displayed by the video system as a spatial map of the collected and analyzed data (b). Any fluorescence power values that were indeterminate between cluster values of $n$ and $n + 1$ were denoted as "$n+$" on the image. Values of 7 or more were denoted as 7+. Bar is 5 $\mu$m.

of development in our laboratory where they can provide useful biological information and the techniques are being applied with the objectives of understanding the constraints on cell-surface receptor diffusion.

## ELECTROPHORESIS OF LDL-R ON J.D. CELLS

Application of electric fields parallel to the surfaces of tissue cells in culture can induce lateral flow of cell-surface proteins. M.-m. Poo and associates [Poo, 1981] have measured the electrophoretic segregation of various receptors on several cell types. Applied fields of $\approx 10$ V/cm can induce rather complete segregation in several hours. Provided segregation is not maintained for too long a time, it relaxes rapidly toward the original

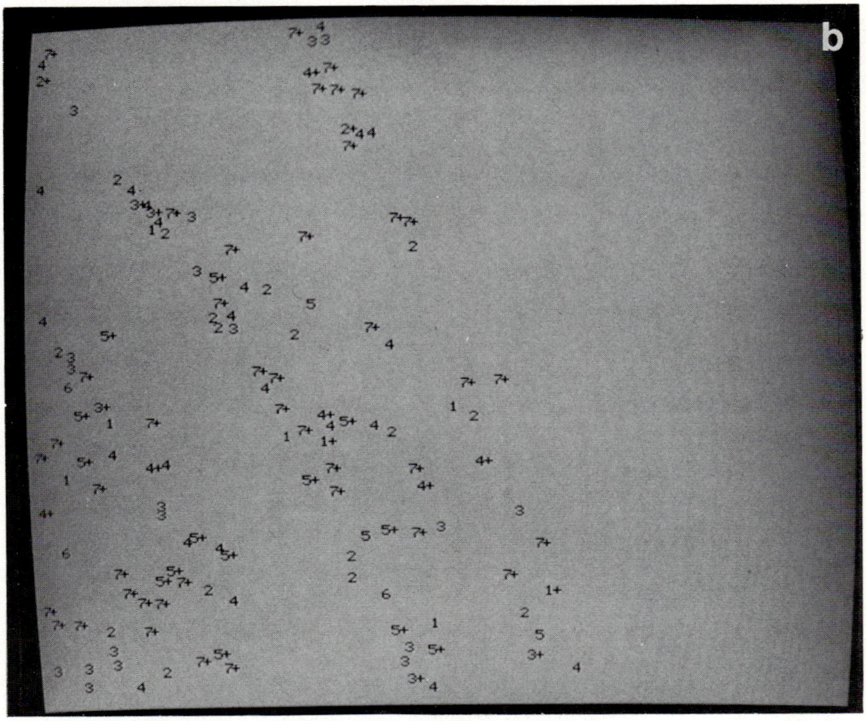


Fig. 6 (continued).

uniform distribution on turning off the field. The relaxation can be analyzed by measuring the time to restore a uniform distribution in order to determine the postfield diffusion coefficient.

In many cases the postfield diffusion is much faster ($D \approx 10^{-9}$ $cm^2$/sec) than prefield diffusion measured by FPR. It thus resembles diffusion on blebs where protein diffusion constraints have been disabled. On the other hand, if the electric field is left on to hold the segregation for several hours, some receptor types are immobilized at the edges of the cells. Several mechanisms can be invoked to account for these still-unexplained changes. We think it likely that electrophoresis disrupts the constraints that ordinarily limit cell-surface protein diffusion and that aging of the segregated population will restore them. Alternatively, the fluorescent label present on receptors during the diffusion measurements in FPR experiments may slow diffusion. Post-electrophoresis diffusive recovery is ordinarily carried out in the absence of the fluorescence label. We are not aware of any postelectrophoresis FPR measurements of diffusion.

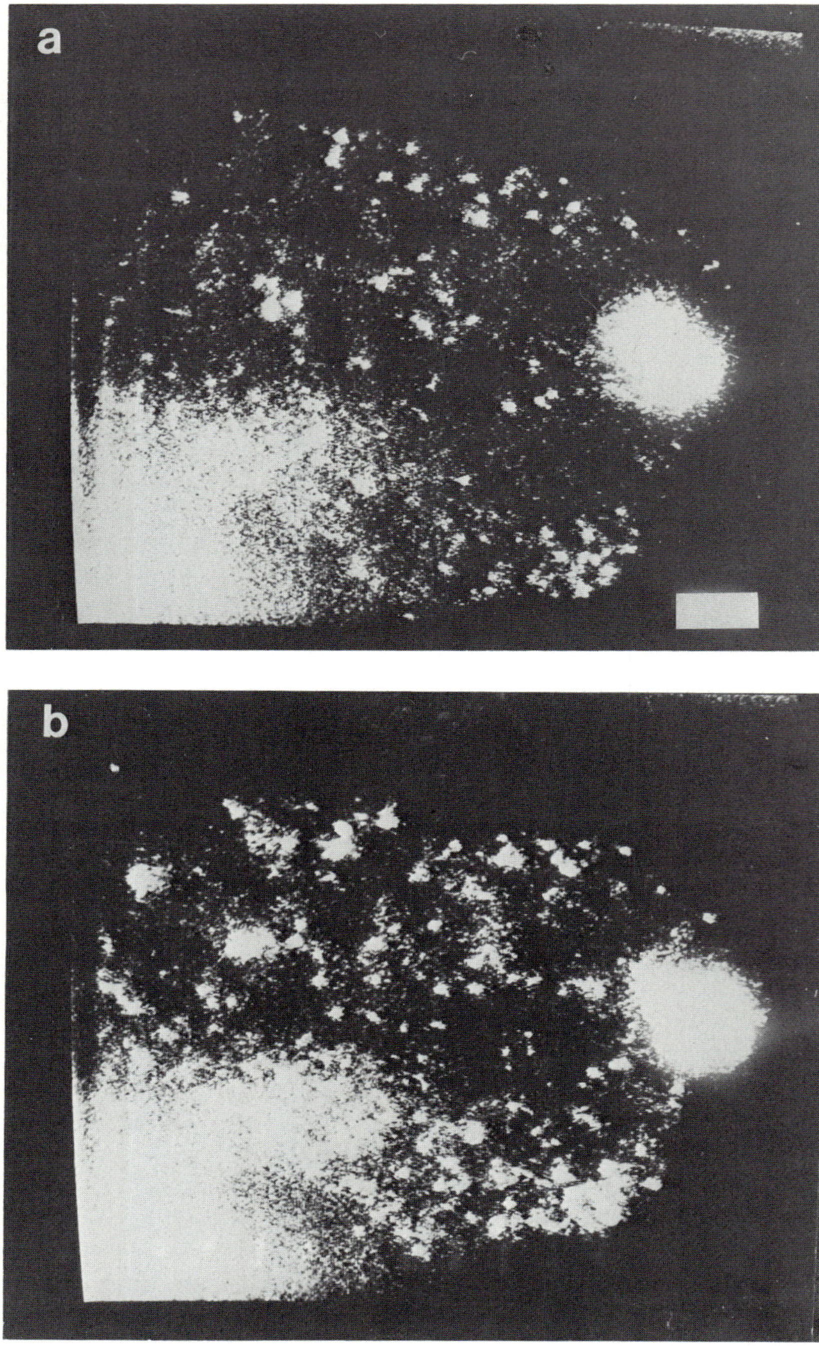


Fig. 7. The mobility of the liganded LDL receptor on J.D. cells. a. Zero-time video image of diI-LDL distribution. b. The distribution after 20 minutes. c. Tracks of several particles at 20-second time resolution. The heavy dots mark the initial position of the particles.

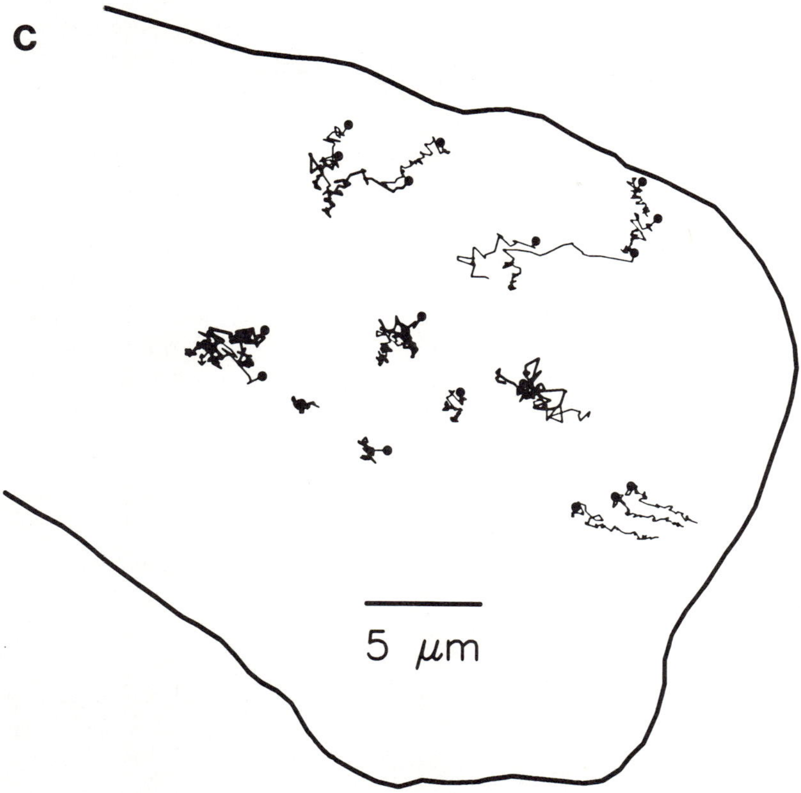


Fig. 7 (continued).

Measurements of electrophoretic segregation of LDL receptors on J.D. cells and subsequent relaxation were carried out in our laboratory by David Tank and William Fredericks several years ago [Tank et al., 1985]. It was found that the unliganded LDL-R diffused faster after electrophoretic segregation than did the liganded LDL-R. Because LDL is so very large it seemed a likely source of hydrodynamic drag in the aqueous phase and glycocalix. Subsequent experiments on glycoprotein diffusion in membranes in the presence of highly viscous aqueous media are being carried out in our laboratory by Patricia Donaldson. They tend to indicate that aqueous drag should be negligible. Changes of the interaction of LDL-R with the J.D. cell surface may be induced by electrophoresis since internalization of the LDL receptor complex—in these internalization-deficient mutants—seems to be enhanced by electrophoresis.

Electrophoretic segregation of the LDL receptor after about 1 hour at 10 V/cm is illustrated by Figure 8. For this figure the cell is fixed and stained with diI-LDL after electrophoresis. The cell morphology, the actin cytoskeleton, and the distribution of clathrin-coated pits are not changed by the electrophoresis.

The dynamics of the diffusive relaxation can be qualitatively analyzed to determine the diffusion coefficients of the LDL receptor as shown in Figure 9 [Tank et al., 1985]. Our analysis of data of the type illustrated in Figure 9 did not take explicit advantage of the capability for resolving and recording individual receptor molecules and clusters. The procedure was based on statistical analysis of measurements on many fixed cells of a go–no go criterion for residual segregation after diffusive relaxation for many times. However, the brilliant fluorescence of the diI-LDL label was necessary in order to detect and measure the inhomogeneous distribution of the sparse population of LDL receptors.

## SUMMARY

The ability to observe the distribution of very small receptor populations on living cell surfaces provides access to critical and elusive biological

Fig. 8. Distribution of LDL-R on a J.D. cell after electrophoresis.

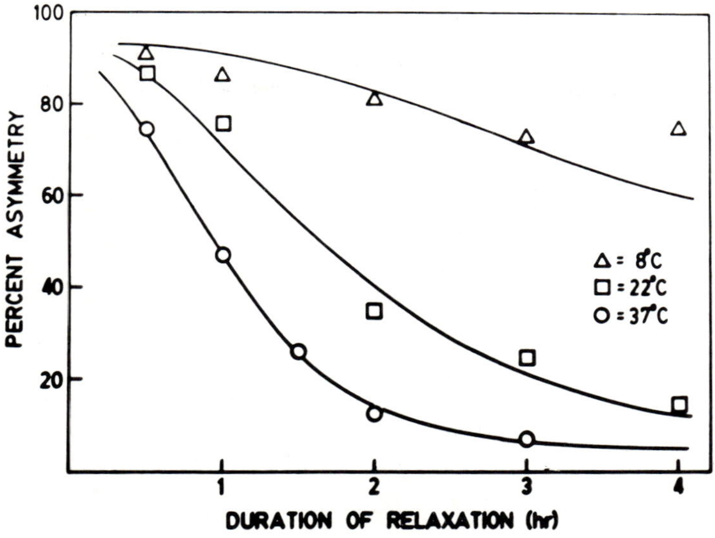


Fig. 9. Analysis of post-electrophoresis receptor distribution relaxation to determine diffusion coefficients.

processes. We continue to study the mobility and internalization of the LDL-R by tracking the behavior of individual receptors. Analogous experiments would help to elucidate the transmembrane signaling mechanisms involved in hormonal responses, particularly of growth hormones whose small populations of receptors appear to be internalized at coated pits. The development of suitable brilliant fluorescent labels that are receptor specific is necessary to follow this promising strategy. Some more abundant cell-surface receptors that aggregate in the course of their response are being tracked without need for resolution of individual molecules. Many such projects are under way in various laboratories using image-intensified video microscopy with digital image analysis. We forecast broad utility for this powerful tool.

## ACKNOWLEDGMENTS

This work was supported by grants from the NSF (PCM 83-03404), NIH GM33028-02, and the Cornell Biotechnology Program.

## REFERENCES

Axelrod D, Koppel DE, Schlessinger J, Elson E, Webb WW (1976): Mobility measurements by analysis of fluorescence photobleaching recovering kinetics. Biophys J 16:1055–1069.

Barak LS, Webb WW (1981): Fluorescent low density lipoprotein for observation of dynamics of individual receptor complexes on cultured human fibroblasts. J Cell Biol 90:595–604.

Barak LS, Webb WW (1982): Diffusion of low density lipoprotein-receptor complex on human fibroblasts. J Cell Biol 95:846–852.

Brown MS, Goldstein JL (1976): Analysis of a mutant strain of human fibroblasts with a defect in the internalization of receptor-bound low density lipoprotein. Cell 9:663–674.

Carley WW, Barak LS, Webb WW (1981): F-actin aggregates in transformed cells. J Cell Biol 90:797–802.

Elson EL, Magde D (1974): Fluorescence correlation spectroscopy. I. Conceptual basis and theory. Biopolymers 13:1–27.

Elson EL, Webb WW (1975): Concentration correlation spectroscopy: A new biophysical probe based on occupation number fluctuations. Ann Rev Biophys Bioeng 4:311–334.

Goldstein B, Griego R, Wofsy C (1984): Diffusion-limited forward rate constants in two dimensions: Application to the trapping of cell surface receptors by coated pits. Biophys J 46:573–586.

Gross D, Webb WW (1983): Time lapse video recording of individual molecular motions of LDL-receptor complex on living human fibroblasts. Biophys J 41:215a.

Gross D, Webb WW (1984): Molecular counting in small clusters of LDL on cell surfaces by fluorescence intensity quantization. Biophys J 45:269a

Jacobson K, Elson E, Koppel D, Webb WW (1982): Fluorescence photobleaching in cell biology. Nature 295: 283–284.

Jacobson K, Elson E, Koppel D, Webb WW (1983): International workshop on the application of fluorescence photobleaching techniques to problems in cell biology. Fed Proc 42:72–79.

Magde D, Elson E, Webb WW (1972): Thermodynamic fluctuations in a reacting system—Measurements by fluorescence correlation spectroscopy. Physiol Rev Lett 29:705–708.

Magde D, Elson EL, Webb WW (1974): Fluorescence correlation spectroscopy. II. An experimental realization. Biopolymers 13:29–61.

Magde D, Webb WW, Elson EL (1978): Fluorescence correlation spectroscopy. III. Uniform translation and laminar flow. Biopolymers 17:361–376.

Poo M-m (1981): In situ electrophoresis of membrane components. Ann Rev Biophys Bioeng 10:245–276.

Poo M-m, Cone RA (1974): Lateral diffusion of rhodopsin in the photoreceptor membrane. Nature 247:438–441.

Tank DW, Fredericks WJ, Barak LS, Webb WW (1985) Electric field-induced redistribution and postfield relaxation of low density lipoprotein receptors on cultured human fibroblasts. J Cell Biol 101:148–157.

Webb WW (1976): Applications of fluorescence correlation spectroscopy. Q Rev Biophys 9:49–68.

Webb WW, Barak LS, Tank DW, Wu E-S (1981): Molecular mobility on the cell surface. Biochem Soc Symp 46:191–205.

*Applications of Fluorescence in the Biomedical Sciences, pages 423–447*


# 19 Optical Measurement of Membrane Potential in Simple and Complex Nervous Systems

*Jill A. London, Dejan Zecevic, Leslie M. Loew, Harry S. Orbach, and Lawrence B. Cohen*

Optical measurements of membrane potentials using a molecular probe might be beneficial in a variety of circumstances. Such a probe could provide a useful technique for measuring membrane potential in systems where, for reasons of scale, topology, or of complexity, the use of electrodes is inconvenient or impossible. The possibility of using optical methods was first suggested in 1968 by the discovery of potential-dependent changes in intrinsic optical properties of squid giant axons [Cohen et al., 1968]. Shortly thereafter, Tasaki et al. [1968] found stimulus-dependent changes in fluorescence of stained axons and in 1971 we began a search for dyes which would give signals that were large enough to be useful for monitoring membrane potential. By now more than 1,000 dyes have been tested for their ability to act as molecular transducers of changes in membrane potential into changes in three types of optical signals: absorption, birefringence, and fluorescence. This screening effort has resulted in the discovery of dyes with a signal-to-noise ratio 100 times larger than was available from any signal in 1971. Several of these dyes (see, for example, Fig. 1) have been used to monitor changes in potential in a variety of preparations. For reviews see Cohen and Salzberg [1978], Waggoner [1979], Freedman and Laris [1981], Salzberg [1983], and

Department of Physiology, Yale University School of Medicine, New Haven, Connecticut 06510 (J.A.L., D.Z., L.B.C.); Department of Physiology, University of Connecticut Health Center, Farmington, Connecticut 06032 (L.M.L.); Division of Biology, California Institute of Technology, Pasadena, California 91125 (H.S.O.).

Grinvald [1985]. The review by Cohen and Lesher [1985] emphasizes experimental methods.

In this chapter we begin with the evidence that has been used to show that optical signals are potential dependent. Then we briefly discuss the mechanisms responsible for the signals. Finally, we discuss two applications, one from gastropod molluscs, where the number of photodetectors is approximately equal to the number of neurons and the individual action potentials of individual neurons are measured, and the second, from mammalian cortex, where light from very many neurons falls on each detector and only the average activity can be measured.

All of the optical signals described in this paper are fast signals as defined in Cohen and Salzberg [1978]. These signals are presumed to arise from membrane-bound dye and follow changes in membrane potential with time courses that are rapid compared to the rise time of an action potential. The use of a second, slower type of potential-dependent signal, the redistribution signals, is not discussed.

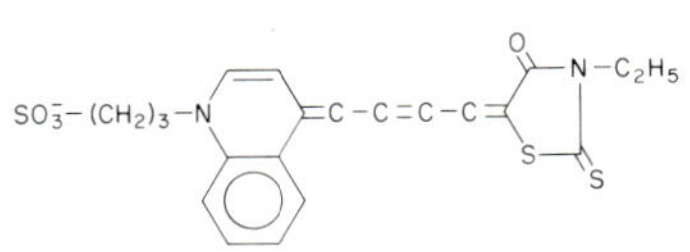

XVII, Merocyanine, Absorption, Birefringence

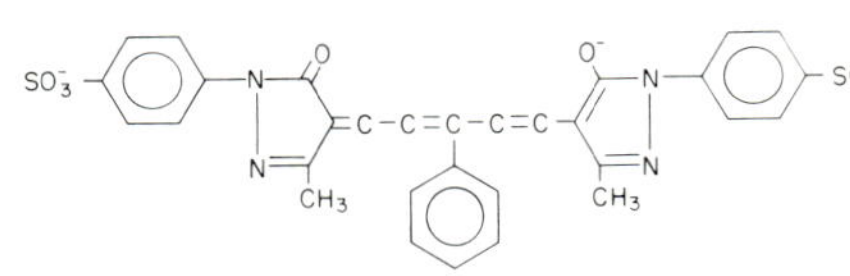

RH155, Oxonol, Absorption

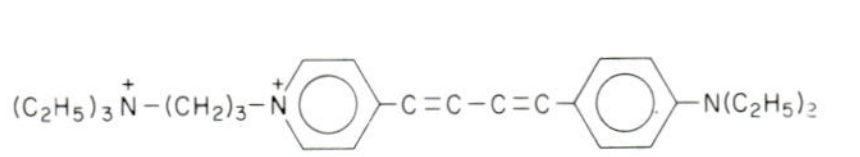

RH414, Styryl, Fluorescence

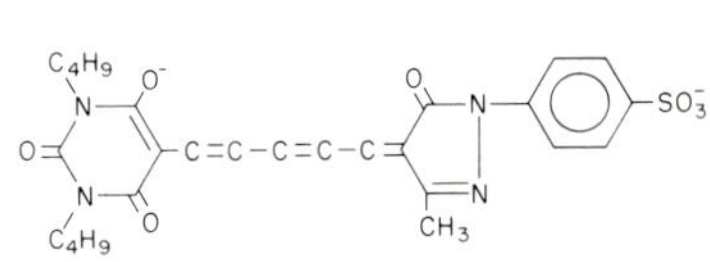

XXV, Oxonol, Fluorescence, Absorption

Fig. 1. Structures of several dyes which have been used to monitor membrane potential. The merocyanine (XVII) was the dye used in the experiments illustrated in Figures 3, and 4; the oxonol, RH155, was used in the experiments illustrated in Figures 10 and 11; and the styryl, RH414, was used in the experiments illustrated in Figures 12–14. Dye XVII and the oxonol XXV were first synthesized by Dr. A.S. Waggoner, Center for Fluorescence, Carnegie Mellon University, 4400 Fifth Ave., Pittsburgh, PA, as WW375 asnd WW781. The oxonol, RH155 and styryl, RH414, were first synthesized by Rina Hildesheim and Amiram Grinvald, Dept. of Neurobiology, Weizmann Institute, Rehovot, Israel. Dye XVII is available as NK2495 and RH155 is available as NK3041 from Nippon Kanko-Shikiso Kenkyusho Co., Ltd., 700 Okayama, Japan. RH414 is available as dye 1112 and dye XXV is available as W-435 from Molecular Probes, Junction City, Oregon.

## SOME OPTICAL SIGNALS ARE POTENTIAL DEPENDENT

The squid giant axon has provided a useful preparation for distinguishing among possible origins for optical signals and for testing new dyes. Figure 2 is a schematic drawing of the apparatus used to measure absorption and fluorescence: a very simple filter spectrophotometer and spectrofluorimeter. The axon is perpendicular to the page and is suspended in a cuvette filled with seawater. The light intensity reaching the photodiode in an absorption measurement is about $10^3$ times larger than in fluorescence. In birefringence measurements the intensity is intermediate between fluorescence and absorption.

Figure 3 shows the results of a measurement of light absorption during an action potential in a squid axon stained with a merocyanine dye (XVII, roman numbers refer to dyes in Figure 1 or in the tables of Ross et al. [1977] or Gupta et al. [1981]. The dotted trace is the light intensity transmitted through the axon at 750 nm; the smooth curve is the potential measured between internal and external electrodes. The axon membrane was space clamped, and the internal potential-sensing electrode was in the illuminated region of the axon, so that the optical and electrical measurements were made at the

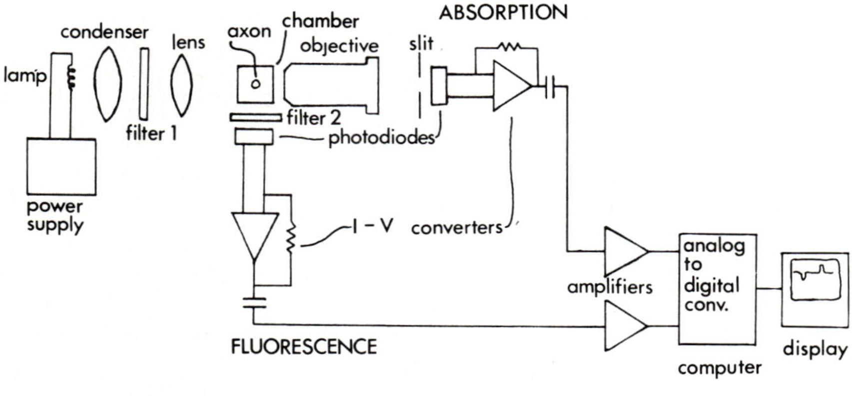


Fig. 2. Schematic diagram of the apparatus used for simultaneous measurements of absorption and fluorescence of a squid axon. Filter 1 was an interference filter whose width at half-height was usually 30 nm. The barrier filters (filter 2) were obtained from Schott Optical Glass, Inc. For the absorption measurements an image of the axon was formed with a 10× objective lens, and slits in the image plane were used to block the light that did not pass through the axon. Birefringence can be measured in the absorption path by inserting two polarizers, one between filter 1 and the axon, and the second between the objective and the photodetector. The polarizers are crossed (oriented perpendicular to each other) and at an angle of 45° to the long axis of the axon [from Ross et al., 1977].

Dye XVII

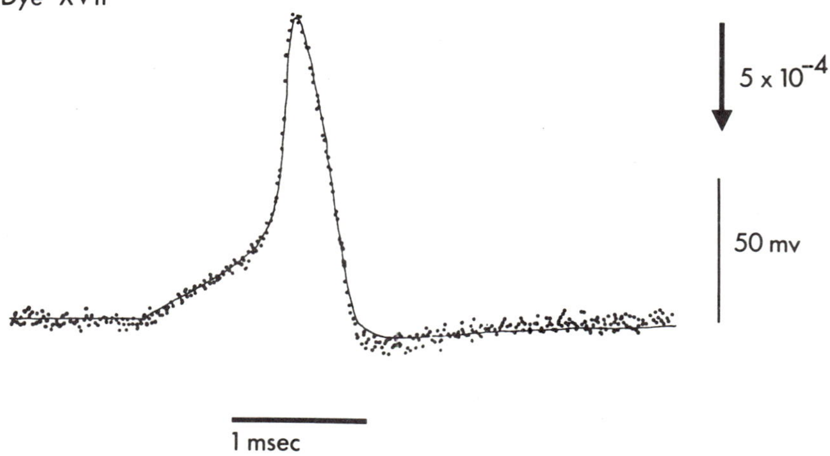


Fig. 3. The change in absorption (dots) of a giant axon stained with dye XVII during a membrane action potential (smooth trace) simultaneously recorded with an internal electrode. The change in absorption and the action potential had the same time course. In this figure and in Figure 4 the direction of the arrow adjacent to the optical trace indicates the direction of an increase in absorption; the size of the arrow represents the stated value of a change in absorption, $A$, in a single sweep divided by the resting absorption due to the dye, $A_r$. Incident light of 750 nm was used; 32 sweeps were averaged. The response time constant of the light measuring system was 5 $\mu$sec [from Ross et al., 1977].

same time and place. Because the two measurements had very similar time courses, it seemed likely that the absorption signal was related to the changes in membrane potential of the action potential and not to the ionic currents or the membrane permeability increases that occur during the action potential. More direct evidence for potential dependence can be obtained from voltage-clamp experiments such as the one illustrated in Figure 4. The top trace is the absorption signal; clearly, it has a time course similar to that of the membrane potential (middle trace) and distinctly different from that of the permeability changes or the ionic currents (bottom trace). This kind of result has been obtained for many dyes, including the four illustrated in Figure 1.

When an experiment was performed by using four potential steps, and the size of the absorption changes versus the size of the potential step was plotted, the absorption change is linearly related to membrane potential over the range ± 100 mV from the resting potential. In subsequent experiments with larger steps carried out with dyes XVII and XXII, the absorption signals were found to be linearly related to membrane potential over the range −130 to +200 mV from the resting potential [Gupta et al., 1981]. A linear

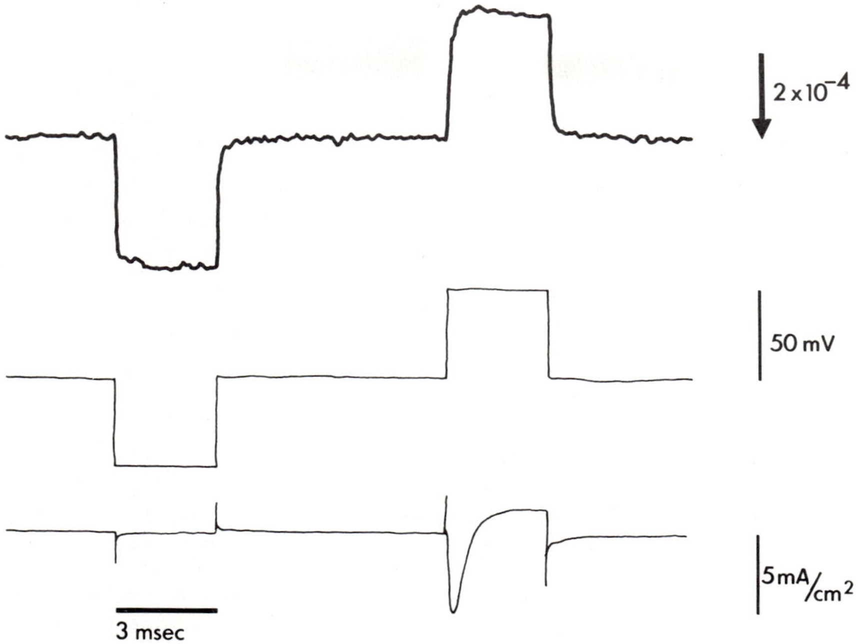


Fig. 4. Changes in absorption of a giant axon stained with dye XVII (top trace) during hyperpolarizing and depolarizing steps (middle trace). The bottom trace is the current density. The absorption changes had the same shape as the potential changes and were insensitive to the large currents and conductance changes that occurred during the depolarizing step. The holding potential was the resting potential, and hyperpolarization is represented downward; inward currents are downward. Incident light of 750 mn was used; 128 sweeps were averaged; the time constant of the light measuring system was 20 $\mu$sec [from Ross et al., 1977].

relationship between optical signal and membrane potential has been obtained with many dyes. Thus, in many instances there is strong evidence that the signals obtained with msec potential steps in squid axons depends in a linear manner on changes in the transmembrane potential [Cohen et al., 1970, 1974; Conti and Tasaki, 1970; Patrick et al., 1971; Davila et al., 1974; Conti, 1975; Ross et al., 1977; Gupta et al., 1981] although there was some earlier disagreement about this conclusion [Conti et al., 1971; Tasaki et al., 1972].

These dye signals can be rapid. Our initial attempts to measure differences between the potential change and the optical signal showed that some of the signals were too fast to measure and thus lagged behind the change in membrane potential by less than 10 $\mu$sec [Ross et al., 1977]. Measurements with a faster apparatus by Salzberg, Bezanilla, and Obaid, showed that the

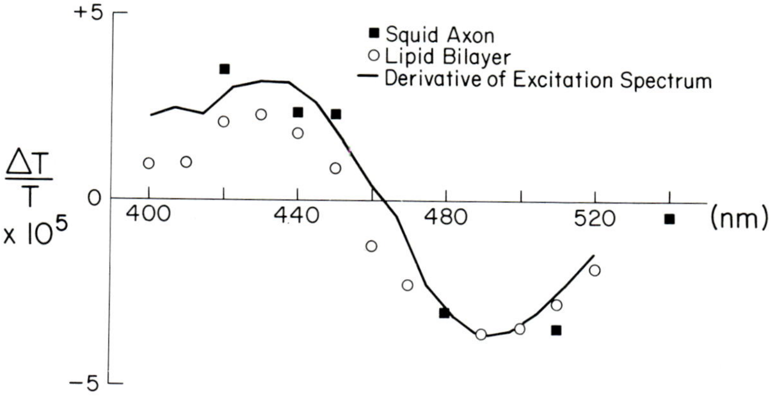


Fig. 5. Spectra of transmission signals obtained with a styryl dye (XXVI, di6ASPPS) on lipid bilayers (open circles) and squid axons (filled squares). The signal spectra are compared with the derivative of the resting excitation spectrum (solid curve) measured on lipid bilayers. The three curves were normalized to the same height at 490 nm. Thus the scale on the ordinate is only approximate. The spectra of the signals are similar to the derivative, as would be expected from an electrochromic mechanism (from Fig. 3 and Fig. 7 of Loew and Simpson [1981] and Fig. 2 of Loew et al. [1985]). △T/T is the change in light transmission over the resting transmission.

signals obtained with dyes XXVI (styryl), XVII, and XXII (merocyanines) were still too rapid to measure and thus lagged behind the change in membrane potential by less than 2 μsec [Loew et al., 1985; B.M. Salzberg, F. Bezanilla, and A.L. Obaid, unpublished results]. This rapid tracking of potential change by optical signal was, however, sometimes not obtained either when relatively low or relatively high concentrations of dye were used; with low concentrations, time constants as slow as 70 μsec were obtained with dye I and, with high concentrations, very slow components have appeared [Ross et al., 1977; Ross and Krauthamer, 1984].

The results in Figures 3 and 4 suggest that when these dye signals are used to measure rapid events the signals represent changes in membrane potential. However, in a number of preparations, much slower signals (100–1000 msec) are obtained [e.g., Orbach and Cohen, 1983; Orbach et al., 1985; Kauer et al., 1984; Lev-Ram and Grinvald, 1984]. Clearly, it would be useful to have independent evidence that such slow signals represent a change in membrane potential. But evidence has not been obtained and thus the origins of slow optical signals must be interpreted with caution.

## MECHANISMS

The spectra of the potential-dependent signals may provide clues about their origins. Qualitatively different spectra have been obtained with different dyes. If the changes in absorption (or fluorescence) resulted from a simple wavelength shift of the spectrum, then the signal spectrum should be similar to the derivative of the resting spectrum. With a pyridinium-styryl dye, the spectrum of absorption signals is in fact well fit by the derivative of the excitation spectrum. The smooth curve in Figure 5 is the derivative of the excitation spectrum. Clearly the experimental data from squid (filled squares) and lipid bilayers (open circles) fall reasonably close to the derivative. This result suggests that the signal results from a simple wavelength shift of absorption. The spectra of the signals were independent of the charge on the dye, and the dye signals followed changes in membrane potential with a time constant of less than 2 $\mu$sec [Loew et al., 1985]. These results are consistent with the possibility that these absorption signals result from electrochromism. The fluorescence spectra are not as simple [Loew et al., 1985] and thus the fluorescence signals probably result from other mechanisms in addition to electrochromism.

The absorption spectrum of dye I on squid axon is shown in Figure 6A. This spectrum has a relatively complicated shape [including the long wavelength sign reversal, not found with the styryl dye (Fig. 5)] but it can be synthesized rather accurately from the difference spectra of monomers and dimers obtained from spectral studies of dye bound to lipid vesicles [Figs. 3–5 of Waggoner and Grinvald, 1977]. Although the extraction of monomer and dimer spectra involves some assumptions, the ability to resynthesize the spectra in Figure 6A encouraged us to think that a potential-dependent, monomer-dimer equilibrium would explain the signals found with this dye [Ross et al., 1974, 1977]. However, it was difficult to explain three additional results by using only this mechanism. Two additional mechanisms, dye rotation and energy transfer from monomer to dimer, have been suggested to explain these and other features of signals found with this dye [Dragsten and Webb, 1977; Ross et al., 1977].

The absorption changes of some membrane-bound dyes were in the same direction at all wavelengths (for example, dye XVII, Fig. 6B), a result that has not yet been satisfactorily explained. Heiny and Vergara [1984] have started sorting out the complex dichroic signals with these dyes. They suggest that both dye reorientation and spectral shifts are involved.

The mechanisms responsible for the fluorescence changes found with aminonaphthalenesulfonates have been investigated in some detail [Conti, 1975]. Even though the fractional changes in fluorescence with this class of

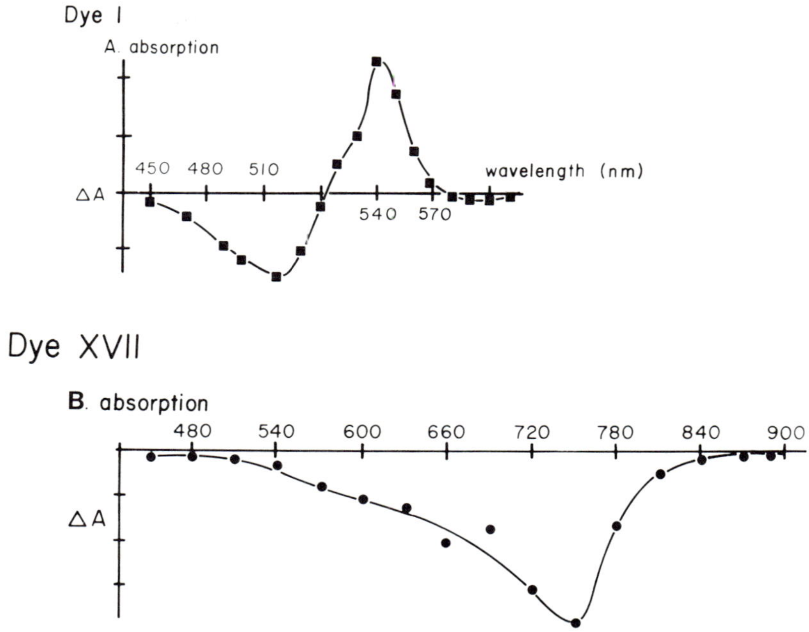


Fig. 6. The wavelength dependence of the absorption changes of dye I and dye XVII in experiments on squid axons. The absorption change found with dye I seemed to be consistent with a potential-dependent monomer-dimer equilibrium, whereas the curve found with dye XVII was qualitatively different [from Ross et al., 1977].

dye are only ten times smaller than those found with dye I, no changes in absorption have been detected [H.V. Davila, B.M. Salzberg, and L.B. Cohen, unpublished observations], indicating yet another mechanism. Conti [1975] suggests that three different phenomena are sufficient to account for all the results obtained with aminonaphthalenesulfonates: (1) a change in dye environment to account for shifts in the emission spectrum, (2) a change in dye binding to account for slow components, and (3) dye rotation to account for the sign of several signals.

## DYES

One of the major difficulties in applying optical methods to the study of neuron function is the small signal-to-noise ratio, a problem that will become evident later in this chapter. The choice of dye is a very important factor in maximizing the signal-to-noise ratio in an optical measurement. In only a few instances, where there is a large density of synchronously active membrane, has it been possible to obtain large signals without testing a fair number of dyes. In addition, in some preparations, photodynamic damage due to illumination of the dye in the presence of oxygen may also affect the choice of dye. Pharmacologic affects and dye bleaching are also considered in this section.

**Screening**

By using squid giant axons more than 1,000 dyes have been tested for signal size in response to changes in membrane potential. This screening was made possible by the heroic synthetic efforts of two laboratories. The group of Alan Waggoner, Jeff Wang, and Ravender Gupta of Amherst College and that of Rina Hildesheim and Amiram Grinvald at the Weizmann Institute have both synthesized more than 300 dyes. In addition, Leslie Loew has rekindled interest in and synthesized a number of styryl dyes; several merocyanine dyes have been made by Nippon Kankoh-Shikiso Kenkyusho Co. Ltd. Included in this synthetic activity were about 100 analogues of each of the four dyes illustrated in Figure 1. In each of these four groups there are ten or 20 dyes which gave identically large signals (within a factor of two of the dye illustrated) on squid axons.

However, a group of dyes which gave nearly identical signals on squid axons had very different signal sizes on other preparations and thus many dyes had to be tested to maximize the signal. Examples of preparations where a number of dyes had to be screened are the *Navanax* and *Aplysia* buccal ganglia [London et al., 1985], rat cortex [Orbach et al., 1985], and tissuecultured neurons [Ross and Reichart, 1979; Grinvald et al., 1981a]. Some of the dyes were unable to penetrate through connective tissue or along intercellular spaces to the membrane of interest. Others appeared to have a relatively low affinity for neuronal versus nonneuronal (connective) tissue. However, in some instances, the dye penetrated well and the staining appeared to be specific but nonetheless the signals were small. Ross and Krauthamer [1984] have reported a case where supraesophageal ganglia from different barnacle species of the same genus had qualitatively different signals.

**Pharmacologic Effects**

In many preparations high concentrations of dye produced pharmacologic effects. However, in some instances the dye concentration needed to obtain the maximum signal size was lower than the concentration at which pharmacologic effects were detected. These include the squid giant axon [Cohen et al., 1974; Ross et al., 1977; Gupta et al., 1981], neuroblastoma cells in tissue culture [Grinvald et al., 1982], heart muscle [Morad and Salama, 1979], skeletal muscle [Nakajima and Gilai, 1980; Baylor et al., 1981], the barnacle supraesophageal ganglion [Salzberg et al., 1977; Grinvald et al., 1981b], and the *Navanax* buccal ganglion [London et al., 1985].

### Photodynamic Damage and Dye Bleaching

In certain experiments, e.g., on neuroblastoma neurons with the styryl dye, RH414 (Fig. 1), photodynamic damage (due to the interaction of light, dye, and oxygen) limited the duration of the experiments [Grinvald et al., 1982]. In others, e.g., on *Navanax* buccal ganglia with the oxonol dye RH155 (Fig. 1), photodynamic damage was not detected [London et al., 1985]. Similarly, dye bleaching has caused difficulties in some preparations but not in others. Grinvald et al. [1982] reported a 3% bleaching from 350 msec of illumination with a styryl dye while we found that bleaching was undetectable after 5 min of illumination with the oxonol [London et al., 1985]. This difference in severity is due, in part, to the difference in dyes that were used; however, higher light intensities were used in the experiments where damage and bleaching were severe.

## RECORDING ACTIVITY OF INDIVIDUAL NEURONS IN A MOLLUSCAN CENTRAL NERVOUS SYSTEM

The activity of individual neurons has been studied with the long-range goal of describing how this activity underlies behavior and how changes in the interactions of these neurons account for simple forms of learning. Techniques which use microelectrodes to accomplish this are limited in that they can observe single-cell activity in only as many cells as one can place electrodes. In principle, optical methods do not have this restriction since there are many devices with a large number of photodetectors. At present we are using 124 elements of a 12 × 12 array. This number of photodetectors corresponds with the number of neurons in the invertebrate ganglia we have been using.

The preparation is placed in a recording chamber which is positioned on the stage of a microscope (Fig. 7). The microscope is equipped with a lamp housing (12-V, 100-W tungsten halogen bulb) and a bright field condenser. The light is passed through a heat filter and an interference filter. The selection of the peak transmission wavelength of the interference filter is dependent on the particular dye used. A long working distance objective forms a magnified real image of the ganglion at the objective image plane; the photodiode array is placed at that plane. Each of the $1.4 \times 1.4$ mm$^2$ active elements of the array are separated by 100 $\mu$m of insulating regions. The output of each photodiode was amplified; the signals were high-frequency filtered and AC coupled (so that only changes in intensity were recorded). The amplifier outputs were passed to an A-D converter and

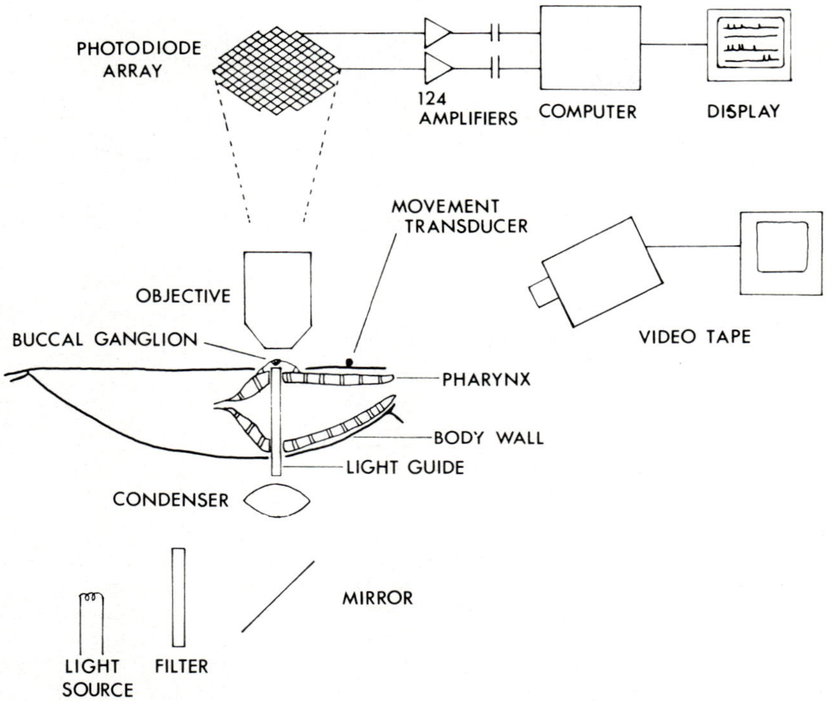


Fig. 7. Schematic drawing of whole animal recording apparatus. The light is made quasi-monochromatic by an interference filter and is focused by a condensor onto the light guide. The light guide was constructed from a 1-mm-clad quartz cylinder of glass encased in a metal tube; at the top of the light guide was a platform which suspended the ganglion over the glass. The image of the ganglion was projected onto the 12 × 12 photodiode array. The signals from individual diodes were fed to individual amplifiers, A-D converted, stored in a computer, and later displayed. A movement transducer was placed on the pharynx to record feeding movements. A video camera was also used to record the behavior during the optical monitoring [from London et al., 1985].

subsequently stored and analyzed in a computer [Grinvald et al., 1981b]. In recordings from invertebrate ganglia, the signal-to-noise ratio in fluorescence measurements was smaller than in absorption measurements; therefore, absorption was measured.

When a single detector receives the light from a single neuron, the changes in light absorption mirror sub- and a suprathreshold events (Fig. 8). On the left, a subthreshold membrane potential change as a result of injection of just-subthreshold depolarizing current is recorded electrically (bottom trace). A small change in the light absorption (top traces) of the cell is recorded

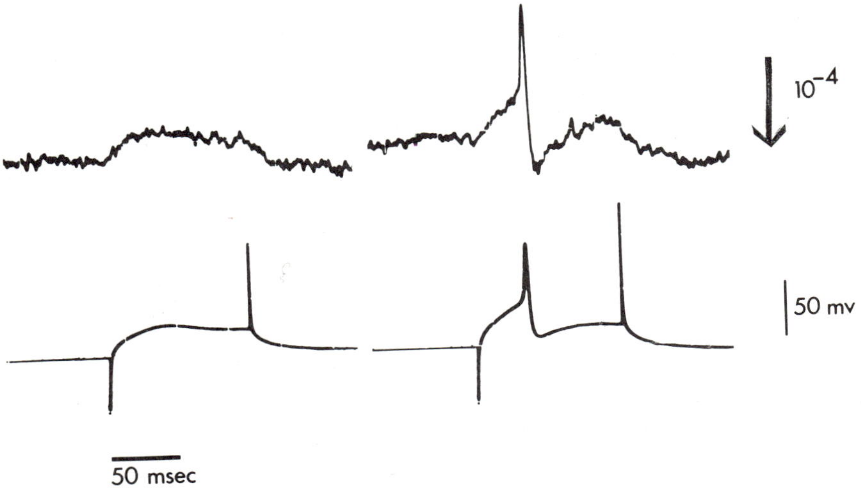


Fig. 8. Comparison of simultaneous changes in light absorption at 720 nm with voltage changes in a stained neuron from a barnacle supraesophageal ganglion. For both subthreshold and action potential activity the optical signals (top traces) have a shape that is essentially identical to the voltage signals (bottom trace) ignoring the stimulation artifacts on the voltage signals. These artifacts occur at the beginning and at the end of the time that current is passed through the electrodes. The merocyanine dye used in this experiment is dye XVII (Fig. 1). The direction of the arrow adjacent to the optical trace indicates the direction of an increase in absorption; the length of the arrow represents $\triangle A/A$, the change in absorption over the resting absorption [from Salzberg et al., 1977].

simultaneously. When the current is increased slightly so that it is just suprathreshold (on the right) there is an action potential in both the electrical and optical recordings. Since this recording is of a single trial (it was not signal averaged) it is clear that with one detector one can follow the action potential activity of one neuron.

However, when one positions many closely spaced and regularly arranged detectors in the image plane in order to measure as much of the light as possible from the ganglion, then the relationship between the detectors and neuron images becomes complex. A schematic drawing of the projection of neurons onto the array is illustrated in Figure 9. Each square represents the surface area of the preparation to which a single photodiode is sensitive. First, the activity of one cell may be recorded on several different detectors and generate duplicate signals. The light of several cells can be seen to extend beyond the boundary of a single detector. These cells' activity will be recorded simultaneously on several detectors. Second, some cells cover the entire area of a diode and therefore the signal recorded from these cells on

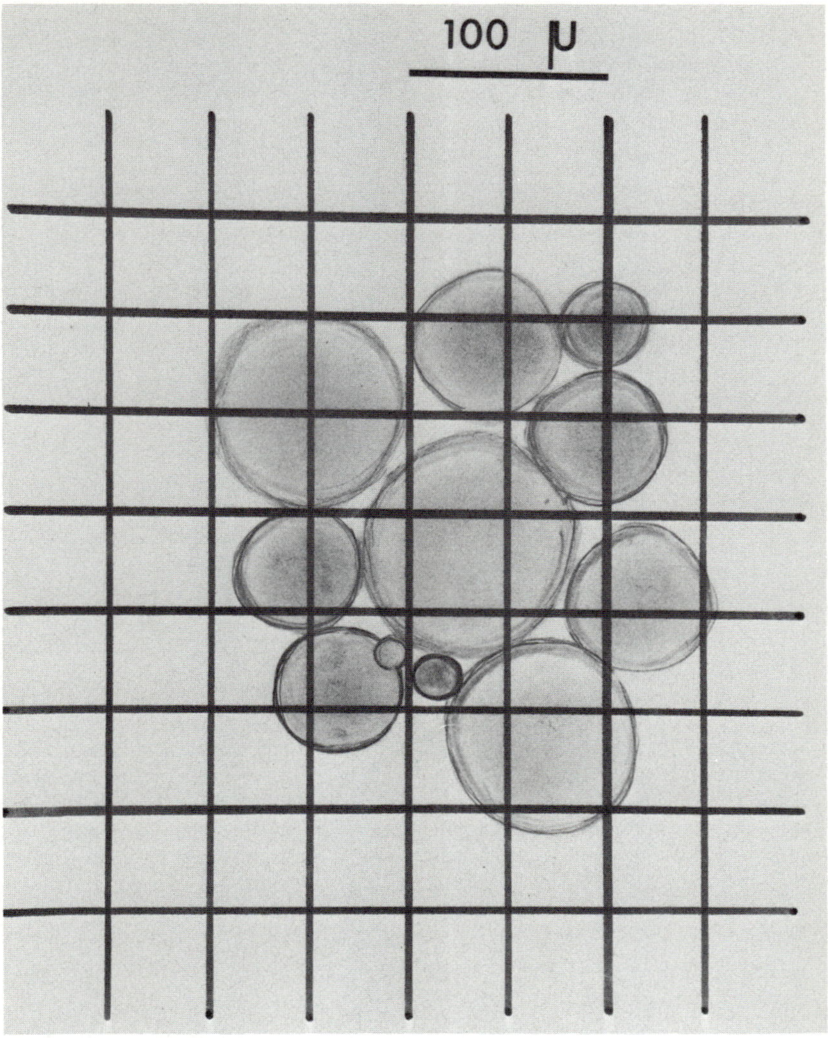


Fig. 9. Drawing of the projected image of an idealized cluster of cells onto the array. Note how various-sized areas of one cell may fall on different detectors, giving rise during a recording to duplicate spike activity of different size on neighboring detectors. Note also how areas of more than one cell project their images onto a single detector, thus one detector may record the activity of several cells. We thank Amiram Grinvald for preparing this figure.

that detector will be relatively large. But other detectors may have only a small part of a cell's image projected onto them, and the signal from these detectors will be small. Third, it can be seen that one detector may record the activity of several cells, for example, when sections of three different cells are projected onto one detector. Such a recording can be called a multicellular recording.

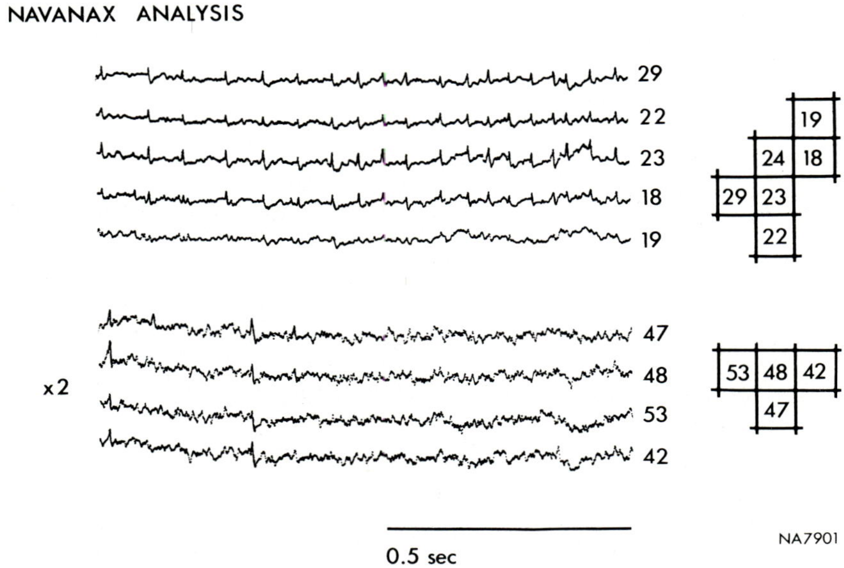


Fig. 10. Optical recordings from the buccal ganglion of *Navanax inermis* using a few elements of a 12 × 12 array. The numbers to the right of each trace identify the detector from which the trace was taken. The drawings to the right represent the relative position of the detectors whose activity is displayed. In the top section, four detectors exhibit simultaneous spike activity, presumably from the same neuron. The bottom section is recording from four detectors (amplified times 2), which indicates some of the difficulties of this technique. There is significant noise which may mask small action potentials. Also there are slow signals occurring at the end of the recordings which are probably due to movement artifacts [from London et al., 1985].

The detection of duplicate signals on adjacent elements of the array, and multicellular recordings, are shown in Figure 10 from a buccal ganglion from the mollusc *Navanax inermis*. In the top section the activity of one cell is recorded on four adjacent detectors, 18, 22, 23, and 29. However, smaller signals from a second, different cell also appear, recorded only on detector 18 and perhaps 19. The first spike from this second cell is clearly seen on detector 18 between the first and second spike of the larger cell. In the bottom section of this figure, a different set of detectors illustrates the same point—two spikes from the small cell occur simultaneously on four adjacent detectors. Two of the problems associated with optical recordings are also illustrated. One problem arises from the signal-to-noise ratio. The high background noise will obscure small action potentials. For example, there may also be spike activity simultaneously on detectors 47 and 48 following the seconds pike, but the signals are not much larger than the noise . A second problem arises from the slow changes in light intensity, which are probably movement artifacts. These slow changes could also obscure some of the neuronal activity.

**Recording From a Behaving Animal**

Recently we have been able to use the diode array to monitor activity from the nervous system of a minimally dissected preparation. In this preparation a nearly normal behavior is preserved. Activity of the buccal ganglion of the marine slug, *Navanax,* was monitored while the animal fed. *Navanax* is a carnivore which feeds on several species of opisthobranchs including other *Navanax,* and some species of fish [Paine, 1963]. Roughly, the feeding movements consist of the protraction of the mouth, prey contact, and grasping with the lips, followed by an explosive expansion of the pharynx. This expansion creates a negative pressure so that the prey is sucked into the pharynx. The prey is subsequently pushed into the esophagus where digestion begins. Neurons controlling the expansion have been described and reside in the buccal ganglion [Wollacott, 1974; Spira et al., 1980]. A movement transducer, consisting of a glass rod connected to a force displacement recorder (Grass Instrument Corp.), was placed on the pharynx to record feeding movements. A video camera and recorder were also used to monitor the behaviors.

The action potential data from these experiments was displayed in a raster diagram (Fig. 11). Each vertical line represents the occurrence of one action potential. The bottom trace is a recording from the movement transducer which had been positioned on the pharynx. An upward deflection indicates expansion. The dotted line represents the position of the transducer, and therefore the state of the pharynx, at rest. The numbers just to the left of the spike traces represent the number of detectors on which this particular neuron's activity was recorded. The numbers to the far left represent the identification number of the detector used to enter this neuron's activity into the computer and to generate the raster trace. The generation of this kind of raster diagram requires several steps. First, original data from 20 detectors are displayed; these detectors are nearest neighbors so that duplicates, i.e., the same cell's activity seen on more than one detector, may be identified in the data and the occurrence of the duplicate may be recorded and then removed from the data. Then, the times of occurrence of the identified action potentials are entered into the computer. Finally, these events are then printed out in the raster display. The results shown here are preliminary. A more detailed identification of the neurons from this data is in progress.

During feeding (Fig. 11) 24 cells were active. As in spontaneous expansions, during the feeding there is a burst of activity reflected as increased

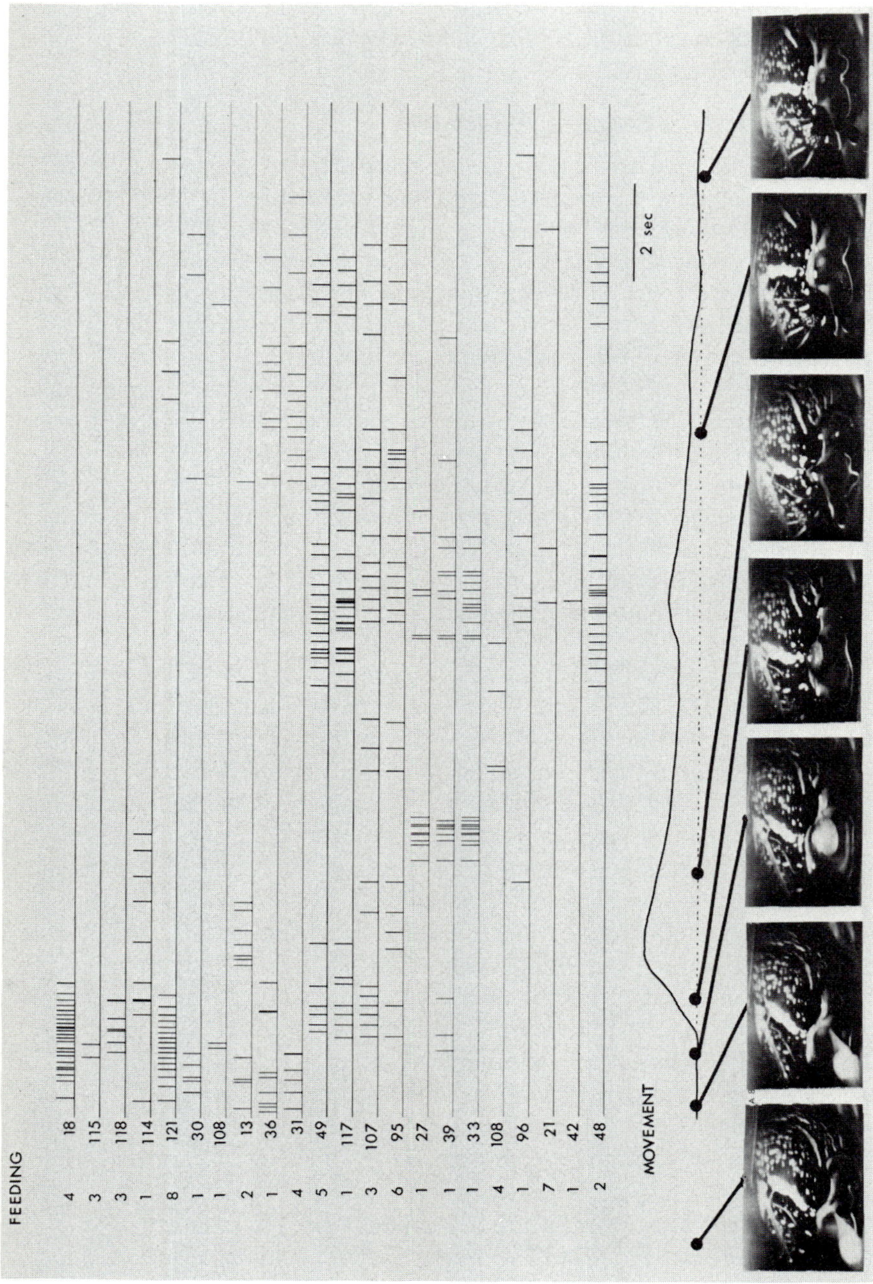
FEEDING
4 18
3 115
3 118
1 114
8 121
1 30
1 108
2 13
1 36
4 31
5 49
1 117
3 107
6 95
1 27
1 39
1 33
4 108
1 96
7 21
1 42
2 48
MOVEMENT
2 sec

spike activity prior to the expansion. Several large-sized cells, on detectors 121 and 95, are active early during the expansion. There are also a number of medium-sized neurons active in this feeding experiment (cell's whose activity is seen on two to five detectors). And a number of the cells are small; their activity was found on only one detector. Photographs of the animal made during the optical recording are shown at the bottom of Figure 11.

We wanted to know whether we were recording from all the cells that were active during a behavior, or only a small fraction of the active cells. To answer this question we compared the number of cell bodies in the ganglion to the number of cells we could identify by their optical activity during an experiment where we tried to activate as many cells as possible in the ganglion. To determine the number of cell bodies, ganglia were stained with methylene blue and the number of cell bodies was counted by using a computer-interfaced microscope [Macagno, 1978; Boyle et al., 1983]. The number of cells we counted in three cases where both the anatomical count and optical measurements were done on the same ganglia is indicated in Table 1. To obtain the number of cells whose activity could be detected optically, the preparation was activated by stimulating the peripheral nerves and connectives with suction electrodes. The number of neurons whose activity could be detected optically is also shown in Table 1. On average we

**TABLE 1. Optically Detectable Neuronal Activity**

| Preparation | No. of cells counted | No. of neurons detected optically |
|---|---|---|
| 1 | 183 | 140 |
| 2 | 162 | 128 |
| 3 | 203 | 112 |

Fig. 11. Raster diagram generated from an optical recording from the buccal ganglion during feeding. All the neurons whose activity was detected are shown. The photographs at the bottom are from the videotape recording made during the optical measurement. The number just to the left of a trace is the number of detectors which recorded this activity, an indication of cell size. The number to the left of the cell size number is the cell's identification number. This number is the number of one detector on which the cell's activity is observed; this detector was used to enter the cell's activity into the computer. Usually, a detector is chosen for a particular cell because it recorded the largest signal for that cell. The bottom trace is a record of the movement of the pharynx; an upward deflection indicates expansion [from London et al., 1985].

detected activity in a little over 70% of the neurons. This percentage may be an underestimate for two reasons. First, we may not correctly separate the activity of small cells. These cells' activities may occur on the same detector and have small signals which appear too similar to allow us to differentiate between them. Second, the stimulation we used may not activate all the neurons.

We now have the capability to monitor the action potential activity of a large fraction of the total number of neurons in ganglia like the buccal ganglia of *Navanax*. We hope that this capability will make it easier to determine the neuronal basis of the behaviors generated by such ganglia. In addition we plan to continue developing better dyes and apparatus to improve the signal-to-noise ratio.

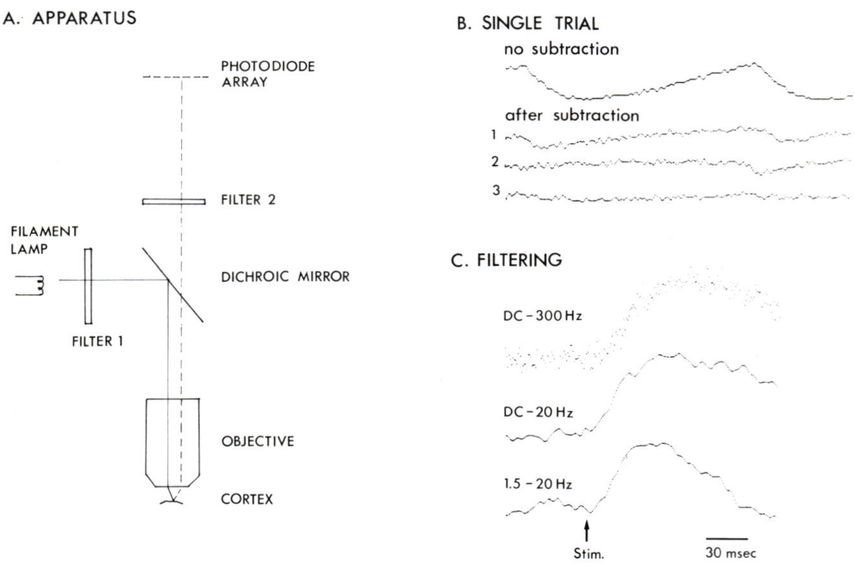


Fig. 12. A. Schematic drawing of optical apparatus for epifluorescence measurements from rat cortex. B. Subtraction reduces noise locked to the electrocardiogram. Four traces are illustrated; the top one is a single recording without subtraction; the bottom three are individual trials where a recording was subtracted from its predecessor. C. Filtering to reduce noise. The top two traces are the same data but at different bandwidths. The third trace was from the subsequent measurement on the same preparation. The middle trace was obtained from the top trace by using ten passes of a 1-2-1 digital filter. For each point in the trace, this filter took twice the value of the center point, added the result to the sum of the two side points, divided the total by four, and replaced the old center point with the result of the division. A heating pad was used to regulate the rectal temperature between 35°C and 38.5°C. Staining was carried out at rectal temperatures between 31°C and 37°C [from Orbach et al., 1985].

## MONITORING ACTIVITY IN VERTEBRATE BRAINS

The same apparatus we used to monitor activity in *Navanax* ganglia can be used to measure activity in vertebrate central nervous systems. Only here, each detector will receive light from many neurons ($10^3$–$10^5$) and processes, and thus the signals are almost certainly not going to represent the activity of individual neurons but some average of the change in membrane potential of all the membranes imaged onto that detector. In addition, in vertebrate experiments, spatial resolution will be limited by the light-scattering properties of the preparations. In salamander olfactory bulb, scattering would smear the light from a small spot to an area of about 200 $\mu$m [Orbach and Cohen, 1983]. Furthermore, vertebrate preparations are relatively thick, and with the simple optical system we have used, signals from planes that are out of focus will be included with the signals that are from in-focus planes.

We investigated the cortical responses to whisker movements in the rat. The upper layers of somatosensory cortex were stained with styryl or oxonol dyes by placing a pool of dye solution on top of the cortex. A schematic drawing of the light path is illustrated in Figure 12A. Epi-illumination was used and fluorescent signals were measured since absorption-dependent light-scattering signals were small in experiments on salamander olfactory bulb. The right-hand side of the figure illustrates two kinds of noise-reduction techniques we used: subtraction to reduce movement artifacts associated with the heart-beat and filtering to remove high-frequency noise.

## CORTICAL RESPONSES TO WHISKER MOVEMENT

Figure 13 illustrates the results of an experiment where 124 adjacent regions of cortex were simultaneously monitored. The traces are arranged so that their relative positions in the figure correspond to relative positions of the areas of the preparation imaged onto the detectors. When the tip of whisker B1 [for nomenclature, see Simons, 1978] was moved by 1 mm, the number of photons reaching the central photodetectors changed by 0.001 of the resting value ($\Delta F/F = 10^{-3}$). The time of the beginning of the whisker movement is given by the bars under the traces in row 6. A 4× objective was used; the centers of neighboring detectors received light from spots on the cortex that are separated by 283 $\mu$m. A decrease in intensity is plotted upward. This direction of plotting was chosen because in other preparations, depolarizations led to a decrease in intensity with styryl dyes of this type at this wavelength [Gupta et al., 1981; Grinvald et al., 1982, 1984; Orbach and Cohen, 1983; Loew et al., 1985]. In a few experiments we were able to detect signals in single trials, but most required averaging of several trials; 32 trials were averaged for Figure 13.

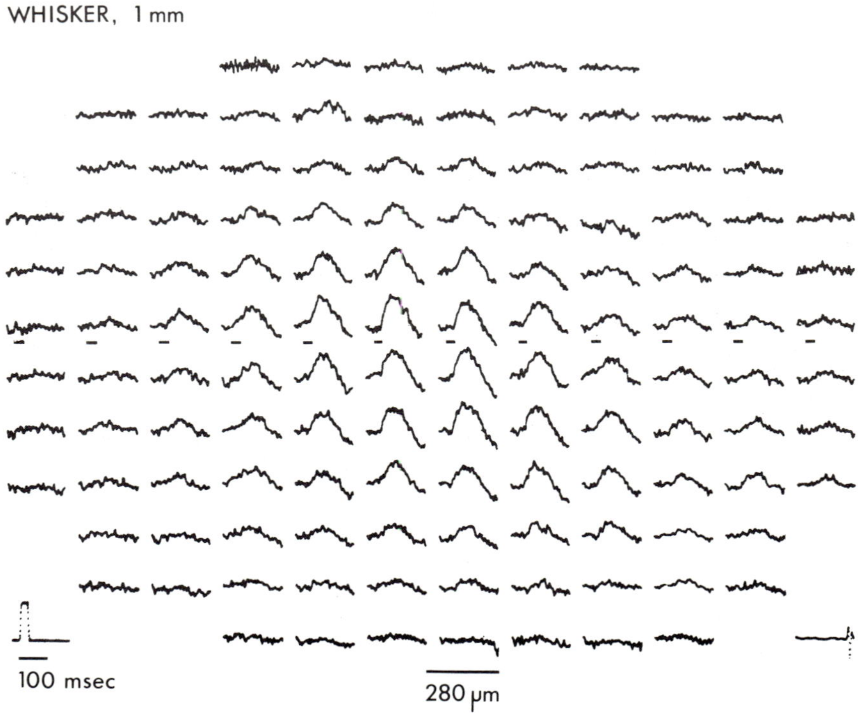


Fig. 13. Optical detection of a localized region of activity. Simultaneous intensity changes measured with 124 photodiodes from rat somatosensory cortex that result from a 1-mm whisker movement. The signals are restricted to the central detectors of the array. The fractional changes in fluorescence ($\triangle F/F$) on the central four detectors are at least 20 times larger than the mean signal on all of the detectors around the very edge of the array. The time of the beginning of the whisker movement is indicated by the bars below the traces in row 6. Top is medial, bottom is lateral, left is posterior, and right is anterior. Before staining, the amount of light reaching the photodiodes was lower by a factor of at least ten. Thus, most of the light in the measurement results from dye-related cortical fluorescence. The records were smoothed with ten passes of the 1-2-1 digital filtering routine. The inset at the lower left shows the timing of the current to the galvanometer-stimulator. The missing traces are the result of malfunctioning amplifiers. The whisker was moved in the caudal to rostral direction; 32 trials were averaged. The interval between stimuli was 9 sec. The cortex was stained with a 2 mg/ml solution of RH414 for 100 min [from Orbach et al., 1985].

In the experiment illustrated in Figure 13, the delay between the start of the whisker movement and the start of the intensity change was measured without digital filtering on the averaged signal of the 16 central detectors and was 7 msec. The intensity change reached a peak about 50 msec after the start of the movement and then returned toward the baseline. To facilitate

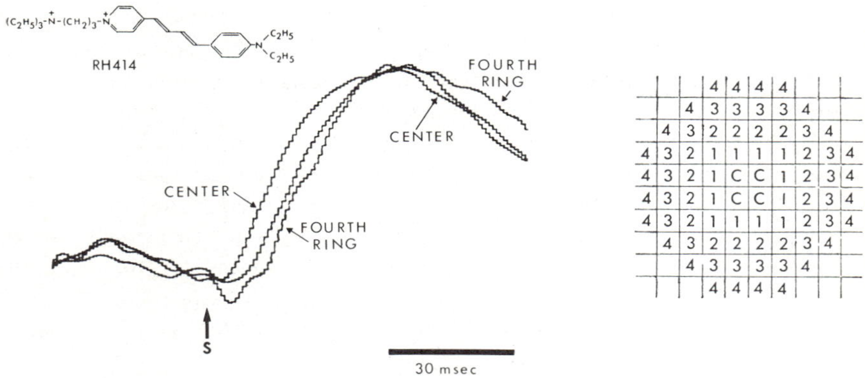


Fig. 14. Time course of the signals from three different parts of the response area: the center, the edge (fourth ring), and an intermediate area (second ring). The signal had a briefer delay at the center and a longer delay at the edge. The inset, on the left, illustrates the arrangement of the central detectors (C) and the four rings of detectors (1–4) which we used. The signals illustrated are the sum of the four central detectors (C), the 16 detectors in ring 2, and the 24 detectors in ring 4. The three signals were normalized to the same height; 64 trials were averaged for the original data. The center is the sum of four detectors, the second ring is the sum of 16 detectors, and the fourth ring is the sum of 24 detectors. Whisker B1 was moved by 6 mm. The time of the beginning of the whisker movement is indicated by the arrow. S represents stimulus. The interval between trials was 5 sec. The cortex was stained with a 4 mg/ml solution of RH414 for 60 min [from Orbach et al., 1985].

comparison of the time course of the signals, we combined the outputs of the central four detectors and the four concentric rings of detectors surrounding the central four. The arrangement of the central detectors (labeled C) and the four rings are shown in the inset of Figure 14. In most of the 22 measurements that were analyzed, the signal at the center either had a shorter delay or had a more rapid rise than the signals from the edge of the response curve. Figure 14 illustrates the results from another preparation where the time courses of the central four, the second ring, and the fourth ring were scaled to the same peak height. The response from the fourth ring is delayed by about 8 msec compared to the response from the central four. The response from the second ring has an intermediate time course. Thus, while the shapes of the signals throughout the response area were not dramatically different, the latency or rise time at the edge was often substantially slower. One interpretation of the difference in latency is that the activity propagates from the central region outward.

In Figure 12C, the signal, filtered at 1.5–20 Hz, had returned to the baseline by the time of the next heartbeat, about 150 msec after the start of the stimulus. This was our average result. However, since much of this decline toward the baseline is caused by the low-frequency cutoff of our amplifiers (compare DC 20 Hz with 1.5–20 Hz), on average, the actual signal had not returned by 150 msec. The signals illustrated in Figure 13 are unusual in that they cross below the original baseline by that time.

We also measured optical signals during an interictal epileptiform discharge elicited by bicuculline.These signals were about five times larger than those illustrated in Figures 13 and 14. Using signal size during the interictal discharge, we can make, with assumptions, a very rough estimate of the percentage of activated neurons necessary to generate the signals obtained in experiments illustrated in Figures 13 and 14. The peak signals in Figure 13 could be accounted for by a 20-mV depolarization in 25% of the neurons or a very high firing rate of a smaller percentage of neurons. Thus, activation of a substantial fraction of the neurons may be necessary to give rise to signals of the size we obtained.

Our experiments demonstrate that optical measurements can simultaneously detect activity-related signals from 124 sites on cortex with a time resolution of milliseconds, and a spatial resolution of a few hundred microns. Dramatic improvements in the signal-to-noise ratio and spatial resolution and a large increase in the number of simultaneously monitored areas may be possible. Preliminary experiments carried out on cat visual cortex suggest that much of the low-frequency noise remaining after the heartbeat subtraction procedure can be removed by temporarily stopping the breathing of an artificially respired animal. Once the movement noise is eliminated, the effects of the remaining noise, dark noise and shot noise, can be reduced by increasing the incident light intensity. In addition, better dyes may be found or designed. Finally, modification of the optics should reduce the effects of scattering and the contribution from out-of-focus layers and thereby improve resolution.

## SUMMARY

Optical measurements of neuron activity have been developed to the point where they can make substantial contributions in a number of areas of neurobiology. Future improvements in signal-to-noise ratio and spatial resolution would increase their utility.

## ACKNOWLEDGMENTS

We wish to thank Rina Hildesheim, Amiram Grinvald, Alan Waggoner, and the Nippon Kankoh-Kenkyusho Co. for supplying us with dyes. This work was supported in part by a Grass Foundation Fellowship to J.A.L., a

Fulbright grant, 83-04765, to D.Z., and Public Health Service grants NS08437, NS07169, and GM35063.

## REFERENCES

Baylor SM, Chandler WK, Marshall MW (1981): Studies in skeletel muscle using optical probes of membrane potential. In Grinnel AD, Brazler MAB (eds): "The Regulation of Muscle Contraction: Excitation-Contraction Coupling." New York: Academic Press, pp 97–130.

Boyle MB, Cohen LB, Macagno ER, Orbach HS (1983): The number and size of neurons in the CNS of gastropod molluses and their suitability for optical recording of activity. Brain Res 266:305–317.

Cohen LB, Hille B, Keynes RD (1968): Light scattering and birefringence changes during nerve activity. Nature 218:438–441.

Cohen LB, Landowne D, Shrivastav BB, Ritchie JM (1970): Changes in fluorescence of squid axons during activity. Biol Bull Mar Biol Lab Woods Hole 139:418–419.

Cohen LB, Lesher S (1985) Optical monitoring of membrane potential: Methods of multi-site optical measurement. In DeWeer P, Salzberg BM (eds): "Optical Methods in Cell Physiology." New York: John Wiley and Son, Inc., in press.

Cohen LB, Salzberg BM, Davila HV, Ross WN, Landowne D, Waggoner AS, Wang CH (1974): Changes in axon fluorescence during activity: Molecular probes of memebrane potential. J Membr Biol 19:1–36.

Cohen LB, Salzberg BM (1978): Optical measurement of membrane potential. Rev Physiol Biochem Pharmacol 83:35–88.

Conti F (1975): Fluorescent probes in nerve membranes. Ann Rev Biophys Bioeng 4:287–310.

Conti F, Tasaki, I (1970): Changes in extrinsic fluorescence in squid axons during voltage clamp. Science 169:1,322–1,324.

Conti F, Tasaki I, Wanke E (1971) Fluorescence signals in ANS-stained squid axons during voltage clamp. Biophysics 8:58–70.

Davila HV, Cohen LB, Salzberg BM, Shrivastav BB (1974): Changes in ANS and TNS fluorescence in giant axons from *Loligo*. J Membr Biol 15:29–46.

Dragsten PR, Webb WW (1978) Mechanism of membrane potential sensitivity of fluorescent membrane probe merocyanine 540. Biochemistry 17:5228–5240.

Freedman JC, Laris PC (1981) Electrophysiology of cells and organelles: Studies with optical potentiometric indicators. Int Rev Cytol Suppl 12:177–246.

Grinvald A (1985) Real-time optical mapping of neuronal activity: From growth cones to the intact mammalian brain. Ann Rev Neurosci (in press).

Grinvald A, Ross WN, Farber IC (1981a): Simultaneous optical measurements of electrical activity from multiple sites on processes of cultured neurons. Proc Natl Acad Sci USA 78:3245–3249.

Grinvald A, Cohen LB, Lesher S, Boyle MB (1981b): Simultaneous optical monitoring of activity of many neurons in invertebrate ganglia using a 124 element photodiode array. J Neurophysiol 45:829–840.

Grinvald A, Hildesheim R, Farber IC, Anglister L (1982): Improved fluorescent probes for the measurement of rapid changes in membrane potential. Biophys J 39:301–308.

Gupta RK, Salzberg BM, Grinvald A, Cohen LB, Kamino K, Lesher S, boyle MB, Waggoner

AS, Wang CH (1981): Improvements in optical methods for measuring rapid changes in membrane potential. J Membr Biol 58:123–137.

Heiny, JA, Vergara J (1984): Dichroic behavior of the absorbance signals from dyes NK2367 and WW375 in skeletal muscle fibers. J Gen Phys 84:805–837.

Kauer JS, Senseman D, Cohen LB (1984): Voltage-sensitive dye recording from the olfactory system of the tiger salamander. Soc Neurosci Abstr 10:846.

Lev-Ram V, Grinvald A (1984): Is there a potassium dependent depolarization of the paranodal region of myelin sheath? Optical studies of rat optic nerve. Soc Neuro 1 Abstr 10:948.

Loew LM, Cohen LB, Salzberg BM, Obaid AL, Benzanilla F (1985): Charge shift probes of membrane potential. Characterization of aminostyrylpyridinium dyes on the squid giant axon. Biohys J 47:71–77.

Loew LM, Simpson LL (1981): Charge shift probes of membrane potential: A probable electrochromic mechanism for p-aminostyrylpyridinium probes on a hemispherical bilayer. biophys J 34:353–365.

London JA, Zecevic D, Cohen LB (1985): Optical recording of action potential activity from individual neurons in invertebrate central nervous systems. In DeWeer P, Salzberg BM (eds): "Optical Methods in Cell Physiology." New York: John Wiley and Son, Inc., in press.

Macagno (1980): Number and distribution of neurons in leech segmental ganglia. J Comp Neurol 190:283–302.

Morad M, Salama G (1979): Optical probes of membrane potential in heart muscle. J Physiol (Lead) 292:267–295.

Nakajima S, Gilai A (1980): Action potentials of isolated single muscle fibers recorded by potential sensitive dyes. J Gen Physiol 76:729–750.

Orbach HS, Cohen LB (1983): Optical monitoring of activity from many areas of the *in vivo* and *in vitro* salamander olfactory bulb: A new method for studying functional organization in the vertebrate central nervous system. J. Neurosci 3:2251–2262.

Orbach HS, Cohen LB, Grinvald A (1985): Optical mapping of electrical activity in rat somatosensory and visual cortex. J Neurosci 5:1886–1895.

Paine, R.T. (1963): Food recognition and predation on opisthobranchs by *Navanax inermis. Veliger* 6:1–9.

Patrick J, Valeur B, Monnerie L, Changeux J-P (1971): Changes in entrinsic fluorescence intensity of the electroplax membrane during electrical excitation. J Membr Biol 5:102–120.

Ross WN, Krauthamer V (1984): Optical measurements of potential changes in axons and processes of neurons of a barnacle ganglion. J Neurosci 4:659–672.

Ross WN, Salzberg BM, Cohen LB, Grinvald A, Davila HV, Waggoner AS, Wang CH (1977): Changes in absorption, fluorescence, dichroism, and birefringence in stained giant axons: optical measurement of membrane potential. J Membr Biol 33:141–183.

Salzberg BM (1983): Optical recording of electrical activity in neurons using molecular probes. In Barker JL, McKelvy JF (eds): "Current Methods in Cellular Neurobiology ." New York: John Wiley, pp 139–187.

Salzberg BM, Grinvald A, Cohen LB, Davila HV, Ross WN (1977): Optical recording of neuronal activity in an invertebrate central nervous system: Simultaneous monitoring of several neurons. J Neurophysiol 40:1281–1291.

Simons DJ (1978): Response properties of vibrissa units in rat SI somatosensory cortex. J Neurophysiol 41:798–820.

Spira ME, Spray DC, Bennett MVL (1980): Synaptic organization of expansion moteneurons of *Navanax inermis.* Brain Res 195:241–269.

Tasaki I, Watanabe A, Hallett A (1972): Fluorescence of squid axon membrane labeled with hydrophobic probes. J Membr Biol 8:109–132.

Tasaki I, Watanabe A, Sandlin R, Carnay L (1968): Changes in fluorescence, turbidity, and birefringence associated with nerve excitation. Proc Natl Acad Sci USA 61:883–888.

Waggoner AS (1979): Dye indicators of membrane potential. Ann Rev Biophys Bioeng 8:47–68.

Waggoner AS, Grinvald A (1977): Mechanisms of rapid optical changes of potential sensitive dyes. Ann NY Acad Sci 303:217–241.

Woolacott M (1974): Patterned neural activity associated with prey capture in *Navanax*. J Comp Physiol 94:69–84.

*Applications of Fluorescence in the Biomedical Sciences, pages 449–459*


# 20 New Aspects of Microscopy in Cell Biology

*Frederick Lanni*

The science of microscopy is in flux. Over the past 20 years or so there have been many advances in the sciences and technologies related to optics, and in the nature of the research questions being asked by the users of microscopes. The utility of the basic brightfield microscope increased greatly in biology with the development of phase-contrast instruments, principally by Zernike [1] and Nomarski [2], and with the development of the fluorescence microscope [3]. More recently, the microscope has been augmented by the laser for illumination and for interferometric positioning [4], by a variety of detectors ranging from photomultipliers to solid-state arrays and intensified video cameras, by active optical devices such as modulators, phase retarders, interferometers, and monochromators, by guided-wave and evanescent wave devices, by the addition of optical scanners and specimen translators, and not least by computerized data acquisition systems. One result is that more and more frequently microscopes perform their most important functions in the absence of an observer! In most cases the eye and brain of the microscopist cannot discriminate or remember the vast quantity of information encoded into one or more microscope images. It is a challenge to the user of a microscope to assemble an instrument and design an experiment in which the images that are produced are maps of only a set of well-defined specimen parameters.

This introduction is followed by four research reports (see in this volume: Axelrod et al., Koppel et al., Gaub et al., and Lanni) in which fluorescence microscopy has been applied to problems in cell biology. The specific research questions addressed in these reports are almost totally unrelated, but

Department of Biological Sciences, Center for Fluorescence Research in Biomedical Sciences, Carnegie-Mellon University, Pittsburgh, Pennsylvania 15213.

in terms of microscopy all share several similarities: all involve fluorescent tracers, image-forming instruments (as opposed to scanning instruments), and quantitative output. All involve relatively recent developments in fluorescence microscopy

It should be noted, however, that developments have occurred beyond the scope of these researches. It may be of some use to attempt to review this broader field because there will undoubtedly be combination and further synthesis in the future.

## IMAGE FIELD AND OBJECT STRUCTURE

Cells, usually the specimens of interest in biological microscopy, are complex optical objects: The refractive index varies with location in the cell volume, and can be anisotropic as well [5,6]. When cells are prepared by staining or by the introduction of fluorescent tracers, there will be an inhomogeneous absorptive component to the index. Emission of radiation by tracer molecules occurs in the same milieu. Fixation and embedding techniques minimize refractive mismatch, but for study of live cells little can be done beyond limited adjustment of the refractive index of the culture medium in contact with the specimen.

What then is the relationship between the optical structure of the specimen and the image field formed by a microscope? This is a central question in quantitative microscopy, and has been approached in several ways. In a recent report Streibl [7], building upon the work of Frieden [8], demonstrates how imaging numerical aperture (NA) and illumination aperture (INA) limits the three-dimensional zone of spatial frequencies that can be transferred to the image field. It is further demonstrated, in a first-order Born approximation, that two three-dimensional transfer functions can be defined which linearly filter the "refractive" and "absorptive" parts of the object scattering potential in the formation of the image field. However, to apply this formalism, it had to be assumed that object refractive index varies slowly over wavelength-sized domains, which is generally not true for cells. A second approach has been taken by Kerker et al. [9] in which scattering from and excitation of fluorescence in a small structured particle has been treated as an electromagnetic boundary-value problem. The image field formed by a microscope is a functional of the long-range scattered or emitted field due to the particle (object). While this approach is currently limited by computational complexity, it can in principle be applied to realistic cellular model objects. An extension of this method, by use of the theory of dipole radiation in a bounded dielectric [10–14], should make possible the calculation of the emission properties of fluorescent dye molecules in inhomogeneous optical environments such as when attached to a large protein molecule or membrane, or as a free dye in a cell.

A third approach to understanding the relationship between object structure and image field has been taken by Axelrod [15], who derived a transfer function that connects object emission dipole distribution to the polarized components of the fluorescence image, accounting for the mixing that occurs due to the high aperture of the microscope. This transform has been used by Burghardt and Thompson [14] and by Thompson et al[16] to determine the angular moments of dye molecules in microscopic regions of bounded planar dielectric specimens of biological interest.

Outside the current realm of fluorescence microscopy (but within the subject of image field and object structure) transmission microscopes operated with highly coherent light sources have been used to record holograms of biological objects [17,18 for review]. A difficulty encountered in coherent image formation at high aperture is the appearance of interference components in the hologram due to scattered waves from out-of-focus parts of the specimen and from optical surfaces of the microscope. Nevertheless, a high-NA holographic microscope remains an attractive goal because living cells are not thin sections but are three-dimensional structures. In fluorescence microscopy, dye molecules in the specimen emit incoherently: stimulated emission or superradiance is normally not a significant factor and no hologram can be made. However, the dye molecules used in fluorescence microscopy and those used in dye lasers are very similar, and it is conceivable that a pulsed laser can be used both to excite dye molecules in a specimen and to pump a dye laser tuned to stimulate emission from the specimen. The refractive structure of the specimen will contribute strongly to the resulting fluorescence image; it may be possible to compensate for this by using only stimulation pulses to generate a second hologram which would represent only the refractive structure of the object.

In many cases it is not necessary to have a complete theory of image formation in three dimensions as a prerequisite to having a quantitative microscope. A time-honored activity in microscopy is the construction of instruments and experiments principally sensitive to one specimen optical property at a time. As pointed out by Taylor et al. [19], the fluorescence microscope is such a device, especially when the specimen (a cell) contains not only the tracer of interest, usually a fluorescent protein analog, but also a spectrally distinguishable compartment marker (such as a fluorescent dextran for cytoplasmic solvent volume). This experimental design has been used effectively to distinguish concentration and volume effects in fluorescence images of live cells [20–22]. A small amount of "outside" information, in this case, knowing that a particular type of marker distributes diffusely in the cytoplasmic compartment, goes a long way toward turning a small set of images into a map of a well-defined parameter—the axially averaged concentration of the tracer of interest. In polarization microscopy, instruments that determine birefringence and dichroism independently have been developed

[23,24], so that maps of molecular orientation and of concentration can be made.

## SCANNING MICROSCOPES

As a class, scanning microscopes are point-source-imaging optical devices. In such an instrument a small zone of excitation is moved relative to the specimen (or vice versa) and the resulting time series is synchronously converted into a spatial array. Compared to the image-forming microscope, the scanning microscope is a relatively recent invention, introduced in 1951 by Young and Roberts [25]. Due to the introduction of laser sources which can be efficiently focused into the diffraction limit of a lens system [26], high scan rates at high signal-to-noise ratio became possible. Of equal importance has been the development of a wide variety of beam deflection devices utilizing servocontrolled mirrors [27,28], rotating mirrors [29], or acousto-optic cells, and the development of precision specimen translation devices [4,30] based on stepping motors, oscillating drivers, and direction-sensitive interferometers.

The principal attractions of scanning microscopy are five: (1) use of a fast, low-noise detector (such as a photomultiplier) having high linearity and high dynamic range, (2) control of detection field and detection depth-of-field [31, 32], (3) reduction of the constraints on objective lens design [29,32], (4) photoelectronic [31] or photoacoustic modes of operation, and (5) autofocusing capability for certain types of specimens [29,31].

A new and different approach to achieve ultrahigh resolution is embodied in the near-field scanning microscope [33,34]. In this instrument an intense light source is coupled to a waveguide having one or more very small apertures or antennae. The waveguide is apposed to and scanned over the specimen. Excitation due to the intense near-field of each aperture occurs in zones of the specimen far smaller than the optical resolution limit of the microscope. When used with a high-precision specimen translator, the resolution limit for the instrument is set by the aperture or antenna size rather than by the optical NA and detected wavelength.

## FLUORESCENCE MICROSCOPY

It is in this area of microscopy that biological applications have been most numerous and varied. A wide range of recent ideas and experiments related to fluorescence microscopy can be found in the research reports published by the New York Academy of Sciences as the proceedings of the Eighth Katzir Conference [35], which will not be further reviewed here.

Three areas of development are related to the research reports that follow this introduction: measurement of tracer translational mobility in and on cells, the use of evanescent wave devices, and fluorescence microtomography.

Fluorescence photobleaching recovery (FPR) or fluorescence redistribution after photobleaching (FRAP) names a family of techniques in which submillisecond irreversible photochemistry is used to generate a tracer gradient in the spatial distribution or angular distribution of fluorescent probe molecules in a specimen. Relaxation of the tracer distribution to equilibrium is determined by subsequent measurements of fluorescence emission (or its polarized components) at low excitation rates, so that no significant additional bleaching occurs. The technique was originally developed for the measurement of lateral molecular motion in cell membranes [36–40], but has been applied to bulk specimens and to tracers in the cytoplasm of living cells [22,41–51]. In addition, it has been recognized that focal, polarized, and distributed FRAP patterns, as well as scanning and imaging detection systems, are complementary so that for a particular type of specimen or for determination of a particular transport coefficient, there may be an optimal instrument mode [52–60]. This particular point is well illustrated in this volume by Koppel et al. in which three different FRAP patterns are used in conjunction with a scanning detector to unambiguously determine the values of diffusion, flow, and boundary permeability coefficients for a membrane protein on the compartmentalized surface of sperm cells.

Evanescent wave excitation, or total internal reflection (TIR) illumination, was introduced as a microscopical darkfield technique by Ambrose [61]. In TIR, a beam of light incident on a high-to-low refractive index boundary at an angle exceeding the critical angle is 100% reflected at the interface. This produces an evanescent field in the low-index material that is intense only within a fractional wavelength distance from the interface, under typical conditions. A refractive inhomogeneity within range of this field will scatter radiation from it. A suitable dye molecule located within range of the field will scatter and absorb energy as well, leading to fluorescence emission. TIR excitation of fluorescence (TIRF) was used by Hirschfeld [62] to detect single fluorescent antibody molecules adsorbed to a glass microscope slide. Axelrod, Burghardt, and Thompson [63–65] have demonstrated that tracer mobility at solid/liquid interfaces can be measured by FRAP or FCS (fluorescence correlation spectroscopy) experiments in the TIRF mode, and have applied TIRF microscopy to the study of cell-substrate contact. TIRF microphotometry has also been used to estimate the distance between the substrate and fluorescent tracer in the cytoplasmic compartment of interphase 3T3 cells [66].

All quantitative studies involving TIR excitation of scatter or fluorescence must also involve consideration of the effects of the refractive boundary on the scatter or emission properties of the specimen. Any molecule (or dipole radiator) that is within range of the evanescent excitation field will also be close enough to the interface so that the directional radiation probability and the radiative lifetime will be perturbed [10–14,16,66]. As pointed out by Lukosz and Kunz [10–12] there are two sources of perturbation: partial reflection of the dipole field at the refractive interface, and refraction of evanescent field components of the dipole. The reflected field shifts the phase of the dipole field at the dipole location thus changing the emission rate, and changes the emission pattern in the low-index half-space due to interference. The refracted field includes Fourier components that are evanescent in the low-index specimen, but become plane-waves in the high-index substrate and therefore carry energy away from the dipole.

In a following report, Axelrod et al. (in this volume) make use of the short range of TIR excitation of fluorescence, and of the perturbation (quenching) of emission dipoles on a dielectric/metal boundary. This combination permits measurement of emission due to tracer molecules adsorbed to a flattened erythrocyte membrane preparation, in the presence of both an equilibrium bulk concentration of the tracer and tracer adsorbed to the substrate. The rate constants for tracer chemical exchange and for diffusion on the flattened membrane are estimated by FRAP experiments in the TIRF mode. In the report by Gaub et al. (in this volume) TIR illumination is used to excite fluorescence emission via intermolecular energy transfer, where the "donor" molecule is a dye linked to antigen molecules recognized by a T-lymphocyte, and the "acceptor" dye molecule is linked to major histocompatability complex (MHC) molecules in a synthetic bilayer supported by the totally reflecting substrate. An experiment is proposed in which TIRF, FRAP, and intermolecular energy transfer are used together to measure the relative proximity and mobility of donor and acceptor molecules within the contact zone between a single T-cell and the bilayer substrate.

## FLUORESCENCE MICROTOMOGRAPHY

As previously stated, living cells are complex optical objects; one source of complexity is that cells are not thin films, and generally have an axial extent that is greater than the depth-of-field of a high-resolution microscope. For study of the internal structures of cells, fluorescent analogs of native intracellular proteins have been developed, as have fluorescent stains specific to a particular cell component such as membrane, DNA, acidic vesicles, mitochondria, etc. These probes open up the possibility of mapping in three dimensions the intracellular distribution of a specific molecule or assembly at the optical resolution limit. This distribution involves not only the lateral (x,y) coordinates mapped directly into a microscope image, but also the axial (z) coordinate which is projected into the image. Any one fluorescence image therefore contains an in-focus component due to the object section located at the geometric object plane of the microscope, and components more and more severely out-of-focus due to object sections progressively more distant from the geometric object plane. Agard and Sedat [67,68], and Fay and Fogarty [69] recognized that a single fluorescence image was related to the true object by a convolution of the object with the point-source transfer function (PSTF) of the microscope, and therefore that a set of images recorded as the object was stepped (or tilted) through the geometric object plane could be processed by using the PSTF to remove from each image the out-of-focus components. The "object" in this case is the three-dimensional distribution of fluorescence emitters within the specimen; the completely in-focus set of processed images is a discrete reconstruction of this object, so that the technique could be called "fluorescence microtomography" (name suggested by the work of Agard and Sedat).

The simple relationship described above, relating the fluorescence image set and the object, holds strictly only if the object and surround are isorefractive (generally not true for living cells), the object is only weakly absorbing (may not be true if high concentrations of tracer accumulate), and if emission from any part of the object is unpolarized. The third restriction derives from the fact that an emitted light wave entering the microscope will not have equal amplitude over a wavefront if the source of that wave is, on the average, an oriented dipole. A general theory of fluorescence polarization microtomography would include a PSTF for each principal orientation of the emission dipole relative to the optical axis of the microscope. This would be an extension of the results of Axelrod [15] on microscopic fluorescence polarization.

In the original applications of fluorescence microtomography, the only identified aberration appearing in computed reconstructions was halation attributed to scattering of emission by dense subcellular structures [67]. It is likely that in these specimens local dipole orientation was well averaged by either "wobble" due to thermal motion, or by orientational disorder at a scale below the optical resolution limit.

Recent developments in fluorescence microtomography include the use of a scanning microscope to generate three-dimensional reconstructions without any required image processing [32], and the introduction of zonal excitation to decompose fluorescence images into Fourier coefficient maps. A more detailed discussion is included in another report by Lanni (in this volume).

## CONCLUSION

It has always been difficult to separate microscopy as an optical science from biological microscopy. Many of the advances discussed herein originated from the ever-more-detailed questions raised by scientists studying cells. However, numerous advances have been made in other areas of microscopy, especially related to materials science and microelectronics. Undoubtedly there will be a flow of ideas and methods both ways.

Fluorescence microscopy in particular has been immensely useful in cell biological research. Along with use has come the realization that dye molecules are not always innocuous probes of a dense, reactive system such as a living cell, due to the photochemistry of excited states produced during the measurement process. For many experiments now contemplated by cell biologists, minimization of total radiation dose to the specimen is thus an important consideration. In some cases it will be a luxury to be able to view the specimen directly, since this involves diversion of an informational photon flux that would otherwise be detected by a measuring instrument. Machine-vision microscopes of the future probably will not resemble the classic compound instrument (scanning microscopes already do not) that is so familiar, but will be optimized to form an image matched to a man-made detector rather than to a retina. It is very likely, then, that the state of flux that microscopy appears to be in will keep the users of microscopes glued to their video terminals if not their oculars.

## REFERENCES


1. Zernike F: Phase contrast, a new method for the microscopic observation of transparent objects. I. Physica 9:686, 974. 1942.
2. Allen RD, David GB, Nomarski G: The Zeiss-Nomarski differential interference equipment for transmitted-light microscopy. Zeit f wiss Mikroscopie u mikr Tech 69:193–221, 1969.
3. Ellinger P, Hirt A: Method and device for microscopically viewing living objects. US Patent 1,891,348, 1932.
4. Ichioka Y, Kobayashi T, Kitagawa H, Suzuki T: Digital scanning laser microscope. Appl Optics 24:691–696, 1985.
5. Ruch F: Birefringence and dichroism of cells and tissues. In Pollister AW (ed): "Physical Techniques in Biological Research, IIIA." 2nd ed, Academic Press, New York: 1966, pp 57–86.
6. Sato H, Ellis GW, Inoué S: Microtubular origin of mitotic spindle form birefringence. J Cell Biol 67:501–517, 1975.
7. Streibl N: Three-dimensional imaging by a microscope. J Optic Soc Am A 2:121–127, 1985.
8. Frieden BR: Optical transfer of the three-dimensional object. J Optic Soc Am 57:56–66, 1967.
9. Kerker M, Chew H, McNulty PJ, Kratohvil JP, Cooke DD, Sculley M, Lee M-P: Light scattering and fluorescence by small particles having internal structure. J Histochem Cytochem 27:250–263, 1979.
10. Lukosz W, Kunz RE: Light emission by magnetic and electric dipoles close to a plane interface. I. Total radiated power. J Optic Soc Am 67:1607–1614, 1977.
11. Lukosz W, Kunz RE: Light emission by magnetic and electric dipoles close to a plane interface. II. Radiation patterns of perpendicular oriented dipoles. J Optic Soc Am 67:1615–1619, 1977,

12. Lukosz W, Kunz RE: Fluorescence lifetime of magnetic and electric dipoles near a dielectric interface. Optics Comm 20:195–199, 1977.
13. Carniglia CK, Mandel L, Drexhage KH: Absorption and emission of evanescent photons. J Optic Soc Am 62:479–486, 1972.
14. Burghardt TP, Thompson NL: Effect of planar dielectric interfaces on fluorescence emission and detection. Biophys J 46:729–737, 1984.
15. Axelrod D: Carbocyanine dye orientation in red cell membrane studied by microscopic fluorescence polarization. Biophys J 26:557–574, 1979.
16. Thompson NL, McConnell HM, Burghardt TP: Order in supported phospholopid monolayers detected by the dichroism of fluorescence excited with polarized evanescent illumination. Biophys J 46:739–747, 1984.
17. van Ligten RF, Osterberg H: Holographic microscopy. Nature 211:282–283, 1966.
18. Smith RW: Holographic methods in microscopy. J Microsc 129:29–47, 1983.
19. Taylor DL, Amato PA, Luby-Phelps K, McNeil P: Fluorescent analog cytochemistry. Trends Biochem Sci 9(3):88–91, 1984.
20. Wang Y-L, Taylor DL: Distribution of fluorescently labeled actin in living sea urchin eggs during early development. J Cell Biol 81:672–679, 1979.
21. Luby-Phelps K, Amato PA, Taylor DL: Selective immunocytochemical detection of fluorescent analogs with antibodies specific for the fluorophore. Cell Motil 4:137–149, 1984.
22. Luby-Phelps K, Lanni F, Taylor DL: The behavior of a fluorescent analog of calmodulin in living 3T3 cells. J Cell Biol 101:1245–1256, 1985.
23. Tayor DL, Allen RD, Benditt EP: Determination of the polarization optical properties of the amyloid-Congo Red complex by phase modulation microspectrophotometry. J Histochem Cytochem 22:1105–1112, 1974.
24. Mickols W, Maestre MF, Tinoco I: Differential microscopy. Biophys J 47:7a, 1985.
25. Young JZ, Roberts F: A flying spot microscope. Nature 167:231, 1951.
26. Kogelnik H, Li T: Laser beams and resonators. Proc IEEE 54:1312–1329, 1966.
27. Koppel DE: Fluorescence redistribution after photobleaching. Biophys J 28:281–292, 1979.
28. Hansen EW, Allen RD, Strohbehn JW, Chaffee MA, Farrington DL, Murray WF, Pillsbury TA, Riley MF: Laser scanning phase modulation microscope. J Microsc (in press).
29. Shack R, Baker R, Buchroeder R, Hillman D, Shoemaker R, Bartels PH: Ultrafast laser scanner microscope. J Histochem Cytochem 27:153–159, 1979.
30. Marsman HJB, Stricker R, Wijnaendts van Resandt RW, Brakenhoff GJ, Blom P: Mechanical scan system for microscopic applications. Rev Sci Instrum 54:1047–1052, 1983.
31. Sheppard CJR: Applications of scanning optical microscopy. In Baker LR (ed): "Microscopy—Techniques and Capabilities." Proc SPIE 368:88–95, 1982.
32. Wijnaendts van Resandt RW, Marsman HJB, Kaplan R, Davoust J, Stelzer EHK, Stricker R: Optical fluorescence microscopy in three diminsions: Microtomoscopy. J Microsc 138:29–34, 1985.
33. Lewis A, Isaacson M, Harootunian A, Murray A: Development of a 500 Å spatial resolution light microscope. Ultramicroscopy 13:227–232, 1984.
34. Wessel J: Surface-enhanced optical microscopy. J Optic Soc Am B 2(9):1538–1541, 1985.
35. Morawetz H, Steinberg IZ (eds): Luminescence from biological and synthetic macromolecules. Eighth Katzir Conference, Ann. New York Acad. Sci. 366. The New York Academy of Sciences, New York, 1981.
36. Peters R, Peters J, Tews KH, Bähr W: A microfluorimetric study of diffusion in erythrocyte membranes. Biochim Biophys Acta 367:282–294, 1974.

37. Axelrod D, Koppel DE, Schlessinger J, Elson E, Webb WW: Mobility measurement by analysis of fluorescence photobleaching recovery kinetics. Biophys J 16:1005–1069, 1976.
38. Edidin M, Zagyansky Y, Lardner TJ: Measurement of membrane protein lateral diffusion in single cells. Science 191:466–468, 1976.
39. Jacobson K, Wu E, Poste G: Measurement of the translational mobility of concanavalin A in glycerol-saline solutions, and on the cell surface by fluorescence recovery after photobleaching. Biochim Biophys Acta 433:215–222, 1976.
40. Koppel DE, Axelrod D, Schlessinger J, Elson EL, Webb WW: Dynamics of fluorescence marker concentration as a probe of mobility. Biophys J 16:1315–1329, 1976..
41. Barisas BG, Leuther MD: Fluorescence photobleaching recovery measurements of protein absolute diffusion constants. Biophys Chem 10:221–229, 1979.
42. Lanni F, Ware BR: Detection and characterization of actin monomers, oligomers, and filaments in solution by measurement of fluorescence photobleaching recovery. Biophys J 46:97–110, 1984.
43. Gorti S, Plank L, Ware BR: Determination of electrolyte friction from measurements of the tracer diffusion coefficients, mutual diffusion coefficients, and electrophoretic mobilities of charged spheres. J Chem Phys 81:909–914, 1984.
44. Wojciesczyn JW, Schlegel RA, Wu E-S, Jacobson KA: Diffusion of injected macromolecules within the cytoplasm of living cells. Proc Natl Acad Sci USA 78:4407–4410, 1981.
45. Wang Y-L, Lanni F, McNeil PL, Ware BR, Taylor DL: Mobility of cytoplasmic and membrane-associated actin in living cells. Proc Nat Acad Sci USA 79:4660–4664, 1982.
46. Kreis TE, Geiger B, Schlessinger J: Mobility of microinjected rhodamine actin within living chicken gizzard cells determined by fluorescence photobleaching recovery. Cell 29:835–845, 1982.
47. Jacobson K, Wojcieszyn J: The translational mobility of substances within the cytoplasmic matrix. Proc Natl Acad Sci USA 81:6747–6751, 1984.
48. Leslie RJ, Saxton WM, Mitchison TJ, Neighbors B, Salmon ED, McIntosh JR: Assembly properties of fluorescein-labeled tubulin in vitro before and after fluorescence bleaching. J Cell Biol 99:2146–2156, 1984.
49. Salmon ED, Saxton WM, Leslie RJ, Karow ML, McIntosh JR: Diffusion coefficient of fluorescein-labeled tubulin in the cytoplasm of embryonic cells of a sea urchin: Video image analysis of fluorescence redistribution after photobleaching. J Cell Biol 99:2157–2164, 1984.
50. Salmon ED, Leslie RJ, Saxton WM, Karow ML, McIntosh JR: Spindle microtubule dynamics in sea urchin embryos: Analysis using a fluorescein-labeled tubulin and measurements of fluorescence redistribution after laser photobleaching. J Cell Biol 99:2165–2174, 1984.
51. Saxton WM, Stemple DJ, Leslie RJ, Salmon ED, Zavortink M, McIntosh JR: Tubulin dynamics in cultured mammalian cells. J Cell Biol 99:2175–2186, 1984.
52. Smith BA, McConnell HM: Determination of molecular motion in membranes using periodic pattern photobleaching. Proc Natl Acad Sci USA 75:2759–2763, 1978.
53. Koppel DE: Fluorescence redistribution after photobleaching. Biophys J 28:281–292, 1979.
54. Koppel DE, Sheetz MP, Schindler M: Lateral diffusion in biological membranes. Biophys J 30:187–192, 1980.
55. Smith LM, McConnell HM, Smith BA, Parce JW: Pattern photobleaching of fluorescent lipid vesicles using polarized laser light. Biophys J 33:139–146, 1981.
56. Lanni F, Ware BR: Modulation detection of fluorescence photobleaching recovery. Rev Sci Instrum 53:905–908, 1982.

57. Davoust J, Devaux PF, Leger L: Fringe pattern photobleaching, a new method for the measurement of transport coefficients of biological macromolecules. EMBO J 1:1233–1238, 1982.
58. Koppel DE, Sheetz MP: A localized pattern photobleaching method for the concurrent analysis of rapid and slow diffusion processes. Biophys J 43:175–181, 1983.
59. McGregor GN, Kapitza H-G, Jacobson KA: Laser-based fluorescence microscopy of living cells. Laser Focus/Electro-Optics, 20(11):85–93, 1984.
60. Wegener WA, Rigler R: Separation of translational and rotational contributions in solution studies using fluorescence photobleaching recovery. Biophys J 46:787–793, 1984.
61. Ambrose EJ: The movements of fibrocytes. Exp cell Res 8(Suppl.):54–73, 1961.
62. Hirschfeld T: Optical microscopic observation of single small molecules. Appl Optics 15:2965–2966, 1976.
63. Axelrod D, Burghardt TP, Thompson NL: Total internal reflection fluorescence. Ann Rev Biophys Bioeng 13:247–268, 1984.
64. Axelrod D, Thompson NL, Burghardt TP:Total internal reflection fluorescent microscopy. J Microsc 129:19–28, 1983.
65. Axelrod D: Cell-substrate contacts illuminated by total internal reflection fluorescence. J Cell Biol 89:141–145, 1981.
66. Lanni F, Waggoner AS, Taylor DL: Structural organization of interphase 3T3 fibroblasts studied by total internal reflection fluorescence microscopy. J Cell Biol 100:1091–1102, 1985.
67. Agard DA, Sedat JW: Three-dimensional architecture of a polytene nucleus. Nature 302:676–681, 1983.
68. Agard DA: Optical sectioning microscopy: Cellular architecture in three dimensions. Ann Rev Biophys Bioeng 13:191–219, 1984.
69. Fay FS, Fogarty K: Computerized 3-dimensional immunocytochemistry in single cells. J Cell Biol 97:294a, 1983.

*Applications of Fluorescence in the Biomedical Sciences, pages 461–476*


# 21 Adsorption Kinetics on Biological Membranes: Measurement by Total Internal Reflection Fluorescence

*Daniel Axelrod, Robert M. Fulbright, and Edward H. Hellen*

The distribution and dynamics of fluorescent-labeled molecules at or near "solid" surfaces (e.g., membranes) is of considerable interest in surface chemistry, surface physics, and cell biology. Unfortunately, the number of fluorophores close enough to interact with the surface may be dwarfed by the number farther away but still excited and detected by a conventional optical system (such as epi-illumination in a microscope). Illumination by the evanescent wave of a totally internally reflected laser beam overcomes this problem by selectively exciting just those molecules within less than a wavelength of the surface [see Axelrod, 1984, for a review]. As an illustration of the selective surface illumination achieved with total internal reflection fluorescence (TIRF) optics, Figure 1 compares epi-illumination with TIRF on crenated whole human red cells labeled with the lipid probe 3,3′-dioctadecylindocarbocyanine (diI-C18-3, or diI).

By combining TIR with fluorescence photobleaching recovery (FPR) or fluorescence correlation spectroscopy (FCS), the chemical kinetic rates of surface adsorption and desorption of the fluorescent labeled molecule can be measured [Thompson et al., 1981]. However, the surfaces and soluble molecules studied with TIR/FCS and TIR/FPR in the past were chosen mainly for convenience in testing the experimental technique rather than for their

Department of Physics, Biophysics Research Division, University of Michigan, Ann Arbor, Michigan 48109.

intrinsic biological relevance. In this chapter, we discuss applications of TIR/FPR to nonspecific binding between a peptide hormone and a cell surface—in particular, fluorescent-labeled epidermal growth factor (EGF) and human erythrocyte ghost membrane. Study of this nonspecific interaction may suggest whether specific reaction rate enhancement by nonspecific adsorption followed by surface diffusion ("reduction of dimensionality" as proposed by Adam and Delbruck [1968]) is likely to occur on a cell surface.

Several theoretical and experimental aspects related to employing TIR/FPR on cell membranes are discussed in some detail here because they are nontrivial and may find general application. These aspects are:

1. Theoretical analysis of the kinetic parameters of nonspecific surface adsorption that allow "reduction of dimensionality" reaction rate enhancement.
2. Experimental techniques for flattening erythrocyte ghosts onto solid surfaces.
3. Suppression of fluorescence of peptides adsorbed directly to the substrate upon which the membrane adheres. This suppression is accomplished by coating the surface with a thin layer of aluminum, which quenches fluorescence in its vicinity.
4. A general discussion of the emission patterns of fluorophores near bare and metal-coated dielectrics.
5. Light-induced artifacts affecting adsorption.
6. TIR/FPR results and their significance with respect to surface chemical kinetics.

Other recent biophysical applications of TIRF in our lab and elsewhere are not discussed in detail here. These include time-lapse microscope video recording of cells in culture in the continuous presence of a fluorescent ligand in solution; fluorescent antibody marking of cytoplasmic filaments in thick cells; quantitation of blood protein adsorption on artificial biomaterials; measurement of cell-to-substrate distances; and spectral changes of adsorbed materials.

## THEORETICAL ANALYSIS OF REDUCTION OF DIMENSIONALITY

The Adam-Delbruck hypothesis is that a ligand diffusing in three-dimensional bulk solution might find its membrane-bound receptor by first nonspecifically adsorbing to the membrane and then surface diffusing in two dimensions toward the receptor. Under certain conditions, this mode of approach should lead to a shorter mean time for capture than a pure three-

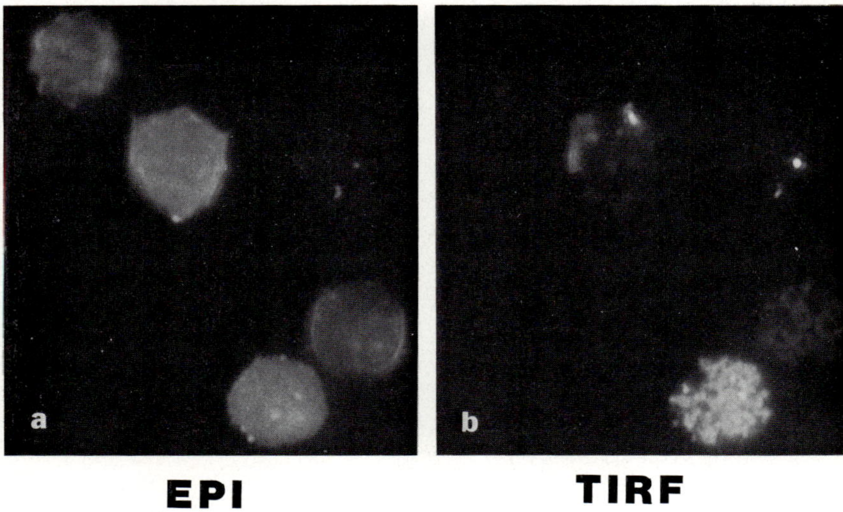


Fig. 1. Crenated whole human red cells labeled by the lipid probe 3,3′dioctadecylindocarbocyanine (diI-C18-3). a. Epi-illumination (EPI). b. Total internal reflection fluorescence (TIRF).

dimensional search for the membrane receptor. Whether this advantage is significant depends on several factors, but in the simplest case of a spherical cell with uniformly spaced receptors on its surface and immersed in an infinite three-dimensional reservoir of ligand, there are only four relevant factors: (1) the number of receptors on the cell surface; (2) the fraction of the cell surface occupied by receptors; (3) the surface diffusion coefficient of the nonspecifically adsorbed ligand; and (4) the mean time the ligand is reversibly adsorbed (i.e., the reciprocal of the kinetic rate of desorption).

The first two factors determine the efficiency by which the cell receptors capture ligands without the aid of surface diffusion. This efficiency is expressed as the ratio of the net flow rate $J$ of the ligand toward the cell to the maximum possible flow rate $J_{\max}$ toward hypothetical cell of the same size whose entire surface is receptive. According to Berg and Purcell [1977],

$$\frac{J}{J_{\max}} = \frac{1}{1 + \dfrac{\pi a}{Ns}} \tag{1}$$

where $s$ = radius of the receptive site of each receptor, $a$ = radius of the cell, $N$ = number of receptors on the cell.

One can show that

$$J_{max} = 4\pi D_3 \; a \; C(\infty) \tag{2}$$

where $D_3$ = three-dimensional bulk diffusion coefficient, $C(\infty)$ = concentration of ligand in solution very far from the cell surface.

For reasonable estimates of these parameters ($s$ = 5 Å and $a$ = 10 $\mu$m), then $J = J_{max}/2$ at a receptor number $N$ = 63,000 (or 16 receptors/$\mu m^2$). This calculation shows that a cell is still quite efficient in capturing ligands at quite low receptor concentrations even without surface diffusion. Note however, that some important receptors (e.g., insulin receptors) are found at even lower concentrations, on the order of 1 receptor/$\mu m^2$. Also, these calculations only apply assuming the receptors are "perfect" sinks. Imperfect sinks, which may result from unfavorable relative orientations, for example, will generally reduce $J/J_{max}$.

If $J/J_{max} \ll 1$, then surface diffusion may significantly increase the flow toward receptors (or, equivalently, the reaction rate). This increase is significant only if the average time spent by a nonspecifically adsorbed ligand on the surface before desorption, $\tau_{off}$, is larger than the average time required for two-dimensional diffusion to the receptor, $\tau_D$.

Berg and Purcell [1977] provide an approximate expression for $\tau_D$ in the case where the radius of a receptive area $s$ is much smaller than the average separation of receptors. The condition under which surface diffusion may significantly enhance reaction rates (provided $J/J_{max} \ll 1$) is

$$\tau_{off} > \frac{2a^2 \ln \dfrac{2}{\sqrt{N}} \dfrac{a}{s}}{DN} = \tau_D \tag{3}$$

where $D$ is the surface diffusion coefficient.

TIR/FPR and TIR/FCS are particularly well suited to direct measurements of $\tau_{off}$ and $\tau_D$. This paper describes the measurement of $\tau_{off}$ for EGF on erythrocyte ghosts.

## FLATTENING GHOST MEMBRANES ON A SOLID SURFACE

Human erythrocytes are adsorbed to a 1″ × 1″ surface of either bare or aluminum vacuum-coated microscope slide glass (see next section) that has been covalently bonded with poly-l-lysine. The procedure for covalent attaching poly-l-lysine, which involves presilanization and succinylation of glass, has been described by Jacobson et al. [1978] for treatment of spherical glass beads.

After the red cells are adsorbed onto the surface, they are hemolyzed by hypoosmotic shock. The ghost membranes spread out in a manner which yields, when visualized by a fluorescent label, a circular disk of diameter 6–8 $\mu$m with a characteristic "crescent" and "bite" configuration as shown in Figure 2.

Figure 3 presents a schematic drawing of the likely morphology of the spread ghost, as deduced from fluorescent labeling with lipid probe diI or with rhodamine-wheat germ agglutinin (R-WGA) either before (pre-label) or after (postlabel) adsorption and spreading of the cells on the surface (Fig. 4). With diI labeling (either pre or post), both the bite and crescent are labeled, with the crescent approximately twice as bright (as viewed under epi-illumination), suggesting a single membrane layer in the bite and a double layer in the crescent (Fig. 4a). With R-WGA postlabeling (Figs. 2b, 4b), only the crescent is labeled. This suggests that (1) the exposed surface of the bite membrane is cytoplasmic with no WGA receptors and (2) the spacing between ghost and the solid substrate is so close that the R-WGA molecules

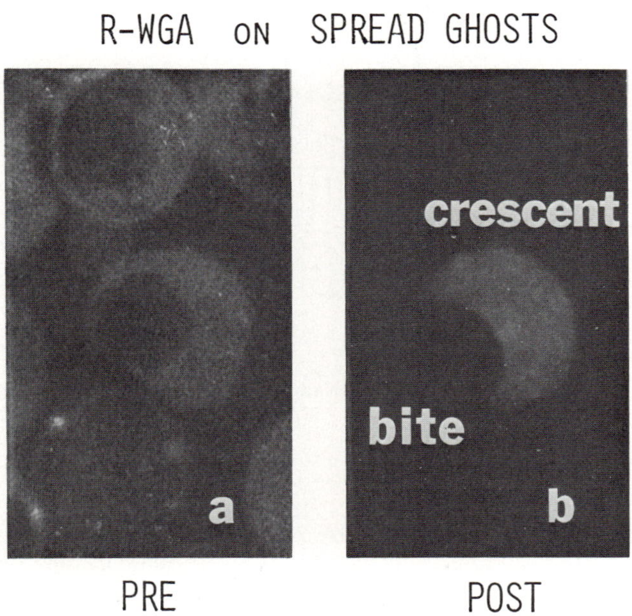


Fig. 2. Erythrocyte ghost membranes spread on poly-l-lysine-coated glass, showing "crescent" and "bite" configuration, with rhodamine–wheat germ agglutinin (R-WGA) labeling. a. Prelabeled (before adhering to glass) b. Postlabeled (after adhering to glass). Epi-illumination with a 50× water immersion N.A. 1.00 objective was used. × 2550.

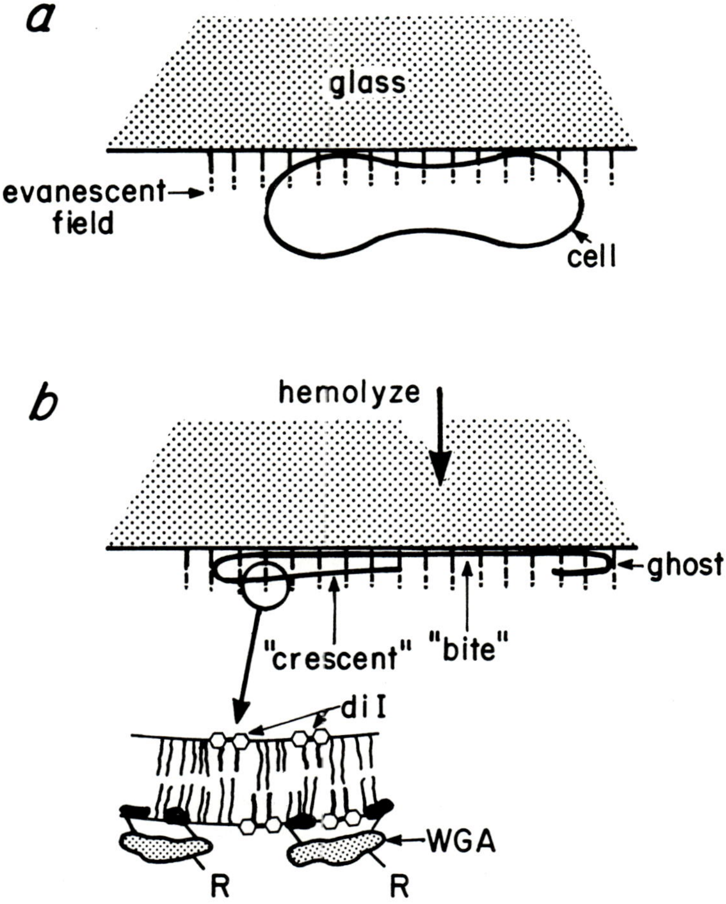


Fig. 3. Schematic drawing of the likely morphology of spread erythrocyte ghosts.

cannot diffuse into it. With R-WGA prelabeling (Fig. 4c), the pattern is somewhat similar to diI labeling, strengthening the view that the bite consists of a single membrane with its outside surface facing the substrate and cytoplasmic surface facing the open solution. Note that under TIR illumination of Figure 4b and c, the pattern is qualitatively similar to epi-illumination for both diI (Fig. 4a) and R-WGA labeling (Fig. 2), suggesting that the membrane is indeed flattened on the glass.

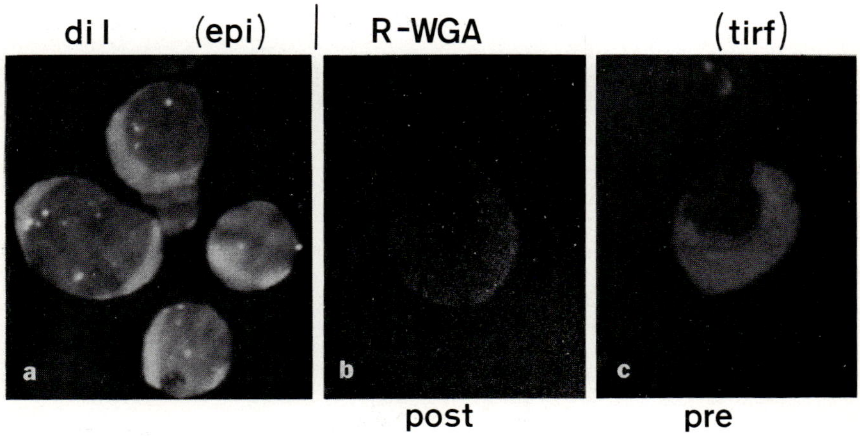


Fig. 4. Spread erythrocytes labeled with (a) diI under epi-illumination: (b) R-WGA (postlabeled) under TIRF; (c) R-WGA (prelabeled) under TIRF. A 50× water immersion N.A. 1.00 objective was used. × 2150.

## SUPPRESSION OF SUBSTRATE-ADSORBED FLUORESCENCE

To study the interaction of EGF with spread ghost membranes, the microscope slide with adsorbed membranes is placed in equilibrium with a solution of fluorescein (F)-EGF in Hank's balanced salt solution without calcium or magnesium (HBSS/Ca-free) in a chamber shown in Figure 5. When viewed with epi-illumination, the fluorescence of F-EGF in bulk solution totally obscures the fluorescence of F-EGF adsorbed to either the ghost membranes or surrounding substrate. But when viewed by TIR, fluorescence from the surface dominates. Unfortunately, however, the concentration of F-EGF adsorption directly onto a poly-lysinated glass substrate is very large compared to that on the ghost membranes and the fluorescent field again appears uniform, at least initially. But as F-EGF bleaches during TIRF observation, the narrow aqueous layer between a ghost membrane and the substrate becomes dark due to a slow diffusional exchange of bleached and unbleached F-EGF. The membrane with its characteristic crescent and bite then becomes discernible as dark disks on a bright background (Fig. 6a).

This overwhelming but relatively bleachable contribution due to F-EGF adsorption to the substrate can be overcome by two possible strategies: (1) designing the chemistry of the substrate to bind less F-EGF while still strongly attracting ghost membranes; or (2) quenching the fluorescence of F-EGF residing very close to the substrate. We have chosen the latter strategy

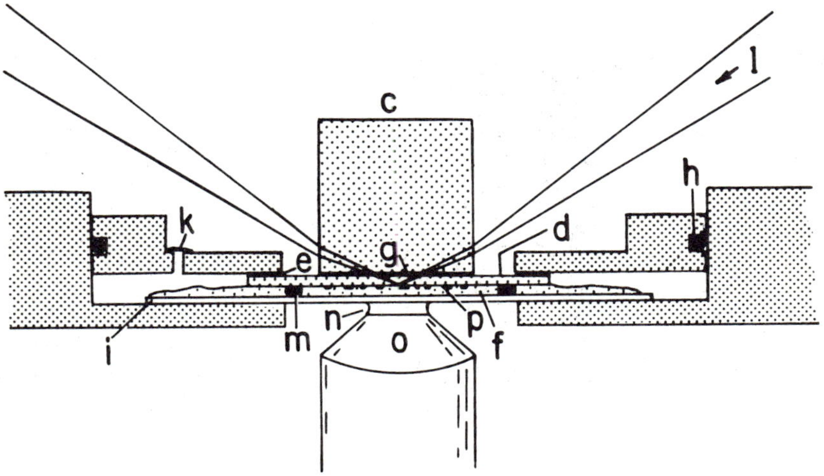


Fig. 5. Airtight sample chamber for TIR/FPR experiments on spread erythrocytes. Abbreviations are (c) fused silica cube, approximately 1 cm$^3$; (d) fused silica (glass) slide, 1″ × 1″ × 1 mm, coated on the lower surface with approximately 200 Å of aluminum and derivatized with poly-l-lysine; (e) silicone grease seal; (f) fluorescein-epidermal growth factor (F-EGF) solution; (g) thin layer of glycerol to maintain optical contact between cube and slide; (h) neoprene O-ring; (i) glass coverslip; (k) sealable vent; (l) incident laser beam, focused at total internal reflection interface; (m) Teflon spacer ring, thickness 0.056 mm; (n) immersion oil or water; (o) microscope objective, either 100× N.A. 1.2 water, or 50× N.A. 1.00 water, or 40× N.A. 1.3 oil. The finely stippled parts of the chamber are Plexiglas; the vertical dimension is somewhat exaggerated for clarity.

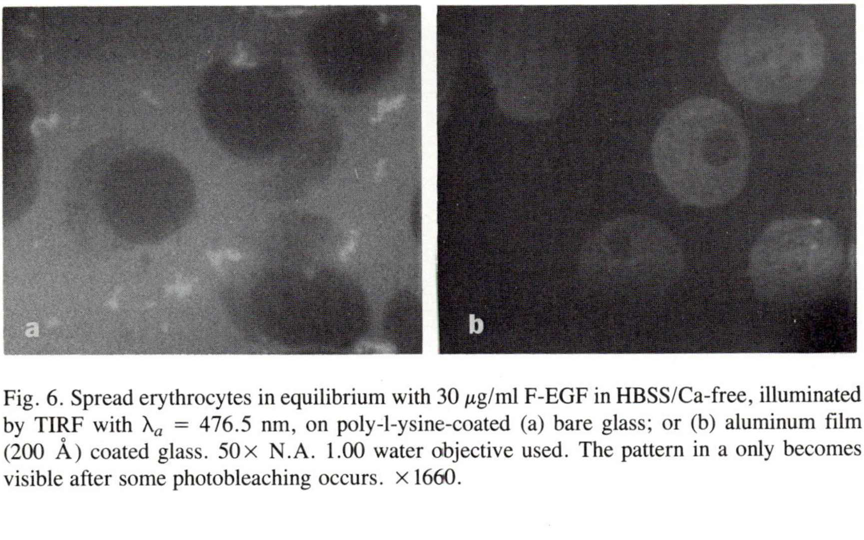


Fig. 6. Spread erythrocytes in equilibrium with 30 $\mu$g/ml F-EGF in HBSS/Ca-free, illuminated by TIRF with $\lambda_a$ = 476.5 nm, on poly-l-ysine-coated (a) bare glass; or (b) aluminum film (200 Å) coated glass. 50× N.A. 1.00 water objective used. The pattern in a only becomes visible after some photobleaching occurs. ×1660.

by first vacuum coating the glass with ~200 Å of aluminum, exposing to air to produce an oxide layer, then subsequently silanating, succinylating, and poly-l-lysinating the surface, and finally adsorbing and hemolyzing erythrocytes in exactly the same manner as on bare glass. When placed in equilibrium with F-EGF and illuminated by TIRF, the ghost membranes on the aluminum coating stand out as bright disks on a dark background (Fig. 6b). Clearly, fluorescence from F-EGF adsorbed to the membrane in the bite and crescent regions is not quenched nearly as strongly as that adsorbed directly onto the derivatized aluminum oxide coating.

## THEORY OF FLUOROPHORE EMISSION NEAR METAL FILMS

In order to better understand the dependence of fluorophore quenching on fluorophore orientation and distance from the surface, we performed extensive computer calculations of the radiation power emitted into the aqueous solution and also through the metal film into the glass, as a fraction of the total power dissipated by the fluorophore. The methods and results of these calculations, to be described in detail elsewhere, are of general interest and are described briefly here.

We initially follow the approach of a recent paper by Ford and Weber [1984]. The fluorophore is modeled classically as an oscillating dipole $\vec{\mu} e^{-i\omega t}$ of some arbitrary orientation placed at $\vec{r} = \hat{z}d$ where $z = 0$ is the metal/liquid interface and $z > 0$ is the liquid region. The power dissipated by the dipole via electromagnetic modes (i.e., not counting intrinsic radiationless decay) is

$$P = \frac{\omega}{2} Im\,(\vec{\mu}^{\,*} \cdot \vec{E}) \tag{4}$$

where $\vec{E}$ is the total electric field at the dipole which arises both from its own near-field and the reflection from the metal surface. Some of this total power $P$ radiates to long distances, but some does not. The "missing" energy becomes trapped in the metal and is ultimately converted to heat.

Both the $\vec{E}$ field at the dipole (used in Eq. 4) and the $\vec{E}$ field at large distances from the dipole (used to calculate radiated power) can be obtained by starting from a plane-wave expansion of the well-known field $\vec{E}(\vec{r})$ of an isolated oscillating dipole [Ford and Weber, 1984]. This expansion is basically a Fourier transform integral of $\vec{E}(\vec{r})$ where the integration is taken over wave vector $k$:

$$\vec{E}(\vec{r}) = \int \vec{E}(\vec{k})\, e^{i\vec{k}\cdot\vec{r}}\, dk_x dk_y dk_z \tag{5}$$

Although the integral is taken over all $\vec{k}$ space, we must impose the restriction that $k = (k_x^2 + k_y^2 + k_z^2)^{1/2} = 2\pi\nu/c$ is a constant since frequency $\nu$ is constant. For the region of integration where $k_x$, $k_y$, and $k_z$ are all real and $\leqslant k$, the plane waves have two-dimensionally infinite wave fronts and simply propagate away from the dipole in all directions. But since $k_x$ and $k_y$ vary from $-\infty$ to $+\infty$ in Eq. 5, $k_z$ necessarily will be imaginary for $(k_x^2 + k_y^2)^{1/2} > k$. Physically, this region corresponds to short-wavelength "evanescent waves" propagating parallel to the surface at $z = 0$ and exponentially decaying in both directions from the $z = d$ plane. The amplitudes and phases of each of these wave types, both infinite and evanescent, are contained in $\vec{E}(\vec{k})$ of Eq. 5. The evanescent waves and those plane waves propagating toward the surface—bare glass or thin metal coating on glass—all produce transmitted and reflected waves with relative amplitudes and phases given by Fresnel coefficients and directions given by Snell's law. For the metal coating, we use Fresnel coefficients derived for dielectric coatings [Born and Wolf, 1975], noting that the real part of the dielectric constant of a metal is negative. The incident plus the reflected waves (for $z > 0$) or the transmitted waves (for $z < 0$) are then integrated over the appropriate $\vec{k}$ range as a function of observation point $\vec{r}$ with the approximation that $|\vec{r}| \gg \lambda$, to give the far field $\vec{E}$. The radiated power is then $c/8\pi \, \epsilon \, \vec{E}^* \cdot \vec{E}$, where $\epsilon$ is the dielectric constant of the medium in which the light is observed (i.e., water or glass). Figure 7 shows the calculated radiation pattern of dipoles near a bare glass or metal-coated surface as a function of observation angle.

Two qualitative physical effects are of interest. In the case of bare glass, for $d < \lambda$, a significant portion of the evanescent waves emitted from the dipole interact with the surface and appear as waves propagating through the glass away from the interface at angles greater than the critical angle for total internal reflection. As a result, fluorescent dipoles in water near a glass surface tend to emit a disproportionate share of their energy into the glass [but that extra energy can only be captured with a very high ($>1.33$) numerical aperture lens]. In the case of metal, much of the emitted energy is lost to heat for $d < \lambda/10$, but a significant portion of the radiation passes *through* the metal into the glass in a hollow cone of light. This occurs because the Fresnel transmission coefficient becomes very large (due to the negative real dielectric constant) for a certain narrow range of $k$ vectors. Ford and Weber [1984], who use a somewhat different mathematical approach in the last stage of these calculations, associate this cone of light with the generation of surface plasmon waves at the metal/water interface which for metal films leak energy into the glass.

Figure 8 shows the power of fluorescent light emitted into a 1.0 numerical aperture objective in the water relative to the total power dissipated by the dipole as a function of $d$ for both bare glass and metal coatings. Figure 8

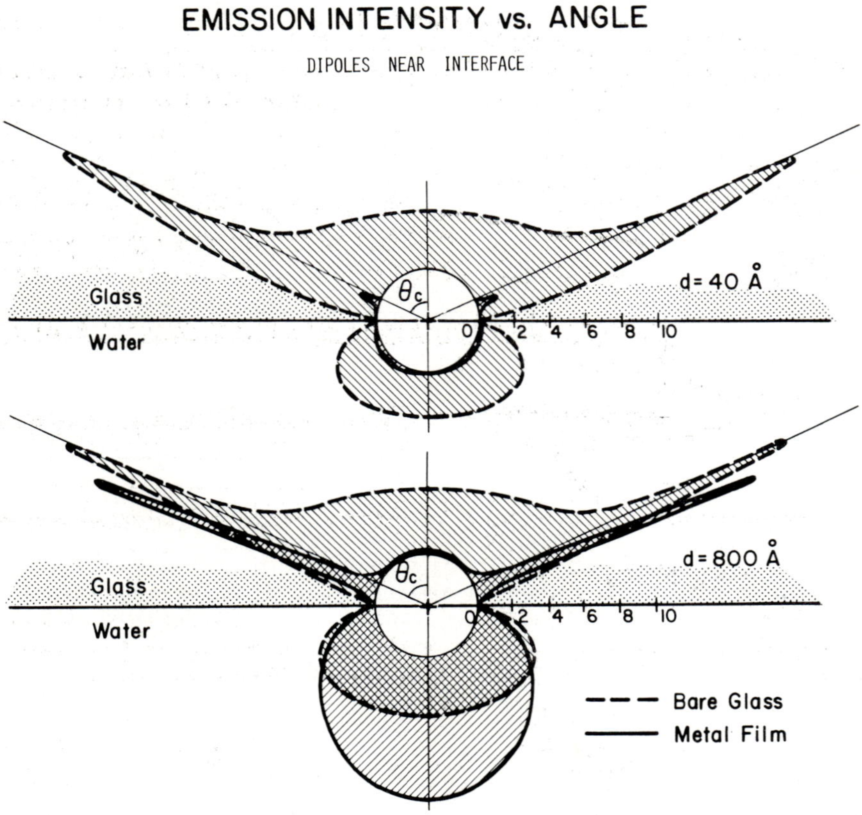


Fig. 7. Normalized radiated power (averaged over azimuthal angle) vs. polar angle $\theta$. Normalization is with respect to the total power dissipated (including non-radiative dissipation) by the dipoles at a distance $d$ from the interface. Equal weighting of results for dipoles parallel and perpendicular to the interface is assumed, as explained in the text. For each $d$, two cases are shown: water/glass (dashed lines) and water/aluminum (solid lines). At each angle, the radial distance from the circle (surrounding the dipole's location) is proportional to the radiated power. The aluminum film is 200 Å thick. The radiative quantum yield for an isolated dipole is assumed to be 0.5. The emission wavelength is 520 nm. The indicies of refraction of the water, glass (fused silica), and aluminum are: $n_w = 1.33$; $n_g = 1.46$; and $\hat{n}_m = 0.73 + i\,5.75$.

clearly shows the effective range of the metal quenching. The periodic oscillations are due to interference between direct and reflected waves; these oscillations decrease in amplitude as $d \rightarrow \infty$.

The exact behavior of dipole radiation near a surface of course depends on the orientation of the dipole. The results depicted in Figures 7 and 8 correspond to an equally weighted average of results for excited molecular

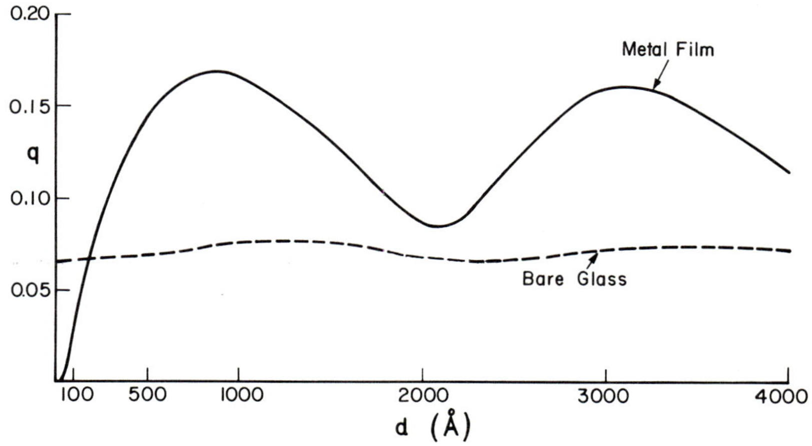


Fig. 8. Effective quantum yield for a 1.00 aperture objective as a function of the distance between the dipoles and the interface. The effective quantum yield is the power of the fluorescent light emitted into the objective divided by the total power (including radiationless decay) emitted by the dipoles. Equal weighting of results for dipoles parallel and perpendicular to the interface is assumed, as explained in the text. Two cases are shown: water/glass (dashed line) and water/metal (200 Å aluminum) (solid line) interfaces. Relevant parameters are the same as in Figure 7.

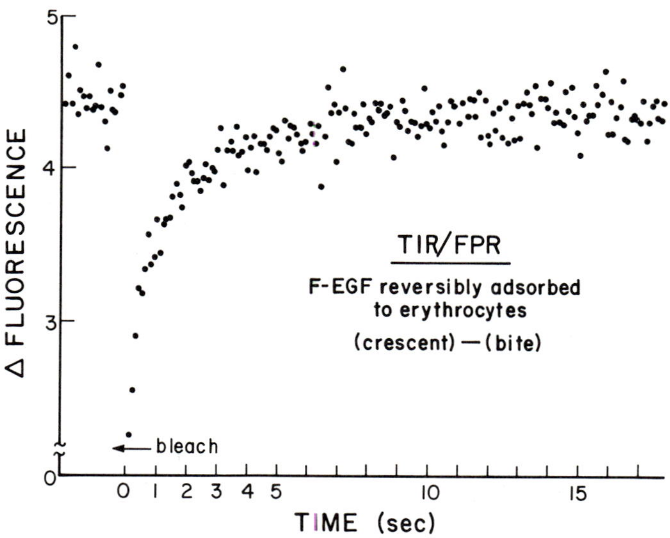


Fig. 9. TIR/fluorescence photobleaching recovery (FPR) results for F-EGF on spread erythrocyte ghosts. The relevant experimental parameters are given in the text. A 40× N.A. 1.3 oil objective was used.

orientations parallel and perpendicular to the surface. This is approximately the situation when randomly oriented dipoles are excited by a TIR evanescent wave polarized in the plane of incidence (p-polarized) with an incidence angle approaching 90°.

All the above considerations apply to the emitted light. The excitation light in our experiments is a TIR evanescent wave which, for the same angle of incidence in the glass, has the same characteristic exponential decay distance at either bare or metal-coated glass interfaces with water. The intensity of the TIR evanescent wave is not discussed in detail here, but it can be derived with the same Fresnel coefficient approach outlined above. A p-polarized TIR evanescent wave has a substantial intensity even for metal coatings several hundred angstroms thick. However, the intensity of an s-polarized evanescent wave is much weaker.

## LIGHT-INDUCED ARTIFACTS AFFECTING ADSORPTION

Upon prolonged TIR illumination, the fluorescence of the spread erythrocyte ghost sample in equilibrium with F-EGF solution gradually and irreversibly increases. The effect occurs in all regions: crescents, bites, and substrate. The kinetics of this increase and the dependence on intensity, time of illumination, or the nature of the soluble protein, fluorophore, or surface were not investigated in detail. At very high light intensities, fluorescence again decreases after the increase, presumably due to the more familiar, if not better understood, photobleaching effect. The increasing fluorescence is most simply interpreted as an excited-state crosslinking of F-EGF to proteins already bound to the surface.

Fortunately, this putative crosslinking effect can be almost completely inhibited by elimination of $O_2$ from the F-EGF solution. This was accomplished by degassing with pure $N_2$ and then adding to the F-EGF solution an enzyme-substrate system which consumes any residual molecular oxygen. This system is 1 $\mu$M protocatechuic deoxygenase and 1 mM protocatechuic acid. The TIRF sample is sealed airtight by a neoprene O-ring and high-vacuum silicone grease (Fig. 5). With adequate degassing and sealing, the enzyme system is not absolutely necessary.

## TIR/FPR EXPERIMENTS ON GHOSTS

The optics, the analysis, and the protocols of TIR/FPR have been described previously [Thompson et al., 1981; Burghardt and Axelrod, 1981]. We discuss here the special features of TIR/FPR experiments on spread erythrocyte ghosts.

Because of the possible heating of the sample due to light absorption in the metal coating, illumination intensity was kept low: 0.01 mW/$\mu$m$^2$ at $\lambda$ =

476.5 nm during the 0.13-sec bleaching flash, and $\sim 0.02\ \mu W/\mu m^2$ during the attenuated observation beam. This low intensity necessitated the use of signal averaging. In our current system, a laboratory minicomputer (Nova 3/12) starts each scan, triggers an acousto-optic modulator to transmit a bright bleaching flash of programmable duration, accumulates the photon count data and adds it to previous scans, pauses for a presettable number of seconds to allow complete recovery, and then starts a new scan. (The low intensity, combined with the very low effective quantum efficiency of fluorophores near a metal surface, also makes the related technique of TIR/FCS difficult in this application).

Although we gather data from only parts of single ghosts, the illuminated region is much larger than a single ghost. We limit the field of view to crescent or bite regions in separate experiments, with an image plane diaphragm of $1.6 \times 1.6\ \mu m$ effective size. The bite data are then subtracted from the crescent data so that the difference, in principle, reflects only molecular dynamics on the bilayer more distal from the substrate.

Figure 9 shows the combined result after a total of 120 runs on two different cells in equilibrium with $\sim 30\ \mu g/ml$ F-EGF in HBSS/Ca-free. The characteristic time of the recovery is about 0.2 sec and the recovery is nearly 100%. This result suggests that the average residence time of F-EGF on the ghost membrane is about 0.2 sec and that almost all the adsorption is reversible. But a firmer interpretation of the recovery time depends on the equilibrium constant of the adsorption.

Preliminary information about the equilibrium constant of adsorption was obtained by a depletion technique. A 0.5-ml solution of F-EGF (8 $\mu g/ml$ in HBSS/Ca-free) was equilibrated with 1 ml of washed and packed human red cells and centrifuged. The resulting depletion of F-EGF from the supernatant was assayed by fluorescence intensity. The depletion is a measure of the maximum amount of F-EGF that might be adsorbed onto the external surface of red cells. A similar experiment was performed on red cell ghosts during mild sonication so that both the external and cytoplasmic membrane reach equilibrium with the solution (as occurs in the TIR/FPR experiments). The measured depletion shows that binding is quite weak: an upper bound for the surface concentration (both external and cytoplasmic) of F-EGF is $\sim 600$ molecules/$\mu m^2$. At the bulk concentration of F-EGF used in TIR/FPR experiments this gives a surface concentration of no more than about 2,300 molecules/$\mu m^2$. The data is too preliminary to report the ratio of concentration on the two surfaces but it appears that substantial concentrations are adsorbed onto *both* surfaces. This tentative conclusion is supported by comparing TIR fluorescence intensity measurements on crescents and bites and interpreting those intensities in view of the known quenching curves (Fig. 8) as a function of distance.

It remains then to interpret the 0.2-sec apparent desorption time. The first question is whether that time represents kinetics in the reaction rate limit (i.e., where the desorption rate determines the recovery rate) or the bulk diffusion limit (where bulk diffusional escape of F-EGF from the solution in the vicinity of the surface determines the recovery rate) [Thompson et al., 1981]. Based on the equilibrium results, one can show that these results are indeed in the reaction limit. The second question concerns whether the recovery is merely due to bulk diffusional flow of unbleached F-EGF from open solution into the constricted region between the overleafed membranes. However, the maximum time for such diffusion is much shorter than 0.2 sec; again, it appears that 0.2 sec is the actual desorption time of F-EGF from erythrocyte membrane.

Referring now to the semitheoretical analysis of "reduced dimensionality" kinetics summarized in Eq. 3, we can now calculate how fast surface diffusion must be to speed reaction kinetics with hypothetical membrane receptors given a $\tau_{\text{off}} = 0.2$ sec. (Of course, the receptor-free ghost membranes used in our experiments serve only as a rough model for receptor-containing living cell membranes.) Assuming $N = 1{,}000$, $a = 10\ \mu\text{m}$, and $s = 5\,\text{Å}$, we calculate $D > 4.8 \times 10^{-9}\ \text{cm}^2/\text{sec}$. This value is certainly reasonable, because it shows that $D$ need be no faster than a surface diffusion coefficient already observed ($5 \times 10^{-9}\ \text{cm}^2/\text{sec}$) [Burghardt and Axelrod, 1981] for one protein (bovine serum albumin) on the surface (glass). Also $D$ need be no more than a small fraction of the diffusion coefficient in the bulk.

TIR/FPR is particularly appropriate for directly measuring $D$; such experiments are now in progress.

## ACKNOWLEDGMENTS

We thank Drs. Nancy L. Thompson and W.H. Weber for very useful discussions about dipole emission patterns near surfaces. The work was supported by the NIH grant 14565.

## REFERENCES

Adam G, Delbruck M (1968): Reduction of dimensionality in biological diffusion processes. In Rich A, Davidson N (eds): "Structural Chemistry and Molecular Biology," WH Freeman Press, pp. 198–215.

Axelrod D (1984): Total internal reflection fluorescence. Ann Rev Biophys Bioeng 13:247–268.

Berg H, Purcell EM (1977): Physics of chemoreception. Biophys J 20:193–219.

Born M, Wolf E (1975): "Principles of Optics," Fifth ed. Oxford: Pergamon Press, pp 61–66.

Burghardt TP, Axelrod D (1981): Total internal reflection/fluorescence photobleaching recovery study of serum albumin adsorption dynamics. Biophys J 33:455–468.

Ford GW, Weber WH (1984): Electromagnetic interactions of molecules with metal surfaces. Phys Rep 113:195–287.

Jacobson BS, Cronin J, Branton D (1978): Coupling polylysine to glass beads for plasma membrane isolation. Biochim Biophys Acta 506:81–96.

Thompson NL, Burghardt TP, Axelrod D (1981): Measuring surface dynamics of biomolecules by total internal reflection fluorescence with photobleaching recovery or correlation spectroscopy. Biophys J 33:435–454.

*Applications of Fluorescence in the Biomedical Sciences, pages 477–497*


# 22 Fluorescence Photobleaching Analysis of Cell Surface Regionalization

*Dennis E. Koppel, Paul Primakoff, and Diana Gold Myles*

In recent years, there has been considerable interest in the study of lateral diffusion in membranes, both in terms of the potential functional significance of the process [for a recent review, see Axelrod, 1983], and in regard to the information on membrane structure and interactions that such studies can provide [e.g., Koppel et al., 1981]. In our laboratories, we have been particularly interested in the ways that cells control the randomizing effect of lateral diffusion in order to produce and maintain specialized functional regions on the cell surface. For this reason, we are using measurements of fluorescence redistribution after photobleaching (FRAP) to characterize the surface dynamics of specific surface antigens of guinea pig sperm cells.

The guinea pig sperm surface has been mapped with monoclonal antibodies and shown to be a dynamic mosaic of distinct regions with different compositions [Myles et al., 1981; Primakoff and Myles, 1983]. The surface protein recognized by antibody PT-1, for example, is confined to the posterior tail region prior to sperm capacitation. During capacitation, in a process apparently requiring active cell swimming, the PT-1 antigen migrates and concentrates onto the anterior tail [Myles and Primakoff, 1984]. FRAP experiments on cells before capacitation have demonstrated that PT-1 antigen exhibits free diffusion within its original domain [Myles et al., 1984], effectively ruling out immobilization as a mechanism for maintaining the original regionalization of this antigen. Ongoing FRAP studies are continuing the elucidation of the mechanisms of localization and migration.

Departments of Biochemistry (D.E.K.) and Physiology (P.P., D.G.M.), University of Connecticut Health Center, Farmington, Connecticut 06032.

In the course of this work, we have employed several distinctly different FRAP experimental techniques and theoretical analyses, each designed to focus optimally on one particular aspect of the problem. Periodic pattern bleaching [Smith and McConnel, 1978; Koppel and Sheetz, 1983] measures dynamics averaged over large areas of one or more regions. Localized line or spot bleaching [Axelrod et al., 1976; Koppel, 1979] yields spatially resolved information, allowing a comparison between the properties of different regions and different locations within the same region. Bleaching whole sections of the surface with a scanning slit (an approach new to this work) yields information on interregional transport. Boundaries between regions can be characterized as barriers of finite permeability to diffusive flux. It is the purpose of this paper to demonstrate the general utility of these methods for the study of regionalized cell surfaces with examples of data for PT-1 antigen, placing particular emphasis on a new method for the determination of interregional permeabilities.

## EXPERIMENTAL METHODS AND RESULTS

### Cell Labeling and Preparation

Cauda epididymal sperm were obtained from mature, male Hartley guinea pigs. Freshly removed sperm were labeled with PT-1 antibody followed by fluorescein- or rhodamine-labeled Fab fragments of goat antimouse IgG, as previously described [Myles et al., 1981]. For the FRAP experiments, labeled cells were immobilized between slid and coverslip in 0.66% agarose in $Mg^{2+}$-HEPES [Myles et al., 1984]. In selected cases, antigen migration was initiated and allowed to proceed during 30–60-min incubations in minimum culture medium (MCM) at 37°C [Myles and Primakoff, 1984] prior to the immobilization in agarose in MCM.

PT-1 antibody binding patterns were observed with immunofluorescence microscopy. On nonincubated sperm (Fig. 1A,B), staining is restricted to the posterior tail region. After incubation in MCM, the PT-1 antigen migrates into the anterior region. Approximately half of the cells show antigen migration after 30 min of incubation [Myles and Primakoff, 1984]. After an extended incubation at 37°C, many cells showed a marked accumulation of PT-1 antigen in the anterior tail region (Fig. 1C,D).

For FRAP experiments, sperm tails were aligned along the laser scan axis (see below) with a rotatable microscope stage. Whenever possible, sperm were chosen that were associated in stacks (rouleaux) which allowed for several aligned cells to be bleached simultaneously. This enhanced the fluorescence signal, and enabled an efficient averaging over the cell population. To insure cell viability, cells were selected that had part of the tail in a nonsolid pocket of the agarose, and therefore continued to beat outside of the area studied.

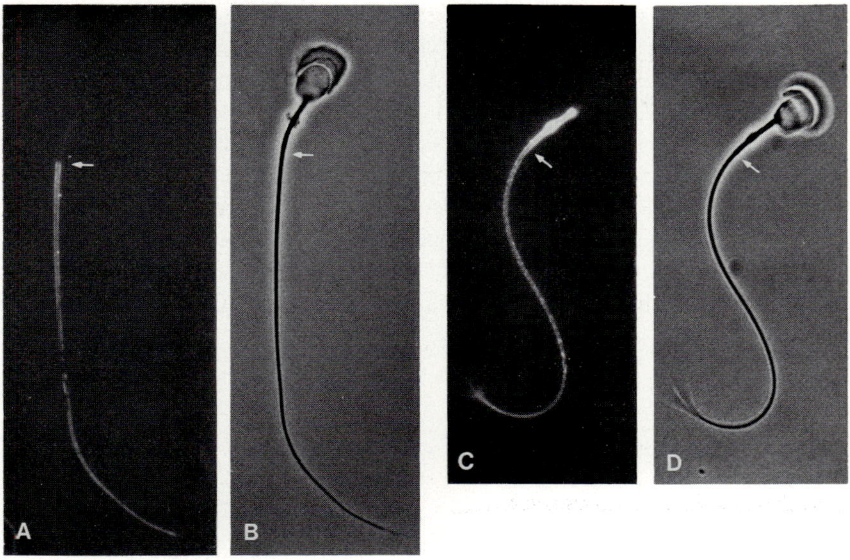


Fig. 1. Binding pattern of PT-1 monoclonal antibody on nonincubated sperm and sperm incubated in $Mg^{2+}$ – minimum culture medium (MCM). Arrows indicate the junction between the posterior tail and anterior tail. A. Antibody binding pattern of PT-1 antibody on fresh cauda epididymal sperm demonstrated by indirect immunofluorescence. Staining is restricted to the posterior tail. B. Phase-contrast micrograph of the same sperm. C,D. PT-1 antibody binding after 90-min incubation at 37°C. PT-1 antigen has accumulated on the anterior tail region, but the posterior tail region is still stained. Reproduced from *The Journal of Cell Biology*, 1984, volume 99, pp. 1634–1641, by copyright permission of the Rockefeller University Press. ×800.

## Fluorescence Redistribution After Photobleaching

The optical apparatus designed to perform FRAP experiments has been described previously [Koppel, 1979]. The main feature which distinguishes this instrument from others of similar construction is an optical scanning mirror immediately preceeding the incident light illuminator of the fluorescence microscope. The scanning capability provided by this mirror is used to produce a measured convolution function combining the distribution of fluorescent molecules and the monitoring laser illumination pattern [Koppel, 1983]. For a localized illumination pattern (e.g., a focused spot, line, or slit image), fluorescence measured as a function of scanning position gives a spatially resolved mapping of the surface. For periodic pattern illumination, on the other hand, scanning allows separate determinations of the amplitude and phase of the modulated fluorescence pattern.

Figure 2 shows the fluorescence bleaching patterns (the cross-hatched areas), relative to sperm cell morphology, that are employed in the three FRAP experimental strategies: periodic pattern, localized line, and scanning slit experiments. These are described, individually, in the sections below.

**Periodic patterns.** Pattern photobleaching has been employed to measure the average diffusion properties of the posterior tail region. For bleaching and monitoring, the cells are illuminated with laser light that has been masked with Ronchi rulings to form an extended pattern of stripes (of repeat distance $\xi_o^{-1}$) oriented perpendicularly to the optical scan axis. A bleaching pulse produces a periodic distribution of unbleached molecules along the sperm

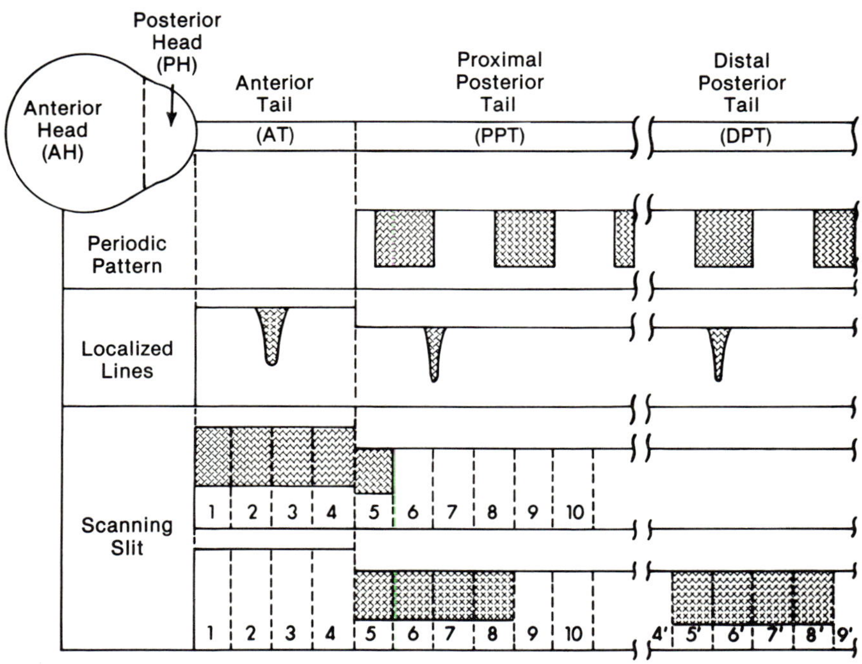


Fig. 2. Schematic diagram of fluorescence bleaching patterns used in three methods of operation (see text). Each horizontal section of the figure represents an assumed initial concentration profile of the fluorescent molecules on the cell surface. The cross-hatched areas indicate the regions bleached. In each case, fluorescence redistribution after bleaching is monitored with successive scans of the attenuated monitoring beam. The guinea pig sperm head and anterior tail are both approximately 11 $\mu$m long. The complete posterior tail is about 91 $\mu$m long.

tails. The modulation amplitude and phase of this distribution is calculated as a function of time after bleaching from a series of fluorescence scans with the attenuated monitoring beam [Koppel and Sheetz, 1983]. Diffusion theory predicts an exponential decay of the fluorescence modulation amplitude ($|A(\xi_o,t)|$) with a decay rate proportional to diffusion coefficient $D$, and inversely proportional to the square of the pattern repeat distance.

Figure 3 presents a set of periodic pattern data on nonincubated cells. The bottom trace shows the measured fluorescence intensity as a function of time before and after bleaching, monitored with a series of 84 twelve-point scans. The top figure graphs the values of the computer-calculated modulation amplitudes for each postbleach scan. The modulation is seen to decay rapidly, and nearly completely, at a rate corresponding to a value of $D = 2.5 \times 10^{-9}$ $cm^2/sec$.

Figure 4 presents the same data again, arranged now as a stack of individual scans with time progressing from top to bottom. In this way, one can see that the modulation of the fluorescence scan decays without evidence of any systematic phase shift which would have indicated the presence of directed, nondiffusive motion.

Figure 5 confirms the dependence of modulation decay rate on pattern repeat distance predicted by diffusion theory. The two sets of data were taken by using Ronchi rulings of 50 (open squares) and 100 (filled squares) lines/inch, giving pattern repeat distances on the cells of 9.5 and 4.75 $\mu m$, respectively. As predicted, the corresponding modulation decay rates differ by a factor of 4.0. This is shown graphically by replotting the data (bottom) with time axes scaled to incorporate the pattern repeat distances.

Figure 6 demonstrates the application of a useful control for this type of experiment. The top two data sets are for cells fixed with 4% formaldehyde for 30 min (circles) and 1.5% glutaraldehyde for 40 min (triangles). the virtual abolition of modulation decay for gluataraldehyde-fixed cells demonstrates that the recovery observed with unifixed cells (squares) corresponds to a redistribution in the membrane and not to a "hopping" of labeled antibody from site to site. Interestingly, even what might be considered extensive fixation with formaldehyde does not completely immobilize sperm surface proteins.

**Localized lines.** We have used localized bleaching and monitoring beams to compare the dynamics of PT-1 antigen on the anterior, proximal posterior, and distal posterior tail regions (see Fig. 2) of cells with and without prior incubation. As with bacterial filaments [Schindler et al., 1980], the geometrical constraints of the cell surface make the use of the conventional focused laser spot configuration difficult to employ and interpret. Consequently we have defocused the laser spot in one dimension with a cylindrical lens to produce a focused "line" of laser light perpendicular to the optical scan axis.

As for pattern bleaching, this reduces the analysis to a simple problem of diffusion in one dimension [Koppel, 1979].

Figure 7 presents a set of raw data photographed directly off the oscilloscope screen as it was recorded. The laser beam was continuously cycled through four positions on a rouleaux of three incubated cells producing the four fluorescence transients shown. The top three traces were recorded from evenly spaced positions, 1.5 $\mu$m apart, centered on the anterior tail regions

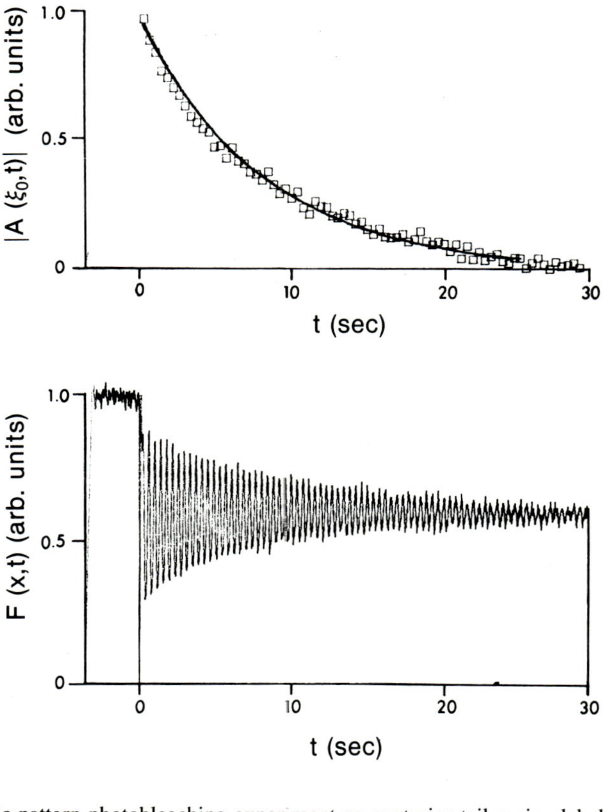


Fig. 3. Periodic pattern photobleaching experiment on posterior tail region labeled with PT-1 followed by fluorescein-conjugated Fab fragments of goat antimouse IgG. The sample was bleached (t = 0) with a pulse of laser light in the form of a pattern of stripes with repeat distance $\xi_o^{-1} = 9.5\ \mu$m. Fluorescence scans with the attenuated stripe pattern were taken every 0.384 seconds before and after bleaching. $F(x,t)$: full display of recorded fluorescence intensities for 84 consecutive 12-point scans. A clear sinusoidal modulation is observed after bleaching. $|A(\xi_o,t)|$: computer-calculated modulation amplitude for each postbleach scan. Solid line is exponential fit to modulation decay. Reproduced from *The Journal of Cell Biology*, 1984, volume 98, pp. 1905–1909, by copyright permission of The Rockefeller University Press.

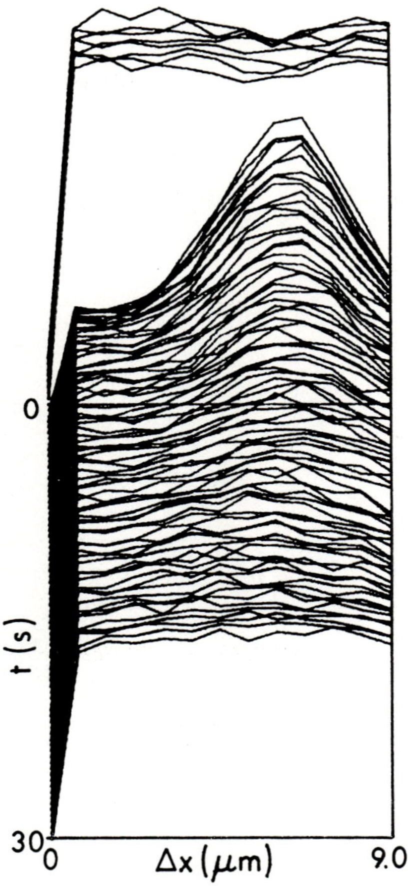


Fig. 4. Display of the data of Figure 3, rearranged as a vertical stack of individual 12-point fluorescence scans with time progressing from top to bottom. The bleach occurred after eight initial prebleach scans. $\Delta x$ is the relative shift of the monitoring illumination pattern along the scan axis.

of the cells. The fourth was recorded from a position 10 μm down onto the proximal posterior tail. From the prebleach data ($t < 0$), it is seen that the intensity on the anterior tail is four times higher than that of the posterior tail for these particular cells. During the scanning cycle at $t = 0$, two bleaching pulses were applied: one at the middle of the three anterior tail positions, and one on the posterior tail. Localized diffusion properties are determined from the rate and extent of fluorescence recovery measured at these positions. The fluorescence recorded from either side of the anterior tail bleach serves two

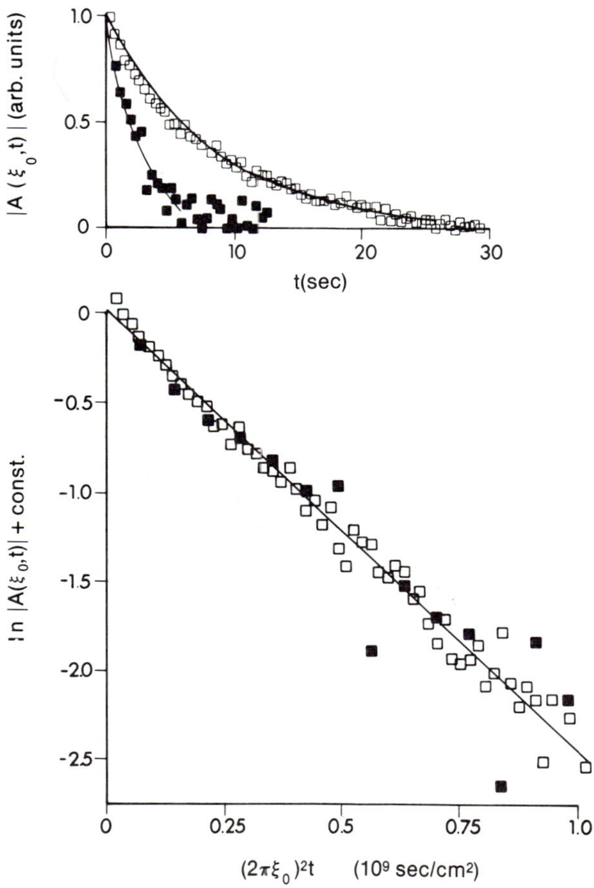


Fig. 5. Periodic pattern photobleaching of posterior tail regions of PT-1-labeled cells. Top. Modulation amplitudes as function of time after bleaching recorded with $\xi_0^{-1} = 9.5\ \mu m$ (□) and 4.25 $\mu m$ (■). Solid lines are least-squares exponential fits. Bottom. Semilog plots of same sets of data with time axes scaled to compensate for the difference in characteristic distance ($\xi_0^{-1}$). Solid line is least-squares straight-line fit with a slope equal to a diffusion coefficient of $2.5 \times 10^{-9}$ $cm^2$/sec. Reproduced from *The Journal of Cell Biology*, 1984, volume 98, pp. 1905–1909, by copyright permission of The Rockefeller University Press.

additional purposes. Immediately after bleaching, they map out the characteristic width of the bleached regions, a quantity needed for the determination of absolute diffusion coefficients [Koppel, 1979]. The time course of fluorescence changes at these positions after bleaching, furthermore, would indicate the presence of any systematic flow or drift. The fact that these curves run parallel, in this case, rather than cross or diverge, indicates that no such motion has occurred.

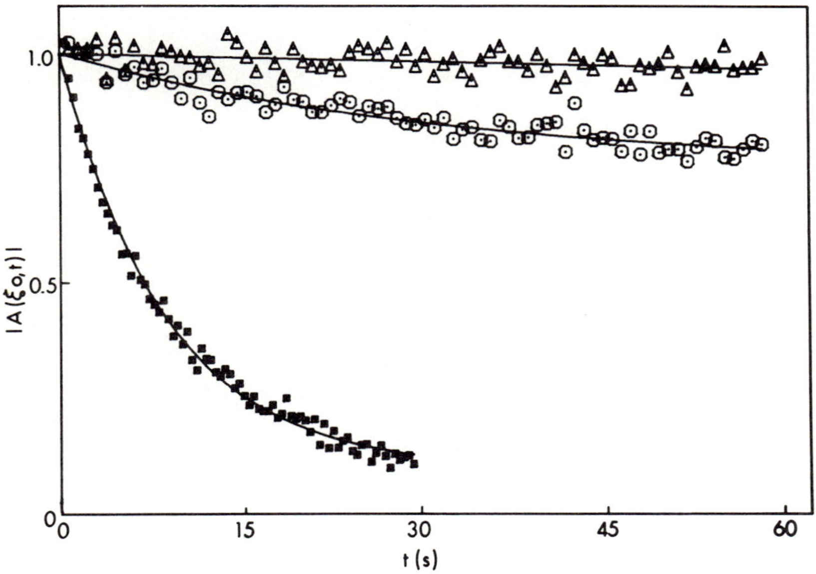


Fig. 6. Periodic pattern photobleaching of posterior tail regions of fixed (Δ,⊙), and unfixed cells (■), recorded with $\xi_o^{-1} = 9.5\ \mu m$. Cells were fixed with 1.5% glutaraldehyde for 40 min (Δ), or 4% formaldehyde for 30 min (⊙), before antibody labeling. Solid curves are least-squares fits of each modulation amplitude ($|A(\xi_o,t)|$) measured as a function of time after bleaching to a single exponential plus a constant.

Figure 8 contains data averaged from many individual experiments of the type shown above, inverted and displayed as the fluorescence depletion per cell (coincident with the bleaching position) as a function of time after bleaching.

**Scanning slits.** In order to assess the nature of interregional diffusion on incubated cells, we have developed a new variation on the FRAP technique. A slit is illuminated with laser light and imaged down onto the cells to illuminate an area of a few microns across. The slit image is cycled through a set of discrete, evenly spaced locations such that the whole area is covered and monitored without gaps or overlaps. Bleaching can be effected at any one or more of these locations. Bleaching patterns are designed to fulfill two criteria: (1) They should maximize the distinction between an absolute barrier to diffusion on the one hand and partial or free diffusion across the boundary on the other. (2) They should contain an internal control of one form or another so that the extent of diffusion across the boundary can be evaluated in a quantitative manner.

Figure 2 shows three useful patterns that we are using. In the first, the bleaching of the entire anterior tail region (sections 1–4) works to fulfill

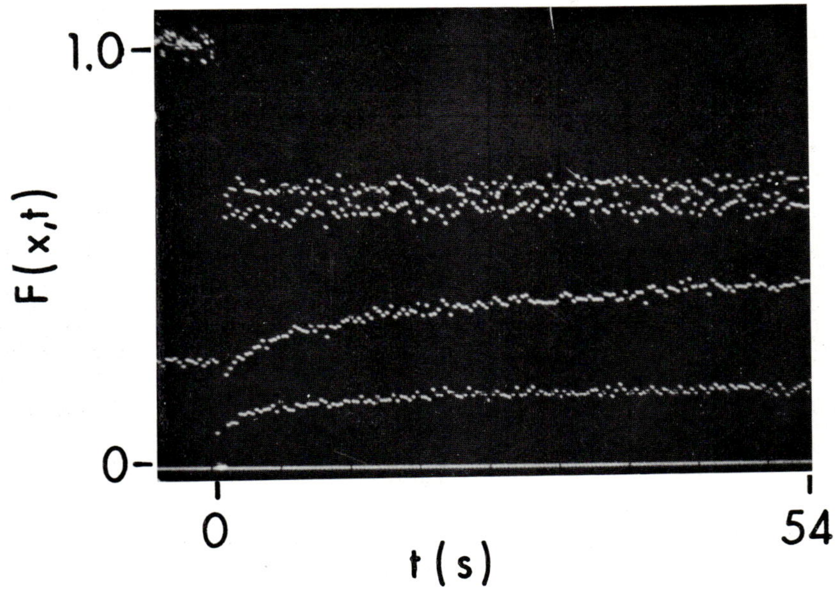


Fig. 7. Localized line fluorescence redistribution after photobleaching (FRAP) data photographed directly off the oscilloscope screen as it was recorded from a rouleaux of three cells fluorescently labeled after a 45-min incubation in MCM at 37°C. The focused laser beam was successively scanned over four discrete positions. The top three traces were recorded with the beam positioned at three evenly spaced locations, 1.5 $\mu$m apart, centered on the anterior tail regions of the cells. The fourth is from a position 10 $\mu$m down on the proximal posterior tail. At $t = 0$, the cells were bleached at two locations: the position on the posterior tail and the middle position of the three on the anterior tail.

criterion #1. Any decay of the depletion in this area indicates that molecules must be able to cross the AT/PT boundary to some extent (at least from the posterior to the anterior side). The additional bleaching of section 5, the first section of the adjacent proximal posterior tail, acts as the internal control specified in criterion #2. The rate and extent of recovery in sections 1–4 can be compared to that observed in section 5.

Bleaching sections 5–8 works in a somewhat different way. Recovery across the section 4/5 boundary is compared to that observed across that of sections 8 and 9. A markedly asymmetric recovery would be indicative of less than free diffusion across the boundary. This configuration complements the one described above in that it measures transport from the anterior to posterior regions. Bleaching sections 5′–8′ on the distal posterior tail of another cell acts as a control for this experiment.

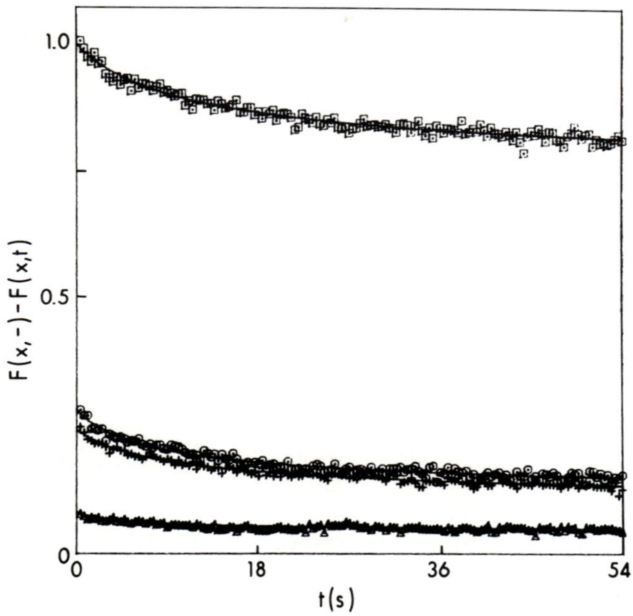


Fig. 8. Localized line FRAP data averaged from many individual experiments of the type shown in Figure 7. Fluorescence depletion per cell at the bleaching positions on the anterior tail regions of cells with incubation (⊡) and without incubation (Δ). Incubations were for 30–45 min in MCM at 37°C before fluorescent labeling. Corresponding data for the bleached position on the proximal posterior tail from the same experiments on the same cells are shown; with incubation (⊙) and without incubation (+).

We have been evaluating experiments of this type using data simulations. The AT/PT boundary is modeled as an interface of finite permeability $h$. $J(t)$, the net flux across the boundary, is proportional to $h$; i.e.,

$$J(t) = h\Delta c(t) \tag{1}$$

where $\Delta c(t)$ is the concentration difference (anterior minus posterior) across the boundary. As $h$ approaches 0 in this model, $J(t)$ approaches 0, and the boundary becomes an impermeable barrier. As $h$ approaches infinity, $\Delta c(t)$ approaches 0, and one has the case of unrestricted diffusion across the boundary. The complete solution for the model, applicable to the bleaching configuration considered above, is presented in the Appendix. Equations for the Laplace transforms of the concentration distributions can be solved analytically. Solutions for the time-dependent functions are produced by numerical integration.

As an example, let us consider the time course of fluorescence depletion, ($\Delta F(x_i,t) = [F(x_i,-)-F(x_i,t)]$, where $F(x_i,-)$ and $F(x_i,t)$ are the fluorescence intensities from section $i$ before and at time $t$ after bleaching, respectively), predicted for the first scanning slit bleaching configuration of Figure 2. As indicated by the cross-hatching, the cells are bleached uniformly on sections 1–5, and monitored on sections 1–10. For simplicity, we will concentrate on the fluorescence transients monitored only on sections 4–7, an area encompassing both the regional boundary and the bleaching boundary. Figure 9 contains simulations for these sections calculated for different values of $hL_2/D$ (where $L_2$ is the total width of the five sections bleached). The discontinuities in the slopes of the curves result from an 11-fold change of time-scale part way through the calculations, duplicating experimental procedures designed to capture both rapid and slow fluorescence changes.

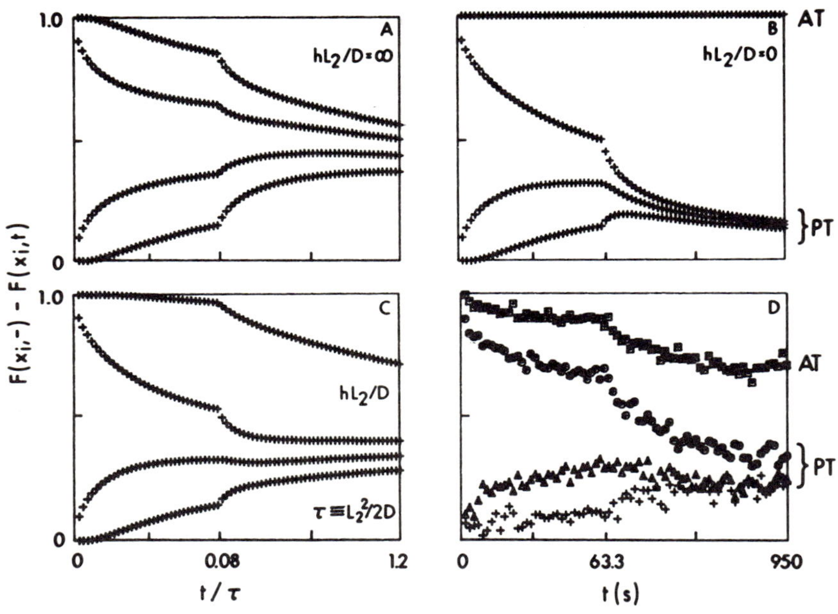


Fig. 9. Simulated and measured fluorescence depletions for a scanning slit FRAP experiment with the bleaching pattern in which sections 1–5 are bleached (see Fig. 2). The four traces from top to bottom in each case correspond to the fluorescence monitored in sections 4–7. The data points are evenly spaced in time before and after the break in slopes marking the imposition of an 11-fold change in time-scale. Data simulations (A–C) were computed according to the theory presented in the Appendix with $\alpha_1 = \alpha_2 = 1.0$, $L_1' = 0.8$, and $h'$ ($= hL_2/D$) equal to $\infty$, 0, and 1.0 in A, B, and C, respectively. Actual data (D) were recorded from a rouleaux of three cells fluorescently labeled after a 30-min incubation in MCM at 37°C.

In the data simulation for unrestricted diffusion across the AT/PT boundary (infinite h, Fig. 9A), the increases and decreases of fluorescence depletion are nearly symmetric about the bleaching boundary. This is as would be expected. The symmetry would be perfect if the posterior tail ended with a reflective barrier after section 10, comparable to the presumed barrier between the anterior tail and head. The slow downward trend of all four curves in Figure 9A seen at long times is evidence of the propagation of depletion beyond the far edge of section 10.

For anything other than unrestricted diffusion across the regional boundary, there is necessarily an additional asymmetry about the bleaching boundary, indicative of a buildup of $\Delta c(t)$ at the regional boundary. For an impermeable barrier (zero $h$, Fig. 9B), one has a maximal asymmetry. The fluorescence depletion on section 4 does not decay at all. For intermediate values of $hL_2/D$ (Fig. 9C), one has intermediate values of asymmetry. A set of actual data is presented in Figure 9D. The general pattern is seen to fall in between that predicted for unrestricted diffusion (Fig. 9A) and an impermeable barrier (Fig. 9B), much as in Figure 9C.

To aid in the determination of the actual value of $h$ in this model most consistent with a particular set of data, we have developed a quantitative analysis based on the observed asymmetry of the fluorescence recoveries, as described above. We define an asymmetry function, designated $P(t)$ in keeping with a functional form analogous to that of a fluorescence polarization:

$$P(t) = \frac{\Delta F_+(t) - \Delta F_-(t)}{\Delta F_+(t) + \Delta F_-(t)} \tag{2}$$

where

$$\Delta F_+(t) = \Delta F(x_4,t) - \Delta F(x_5,t)$$
$$\Delta F_-(t) = \Delta F(x_6,t) - \Delta F(x_7,t)$$

Figure 10 presents curves of $P(t)$ calculated from data simulated for each of several values of $hL_2/D$. It is seen that with the possible exception of the trace for $h = 0$, each reaches a well-defined maximum deviation from zero ($P_{max}$) during the range of characteristic times covered by experiment. These are plotted in Figure 11 in order to allow a graphical estimation of $h$ from experimentally determined values of $P_{max}$. For the data of Figure 9D, $P_{max}$ is estimated at 0.87 (see Fig. 12), giving a value of $hL_2/D$ of about 0.4.

## DISCUSSION

Our data illustrate the ways that various types of FRAP experiments can be used to characterize the properties of regionalized cell surfaces and the

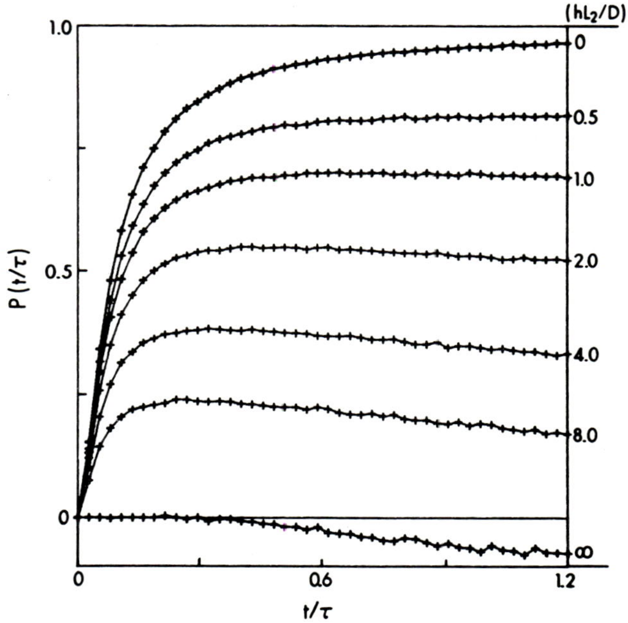


Fig. 10. Calculated values of $P(t/\tau)$, a measure of the asymmetry of fluorescence recovery of scanning slit experiments, as defined in the text. The points were calculated from simulated data of the type shown in Figure 9, for each of the values of $h'$ ($= hL_2/D$) indicated. The scatter in these curves results from errors introduced in the numerical inverse Laplace transformation.

phenomenon of interregional surface migrations. The same basic approaches can be applied to study the movement of molecules between regionalized intracellular compartments, such as the cytoplasm and the nucleus.

Measured diffusion characteristics can be used to eliminate some hypothetical mechanisms for producing and maintaining regional distributions, thereby supporting others. The demonstration of long-range mobility for components localized to a limited domain, for example, rules out immobilization as the sole means of maintaining that localization. The data for PT-1 antigen is particularly striking in this regard. On non-incubated cells, diffusion is observed at a rate near the theoretical upper limit for a freely diffusing membrane protein [Myles et al., 1984]. This result is consistent with the idea that the annulus, observed by electron microscopy at the anterior tail/posterior tail junction [see Figs. 20, 21 of Friend, 1984], acts as a barrier to diffusion for this surface antigen.

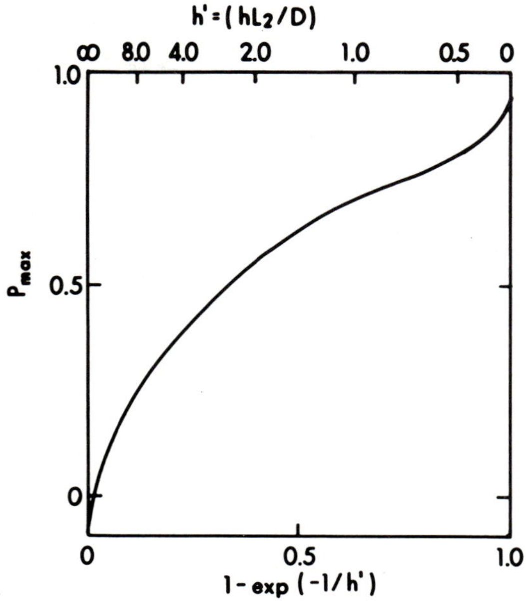


Fig. 11. Values of $P_{max}$, the maximum deviations of $P(t)$ from zero, calculated from data simulations of the type shown in Figure 10. The ordinate, $1 - \exp(-1/h')$, was chosen to present the whole range of $h'$, from zero to infinity, in a convenient way.

We have developed the scanning slit approach for examining the nature of interregional junctions. Previously, Wolf and Voglmayr [1984] used qualitative image analyses to test if fluorescent lipid probes added to the membranes of ram sperm could diffuse between regions. Our approach extends this idea to quantitate the process of interregional diffusion. The interregional boundary is modeled as a barrier of finite permeability $h$. The value of $(hL_2/D)$ is determined from the asymmetry of the fluorescence recovery pattern by using an empirical calibration curve (Fig. 11) derived from simulated data. Direct nonlinear fits of fluorescence data to the theoretical functions are not possible in reasonable periods of time because of the considerable complexity of the theoretical computation.

Physically, barriers between regions can be modeled as a narrow bands in which the effective diffusion coefficient $(D')$ is markedly reduced. By the simple partition-diffusion model, derived for the analysis of basal diffusion across cell membranes [Lieb and Stein, 1969], $h$ has the following form:

$$h = KD'/\lambda \qquad (3)$$

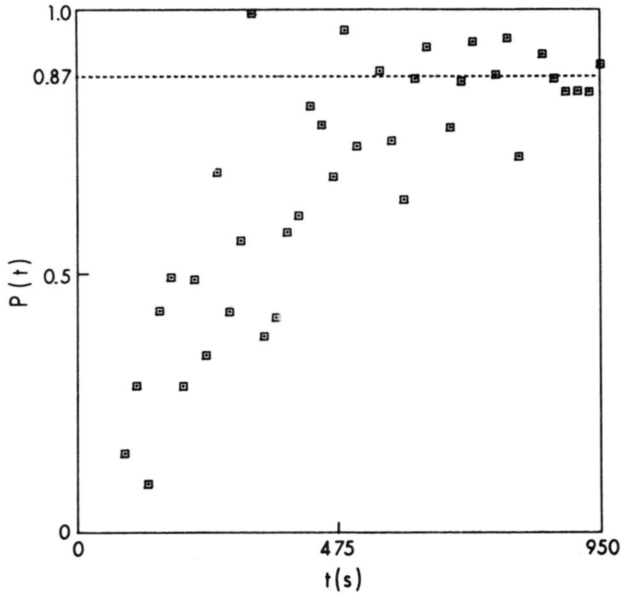


Fig. 12. Values of $P(t)$ calculated for the data of Figure 9D.

where $K$ is the partition coefficient of the diffusing molecules in the barrier region, and $\lambda$ is the thickness of the barrier, so that

$$(hL_2/D) = K(D'/D)(L_2/\lambda) \tag{4}$$

Whatever the actual reality of this formulation, it serves at least to give a useful physical picture for an otherwise abstract parameter.

We are currently exploring the idea that the migration of PT-1 antigen on the anterior tail, observed during cell capacitation, is a diffusion-mediated process. In general, diffusion measurements can provide information of two types in this regard. Is diffusion rapid enough to support observed rates of migration in the first place? If not, then diffusion-mediated mechanisms can be definitively ruled out. If yes, then diffusion-mediated processes remain viable hypotheses. Preliminary analysis supports the latter for the case of PT-1 antigen migration. In either event, measurements of localized diffusion characteristics can help clarify the nature of the processes that maintain the evolving distributions.

## ACKNOWLEDGMENTS

This investigation was supported by research grants GM23585 (D.E.K.) and HD16580 (D.G.M.) from the U.S. Public Health Services. P.P. held an American Heart Established Investigatorship Award.

## REFERENCES

Axelrod D (1983): Lateral motion of membrane proteins and biological function. J Membr Biol 75:1–10.

Axelrod D, Koppel DE, Schlessinger J, Elson EL, Webb WW (1976): Mobility measured by fluorescence photobleaching recovery kinetics. Biophys J 16:1055–1069.

Crank J (1956): "The Mathematics of Diffusion." London: Oxford University Press.

Friend DS (1984): Membrane organization and differentiation in the guinea-pig spermatozoon. In Van Blerkom J, Motta PM (eds): "Ultrastructure of Reproduction." Boston: Martinus Nijhoff Publishers, pp 75–85.

Koppel DE (1979): Fluorescence redistribution after photobleaching: A new multi-point analysis of membrane translational dynamics. Biophys J 28:281–291.

Koppel DE (1983): Fluorescence photobleaching as a probe of translational and rotational motions. In Sha'afi RI, Fernandez SM (eds): "Fast Methods in Physical Biochemistry and Cell Biology." Amsterdam: Biomedical Press, pp 339–368.

Koppel DE, Sheetz MP (1983): A localized pattern photobleaching method for the concurrent analysis of rapid and slow diffusion processes. Biophys J 43:175–181.

Koppel DE, Scheetz MP, Schindler M (1981): Matrix control of protein diffusion in biological membranes. Proc Natl Acad Sci (USA) 78:3576–3580.

Lieb WR, Stein WD (1969): Biological membranes behave as non-porous polymeric sheets with respect to the diffusion of non-electrolytes. Nature 224:240–243.

Myles DG, Primakoff P (1984): Localized surface antigens of guinea pig sperm migrate to new regions prior to fertilization. J Cell Biol 99:1634–1641.

Myles DG, Primakoff P, Bellve AR (1981): Surface domains of the guinea pig sperm defined by monoclonal antibodies. Cell 23:433–439.

Myles DG, Primakoff P, Koppel DE (1984): A localized surface protein of guinea pig sperm exhibits free diffusion in its domain. J Cell Biol 98:1905–1909.

Primakoff P, Myles DG (1983): A map of the guinea pig sperm surface constructed with monoclonal antibodies. Dev Biol 98:417–428.

Schindler M, Osborn MJ, Koppel DE (1980): Lateral diffusion of lipopolysaccharide in the outer membrane of Salmonella typhimurium. Nature 285:261–263.

Smith BA, McConnell HM (1978): Determination of molecular motion in membranes using periodic pattern photobleaching. Proc Natl Acad Sci USA 75:2759–2763.

Wolf DE, Voglmayr JK (1984): Diffusion and regionalization in membranes of maturing ram spermatozoa. J Cell Biol 98:1678–1684.

## APPENDIX: RESTRICTED DIFFUSION AT A REGIONAL BOUNDARY

We seek solutions to the diffusion equation in one dimension,

$$\frac{\partial c}{\partial t} = D \frac{\partial^2 c}{\partial x^2} \tag{A1}$$

for each of the three regions along the x-axis: (1) $c_1(x,t)$ for $0 \leqslant x \leqslant L_1$, the region representing the anterior tail; (2) $c_2(x,t)$ for $L_1 < x \leqslant L_2$, the region representing an adjacent segment of posterior tail; and, (3) $c_3(x,t)$ for $x > L_2$, the region encompassing the rest of the posterior tail. Region 2 is separated out from region 3 to allow for a selective bleaching of part of the posterior tail, as in Figure 2. In keeping with our usual data analysis in terms of fluorescence depletions, we define the $c_i(x,t)$'s as photoinduced concentration depletions. For initial conditions, we specify

$$c_1(x,0) = \alpha_1 c_0 \tag{A2a}$$

$$c_2(x,0) = \alpha_2 c_0 \tag{A2b}$$

$$c_3(x,0) = 0 \tag{A2c}$$

where $\alpha_1$ and $\alpha_2$ can be set to either 0 or 1, depending upon the bleaching pattern selected.

We proceed by the method of Laplace transformation [Crank, 1956]. Multiplying both sides of Eq. (A1) by $e^{-st}$, and integrating by parts over $t$ in the range $0 \leqslant t < \infty$, we have (incorporating initial conditions A2):

$$D \frac{d^2\bar{c}_1}{dx^2} = s\bar{c}_1 - \alpha_1 c_0 \tag{A3a}$$

$$D \frac{d^2\bar{c}_2}{dx^2} = s\bar{c}_2 - \alpha_2 c_0 \tag{A3b}$$

$$D \frac{d^2\bar{c}_3}{dx^2} = s\bar{c}_3 \tag{A3c}$$

where, in each case, $\bar{c}_i$ is the Laplace transform of $c_i$, i.e.,

$$\bar{c}_i\,(x,s) = \int_0^\infty c_i(x,t)\, e^{-st} dt.$$

Eqs. A3 are readily solved in terms of 6 constants, giving

$$\bar{c}_1 = a_1 e^{qx} + b_1 e^{-qx} + \alpha_1 c_0/s \tag{A4a}$$

$$\bar{c}_2 = a_2 e^{qx} + b_2 e^{-qx} + \alpha_2 c_0/s \tag{A4b}$$

$$\bar{c}_3 = a_3 e^{qx} + b_3 e^{-qx} \tag{A4c}$$

where

$$q = (s/D)^{1/2}$$

We can solve for each of the unknown constants by applying the appropriate boundary conditions to the problem.

At $x = 0$, we have

$$\frac{\partial c_1}{\partial x} = 0 \tag{A5}$$

formally establishing the anterior tail/posterior head boundary as impermeable to diffusion. At $x = L_1$, for all $t > 0$,

$$\frac{\partial c_1}{\partial x} = \frac{\partial c_2}{\partial x} \tag{A6}$$

$$D \frac{\partial c_1}{\partial x} = h(c_2 - c_1) \tag{A7}$$

The former condition precludes a buildup or depletion on the interface between regions. The latter establishes the AT/PT boundary as a barrier of finite permeability, as discussed in the text above. At $x = L_2$, for $t > 0$,

$$\frac{\partial c_2}{\partial x} = \frac{\partial c_3}{\partial x} \tag{A8}$$

analogous to Eq. A6, and

$$c_2 = c_3 \tag{A9}$$

since, in the absence of a diffusion barrier, concentration discontinuities are not maintained. Finally, $c_3$ must stay finite as x approaches infinity, so that $a_3 = 0$. The actual tip of the tail is sufficiently far away that, for all practical purposes, at all but extremely long times, we can consider the somewhat simplified problem of diffusion in semiinfinite space. Combining Eqs. A4 through A9, we have:

$$a_2 = -\frac{1}{2}\alpha_2 c_0\, e^{-qL_2} \tag{A10a}$$

$$a_1 = b_1 \tag{A10b}$$

$$= \left[\frac{1 + (1 - \frac{\alpha_1}{\alpha_2})\, e^{q(L_2 - L_1)}}{1 + \frac{1}{2}(Dq/h)(1 - e^{-2qL_1})}\right] a_2$$

$$b_2 = (1 - e^{2qL_1})a_1 + e^{2qL_1}a_2 \tag{A10c}$$

$$b_3 = -e^{2qL_1}a_2 + b_2 \tag{A10d}$$

For the purpose of general calculation, the $\bar{c}_i$'s are recast as functions of a number of normalized, dimensionless parameters:

$$s' = s\tau \tag{A11a}$$

$$x' = x/L_2 \tag{A11b}$$

$$L'_1 = L_1/L_2 \tag{A11c}$$

and

$$h' = hL_2/D \tag{A11d}$$

where

$$\tau = L_2^2/2D \tag{A11e}$$

In this notation (see Eqs. 4 and 10),

$$qx = x'\sqrt{2s'} \tag{A12a}$$

$$qL_1 = L_1'\sqrt{2s'} \tag{A12b}$$

$$qL_2 = \sqrt{2s'} \qquad \text{(A12c)}$$

and

$$Dq/h = \sqrt{2s'}/h' \qquad \text{(A12d)}$$

Eqs. 12 complete the derivation of the $\bar{c}_i$'s. Two final operations are needed to convert these $\bar{c}_i$'s to the desired fluorescence intensity depletions. Integration over segments of $x'$, an operation duplicating the effect of uniform slit illumination, is easily performed on Eq. 4 analytically. The inverse Laplace transform producing the final function of $t/\tau$ must be computed numerically.

*Applications of Fluorescence in the Biomedical Sciences, pages 499–503*


# 23 Total Internal Reflection Fluorimetry on Single T-Lymphocytes in Contact With Supported Lipid Bilayers

*H.E. Gaub, A.A. Brian, T.H. Watts, and H.M. McConnell*

Cell-cell communication is essential for the development and maintenance of the complex integrity and structure of higher organisms. Prominent examples, where direct cell-cell contact has been shown to be required, are cell differentiation [1] and the triggering of cellular immune responses [2]. In the latter case, the transfer of information from non-specific phagocytic cells to highly specific lymphocytes has to occur. In that process, the antigen is presented along with a molecule of the major histocompatibility complex (MHC) on the membrane of an antigen-presenting cell (APC). The "self" and the "foreign" molecule are then recognized in conjunction by the T-cell receptor, triggering proliferation of the corresponding clone [3]. Recent cloning of the T-cell receptor [4,5] has led to a better insight into possible mechanisms of the complicated recognition process. However, there is the need to develop experimental methods and techniques to separate single events in the complex cell interaction by modeling artificial systems and testing the cellular response on well-defined reconstituted systems.

The recognition of the antigen and MHC molecule takes place within the small contact area between the APC and the T-cell. Any kind of direct observation of that process on the molecular level will, when performed in suspension, always measure the interacting molecules against the background of molecules that are not involved in that contact. The low ratio of the contact areas to the whole cell surfaces will thus make straightforward investigation very difficult.

However, methods have been developed recently to replace the membrane of the APC by artificial reconstituted systems [6], especially by supported

Department of Chemistry, Stanford University, Stanford, California 94305 (H.E.G., T.H.W., H.M.M.), Department of Chemistry, University of California at San Diego, La Jolla, CA 92093 (A.A.B.).

planar phospholipid bilayers containing MHC molecules [7,8]. Thus the interacting membrane is restricted to a substrate and can be selectively observed by surface sensitive techniques without background from the cell membrane areas not involved in that interaction, or the surrounding solution. The technique presented in this article accomplishes that by fluorescence excitation of the contact area with total internal reflection [9]. The penetration depth of the evanescent wave is restricted, in this case to a range between 500 and 1000 Å, depending on the incidence angle [10]. As now only a layer with a thickness of the order of molecular dimensions is measured, fluorescence spectroscopy is the method of choice to detect environmental changes and interactions of specifically labeled molecules that are involved in the recognition process.

The reduction of the measured volume makes the use of energy transfer techniques especially suitable to investigate the intermolecular distances of acceptor-labeled MHC molecules in the supported bilayer and donor-labeled antigen in solution, as well as their redistribution, induced by the binding of the T-cell receptor (Fig. 1). With this approach, it should be possible to distinguish between the various proposed configurations of this ternary complex [11].

In order to obtain the fluorescence spectrum from the contact area of a single cell, a microfluorometer (schematically shown in Fig. 2) has been

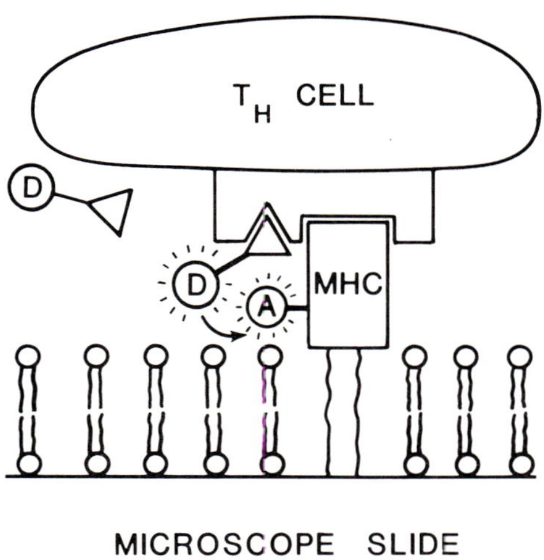


Fig. 1. Schematic representation of the T-cell binding to a supported planar lipid bilayer containing major histocompatability complex (MHC) molecules. The water-soluble antigen is donor-labeled (D) and is excited only when near the interface. Energy transfer can occur if by binding to the receptor, the donor gets close enough to the acceptor-labeled MHC molecule (A).

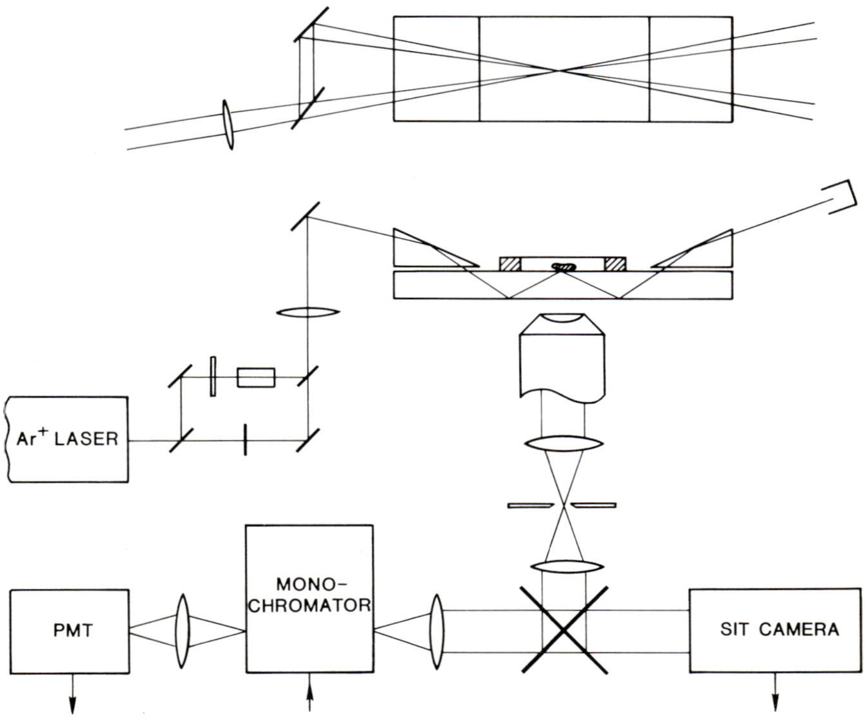


Fig. 2. The microfluorometer is based on an (Nikon Diaphot TMD) inverted microscope. For the practical use of TIR excitation, a fixed stage (with respect to the vertical axis) is required. The focused beam (f = 60 cm) of an argon ion laser (Spectra Physics Ar 2000) is coupled by a high refractive index prism (n = 1.8) into a quartz microscope slide (using glycerol as an index matching medium). To avoid scattering, p-polarization is chosen. In the object plane, the focused beam has a spot size of 200 $\mu$m. The incidence angle can be varied, and, by means of that, the penetration depth of the evanescent wave can be chosen between 500 and 1000 Å (typically 500 Å). The pattern for photobleaching is generated by splitting the beam in two halves and superimposing them in the object plane. The resulting stripe interference pattern has a $\sin^2$ (kx) profile with a periodicity of 2–10 $\mu$m depending on the in-plane angle of the two beams. A field iris diaphragm in the intermediate image plane is used to limit the area from which fluorescence is collected. The fluorescent image can either be monitored with an SIT camera (COHU 5000) or switched to the monochromator (Spex 1681). In this case the image plane is focused with a lens f = 60 mm on the entrance slit and fills without loss the area of the grating. The output slit is threefold enlarged and projected on the cathode of an RCA 31034 photomultiplier (PMT; cooled housing −30°C). Photon counting is done by a Le Croy 3500 with an EG&G 436 discriminator, and EG&G 535 preamp. modules. The monochromator is driven by a Spex Compudrive. Bleaching and observation beams are produced by a system of 90:10 beam splitters and mirrors. The observation beam can be further attenuated by a quartz half-wave plate in a rotation stage in front of a fixed polarizing beam splitter. Selection between the bleach and observation beams is made by electromechanical shutters.

developed based on an inverted microscope. To avoid loss in fluorescence intensity, special care has been taken to match the aperture angle of the microscope output with that of the monochromator input. Spectra with 100 counts per second can easily be obtained from $10^4$ molecules in the contact area without observable bleaching during the measurement (Fig. 3). Additional information about the redistribution of the involved molecules can be obtained, in principle, by photobleaching one of the components with an evanescent wave interference pattern and observing the time dependence of energy transfer (Fig. 2). This can be done either by pulsed or continuous bleaching.

A preliminary result obtained with this setup is given in Figure 3. It shows the fluorescence spectrum of the fluoresceinated antigen in solution and the Texas Red–labeled MHC molecules in the planar membrane in presence of the T-cell. Although the specificity of the interaction has yet to be proven, the spectrum shows that the device described above is sensitive enough to monitor molecular events in the contact area between single cells and supported planar membranes. Further experiments will clarify the question of whether the antigen and the I-A form a complex that is recognized by the T-cell receptor, or, whether the receptor has two independent binding sites (Fig. 1). For recent results see reference [12].

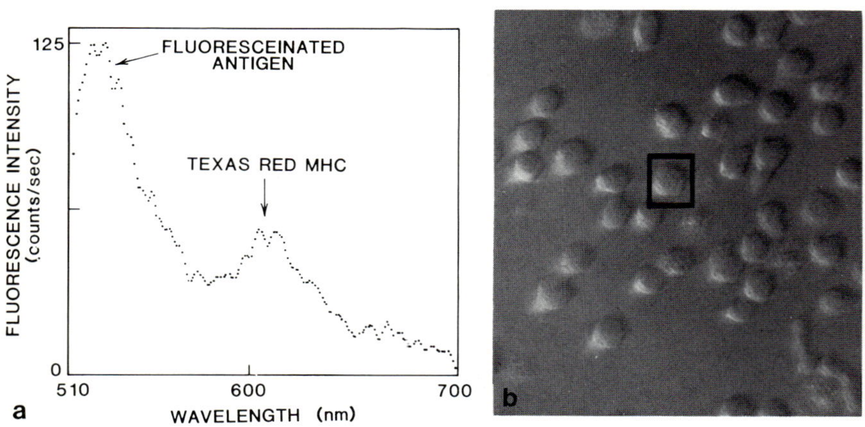


Fig. 3. Fluorescence spectrum of the contact area of the T-cell indicated in the picture. The MHC molecule (in this case I-$A^d$) was isolated from A20-1.11 cells and Texas Red labeled to a fluorophore/protein ratio of 0.7/1. It was reconstituted in a planar Egg-PC-Cholesterol (7/2 mole ratio) bilayer at a lipid-to-protein ratio of 25:1 by weight. It was incubated with T-cells (3DO-54.8) along with the water-soluble fluoresceinated antigen on the microscope stage for 15 min (exc 488 nm/scan speed 1 nm/sec/bandwith $\pm$ 5 nm).

## ACKNOWLEDGMENTS

This work was supported by NIH grant 5R01 AI13587 (H.M.M.), the Medical Research Council of Canada (T.H.W.), and the Deutsche Forschungsgemeinschaft (H.E.G.).

## REFERENCES

1. Edelman GM (1983): Cell Adhesion Molecules. Science 219:450–457.
2. Inaba K, Witmer MD, Steinman RM (1984): Clustering of dendritic cells, helper T lymphocytes, and histocompatible B cells during primary antibody responses in-vitro. J Exp Med 160:858–876.
3. Kappler JW, Skidmore B, White J, Marrack P (1981): Antigen-inducible, H-2-restricted, interleukin-2-producing T cell hybridomas. J Exp Med 153:1198–1214.
4. Hedrick SM, Nielsen EA, Kavaler J, Cohen DI, Davis MM (1984): Sequence relationship between putative T-cell receptors, polypeptides and immunoglobulins. Nature 308:153–158.
5. Saito H, Krantz DM, Tagagaki Y, Hayday AC, Eisen HN, Tonegawa S (1984): Complete primary structure of a heterodimeric T-cell receptor deduced from C-DNA sequences. Nature 303:757–762.
6. Walden P, Nagy ZA, Klein J (1985): Induction of regulatory T-lymphocyte response by liposomes carrying major histocompatibility complex molecules and foreign antigen. Nature 315:327–329.
7. Watts TH, Brian AA, Kappler JW, Marrack P, McConnell HM (1984): Antigen presentation by supported planar membranes containing affinity-purified I-A. Proc Natl Acad Sci USA 81:7564–7568.
8. Brian AA, McConnell HM (1984): Allogenic stimulation of cytotoxic T-cells by supported planar membranes. Proc Natl Acad Sci USA 81:6159–6163.
9. Weis RM, Balakrishnan K, Smith BA, McConnell HM (1982): Stimulation of fluorescence in a small contact region between rat basophil leukemia cells and planar lipid membrane targets by coherent evanescent radiation. J Biol Chem 257:6440.
10. Axelrod D, Burghardt TP, Thompson NL (1984): Total internal relection fluorescence. Ann Rev Biophys Bioeng 13:247–268.
11. Pernis B, Axel R (1985): A one and a half receptor model for MHC-restricted antigen recognition by T-lymphocytes. Cell 41:13–16.
12. Watts TH, Gaub HE, McConnell HM (1986): T cell mediated association of peptide antigen and major histocompatibility complex protein detected by energy transfer in an evanescent wave-field. Nature (in press).

*Applications of Fluorescence in the Biomedical Sciences, pages 505–521*


# 24 Standing-Wave Fluorescence Microscopy

*Frederick Lanni*

The central problem of the science of microscopy is the formation of images that magnify an object and accurately reproduce features up to a limit of resolution set by the laws of physical optics. A useful image or object reconstruction is a map of a particular attribute of the specimen; ideally, an amplitude-contrast image is a map of optical absorbance, a phase-contrast image is a map of optical path length, and a fluorescence image is a map of local luminescence emission rate, for example.

In the usual case, an image is a two-dimensional record of intensity as a function of two lateral coordinates. For an ideal plane object, these coordinates provide a unique mapping of the resolvable features of the specimen into the image. More commonly, however, the image is a two-dimensional record that is an axial projection of a three-dimensional object. In high-resolution microscopy, this projection will also be spatially filtered due to the limited depth-of-field of the image-forming optics. For these two reasons the relative axial position of object features is not encoded uniquely into a single image; a three-dimensional reconstruction of the object requires more data.

In the study of living cells by microscopy at the resolution limit one has, in essence, a problem in microtomography. This is because the features of interest are typically much smaller than the axial dimension of the specimen, and because the depth-of-field is usually smaller as well. Lans Taylor and I became interested in fluorescence microtomography due to our desire to understand the disposition of the cytoskeleton within the cytoplasmic volume and the redistribution of cytoskeletal elements during motile processes. The

Center for Fluorescence Research in Biomedical Sciences, Carnegie-Mellon University, Pittsburgh, Pennsylvania 15213.

use of fluorescent analogues of cytoplasmic proteins such as actin, and of free-volume tracers such as fluorescein-dextran, opens up the possibility of accurate three-dimensional mapping of specific molecular distributions within single living cells [1–3].

The classical method for encoding depth-in-field into image records is stereography (Table 1), for which a minimum of two images is required. Stereography is generally applicable in the low aperture or large depth-of-field limit typical of macrophotography or electron microscopy. In the case where highly coherent light is used to illuminate an object (or where the emission from the object is coherent) the hologram can be recorded. In this particular case a single two-dimensional "image" of the object contains enough information for extensive object reconstruction in three dimensions [4,5]. Recent developments in the general relation between object structure and microscope image have been made by Streibl [6].

In the current practice of fluorescence or phosphorescence microscopy, the image formed by the microscope is due to spatially incoherent emission of light from structures within the object. Under these circumstances the hologram of the emitter distribution can only be made by homodyne methods. State-of-the-art instrumentation for high-resolution fluorescence microtomography, developed by Agard and Sedat [7,8], and by Fay and Fogarty [9], is based upon the acquisition of serial images of a specimen as it is precisely tilted or stepped through the object focal plane of a microscope. For objects that are compact enough to nearly fit within the focal field of the microscope, serial tilt images alone can be used to compute a tomographic reconstruction. This is the microscopic equivalent of stereography. For larger objects such as single cells which have an axial extent greater than the depth-of-field of the microscope, serial focal plane images are recorded. The resulting three-dimensional array of "stacked" image data must then be processed to remove from each image the out-of-focus contribution of parts of the specimen not near that particular plane in the object. From a theoretical standpoint this is a deconvolution operation for which the three-dimensional point-source image function (point spread function, PSF) must be separately determined for

**TABLE 1. Methods for Encoding Depth-in-Field Into a Set of One or More Image Records**

| |
|---|
| Stereography |
| Holography |
| Serial focal-plane images |
| Serial tilt-plane images |
| Focal volume scanning |
| Zonal illumination |

the instrument. It has been demonstrated that these methods lead to good reconstructions of sparsely punctate fluorescent specimens. In the general case of an arbitrary emitter distribution within the specimen and finite noise in the data, convergence of a deconvolution algorithm to an optimum reconstruction can be far from robust, so that an alternative means for encoding axial position into fluorescence or phosphorescence images would be very useful. One route to a more generally invertible encoding scheme is a focal volume-scanning microscope such as the confocal instrument developed by R. Wijnaendts et al. [10]. In this nonimaging device, a high-aperture focal beam excites emission at a location in the specimen conjugate to the detector aperture during scanning. The two important aspects of this device are that the excitation field is weighted by focusing to produce a maximum emission rate in a particular specimen volume element, and that detection is spatially selected to be maximally sensitive to emission from that same volume element. Addition of annular multichannel detection to this instrument would make possible processing equivalent to deconvolution to remove from each "image" volume element photocounts due to emission from neighboring conjugate volume elements in the specimen.

An ideal form of illumination for luminescence microscopy would be zonal excitation by a thin "plane" of light that could be made coincident with the focal plane of the microscope. Then emission would be excited only from in-focus parts of the specimen, and no deconvolution would be necessary, even with full-field image detection. The wave properties of light, however, make formation of such a field essentially impossible in the absence of a waveguide structure.

An alternative to highly localized illumination is zonal excitation by a spatially distributed field having a simple, well-defined intensity profile. The electromagnetic standing-wave field formed by intersecting two coherent light beams can have the necessary characteristics under certain conditions. If the intersecting beams are coherent (laterally and longitudinally), collimated, s-polarized, and of equal intensity, the standing-wave field in the overlap region consists of a periodic array of planar zones in which stationary nodes of zero electric intensity separate zones of stationary high intensity. In fluorescence microscopy, the excitation field induces electric dipole transitions in dye molecules. Therefore, emission from a fluorescent specimen located in a nodal standing-wave field will be maximal at the antinodes of the field and zero where nodal planes intersect the specimen. In the simple arrangement of Figure 1, two beams of light of the proper wavelength to excite dye molecules in the specimen cross so as to produce a set of nodal planes normal to the optical axis of a fluorescence microscope. The image formed by the microscope will have overlapping contributions from the several antinodal zones that span the specimen. By controlling the node

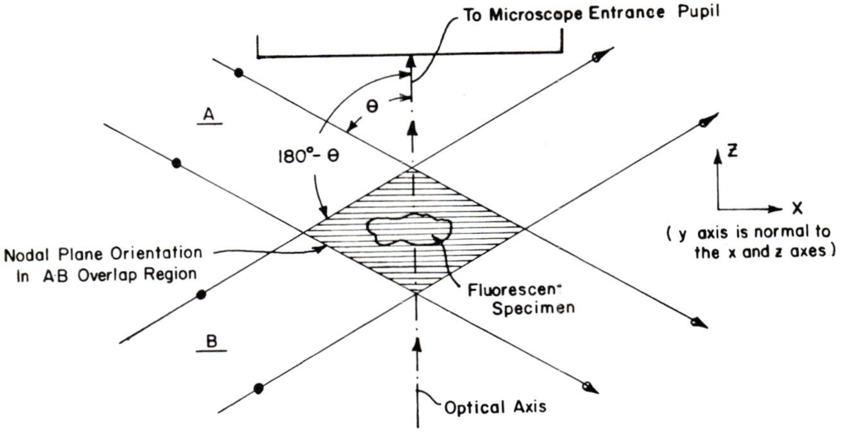


Fig. 1. Basic arrangement for standing-wave fluorescence (SWF) microscopy. Two coherent plane-wave fields, A and B, cross at the location of the specimen on the optical axis of a microscope, and generate a standing-wave field. The axis and the beams define the plane of incidence. When the beams make complementary angles with respect to the axis, the nodal planes of the standing-wave field are normal to the axis. If the beams are s-polarized the electric field has nodes. If the beams are p-polarized the magnetic field has nodes. A fluorescent specimen will emit at the locations of the antinodes of the field, and minimally about the nodes.

spacing of the standing-wave field and its phase relative to a specimen-fixed reference point, a set of fluorescence images can be recorded which are decompositions of the "full" fluorescence image. It should be possible to extract structural detail from this set of images readily because the spatial averaging is less severe than in the usual case: the information specifying the object structure is distributed among a set of records rather than being summed into one. We call this arrangement standing-wave fluorescence (SWF) microscopy.

In principle, the type of distribution of emitters within the specimen—diffuse, punctate, filamentous, or complex—will have little effect on the quality of a three-dimensional reconstruction synthesized from a set of zonally excited fluorescence images. One reason for this is that in-focus diffuse components of the object can be distinguished from out-of-focus punctate components by the different weighting each zone will give to different SWF images recorded at constant focal plane position. This makes the development of a practical SWF microscope attractive for the study of living cells by fluorescent analogue cytochemistry, since diffuse and complex emitter distributions are often encountered.

More important in SWF microscopy than emitter distribution type is the amount of phase and amplitude distortion suffered by wavefronts of the excitation field as they traverse the specimen. The result of such distortion is complication of the pattern of fluorescence excitation in the specimen. In the following discussion I have assumed that the specimen is a weak phase object and a weak absorbance object. The condition of low phase distortion is closely obtained in embedded cytological specimens, and in fixed wet-mount specimens in which the mounting liquid can be much more refractive than water. For living mammalian cells on slides in culture medium, the average refractive index mismatch between cytoplasm (average $n = 1.36$) and surround ($n \cong 1.34$) is about 0.02. Within single cells, the index may vary from location to location, and may be as high as 1.40 for dense protein fibers or for dense vesicles. Our experiments in which fixed cells have been used while immersed in saline solution ($n = 1.336$) closely model the optical properties of live cells, and show that distortion of the standing-wave field is apparently not severe. A more definitive test will be the quality of the reconstructions of biological objects having sharply defined features, such as the fluorescently stained nucleus of a living cell.

## THEORY

The pattern and intensity of a standing-wave field produced by crossing two plane-wave fields is a function of six parameters: wavelength ($\lambda_o$), refractive index ($n$), amplitude, angle of intersection, polarization, and relative phase [11]. For zonal excitation of dye molecules, a nodal electric field pattern must be generated, and this sets the requirements that the two plane-wave fields be of equal amplitude and s-polarized [electric field vector normal to the plane of incidence (Fig. 1)]. As shown in Figure 1, plane-wave field A propagates at an angle $\theta$ relative to the microscope optical axis. Field B propagates at an angle of $180° - \theta$. The electric amplitudes of A and B are

$$\underline{E}_A = \hat{y}E_o \sin(\underline{k}_A \cdot \underline{r} - \omega t + u)$$

$$\underline{E}_B = \hat{y}E_o \sin(\underline{k}_B \cdot \underline{r} - \omega t + v)$$

where $\underline{k}_A = k_1\hat{x} - k_2\hat{z}$, $\underline{k}_B = k_1\hat{x} + k_2\hat{z}$, $k_1 = (2\pi n/\lambda_o)\cos\theta$, $k_2 = (2\pi n/\lambda_o)\sin\theta$, and $u$ and $v$ are unspecified phase angles.

The electric intensity is related to $\underline{E}_A$ and $\underline{E}_B$ by

$$I_e = (\tfrac{1}{2})cn\epsilon_o \langle |\underline{E}_A + \underline{E}_B|^2 \rangle$$

The result of time-averaging $|\underline{E}_A + \underline{E}_B|^2$ is that $I_e$ is a periodic function only of the axial coordinate, z, and represents a nodal standing-wave field:

$$I_e = I_o[1 - \cos(Kz + \varphi)] \tag{1}$$

where $K = |2k_2| = 4\pi n \cos\theta/\lambda_o$. The period of the field, which is the node spacing ($\Delta s$), is given by $2\pi/K$:

$$\Delta s = \lambda_o/2n \cos\theta \tag{2}$$

The parameter $\phi$, which is a function of the original phase angles $u$ and $v$, is the phase of $I_e(z)$ and specifies the position of the nodes relative to a laboratory-fixed reference plane. This is illustrated in Figure 2, which shows graphically how a 90° shift in field phase results in excitation of different zones of a specimen along the axial coordinate. Lateral ($x,y$) resolution in SWF microscopy is obtained through the usual image-forming function of the microscope; control of standing-wave field phase and node spacing gives the microscope a form of axial resolution.

The minimum node spacing is obtained when the intersecting fields A and B are counterpropagating ($\theta = 0°$): then

$$\Delta s_{\text{min}} = \lambda_o/2n$$

The axial resolution limit can be estimated by noting that two small objects will be differentially excited by a 180° shift in the field if their axial separation is $\Delta s_{\text{min}}/2 = \lambda_o/4n$. If ultraviolet light of wavelength equal to 365 nm is used, the axial resolution limit is then 68 nm. This axial superresolution is obtainable because it is due to the excitation field rather than the emission field which is diffracted by the microscope during the image-forming process. Lateral and axial superresolution may be obtainable together by operating a confocal scanning microscope [10] in the standing-wave excitation mode, by use of counterpropagating focal beams from the same laser, or by not constraining the nodal planes to be normal to the optical axis in an image-forming microscope.

The fluorescence emission rate from any specimen volume element $dV(\underline{r})$ will be proportional to the number density of dye molecules $p(\underline{r})$ in that element, and to the local intensity of the excitation field:

$$df(\underline{r};K,\phi) = a\, p(\underline{r})\, I(\underline{r};K,\phi)\, dV$$

The emitter distribution $p(\underline{r})$ is the three-dimensional structural map of the specimen, and is the object of interest here: for simplicity it has been assumed

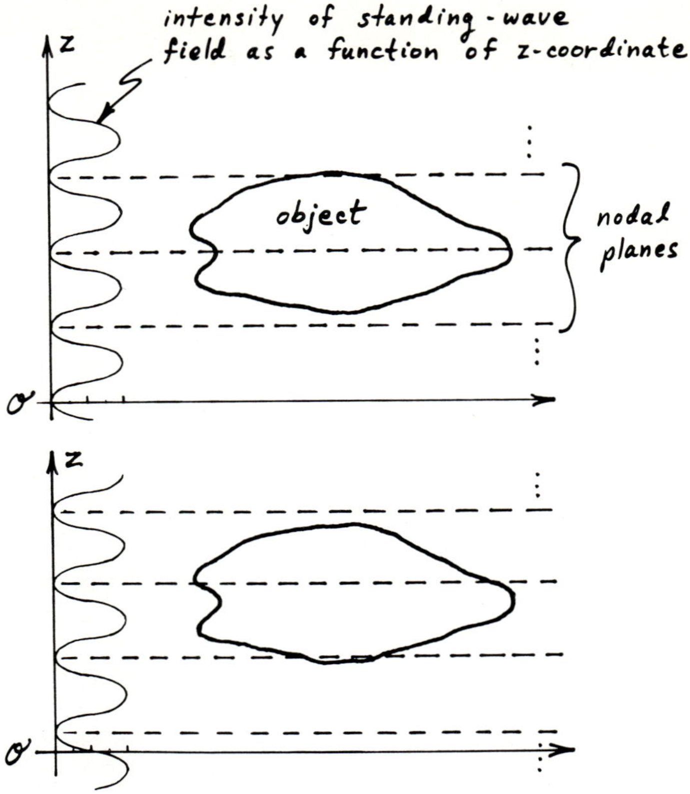


Fig. 2. Location of nodal planes relative to a specimen-fixed origin for two different phases of the standing-wave field.

that the dye molecules are randomly tumbling so that $p(\underline{r})$ is a scalar. In the large depth-of-field case, the SWF image $F(x,y;K,\phi)$ is the axial integral of all contributions $df(\underline{r})$:

$$
\begin{aligned}
F(x,y;K,\phi) &= A' \int p(\underline{r})\, I_e(z;K,\phi)\, dz \\
&= A \int p(\underline{r})\, dz - A \cos\phi \int p(\underline{r}) \cos Kz\, dz \qquad (3) \\
&\quad + A \sin\phi \int p(\underline{r}) \sin Kz\, dz
\end{aligned}
$$

Setting $\phi = 0°, 90°$, and $180°$ with respect to a specimen-fixed reference plane defines three linearly independent image fields:

$$Q(x,y;\ K) = F(x,y;\ K,\ 0°)$$
$$R(x,y;\ K) = F(x,y;\ K,\ 90°)$$
$$S(x,y;\ K) = F(x,y;\ K,\ 180°)$$

Then

$$\tfrac{1}{2}S + \tfrac{1}{2}Q = A \int p(\underline{r})\, dz \tag{4}$$

$$\tfrac{1}{2}S - \tfrac{1}{2}Q = A \int p(\underline{r}) \cos Kz\, dz \tag{5}$$
$$= F_c(x,y;\ K)$$

$$R - \tfrac{1}{2}S - \tfrac{1}{2}Q = A \int p(\underline{r}) \sin Kz\, dz \tag{6}$$
$$= F_s(x,y;\ K)$$

Pixel-by-pixel summing and differencing of the images Q, R, and S therefore yields the "DC", cosine, and sine maps of the Fourier coefficient at spatial frequency $K$ of the axial transform of $p(\underline{r})$. In principle, this provides a way to synthesize a three-dimensional reconstruction of $p(\underline{r})$ from arrays of its Fourier components: a data set would consist of triplets of images $Q(x,y;\ K_i)$, $R(x,y;\ K_i)$, and $S(x,y;\ K_i)$ recorded for a set of excitation spatial frequencies $K_1$, $K_2$, etc. An important consideration is whether the range of accessible spatial frequencies (or the range of accessible nodal plane spacings, $\Delta s$) allows for sampling of the axial Fourier transform of $p(\underline{r})$ at enough points to permit a useful reconstruction to be synthesized.

## ENGINEERING CONSIDERATIONS

A practical standing-wave illuminator will include means for superposing two coherent plane-wave fields at the location of the specimen in a microscope, and will allow for control of node spacing and node position relative to a microscope-fixed reference point. Our prototype device (Fig. 3) utilizes total internal reflection (TIR) of an incident beam so that the specimen is located in the overlap region of two equal-amplitude beams [12]. The nodal planes in this volume are parallel to the glass plate between the specimen and the microscope objective lens. Total reflection of the incident beam occurs at the glass:air interface, but because the water:glass interface between specimen and plate is also quite reflective at high angle, the plate must be an etalon for the best results. Nodal plane spacing is set by the angle of the incident s-polarized beam. Nodal plane position is set by raising or lowering the

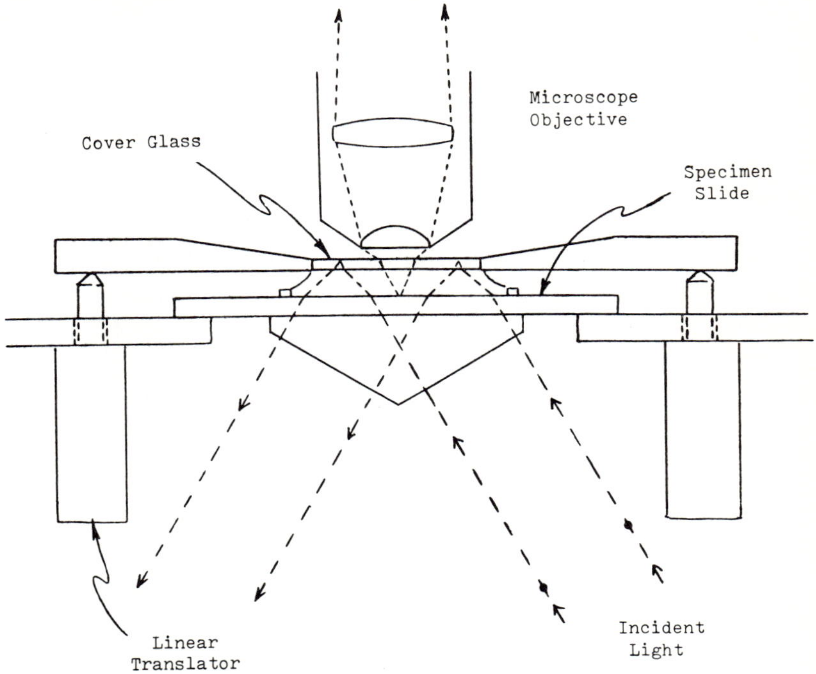


Fig. 3. Standing-wave illuminator. A collimated (plane-wave) beam of light of the proper wavelength to excite fluorescence in the specimen is coupled to the specimen slide via a prism. The angle of incidence of the waves at the slide:specimen interface is below the critical angle, so that transmitted waves pass through the specimen at an angle $\theta_i$. The beam is partially reflected at the lower surface of the cover glass and totally reflected at the upper surface of the cover glass. The reflected plane-wave field passes through the specimen at the complementary angle to $\theta_i$ if cover glass and slide are parallel. The node spacing of the standing-wave field formed by the superposition of incident and reflected beams is $\lambda_o/2n \cos \theta_i$, where $n$ is the specimen refractive index. It should be noted that there will always be partial reflections at the specimen:glass surfaces.

etalon relative to the specimen. Piezoelectric elements can in principle position the etalon with sufficient precision ($\sim\lambda/80$) and accuracy for the highest-resolution work.

Two engineering weaknesses are evident in the prototype standing-wave illuminator of Figure 3. One is that total reflection at the etalon would not occur if an immersion lens was coupled to its upper surface. This limits use to "air" objectives and therefore to numerical apertures (NA) less than about 0.7. High lateral resolution obtainable with NA in the range 1.0–1.4 is therefore not possible with this design. The second problem is that total reflection of the incident beam does not occur when the incidence angle of the beam at the glass:air interface is below the glass:air critical angle. This means that the minimum node spacing ($\lambda_o/2n$) cannot be obtained in this device. The minimum attainable node spacing is about $(0.54)\lambda_o$ rather than $(0.37)\lambda_o$ for normal incidence. The maximum useful node spacing is roughly twice the specimen thickness, or about 10 $\mu$m for cellular specimens. This requires that the incidence angle in the specimen be nearly 89°, which is difficult to achieve in practice without severe wavefront aberration. Alternative designs exist for illuminators which do not rely on TIR for beam folding. In one design, the first element of the objective lens would be a removable multilayer reflector that has peak reflectance for the excitation wavelength at a particular incidence angle. A side effect of this multilayer is that it would lead to multiple-beam interference of emitted waves entering the microscope. In a third possible design, one beam exits from the microscope into the specimen, and a second beam passes through the specimen into the microscope. In this design the intensity, phase, and angle of each beam can be independently adjusted. Nodal plane position can be set easily by changing the phase of either beam by use of a variable retarder.

## SWF IMAGES OF CELLULAR SPECIMENS

For our first test of the microscope, 3T3 cells that had been fixed, stained for F-actin with rhodamine-phalloidin, and embedded in index-matching medium (Flo-Texx) were used. The upper cover glass was kept in a fixed position relative to the specimen, and the specimen was held at one location relative to the focal plane of the microscope. Because of the high index of the specimen, an immersion objective (Zeiss 40× 0.75 NA) was coupled to the cover glass with water. The incidence angle of the excitation beam (514 nm, from an argon-ion laser source) was set to be greater than the glass-water critical angle, creating a node spacing of 1.2 $\mu$m in the specimen. Nodal plane position was changed by varying the incidence angle slightly. Figure 4 shows the fluorescence image of a group of three cells, followed by the SWF images for three different phases. It can be seen that features that are confused in the fluorescence image show up clearly in different SWF images. In this simple case, only about one cycle of the nodal field fits within the axial range of the specimen, probably due to dehydration of the specimen during preparation.

For a more critical test of the SWF microscope, 3T3 cells were fixed and stained with rhodamine phalloidin and then examined as a wet-mount preparation in phosphate-buffered saline. The refractive properties of this specimen should be essentially the same as those of live cells. The prototype illuminator of Figure 3 was used with a 40× 0.65 NA "dry" objective. The

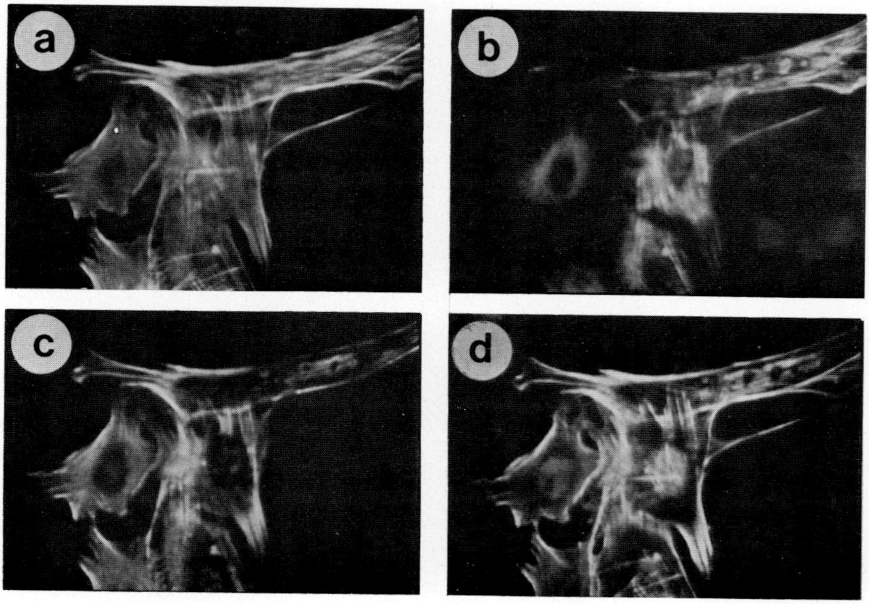


Fig. 4. Fluorescence and SWF images of 3T3 cells. Cells grown on a glass slide were fixed, stained with rhodamine-phalloidin, embedded in liquid mounting medium (Flo-Texx), and covered with glass. a. Fluorescence image (> 590 nm) excited at 546 nm by using standard episcopic illumination. b–d. Fluorescence images of the same group of cells excited at 514 nm by a nodal standing-wave field of axial period equal to 1.2 $\mu$m. The position of the nodes relative to the specimen changes in each frame. Photographs were made by using a RCA SIT camera and video monitor for image intensification. ×300.

fluorescence image of a single cell is shown in Figure 5, followed by pairs of SWF images recorded at node spacings of 0.89, 0.50, and 0.35 $\mu$m (Figs. 6–8). In each of the paired images the phase of the excitation field has been shifted by roughly 180°. It can be seen that the number of antinodal zones that intersect the cell increases as the spacing is decreased.

## FLUORESCENCE MICROTOMOGRAPHY

Under what circumstances can a set of SWF images be used to synthesize a three-dimensional reconstruction of a specimen? The photographs presented here are qualitatively useful, but in general do not represent a data set from which a Fourier synthesis of $p(\underline{r})$ can be made. An immediately apparent problem is that the accessible axial spatial frequency range of the standing-wave field is quite limited (0.1–3.8 $\mu m^{-1}$ for $\lambda_o$ = 488 nm); this means that

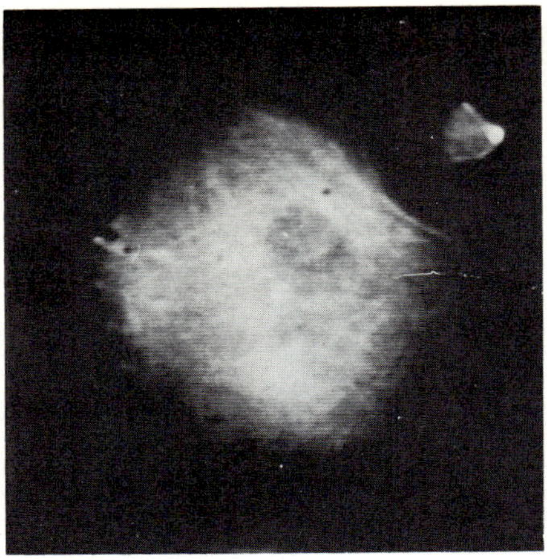

Fig. 5. Fluorescence image of 3T3 cell fixed and stained with rhodamine-phalloidin. Wet mount in phosphate-buffered saline. ×600

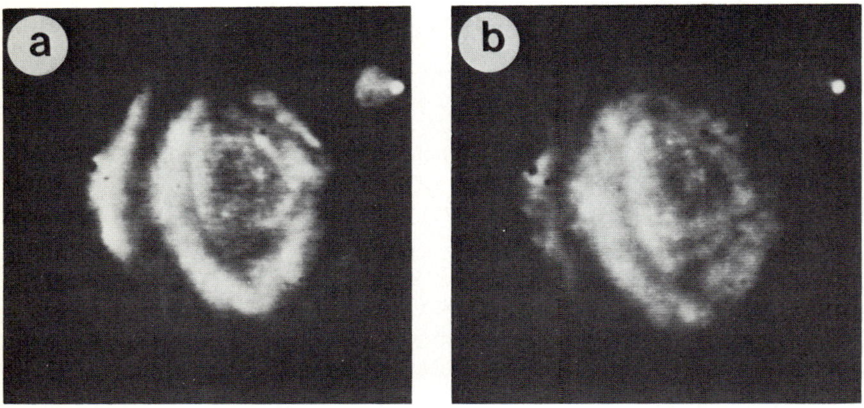


Fig. 6. SWF images of cell of Figure 5. Excitation wavelength = 514 nm. Axial period (node spacing) = 0.89 $\mu$m. Field phase in a and b differs by about 180°. ×600.

most of the Fourier transform of a "general" object of indefinite axial extent would be inaccesible to sampling. The finite axial extent of the specimens of interest, however, reduces the sampling requirements. This can be developed as follows:

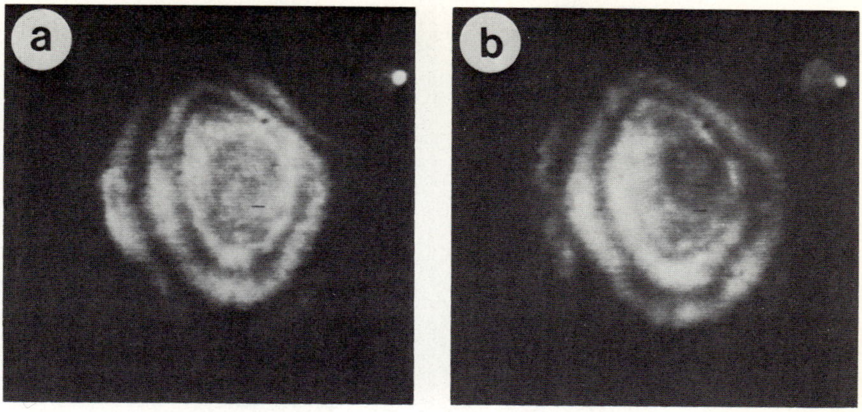


Fig. 7. SWF images of cell of Figure 5. Axial period = 0.50 μm. Field phase in a and b differs by about 180°. ×600.

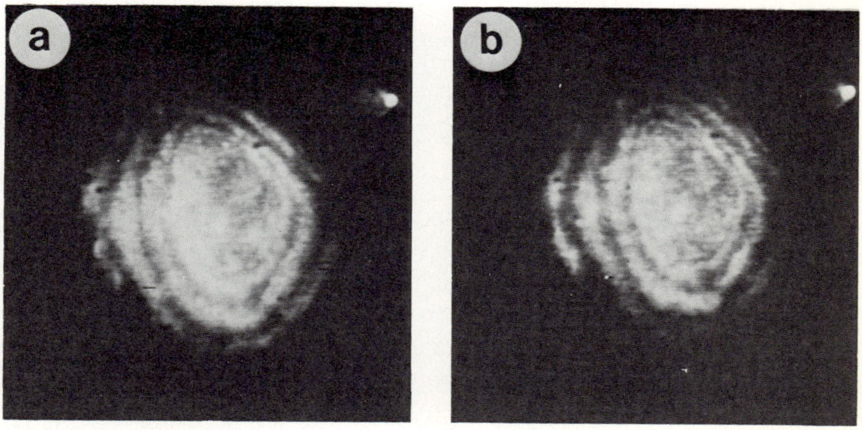


Fig. 8. SWF images of cell of Figure 5. Axial period = 0.35 μm. Field phase in a and b differs by about 180°. ×600.

Suppose $P$ is a periodic function of the axial coordinate $z$. Then $P(z)$ can be expressed as a Fourier series of fundamental spatial frequency $1/z_o$;

$$P(z) = \sum_{-\infty}^{+\infty} c_n e^{i2\pi nz/z_o}$$

where

$$Re(c_n) = Re(c_{-n})$$

and

$$-Im(c_n) = Im(c_{-n})$$

so that $P(z)$ is real-valued. A spatially truncated function $p(z)$ representing a microscope specimen of axial range $z_o$ can be represented as the product of $P(z)$ with a window function $W(z)$:

$$p(z) = P(z)\ W(z)$$

where

$$W(z) = \begin{cases} 1, & -z_o/2 \leqslant z \leqslant z_o/2 \\ 0, & \text{elsewhere} \end{cases}$$

The transform of $W(z)$ is the sampling function $\hat{W}(k) = \sin(\tfrac{1}{2}\ kz_o)/(\tfrac{1}{2}k)$ shown in Figure 9. The transform of $p(z)$ is an equispaced series of such functions weighted by the Fourier coefficients of $P(z)$:

$$\hat{p}(k) = z_o \sum_{-\infty}^{+\infty} c_n \frac{\sin(\tfrac{1}{2}kz_o + \pi n)}{\tfrac{1}{2}kz_o + \pi n} \tag{6}$$

The important result of this development of $\hat{p}(k)$ is that per unit spatial frequency, $\hat{p}(k)$ has only a finite and constant number of degrees of freedom. This result is in accord with the Sampling Theorem [13]. A second result is that because $\hat{W}(k)$ extends over all frequencies, coefficients $c_j$ outside the directly sampled spatial frequency range contribute to measured values of $\hat{p}(k)$. As outlined in the Theory section, a triplet of SWF images can be used to compute $F_s(x,y;\ K)$ and $F_c(x,y;\ K)$, the sine and cosine Fourier maps of the specimen at spatial frequency $K/2\pi$. The connection between the above development of $\hat{p}(k)$ and SWF images is therefore

$$F_c(x,y;\ K) + i\ F_s(x,y;\ K) = \hat{p}(K) \tag{7}$$

A band-limited reconstruction of the object $p(z)$ is completely specified by $2N + 2$ parameters: the set of $2N + 1$ coefficients between (and including) $c_{-N}$ and $c_N$, and the parameter $z_o$. The relationship between these coefficients and the SWF data set can be expressed by the vector-matrix equation

$$\underline{F} = [S]\ \underline{c} \tag{8}$$

or

$$\left( F_c(x,y;\ K_j) + iF_s(x,y;\ K_j) \right) = \left[ S_{jk}(K_j,z_o,k) \right] \left( c_k(x,y) \right)$$

and the equation

$$F_{DC}(x,y;\ 0) = z_o \sum_{-N}^{N} c_n(x,y)$$

where the elements of the sampling matrix [*S*] are

$$S_{jk} = \frac{\sin(\tfrac{1}{2}K_j z_o + \pi k)}{\tfrac{1}{2}K_j z_o + \pi k}$$

The robustness of any computation of the coefficients $c_k$ will depend upon the chosen set of spatial frequencies $K_j$ and the noise level of the imaging data.

## THEORY FOR HIGH-APERTURE INSTRUMENT

The foregoing theory applies strictly only to the large depth-of-field case. Since depth-of-field and numerical aperture vary inversely, the use of high-NA objective lenses to obtain high lateral resolution complicates the theory. In the general case any one SWF image will contain contributions from an in-focus zone and out-of-focus zones. For a single spatial frequency $K_j$ then, a set of SWF image triplets can be recorded in which the focal plane of the microscope is stepped to a new location for each triplet. This set of images $\{Q_m(K_j),\ R_m(K_j),\ S_m(K_j)\}$ can then be processed by using the PSF of the instrument to compute a completely in-focus triplet of images $Q(K_j)$, $R(K_j)$ $S(K_j)$. This stage of processing is the same as that developed by Agard and Sedat, and by Fay and Fogarty. The in-focus SWF image triplets then can be processed as outlined for the large depth-of-field case. It can be seen that SWF microscopy is an extension of current instrumentation and that an "ideal" instrument will control both focal plane and nodal plane position.

## APPLICATION

Microtomography has been the recurring theme of this outline. If one goes back to Table 1, it can be now be seen that the listed methods for encoding

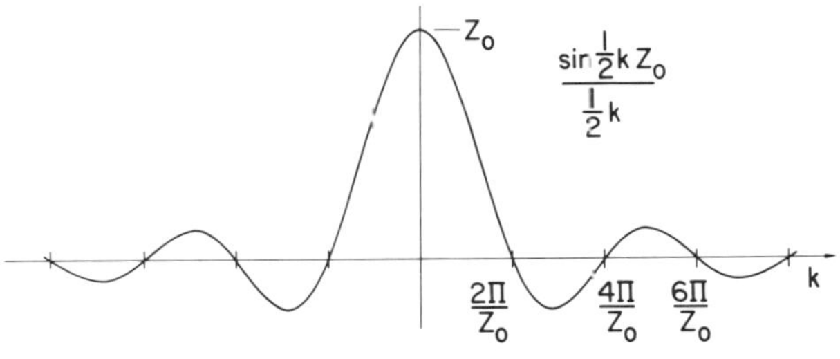


Fig. 9. Fourier transform of the window function *W(z)*.

axial information are not mutually exclusive and in fact may yield a more powerful instrument in combination.

Another application of SWF microscopy should be in microspectrofluorimetry, in which three-dimensional spectral maps could be made. If the distribution function $p(\underline{r})$ were to be treated as a time-dependent quantity $p(\underline{r},t)$, or as a vector or tensor quantity $\underline{p}(\underline{r})$ or $\underline{p}(\underline{r},t)$, there could be an application of this technique in fluorescence photobleaching recovery experiments [14,15] or in the mapping of fluorescence polarization, respectively. Because of the nanosecond lifetime of the excited state that gives rise to fluorescence emission, it should be possible to excite the specimen with a pulsed standing-wave field and record image photocounts subsequently. This makes practical instrument designs in which one beam is propagating into the microscope, a situation that would otherwise lead to high background photocounts. Outside of the field of fluorescence microscopy, standing-wave excitation may be useful in Raman microscopy of biological specimens, or in coherent microscopy of weak phase objects in which simplification of the hologram is necessary.

## ACKNOWLEDGMENTS

I thank colleagues Drs. D. Lansing Taylor and Alan S. Waggoner for critical observations made early in this project, Mr. Joseph Hughes for discussions held many years ago, Dr. Philip A. Amato for setting aside valuable specimens for my use, Mr. William P. George for expert construction of the prototype illuminator, and Dr. C. Roy Worthington for critical reading of the manuscript. This research was supported by a grant from the National Institutes of Health, GM34639-01.

## REFERENCES

1. Taylor DL, Amato PA, Luby-Phelps K, McNeil P: Fluorescent analog cytochemistry. Trends Biochem Sci 9(3):88–91, 1984.
2. Taylor DL, Wang Y-L: Molecular cytochemistry: Incorporation of fluorescently labeled actin into living cells. Proc Natl Acad Sci USA 75:857–861, 1978.
3. Luby-Phelps K, Lanni F, Taylor DL: The behavior of a fluorescent analog of calmodulin in living 3T3 cells. J Cell Biol 101:1245–1256, 1985.
4. Smith RW: Holographic methods in microscopy. J Microsc 12:29–47, 1983.
5. Smith RW, Empson TR: Holography applied to stereo-microscopy. In Baker LR (ed): "Microscopy—Techniques and Capabilities." Proc SPIE 368:104–109, 1982.
6. Streibl N: Three-dimensional imaging by a microscope. J Opt Soc Am A 2:121–127, 1985.
7. Agard DA, Sedat JW: Three-dimensional architecture of a polytene nucleus. Nature 302:676–681, 1983.
8. Agard DA: Optical sectioning microscopy: Cellular architecture in three dimensions. Ann Rev Biophys Bioeng 13:191–219, 1984.
9. Fay FS, Fogarty K: Computerized 3-dimensional immunocytochemistry in single cells. J Cell Biol 97:294a, 1983.
10. Wijnaendts van Resandt RW, Marsman HJB, Kaplan R, Davoust J, Stelzer EHK, Stricker R. Optical fluorescence microscopy in three dimensions: Microtomoscopy. J Microsc 138:29–34, 1985.
11. Longhurst RS: "Geometrical and Physical Optics," 3rd Ed. London: Longman Group Ltd., 1973.
12. Temple PA: Total internal reflection microscopy: A surface inspection technique. Applied Optics 20:2656–2664.
13. Goldman S: "Information Theory." New York: Dover Publications, Inc., 1968.
14. Jacobson K, Elson E, Koppel D, Webb W: International workshop on the application of fluorescence photobleaching techniques to problems in cell biology. Fed Proc 42:72–79.
15. Ware BR: Fluorescence photobleaching recovery. Am Lab 16(4):16, 1984.

# PART III
# Flow Cytometry

*Applications of Fluorescence in the Biomedical Sciences*, pages 525–530


# 25 Flow Cytometry in Cell Biology

*Robert F. Murphy*

Flow cytometry is an analytical and preparative technique which is characterized by the use of selected wavelengths of light to interrogate cells and other particles one at a time. The basic principles of flow cytometry and sorting have been described in a number of excellent review articles [Horan and Wheeless, 1977; Arndt-Jovin and Jovin, 1978; Shapiro, 1983] and books [Melamed et al., 1979; Laerum et al., 1980; Shapiro, 1985; Van Dilla et al., 1985; Melamed et al., 1986]. During the past 10 years, the number of applications of flow cytometry in biology and medicine has increased dramatically. This growth can be attributed to the fact that flow cytometry provides a unique combination of advantages for use in research: it provides statistically significant correlated measurements of multiple parameters (or variables) on heterogeneous populations of fixed or live cells with excellent temporal resolution. Table 1 lists the major cell-sorter manufacturers and summarizes the major features of commercially available instruments. The intent of this article is to provide a brief introduction to flow cytometry, and to describe a number of the ways in which it has been applied recently to problems in cell biology.

## FUNDAMENTAL PARAMETERS OF FLOW CYTOMETRY

The vast majority of applications of flow cytometry take advantage of the properties of fluorescent dyes to measure the level of specific macromolecules which are present in or associated with individual cells. Two major light sources are used to provide the required light intensity at the appropriate excitation wavelengths: arc lamps and continuous wave lasers. Fluorescent light emitted from the cell after entering the exciting beam is collected by using lenses and optical filters and measured using one or more photomultiplier tubes. Two chapters from the book by Van Dilla et al. [1985] are of

Department of Biological Sciences, Center for Fluorescence Research in Biomedical Sciences, Carnegie-Mellon University, Pittsburgh, Pennsylvania 15213.

**TABLE 1. Characteristics of Commercial Cell Sorters**

| Manufacturer | Models | Major features[4] |
|---|---|---|
| Becton Dickinson Immunocytometry Systems<br>2375 Garcia Avenue<br>Mountain View, CA 94086 | FACS 440<br>FACS Star[1]<br>FACS Analyzer[2,3] | a,c,f,g<br>c,e,f<br>f,h |
| Coulter Electronics, Inc.<br>590 West 20th Street<br>Hialeah, FL 33010 | EPICS V<br>EPICS C[1] | b,d,e,f,h<br>b,e,f,h |
| Ortho Diagnostic Systems<br>410 University Ave.<br>Westwood, MA 02090 | Cytofluorograph II,[2] $II_s$<br>Spectrum III[1,2]<br>ICP 22[2,3] | c,d,f<br>e |

[1]Single laser only.
[2]Nonsorting flow cytometer.
[3]Arc-lamp illumination.
[4]a, Open optical bench; b, Straight through optics; c, Image forming optics; d, Pulse shape analysis; e, Computer-controlled instrument setup; f, List-mode data acquisition and storage; g, Multiuser, multiinstrument computer system; h, Electronic volume measurement.

particular relevance here: the factors influencing the choice of light source and light collection optics are clearly described by Wheeless and Kay [1985], and an excellent summary of the electronics used in flow cytometers and sorters is contained in Hiebert and Sweet [1985].

Perhaps the most important factor in making quantitative measurements by using flow cytometry, as in spectrofluorometry or fluorescence microscopy, is the choice of appropriate fluorescent probes (or combinations of probes). For a thorough discussion of currently available probes the reader is referred to the chapter by Waggoner in this volume.

## AN OVERVIEW OF TYPICAL APPLICATIONS

The analysis of cell cycle kinetics by using fluorescent probes specific for DNA, RNA, and protein is one of the earliest applications of flow cytometry. This subject has been reviewed by Darzynkiewicz et al. [1980] and is also the subject of the chapter by Stewart et al. in this book.

The second major application which has dominated the field of flow cytometry for many years is quantitative immunofluorescence measurements of cell-surface antigens on lymphocytes and other cells involved in immune responses. This field developed from the pioneering work of the Herzenberg group at Stanford; two reviews of note are Herzenberg and Herzenberg [1978] and Loken and Stall [1982]. Much of the usefulness of flow cytometry for the analysis of peripheral blood cells stems from the differential light scattering by live and dead cells [Loken and Herzenberg, 1975] and by different cell types [Salzman et al., 1975]. The use of fluorescent probes for the analysis of platelets, a blood cell type which was until recently infrequently studied, is the subject of a chapter by Corash et al. in this volume.

An application with significant potential both in the research and clinical arenas is the use of flow cytometry to analyze and sort individual chromosomes. The combination of flow sorting and recombinant DNA techniques has permitted the creation of libraries containing DNA sequences from specific chromosomes [Davies et al., 1981; Krumlauf et al., 1982; Kunkel et al., 1982], and flow karyotypes of cultured hamster, mouse, and human cell samples have been obtained by using combinations of base-specific fluorochromes [e.g., Langlois et al., 1982].

The advantages of flow cytometry for the study of ligand binding to cell surfaces have been recognized for some time [Kraemer et al., 1973; Steinkamp and Kraemer, 1974; Bohn, 1976]; this application has recently been extended to analysis of the events which follow ligand binding [Murphy et al., 1982a,b; Finney and Sklar, 1983] and to hormones which have intracellular receptors, such as steroids [Van et al., 1984]. A number of papers describe the practical aspects of this application [Steinkamp and Kraemer, 1979; Bohn, 1980; Sklar et al., 1984; Murphy, 1986]; the chapter by Murphy and Roederer in this volume provides an overview of some recent developments.

Since phagocytosis plays a crucial role in combating infections, an accurate method of quantitating the phagocytic capacity of individual cells in heterogeneous samples has potential utility both for research into the mechanism of phagocytosis and for detecting defective cells. A number of laboratories have described the use of flow cytometry for measuring the uptake of fluorescent particles [Bassoe et al., 1980; Valet et al., 1981; Steinkamp et al., 1982].

As discussed above, light scattering provides an indication of cell viability. It may also be used to measure changes in cell shape resulting from surface stimulation. A number of changes in light scattering have been observed following presentation of chemotactic factors to neutrophils [Sklar et al., 1984; McNeil et al., 1985; Fletcher and Seligmann, 1985]. These reflect ruffling of cell membranes, degranulation (release of vesicles containing degradative enzymes), and cell elongation.

Another means by which cell responses to stimulation may be measured is with the use of fluorescent probes which reflect the ionic events, such as changes in membrane potential and calcium concentration, which may cause or result from cell activation. The chapter by Chused et al. describes how these techniques can be applied to the study of leukocyte physiology.

As discussed in the chapter by Taylor et al. in this volume, incorporation of fluorescent analogs of cellular macromolecules into viable cells provides very powerful tool for determining the mechanisms of a number of basic cellular processes. The success of this approach is critically dependent on the methods used for introducing these labeled macromolecules into living cells. Flow cytometry has been used to provide a rapid and accurate method for assaying cell viability and extent of cell loading [McNeil et al., 1984; Fechheimer et al., 1984].

## CONCLUSIONS

While the ability to measure more than one parameter simultaneously is clearly one of the major advantages of flow cytometry, the majority of uses to date have relied primarily on only one or two fluorescent parameters. As more varied fluorescent dyes become available, the number of parameters which can be measured simultaneously will increase; this process has begun already. The complexity of data collected by using three or more fluorescent probes will increase the need for automated statistical analysis of flow cytometric data [e.g., Murphy, 1985]. The future of flow cytometry in cell biology lies in increasing the information that can be obtained from a single sample, and in performing real-time analysis of complex cell biological processes, both on live cells (e.g., ligand binding, redistribution, internalization, and processing; generation of second messengers, cell responses such as shape changes) and on cell fractions (e.g., chromosome analysis, biochemical characterization of organelles, vesicle fusion).

## REFERENCES

Arndt-Jovin DJ, Jovin TM (1978): Automated cell sorting with flow systems. Ann Rev Biophys Bioeng 7:527–558.

Bassoe C-F, Solsvik J, Laerum OD (1980): Quantitation of single cell phagocytic capacity by flow cytometry. In Laerum OD, Lindmo T, Thorud E (eds): "Flow Cytometry IV." Oslo: Universitetsforlaget, pp 170–174.

Bohn B (1976): High-sensitivity cytofluorometric quantitation of lectin and hormone binding to surfaces of living cells. Exp Cell Res 103:39–46.

Bohn B (1980): Flow cytometry: A novel approach for the quantitative analysis of receptor-ligand interactions on surfaces of living cells. Mol Cell Endocrinol 20:1–15.

Darzynkiewicz Z, Traganos F, Melamed MR (1980): New cell cycle compartments identified by multiparameter flow cytometry. Cytometry 1:98–108.

Davies KE, Young BD, Elles RG, Hill ME, Williamson R (1981): Cloning of a representative genomic library of the human chromosome X after sorting by flow cytometry. Nature 293:374–376.

Fechheimer M, Denny C, Murphy RF, Taylor DL (1984): Use of a new method for loading macromolecules into living cells to measure the cytoplasmic pH of *Dictyostelium discoideum* amoebae. J Cell Biol 99:60a.

Finney DA, Sklar LA (1983): Ligand/receptor internalization: A kinetic, flow cytometric analysis of the internalization of N-formyl peptides by human neutrophils. Cytometry 4:54–60.

Fletcher MP, Seligmann BE (1985): Monitoring human neutrophil granule secretion by flow cytometry: Secretion and membrane potential changes assessed by light scatter and a fluorescent probe of membrane potential. J Leukocyte Biol 37:431–447.

Herzenberg LA, Herzenberg LA (1978): In Weir DM (ed): "Handbook of Experimental Immunology." Oxford: Blackwell Scientific Publications, Chapter 12.

Hiebert RD, Sweet RG (1985): Electronics for flow cytometers and sorters. In Van Dilla MA, Dean PN, Laerum OD, Melamed MR (eds): "Flow Cytometry: Instrumentation and Data Analysis." New York: Academic Press, pp 129–162.

Horan PK, Wheeless LL (1977): Quantitative single cell analysis and sorting. Science 198:149–157.

Kraemer PM, Tobey RA, Van Dilla MA (1973): Flow microfluorometric studies of lectin binding to mammalian cells. I. General features. J Cell Physiol 81:305–314.

Krumlauf R, Jeanpierre M, Young BD (1982): Construction and characterization of genomic libraries from specific human chromosomes. Proc Natl Acad Sci USA 79:2971–2975.

Kunkel LM, Tantravahi U, Eisenhard M, Latt SA (1982): Regional localization on the human X of DNA segments cloned from flow sorted chromosomes. Nucleic Acids Res 10:1557–1578.

Laerum OD, Lindmo T, Thorud E (1980): "Flow Cytometry IV." Oslo: Universitetsforlaget.

Langlois RG, Yu LC, Gray JW, Carrano AV (1982): Quantitative karyotyping of human chromosomes by dual beam flow cytometry. Proc Natl Acad Sci USA 79:7876–7880.

Loken MR, Herzenberg LA (1975): Analysis of cell populations with a fluoresence-activated cell sorter. Ann NY Acad Sci 254:163–171.

Loken MR, Stall AM (1982): Flow cytometry as an analytical and preparative tool in immunology. J Immunol Methods 50:R85–R112.

McNeil PL, Murphy RF, Lanni F, Taylor DL (1984): A method for incorporating macromolecules into adherent cells. J Cell Biol 98:1556–1564.

McNeil PL, Kennedy AL, Waggoner AS, Taylor DL, Murphy RF (1985): Light-scattering changes during chemotactic stimulation of human neutrophils: Kinetics followed by flow cytometry. Cytometry 6:7–12.

Melamed MR, Mullaney PF, Mendelsohn ML (1979): "Flow Cytometry and Sorting." New York: John Wiley and Sons.

Melamed MR, Lindmo T, Mendelsohn ML (1986): "Flow Cytometry and Sorting." Second edition. New York: John Wiley and Sons, in press.

Murphy RF (1985): Automated identification of subpopulations in flow cytometric list mode data using cluster analysis. Cytometry 6:302–306.

Murphy RF (1986): Ligand binding, endocytosis and processing. In Melamed MR, Lindmo T, Mendelsohn ML (eds): "Flow Cytometry and Sorting." Second edition. New York: John Wiley and Sons, in press.

Murphy RF, Jorgensen ED, Cantor CR (1982a): Kinetics of histone endocytosis in chinese hamster ovary cells. A flow cytofluorometric analysis. J Biol Chem 257:1695–1701.

Murphy RF, Powers S, Verderame M, Cantor CR, Pollack R (1982b): Flow cytofluorometric analysis of insulin binding and internalization by Swiss 3T3 cells. Cytometry 2:402–406.

Salzman GC, Crowell JM, Martin JC, Trujillo TT, Romero A, Mullaney PF, LaBauve PM (1975): Cell identification by laser light scattering: Identification and separation of unstained leukocytes. Acta Cytol 19:374–377.

Shapiro HM (1983): Multistation multiparameter flow cytometry: A critical review and rationale. Cytometry 3:227–243.

Shapiro HM (1985): "Practical Flow Cytometry." New York: Alan R. Liss, Inc.

Sklar LA, Finney DA, Oades ZG, Jesaitis AJ, Painter RG, Cochrane CG (1984): The dynamics of ligand-receptor interactions. Real-time analysis of association, dissociation, and internalization of an N-formyl peptide and its receptors on the human neutrophil. J Biol Chem 259:5661–5669.

Sklar LA, Oades ZG, Finney DA (1984): Neutrophil degranulation detected by right angle light scattering: Spectroscopic methods suitable for simultaneous analyses of degranulation or shape change, elastase release, and cell aggregation. J Immunol 133:1483–1487.
Steinkamp JA, Kraemer PM (1974): Flow microfluorometric studies of lectin binding to mammalian cells. II. Estimation of the surface density of receptor sites by multiparameter analysis. J Cell Physiol 84:197–204.
Steinkamp JA, Kraemer PM (1979): Quantitation of lectin binding by cells. In Melamed MR, Mullaney PF, Mendelsohn ML (eds): "Flow Cytometry and Sorting." New York: John Wiley and Sons, pp 497–504.
Steinkamp JA, Wilson JS, Saunders GC, Stewart CC (1982): Phagocytosis: Flow cytometric quantitation with fluorescent microspheres. Science 215:64–66.
Valet G, Jenssen HL, Krefft M, Ruhenstroth-Bauer G (1981): Flow-cytometric measurements of the transmembrane potential, the surface charge density and the phagocytic activity of guinea pig macrophages after incubation with lymphokines. Blut 42:379–382.
Van NT, Raber M, Barrows GH, Barlogie B (1984): Estrogen receptor analysis by flow cytometry. Science 224:876–879.
Van Dilla MA, Dean PN, Laerum OD, Melamed MR (1985): "Flow Cytometry: Instrumentation and Data Analysis." New York: Academic Press.
Wheeless LL, Kay DB (1985): Optics, light sources, filters, and optical systems. In Van Dilla MA, Dean PN, Laerum OD, Melamed MR (eds): "Flow Cytometry: Instrumentation and Data Analysis." New York: Academic Press, pp 21–76.

*Applications of Fluorescence in the Biomedical Sciences, pages 531–544*


# 26 Probes for Use in the Study of Leukocyte Physiology by Flow Cytometry

*Thomas M. Chused, H. Alexander Wilson, Bruce E. Seligmann, and Roger Tsien*

The measurement of fluorescence by flow cytometry was developed in an immunology laboratory [Loken and Herzenberg, 1975], impelled by the desire to analyze the heterogeneity of lymphocyte surface differentiation antigens and determine their relationship to immune function. This technological advance has been very rewarding: flow cytometry is now an indispensable tool of the cellular immunologist. It has enabled us to define subsets of T and B lymphocytes, monocytes, and granulocytes by their surface antigens and to purify them for determination of their function. Flow cytometry is responsible for our current appreciation of the strict relationships between surface differentiation antigen phenotype and functional capacity of leukocyte subsets. There has been considerable "synergy" between flow cytometry and another recent technical advance—monoclonal antibodies. Flow cytometry is regularly used to screen for desired antibody specificities and to select desired hybridoma cells by sorting [Parks et al., 1979]. Because of their well-defined and exquisite specificity monoclonal antibody reagents, in turn, have become invaluable reagents for the highly quantitative technique of flow cytometry.

Some of the important questions now facing immunologists regard the mechanisms of immunoregulation—how lymphocytes sense relevant ligands in their environment and how this information is transduced across the cell membrane to control gene translation. This process of signal transduction,

Laboratory of Microbial Immunity (T.M.C., H.A.W.), Laboratory of Clinical Investigation (B.E.S.), National Institute of Allergy and Infectious Diseases, National Institutes of Health, Bethesda, Maryland 20205; University of California, Berkeley, California 94720 (R.T.).

which is actually of general interest to cell physiologists, has, until now, been approached by conventional biochemical analysis of cell lysates and by single-cell micropuncture or patch-clamp techniques. However, the recent development of probes that report a physiologic state of the cell in terms of fluorescence is rapidly expanding our understanding of cellular responses. The quantitative method of flow cytometry is ideally suited to the use of these probe reagents, since it provides both resolution at the level of individual cells and the ability to identify subsets in complex cell mixtures by simultaneously labeling them with fluorescent monoclonal antibodies emitting at a different wavelength than the physiologic probe. This paper will focus on our recent work with probes sensitive to membrane potential and intracellular calcium.

## RESULTS

### Cyanine Dyes

Previous experimental attempts to measure membrane potential in terms of fluorescence have utilized cyanine dyes—the first dyes used for this purpose [Sims et al., 1974; Hoffman and Laris, 1974]—almost exclusively. These molecules are symmetric, fluorescent cations with a single charge delocalized over an extensive pi-electron system (Fig. 1). Due to this delocalization, these charged molecules are lipophilic and exhibit various degrees of membrane permeability depending upon the side chain length. The particular

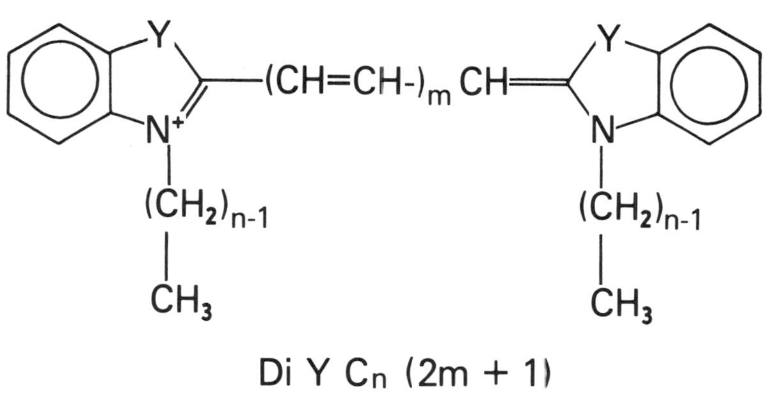


Fig. 1. Structure of cyanine dyes, illustrating single positive charge and extensive pi-electron system. The "Y" member of the ring may be oxygen (O), isopropyl (I), or sulfur (S). The number of methene groups, "m," effects the absorption and emission wavelengths. The alkyl side chains, of length "n," influence the lipid solubility.

molecules used for membrane potential studies readily cross the plasma membrane of cells. Because of these properties the equilibrium distribution of the dyes across cell membranes relfects, in part, the transmembrane electrical potential. Thus, they have found widespread use as indirect indicators of electrical potential difference in cells and organelles [Waggoner, 1979; Azzone et al., 1984]. Since manipulation of the dye structure alters their spectral and physiochemical properties, it is possible to select cyanine dyes whose excitation spectra match the laser lines available in the flow cytometer. The two cyanine dyes we have examined, di-O-$C_5$(3) and di-I-$C_5$(3) (kindly provided by Dr. A. Waggoner), exhibit a number of properties appropriate to use in the flow cytometer: increased fluorescence efficiency in low-polarity environments, high lipid/water partition coefficients [Sims et al., 1974], and less of a tendency to form molecular aggregates than the prototype dye, di-S-$C_5$(3) [Freedman and Hoffman, 1979; Seligmann et al., 1980; Guillet and Kimmisch, 1981].

**Basis of membrance potential.** The resting potential across cell membranes depends on the concentration gradient of $K^+$ ions, which are concentrated in the cells by Na-K ATPase, and on the relative permeability of the plasma membrane to $K^+$ ions being higher than that of other ions. Under these conditions, as $K^+$ ions selectively leak out, a countervailing electron gradient is established leaving the cytoplasm at a negative potential with respect to the medium. By convention, membrane depolarization is taken to mean a reduction in the electron gradient and thus a less negative internal potential, while hyperpolarization implies a more negative potential. Thus, if a cell hyperpolarizes the equilibrium distribution of the positively charged cyanine dyes shifts toward a greater concentration of intracellular dye. Conversely, depolarization induces a loss of dye from the cell. Initial studies with the less fluorescent cyanine dye, di-S-$C_5$(3), relied on the fluorescence quenching which occurs at high dye concentrations due to aggregation of the probe molecules. Under these conditions, fluorescence decreases as the intracellular dye concentration increases. Unlike these experiments with di-S-$C_5$(3), adequate fluorescence signals can be obtained at concentrations of di-O-$C_5$(3) and di-I-$C_5$(3) below $10^{-7}$ M. One unresolved issue concerning the use of cyanine dyes is the assumption that in lymphocytes [Shapiro et al., 1980; Monroe and Cambier, 1983] and in other cell types [Johnson et al., 1981], these concentrations are low enough to avoid the well-documented plasma membrane [Simons, 1979; Rink et al., 1980] and mitochondrial [Waggoner, 1976; Kinnally et al., 1978] toxicity of these agents. Cyanine dyes, like other cationic probes, uncouple oxidative phosphorylation [Azzone et al., 1984].

**Limitations affecting measurement of membrane potential with cyanine dyes.** A second problem in the use of cyanine dyes with intact cells is the substantial, proton-driven membrane potential across the inner mitochondrial membrane [Rottenberg, 1979; Nichols, 1982]. There is disagreement over whether cyanine dyes measure mitochondrial potentials [Johnson et al., 1981; Levenson et al., 1982] or plasma membrane potential [Shapiro et al., 1980]. Further, several reports have suggested that cyanine dyes indicate plasma membrane voltage changes within minutes after exposure of T cells to mitogenic lectins [Shapiro, 1981] or B cells to anti-immunoglobulin or lipopolysaccharide and dextran sulfate [Monroe and Cambier, 1983]. However, studies with radiolabeled cation probes in mitochondria-poisoned lymphocytes did not confirm the T cell results [Felber and Brand, 1982]. To resolve these issues and to determine the utility of cyanine dyes for the measurement of membrane potential in lymphocytes, which have abundant mitochondria, we have carefully analyzed (1) the source of cyanine dye fluorescence from lymphocytes, (2) the factors determining the distribution of cyanine dye fluorescence within cells, and (3) the toxicity of cyanine dyes at low dye concentrations [Wilson et al., 1985].

These experiments were done by adding known amounts of cyanine dye to tubes of lymphocytes maintained in serum-free tissue culture medium at 37°C. At the initiation of these studies, we observed that dye adsorbed to the walls of the Teflon tubing used in the fluidics system of the flow cytometer. The fraction of dye adsorbed depends on the rate of sample flow, which must be held steady to obtain reproducible steady-state dye partitioning between the cells, medium, and sample tubing. At the nanomolar concentrations of cyanine dyes used, saturation of the tubing required prolonged exposure. Because this was impractical we adopted the method of sequentially flushing the tubing between samples with household bleach and ethanol (which removed adsorbed dye as well as lipophilic ionophores such as valinomycin). By adding dye to sample as it was placed on the cytometer and making measurements at a fixed time afterward, highly reproducible fluorescence values were obtained.

The fluorescence histogram of murine splenic lymphocytes in physiologic 5 mM $[K^+]_o$ medium and 20 nM di-O-$C_5$(3) is shown in Figure 2a. Increasing $[K^+]_o$ to 60 mM depolarizes the cells and reduces the fluorescence, as indicated in Figure 2b. In contrast to the decrease observed in high $[K^+]_o$ medium, alteration of extracellular $Na^+$, $Cl^-$, or $Ca^{++}$ concentrations or treatment with the Na-K ATPase inhibitor, ouabain, had no effect on cell fluorescence, suggesting that the resting plasma membrane potential is primarily due to $K^+$ ion. Thus in high $K^+$, the potential across the plasma membrane is zero [Grinstein et al., 1982]. However, when 0.1 $\mu$M valinomycin was added to the high $[K^+]_o$ suspension fluorescence was additionally decreased (Fig. 2d). Valinomycin is both a potassium ionophore and an uncoupler of mitochondrial oxidative phosphorylation [Rottenberg, 1970].

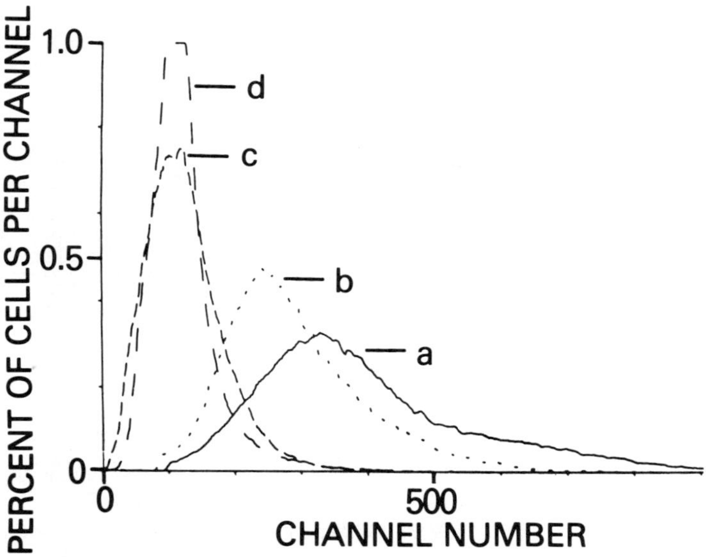


Fig. 2. Effect of high potassium and azide on di-O-$C_5$(3) lymphocyte fluorescence. Murine splenocytes (2 × $10^6$ cells/ml) were stained with an initial dye concentration of 20 nM. Fluorescence profiles were obtained under the following conditions: A, 5.4 mM $K^+]_o$; B, 60 mM $[K^+]_o$; C, 60 mM $[K^+]_o$ + 30 mM azide; D, 60 mM $[k^+]_o$ + 0.2 μM valinomycin.

Because cellular $K^+$, determined by flame photometry, was not significantly altered by exposure to 0.2 μM valinomycin for 30 minutes, and the fluorescence profile induced by the respiratory chain inhibitor, azide (Fig. 2c), was nearly identical to that produced by valinomycin, we concluded that mitochondrial uncoupling caused the additional loss of fluorescence.

**Partition of cyanine dyes.** This result presented us with a method to measure separately the contributions of the plasma membrane potential and the mitochondrial membrane potential to the fluorescence signal from cyanine dyes. The fluorescence difference between 5 and 120 mM $[K^+]_o$ suspensions is attributable to the change in plasma membrane potential from −60 (the resting value) to 0 mV and the additional shift after adding valinomycin is due to changing mitochondrial potential from −150 (normal physiologic value) to 0 mV. These determinations were made for varying concentrations of cells at a fixed di-O-$C_5$(3) concentration of 20 nM. The result was that a low cell density (2.5 × $10^5$/ml), which gives a high ratio of extracellular dye to cells, cyanine fluorescence reported principally the plasma membrane potential, while at high cell density (100 × $10^5$/ml), corresponding to a low amount of extracellular dye/cell, it reflected mitochondrial potential. As expected, the mean cell fluorescence decreased as the number of cells increased. In addition, the fraction of fluorescence that was insensitive to both plasma membrane and mitochondrial potential increased from 25% at low cell density to 60% at high cell density.

**Inadequate buffering of cyanine dyes.** To have an adequate dynamic range for potential measurements, only a small fraction of the indicator dye should be taken up by cells. The amount of di-O-$C_5$(3) remaining in the supernatant was determined as a function of $[K^+]_o$ at 20 × $10^5$ cells/ml. In the absence of valinomycin only 15% of the dye was recovered in the supernatant at 5 mM $[K^+]_o$, increasing to 22% at 80 mM $[K^+]_o$. With valinomycin the corresponding values were 17 and 64%. Thus the loss of dye from the cells when both mitochondria and plasma membrane are depolarized is much larger than when only the plasma membrane is affected. While these results show that the cyanine dyes respond strongly to changes in potential, they indicate that the cyanine-cell system is very poorly buffered. This introduces the possibility that dye extruded by one cell population because of depolarization may be taken up by another cell population (which has not changed its membrane potential) simply because of the increase in available extracellular dye. In addition, this makes it impossible to calibrate cyanine dye fluorescence in terms of absolute membrane potential.

**Nonaqueous sequestration and self-quenching of cyanine dyes.** By using the accepted values of −60 mV for plasma membrane and −150 mV for mitochondrial potential, the Nernst equation gives the ratio of free dye concentrations in medium:cytosol:mitochondria as 1:10:3,000. By using these values and assuming that lymphocyte cell water volume is 1–2 $\mu l/10^7$ cells [Deutsch et al., 1979; Segal et al., 1981] and that mitochondrial water volume is 5% of cytosol volume [Deutsch et al., 1979; Brand and Felber, 1984], the cellular accumulation of free dye by 2 × $10^6$ cells/ml should reduce the extracellular dye concentration <5%. However, we observed that under these conditions the supernatant dye concentration declines by >80%. Thus, as in other cell types [Hladky and Rink, 1976], most of the cell-associated dye is bound to cellular constituents rather than free in the aqueous phase. We have shown [Wilson et al., 1985] that binding of dye to nonaqueous components saturates in the mitochondria before it does in the cytosol. Further, the fluorescence efficiency of cytosol-associated dye is five times greater than that of mitochondria-associated dye, suggesting that significant fluorescence quenching occurs in the mitochondrial compartment.

We conclude from our results that there are a number of difficulties in the application of cyanine dyes to the measurement of plasma membrane potential. Cyanine dyes are distributed in "series"—first between medium and cytosol and then between cytosol and mitochondria. Because effective mitochondrial dye binding and fluorescence efficiency are not constant, separation of the fluorescence signals produced by the plasma membrane and mitochondrial potentials is not possible. A conceivable response to this dilemma would be to depolarize mitochondrial membranes with mitochondrial toxins and examine alterations of the plasma membrane potential in isolation. Unfortunately, all mitochondrial toxins we have tested also rapidly depolarize the cell membrane.

Two additional phenomena which we observed provided a further cautionary note concerning the use of cyanine dyes. First, cyanine dyes are selectively toxic for murine B lymphocytes. Initial concentrations of 50–200 nM di-O-$C_5$(3) caused selective depolarization of B cells (demonstrated by using simultaneous two-color flow cytometry). This could be caused by cyanine inhibiting $K^+$ efflux. Second, because the concentration of cyanine dye is so poorly buffered, manipulations resulting in the selective loss of dye from one subpopulation in a cell mixture cause an artifactual increase in the fluorescence of the unaffected cells.

In conclusion, cyanine dyes are not suitable for determination of membrane potential in intact, mitochondria-containing cells. We should emphasize, however, that in cells with few or no mitochondria, e.g., neutrophils, membrane potential changes indicated by cyanine dyes are reported correctly [Seligmann and Gallin, 1983].

## Oxonol Dyes

**Characteristics of oxonol dyes.** Oxonol dyes, like the cyanines, are symmetric, membrane-permeant, fluorescent molecules with a delocalized charge. Functionally, they differ from the cyanine probes only in that they bear a negative charge. Thus, oxonol probes are excluded from cells as they hyperpolarize but are taken up as they depolarize—the opposite of the cyanine dyes. Although occasionally used [Rink et al., 1980; Tsien et al., 1982], most oxonol dyes have been found to have only limited utility for bulk cell measurements in a cuvette because only a small fraction of the dye is cell-associated and because excitation and emission spectra overlap. In this situation only a small part of the fluorescence signal emanates from the cell-bound dye, even though the fluorescence of oxonol dyes, like cyanines, is enhanced in nonpolar environments. An exception to this is one probe, bis-(1,3-diisopropylbarbiturate) trimethine oxonol [di-iso-BA-$C_3$(3)] (kindly provided by Dr. A. Waggoner), which we have used in a variety of cell types including neutrophils, endothelial cells, and a mast cell line (Fig. 3; unpublished observations). Several of the properties of most oxonol dyes, however, make them highly suitable for use in the flow cytometer: (1) Because a small fraction of dye is taken up by cells, fluctuation of extracellular dye concentration is minimized; (2) as cells pass through the nozzle of the flow cytometer, a few microseconds before passing the laser beam, they are

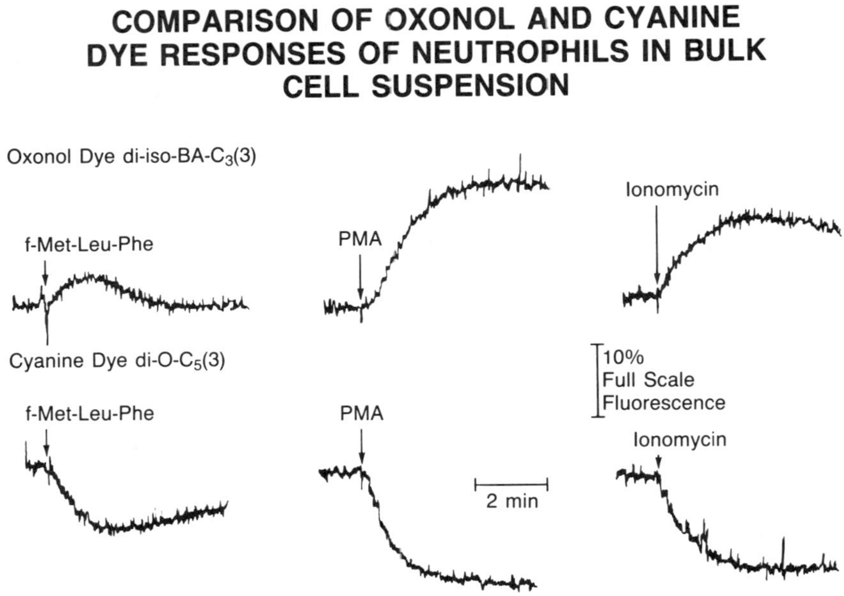


Fig. 3. Use of oxonol dye in bulk cell suspensions. The upper panel shows fluorometer tracings of human granulocytes at $2 \times 10^6$ cells/ml were equilibrated with $10^{-6}$ M of the oxonol dye, di-iso-$C_3$-BA(3), and then stimulated with the indicated agent: N-formyl methionyl leucyl phenylalanine (f-Met-Leu-Phe), $10^{-6}$ M; phorbolmyristate acetate (PMA), 20 ng/ml; and the calcium ionophore, ionomycin, $10^{-6}$ M. The lower panel shows granulocytes at $0.5 \times 10^6$ cells/ml equilibrated with 25 nM of the cyanine dye, di-O-$C_5$(3), and exposed to the same stimulants. The oxonol and cyanine dyes indicate qualitatively similar changes in electrical potential.

almost completely separated from dye in the external medium, substantially enhancing the signal:noise ratio; and (3) because of their negative charge, they are strongly excluded from mitochondria, which therefore contribute minimally to the fluorescence signal.

**Advantages of using a flow cytometer with oxonol dyes.** While superior to cyanine dyes, oxonol reagents also have certain limitations. First, because small amounts of dye are taken up by the cells, fluorescence signals are much weaker than with cyanine dyes. This can be overcome by the use of a sensitive flow cytometer. Second, at sufficiently negative membrane potentials the force driving dye extrusion may not overcome cellular dye binding forces, causing a flattening of the fluorescence versus voltage curve. Of a number of oxonol dyes tested (kindly provided by Dr. A. Waggoner), we found bis-(1,3-dibutylbarbiturate) trimethine oxonol [di-Ba-$C_4$(3)], at concentrations of 100–200 nM, to be the most satisfactory for green fluorescence and bis-(1,3-dibutylbarbiturate) pentamethine oxonol [di-BA-$C_4$(5)] most useful for red fluorescence. These reagents gave linear fluorescence-to-voltage relationships for lymphocytes over the wide range of 0 to $-100$ mV. Finally, the fluorescence:voltage relationship of oxonol dyes cannot be calibrated in the conventional method with the potassium ionophore, valinomycin, because negatively charged dyes interact with valinomycin-$K^+$ complexes [Gains and Dawson, 1975; Haynes and Simkowitz, 1977].

**Calibration.** We found that di-BA-$C_4$(3) fluorescence can be calibrated by activating a calcium-sensitive potassium channel [Tsien et al., 1982; Wilson and Chused, 1985]. The calcium ionophore, A23187, caused T lymphocytes to hyperpolarize with the effect maximal at a dose of 50 nM. Interestingly, B cells were unaffected whereas all murine thymocytes and 70% of human peripheral blood lymphocytes hyperpolarized. Thus, the calcium-sensitive $K^+$ channel is restricted to T lymphocytes. The di-BA-$C_4$(3) fluorescence of A23187-treated T cells varies linearly with the log of $[K^+]_o$ between 2.5 and 120 mM. At $[K^+]_o < 2.5$ mM the curve flattens considerably. Since unmanipulated T and B cells have the same di-BA-$C_4$(3) fluorescence, assuming an intracellular $K^+$ value of 120–140 mM, the calibration data indicates that lymphocytes have a resting membrane potential of $-66$ to $-71$ mV.

**Rectification of membrane potential.** Lymphocytes behave as potassium electrodes only at $[K^+]_o$ values above the physiologic range. Between 2.5 land 10 mM $[K^+]_o$ lymphocyte membrane potential does not change (Fig. 4) and is lower (less negative) than the $K^+$ diffusion potential, indicating significant permeability to another ion. To identify the ionic component involved in this rectification process, ion manipulation experiments were performed. Alterations of $[Cl^-]$ independently of $[K^+]_o$ showed that chloride permeability was negligible. Removal of sodium, however, produced fluorescence values similar to those observed for A23187-treated cells (Fig. 4). The rectification mechanism, then, requires the presence of sodium. Either a fixed sodium leak current and a voltage-sensitive potassium channel or a voltage-sensitive sodium channel could account for these results.

In contrast to lymphocytes, monocytes exhibited significant permeability to both $Cl^-$ and $Na^+$ and, like B cells, did not hyperpolarize in response to A23187. Direct recordings from adherent monocytes have identified a barium-sensitive $K^+$ channel [Gallin, 1981]. In agreement with this finding, addition of 3 mM $BaCl_2$ to a mixture of peripheral blood lymphocytes and monocytes caused depolarization of the monocytes but had no effect on the lymphocytes. Like lymphocytes, monocytes exhibit insensitivity of membrane potential to variation of $[K^+]_o$ between 2 and 10 mM. Neutrophils, however, do not rectify.

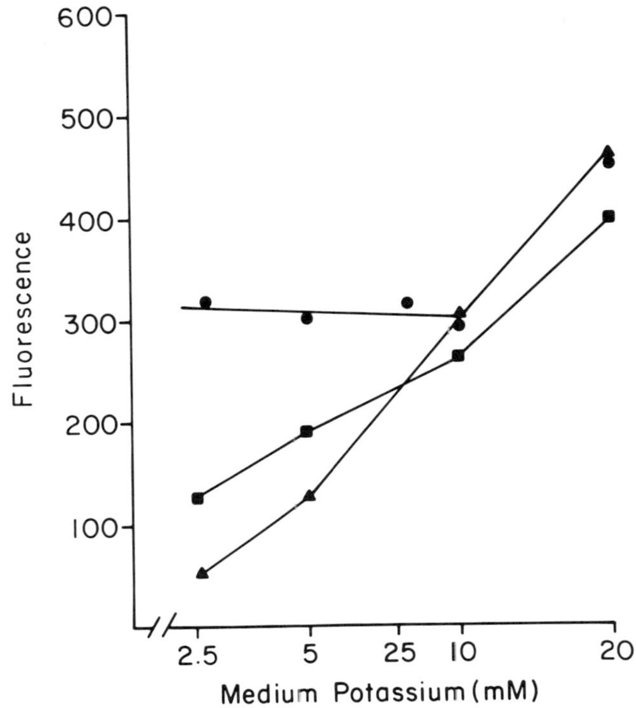


Fig. 4. Dependence of membrane potential on extracellular $Na^+$ and $Cl^-$ at low extracellular $K^+$ concentration. Human peripheral blood lymphocytes (PBL) ($10^6$ cells/ml) were equilibrated with di-BA-$C_4$(3) in media of various ionic composition and mean fluorescence values determined. Extracellular $[k^+]_o$ is noted on the ordinate for all determinations. Circles indicate fluorescence values in media where the product $[k^+]_o \times [Cl^-]_o$ remained constant at 300 mM by substitution of sodium isethionate for NaCl and potassium gluconate for KCl where $[k^+]_o$ values are greater than 2.5 mM. Under these conditions, these T lymphocytes maintain a constant membrane potential with varying $[k^+]_o$. Triangles indicate cells exposed to the calcium ionophore, A23187, while the sum of $[k^+]_o$ and $[Na^+]_o$ remained constant. The increase in intracellular $[Ca^{++}]$ activates a calcium-sensitive potassium channel which overrides the ionic mechanism that buffers membrane potential. Squares indicate determinations in < mM $Na^+$ medium in the absence of A23187. Removal of sodium prevents the buffering of membrane potential, indicating that $Na^+$ ion participates in this process.

**Pharmacologic effects on T cell calcium-sensitive potassium channel.** Flux rates of the potassium analogue rubidium are increased by A23187 and this response is blocked by quinine, an inhibitor of calcium-sensitive $K^+$ channels in other tissues, and by trifluperazine and other calmodulin antagonists [Grinstein et al., 1982]. Since the membrane potentials or unmanipulated T and B cells are identical but only T cells possess the calcium-sensitive potassium channel, it is unlikely that it contributes to the determination of resting potential. We predicted that it would be necessary to depolarize T cells in order to determine the pharmacologic sensitivity of this channel. Using this approach, we found that quinine blocked the repolarization of T cells depolarized by A23187. On the other hand, 4-aminopyrine, which effects voltage-sensitive $K^+$ channels rather than calcium-sensitive $K^+$ channels [Schwarz and Passow, 1983], did not prevent the repolarization of T cells depolarized by A23187. As expected from the rubidium flux data, trifluperazine blocked the hyperpolarization induced by A23187 without affecting resting membrane potential.

**Utility of oxonol dyes for indirect measurement of membrane potential.** In conclusion, indirect measurement of membrane potential in mitochondria-containing cells can be accurately performed by using oxonol dyes in the flow cytometer. Their negative charge prevents significant mitochondrial accumulation. Since only 3% of the di-BA-$C_4$(3) is taken up by cells, the probe concentration is highly buffered. Because most of the extracellular dye is removed by the cytometer fluidics, adequate and reproducible signal:noise ratios are obtained. By the use of this system we have shown that T, but not B, lymphocytes possess a calcium-sensitive $K^+$ channel; and that lymphocytes and monocytes, but not neutrophils, rectify their membrane potential at physiologic $[K^+]_o$ values.

## Measurement of Intracellular Free Calcium

Calcium participates in signal transduction across the plasma membrane in a wide variety of cells. Because the resting level of $[Ca^{++}]_i$ is approximately 100 nM, direct measurement is difficult. The development of quin2, which reports $[Ca^{++}]_i$ as fluorescence magnitude, has permitted a series of studies which delineate the significance of calcium's role in intracellular signaling. Becaues the laser wavelengths available are not appropriate and the fluorescence output of quin2 at acceptable levels of cell loading is too low for use in the cytometer, these experiments have been performed on bulk cell suspensions in a fluorometer. It would clearly be advantageous to obtain population distributions by making single cell measurements and to be able to correlate fluorescent $[Ca^{++}]_i$ responses with cell surface differentiation antigens.

## Indo-1

A family of "second-generation" calcium-sensing fluorescent probes with greatly improved fluorescence properties have recently been synthesized [Grynkiewicz et al., 1985]. These new dyes have up to 30-fold brighter fluorescence, permitting both greatly reduced loading levels and sufficiently strong signals to be detected at the single-cell level. One of these new reagents, indo-1, has the unique feature that the wavelength of the emission

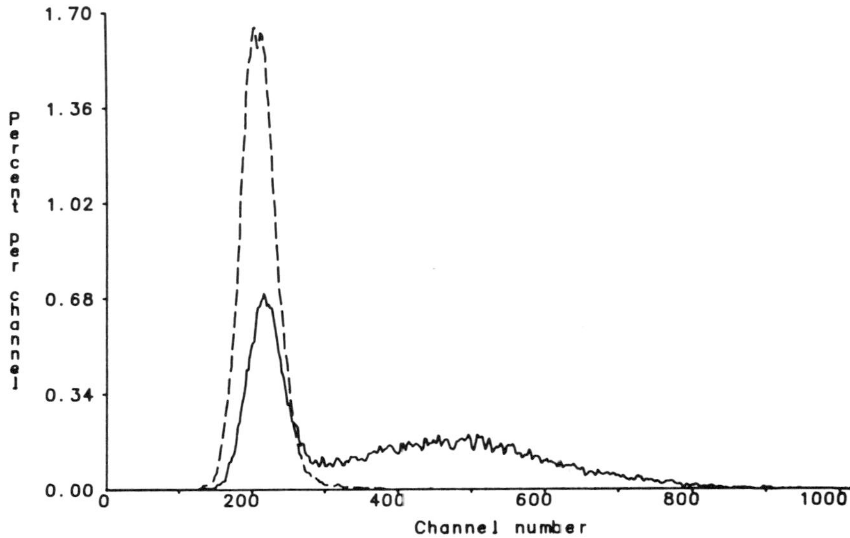


Fig. 5. Detection of changes in intracellular free $Ca^{++}$ by indo-1. Mouse spleen cells were loated $2 \times 10^7$ cells/ml with 5 $\mu$M of the acetoxymethyl ester of indo-1 at 37°C for 45 minutes and then diluted to $2 \times 10^6$ cells/ml. The fluorescence at 404 nm was electronically divided by that at 485 nm. The solid tracing indicates $[Ca^{++}]_i$ 30 seconds after exposure to antimouse IgG. $[Ca^{++}]_i$ increases in the B cells to form a peak shifted to the right. The nonresponding T cells remain in a peak that overlaps the unstimulated control (dashed line).

maxima shifts to lower wavelengths as the calcium concentration increases. By determining the ratio of fluorescence emitted at two appropriate wavelengths by cells loaded with the acetoxymethyl ester of indo-1, $[Ca^{++}]_i$ can be determined independently of the intracellular dye concentration. Dividing the fluorescence at 404 nm by that at 485 produces the highly restricted distribution $[Ca^{++}]_i$ shown in Figure 5. Twenty seconds after exposure of this cell suspension to rabbit antimouse immunoglobulin, the sharp, clearly defined rise in B lymphocyte $[Ca^{++}]_i$ indicated in Figure 5 was observed.

## SUMMARY

This review describes the successful utilization of two probes which report important aspects of a cell's physiological state in terms of the highly sensitive physical modality—fluorescence. The oxonol dye di-BA-$C_4$(3) accurately measures membrane potential and indo-1 similarly determines $[Ca^{++}]_i$. Additional probes are available for other parameters, e.g., intracellular pH [Valet et al., 1984]. By utilizing these reagents, we will be able to determine the immediate consequences of interactions of lymphocyte receptors with ligands. We anticipate that enhanced knowledge of these reactions will improve our understanding of the process of immunoregulation. In addition, these techniques should be of considerable value for the physiologic analysis of many other types of cells.

## ACKNOWLEDGMENTS

We thank Linette J. Edison for flow cytometry support and Elinor M. Brown for technical assistance. H.A.W. conducted these studies while an Investigator of the Arthritis Foundation.

## REFERENCES

Azzone GF, Peitroban D, Zoratti M (1984): Determination of the proton electrochemical gradient across biological membranes. Curr Top Biophys Bioeng 13:1–77.

Brand MD, Felber SM (1984): Membrane potential of mitochondria in intact lymphocytes during early mitogenic stimulation. Biochem J 217:453–459.

Deutsch CJ, Holian A, Holian SK, Daniele RP, Wilson DF (1979): Transmembrane electrical and pH gradients across human erythrocytes and human peripheral lymphocytes. J Cell Physiol 99:79–94.

Felber SM, Brand MD (1982): Factors determining the plasma-membrane potential of lymphocytes. Biochem J 204:577–585.

Freedman JD, Hoffman JF (1979): The relation between dicarbocyanine dye fluorescence and the membrane potential of human red blood cells set at varying Donnan equilibria. J Gen Physiol 74:187–212.

Gains N, Dawson AP (1975): Transmembrane electrophoresis of 8-anilino-1-naphthalenesulfonate through egg lecithin liposome membranes. J Membr Biol 24:237–248.

Gallin EK (1981): Voltage clamp studies in macrophages from mouse spleen cultures. Science 214:458–460.

Grinstein S, Clarke CA, Dupre A, Rothstein A (1982): Volume induced increase in anion permeability in human lymphocytes. J Gen Physiol 80:801–823.

Grynkiewicz G, Poenie M, Tsien RY (1985): A new generation of $Ca^{2+}$ indicators with greatly improved fluorescence properties. J Biol Chem 260:3440–3450.

Guillet EG, Kimmich S (1981): Di-O- and di-S: Interactions with red blood cells, ghosts and phospholipid vesicles. J Membr Biol 59:1–11.

Haynes DH, Simkowitz P (1977): 8-anilino-1-naphthalene sulfonate: A fluorescent probe of ion and ionophore transport kinetics and transport asymmetry. J Membr Biol 33:63–108.

Hladky SB, Rink TJ (1976): Potential difference and the distribution of ions across the human red blood cell membrane. A study of the mechanism by which the fluorescent cation, di-S, reports membrane potential. J Physiol (Lond) 263:287–310.

Hoffman JF, Laris, PC (1974): Determinations of membrane potential in human and amphiuma red blood cells by means of a fluorescent probe. J Physiol (Lond) 239:519–552.

Johnson LV, Walsh ML, Bockus BJ, Chen LB (1981): Monitoring of relative mitochondrial membrane potential in living cells by fluorescence microsocopy. J Cell Biol 88:526–535.

Kinnally KW, Tedeschi H, Maloff BL (1978): Use of dyes to estimate the electrical potential of the mitochondrial membrane. Biochemistry 17:3419–3428.

Levenson R, Macara IG, Smith RL, Cantly L, Housman D (1982): Role of mitochondrial membrane potential in the regulation of murine erythroleukemia cell differentiation. Cell 28:855–863.

Loken MR, Herzenberg LA (1975): Analysis of cell populations with a fluorescence-activated cel sorter. Ann NY Acad Sci 254:163–171.

Monroe JG, Cambier JD (1983): B cell activation. I. Anti-immunoglobulin-induced receptor cross linking results in a decrease in plasma membrane potential of murine B lymphocytes. J. Exp Med 157:2073–2086.

Nichols DG (1982): In: "Bioenergetics." New York, NY: Academic Press, pp 65–91.

Parks DR, Bryan VM, Oi VT, Herzenberg LA (1979): Antigen-specific identification and cloning of hybridomas with a fluorescence-activated cell sorter. Proc Natl Acad Sci USA 76:1962–1966.

Rink TJ, Montecucco C, Hesketh TR, Tsien RY (1980): Lymphocyte membrane potential assessed with fluorescent probes. Biochim Biophys Acta 595:15–30.

Rottenberg H (1979): ATP synthesis and electrical membrane potential in mitochondria. Eur J Biochem 15:22–28.

Schwarz W, Passow H (1983): $Ca^{+2}$-activated-$K^{+}$ channels in erythrocytes and excitable cells. Ann Rev Physiol 43:359–374.

Segal GB, Cokelet GR, Lichtman MA (1981): The measurement of lymphocyte volume: Importance of reference particle deformability and counting solution tonicity. Blood 57:849–99.

Seligmann BE, Gallin JI (1983): Comparison of indirect probes of membrane potential utilized in studies of human neutrophils. J Cell Physiol 115:105–115.

Seligmann BE, Gallin EK, Martin DL, Shain W, and Gallin JI (1980): Interaction of chemotactic factors with human polymorphonuclear leukocytes: Studies using a membrane potential-sensitive cyanine dye. J Membr Biol 52:257–272.

Shapiro HM (1981): Flow cytometric probes of early events in cell activation. Cytometry 1:301–312.

Shapiro HM, Natale PJ, Kamentsky LA (1980): Estimation of membrane potentials of individual lymphocytes by flow cytometry. Proc Natl Acad Sci USA 76:5728–5730.

Simons TJB (1979): Actions of a carbocyanine dye on calcium-dependent potassium transport in human red cell ghosts. J Physiol (Lond) 288:481–507.

Sims PJ, Waggoner AS, Wang CH, Hoffman JS (1974): Studies on the mechanism by which cyanine dyes measure membrane potential in red blood cells and phosphatidyl cholinic vesicles. Biochemistry 13:3315–3330.

Tsien RY, Pozzan T, Rink TJ (1982): T-cell mitogens cause early changes in cytoplasmic free $Ca^{2+}$ and membrane potential in lymphocytes. Nature 295:68–70.

Valet G, Warnecke HH, Kahle H (1984): New possibilities of cytostatic drug testing on patient tumor cells by flow cytometry. Blut 49:37–43.

Waggoner AS (1976): Optical probes of membrane potential. J Membr Biol 27:317–334.

Waggoner AS (1979): Dye indicators of membrane potential. Ann Rev Biophys Bioeng 8:847–868.

Wilson HA, Chused TM (1985): Lymphocyte potential and $Ca^{++}$-sensitive potassium channels described by oxonol dye fluorescence measurements. J Cell Physiol 125:72–81.

Wilson HA, Seligmann BE, Chused TM (1985): Voltage sensitive cyanine dye fluorescence signals in lymphocytes: Plasma membrane and mitochondrial components. J Cell Physiol 125:61–71.

*Applications of Fluorescence in the Biomedical Sciences, pages 545–566*


# 27 Flow Cytometric Analysis of Endocytic Pathways

*Robert F. Murphy and Mario Roederer*

Receptor-mediated endocytosis is responsible for the uptake of many cellular proteins, hormones, toxins, and viruses [for reviews see Goldstein et al., 1979; Helenius et al., 1983; Pastan and Willingham, 1983]. Although there is evidence that all follow a common pathway into the cell [e.g., Maxfield et al., 1978], subsequent processing steps result in the differential sorting, degradation, and recycling of ligands and receptors. A number of studies on virus [Helenius et al., 1980; Miller and Lenard, 1980; Yoshimura et al., 1982; Marsh et al., 1983] and toxin [Sandvig and Olsnes, 1980; Draper and Simon, 1980; Marnell et al., 1982; Boquet and Duflot, 1982] endocytosis demonstrate a crucial role of the acidification of these ligands in their infectious mechanisms. Endosome acidification has also been demonstrated to play a role in recycling of some receptors [Tietze et al., 1980; Gonzalez-Noriega et al., 1980; Van Leuven et al., 1980].

Fluorescent measurements of the pH of compartments involved in endocytosis were first performed by Ohkuma and Poole [1978]. Using a fluorometer, they measured the excitation spectrum of cells containing fluorescein isothiocyanate (FITC)-dextran, and showed that the ratio of fluorescence emission resulting from excitation at 450 nm and 490 nm was proportional to pH. We have used fluorescein-labeled ligands and flow cytometry to demonstrate that the acidification process occurs much more rapidly than had previously been thought [Murphy et al., 1982a,b]. Tycko and Maxfield

Department of Biological Sciences, Center for Fluorescence Research in Biomedical Sciences, Carnegie-Mellon University, Pittsburgh, Pennsylvania 15213.

presented evidence that alpha-2-Macroglobulin is acidified before it enters lysosomes [Tycko and Maxfield, 1982].

In contrast to many ligands which are rapidly degraded in lysosomes, transferrin is not appreciably degraded by some cell types. Van Renswoude et al. [1982] demonstrated that transferrin-containing vesicles are nonetheless mildly acidified. Using dual fluorescence flow cytometry, we have demonstrated two phases of acidification of insulin by 3T3 cells [Murphy et al., 1984a]; these may correspond to entry into the mildly acidic, transferrin-containing compartment followed by transfer to lysosomes. Epidermal growth factor (EGF) has been demonstrated to be degraded in lysosomes [Carpenter and Cohen, 1979]; however, at least some EGF is internalized into nonlysosomal vesicles [Miskimmins and Shimizu, 1984]. Mellman and co-workers [Mellman et al., 1984; Mellman and Plutner, 1984] have demonstrated that complexes formed by monovalent, but not polyvalent, antibodies against the macrophage Fc receptor are recycled. This recycling may occur before acidification takes place.

Taken together, these data on whole cells for a variety of cell types and a number of ligands indicate that at least two types of endocytic compartments may be acidic, and that some receptor-ligand complexes (e.g., those involving transferrin and monovalent antibodies to Fc receptor) may be endocytosed and recycled without encountering lysosomes. This separation into at least two compartments is strongly supported by the data of Galloway et al., [1983], who found that two distinct, FITC-dextran-containing fractions isolated on Percoll gradients were capable of acidifying their contents in vitro in the presence of ATP. To complicate the acidification picture, both fractions of clathrin-coated vesicles [Forgac et al., 1983] and Golgi vesicles [Glickman et al., 1983] have also been demonstrated to contain proton pumps. In addition, at least two types of endosomes may exist: "light" endosomes (primary endocytic vesicles?) and "heavy" endosomes [Merion and Sly, 1983].

Despite significant progress during the past 5 years, a number of questions regarding endocytosis have been only partially answered. What is the pH of the initial compartment? What are the kinetics of acidification at different temperatures? Is the initial acidification process passive (ion antiport) or active (proton pump)? Do all ligands follow the same initial pathway of internalization, and at what point do they diverge?

Flow cytometry is ideally suited for answering these questions, for several reasons: (1) It allows measurement of more than one parameter (variable) simultaneously. These include intrinsic (cell volume, light scattering at various angles) and extrinsic (fluorescence of labeled probes) parameters. (2) Measurements can be made on a cell-by-cell basis on viable cells, so that physiological variables such as pH can be measured and the correlations between measured parameters can be determined. (3) Data for large numbers of cells can be collected rapidly. Thus, statistically significant results may be obtained even on very heterogeneous samples. (4) Subpopulations of cells may be physically isolated by flow sorting for further microscopic or biochemical analysis.

We have undertaken a series of experiments to address the questions listed above, and in the process have developed new techniques for the analysis of cell systems by flow cytometry. In particular, we have devised methods for (1) correcting for autofluorescence in low signal-to-noise systems, (2) acquiring temporally resolved data yielding high-resolution kinetics of processes such as binding, internalization, and acidification of a fluorescent ligand, (3) obtaining the temperature dependence of these processes, and (4) analyzing subcellular components, such as endosomes and lysosomes, in unfractionated cell lysates.

## AUTOFLUORESCENCE CORRECTION

Analysis of peptide receptors by using fluorescent techniques is often complicated by high levels of autofluorescence from unlabeled cells, most of which can be attributed to pyridine and flavin nucleotides [Aubin, 1979; Benson et al., 1979.] The amount of autofluorescence of a typical 3T3 cell is equivalent to 35,000 molecules of fluorescein; thus, when working with ligands such as FITC-EGF, of which only 21,000 bind to a cell, autofluorescence is a major portion of the signal. A wide variation in cell size is often observed in cultured cells such as fibroblasts. This variation can affect both the number of receptors and the autofluorescence, and can be large enough that a small cell with a number of ligands bound can show less fluorescence than a large unstained cell. We have devised a method for reducing the error due to autofluorescence by using either light scattering or a second wavelength "autofluorescence" emission to normalize for cell size. The basic procedure is illustrated in Figure 1.

Monolayers (60 mm) of 3T3 cells were washed twice with phosphate-buffered saline (PBS); 0.5 ml of Dulbecco's modified Eagle's medium (DMEM) was added. After 10 min, FITC-EGF was added to one plate to a concentration of 10 nM, to another, it was added to 10 nM along with 325 nM unlabeled EGF. This concentration of unlabeled EGF should reduce the binding of labeled EGF by 94%. After 30 min at 37°C, the cells were washed twice with PBS and trypsinized for 5 min at 37°C. These samples were then run on the fluorescence-activated sorter (FACS) and analyzed for forward scatter, right-angle scatter, and fluorescence at 530 nm [*F*(488,530), FITC] and 625 nm [*F*(488,625), autofluorescence] when excited by an argon laser

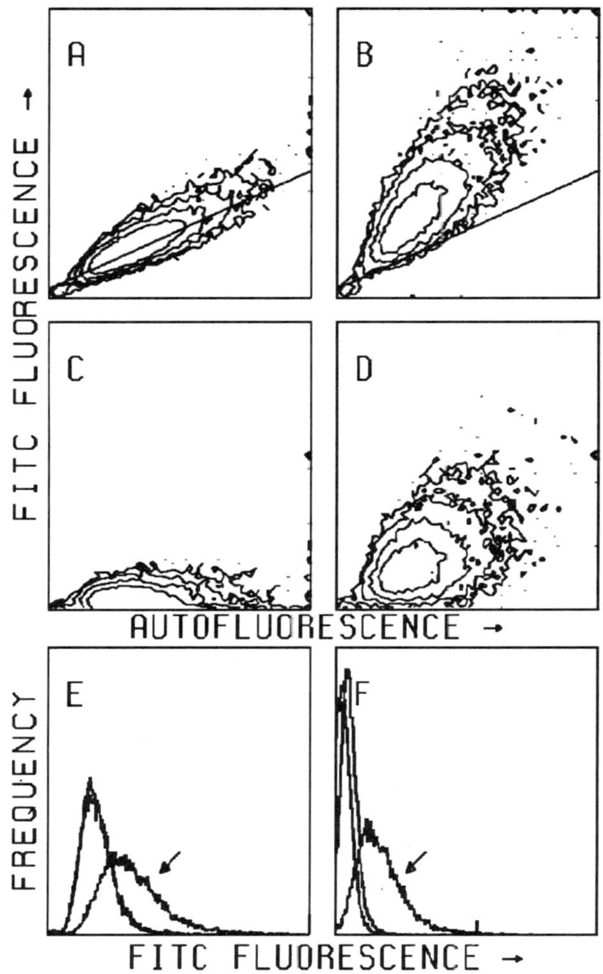


Fig. 1. Autofluorescence correction method. A–D. Dual parameter histograms are shown for two samples: control cells (A,C) and cells incubated with FITC-EGF as described in the text (B,D). Contours are drawn at 2, 6, 18, and 48 cells. Panel A shows the correlation between autofluorescence at 530 nm and 625 nm for the unstained cells. The line shows the best-fit line relating these parameters. The slope of this line predicts the proportion of 625 fluorescence which will be present in the 530 signal, regardless of whether "real" 530 fluorescence is present. Panel B demonstrates that most of the labeled cells are above this line. By correcting for autofluorescence (see text), the control distribution is shifted down to the horizontal axis, and the vertical axis becomes "real" 530 fluorescence (C,D). E–F. Single-parameter histograms are shown for the uncorrected (E) and corrected (F) F(488,530) of three samples each. The three histograms are: control cells, cells plus FITC-EGF (arrows), and cells plus FITC-EGF plus a 32-fold excess of unlabeled EGF (30-min incubations at 37°C). (The histograms for the control and blocked samples are almost identical.) Note the much smaller overlap between the control and the labeled samples in the corrected (right) histograms compared with the uncorrected histograms.

at 488 nm; 10,000 live cells (selected by thresholding on forward scatter) were analyzed for each sample.

Because autofluorescence of cultured cells shows wide excitation and emission spectra [Aubin, 1979; Benson et al., 1979], it is possible, with appropriate optical filters, to analyze a sample of cells and collect a parameter containing only autofluorescence. In this example, in which the probe is FITC-EGF the autofluorescence is detected by a bandpass filter at 625 nm with a bandwidth of 35 nm. There is little detectable fluorescein fluorescence within this band, so all signal is due to autofluorescence (data not shown).

In order to correct for the amount of autofluorescence in the 530-nm-band signal, therefore, only one assumption need be made: that the autofluorescence at 625 nm is proportional to the autofluorescence at 530 nm. To obtain this constant of proportionality, a control cell sample (with no FITC-EGF) is analyzed (Fig. 1a). It is observed that a straight line with a slope of 0.454 and a correlation coefficient of 0.864 relates the emission at 530 nm to that at 625 nm, indicating that this assumption is valid. This slope was then used in subsequent experiments with FITC-EGF to correct for the autofluorescence. In particular, the amount of signal in the autofluorescence parameter was used to gauge the amount of autofluorescence in the fluorescence parameter:

$$F_{\mathrm{real}} = F(488{,}530) - F(488{,}625) \times S_{625,530}$$

where $F_{\mathrm{real}}$ is the amount FITC fluorescence which would have been observed in the absence of any autofluorescence, $F(488{,}530)$ and $F(488{,}625)$ are the measured fluorescence values as described above, and $S_{625,530}$ is the spillover between them calculated from the control (in the example $S_{625,530} = 0.454$). This calculation was done on a cell-by-cell basis (on list-mode data using the CALC4 program we have developed) for the samples containing FITC-EGF; the results for all cells were combined to obtain corrected histograms (Fig. 1E,F), means, and percent positives (Table 1). The success of the correction is shown by the reduced overlap between the histograms and the increase in percent positive from 48% to 84%.

The degree to which noise is reduced should be related to the correlation between the actual autofluorescence at 530 nm and the parameter being used to estimate it in the control sample. For instance, if we compare forward-angle scatter, right-angle scatter, and $F(488{,}625)$ as a measure of the autofluorescence at 530 nm, then we expect the correction to be the best using $F(488{,}625)$, followed by side scatter, with the least correction being obtained with forward scatter (see Table 1 for correlation coefficients). The order of the latter two comes from the expected correlation of autofluorescence with

**TABLE 1. Autofluorescence Correction**

| Parameter used for correction | Correlation with *F* (488,530) | % positive[1] | Coefficient of variation (%) | |
|---|---|---|---|---|
| | | | Control | FITC-EGF[2] |
| None | N/A | 48.5 | 38 | 41 |
| Forward scatter | 0.236 | 58.7 | 33 | 37 |
| Side scatter | 0.452 | 70.0 | 20 | 30 |
| *F* (488,625) | 0.864 | 84.4 | 29 | 35 |

[1]Number of cells which exceed a threshold selected to eliminate 95% of the appropriate unlabeled control sample.
[2]FITC, fluorescein isothiocyanate; EGF, epidermal growth factor.

cell size, which is better reflected by side scatter. In Table 1, we see that this order is reflected in the percent positive after correction.

## ON-LINE KINETICS

Until recently, the data obtained by flow cytometry normally was from static populations; any active processes were stopped in order to acquire statistically significant data. Martin and Swartzendruber [1980] first included a measure of elapsed time as another parameter in flow cytometry. Previously, kinetic information was obtained only by multiple discrete measurements. These authors discuss the disadvantages of this technique, and demonstrate the increased accuracy in the determination of rapid kinetics by including time as a parameter. As opposed to making time an actual parameter for every cell, we have developed a system which continuously acquires data from the cytometer, on a cell-by-cell basis, and also records the number of cells analyzed during fixed intervals from 1/60 sec to 60 min [McNeil et al., 1985; Roederer and Murphy, in preparation]. Usually, intervals of 1 sec are used. Here we show two illustrations of the use of this system. First, we have used the method to ascertain the kinetics of the acidification of internalized insulin. Secondly, we have measured the kinetics with which endocytosed fluid is delivered to proteolytically active compartments.

### Insulin Acidification Kinetics

Figure 2 shows the binding, internalization, and acidification of FITC- and tetramethyl rhodamine isothiocyanate (TRITC) insulin by 3T3 fibroblasts. Cells were suspended by scraping and additions were made directly to the cell suspension (see legend). Because the sample is being run continuously on the sorter, the amount of fluorescence per cell is measured as a function of time for the whole population. Thus, by averaging the measured

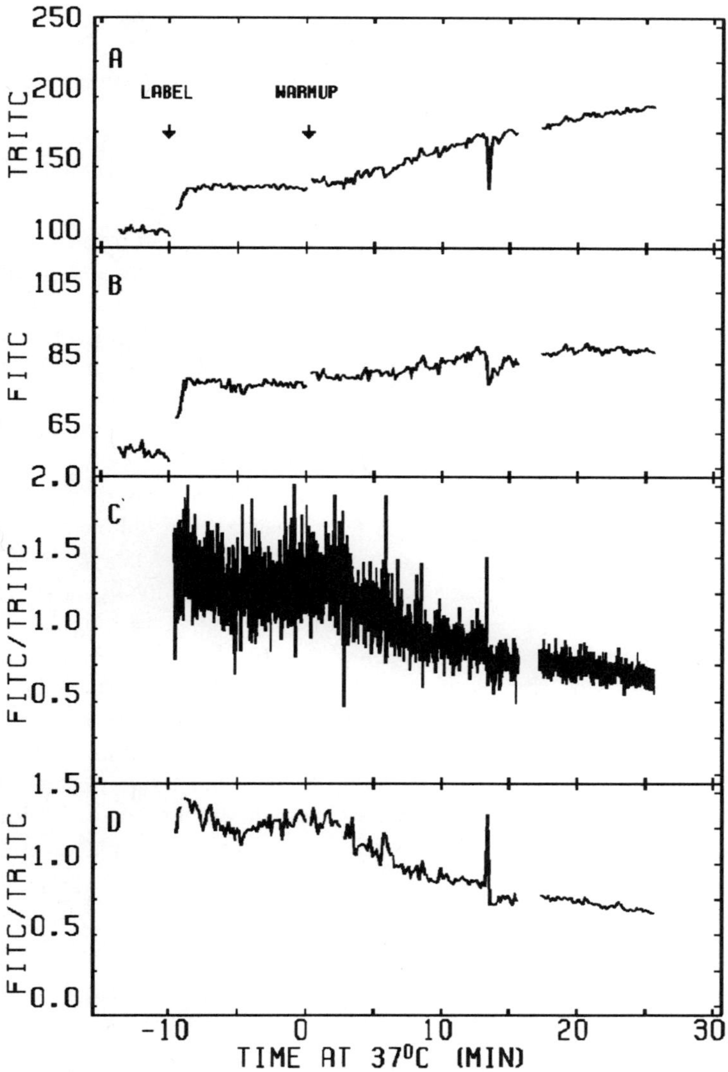


Fig. 2. Binding, internalization, and acidification of insulin. 60-mm monolayers were washed twice with PBS and then scraped into PBS. After 10 min on ice, they were run on the FACS to obtain autofluorescence levels for both red and green. FITC- and TRITC-insulin were added to the sample while it was still on ice [time of addition marked with an arrow in (A)]. After the amount of binding had stabilized, the cells were warmed to 37°C (second arrow). A,B. Uncorrected fluorescent signals for FITC-insulin (A) and TRITC-insulin (B) as a function of time. C,D. These plots were obtained by averaging the two fluorescent signals over 1 (C)- or 10 (D)-sec periods, subtracting out the average autofluorescences, and taking the ratio of the resulting FITC value to the TRITC value.

parameters over short periods of time (between 1 and 10 sec, or about 200 and 2,000 cells per period, respectively) to eliminate a large part of statistical noise, the internalization of the fluorescent ligand is observed as a function of time on a resolution of 1 or 10 seconds.

By subtracting the autofluorescence and then taking the ratio of the FITC and TRITC signals, a value proportional to the pH in which the ligand is found is obtained. Several features are evident in Figure 2: (1) binding is complete within 1 to 2 min of labeling; (2) a continuous and linear uptake of TRITC-insulin is observed after warmup; (3) a small increase in the FITC signal is observed, resulting in a ratio which steadily declines from roughly 1.5 to 0.7 over 30 min. This corresponds in a pH change from 7.4 to about 6.1. Since the amount of surface-bound ligand remains constant (confirmed by trypsinization; data not shown), we expect only a small increase in the FITC signal after warmup, since the fluorescence which would result from internalized ligand is rapidly quenched by acidification.

As was originally described by Bohn [1976] and discussed in detail by Murphy [1986], in flow cytometry the fluorescence from unbound ligand is usually negligible compared to that observed from the labeled cell (for appropriate numbers of receptors and high enough affinity ligands). However, two problems arise in analyzing steps subsequent to binding. First, there can be as much as one surface complement of ligand on the membrane which is not acidified. Thus, to obtain internal pH, corrections must be made for the amount of surface-bound ligand. This amount can be determined from parallel samples at fixed time points by using either trypsinization or pH shifts (for fluorescein-conjugated ligands; Finney and Sklar [1983]). With this information, the total fluorescence observed can be corrected for the surface complement. Second, there is also a distribution of ligand throughout the endocytic compartments, and a calculated pH will be the average of all compartments in which the ligand is found throughout the cell. Both of these problems can be partially overcome by using a pulse-chase protocol in which the processing of ligand can be expected to be synchronized.

## MNA-Peptide Hydrolysis

Another example of the acquisition of on-line kinetics data is shown in Figure 3. In order to measure the kinetics with which endocytosed material is degraded, a fluorogenic enzyme subtrate, N-carbobenzyloxy-ala-arg-arg-4-methoxy-$\beta$-naphthylamine (MNA-peptide), was used [Murphy, Bowser, and Roederer, in preparation]. The basis for the assay has been described by Dolbeare and Smith [1977] and Dolbeare and Vanderlaan [1979]. Briefly, the free MNA resulting from enzymatic digestion of the MNA-peptide reacts with 5-nitrosalicylaldehyde (NSA) to form an insoluble product with an emission maximum near 530 nm. By measuring MNA-NSA fluorescence as a function of time after additon of MNA-peptide, we can determine the earliest point at which endocytosed fluid reaches a compartment in which proteolytic enzymes are found.

A subconfluent 60-mm monolayer culture was used for each sample. The medium was removed, the cells were quickly washed twice with PBS, and the cells were removed by scraping with a rubber policeman in 2.9 ml of PBS. These unlabeled cells were then analyzed on the FACS for 2–5 min (at the appropriate temperature, see below), to determine an autofluorescence "baseline." The sample was removed from the FACS, MNA-peptide (0.5 mM) and NSA (1 mM) were added (in the presence or absence of inhibitors as described below), and analysis was continued. The sample temperature was maintained by a circulating water bath during acquisition. To control for nonspecific hydrolysis of the substrate and fluorescence resulting from NSA alone, proteinases were inhibited by the addition of leupeptin immediately before the addition of MNA-peptide and NSA. To control for uptake and/or hydrolysis by nonendocytic processes (e.g., secretion of enzymes into media, cell lysis, or diffusion through the plasma membrane), endocytosis was inhibited by using cytochalasin B [Wagner et al., 1971]. In this case, cells were incubated in cytochalasin B for 20 min at 37°C before scraping and addition of NSA and MNA-peptide.

As is shown in Figure 3, hydrolysis of MNA-peptide begins almost immediately after its addition to cells, and continues linearly for at least 35 min. The vast majority of this hydrolysis is due to enzymatic cleavage, since 95% can be inhibited by coincubation with leupeptin; most is also dependent on endocytosis, since 86% is inhibited when cells are preincubated with cytochalasin B. The kinetics of appearance of the MNA-NSA fluorescence thus provides an excellent reflection of the kinetics with which endocytosed fluid reaches a compartment containing protease.

Since these measurements were made in the continual presence of 1 mM NSA, the rate of cleavage observed may not reflect the rate which would be observed in the absence of NSA. We are currently investigating the possibility of using direct UV excitation of the liberated MNA to eliminate this possibility.

## COMPARISON OF ACIDIFICATION KINETICS FOR DIFFERENT LIGANDS

There is some disparity in the reports of the pH to which various ligands are internalized. To determine whether the initial kinetics of acidification are biphasic [Murphy et al., 1984a], the acidification of three fluorescent conjugates were analyzed by different methods. A summary of these results is shown in Figure 4. For EGF, FITC-EGF was added to 10 nM; the cells were incubated for various periods, trypsinized, and analyzed by using the amine-

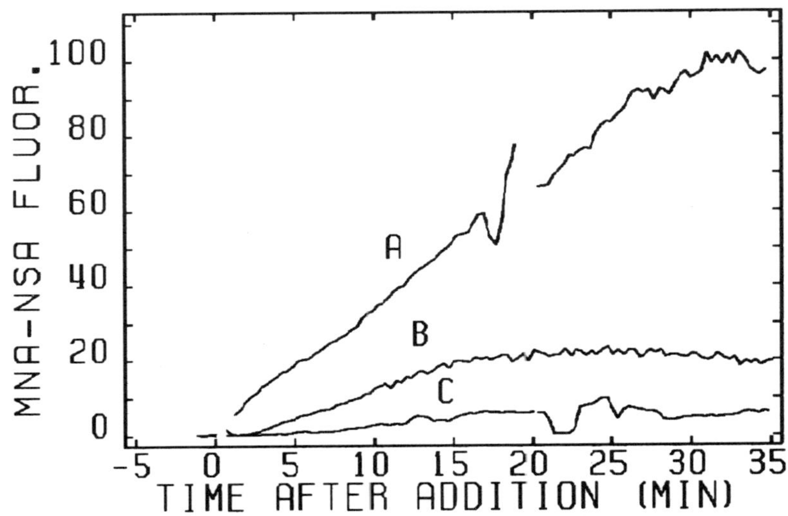


Fig. 3. MNA-peptide hydrolysis kinetics. Cells were scraped from tissue culture dishes at 0°C and run on the FACS at 37°C to obtain the background autofluorescence. MNA-peptide (Research Plus), NSA (Eastman Kodak), and either cytochalasin B (endocytosis inhibitor) or leupeptin (proteolysis inhibitor) or no inhibitor was added and the cells were analyzed during continuous incubation at 37°C. The fluorescence resulting from excitation at 488 nm was measured for each cell by using a 530-nm bandpass filter (30-nm bandwidth). A. Cells plus MNA-peptide and NSA. B. Cells incubated with 20 μg/ml cytochalasin B for 20 min, scraped, and run in MNA-peptide and NSA as in A. C. Cells plus MNA-peptide, NSA, and 100 μg/ml leupeptin; 86% of the hydrolysis is due to postendocytic enzymatic cleavage, since 5% is not blocked by leupeptin and an additional 9% (a total of 14%) is not blocked by cytochalasin B.

ratio method [Murphy et al., 1982a]. The EGF went through only one round of internalization as demonstrated by a constant level of signal (after methylamine addition) independent of incubation time (data not shown); thus, it enters the endocytic pathway synchronously and provides a signal equivalent to a pulse-chase experiment. For insulin, both FITC- and TRITC-insulin were added simultaneously to a concentration of 100 nM as shown above in Figure 2. The labeled insulins were taken up linearly with time (data not shown). Finally, FITC-and XRITC-dextran were added simultaneously to monolayers to a concentration of 1 mg/ml. After a 5-min incubation at 37°C, the cells were washed 8 times with PBS, and incubation continued for various periods of time in media at 37°C. pH calculations were done as follows: for EGF, autofluorescence was subtracted from the average fluorescein signal before and after the addition of methylamine. This ratio was compared to a pH calibration performed on a fluorometer to obtain a value corresponding to the pH in which the FITC-EGF was found. The background for insulin was determined by binding insulin to cells at 0°C. This represents autofluorescence plus the signal due to surface-bound insulin. This value was subtracted from the subsequent incubation samples to obtain a value for internal fluorescence. The ratio of the background-corrected fluorescein and rhodamine signals was compared to a pH calibration curve to obtain an average pH of internal insulin. Because ligand was continuously present, the resulting pH is an average for ligand throughout the endocytic pathway, and is expected to be higher than that for ligand that has synchronously entered the cells. The dextran pH was calculated by subtracting autofluorescence, taking the ratio of the fluorescein and rhodamine signals, and comparing to a dextran calibration curve for that experiment.

Figure 4 shows the close correlation between the pH calculated for dextran and EGF and the slightly higher values obtained for insulin. All three ligands demonstrate a rapid acidification to approximately pH 6.0 (within 5 min of labeling), followed by a slow acidification to a pH as low as 5.2 (50 min with EGF). A previous report comparing the pH to which alpha-2-macroglobulin and transferrin are acidified [Yamashiro et al., 1984] suggests that

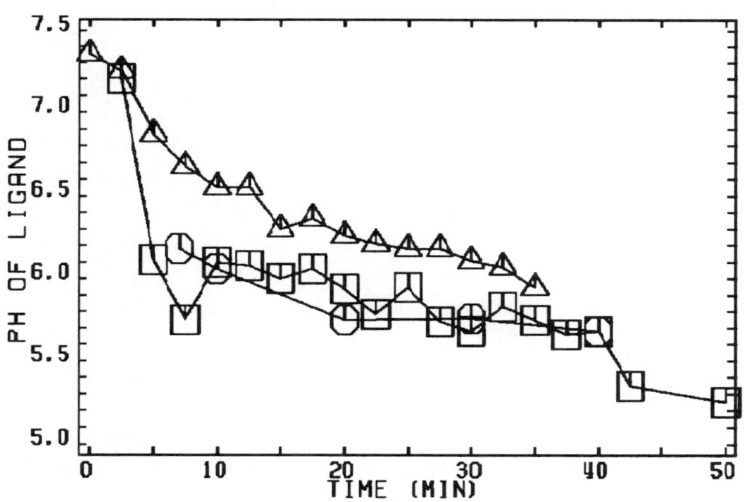


Fig. 4. Acidification kinetics for dextran, insulin, and EGF. Swiss 3T3 cells were incubated continuously with either 10 nM FITC-EGF (□) or 100 nM FITC- and TRITC-insulin (Δ), or pulsed for 5 min with 1 mg/ml FITC- and XRITC-dextran and chased (○). The average pH at which ligand is found is plotted as a function of time after addition of the label. Note the close overlap between the EGF and the dextran curves. The insulin curve shows a slightly higher average pH, since the continuous incubation procedure yields an average over all compartments. Although EGF was also present in a continuous incubation, the ligand was synchronously endocytosed in the time frame of this experiment (see text), and thus resembles a pulse-chase experiment.

the initial pH is ligand-dependent. It was reported that FITC-transferrin was acidified to pH 6.4 within 10 min, and that it was not acidified to a lower pH. FITC-$\alpha_2$-macroglobulin, on the other hand, was rapidly acidified to either pH 5.4 ± 0.1 [Yamashiro et al., 1984] or pH 5.0 ± 0.2 [Tycko and Maxfield, 1982] and was not further acidified, even when delivered to lysosomes. However, we found that the initial pH to which insulin was acidified was higher (approximately pH 6) than that which it eventually reached (approximately pH 5) [Murphy et al., 1984a]. Biphasic acidification kinetics were also observed for antibodies directed against H-2K [Murphy et al., 1984b]. The evidence in Figure 4 demonstrates that at least these three ligands are intially exposed to pH 6 and then (possibly) later to a lower pH. This explains the observation (A. Helenius, personal communication) that the Semliki Forest virus (SFV) fus-1 mutant (which requires a pH of 5.5 for infection [Kielian et al., 1984]) requires at least 45 min at 20°C to infect while the wild-type virus can infect in less than 5 min.

## TEMPERATURE DEPENDENCE OF ACIDIFICATION

From the in vitro data of Galloway et al. [1983], the mechanism of endosome acidification in vivo would appear to be through an ATP-requiring proton pump. This conclusion is supported by the data of Maxfield [1982] and Yamashiro et al. [1983]. These data are inconclusive, however, since (1) no evidence is presented in either paper that the probe is not contained in lysosomes; (2) the fluorescence values are not calibrated to allow accurate comparison either internally or with other papers (e.g., Fig. 2 of Maxfield [1982] and Fig. 1 of Yamashiro et al. [1983]); (3) the digitonin-"permeabilized" cells may not reflect the in vivo system. Additional experiments are required to firmly establish the active nature of this initial acidification step.

One means of examining the energy requirements for this initial acidification is to determine its temperature dependence. The temperature dependence of some of the steps in endocytosis has already been addressed by several groups. Some conflict exists, however, and no reliable information is available about the exact pH to which ligands are acidified at temperatures below 37°C. Internalization of ligand begins at approximately 10°C. For example, Pesonen et al. [1984] observed no uptake of ligand at 10°C, whereas Marsh et al. [1983] observed internalization of roughly 10% of the amount taken up at 37°C. By 15°C, 40% of the 37°C amount of ligand was internalized. Virus is capable of infecting cells at 15°C, demonstrating acidification at this temperature. Pesonen et al. [1984] found no transcytosis of G protein at 15°C, indicating that sorting may be blocked at this temperature. Oka and Weigel [1983] demonstrated that at 18°C an asialoglycoprotein is found in two distinct compartments: one in which dissociation occurs, and one in which the the ligand remains complexed to the receptor. It was not ascertained whether this was due to sorting after internalization, or internalization of ligand into two different compartments.

Transcytosis of G protein does occur at 20°C, indicating that sorting of ligands occurs at this temperature. Galloway et al. [1983] demonstrated that FITC-dextran internalized at this temperature is found in endosomes that are capable of acidifying in vitro. Matlin and Simons [1983] found that hemagglutinin protein is not exocytosed at 20°C even 2 hours after its synthesis. There is evidence that part of the endocytic and exocytic pathways converge at the Golgi complex, and this experiment would indicate therefore that there is still a processing step blocked at 20°C. Hopkins and Trowbridge [1983] showed that, at 22°C, transferrin receptor-antibody complexes traverse several compartments to arrive in juxtanuclear multivesicular bodies. However, no degradation was observed at this temperature. Upon warming to 37°C, the receptor was immediately transferred to lysosomal compartments and degraded. Finally, at 25°C, degradation was observed for G protein [Pesonen et al., 1984] and viral proteins [Marsh et al., 1983], but not the EGF receptor [Stoscheck and Carpenter, 1984]. The EGF receptor is degraded at 37°C.

We have addressed the question of the temperature dependence of various steps in the endocytic pathway by labeling cells with FITC- and XRITC-dextran at various temperatures. Figure 5 shows typical results from these experiments [Roederer and Murphy, in preparation]. At 11°C, acidification does not occur until after 20 min, at which point it proceeds to pH 6.1. At temperatures between 17°C and 21°C, acidification to pH 6.1 proceeds fairly rapidly but does not proceed beyond that point, at least up to 40 min. At 25°C and 37°C, dextran is already at pH 6.1 at the earliest time point (2-min chase following a 5-min pulse); however, acidification continues to almost pH 5.5 at 40 min.

Figure 5 also shows the temperature dependence of exocytosis of the internalized dextrans. At 11°C, the amount of dextran remaining slowly decreases to 60%. At temperatures up to 21°C, this process occurs much more rapidly, but remains at 60% even at 40 min. At 25°C, however, the amount remaining begins to decrease below 60%, and at 37°C almost 70% of the internalized ligand is lost. This correlates well with the temperature dependence of ligand processing and sorting. Our conclusion is that at temperatures up to 21°C, internalized ligand reaches pH 6.1 within 30 min, while approximately 45% is lost by "reversible endocytosis" [cf. Storrie et al., 1984]. At higher temperatures, acidification proceeds rapidly to pH 6, and more slowly to pH 5.5 or below; more dextran is slowly lost due to a recycling pathway that depends on a fusion or processing event, possibly involving lysosomes.

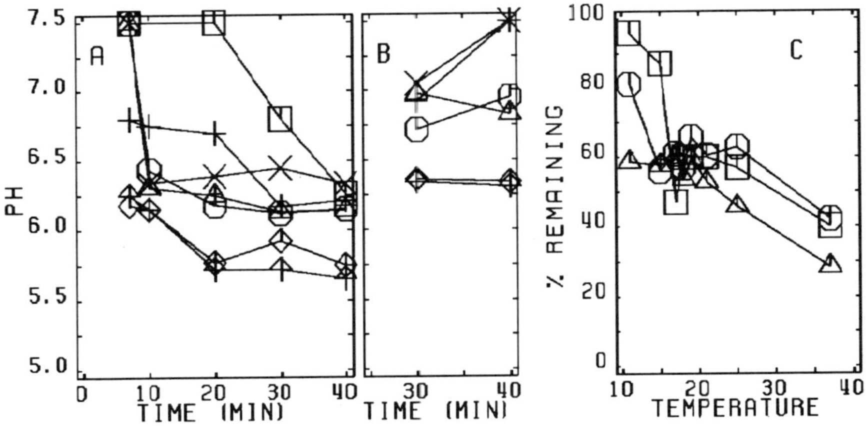


Fig. 5. Temperature dependence of dextran acidification and exocytosis. FITC- and XRITC-dextran were added to cells on tissue culture plates at a concentration of 1 mg/ml for 5 min; the cells were washed extensively with PBS and medium was added for the duration of the chase. All operations were carried out with solutions and cells equilibrated in a refrigerated water bath with the temperature held to within 0.1°C. A. The pH of the compartments containing dextran as calculated according to the text is plotted as a function of time after addition of dextran for cells incubated at 11°C (□), 17°C (○), 18°C (Δ), 19°C (+), 21°C (×), 25°C (◇), and 37°C (↑). B. The pH of the cellular compartments after neutralization with 200 mM methylamine for 5 min at 24°C (symbols as in A). C. The amount of XRITC-dextran remaining in the cell after 20 min (□), 30 min (○), and 40 min (Δ) after addition of dextran, relative to the amount present at the earliest time point (5-min pulse, 2-min chase) is shown as a function of temperature of incubation. At low temperatures, there is a slow loss of ligand to 60% of the initial value. The intermediate temperatures show a rapid loss to 60%, while the warm temperatures show an additional loss to less than 40% of the initial amount.

A distinction between the endosomal and lysosomal compartments can also be made by examining the extent of neutralization by 200 mM methylamine of these compartments (Fig. 5B). The low-temperature samples were neutralized to pH 7.0–7.5, whereas the 25°C and 37°C samples were only neutralized to pH 6.5. Because the neutralization was done at the same temperature (24°C) for all samples, the difference in final pH is not a consequence of temperature sensitivity of the acidification mechanism but rather of the fact that at the lower temperatures, ligand is not delivered to compartments that acidify as actively.

From Arrhenius plots (the natural logarithm of the amount of dextran taken up at the earliest time point versus inverse absolute temperature), we can obtain an energy of activation for the initial internalization step. By linear regression, we obtain an energy of activation of 8.8 kcal/mol (correlation coefficient 0.947), a value less than half that observed previously. Measurements of 17–25 kcal/mol [Mahoney et al., 1977] and 19.5 kcal/mol [Kaplan and Nielsen, 1979] were obtained for the energy of activation of internalization by macrophages, 26.5 kcal/mol was obtained for perfused rat liver cells [Dunn et al., 1980], and 17.6 kcal/mol obtained for fibroblast internalization [Steinman et al., 1974]. Since previous measurements have used long (at least 30 min) incubations or warmup protocols which are not appropriate for this analysis, the value of 8.8 kcal/mol should be more accurate reflection of the activation energy of the initial endocytic event.

## FLOW CYTOMETRIC VESICLE ANALYSIS

As discussed above, a great deal of information about the pathways of endocytosis has been obtained by using a combination of electron microscopy and traditional biochemical techniques. However, some of the details of these pathways are not yet known due to the limitations of the methods used. For example, morphological studies by electron microscopy have the advantage that spatial and structural information may be obtained. However, the method is inherently static, sample sizes are ordinarily limited, and it is difficult to quantitate enzymatic and biochemical properties. Density gradient centrifugation has been useful due to the possibility of performing biochemical analyses on the resulting fractions. However, the length of time required for fractionation introduces the possibility or changes in vesicle properties, as might be expected from proteolysis or loss of ion gradients. In addition, bulk fractionation techniques are unable to address questions which require particle-by-particle analysis, such as whether different endocytic probes or different enzymes are contained in the same compartment.

While the usefulness of flow cytometry for the study of endocytosis has been demonstrated previously [Murphy et al., 1982a, 1984a,b; Finney and Sklar, 1983], it has heretofore been limited mainly to measurements on whole cells. Geisow et al. [1982] have described a flow cytometric assay for extracellular fusion of lysosomes and phagosomes containing 5-$\mu$m latex spheres (phagosomes were detected by light scattering from the incorporated sphere; no attempt was made to detect the unfused lysosomes directly). Flow microfluorometric analysis of synthetic phosphatidylcholine/cholesterol vesicles has been described recently by Allen et al. [1984]. We have recently demonstrated the feasibility of using flow cytometry and sorting to characterize unfractionated cell lysates after labeling with endocytic and enzymatic probes [Murphy, 1985]. This technique offers the advantages of rapid measurement of large numbers of individual vesicles, the ability to simultane-

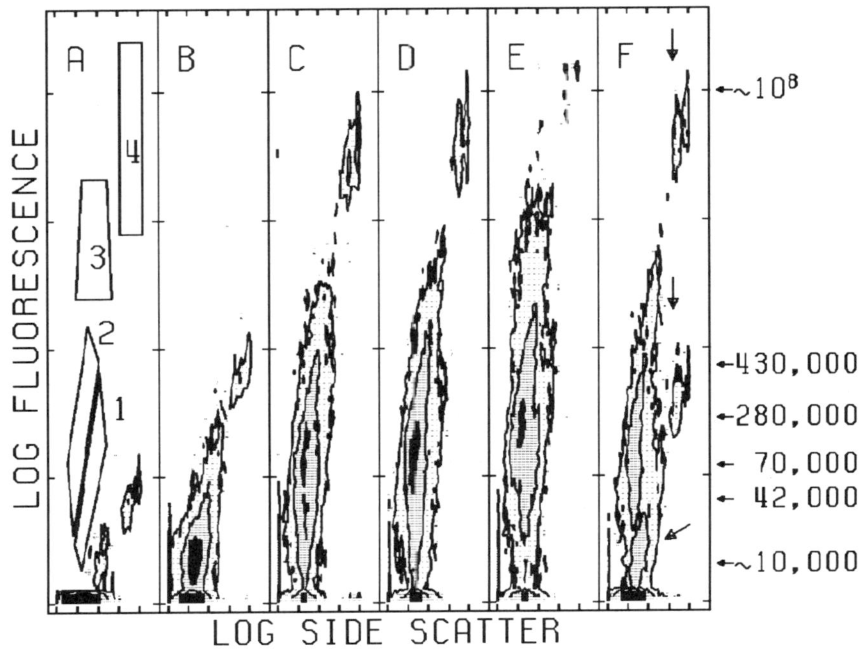


Fig. 6. Detection of endocytic vesicles by using flow cytometry. Crude cell lysates were prepared from control cells which did not receive any FITC-dextran (A), and from cells which were incubated with 50 mg/ml FITC-dextran for 5 sec (B), 20 min (C), 60 min (D), and 180 min (E). Each panel represents 50,000 events. The polygons in A show the regions in C–E where vesicles which show weak (1), intermediate (2), and bright (3) fluorescence are found. The position of intact fluorescent cells (4) is also shown. The arrows show the average fluorescence of four types of calibration microspheres, and two extrapolated values. A mixture of equal volumes of the samples in A and D was also analyzed (F). Note the presence of a population of unlabeled vesicles and two populations of intact cells in this mixture (large arrows). Contours are drawn at 3, 20, and 80 events per bin. Tics are placed at approximately one-log intervals.

ously measure more than one vesicle property, and the possibility of sorting individual vesicles for further analysis.

Primary effort in the area of vesicle analysis has thus far been directed to developing a flow cytometric procedure for analyzing individual endocytic vesicles in cell lysates [Murphy, 1985]. Some of the results of this analysis are shown in Figures 6 and 7. The results indicate that at least three kinetically distinct compartments may be identified by flow cytometry; these may correspond to light endosomes, heavy endosomes, and lysosomes. Figure 7 demonstrates the increased resolution of these populations obtained by normalizing the data by using light scatter.

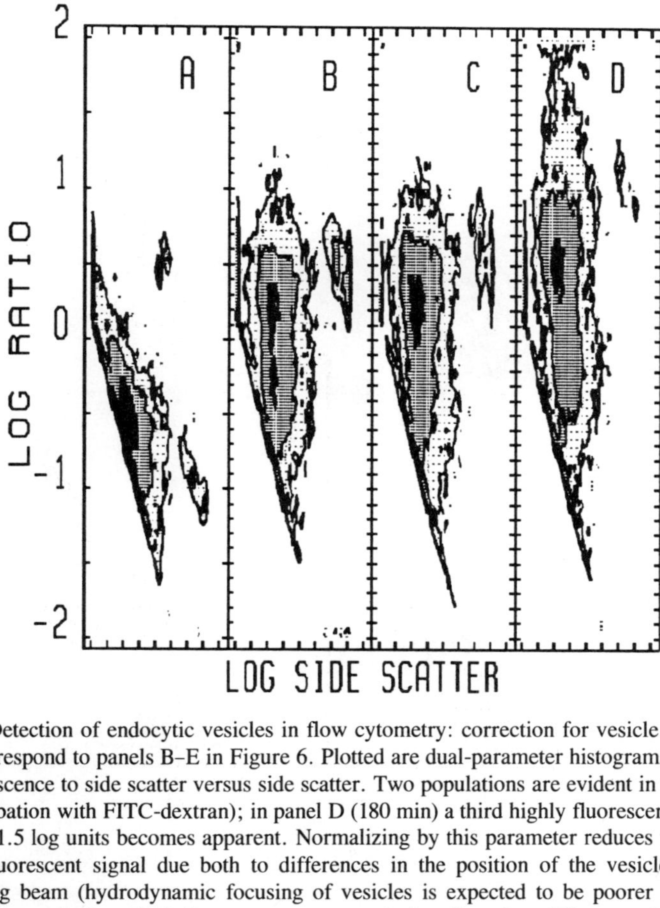


Fig. 7. Detection of endocytic vesicles in flow cytometry: correction for vesicle size. Panels A–D correspond to panels B–E in Figure 6. Plotted are dual-parameter histograms of the ratio of fluorescence to side scatter versus side scatter. Two populations are evident in panel B (20-min incubation with FITC-dextran); in panel D (180 min) a third highly fluorescent population at about 1.5 log units becomes apparent. Normalizing by this parameter reduces the variation in the fluorescent signal due both to differences in the position of the vesicle within the measuring beam (hydrodynamic focusing of vesicles is expected to be poorer than for the much larger cells for which flow cytometers are designed) and to differences in size of the individual vesicles; the resulting populations are more homogeneous and better separated from each other.

A 20-min incubation with FITC-dextran resulted in two distinct labeled populations. Merion et al. [1983] have demonstrated a similar resolution of at least two endocytic compartments after a 15-min incubation with FITC-dextran with Percoll gradients. As incubation time increases, a highly fluorescent population appears. The kinetics of the appearance of this population suggest that it may consist of secondary lysosomes and multivesicular bodies.

This approach has been combined with the use of the MNA-conjugated fluorogenic cathepsin B substrate described above to permit detection of compartments containing this proteolytic enzyme (Fig. 8). Correlation of the

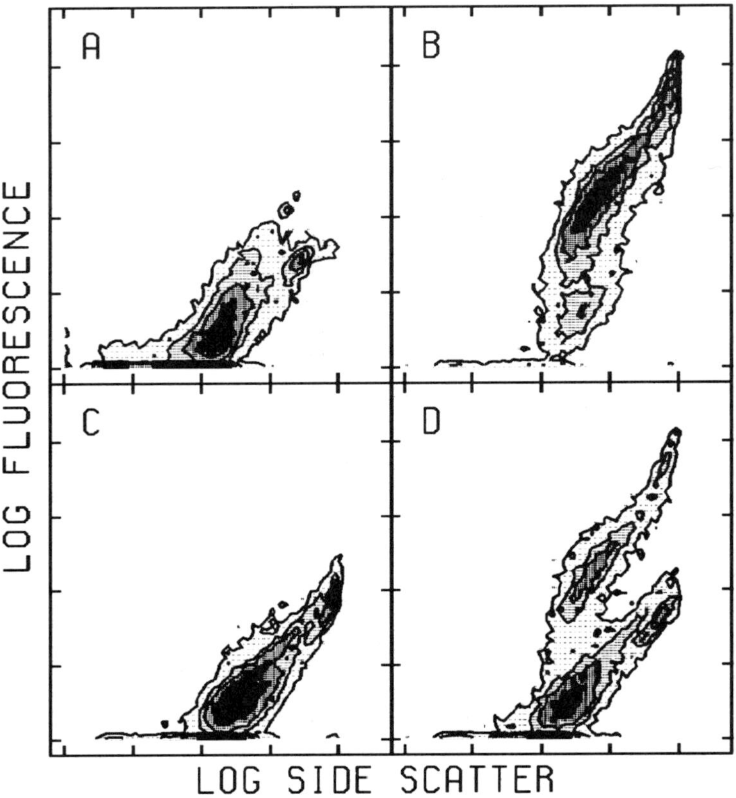


Fig. 8. Detection of vesicles containing cathepsin B activity. Flow cytometric analysis of lysates from cells incubated with (B,C) or without (A) fluorogenic substrate and with (C) or without (A,B) 100 $\mu$g/ml leupeptin. A mixture of equal volumes of the samples shown in B and C was also analyzed (D). The presence of both labeled and unlabeled populations in D demonstrates that the fluorescence is intravesicular. Contours are drawn at 5, 20, 40, and 80 events per bin. Tics are placed at approximately one-log intervals.

presence of endocytic markers with the presence of lysosomal enzymes should be possible by using this technique.

## CONCLUSION

We have developed several new flow cytometric techniques for the analysis of endocytosis. First, by correcting for autofluorescence by using correlated parameters, the statistical significance of results with low signal-to-noise ratios is increased dramatically. This is crucial in many ligand systems where less than 30,000 ligands bind to a cell. We have also demonstrated the usefulness of acquisition of real-time data on the flow cytometer. This allows the observation of kinetic processes with half-times on the order of seconds or minutes. We have used this system to analyze the kinetics with which pinocytosed ligand is acidified and delivered to degradatory compartments.

Additional experiments on other ligands demonstrate that ligand acidification is biphasic; in particular, both EGF and dextran are acidified to pH 6 within 5 min of internalization, followed by a slower acidification to pH 5.2 over the next 50 min. By varying the temperature of incubation, we have demonstrated that the second phase of this acidification is blocked below 21 °C, but the first phase occurs even at 11 °C.

Finally, we have used flow cytometry to analyze endocytic vesicles in unfractionated cell lysates. By analyzing large populations of vesicles labeled with fluorescent dextran, we have shown that there are three kinetically distinct compartments involved in endocytosis. Future experiments using these techniques should permit further elucidation of the biochemical properties of the compartments involved in endocytosis.

## ACKNOWLEDGMENTS

We thank Amy Kennedy, Greg LaRocca, Robert Mays, and Daniel Shmorhun for technical assistance, and Robert Bowser, David Sipe, D. Lansing Taylor, and Alan Waggoner for helpful discussions. This work was supported in part by National Institutes of Health grant GM32508, and Presidential Young Investigator Award DCB-8351364 from the National Science Foundation, with matching funds from Becton Dickinson Monoclonal Center, Inc.

## REFERENCES

Allen JK, Dennison DK, Schmitz KS, Morrisett JD (1984): Direct observation of the distribution of fluorescent probes in phosphatidylcholine/cholesterol vesicles using flow microfluorometry. Anal Biochem 140:409–416.

Aubin JE (1979): Autofluorescence of viable cultured mammalian cells. J Histochem Cytochem 27:36–43.

Benson RC, Meyer RA, Zaruba ME, McKhann GM (1979): Cellular autofluorescence—is it due to flavins? J Histochem Cytochem 27:44–48.

Bohn B (1976): High-sensitivity cytofluorometric quantitation of lectin and hormone binding to surfaces of living cells. Exp Cell Res 103:39–46.

Boquet P, Duflot E (1982): Tetanus toxin fragment forms channels in lipid vesicles at low pH. Proc Natl Acad Sci USA 79:7614:7618.

Carpenter G, Cohen S (1979): Epidermal growth factor. Ann Rev Biochem 48:193–216.

Dolbeare FA, Smith RE (1977): Flow cytometric measurement of peptidases with use of 5-nitrosalicylaldehyde and 4-methoxy-$\beta$-naphthylamine derivatives. Clin Chem 23:1485–1491.

Dolbeare F, Vanderlaan M (1979): A fluorescent assay of proteinases in cultured mammalian cells. J Histochem Cytochem 27:1493–1495.
Draper RK, Simon MI (1980): The entry of diphtheria toxin into the mammalian cell cytoplasm: Evidence for lysosomal involvement. J Cell Biol 87:849–854.
Dunn WA, Hubbard AL, Aronson NN (1980): Low temperature selectively inhibits fusion between pinocytic vesicles and lysosomes during heterophagy of $^{125}$I-asialofetuin by the perfused rat liver. J Biol Chem 255:5971–5978.
Finney DA, Sklar LA (1983): Ligand/receptor internalization: A kinetic, flow cytometric analysis of the internalization of N-formyl peptides by human neutrophils. Cytometry 4:54–60.
Forgac M, Cantley L, Wiedenmann B, Altstiel L, Branton D (1983): Clathrin-coated vesicles contain an ATP-dependent proton pump. Proc Natl Acad Sci USA 80:1300–1303.
Galloway CJ, Dean GE, Marsh M, Rudnick G, Mellman I (1983): Acidification of macrophage and fibroblast endocytic vesicles *in vitro*. Proc Natl Acad Sci USA 80:3334–3338.
Geisow M, D'arcy Hart P, Young MR (1982): Extracellular fusion of macrophage phagosomes with lysosomes. Cell Biol Int Rep 6:361–367.
Glickman J, Croen K, Kelly S, Al-Awqati Q (1983): Golgi membranes contain an electrogenic $H^+$ pump in parallel to a Chloride conductance. J Cell Biol 97:1303–1308.
Goldstein JL, Anderson RGW, Brown MS (1979): Coated pits, coated vesicles, and receptor-mediated endocytosis. Nature 279:679.
Gonzalez-Noriega A, Grubb JH, Talkad V, Sly WS (1980): Chloroquine inhibits lysosomal enzyme pinocytosis and enhances lysosomal enzyme secretion by impairing receptor recycling. J Cell Biol 85:839–852.
Helenius A, Kartenbeck J, Simons K, Fries E (1980): On the entry of semliki forest virus into BHK-21 cells. J Cell Biol 84:404–420.
Helenius A, Mellman I, Wall D, Hubbard A (1983): Endosomes. Trends Biochem Sci 8:245–249.
Hopkins CR, Trowbridge IS (1983): Internalization and processing of transferrin and the transferrin receptor in human carcinoma A431 cells. J Cell Biol 97:508–521.
Kaplan J, Nielsen ML (1979): Analysis of macrophage surface receptors. J Biol Chem 254:7329–7335.
Kielian MC, Keranen S, Kaariainen L, Helenius A (1984): Membrane fusion mutants of semliki forest virus. J Cell Biol 98:139–145.
Mahoney EM, Hamill AL, Scott WA, Cohn ZA (1977): Response of endocytosis to altered fatty acyl composition of macrophage phospholipids. Proc Natl Acad Sci USA 74:4895–4899.
Marnell MH, Stookey M, Draper RK (1982): Monensin blocks the transport of diphtheria toxin to the cell cytoplasm. J Cell Biol 93:57–62.
Marsh M, Bolzau E, Helenius A (1983): Penetration of semliki forest virus from acidic prelysosomal vacuoles. Cell 32:931–940.
Martin JC, Swartzendruber DE (1980): Time: A new parameter for kinetic measurements in flow cytometry. Science 207:199–201.
Matlin KS, Simons K (1983): Reduced temperature prevents transfer of a membrane glycoprotein to the cell surface but does not prevent terminal glycosylation. Cell 34:233–243.
Maxfield FR (1982): Weak bases and ionophores rapidly and reversibly raise the pH of endocytic vesicles in cultured mouse fibroblasts. J Cell Biol 95:676–681.
Maxfield FR, Schlessinger J, Shecter Y, Pastan I, Willingham MC (1978): Collection of insulin, EGF, and $\alpha_2$-macroglobulin in the same patches on the surface of cultured fibroblasts and common internalization. Cell 14:805–810.
McNeil PL, Kennedy AL, Waggoner AS, Taylor DL, Murphy RF (1985): Light-scattering changes during chemotactic stimulation of human neutrophils: Kinetics followed by flow cytometry. Cytometry 6:7–12.

Mellman I, Plutner H (1984): Internalization and degradation of macrophage Fc receptors bound to polyvalent immune complexes. J Cell Biol 98:1170–1177.
Mellman I, Plutner H, Ukkonen P (1984): Internalization and rapid recycling of macrophage fc receptors tagged with monovalent antireceptor antibody: Possible role of a prelysosomal compartment. J Cell Biol 98:1163–1169.
Merion M, Sly WS (1983): The role of intermediate vesicles in the adsorptive endocytosis and transport of ligand to lysosomes by humand fibroblasts. J Cell Biol 96:644–650.
Merion M, Schlesinger P, Brooks RM, Moehring JM, Moehring TJ, Sly WS (1983): Defective acidification of endosomes in chinese hamster ovary cell mutans "cross-resistant" to toxins and viruses. Proc Natl Acad Sci USA 80:5315–5319.
Miller DK, Lenard J (1980): Inhibition of vesicular stomatitis virus infection by spike glycoprotein. J Cell Biol 84:430–437.
Miskimins WK, Shimizu N (1984): Uptake of epidermal growth factor into a lysosomal enzyme-deficient organelle: Correlation with cell's mitogenic response and evidence for ubiquitous existence in fibroblasts. J Cell Physiol 118:305–316.
Murphy RF (1985): Analysis and isolation of endocytic vesicles by flow cytometry and sorting: Demonstration of three kinetically distinct compartments involved in fluid-phase endocytosis. Proc Natl Acad Sci USA 82:8523–8526.
Murphy RF (1986): Ligand binding, endocytosis and processing. In Melamed MR, Lindmo T, Mendelsohn ML (eds): "Flow Cytometry and Sorting," second edition. New York: John Wiley and Sons, in press.
Murphy RF, Jorgensen ED, Cantor CR (1982a): Kinetics of histone endocytosis in chinese hamster ovary cells. J Biol Chem 257:1695–1701.
Murphy RF, Powers S, Verderame M, Cantor CR, Pollack R (1982b): Flow cytofluorometric analysis of insulin binding and internalization by swiss 3T3 cells. Cytometry 2:402–406.
Murphy RF, Powers S, Cantor CR (1984a): Endosome pH measured in single cells by dual fluorescence flow cytometry: Rapid acidification of insulin to pH 6. J Cell Biol 98:1757–1762.
Murphy RF, Tse DB, Cantor CR, Pernis B (1984b): Acidification of internalized class I major histocompatibility complex antigen by T lymphoblasts. Cell Immunol 88:336–342.
Ohkuma S, Poole B (1978): Fluorescence probe measurement of the intralysosomal pH in living cells and the perturbation of pH by various agents. Proc Natl Acad Sci USA 75:3327–3331.
Oka JA, Weigel PH (1983): Recycling of the asialoglycoprotein receptor in isolated rat hepatocytes. J Biol Chem 258:10253–10262.
Pastan I, Willingham MC (1983): Receptor-mediated endocytosis: Coated pits, receptosomes, and the golgi. Trends Biochem Sci 8:250–252.
Pesonen M, Ansorge W, Simons K (1984): Transcytosis of the G protein of vesicular stomatitis virus after implantation into the apical plasma membrane of madin-darby canine kidney cells. I. Involvement of endosomes and lysosomes. J Cell Biol 99:796–802.
Sandvig K, Olsnes S (1980): Diphtheria toxin entry into cells is facilitated by low pH. J Cell Biol 87:828–832.
Steinman RM, Silver JM, Cohn ZA (1974): Pinocytosis in fibroblasts. J Cell Biol 63:949–969.
Storrie B, Pool RR, Sachdeva M, Maurey KM, Oliver C (1984): Evidence for both prelysosomal and lysosomal intermediates in endocytic pathways. J Cell Biol 98:108–115.
Stoscheck CM, Carpenter G (1984): Down regulation of epidermal growth factor receptors: Direct demonstration of receptor degradation in human fibroblasts. J Cell Biol 98:1048–1053.
Tietze C, Schlesinger P, Stahl P (1980): Chloroquine and ammonium ion inhibit receptor-mediated endocytosis of mannose-glycoconjugates by macrophages: Apparent inhibition of receptor recycling. Biochem Biophys Res Comm 93:1–8.

Tycko B, Maxfield FR (1982): Rapid acidification of endocytic vesicles containing alpha-2-macroglobulin. Cell 28:643–651.

Van Leuven F, Cassiman J-J, Van Den Berghe H (1980): Primary amines inhibit recycling of $\alpha_2$-macroglobuline receptors in fibroblasts. Cell 20:37–43.

Van Renswoude J, Bridges KR, Harford JB, Klausner RD (1982): Receptor-mediated endocytosis of transferrin and the uptake of Fe in K562 cells: Identification of a nonlysosomal acidic compartment. Proc Natl Acad Sci USA 79:6186–6190.

Wagner R, Rosenberg M, Estensen R (1971): Endocytosis in chang liver cells. J Cell Biol 50:804–817.

Yamashiro DJ, Fluss SR, Maxfield FR (1983): Acidification of endocytic vesicles by an ATP-dependent proton jump. J Cell Biol 97:929–934.

Yamashiro DF, Tycko B, Fluss SR, Maxfield FR (1984): Segregation of transferrin to a mildly acidic (pH 6.5) para-golgi compartment in the recycling pathway. Cell 37:789–800.

Yoshimura A, Kuroda K, Kawasaki K, Yamashina S, Maeda T, Ohnishi S (1982): Infectious cell entry mechanism of influenza virus. J Virol 43:284–293.

*Applications of Fluorescence in the Biomedical Sciences, pages 567–584*


# 28 Analysis of Platelet Function Using Subcellular and Surface-Associated Fluorescent Probes

*Laurence Corash, Yuen Mok, and Margaret Rheinschmidt*

Normal platelet hemostatic function requires the complex interplay of plasma membrane, subcellular organelles, and the endoperoxide metabolic pathway [Weiss, 1982]. Hereditary or acquired disorders of platelet function may occur as the result of a defect at each of the preceding critical steps. While understanding of the specific biochemical defect for failure at these three functional points has expanded considerably over the past decade [Bellucci, 1983], there has been little advance in clinical laboratory testing to identify these specific lesions in a clinical setting. Moreover, concurrent thrombocytopenia frequently prevents the application of existing tests. Fluorescence-activated flow cytometry offers the potential to develop new tests with improved sensitivity and specificity to diagnose and to monitor platelet function disorders. The purpose of the present study is to demonstrate that platelet function both at the subcellular level and at the plasma membrane level can be examined by this approach and that with the development of additional probes a panel of assays to fully assess platelet function should become available.

## SUBCELLULAR PROBES OF PLATELET FUNCTION

Platelets contain three classes of subcellular organelles which are directly involved in platelet function: alpha granules, dense bodies (delta granules), and lysosomes. These organelles are secreted in response to a wide variety

Department of Laboratory Medicine, University of California, San Francisco 94143.

of platelet agonists as an integral part of normal platelet function [Weiss, 1982]. Congenital absence of at least two granule organelles, dense bodies or alpha granules, is associated with qualitative hemostatic defects characterized by prolonged bleeding times and clinically significant hemorrhage [Bellucci et al., 1983; Weiss et al., 1979]. Additional studies have also demonstrated that these defects may be acquired due to in vivo platelet activation, for example, during cardiopulmonary bypass procedures or in a variety of diseases [Malpass and Harker, 1980; Rosove et al., 1980]. Patients may also develop clinically qualitative platelet dysfunction due to the presence of platelet-associated immunoglobulin on the platelet surface [Karpatkin and Lackner, 1975]. This latter condition may result either in inappropriate granule release or decreased sensitivity to normal platelet agonists. Recent evidence also indicates that normal platelets may continuously secrete subcellular organelles as part of the "normal in vivo platelet aging process" [Kaplan et al., 1979]. This low-level secretion may be accelerated in pathologic disorders associated with suspected platelet activation such as stroke, myocardial infarction, or peripheral vascular disease [Doyle et al., 1980].

Current techniques to assess the content and function of these subcellular compartments are cumbersome and yield limited information. They require large numbers of cells and cannot be performed repeatedly in order to follow changes during a patient's clinical course or after in vitro manipulations. One purpose of this project is to explore the use of fluorescent probes to visualize these platelet subcellular compartments and to correlate these measurements with platelet function. An initial objective was to develop a system which could use small numbers of cells so the studies could be performed in severely thrombocytopenic subjects. In addition, the techniques should be rapid and require small blood samples to permit frequent repetition. The first probe to be used was mepacrine. It has been well demonstrated that this compound is selectively concentrated in platelet dense bodies [Lorez et al., 1977].

This present experiment was designed to answer the following questions.

1. Can platelets be labeled and visualized with the fluorescence-activated flow cytometer?
2. Is the mepacrine label specific for dense body detection?
3. Can an assay be designed to test platelet function?
4. Are these assays achievable in patient samples with severe thrombocytopenia?

## Methods

**Platelet isolation.** Platelets were isolated from whole blood anticoagulated with either 3.8% sodium citrate or a combination anticoagulant containing citrate (13 mM), ethylenediaminetetraacetic acid (EDTA, 1 mM), PGE-1 (1 μg/ml), and theophylline (1.8 mg/ml). This latter anticoagulant has been shown by our laboratory to prevent platelet activation through all of the described isolation procedures [Corash et al., 1977]. Purified whole blood platelets were isolated free of plasma proteins and all other cell classes by using discontinuous arabinogalactan gradients as previously described [Corash et al., 1977]. This technique requires about 60 min of preparative time and can yield adequate numbers of platelets from blood samples with platelet counts as low as 5,000/μl (normal range 150,000–350,000/μl).

**Mepacrine staining.** Total whole blood platelet populations suspended in plasma diluted with isosmolar phosphate-buffered saline containing glucose (5 mM) and citrate (13 mM) were incubated at 37°C for 30 min with freshly prepared mepacrine at final concentrations ranging from 1 μM to 100 μM. Postincubation aliquots of platelet suspension (100 μl) were layered over micro-arabinogalactan density gradients prepared in soft polypropylene microfuge tubes. The gradients were centrifuged for 5 min and the platelet band was cut out of the tube and dropped into isotonic counting fluid for volume and fluorescence analysis. This procedure concentrated the cells and removed unbound dye as well as plasma proteins.

**Size analysis.** Platelet volume distribution and mean volume were measured on a Particle Data, Inc., analyzer equipped with a PDP 11A computer and logarithmic amplifier as previously described [Corash et al., 1977].

**Fluorescence analysis.** Platelet dense body analysis was performed on the Becton-Dickinson FACS Research Analyzer (Becton-Dickinson, Mountain View, CA) equipped with a 50-μm orifice using a 435-nm excitation filter with a 10-nm band pass, a 462-nm first dichroic filter, and a 480-nm SP emission filter. Instrument settings were as follows: volume—current 0.5 mA, log amplification, gain +3 position, threshold +1 position; fluorescence—PMT 450 mV, log amplification, gain 4+ position, threshold off; RTO: FL-1/Vol. Lamp focus was performed with well sonicated 9.5-μm fluorescein alignment beads. Volume gates of 35–255 were used to reduce electrical noise and fluorescence gates of ten to 256 were used to reduce low-level nonspecific fluorescence.

**Platelet function analysis.** Isolated platelets were exposed to either gamma thrombin or bovine thrombin or the calcium ionophore A23187 before and after mepacrine staining. Thrombin was used over a dose range from 1.0–0.025 U/ml and A23187 over a range of from 25 to 0.8 μm.

**Data analysis.** The DIHSAN program supplied with the Hewlett-Packard (HP-85) computer was used to perform data analysis. For the purposes of this study three different parameters of platelet fluorescence were measured: (1) the shift in mean fluorescence due to mepacrine labeling was determined in comparison to that of control platelets labeled with buffer alone; (2) the shift in mean fluorescence corrected for volume differences using the ratio of fluorescence to volume (FL-1/Vol) was determined; and (3) the percentage of cells with fluorescence greater than that of unlabeled cells was determined after normalization for volume differences.

## Results

**Mepacrine staining of cells.** Unlabeled platelets exhibit autofluorescence under the present operating conditions. Flow rates were generally adjusted to be between 200 and 400 cells per sec and the analysis was terminated when 10,000 volume-triggered events were analyzed. This usually required 30–45 sec and negligible background fluorescence was observed during equivalent time periods sampling only diluting fluid. Thus, "platelet autofluorescence" did not appear to be nonspecific fluorescent noise but was true platelet autofluorescence. This observation was also supported by the finding that a series of band-pass filters did not eliminate the unlabeled cell fluorescence which would have occurred if the "autofluorescence" were due to backscatter. In addition, autofluorescence was dependent upon platelet volume. Larger platelets demonstrated a greater mean autofluorescence. Platelets labeled with 50 $\mu$M mepacrine and then removed from unbound dye demonstrated a significant shift in the fluorescence distribution without an accompanying volume change (Fig. 1). Under normal conditions platelet volume is symmetrically distributed in a log normal pattern with a mean volume of 6.6 $\pm$ 0.82 $\mu m^3$. Observation of this volume distribution during analysis is important to insure that the left end of the curve, containing the smallest cells, is adequately defined. Mean platelet volume is also an index of platelet activation. If the latter occurs, mean volume usually shifts to a larger size and there is a concomitant increase in autofluorescence.

Because unlabeled platelets exhibit autofluorescence, the measurement of platelet fluorescence after staining was initially expressed as the shift in mean

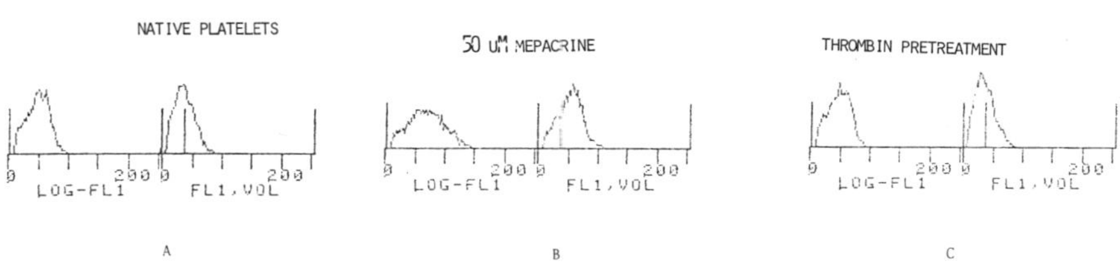


Fig. 1. The fluorescence distribution for platelets labeled with buffer (A) and for platelets labeled with 50 $\mu$M mepacrine (B) are shown as the logarithm of the fluorescence distribution (LOG-FL 1) and as the ratio of fluorescence to volume (FL1,VOL). Pretreatment of platelets with thrombin (C) prior to mepacrine exposure results in secretion of dense bodies and markedly reduced mepacrine labeling. The vertical marker represents the median of the native platelet fluorescence distribution for comparison in each panel.

**TABLE 1. Mepacrine-Induced Fluorescence of Normal Platelets**[1]

| Buffer | Mepacrine (50 $\mu$M) | Fluorescence shift |
|---|---|---|
| 48.1 ± 3.7 | 81.5 ± 5.9 | 33.4 ± 3.6 |

[1]Platelet fluorescence is expressed as the mean channel of the platelet fluorescence distribution (FL-1) after exposure to either buffer or mepacrine. The shift in platelet fluorescence is expressed as the increment in the mean channel of the platelet fluorescence distribution after exposure to mepacrine. These data represent the mean and standard deviation of nine experiments. Incubation with buffer caused no shift in mean fluorescence compared to platelets suspended in arabinogalactan media alone. These data are not normalized for volume differences.

fluorescence as compared to that of an unlabeled control sample. This provided a discrete measurement of the magnitude of platelet fluorescence. This measurement was made on platelets from nine normal subjects and demonstrated a consistent shift in mean fluorescence after mepacrine staining (Table 1).

A dose-response curve was performed with 1 $\mu$M, 50 $\mu$M, and 100 $\mu$M mepacrine; 50 $\mu$M mepacrine was found to be the lowest concentration to give maximal fluorescence. Red cell contamination of platelet suspensions resulted in reduced fluorescence; thus the samples must be substantially red cell free. This was achievable by using the described preparative techniques.

**Specificity of the mepacrine label.** Previous studies using fluorescence microscopy have demonstrated that mepacrine is selectively concentrated in platelet dense bodies in a punctate pattern. In order to determine that the label was in the dense body compartment and not nonspecifically in the cytoplasm, the following studies were performed. Platelets were treated with either A23187 (25 $\mu$M) or thrombin (1.0 U/ml) prior to the addition of mepacrine to cause secretion of dense bodies. This should result in depletion of mepacrine storage sites and there should be no platelet fluorescence if the label is dense body specific. This experiment is complicated by the observation that both A23187 and thrombin cause an increase in platelet volume and a nonspecific increase of platelet autofluorescence. This artifact can be controlled for by comparison with a treated unlabeled control sample or by the use of the ratio mode with fluorescence divided by volume (FL-1/Vol) in order to normalize for volume differences. Under these conditions, there is no evidence of platelet labeling after dense body secretion.

Treatment of previously labeled intact cells with either A23187 or thrombin also causes secretion of all the labeled storage sites and disappearance of all mepacrine-associated platelet fluorescence (Fig. 2). Use of the ratio mode to monitor dense body secretion is a convenient way to normalize for the

Fig. 2. The fluorescence distribution (LOG-FL1) and the ratio of fluorescence to volume (FL1,VOL) distribution are shown for platelets in the native state after labeling with 50 $\mu$M mepacrine (A) and after treatment with the calcium ionophore A23187 (B). The vertical marker indicates the median of native, unlabeled platelets as in Figure 1A. Note that A23187 causes a marked reduction in fluorescence indicative of dense body secretion.

volume-induced increased autofluorescence due to the agonist and it also normalizes for intra-individual differences in native platelet volumes.

**Correlation with previous studies.** Previous studies from this laboratory have shown that platelet subcellular content is density dependent [Corash et al., 1977]. High-density platelets have a greater number of dense bodies and alpha granules compared to low-density platelets [Corash et al., 1984]. Density cohort radiolabel studies from this laboratory and others now clearly demonstrate that differences in platelet density can be partially attributed to platelet aging. High-density platelets are young platelets which gradually decrease in density and organelle content during aging in the peripheral circulation [Corash et al., 1978; Rand et al., 1981]. Previous studies using whole mount electron micrographs have shown a density-dependent relationship for dense body content and endogeneous serotonin content, a major dense body component [Corash et al., 1977, 1984]. Mepacrine labeling of platelet-density-derived fractions demonstrates a parallel relationship (Table 2). Thus it appears that the relative dense body content of old and young cells can also be determined with this technique.

**TABLE 2. Mepacrine Labelling of Density-Dependent Platelet Subpopulations**

| Density fraction[1] | Fluorescence shift[2] |
|---|---|
| 1S | 5.6 |
| 2S | 39.5 |
| 3S | 49.4 |
| 4S | 54.7 |

[1]1S, low density, old cells; 4S, high density, young cells.
[2]Shift in mean fluorescence compared to unstained control cells, as in Table 1. Data are not normalized for volume differences.

**TABLE 3. Dose Response to A23187 in Two Subjects**

| Agonist dose ($\mu$M) | Mean fluorescence:volume |
|---|---|
| Subject No. 1 | |
| Buffer | 83.8 |
| 25 | 32.0 |
| 12.5 | 30.4 |
| 6.25 | 31.4 |
| 3.1 | 39.6 |
| Subject No. 2 | |
| Buffer | 89.0 |
| 6.25 | 63.5 |
| 3.1 | 71.5 |
| 1.6 | 80.7 |
| 0.8 | 87.8 |

Cells from two subjects were labeled with 50 $\mu$M mepacrine and stimulated with the agonist dose indicated. Fluorescence was measured as the ratio of fluorescence to volume (FL-1:VOL) and is expressed as the percentage of fluorescent cells compared to unlabeled, agonist-treated control cells. Thus maximum fluorescence occurs with a buffer agonist where 83.8 and 89% of the platelets respectively are fluorescent. Note that the dosage range varies for the two subjects.

Platelets prepared in the absence of fibrinogen can be stimulated by a variety of agonists to cause secretion. This would permit a functional evaluation of platelet secretory activity in addition to the quantitative assessment of the secretory granules. It may also be possible to construct a dose-response curve for individual subject samples to quantitatively determine hyperactive and hypoactive secretory states. To date no adequate tests exist to conveniently evaluate these possibilities of altered platelet function and such a test could be useful in patients with suspected increased platelet activation and associated thrombotic diseases.

We have used two platelet agonists, A23187 and thrombin, to demonstrate the feasibility of a dose-response curve secretory assay (Table 3). Platelets were first purified, then labeled with the fluorescent probe, and exposed to variable concentrations of the secretory agonist A23187. This agonist has two effects: shape change as measured by volume, and secretion as measured by the disappearance of the labeled secretory granules. Subject 1 appears more sensitive to the agonist than does subject 2, since the former undergoes complete secretion at 6.25 $\mu$M and 3.1 $\mu$M A23187 while the latter only secretes partially at this dose. Subject 3 was also more sensitive to thrombin than subject 4 (Table 4). The data in both of these experiments were expressed

**TABLE 4. Dose Response to Thrombin in Two Subjects**

| Agonist dose (U/ml) | Mean fluorescence:volume[1] |
|---|---|
| Subject No. 3 | |
| Buffer | 51.8 |
| 1.0 | 14.7 |
| 0.5 | 25.8 |
| 0.1 | 43.6 |
| 0.05 | 45.4 |
| Subject No. 4 | |
| Buffer | 57 |
| 1.0 | 23.6 |
| 0.1 | 54.9 |
| 0.05 | 52.6 |
| 0.025 | 53.8 |

[1]The change in the percentage of fluorescent cells normalized for volume in response to agonist is expressed as in Table 3. Note that maximum fluorescence occurs with the control buffer agonist.

as the percentage of fluorescent cells in comparison to a nonfluorescent control sample.

**Patient studies.** Two patients with acquired platelet dysfunction were studied. Patient BL had a malignant lymphoproliferative disorder (leukemic reticuloendotheliosis) characterized by splenic enlargement and mild thrombocytopenia (platelet count 120,000/$\mu$l; normal 150,000–350,000/$\mu$l). Patient BL was studied pre- and postsplenectomy with fluorescent probes and electromicrographs. This patient had a bleeding time test greater than 30 min (normal, less than 10 min) with only a modest reduction in absolute platelet count. The dissociation of quantitative and qualitative platelet properties suggested an acquired qualitative defect. Fluorescence analysis showed poor mepacrine labeling of the patient's platelets which improved dramatically after splenectomy (Table 5). The patient's hemostatic functional status also improved postoperatively with correction of the bleeding time to normal. Electron micrographs demonstrated that preoperatively few storage granules were present and that postoperatively platelet ultrastructure had returned to normal. The patient's qualitative disorder was felt to be due to platelet activation in his enlarged spleen. Based on the preoperative observations it was decided that platelet transfusion would not be useful since the infused platelets could also acquire the storage pool lesion and thus not improve

**TABLE 5. Patient BL: Acquired Storage Pool Disease**

| Status | Shift in Mean Fluorescence:Volume Ratio[1] |
|---|---|
| Patient preoperative, buffer | 0 |
| Patient preoperative, 50 μM mepacrine | 6.4 |
| Control, buffer | 0 |
| Control, 50 μM mepacrine | 29.6 |
| Patient postoperative, buffer | 0 |
| Patient postoperative, 50 μM mepacrine | 28.8 |
| Control, buffer | 0 |
| Control, 50 μM mepacrine | 27.7 |

[1]Shift in mean fluorescence expressed using the ratio of fluorescence to volume (FL-1/Vol) after buffer and mepacrine labeling.

hemostasis. Platelet transfusion was utilized only after the splenic circulation had been isolated intra-operatively.

A second patient, TB, was a young child with suspected malignancy and a markedly prolonged bleeding time (>30 min) but with normal platelet count. The patient also demonstrated massive splenic enlargement. The patient's platelets were hypoaggregable to standard platelet agonists with platelet aggregation testing, but this technique did not reveal the etiology of the platelet lesion. Mepacrine fluorescent staining analysis confirmed the absence of platelet dense bodies (Table 6).

**TABLE 6. Mepacrine Dense Body Staining in Patient TB With Suspected Qualitative Platelet Dysfunction**

| Sample | Shift in mean fluorescence:volume ratio[1] |
|---|---|
| Patient | |
| Platelets with buffer | 0 |
| Platelets with 50 μM mepacrine | 1.4 |
| Control | |
| Platelets plus buffer | 0 |
| Platelets plus 50 μM mepacrine | 34.1 |

[1]Shift in mean fluorescence expressed using the ratio of fluorescence to volume (FL-1/VOL) after buffer and mepacrine labeling.

## SURFACE-ASSOCIATED PROBES OF PLATELET FUNCTION

Thrombocytopenia may occur as the sole manifestation of disease or secondary to another disorder. Management of either primary or secondary thrombocytopenia will depend upon correct identification of the causal pathophysiologic processes. Over the past decade a number of research laboratories have provided data to strongly support the hypothesis that primary isolated thrombocytopenia in adults, idiopathic thrombocytopenia purpura (ITP), is immune mediated and in many cases may be more correctly termed auto-immune thrombocytopenia purpura (ATP) [Karpatkin, 1980]. The development of assays to directly measure platelet-associated immunoglobulin (PAIg) corroborated earlier studies that suggested that this disorder is due to the deposition of immunoglobulin (Ig) on the platelet surface leading to abnormal clearance of sensitized platelets. The ability to directly measure PAIg, as opposed to the earlier tests for the presence of serum antiplatelet factors, increased the sensitivity to detect patients with suspected platelet immune destruction. However, the increase in sensitivity also raised new questions about the specificity of PAIg [Kelton et al., 1982]. It is now well recognized that increased PAIg can be seen in thrombocytopenia associated with a wide variety of disorders, including systemic lupus erythematosis [Kelton et al., 1982], lymphoma [Kelton et al., 1982], carcinoma [Schwartz et al., 1982], Hodgkin's disease [Hassidim et al., 1979], acquired immuno deficiency syndrome (AIDS)[Morris et al., 1982], and Hashimoto's thyroiditis [Hymes et al., 1981]. Whether the PAIg observed in either ATP or these secondary disorders is specifically directed against a platelet antigen or is nonspecifically deposited on the platelet surface as part of an immune complex remains unknown in most cases. Several recent reports using immunoblot techniques have identified PAIg directed against integral platelet proteins: glycoprotein Ib [Woods et al., 1984a] and the glycoprotein IIb/IIIa complex [Woods et al., 1984b]. Other studies using sensitive techniques for the detection of cell-associated Ig have also provided renewed evidence that in patients with classical immune thrombocytopenia, the elevated PAIg is specific for immune-mediated platelet destruction and that it is not found in patients with nonimmune thrombocytopenia [LoBuglio et al., 1983]. The pathophysiology of immune-mediated platelet destruction is probably not the same in all disorders; for example, the AIDS-related thrombocytopenia may be due to immune complex disease [Walsh et al., 1984].

A number of different techniques to measure platelet-associated immunoglobulin have been described. Each of these techniques requires platelet isolation from whole blood which may be difficult for patients with severe thrombocytopenia. The ideal assay technique should be sensitive enough to allow the use of small numbers of platelets, be simple to perform in a routine laboratory, permit storage of samples so that assays can be performed in a batch mode, and yet yield data with sensitivity and specificity equivalent to that of the best technique available. The assay described in this report appears to meet many of these criteria in that it requires small numbers of cells, avoids the use of radioisotopes, can be performed in batch mode with sample storage, and utilizes instrumentation with which a multiplicity of different tests can be accomplished. It also exhibits adequate sensitivity and specificity for detection of PAIg.

Measurement of platelet-associated immunoglobulin is useful for both diagnostic and therapeutic purposes. It is indicated for patients who have isolated thrombocytopenia without an antecedent history of a hematologic disorder and in whom physical examination does not indicate the presence of any other systemic disorders. In the appropriate setting, patients with the classical ATP clinical presentation may not require a bone marrow examination if the PAIg test is positive. Measurement of PAIg is indicated for thrombocytopenia during pregnancy since maternal auto-immune thrombocytopenia may lead to neonatal thrombocytopenia if the antibody is IgG [Karpatkin, 1980]. Measurement of PAIg is indicated for patients with systemic lupus erythematosis who develop thrombocytopenia or qualitative platelet function defects during their disease course. Both of these situations may require specific diagnoses to determine alternative therapeutic approaches. The onset of thrombocytopenia which cannot be explained by decreased platelet production in patients with malignant lymphoproliferative disorders [Hassidim et al., 1979; Kelton et al., 1982] or carcinoma [Schwartz et al., 1982] is an indication for measurement of PAIg because it may direct a different therapeutic strategy. Patients with other auto-immune disorders, such as Hashiomoto's thyroiditis, may benefit from measurement of PAIg especially when symptoms of platelet dysfunction arise [Hymes et al., 1981]. Patients who have qualitative platelet dysfunction but a normal platelet count may require measurement of PAIg to document the etiology of their hemostatic disorder. More recently patients with thrombocytopenia during the course of AIDS and the AIDS-related disorders should have PAIg determined to document the presence of immune platelet destruction.

PAIg can also be measured by using an indirect test composed of the patient's serum and a normal donor target platelet. This technique may be useful to detect the presence of drug-dependent platelet antibodies in order to decide which pharmacologic agents are safe in patients with suspected drug-induced thrombocytopenia. The indirect assay for PAIg is also useful to detect maternal antibodies directed against fetal paternal platelet antigens in neonatal alloimmunization. A similar approach may also be used to select platelet donors for patients who are refractory to random donor platelet support.

## Methods

Cell-associated Ig is detected on the surface of intact, washed, fixed platelets isolated from the patient's blood. Normal platelets contain 100–200 molecules of IgG per platelet; this increases severalfold in patients with immune-mediated platelet destruction [LoBuglio et al., 1983]. Cell-surface Ig is detected by using a biotinated goat antihuman immunoglobulin antibody Fab-2 fragment directed against three heavy chain classes (G, M, and A). The Fab-2 fragment is used to prevent nonspecific sticking of the antihuman Ig Fc portion and the platelet Fc receptor. The detection system consists of avidin labeled with fluorescein isothiocyanate (A-FITC) and fluorescence is measured using a fluorescence-activated flow cytometer (FACS Analyzer, Becton-Dickinson Inc., Mountain View, CA). A paired negative control using biotinated goat antirabbit immunoglobulin Fab-2 fragments may be analyzed for each sample; however, based on our experience that this control was consistently negative it was eliminated from the test rountine. Appropriate positive and negative control samples are analyzed with each assay run and must meet specific criteria. The criteria for a positive result are based upon a large normal control population study and the specificity and sensitivity for the assay were determined by a prospective study of patients with clinically documented diagnoses of either immune or nonimmune thrombocytopenia.

This assay may also be used to detect serum antiplatelet antibodies by using an indirect test. Washed, fixed normal donor target platelets are preincubated with the patient test serum, washed free of unbound serum, and analyzed by the same procedure as the direct assay. Parallel control sera with and without known antiplatelet antibodies are analyzed simultaneously to serve as positive and negative controls.

**Specimen collection and platelet isolation.** Whole blood is collected into acid citrate dextrose (ACD) anticoagulant vacuum tubes (#4606, Vacutainer ACD, Vacutainer Systems, Rutherford, NJ). The specimen is stable at room temperature for 24 hours. If the patient's platelet count is greater than 50,000/$\mu$l, one tube (8.5 ml) is adequate. If the platelet count is below 50,000/$\mu$l then 17.0 ml of whole blood is drawn to insure adequate cell numbers. A normal control is drawn and processed in parallel to the patient sample.

Five milliliters of ACD blood is transferred to a polypropylene plastic tube (#2059 Falcon Plastics, Oxnard, CA) to which 2 ml of buffered saline glucose containing 13 mM sodium citrate and 10 mM EDTA (BSG-Citrate-EDTA) is added. The tube is mixed by gentle inversion and centrifuged at 600g for 3 min at 20°C. After centrifugation, the supernatant consisting of diluted platelet-rich plasma (PRP) is removed by using a plastic pipette without disturbing the red cell interface and the buffy coat, which is rich in lymphocytes. The diluted PRP is conserved in a second Falcon 2059 tube. The volume of the residual red cell pellet is restored to 7.0 ml and the tube is mixed and centrifuged again under the same conditions. The second supernatant is removed as before and pooled with the first supernatant. The procedure is repeated one more time and all the supernatant fractions are pooled.

The pooled PRP is centrifuged at 2,000g for 10 min at 20°C to produce a platelet pellet. The platelet-poor supernatant is discarded and the pellet is resuspended in 1.0 ml of 1% ammonium oxalate. A 1-ml Eppendorf pipette is used to gently resuspend the platelet pellet and then 4.0 ml of additional ammonium oxalate is added to the resuspended platelet. The platelets are incubated for 5 min at room temperature to lyse red cells and then pelleted at 2,000g for 10 min. The platelets are resuspended in 1.0 ml of 1% paraformaldehyde solution (PFA), dispersed by gently pipetting with an Eppendorf tip, and brought to a final volume of 3 ml with 1% PFA and incubated at room temperature for 5 min. The platelets are pelleted at 2,000g for 10 min; the supernatant is discarded; the pellet is dispersed in 1 ml Tyrodes buffer containing 10 mM EDTA and 0.5% bovine serum albumin (Tyrodes-EDTA-BSA) and brought to a final volume of 5 ml with the same buffer. The platelets are then pelleted and resuspended in Tyrodes-EDTA-BSA two more times under the same conditions as before. After the final centrifugation, the platelets are resuspended in 1 ml of Tyrodes-EDTA-BSA buffer and stored at 4°C until analyzed. PFA fixed, washed platelets are stable at 4°C for 30 days.

**Preparation of platelets for analysis in the direct test.** The platelet concentration of both the control and the patient sample is adjusted to 200,000 cells/μl with Tyrodes-EDTA-BSA buffer. Twenty-five microliters of platelet suspension is placed into two polystyrene 1-ml centrifuge tubes and 50 μl of goat antihuman Ig-Biotin diluted 1:20 in Tyrodes-EDTA-BSA is added. The dilution of this reagent may vary with different Ig lots. Each new lot is assayed in comparison to the previous lot and the dilution is adjusted based on the positive control to provide a consistent result. After mixing, the tubes are incubated on ice for 30 min and then 1.0 ml of Tyrodes-EDTA-BSA is added to each tube followed by centrifugation at 2,000g for 5 min. The supernatant fraction is discarded and the cells are washed twice more by using Tyrodes-EDTA-BSA. After the second wash 50 μl of Avidin-Fluorescein, diluted 1:20 in Tyrodes-EDTA-BSA, is added to the resuspended platelet pellet and incubated on ice for 30 min. This step is followed by three washes under the same conditions as the first label step. At the end of the final wash the platelets are resuspended in 1.0 ml of Isoton II (Coulter Electronics, Hialeah, FL) and transferred to a plastic tube. Platelet suspensions are stored on ice until analyzed. Samples must be analyzed on the same day they are labeled with fluorescein but previously washed fixed platelet suspensions are useable for 30 days.

**Preparation of control platelets for analysis.** The negative control sample consists of a platelet suspension prepared from a normal donor and processed in parallel to the patient sample. The positive control sample is composed of normal platelets which have been sensitized by incubation with serum containing multiple human leukocyte antigen (HLA) antibodies from patients who have received long-term platelet support. The same procedure as for the indirect test may be used. Positive control platelets are prepared in a batch mode and can be used for up to 5 months with stable results. This permits comparison of daily positive controls over a 1-month period to evaluate assay sensitivity.

**Operating conditions for the Becton Dickinson FACS Research Analyzer.** The PAIg assay is performed by using the 50-$\mu$m orifice. Instrument settings are as follows: volume, current 0.5, log mode, threshold minimum; fluorescence, PMT 600, log mode, threshold off. The alignment is adjusted with glutaraldehyde-fixed chicken red blood cells to obtain maximum sensitivity. The variable gain of the FL-1 channel is adjusted by using 2.0-$\mu$m-diameter beads (Duke Scientific Corp., Palo Alto, CA) so that the fluorescent mean lies within channels 43–47 and the fluorescent mean of the chicken red blood cells lies above channel 200. Volume gates are set at channel 20 and channel 255. This excludes debris below the normal platelet size distribution. A flow rate of 300–500 events per second is maintained and 10,000 events are analyzed in single-parameter mode. FL-1 markers are set at channel 0 and channel 255. The FL-1 mean is recorded and a sample with a mean channel which lies two standard deviations beyond the normal control mean (N = 100) is considered positive. One positive control and at least one negative control are run in duplicate with each assay run. The fluorescent mean of the negative control must lie between channels 58 and 71 and the fluorescent mean of the positive control must lie between channels 110 and 126. This may vary with different lots of reagents but has been constant in our laboratory for the past year. The volume distribution curve for each patient sample must be symmetrical and free of debris with a well-defined left side. If the controls or the volume distribution curve are not acceptable then the test must be repeated.

**Preparation of platelet samples in the indirect test.** Platelets are isolated and prepared as for the direct test from a normal subject with O blood type. The platelet count is adjusted to 200,000 per microliter by using Tyrodes-EDTA-BSA buffer. The target platelets (0.025 ml) are resuspended in 0.1 ml of test serum and incubated at room temperature for 30 min. After incubation, 1.0 ml of Tyrodes-EDTA-BSA buffer is added and the sample is centrifuged at 2,000g for 5 min. The supernatant is discarded. The platelets are washed in Tyrodes-EDTA-BSA one additional time. A negative control is run by using the serum sample from a pool of 20 AB untransfused male donors previously screened to rule out cytotoxic antibodies (Medical Specialty Labs, Boston, MA). The assay is performed as for the direct test.

**Preparation of samples for the indirect test to detect drug-dependent antibodies.** Target platelets are isolated as described above, fixed, and stored at 4°C until needed. The suspected drug must be obtained in soluble form and is diluted to a 2 mM concentration in normal saline. Patient and control sera are serially diluted by using either normal saline or normal saline containing 2 mM drug so that the final concentration of the drug is 1 mM. Target platelets are incubated with the patient and control sera containing either normal saline or normal saline plus drug and analyzed as for the indirect test. A positive result is defined by the same criteria as for the direct assay.

## Results

The diagnostic efficacy of this technique has been evaluated by performing a prospective study of 171 patients who had a laboratory finding of either thrombocytopenia or a clinical suspicion of a qualitative platelet defect based upon a prolonged template bleeding time. The clinical diagnosis was established for each patient independent of the determination of platelet-associated immunoglobulin. The clinical diagnosis for immune-related thrombocytopenia was based on bone marrow examination and subsequent clinical course in response to therapy. The diagnosis of qualitative platelet dysfunction was based on an abnormal template bleeding time and platelet aggregation studies. The results of this study are summarized in Table 7. This study indicates that the sensitivity and specificity of this test are both acceptable. A parallel study to evaluate the indirect test has not been performed as of this time.

It must be kept in mind that this test does not provide specific information as to the nature of the immunoglobulin deposited on the platelet surface. PAIg may represent antibody directed against specific platelet antigens [Woods et al., 1984a,b] or immune complexes nonspecifically absorbed to the platelet surface or an Fc receptor site [Walsh et al., 1984]. However, while the results of this technique are nonspecific, the technique does provide the possibility to analyze eluates [Stricker et al., 1984] obtained from positive platelet preparations by using the indirect test procedure in order to further define the specificity of the cell-associated immunoglobulin. Fab fragments can also be prepared from these platelet Ig eluates and then tested in the indirect assay to discriminate between Fab-dependent and Fc-dependent binding. Thus, this technique offers the potential to serve as a sensitive screening test and as a tool to subsequently define antibody specificity.

**TABLE 7. Sensitivity and Specificity of the FACS Assay for Detection of Platelet Associated Ig[1]**

| Patient population clinical diagnosis | PAIg positive | PAIg negative | Total |
|---|---|---|---|
| Clinically documented immune thrombocytopenia | 92 | 6 | 98 |
| Clinically documented nonimmune thrombocytopenia | 4 | 69 | 73 |
| Totals | 96 | 75 | 171 |

[1]Samples were submitted from patients with a positive history of thrombocytopenia or qualitative platelet dysfunction. Clinical diagnosis is based on bone marrow aspirate, clinical course, and response to therapy.
FACS, fluorescence-activated cell sorter; Ig, immunoglobulin; PAIg, platelet-associated immunoglobulin.
Sensitivity = 93.8%.
Specificity = 94.5%.
Prevalence = 57.3%.
Predictive value + = 95.8%.
Predictive value − = 92.0%.

## CONCLUSIONS

The preceding techniques demonstrate the application of fluorescent-activated flow cytometry to the study of platelet function. This approach offers the advantage of using small volumes of blood and the possibility of frequent repetition. The use of mepacrine as a specific subcellular probe is especially attractive for the diagnosis of a selected platelet lesion. Unfortunately at present there are no other selective probes for detecting alpha granules although preliminary studies from this laboratory suggest that acridine orange may be a useful probe for platelet lysosomes. The addition of new probes should serve to further expand the utility of this general approach to study platelet function.

The surface-associated probes are more fully developed at present and can be applied to intrinsic platelet membrane proteins such as the glycoproteins Ib, IIb, and IIIa. This may serve to facilitate the diagnosis of specific platelet disorders such as the Bernard Soulier syndrome or Glanzman's thrombasthenia. Fluorescent techniques to detect abnormal levels of platelet-associated immunoglobulin offer the advantage that the immunoglobulins can be eluted and modified by subsequent chemical procedures to assist in further identifying both the nature of the binding site and the ligand itself. This approach has been preliminarily demonstrated in the present review and current studies are underway [Stricker et al., 1984] to demonstrate that a combined use of fluorescent-activated flow cytometry and Western blotting can be used to define antigen binding sites.

## ACKNOWLEDGMENTS

The authors wish to acknowledge the cooperation of Dr. Mack Fulwyler from the Laboratory for Cell Analysis in the Department of Laboratory Medicine for his guidance and support in the performance of these studies. We also are appreciative to Eric Chase, Kit Snow, and Christopher English of Becton Dickinson Immunocytometry Systems, Mountain View, California, for instrumentation and reagent support. We wish to thank Elizabeth Brew for typing and editing the manuscript.

## REFERENCES

Bellucci S, Tobelein G, Caen J (1983): Inherited platelet disorders. Prog Hematol 13:223–264.

Corash L, Tan H, Gralnick H (1977): Heterogeneity of human whole blood platelet subpopulations. I. Relationship between buoyant density, cell volume, and ultrastructure. Blood 49:71.

Corash L, Shafer B, Perlow M (1978): Heterogeneity of human whole blood platelet subpopulations. II. Use of a subhuman primate model to analyze the relationship between density and platelet age. Blood 52:726.

Corash L, Costa J, Shafer B, Donlon J, Murphy D (1984): Heterogeneity of human whole blood platelet subpopulations. III. Density dependent differences in subcellular constituents and the relationship to platelet aging. Blood 64:185–193.

Doyle D, Chesterman C, Cade J, McCready J, Rennie G, Morgan J (1980): Plasma concentrations of platelet specific proteins correlated with platelet survival. Blood 55:82.

Hassidim K, McMillan R, Conjalka MS, Morrison J (1979): Immune thrombocytopenia purpura in Hodgkin's Disease. Am J Hematol 6:149–153.

Hymes K, Blum M, Lackner H, Karpatkin S (1981): Easy bruising, thrombocytopenia, and elevated platelet immunoglobulin G in Graves's disease and Hashimoto's thyroiditis. Ann Intern Med 94:27–30.

Kaplan K, Broekman M, Chernoff A, Lesznik G, Drillings M (1979): Platelet alpha-granule proteins: Studies on release and subcellular localization. Blood 53:604.

Karpatkin S (1980): Autoimmune thrombocytopenia purpura. Blood 56:329–343.

Karpatkin S, Lackner H (1975): Association of antiplatelet antibody with functional platelet disorders. Am Med 59:599.

Kelton JG, Powers PJ, Carter CJ (1982): A prospective study of the usefulness of the measurement of platelet associated IgG for the diagnosis of idiopathic thrombocytopenia purpura. Blood 60:1050–1053.

LoBuglio AF, Court WS, Vinocur L, Maglott G, Shaw GM (1983): Immune thrombocytopenia purpura. Use of a 125-I-labeled antihuman IgG monoclonal antibody to quantify platelet-bound IgG. N Engl J Med 309:459–463.

Lorez H, DaPrada M, Rendu F, Pletscher A (1977): Mepacrine, a tool for investigating the 5-hydroxytryptonine organelles of blood platelets by fluorescence microscopy. J Lab Clin Med 89:200.

Malpass T, Harker L (1980): Acquired disorders of platelet function. Semin Hematol 17:242.
Morris L, Distenfeld A, Amorosi E, Karpatkin S (1982): Autoimmune thrombocytopenic purpura in homosexual men. Ann Intern Med 96:714–717.
Rand M, Greenberg J, Packman M, Mustard J (1981): Density subpopulations of rabbit platelets: Size, protein and sialic acid content, and specific radioactivity changes following labelling with 35-sulphate in vivo. Blood 57:741.
Rosove M, Nalin F, Hawig S, Zighelboim J (1980): Severe platelet dysfunction in hairy cell leukemia with improvement after splenectomy. Blood 55:903.
Schwartz KA, Slichter SJ, Harker LA (1982): Immune mediated platelet destruction and thrombocytopenia in patients with solid tumors. Br J Haematol 51:17–24.
Stricker RB, Abrams DI, Corash L, Shuman MA (1984): Target platelet antigen in AIDS-related immune thrombocytopenia. Blood 64:241a.
Walsh CM, Nardi MA, Karpatkin S (1984): On the mechanism of thrombocytopenic purpura in sexually active homosexual men. N Engl J Med 311:635–639.
Weiss H, Witte L, Kaplan K, Lages B, Chernoff A, Nossel H, Goodman D, Baumgartner (1979): Heterogeneity in storage pool deficiency: Studies on granule bound substances in 18 patients including variants deficient in alpha-granules, platelet factor 4, beta-thromboglobulin and platelet derived growth factor. Blood 54:1296.
Weiss HJ (1982): Biological properties of platelets. In Weiss HJ (ed): "Platelets: Pathophysiology and Antiplatelet Drug Therapy." New York: Alan R Liss, pp 29–43.
Woods VL, Kurata Y, Montgomery RR, Tani P, Mason D, Oh EH, McMillan R (1984a): Autoantibodies against platelet glycoprotein Ib in patients with chronic immune thrombocytopenia purpura. Blood 64:156–160.
Woods VL, Oh EH, Mason D, McMillan R (1984b): Autoantibodies against the platelet glycoprotein IIb/IIIa complex in patients with chronic ITP. Blood 63:368–375.

*Applications of Fluorescence in the Biomedical Sciences, pages 585–597*


# 29 Characterization of Cell Populations in Tumors by Multiparameter Flow Cytometry

*Carleton C. Stewart, Anita P. Stevenson, and John A. Steinkamp*

Our understanding of the response of lymphocytes and macrophages to neoplasia has largely come from studies using cells from peripheral tissue like the spleen [Gorelik et al., 1981; Gerson et al., 1980] or the peritoneum [Dingermens et al., 1983; Hammerstrom, 1980] and not from cells derived from the tumor. This has largely been because of the ease with which peripheral cells can be obtained compared to their isolation in relatively pure form from the neoplastic lesion itself. The demonstrated tumoricidal activity of these peripheral cells has been taken as evidence that the ones within the tumor itself must also be tumoricidal. Studies using tumor-derived cells have confirmed the presence of host cells in tumors [Luder et al., 1977; Russell et al., 1976; Evans, 1979; Stewart and Beetham, 1978; Eremin et al., 1982]. These populations have been enriched by various methods and their tumoricidal activity was established in a few instances [Key and Haskill, 1981; Stewart and Beetham, 1978; Russell and McIntosh, 1977; Russell et al., 1977].

Flow cytometry in conjunction with monoclonal antibodies to cell differentiation antigens provide us a means by which a mixture of normal and neoplastic cells can be explicitly resolved and sorted in high purity for functional analysis. In this study, we describe our experience in resolving lymphocytes, macrophages, and tumor cells from tumors.

Experimental Pathology Group, M888, Los Alamos National Laboratory, Los Alamos, New Mexico 87545.

## METHODS

Cell size was measured by using either 90° light scatter or electronic cell volume. Simultaneous dual-laser analysis of blue fluorescence intensity by Hoechst 33342 staining (proportional to DNA content) and green fluorescence intensity of fluorescein-labeled antibodies to the MAC1 antigen (found on macrophages and granulocytes), Thy 1.2 antigen (found on T-lymphocytes and macrophages), or surface immunoglobulin (found on B-lymphocytes and macrophages) was performed. In addition, fluorescence emission greater than 610 nm [corresponding to propidium iodide (PI)-stained nonviable cells] was also recorded. Data was acquired in correlated list mode and stored by using a PDP-11/23 computer [Hiebert et al., 1981; Salzman et al., 1981].

In order to demonstrate the ability of the technique to resolve heterogeneous populations of cells which might be present in a tumor, we analyzed spleen cells, peritoneal cells, and EMT6 tumor cells both individually and mixed in varying but known proportions. Spleens were removed from BALB/c mice (6–12 weeks old, Jackson Laboratory, Bar Harbor, ME) and cells were obtained by extruding them through a 380-$\mu$m stainless mesh with the rubber plunger of a 1 ml syringe. Peritoneal exudate cells (PEC) from BALB/c mice injected with 1.5 ml Brewer's thioglycollate medium 3 days prior to harvest were collected after i.p. injection of 10 ml $\alpha$MEM prewarmed to 37°C and containing 5 units of heparin per ml. These procedures have been previously described in detail [Stewart, 1985]. Stocks of attached EMT6 cells (a gift from Dr. Sara Rockwell) were maintained in 75-cm$^2$ flasks (Falcon) in culture medium [$\alpha$MEM supplemented with 10% fetal calf serum (FCS)]. Cells were maintained free of contamination with *Mycoplasma*. EMT6 cells were removed by incubation for 10 min with 0.125% trypsin containing 0.5 mM EDTA.

Progressing EMT6 tumors implanted into BALB/c mice 7–10 days earlier were removed by dissection. Tumors were minced and suspended in a dispersal solution. The solution consisted of 25 ml saline G (composition in grams/liter: glucose 1.1 g, NaCl 8.0 g, KCl 0.4 g, $Na_2HPO_4 \cdot 7H_2O$ 0.29 g, $KH_2PO_4$ 0.15 g, Phenol red 0.005 g, $MgSO_4 \cdot 7H_2O$ 0.154 g, $CaCl_2 \cdot 2H_2O$ 0.016 g) containing 50 mg collagenase (Worthington Biochemicals, Freehold, NJ) and 500 mg bovine serum albumin (Fraction V, Calbiochem, La Jolla, CA). After incubation for 2 hr at 37°C with constant stirring, the free cells were passed through a 62-$\mu$m nylon filter to remove aggregates and undigested tissue. The yield (based on several hundred tumors) was $2.08 \pm 1.15 \times 10^8$ cells/g with a viability in the order of 80%.

The cell concentration of the various suspensions was determined by using an electronic particle counter (Coulter Electronics, Hialeah, FL). After centrifugation of the cell suspensions at 200g for 10 min, the pellets were resuspended in culture medium and adjusted to $2 \times 10^7$ cells per ml in culture medium containing a final concentration of 10 $\mu$g/ml of Hoechst 33342. This dye can be used to stain the DNA while maintaining good cell viability. Cells were incubated for 45 min at 37°C in a shaker water bath.

The suspensions were then chilled on ice and 50-$\mu$l aliquots were distributed into $10 \times 75$ mm polyethylene tubes and incubated for 15 min with (1) 3 $\mu$l fluorescein isothiocyanate (FITC)-conjugated $F(ab)'_2$ fragment of a goat antibody to mouse heavy and light chains (FGAM) or (2) 10 $\mu$l of the MAC 1 or (3) 10 $\mu$l of the Thy 1.2 monoclonal antibodies (MAbs). Cells were washed in 4 ml phosphate-buffered saline with 0.1% sodium azide and 0.5% bovine albumin (PAB). For the cells that had been incubated with MAC-1, the cells were incubated an additional 15 min with 3 $\mu$l of a fluoresceinated goat antibody to rat Ig. For the cells that had been incubated with Thy 1-biotin, the cells were incubated an additional 15 min with 10 $\mu$l of fluoresceinated Avidin. At the end of the incubation each suspension was diluted by adding 2 ml PAB containing 2 $\mu$g/ml Hoechst dye. Cell suspensions were filtered through a 72-$\mu$m nylon filter and kept on ice until analysis. Propidium iodide was added to each sample (final concentration 2 $\mu$g/ml) at least 3 min before analysis as an indicator of nonviable cells. The cell suspensions were analyzed by using a flow cytometer as individual suspensions or as mixtures in known ratios.

## RESULTS

### Unmixed Cells

In Figure 1, the flow cytometric distributions for unmixed spleen cells, peritoneal exudate cells, and cultured EMT6 tumor cells are shown. The cell size distribution using 90° light scatter is shown in column 1, the blue fluorescence corresponding to DNA is shown in column 2, and the log of green fluorescence is shown in column 3 for the binding of the FGAM. (We will illustrate the use of electronic cell volume for cell size discrimination in a later example.) While not shown, dead cells stained with propidium iodide have been excluded from the analysis.

A single broad size distribution is shown for spleen cells. This distribution does not include erythrocytes because only cells with blue fluorescence (DNA) were analyzed. Three DNA distributions are resolvable: One distribution with a peak in channel 32 (<2c), a second in channel 48 (2c), and a very minor population in channel 96 (4c). The latter population represents a small proportion of aggregated cells. In the last column, a negatively stained population and a broad distribution of positively stained cell can be resolved by using the fluoresceinated goat antimouse Ig reagent.

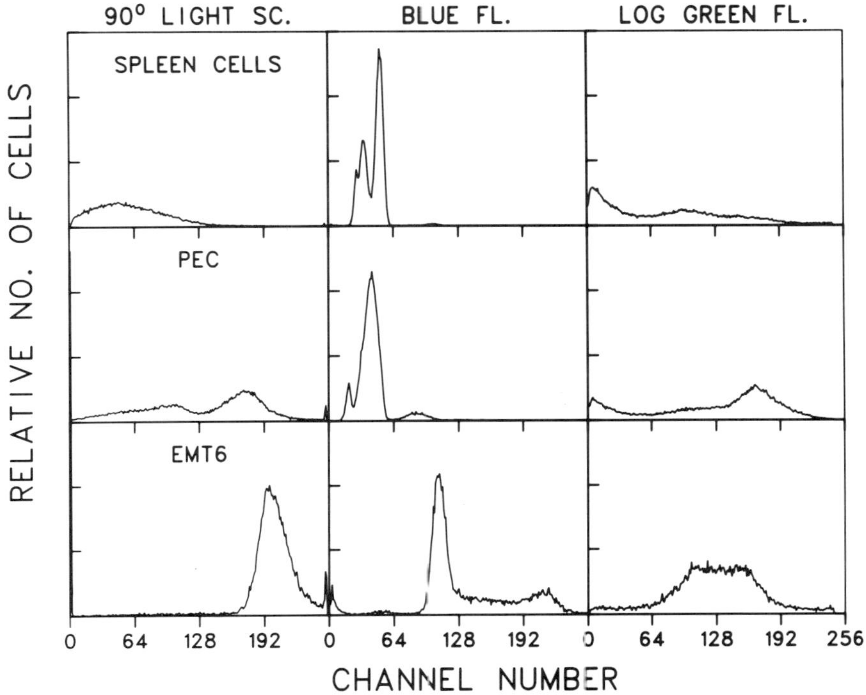


Fig. 1. Unmixed cells. Spleen, peritoneal exudate cells (PEC) and tumor cells were each stained with Hoechst 33342 and FGAM and analyzed by using a dual beam flow cytometer. Dead cells containing propidium iodide have been removed by computer reprocessing.

We investigated the nature of the two major DNA distributions found for the spleen cells by reprocessing the correlated list mode data on each peak and determining where the Ig-bearing B cells were located. As shown in Figure 2, the PI-stained dead cells are shown in the last channel of the log green fluorescence distribution. After reprocessing the data, the dead cells have been eliminated (Fig. 2B–D). As shown in Figure 2C, all of the cells with the lesser DNA content have no surface Ig and are the T cells. Some unlabeled cells, however, are also found in the higher DNA content distribution (Fig. 2D). These cells are presumed to be the "activated" T cells as suggested by Loken, who previously reported that the stoichemetry of Hoechst binding by T cells is different from that of B cells [Loken, 1980].

Returning to Figure 1 the second row contains the distributions of peritoneal exudate cells. The size distribution suggests the presence of three cell populations. The first population contains lymphocytes of the size found in

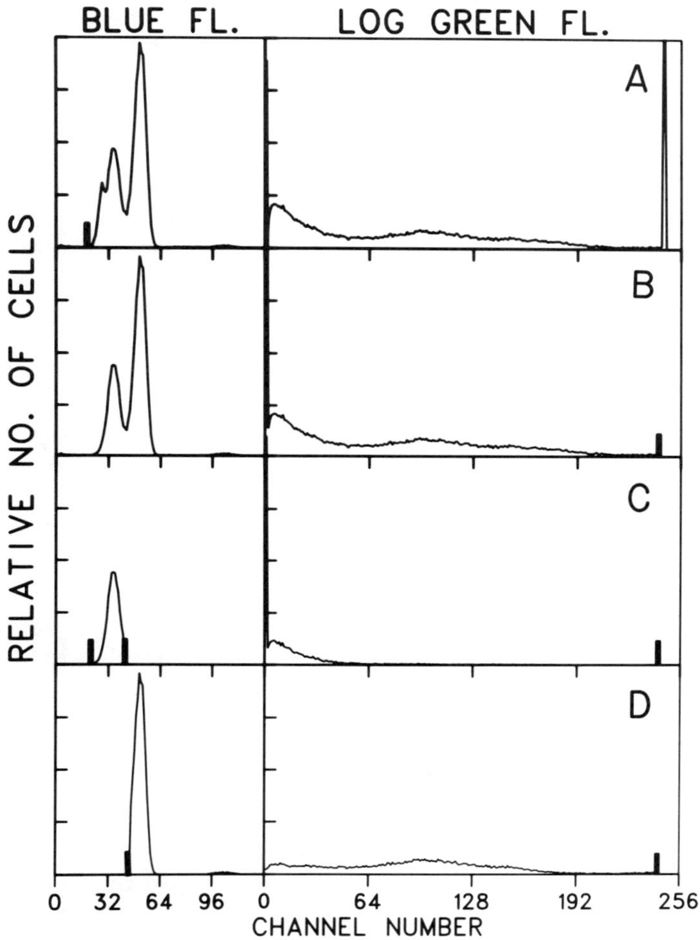


Fig. 2. DNA distribution of spleen cells. The spleen cell distributions shown in Figure 1 were further reprocessed. The Hoechst distribution is shown in the first column and the distribution of cells labeled with FGAM in the second column. The second row shows the resulting distribution after reprocessing on viable cells. The third and fourth rows show the DNA distributions of lower and higher DNA content cells respectively.

the spleen, the second population is made up of small macrophages, while the third population is comprised of large macrophages. The DNA distribution shown in column 2 is broad with the peak in channel 40. The distribution whose peak is in channel 80 represents aggregates. In the last column, three overlapping populations of surface immunoglobulin (sIg)-positive cells are found. When these data were reprocessed (channels 0–64) to show only unstained cells, these unstained cells were found to be both small and large in size and exhibited both <2c and 2c DNA contents consistent with there being T cells and macrophages in this population (distributions not shown). The intermediate stained population (channels 64–128) of cells contained all small cells, and they exhibited a normal DNA content consistent with their being B cells. The brightest cells (channels 128–200) were all large cells and represented macrophages with membrane-bound cytophilic antibody.

The final row in Figure 1 shows the distribution for cultured EMT6 tumor cells. These cells are shown as a homogeneous population of large cells when 90° light scatter is employed. The distribution of tumor cells overlaps the largest cells found in the peritoneal exudate suspension. It is possible to resolve the tumor cells from the large cells found in peritoneal exudates when DNA content is also utilized because the tumor cells have twice the DNA content (4c) of the normal cells. The log green fluorescence of tumor cells was the same whether the cells were labeled with anti-Ig or not (data not shown). Thus, the distribution represents autofluorescence of the cultured tumor cells.

This analysis has allowed us to establish windows for the resolution of each cell type. The rationale for selecting the gating parameters was as follows: Lymphocytes are small cells (90° light scatter channels 10–70) with a <2c or 2c DNA content (blue fluorescence channels 20–60). T cells can be resolved from B cells by whether they have surface Ig (log green fluorescence channels greater than 60). Peritoneal lymphocytes are resolved in the same way, but the macrophages are resolved from tumor cells based on both size and DNA content (90° light scatter channels 71–255) and blue fluorescence (channels 20–60). The autofluorescent tumor cells are resolved from host defense cells on the basis of cell size (90° light scatter channels 150–255) and blue fluorescence (channels 90–255).

## Mixture of Cells

In order to determine the efficacy of resolving lymphocytes, macrophages, and tumor cells, known proportions of spleen cells, peritoneal exudate cells, and EMT6 tumor cells were mixed together, analyzed, and each individual population was resolved by using the windows determined above. The histograms of the mixture and each resolved population are shown in Figure 3. In row 1, the mixture is shown prior to reprocessing the list mode data. In the second row, lymphocytes are shown. In row 3, the negative and Ig-positive macrophages are clearly resolved based on their log green fluorescence. Ig-positive B cells can be resolved based on their log green fluorescence. The autofluorescent EMT6 cells are shown in row 4. Each of the individual

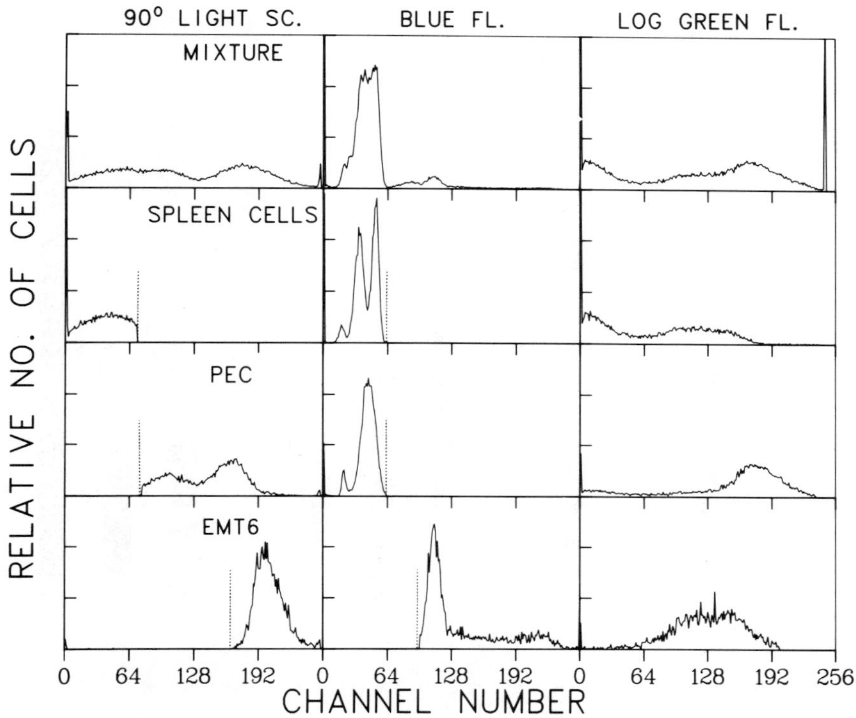


Fig. 3. Resolution of mixed cells. Cells were mixed together so that 90% were host cells and 10% were tumor cells. Row 1 shows the mixture prior to reprocessing. Row 2 shows the lymphocytes, row 3 the macrophages, and row 4 the tumor cells. For each row, the full scale has been adjusted according to the height of the DNA distribution, for illustration purposes, and the areas are not proportional to the number of each cell type.

populations in the mixture were resolved explicitly (compare to Fig. 1 where each individual population was analyzed prior to mixing).

Three separate mixtures of spleen cells, peritoneal cells, and EMT6 cells were analyzed by flow cytometry, and the comparisons of the known percentage to the percentage of each population as determined by flow cytometry are shown in Figure 4. Since lymphocytes were present in the peritoneal exudate and flow cytometry distinguishes them regardless of source, their contribution was added to the known percentage of the cells in the mixture that were spleen lymphocytes. For the different mixtures, flow cytometry was able to resolve each of the populations consistent with the actual amounts mixed together.

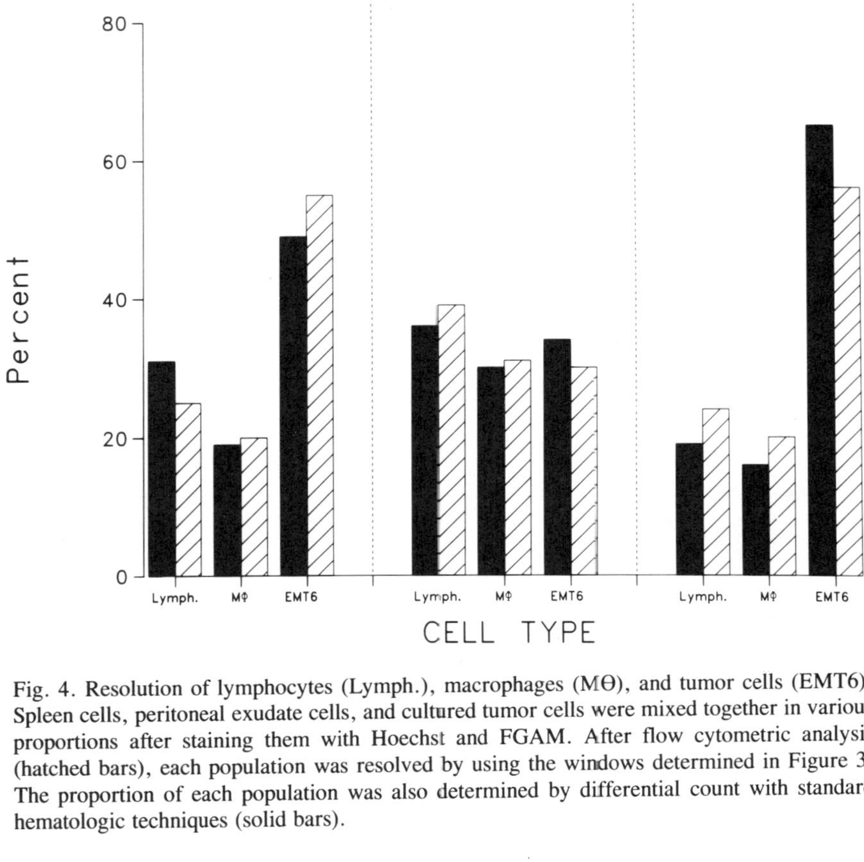


Fig. 4. Resolution of lymphocytes (Lymph.), macrophages (MΘ), and tumor cells (EMT6). Spleen cells, peritoneal exudate cells, and cultured tumor cells were mixed together in various proportions after staining them with Hoechst and FGAM. After flow cytometric analysis (hatched bars), each population was resolved by using the windows determined in Figure 3. The proportion of each population was also determined by differential count with standard hematologic techniques (solid bars).

## The Real Tumor

The next task was to determine whether this procedure could be applied to resolving host cells in a progressing tumor. For this study, electronic cell (Coulter principle) volume was used as the cell size parameter. A typical histogram is shown in Figure 5. Unlike 90° light scatter, where all the cell populations are found within the region of display, the largest cells are off scale and accumulate in channel 255 when Coulter volume is used. While blue fluorescence was in coincidence with Coulter volume to reduce the analysis of debris, there is a considerable fraction of dead cells (propidium-iodide-stained dead cells shown in the log red fluorescence channel). The Hoechst in the dead cells is quenched by the PI, causing what appears to be a poor blue fluorescence distribution. As shown in Figure 6, after reprocessing the data to exclude dead cells and to separate host and tumor cells based

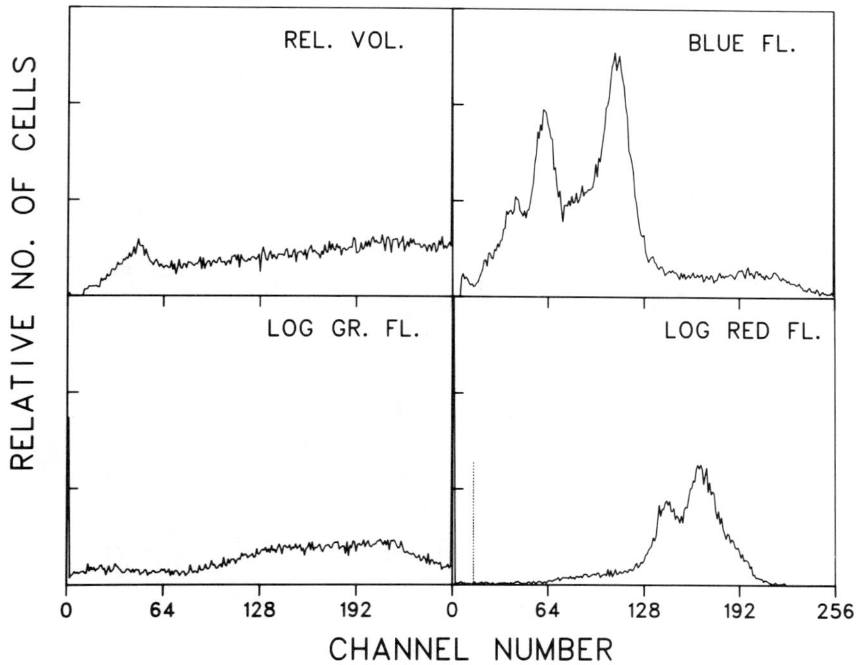


Fig. 5. Distribution of cells from a growing tumor. The first row shows the histograms for electronic cell volume, and Hoechst fluorescence, the second row shows green fluorescence and PI fluorescence from dead cells.

on DNA content, the DNA distributions are clearly improved and the autofluorescent tumor cells are resolved from the unlabeled host cells.

Aliquots of the tumor were labeled with the anti-Ig antibody, or the MAC1 or Thy 1.2 monoclonal antibodies. After resolving the host cells by reprocessing the data on the DNA content of viable cells, the volume and DNA distribution of the labeled cells are shown in Figure 7. The results of analyzing several different individual tumors ranging in weight from 0.1 to 0.5 g are summarized in Table 1.

Sorting the small cells in the volume region represented by channels 20–60 showed them to be mostly lymphocytes and those in the region 70–255 were mostly macrophages. Both sorts, however, contained granulocytes. The percentage of labeled cells was determined in each volume region by reprocessing the data in the two volume regions. Cells in the lymphocyte volume region represented 42% of all the cells. As shown in Table 1 only 22–32%

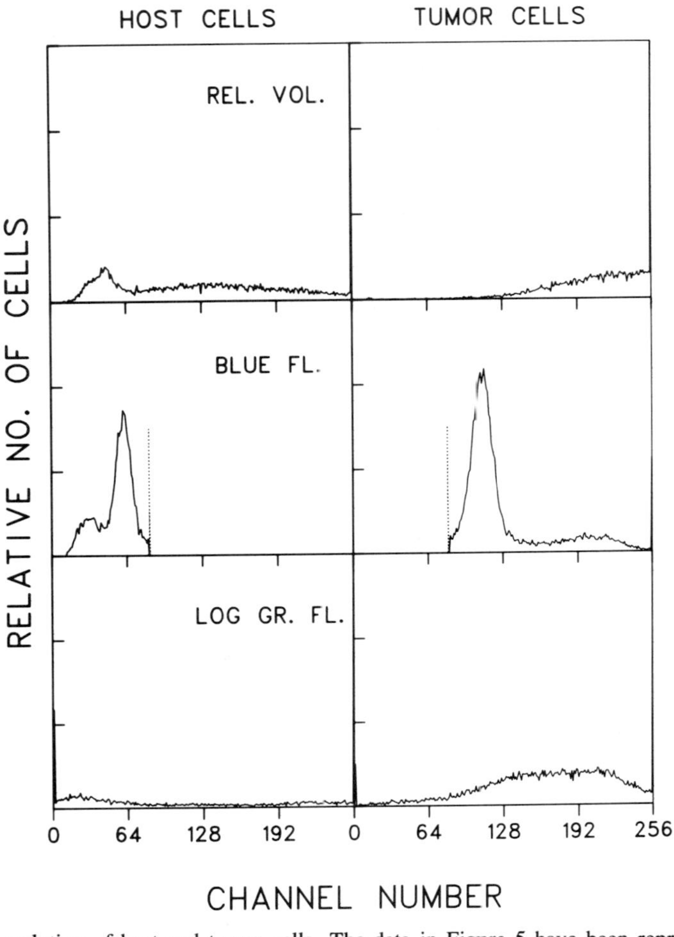


Fig. 6. Resolution of host and tumor cells. The data in Figure 5 have been reprocessed to eliminate dead cells. Host cells have been resolved from tumor cells based on their difference in DNA content in the blue fluorescence histograms.

of the cells in the lymphocyte region labeled with an antibody. In contrast, 70–87% of cells in the macrophage region were labeled. It still needs to be determined whether any of the cells labeled in the lymphocyte region are macrophages or granulocytes and visa versa by actually sorting the labeled cells.

## DISCUSSION

In this study we were able to demonstrate that flow cytometry can be used to resolve lymphocytes, macrophages, and tumor cells with good precision.

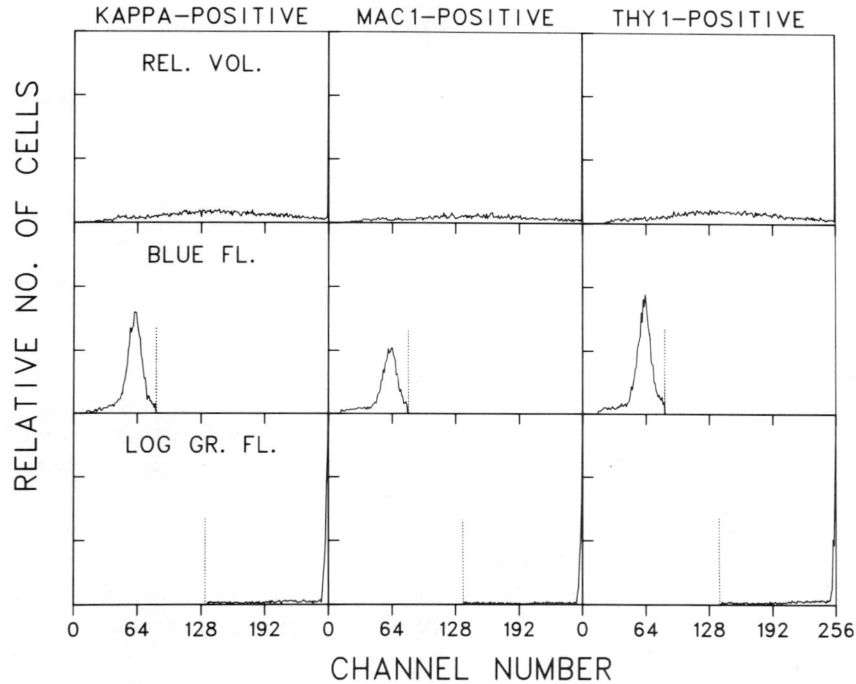


Fig. 7. Subsets of host cells in progressing tumors. Cells from tumors were labeled with three antibodies to identify the various major host leukocytes. After resolving host cells as shown in Figure 6, the electronic volume of the labeled cells was obtaining by using the reprocessing windows shown in the last row.

The findings, however, show that the three antibodies selected do not allow for the explicit identification of T-cells, B-cells, and macrophages and a more complex system for their identification needs to be developed. The procedures, however, represent a rapid means for identification and sorting of host cells from tumors. Other antibodies such as Lyt 1 and Lyt 2 for helper and suppressor/cytotoxic T lymphocytes, respectively, could also be used.

The dispersal technique provides for a high recovery of viable cells without a loss of important surface markers. Dispersion using trypsin and other proteases not only provided a much lower yield of cells but also increases the possibility that surface epitopes and Ig could also be removed. Tumors often contain massive amounts of debris, erythrocytes, and dead cells; in this study these were eliminated by excluding PI-stained cells and by requiring coincidence between size and DNA content. Therefore only viable nucleated cells were analyzed. We found 90° light scatter or electronic

**TABLE 1. Host Cells Infiltrating Tumors**

| | Percent positive cells[1] | |
|---|---|---|
| Label | Lymphocyte region | Macrophage region |
| FGAM | 22 | 83 |
| Thy 1.2 | 32 | 87 |
| MAC-1 | 31 | 72 |

[1]42 ± 3.5% of all the cells were in the lymphocyte region.

volume was better than forward angle light scatter for resolving lymphocytes, macrophages, and tumor cells.

In this study, several insights have been acquired. Few B cells were found in the tumors as nearly all Ig-bearing cells were macrophages. Similarly, two-thirds of the lymphocytelike cells were Thy 1.2 negative; they might be NK cells. The nature of cytophilic antibody on macrophages in the tumor is not known but may be directed against the tumor cells. Most of the macrophages within the tumor also bound Thy 1.2. The existence of Thy 1.2 on macrophages has been previously reported. Double label studies using antibody markers with two different color fluorochromes simultaneously in conjunction with cell sorting will be necessary to explicitly identify these cell populations.

The methods and results reported here suggest that host defense cell content of tumors can be measured rapidly and easily by using flow cytometry. The identification and function of resolved cells can be determined after sorting. Tumors can be analyzed for changes that might occur in the host and tumor cell populations as a result of various therapeutic regimens. Since the procedure is rapid, individual tumors can be analyzed so variation in response among mice can be established. Since the tumor itself represents the most important area for obtaining cells to determine their antitumor activity, application of these procedures should add to our knowledge of the host defense to cancer.

## ACKNOWLEDGMENTS

This work was performed under the auspices of the Department of Energy and the National Flow Cytometry and Sorting Research Resource (grant RR01315 from the National Institutes of Health) and it was supported by grant #CA 33593 from National Institutes of Health, DHS. We wish to thank Ruby Archuleta for typing the manuscript.

## REFERENCES

Dingermans KP, Pels E, van Dongen G, den Otter W (1983): Destruction of murine lymphoma cells by allogeneic peritoneal macrophages *in vitro:* Influence of antiserum. JNCI 70:181.

Eremin D, Coombs RA, Prospero TD, Plumb D (1982): Lymphocyte and B-lymphocyte subpopulations infiltrating human mammary carcinomas. JNCI 69:1.

Evans R (1979): Host cells in transplanted murine tumors and their possible relevance to tumor growth. J Res 26:427.

Gerson JM, Holden HT, Heberman RB (1980): Systemic and in situ natural activity in C57BL/6N and CBA/N mice bearing murine sarcoma virus-induced regressor tumors. JNCI 65:905.

Gorelik E, Kedar E, Sredani B, Herberman R (1981): *In Vivo* anti-tumor effects of local adaptive transfer of mouse and human cultured lymphocytes. Int J Cancer 28:157.

Hammerstrom J (1980): Structure and function of human effusion macrophages from patients with malignant and benign disease. 2. *In Vitro* cytostatic and cytolytic effect on human tumor cell lines. Acta Pathol Microbiol Scand Section C 88:201.

Hiebert RD, Jett JH, Salzman GC (1981): Modular electronics for flow cytometry and sorting: The LACEL System. Cytometry 1:337–341.

Key M, Haskill S (1981): Macrophage-mediated, antibody-dependent destruction of tumor cells in DNA/s mice: In vitro identification of an in situ mechanism. JNCI 66:103.

Loken MR (1980): Separation of viable T and B lymphocytes using a cytochemical stain. Hoechst 33342 28:36–39.

Luder L, Sherne W, Stewart J, Sainsbury R (1977): Macrophage infiltration of breast tumors, a prospective study. J Clin Pathol 30:563.

Russell SW, McIntosh AT (1977): Macrophages isolated from regressing moloney sarcomas are more cytotoxic than those recovered from progressing sarcomas. Nature 268:69.

Russell SW, Doe WF, Hoskin RG, Cochrine CG (1976): Inflammatory cells in solid murine neoplasms. I. Tumor disaggregation and identification of constituent inflammatory cells. Int J Cancer 18:322.

Russell SW, Gillespie GY, McIntosh AT (1977): Inflammatory cells in solid murine neoplasms. III. Cytotoxicity in vitro by macrophages recovered from disaggregated regressing moloney sarcomas. J Immunol 118:1574.

Salzman GC, Wilkins SF, Whitfill JA (1981): Modular computer programs for flow cytometry and sorting: The LACEL system. Cytometry 1:325–336.

Stewart CC (1981): Methods for studying the ontogeny of mononuclear phagocytes. Hndb Exp Immunol (4th Ed) 46:252–266.

Stewart CC, Beetham KL (1978): Cytocidal activity and proliferative ability of macrophages infiltrating the EMT6 tumor. Int J Cancer 22:152.

# PART IV
# Prospectus

*Applications of Fluorescence in the Biomedical Sciences, pages 601–615*


# 30 Solution Spectroscopy and Image Spectroscopy

*Gregorio Weber*

Fluorescence spectroscopy has developed in rapid fashion during the last 30 years and the most remarkable developments have been those related to the purely methodological aspects [1]: For emissions in the ultraviolet and visible regions of the spectrum we can count reliably and efficiently the number of emitted photons, thus closely approaching the theoretical limit in sensitivity of detection, and the numerical manipulation of the data can be accomplished in many cases in parallel with the acquisition. We are thus able to collect and readily digest all the information available in the fluorescence emission.

The application of the methods at hand ought to lead in the near future to a rapid increase in the knowledge of the molecular properties that determine the character of the fluorescence emission but at present there are still serious deficiencies in this aspect of optical spectroscopy. The biologist that wishes to apply the methods of fluorescence spectroscopy as a tool in his own investigations soon finds that in the majority of cases the best way to proceed is to use the data to exhibit and measure small differences between cases and that only much less often can one use the fluorescent properties as an exact indicator of the molecular properties of the biological material. In this task of exhibiting and measuring small differences we posess no physical methods that can compare with fluorescence emission: For all practical purposes the sensitivity of detection is not limited by the ratio of signal to noise but rather by the presence, virtually unavoidable, of fluorescent contaminants present in solvents and chemicals, even of those of highest purity, in dust particles, in the secretions from human skin, etc.

The fluorescence spectroscopist that wishes to carry out a complete analysis of the emission must examine the fluorescence excitation spectrum, the fluorescence emission spectrum, the quantum yield, the fluorescence lifetime, and the state of polarization of the emission. This is a formidable list

Department of Biochemistry, School of Chemical Sciences, University of Illinois, Urbana, Illinois 61801.

of options which is seldom—if ever—systematically followed in investigations of the fluorescence of solutions. The information obtained through these measurements, even when they fall short of the highest precision, greatly exceeds in many cases what is required for a decision among competing interpretations, or what can be usefully employed in practice. The spectrum of fluorophores in solution, particularly in water solution, seldom exhibits fine structure. The majority of such spectra can be described by giving the parameters of a skewed Gaussian [2] and a few measurements of intensity either through filters or through a monochromator with bandwidths of 10–20 nm are sufficient for their definition. Nevertheless when the determination of the fluorescence spectrum is carried out to establish small differences in the average energy of the emission, for example to exhibit variations in the strength of electronic interactions between fluorophore and surroundings, an emission spectrum with a resolution of 1–3 nm may be required. These considerations are particularly important when applying the methods of the fluorescence spectroscopy of solutions to the study of fluorescent images. In this latter case space, time, and spectral resolutions have to be considered. We must be mindful that the demands of spatial resolution and photochemical stability will often conflict with those of a complete spectroscopic analysis involving both temporal and spectral resolution. It is therefore the more necessary in these cases to evaluate carefully the amount and nature of the information to be obtained from each procedure and how best to obtain it in an efficient fashion.*

## FLUORESCENCE SPECTRA

In examining the fluorescence spectrum of pure compounds in solution we are presented with homogeneous samples, both as regards spatial distribution and temporal behaviour. When—as often happens—our solution contains a heterogeneous population of fluorophores we are still operating under the greatly simplifying condition that the heterogeneity is equally distributed throughout the sample and that—barring the most photochemically sensitive cases—the spectra will be practically time independent. This homogeneity and stability are unlikely to be present in the fluorescence from tissues and cells. We must expect regions of high and low concentration of fluorescent material to exhibit different spectra not only as a result of trivial filter effects arising in differences in absorption or scattering among the cell regions but also because of local differences in chemical quenching, in the amount and

*The literature of optical spectroscopy is particularly rich in examples in which resolution is carried to the limits of the art without regard for its possible significance. The results are pages of "dead information" waiting to be brought to life in the future, but having for the present no perceptible effects besides qualifying the authors as careful and dedicated scientists.

character of chemical modification of the fluorophore by the cell metabolism, and in the transfer of electronic energy to nearby chromophores. In fact we expect local variations of fluorescence spectrum to provide an important source of information in the fluorescence spectroscopy of images.

It is customary to distinguish between emissions arising in intrinsic constituents of the biological environment and those introduced from the outside. These two kinds originate from what are respectively called intrinsic and extrinsic fluorescent probes. We expect that the most prevalent intrinsic fluorescence will be that of the cell proteins. The fluorescence of proteins in solution has been extensively studied for the last 30 years. Cellular fluorescence has been photographed [3] and in some instances measured, for example, in muscle fibers in different physiological conditions [4,5], but to my knowledge no consistent attempts have been made to quantitate the images. We foresee that both the unequal distribution of proteins throughout the cell and the very different quantum yields, observed in the isolated proteins in solution, will come into play and result in images with important information content.

In experiments in solution the illuminated portion from which fluorescence can reach the detector commonly has dimensions of ca. 0.5 cm length and 0.04 $cm^2$ cross section, thus filling a volume of $2 \cdot 10^{10}$ $\mu m^3$. The volume of interest illuminated in the microscope in the epifluorescence mode is expected to be confined to perhaps $10^3$ $\mu m^3$. All other conditions being the same as in the solution case, the fluorescence signal from the sample in the microscope will be down by a factor of $10^6$ with respect to the signal from the solution. Fortunately both the flux of the exciting light and the efficiency of collection of the fluorescence are much higher in the microscope. In the microscope the cross section of the exciting beam is five to ten times smaller, and the probability of excitation correspondingly greater, than in the solution. The efficiency of fluorescence collection is notoriously poor in spectrofluorimeters and in the microscope it may improve by a factor as large as 50. As the absorption band of the tryptophan in the protein has a half-width of some 60 nm we can use a much larger bandwidth of excitation and increase in this way the intensity of illumination by a factor of five to ten. Similarly we can gain an extra factor of five, in many cases without serious loss in resolution, by increase in the bandwidth of fluorescence observation. We have no difficulty in recording fluorescence spectra of protein solutions at micromolar concentrations (approximately 0.003 gm %) when the bandwidths of excitation and emission are ca. 3 nm. Considering the improvements in efficiency we can expect to record the fluorescence spectrum of cellular protein if it exists in the illuminated volume at a concentration of mM or higher. Evidently this will often be the case for a typical cellular protein concentration of 20%. In comparing the two types of observations we implicitly assumed that equal times were allowed for both image and solution measurement. Additionally we neglected consideration of the benefits to be obtained by comparison of the intensity emitted by different regions of the fluorescent image. This is an aspect that finds no counterpart in the case of solutions because of the spatially homogeneous character of their fluorescence emission.

Besides the protein fluorescence, which is confined in both excitation and emission spectra to the ultraviolet region, there are in the cell a number of intrinsic fluorophores that emit and often absorb in the visible region. Their emissions have been the object of extensive studies as indicators or as kinetic markers of metabolic events by Britton Chance and his collaborators [6] among others. The most conspicuous of these fluorophores are the reduced nicotinamide nucleotides, the flavin nucleotides, reduced quinones, and one of the few strongly fluorescent flavoproteins, lipoamide dehydrogenase.

Although often measured in cell suspensions or homogenates and occasionally in isolated organs and in tissue slices, no consistent attempts have apparently been made to image their fluorescence emission in the cell.

## EXTRINSIC FLUORESCENT PROBES

The earliest observations of fluorescence microscopy appear to have been done in imitation of cytological stain procedures employing fluorophores instead of the more usual microscopic stains. More recently the sensitivity of fluorescence has been exploited in more sophisticated fashion, by covalently attaching fluorophores to antibodies [7] or by utilizing especially synthesized organic compounds endowed with both specific binding properties and recognizable fluorescence characteristics [8–12]. Fluorophores with binding affinity directed to prevalent chemical constituents like lipids (diphenylhexatriene [13]) or DNA (diamidino phenylindole [10], ethydium bromide [14]) have also been the object of a number of microscopic observations. In all these studies the analysis of the images has remained qualitative and methods like lifetime measurement, state of polarization, and even fluorescence spectral characteristics have not been used to any large extent. In particular no study is known to me of the dependence of the "staining equilibrium" upon the concentration of the fluorophores. Such studies are worth making for their own sake, quite apart from the interest that one may have in mind in each particular case. Compartmentalization, dynamic partition of the fluorophore, and differences in the location of specific binding agents will all play a part in determining the distribution of the fluorophore in the cell. The studies of Knopp and Longmuir [15,16] employing the extrinsic fluorescent probe pyrene butyrate show that even a molecule as diffusible as oxygen exhibits notable differences in concentration in the different cell regions. Any attempt to separate the various influences that determine the distribution must begin with a study of its dependence upon the concentration of the fluorophore to which the cell is exposed.

By collecting all the information available in the fluorescence emission from a specific probe a great deal more can be learned about the structures present in the cell and about their dynamics than what is commonly obtained today by qualitative, mainly visual, observations. This is a complex problem and every case needs to be independently examined. However, some general considerations regarding the problems involved may be useful [17]: If we wish to locate a specific structure by means of a fluorescent probe the demands are straightforward: Apart from desirable spectral characteristics, like the regions of strong absorption and emission, it should have the highest possible affinity for the receptor molecule which it is intended to locate. This is the most common and up to now the most useful way in which fluorescent probes have been employed. Some have tried a more ambitious, and interesting, approach: To infer the course of molecular events by observation of changes in the probe fluorescence. A fluorophore designed for this task must have two indispensable properties: First, it must bind to the specific structure without producing irreversible or even important physiological changes. Second, it must give notice of changes in its environment. These two are evidently contradictory conditions that may be difficult or impossible to fulfill at the same time. A fluorophore that fills a very specific binding site must—to a smaller or larger extent—interfere with the normal functions and if it binds at a place some distance away it is not likely to register changes taking place at the active site. Perhaps if the binding affinity is large enough to ensure a specific location for the fluorophore at a place *close to but not within* the active site it may not *unduly* interfere with the normal functions of the latter and yet show changes in fluorescence during the sequence of events in which the active site takes part. The ideal probe of the dynamics must be a witness of but not an actor in the physiological drama.

A change in the environment may register as a change in fluorescence spectrum, fluorescence yield, or polarization. The fluorescence spectrum is the most sensitive of these properties. With the resolution currently possible with fluorophores in solution a shift in the center of mass[†] of the emission spectrum of 5–10 $cm^{-1}$ is readily detectable [18].

The displacement of the center of mass of the emission toward the blue or red is brought about by any change in the molecular interaction of probe and surroundings that affects unequally the energies of the ground and fluores-

---

[†]The center of mass of the fluorescence spectrum, $\langle \nu_g \rangle$ is calculated employing the relation $\langle \nu_g \rangle$ = Sum of $\{\nu_i \times F(\nu_i)\}$/Sum of $F(\nu_i)$ where $F(\nu_i)$ is the fluorescence intensity per wave number interval emitted at wave number $\nu_i$ and the sum covers the range of wave numbers over which $F(\nu_i)$ has measurable values.

cence states of the probe. There is strict energy conservation in the sense that the change in the *total* energy of the emission equals the change in the energy of interaction of fluorophore and environment. Hence the need to use the center of mass, not the commonly used change in the emission maximum, for an exact determination of the energy change. It will be noticed that a measured spectral shift of 10 $cm^{-1}$ corresponds to 1/20th of the thermal energy at room temperature and that this latter energy corresponds to a shift of approximately 5 nm for a fluorescence emission with center of mass at 500 nm. It follows that the displacement of the center of mass of the fluorescence has all the sensitivity required to display the most minute changes in the interaction of probe and environment.

The molecular interactions that produce the larger spectral effects are those between permanent dipoles. As a result, the extrinsic probes most widely used to characterize the environment have been polar fluorophores, probes that undergo large changes in electronic distribution upon excitation. Studies of polar fluorophores have been of two kinds: Physical chemists have attempted a consistent theory of the spectral effects as a function of the molecular parameters and biochemists have used them as indicators of the polar character of the medium in immediate contact with the probe. As this has been one of the more successful applications of fluorescent probes, and will very likely continue to be much used in investigations in solution and in fluorescence microscopy, a closer examination is warranted. Theories of the effects of solvent polarity on the emission spectrum of fluorophores were proposed by Lippert [19] and others who characterized the environment by its macroscopic properties, refractive index, and dielectric constant, following much earlier work of Onsager [20] on the dielectric properties of liquids. Though these theories provide a reasonable approximation to the overall spectral displacements observed in pure solvents they fail to recognize the heterogeneity of the emitting population that arises because of the plurality of modes of interaction of the fluorophore with the surrounding dipoles. This is an important omission because such differing modes of interaction result in a dependence of the fluorescence spectrum upon the wavelength of excitation. More recently Macgregor and Weber [18,21] have proposed a detailed theory in which the monopole charges of the fluorophore interact with solvent dipoles on a one-to-one basis. This theory, which has been considerably refined since the first publication, accounts well for the shift and broadening of the absorption and fluorescence in media of different polarity as well as the dependence of the fluorescence upon the exciting wavelength. It describes the effects as arising in detailed electrostatic interactions and it can therefore be applied to fluorophores placed in heterogeneous media. Recently it has been used to demonstrate that, contrary to previously held views, the heme pocket of myoglobin is a region of considerable polarity, as large in fact as that of dimethylformamide [22]. In this study the specifically designed probe DANCA (2′-dimethylamino 6′-naphthoyl 4-*trans* cyclohexane carboxylic acid) that forms a 1:1 complex with apomyoglobin was used. The complex disappears upon stoichiometric addition of a mole of hemin. Absorption and low-temperature (−74°C) fluorescence of the bound probe are displaced approximately 1,700 $cm^{-1}$ to the red with respect to the spectra of the free probe in cyclohexane. A monopole calculation of the sum of the electrostatic interactions of the amide dipoles in the protein with the fluorophore monopoles placed in the plane of the heme predicts a spectral displacement of similar magnitude.

There are many examples of the use of fluorescent probes to determine the binding properties of macromolecules in solution and quite a few relating to the presence, amount, or structural disposition of binding structures in the cell. Observation of dynamic behaviour by the use of extrinsic probes has been confined to the determination of rotational or translational motion in membranes and in a few cases to the measurement of distances by observation of electronic energy transfer [23–25]. The possibility of using two different fluorescent probes attached to the same macromolecule in order to observe correlation effects dependent upon the molecular dynamics has often been considered, but it has not met with any success apart from the special case of the estimation of distances between chromophores by measurements of the efficiency of electronic energy transfer. The reason for the failure in the former cases, and the success in the latter, is that, under the normal intensities of illumination, the fraction of the population in which both chromophores are simultaneously excited is vanishingly small. It is important to notice that this limitation does not apply when one wishes to establish, by means of the same method, the existence of an intermolecular—as opposed to intramolecular—correlation. In this case the only additional condition required is that each of the two fluorochrome-labeled species behave very much like the unlabeled population. Double-labeling procedures should find important use in image fluorescence spectroscopy to display correlations between different molecular processes in the cell.

## FLUORESCENCE EXCITATION SPECTRUM

With exceptions, the most important of which is discussed above, the fluorescence spectrum and yield of fluorophores dissolved in a pure solvent are independent of the exciting wavelength. This conclusion can be reached most simply by the rank analysis of a matrix of photometric measurements the elements of which are the fluorescence emissions at several nonoverlapping spectral regions elicited by excitation at a series of wavelengths [26, 27]. Evidently this method can be applied equally well to image and solution fluorescence.

It is unlikely that independence of the fluorescence spectrum from the wavelength of excitation will be the common case for fluorophores introduced in the cell. If heterogeneous binding is present a dependence of the fluorescence upon the exciting wavelength will be detected and this will constitute an important means to determine the existence of such heterogeneous binding. Notice however that energy transfer from nearby absorbing structures can contribute to the excitation of the fluorescence through electronic energy transfer from chromophores located as far as 50 Å from the emitter [28] and that the proteins in the cell will be the most likely contributors to this effect when excitation is carried out at wavelengths of 300 nm and shorter. Because of the very low yield of fluorescence of the nucleic acids their possible contribution to the fluorescence excitation spectra is limited to distances that imply direct contact with the emitting molecule. A well-known case of this kind occurs in NADH where the excitation spectrum shows a strong contribution from the—virtually nonfluorescent—adenine. From such an excitation spectrum the existence of an internal stacking complex involving adenine and nicotinamide was deduced [29].

## FLUORESCENCE LIFETIME AND YIELD

Fluorescence yield and lifetime are almost always simultaneously affected. The fluorescence yield can be changed by processes of two different kinds: Those that depend upon the state of the fluorophore at the time of the excitation and those that can occur, in competition with the emission, during the life of the excited state. The former predate the emission and may be presumed to have reached equilibrium by the time of excitation. They can result in the appearance of molecular species with quantum yield, fluorescence lifetime, and other emission properties different from those observed for the fluorophore dissolved in pure solvents. The processes that occur during the fluorescence life usually change lifetime and yield in the same proportion. The observed fluorescence lifetime is an average of the lifetimes of the molecular species present, weighted according to the contribution of each to the intensity of the emission and to the peculiarities of the method of measurement, pulse [30] or phase [31] fluorometry. In the last 10 years the determination of fluorescence lifetimes with precision of a tenth of a nanosecond or better have become standard in the fluorescence spectroscopy of solutions. The resolution of complex fluorescent decays into components has received much attention: With present-day methodology commonly two and sometimes three concurrent exponential decays can be isolated by employing either pulse or phase measurements [32]. Neither of these two methods of observation has been employed in the analysis of fluorescence images to a significant extent, although one would not expect to encounter in their application difficulties that have not been seen and circumvented in work with solutions. In contrast, the methods of measurement of the quantum yield of fluorescence in solution do not appear to be readily applicable to microscopic images and thus present the fluorescence microscopist with an important and challenging problem. Without such measurements the information obtainable from the determinations of fluorescence lifetime is necessarily incomplete. Despite this limitation the intrinsic information content, the sensitivity, and the relative freedom from artifacts of the measurements of fluorescence lifetime will doubtlessly give wide application to this technique in future analyses of microscopic images.

## FLUORESCENCE POLARIZATION

In solution the polarization of the emitted fluorescence can never exceed 50% [33] because the absorbing chromophores are distributed at random at excitation, but this limit no longer applies when the sample contains preferentially oriented absorbing elements that retain their orientations over periods that are long compared to the fluorescence lifetime. In fact we confidently expect that the fluorescence polarization that results from preferential stable orientation of cell or tissue structures will provide some of the most interesting information to come from fluorescence microscopy. The same rules apply to extrinsic fluorescent probes and in these cases the observation of fluorescence polarizations that exceed 50% requires the binding of the extrinsic fluorophore by preferentially oriented structures. The presence of a small proportion of oriented structures superimposed on a background of randomly oriented emitters may be also demonstrated by excitation of the sample by unpolarized light [34] and measurement of the linear polarization of the light emitted in the direction of excitation: Strict random orientation of the absorbing oscillators precludes the observation of any polarization. When attempting measurements of polarization one must remember the very different conditions prevalent in the measurements in solution and in the microscope as regards the angular aperture of the detection system. In the polarization photometry of solutions the fluorescence cone entering the polarizer makes a solid angle of very few degrees, and the corrections owing to this finite aperture of the beam are in all cases much smaller than the standard error of the determinations [35]. In microscopic observations employing objectives of high numerical aperture this correction may become appreciable [11] though it cannot be large enough to lead to qualitative errors.

Technically the measurements of polarization may appear to be more difficult than in solution, but this need not be the case. In polarization measurements the most likely causes of error arise in the turbidity of the medium, in the existence of parasitic fluorescence of various origins, and in the high local concentrations of fluorophore required in the observation of microscopic images. Turbidity results in scattering of both exciting and fluorescent light; if the return of the exciting radiation to the detector is completely eliminated by filtering, the scattering of both excitation and fluorescence leads to a systematic decrease in the observed polarization. Because of the very small thickness of the scattering layers involved we do not expect this to be an important cause of error. The parasitic fluorescence from cells corresponds largely to their intrinsic fluorescence, or rather to that part of the intrinsic fluorescence that contaminates the emission of interest. It is likely to be maximal for excitation in the ultraviolet region and to decrease with increase in exciting wavelength, to the advantage of those extrinsic fluorophores that absorb in the visible region of the spectrum. The reabsorption of the emitted fluorescence and its reemission as secondary fluorescence results in conspicuous lengthening of the fluorescence lifetime [36,37] and leads also to depolarization, well known in the study of solutions. As "radiative transfer" it stands in contrast to "radiationless transfer" [28] owing to the coupling of electromagnetic oscillators. It depends upon the fraction of fluorescent light reabsorbed within the "observed" volume, that is, the volume of solution from which secondary fluorescent light may reach the detector [38]. It is manifest only in those cases in which there is an important overlap of the absorption and emission spectra of the fluorophore. These cases are the ones that have the larger critical distances for radiationless energy transfer and this circumstance makes it difficult to separate radiative and radiationless effects when one studies the dependence of the polarization upon the concentration of fluorophores in solution. In the study of solutions the method of choice to minimize the radiative effects [39] consists in reducing the layer of solution to the thickness of a few microns. It follows that in the microscopic observation of fluorophores incorporated in cells, where the observed volume has a thickness of this order, one need not expect an important depolarization by radiative transfer, thus greatly facilitating the analysis of both intrinsic polarization and radiationless energy transfer among identical chromophores.

The widest use of fluorescence polarization measurements in solutions has been with the object of determining the rotational diffusion constant of fluorophores [40]. The volume of the probe and the nature of the bonds with the surrounding molecules [41] determine the frictional coefficient for rotation of the probe in the unknown medium. From the latter value one can derive the viscosity of the medium in which the fluorophore moves. In fluorophores dissolved in pure solvents the viscosity thus derived is found to correspond closely to the viscosity determined by the classical flow methods. If a fluorophore is incorporated in a gelatin gel—an experiment carried out nearly 40 years ago [40]—the two values of the viscosity thus determined differ sharply, the polarization indicating a high fluidity and the flow a high viscosity. The two observations are reconciled by noticing that the polarization registers the resistance to rotation produced by the medium in the immediate vicinity of the fluorophore, which is nearly that of the pure solvent, while flow additionally registers the resistance opposed by the multiply connected network of gelatin molecules. These two viscosities may be respectively named the macroscopic and microscopic viscosities of the system. Assignment of microscopic viscosity values to natural environments by means of the polarization of the fluorescence of included probes [42–44] is further complicated by the likely existence of binding equilibria of the fluorophore with biological macromolecules. Nevertheless the assignment of a microscopic viscosity to the medium often permits a straightforward interpretation of the result, at least in qualitative terms, and serves in the comparison between otherwise similar environments.

A very large number of measurements of the microscopic viscosity of biological membranes have been made employing the polarization of fluorescence of included lipophylic probes like perylene and diphenylhexatriene. In contrast there has been only one study, by Udenfriend and co-workers [45], of the viscosity of the cytosol employing similar methods: The fluorescence polarization of an amino acid derivative, N-naphthylalanine, is very low when actively accumulated by cell suspensions, indicating that the viscosity of the cytosol is not appreciably different from that of water. With the advent of methods that permit the introduction of macromolecules, as well as small molecules, in the cytoplasm we can look forward to additional information on the distribution of microscopic viscosities throughout the cell.

Measurements of fluorescence polarization provide rapid, accurate, and convenient means to determine the rotational rate of a fluorophore, but they are not exempt from ambiguity. This arises when the fluorophores constitute, as regards the rate of rotation, a very heterogeneous population. Suppose that the rotor population comprises only two kinds: fast rotors, that when present alone generate a polarization close to zero, and slow rotors that emit light with polarization close to the limiting value for the fluorophore. Depending on the relative numbers of both kinds one will record a polarization of intermediate value, from which a totally spurious rate of rotation of the fluorophore will be deduced if the polarization was assumed to originate in a homogeneous population. It is in these cases that the methods of measurement of polarization in real time present a distinct advantage. Both the decay of the polarization in real time [44] and the differential phase between polarized components [41] determine the rotational rate through a measurement of the *change* in polarization during the fluorescence lifetime. The very fast rotations that occur at times much shorter than the fluorescence lifetime effectively reduce the initial polarization to a lower "zero time" value. The very slow rotations, accomplished in times much longer than the fluorescence lifetime, do not diminish the polarization beyond this initial value; therefore no appreciable polarization changes take place in this case during the fluorescence lifetime and this can only be revealed by the use of the time-resolved measurements. The application of real-time polarization measurements to fluorescence images does not appear to present difficulties besides those experienced in the measurement of the stationary fluorescence polarization.

## CIRCULAR POLARIZATION

Already in 1926 Perrin had reported that he was unable to detect traces of circular polarization in the emission from fluorescein and other dyes. The observation of limiting linear polarizations that closely approach for some fluorophores the value of 0.5 characteristic of a wholly anisotropic linear oscillator is also in agreement with the initial observation of Perrin. In recent years Steinberg and his co-workers [46] have detected the presence of small amounts of circularity, on the order of a few parts in 10,000, in the emission of optically active fluorophores. The differences in absorption of circularly polarized light by optical isomers can reach several percent so that absorption is much more sensitive than emission when it comes to distinguishing between optical isomers. Basing themselves on this premise Turner et al. [47] proposed the use of the differences in *intensity* of the fluorescence alternatively excited by left-handed and right-handed circularly polarized light to detect and measure the optical activity of fluorophores. There is little advantage in the use of this method over the direct measurement of circular dichroic absorption in observations on pure compounds. However, the fluorescence detection of circular dichroism comes into its own when one attempts to measure the circular dichroism of a fluorophore present in a mixture of optically active compounds that do not emit fluorescence or do so in a different region of the spectrum. This is the situation that must inevitably obtain in observations of fluorescence under the microscope. We may recall that many symmetric chromophores can acquire optical activity when they are placed in an asymmetric, optically active, environment. This "extrinsic Cotton effect" observed upon binding of optically inactive chromophores to proteins and nucleic acids has been amply documented [48]. It seems that we have here an important method for the demonstration of the chiral properties of the macromolecules in the cell through observations of the intensity of the emission of optically inactive bound fluorophores excited by circularly polarized light.

## REFERENCES

1. Weber G (1983): Old and new developments in fluorescence spectroscopy. In Cundall RB, Dale RE (eds): "Time Resolved Fluorescence Spectroscopy." New York & London: Plenum Press, pp 1–20.

2. Foerster Th (1951): "Fluoreszenz organischer Verbindungen." Goettingen: Vandenhoeck. & Ruprecht, pp 123–140.
3. Barenboim GM, Domanski AN, Turoverov KK (1974): "Luminescence of Biopolymers and Cells." Eng Trans R.F. Chen. New York & London: Plenum Press, pp 64–75.
4. Borovikov YS, Chernogradskaya NA (1979): Polarized ultraviolet fluorescence microscopy of F-actin. Microsc Acta 81:283–292.
5. Dos Remedios CG, Millikan RGC, Morales MF (1972): Polarization of the tryptophan fluorescence from single striated muscle fibers. J Gen Physiol 59:103–120.
6. Chance B, Schoener B, Oshino R, Itshak F, Nakase Y (1979): Oxidation-reduction ratio studies of mitochondria in freeze-trapped samples. NADH and flavoprotein fluorescence signals. J Biol Chem 254:4764–4771.
7. Frank ED, Warren L (1981): Aortic smooth muscle cells contain vimentin instead of desmin. Proc Natl Acad Sci USA 78:3020–3024.
8. Halpern J, Hinkle PM (1981): Direct visualization of receptors for thyrotropin releasing hormone with a fluorescein-labeled analogue. Proc Natl Acad Sci USA 78:587–591.
9. Hazum E, Nimrod A (1982): Photoaffinity labeling and fluorescence distribution studies of gonadotropin-releasing hormone receptors in ovarian granulosa cells. Proc Natl Acad Sci USA 79:1747–1750.
10. Matsumoto S, Morikawa K, Yamagida M (1981): Light microscopic structure of DNA in solution studied by the 4′ 6-diamidino 2-phenylindole staining method. J Mol Biol 152:501–516.
11. Basu S (1967): Molecular organization of DNA in chromosomes. Biopolymers 16:2315–2328.
12. Bauman JGJ, Wiegant J, Van Duijn G, Luben NH, Sondermeijer PJA, Hennig W, Kubli E (1981): Rapid and high resolution hybridization of polythene chromosomes using fluorochrome-labelled RNA. Chromosoma (Brl) 84:1–18.
13. Shinitzky M, Barenholz Y (1974): Dynamics of the hydrocarbon layer in liposomes of lecithin and sphingomyelin containing dicetylphosphate. J Biol Chem 249:2642–2657.
14. Le Pecq JB, Paoletti C (1967): A fluorescent complex between ethydium bromide and nucleic acids. J Mol Biol 27:87–106.
15. Knopp JA, Longmuir IS (1972): Intracellular measurement of oxygen by quenching of fluorescence of pyrenebutyric acid. Biochim Biophys Acta 279:393–397.
16. Benson DM, Knopp JA, Longmuir IS (1980): Intracellular oxygen measurements of mouse liver cells using quantitative video-fluorescence microscopy. Biochim Biophys Acta 591:187–97.
17. Weber G (1975): Practical applications and philosophy of optical spectroscopic probes. Horiz Biochem Biophys 2:163–198.
18. Macgregor RB, Weber G (1981): Fluorophores in polar media: Spectral effects of the Langevin distribution of electrostatic interactions. Ann NY Acad Sci 366:140–154.
19. Lippert E (1957): Spektroskopische Bestimmung des Dipolmomentes aromatischer Verbindungen in erster angeregten Singulettzustand. Z Elektrochem 61:962–975.
20. Onsager L (1936): Electric moments of molecules in liquids. J Am Chem Soc 58:1486–1493.
21. Weber G (1985): Polar fluorophores in polar solvents. In: "Proceedings of NATO ASI meeting on Excited state probes in biochemistry and biology." September, 1984. Acireale, Italy: Plenum Press, in publication.
22. Macgregor RB, Weber G (1986): Estimation of the polarity of the protein interior by optical spectroscopy. Nature 319:70–73.
23. Wu CW, Stryer L (1972): Proximity relations in rhodopsin. Proc Natl Acad Sci USA 69:1104–1108.
24. Slade Gennis L, Gennis RB, Cantor C (1972): Singlet energy transfer studies in associat-

ing protein systems. Distance measurements on trypsin, $\gamma$-chymotrypsin and their inhibitors. Biochemistry 11:2517–2524.
25. Herman AB, Fernandez SM (1982): Dynamics and topographical distribution of surface glycoproteins during myoblast fusion: A resonance energy transfer study. Biochemistry 21:3275–3283.
26. Weber G (1961): Enumeration of components in complex systems by fluorescence spectrophotometry. Nature 190:27–29.
27. Warner IM, Christian GD, Davidson ER, Callis JB (1979): Analysis of multicomponent fluorescence data. Anal Chem 49:564–573.
28. Foerster Th (1959): Transfer mechanisms of electronic excitation. Disc Faraday Soc 27:1–17.
29. Weber G (1957): Intramolecular transfer of electronic energy in dihydro-diphospho pyridine nucleotide. Nature 180:1409.
30. Ware WR (1983): Technique of pulse fluorometry. In Cundall RB, Dale RE (eds): "Time Resolved Fluorescence Spectroscopy." New York & London: Plenum Press, pp 23–58.
31. Teale FJW (1983): Phase and modulation fluorometry. In Cundall RB, Dale RE (eds): "Time Resolved Fluorescence Spectroscopy." New York & London: Plenum Press, pp 59–79.
32. Gratton E, Jameson DM, Hall RD (1984): Multifrequency phase and modulation fluorometry. Ann Rev Biophys Bioeng 13:105–124.
33. Perrin F (1926): Polarisation de la lumiere de fluorescence. Vie moyenne des molecules dans l'etat excite. J Physique Rad 7:390–401.
34. Weber G (1972): Uses of fluorescence in biophysics. Some recent developments. Ann Rev Biophys Bioeng 1:553–570.
35. Weber G (1952): Polarization of the fluorescence of macromolecules. I. Theory and experimental method. Biochem J 51:145–155.
36. Bailey EA, Rollefson GK (1953): The determination of the fluorescence lifetime of dissolved substances by a phase shift method. J Chem Phys 21:1315–1322.
37. Birks JB, Little WA (1953): Photofluorescence decay times of organic phosphors. Proc Phys Soc Lond (A) 66:921–928.
38. Weber G (1956): Photoelectric method for the measurement of the polarization of fluorescence of solutions. J Opt Soc Am 46:962–970.
39. Weber G (1960): Fluorescence polarization spectrum and electronic energy transfer in solutions of tryptophan, tyrosine and related compounds. Biochem J 75:335–345.
40. Weber G (1953): Rotational brownian motion and polarization of the fluorescence of solutions. Adv Protein Chem 8:415–459.
41. Mantulin WW, Weber G (1977): Rotational anisotropy and solvent-fluorophore bonds: An investigation by polarized differential phase fluorometry. J Chem Phys 66:4092–4099.
42. Shinitzky M, Dianoux AM, Gitler C, Weber G (1971): Microviscosity and order in the hydrocarbon region of micelles and membranes determined with fluorescent probes. Biochemistry 10:2106–2113.
43. Shinitzky M, Barenholz Y (1974): Dynamics of the hydrocarbon layer in liposomes of lecithin and sphingomyelin containing dicetylphosphate. J Biol Chem 249:2652–2657.
44. Chen LA, Dale RE, Roth S, Brand L (1977): Nanosecond time-dependent fluorescence depolarization of diphenylhexatriene in dimirystoyl lecithin vesicles and the determination of "microviscosity." J Biol Chem 252:2163–2169.
45. Udenfriend S, Zaltzman-Nirenberg P, Guroff G (1966): A study of cellular transport with the fluorescent amino acid aminonaphthyl alanine. Arch Biochem Biophys 116:261–270.
46. Steinberg IZ (1978): Circular polarization of luminescence. Ann Rev Biophys Bioeng 7:113–137.

47. Turner DH, Tinoco I, Maestre MF (1974): Fluorescence detected circular dichroism. J Am Chem Soc 96:4340–4342.
48. Daniel E, Yang JT (1973): Analysis of the circular dichroism of the complexes of 8-anilino 1-naphthalene sulfonate with bovine serum albumin. Biochemistry 12:508–512.

# Index